T3-BOP-827

ENGINEERING GRAPHICS FUNDAMENTALS

Betsy Gillham

As an additional learning tool, McGraw-Hill also publishes a problems book to supplement your understanding of this textbook. Here is the information your bookstore manager will need to order it for you: 19129-8 PROBLEMS BOOK TO ACCOMPANY ENGINEERING GRAPHICS FUNDAMENTALS

ENGINEERING GRAPHICS FUNDAMENTALS

Arvid R. Eide
Roland D. Jenison
Lane H. Mashaw
Larry L. Northup
C. Gordon Sanders

Professors of Engineering
Iowa State University

McGraw-Hill Book Company
New York St. Louis San Francisco Auckland Bogotá Hamburg
Johannesburg London Madrid Mexico Montreal New Delhi Panama Paris
São Paulo Singapore Sydney Tokyo Toronto

ENGINEERING GRAPHICS FUNDAMENTALS

234567890HALHAL898765

ISBN 0-07-019126-3

Library of Congress Cataloging in Publication Data
Main entry under title:

Engineering graphics fundamentals.

Includes index.
1. Engineering graphics. I. Eide, Arvid R.
T353.E65 1985 604.2'4 84-21326
ISBN 0-07-019126-3

Cover Photograph Credits

Universal joint and mechanical part by Phototake.
Brackets courtesy of MAGI—CAD/CAM Division.

This book was set in Baskerville by Black Dot, Inc.
The editors were Kiran Verma and J. W. Maisel;
the designer was Merrill Haber;
the production supervisor was Diane Renda.
The drawings were done by Felix Cooper.
Halliday Lithograph Corporation was printer and binder.

Contents

PART FOUR

Introduction to Computer Graphics

Preface

TO THE STUDENT

Engineering graphics is a cornerstone of engineering. The essence of engineering—that is, design—requires graphics as the means of communication within the design process. Graphics serves as the common thread between design and the manufacturing and construction processes.

Study of the fundamentals of engineering graphics is a key to your success as an engineer. Being able to describe an idea with a sketch is a prerequisite of the engineering profession. The ability to put forth a three-dimensional geometry in a form that can be communicated to other engineers, scientists, technicians, craftspersons, and nontechnical personnel is a valuable asset. Of equal importance is knowing how to read and understand the graphics prepared by others.

Your study of engineering graphics begins in a period of rapidly changing graphics technology. The traditional tools of graphics, such as the T square, compass, and drafting machines are being displaced by computer graphics terminals. You are in an exciting era in which you will experience the transition from ink pens, triangles, and dividers to a computer keyboard and from blueprints to databases. The computer will enhance your ability to communicate graphically, but it will not do the engineering for you. You must therefore learn the fundamentals of graphics, develop a measure of expertise for applying these fundamentals to your upcoming engineering analysis and design courses, and, finally, use your graphics capabilities to advantage as a practicing engineer.

The material coverage and organization of the topics in this book are designed to help you effectively develop your graphical communication abilities during your brief introductory study of engineering graphics. The skills and fundamental knowledge acquired during your initial study of engineering graphics will serve you well as you pursue a career in engineering.

TO THE INSTRUCTOR

Engineering graphics, along with most engineering subjects, is undergoing changes as a result of the impact of the computer on technology. Methods of approaching and solving engineering problems are changing. Techniques heretofore shunned because of prohibitive computational time to obtain a reasonable solution are being resurrected. New methods that solve old problems with better accuracy are rapidly being discovered. Generation and verification of geometry by means of computer graphics is now commonplace. Designers are beginning to communicate with analysis groups and the manufacturing function through a computer database rather than a set of design drawings.

Before the impact of computer graphics technology, engineering graphics instruction had been reduced considerably during the 1960s and 1970s. At one time many of us had the luxury of six or more semester credits with the possibility of one of these credits being devoted to an introductory design component. A few may have the luxury of a graphics and design course followed by a descriptive geometry course, but such is more the exception rather than the rule.

The decrease in available time for graphics instruction combined with an increased use of computer graphics might seem at first to signify the end of graphics in engineering curriculums. Nothing could be further from the truth. The computer is a blessing in disguise. Used as a teaching tool, it can enhance learning. Specifically, in graphics, the computer can perform much of the drudgery of drawing, thus permitting more graphics fundamentals to be covered in a fixed time frame. Used as an interactive device, the computer can assist in problem solving by generating more alternatives and permitting more problem variables to be investigated.

We believe strongly that the future for engineering graphics is brighter than ever. Because of the computer, the engineer is a more active participant in the total product cycle from conception to implementation by the consumer. The new engineer must have a global understanding of graphics fundamentals and the ability to apply these fundamentals in a variety of design cycle tasks and to understand the link between design and manufacturing. This book merges the traditional graphics with the new graphics technology and allows the instructor to put together a sound first course in graphics which can satisfy a wide variety of objectives.

The text is organized into four parts. By selecting appropriate coverage for the stated objectives, the instructor can build a two-to four-semester credit course. It would also be possible to divide the material into two 2-credit courses.

Part One introduces the tools of graphics and presents the fundamentals of orthographic projection. Chapter 1 is a brief historical perspective of the development and applications of graphics. The objectives of the text are also presented here. A discussion of the change of graphics technology to computer-based procedures is included. Chapter 2 describes the skills that are needed to communicate graphically. Emphasis is on sketching techniques. Chapter 3 introduces orthographic projection by first developing the concept of a multiview drawing; then introducing the three-plane relationship, common measurements, and auxiliary views; and concluding with the application of orthographic projection to the solution of geometry problems. The three basic problems of space geometry, true length, point view, and edge view of a plane are introduced. Understanding a problem and having the facility to solve geometry problems are enhanced with the use of solution view and solution procedures. By visualizing the solution view, the student can then logically proceed to the correct solution orthographically. The solution view and solution procedure concept are carried through the descriptive geometry material in Part Two.

Part Two, on descriptive geometry, begins with a formal study of lines and planes in Chapters 4 and 5. Chapter 6 develops the point, line, and plane relationships in space with an emphasis on clearances and connectors. Chapter 7 discusses surfaces and procedures for the generation and development of surfaces by standard graphical means. Chapter 8 introduces intersections. Numerous example problems are shown to illustrate the intersection of plane and curved surfaces.

Part Three on applied graphics, covers the topics needed for the communication of geometry within a design loop. Material in this part may be undertaken upon completion of Part One. Chapter 9, on solids, is a more detailed discussion of the correct procedures for construction of multiview drawings. Chapter 10 presents a theoretical discussion of pictorials produced with axonometric and oblique drawing procedures. Sectional views are introduced in Chapter 11. Chapter 12 presents the fundamentals of dimensioning, including production dimensioning and geometric dimensioning and tolerancing. Chapter 13 describes the types of fasteners available to the engineering designer with particular emphasis on threaded fasteners. Extensive appendixes include standard fastener specifications for a wide range of sizes and types. The concepts from Chapters 9 to 13 are merged in Chapter 14 to illustrate the preparation of design drawings. A portion of a design effort is carried through from concept to final documentation to illustrate the interrelationship of the design drawing package.

At the conclusion of each chapter in Parts One to Three, there is a section relating the material just covered to computer graphics. This will give the student an appreciation of how technology is changing. If computer-graphics hardware is available, demonstration of some computer-graphics concepts at the same time the material is covered in Chapters 1 to 14 would be beneficial.

For those who wish to pursue computer graphics and its applications to engineering, Part Four will serve your needs. Chapter 15 surveys the arena of computer graphics by describing some of the principles and the hardware and software necessary to have a computer-graphics system. Chapter 16 introduces 2-D modeling with computer graphics. Chapter 17 extends the material of Chapter 16 to 3-D. Elements of vector algebra and analytic geometry are introduced to support the discussion. Chapter 18 is an introduction to modern design incorporating computer graphics. Conceptual design is discussed along with examples of design by repeated analysis. The chapter concludes with a brief discussion of the preparation of written and oral reports for defense of a proposed design.

If you offer a 2- or 3-credit graphics course, we suggest that coverage includes Parts One and Three, with varying amounts of Part Four as available equipment permits. The material in Part Four can be introduced at any time in the course but would be most effective if introduced parallel to or following coverage of Chapter 3.

For a descriptive geometry course, Parts One and Two would be included with Part Four, particularly Chapter 17, introduced as available equipment permits.

For those with a design component as part of a graphics course, Parts One, Three, and Four with emphasis on Chapter 18 would provide a strong basis for more advanced analysis and design courses.

SI and English units are both used throughout the text. The appendixes include standards for both, and chapter problems include both units.

A coordinated problems book is available that covers all aspects of the text.

Although the text is written with the assumption that the student has had no experience in graphics, a background in algebra and trigonometry is strongly recommended, particularly for the computer-graphics and design aspects. As we learn the potential of computer graphics for engineering, the use of basic mathematical concepts such as analytic geometry and trigonometry will become more prevalent in engineering graphics. This early motivation to apply the mathematics will enhance the student's education.

ACKNOWLEDGEMENTS

The authors wish to express their gratitude to many who supported the concept of this textbook and who contributed to its development. We thank our graphics teachers who, during our formal education, developed within us an appreciation for graphics and its importance in engineering. Our concept of effectively integrating the computer with graphics at the introductory level was greatly influenced by former colleagues Robert J. Bernhard, Steven J. Hooper, M. Gawad Nagati, and Jerald M. Vogel. They also critiqued the manuscript and developed a great deal of software while teaching the experimental graphics–computer graphics course. Many undergraduate teaching assistants are due thanks for their efforts in the mammoth task of converting our ideas and suggestions to usable computer programs. We are grateful to Vicky Eide for her accurate typing of the manuscript. We are especially grateful for the support of our families who have graciously allowed us to take the many extra hours needed to produce this textbook.

Arvid R. Eide
Roland D. Jenison
Lane H. Mashaw
Larry L. Northup
C. Gordon Sanders

ENGINEERING GRAPHICS FUNDAMENTALS

PART ONE

GRAPHIC LANGUAGE AND SKILLS

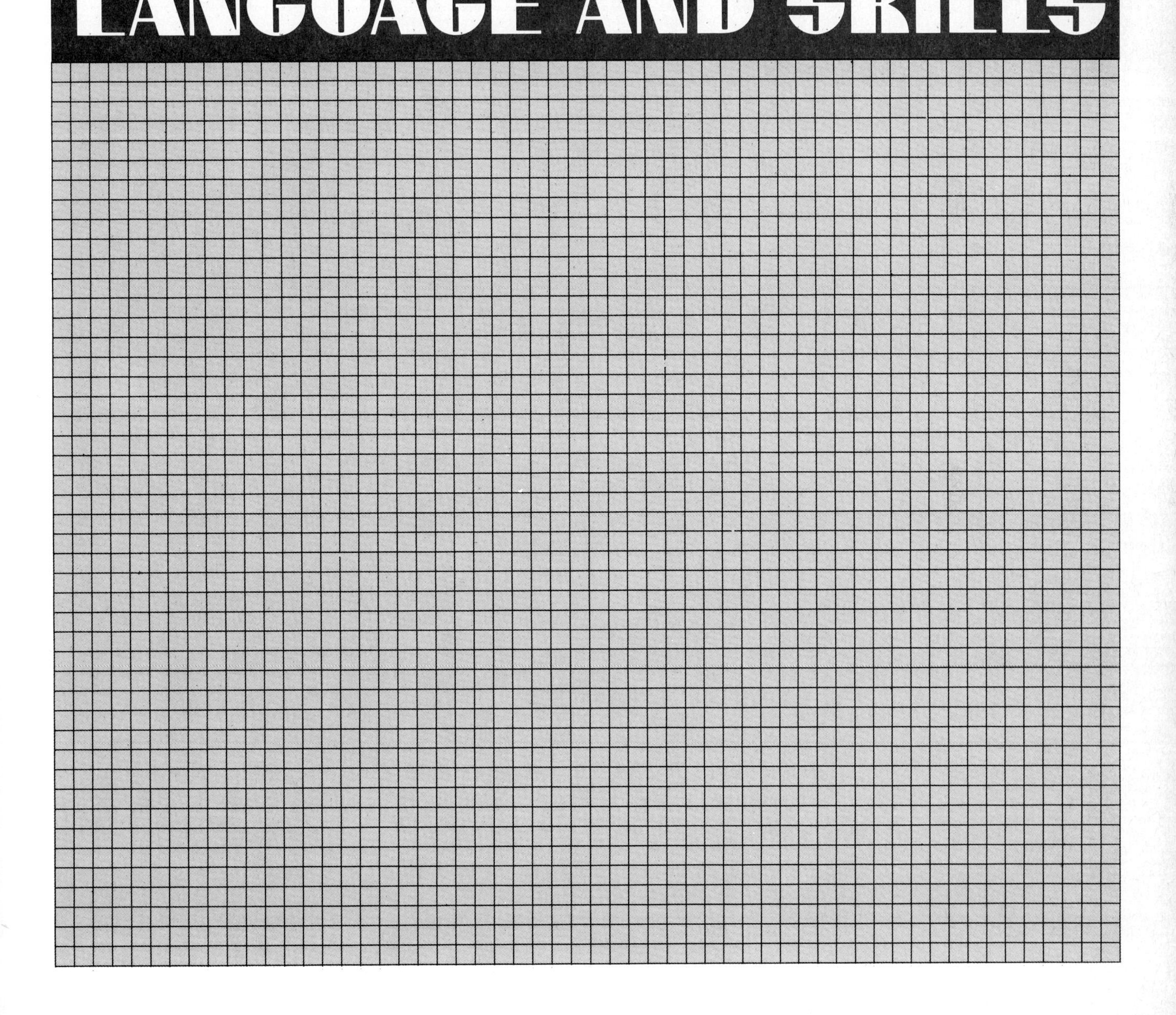

chapter 1

A Language for Engineers

FIG. 1.1 Computer graphics is a new and powerful graphic communication tool. (*Courtesy of Tektronix, Inc.*)

1.1 INTRODUCTION

The ability to communicate is the key to success for a practicing engineer. Graphic communication, along with written and oral communication, constitutes an important part of studying to become an engineer. The fundamentals of graphic language are universal in the industrialized world, an advantage not afforded by the written and spoken language. Thus, graphics may be said to be "the language of engineers."

1.2 HISTORICAL ASPECTS

Various kinds of drawings have been used for thousands of years to convey ideas or represent designs of objects to be constructed. The earliest humans made drawings in the dust with sticks or

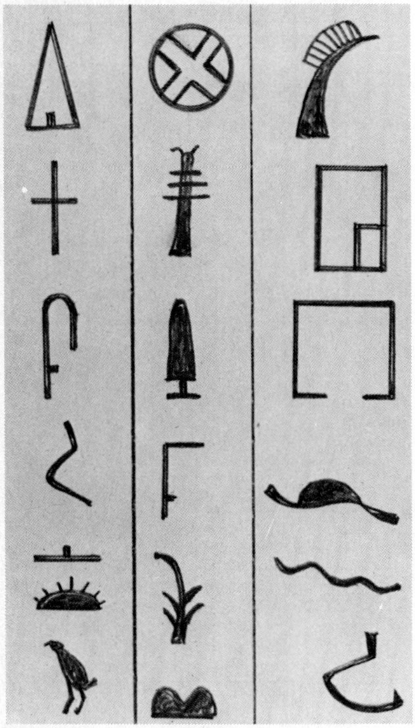

FIG. 1.2 Characters and objects taken from an ancient wall of hieroglyphic writing.

marked cave walls with sharp rocks or instruments. In Egypt around 3400 B.C., hieroglyphics were developed. This graphic language used pictograms to represent objects and phonoglyphs to represent the sounds of the words that described the objects. With the development of parchment and papyrus, more elaborate drawings were made, and the transfer of knowledge from place to place via drawings became possible.

No single event in the history advanced the ability to convey ideas to large numbers of persons more than the invention of the printing press by Johann Gutenberg in about 1450. This was also near the time of birth of Leonardo da Vinci, the brilliant sculptor, artist, scientist, and engineer. Da Vinci's use of colors, light, and shading led to the production of great masterpieces. His drawings of mechanical devices and inventions reflect the use of pictorial drawings (Fig. 1.4 and 1.5).

About 200 years after da Vinci, a French mathematician, Gaspard Monge, published a book entitled *La Géométrie Descriptive,* which generally is considered to be the first published documentation of the principles of graphic geometry. With the nineteenth century came the Industrial Revolution, and technical drawing became the thread tying together the concept of a design and the final manufactured product. Knowledge of graphics became the key requirement for the communication of technology, and this requirement became an integral part of technical education, particularly engineering.

FIG. 1.3 An early printing press now located at the Smithsonian Institution. (*Courtesy of the Smithsonian Institution.*)

FIG. 1.4 A mechanical design by Leonardo da Vinci.

1.3 MODERN TECHNOLOGY

Although the language of graphics has been available for many centuries, its effectiveness in modern technology is due to a gradual evolution toward uniformity of standards. Standardized practices and procedures for graphics came about through the efforts of such organizations as the American National Standards Institute (ANSI), the Society of Automotive Engineers (SAE), the American Society of Mechanical Engineers (ASME), and the American Society for Engineering Education (ASEE). Since the decision of the United States to join the majority of countries in the adoption and use of the Système International d'Unites (SI), most corporations and businesses have supported the sponsoring agencies in the development and implementation of appropriate metric standards.

The fundamentals of graphic projection theory have remained relatively unchanged since they were introduced by Gaspard Monge. In contrast, the methods of doing and presenting technical drawings have changed considerably. The engineer is becoming more and more dependent on the computer to perform the time-consuming drafting function. The modern engineer must be thoroughly grounded in the fundamentals of graphics and the methods of preparing drawings in order to function in the technological environment. Most engineers today are not involved in the actual preparation of drawings by conventional methods but

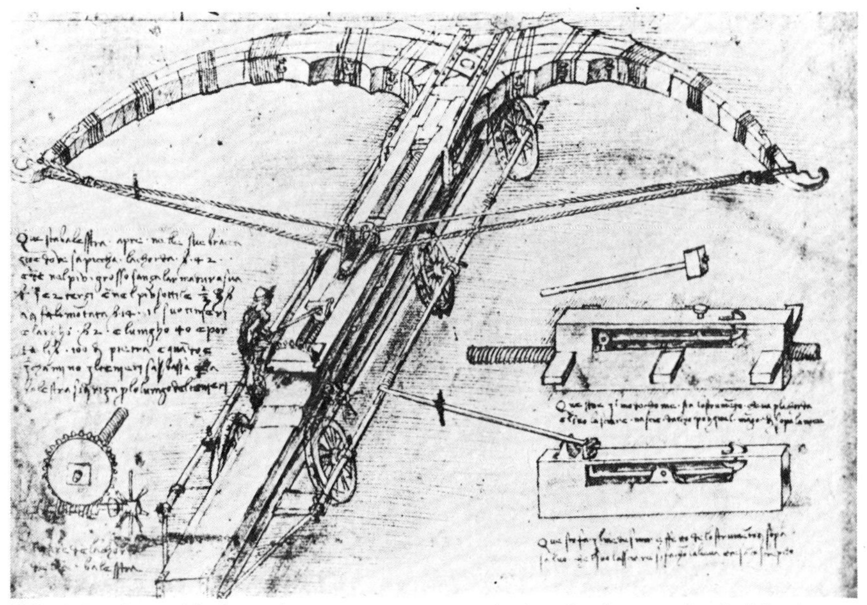

FIG. 1.5 A sophisticated weapons system design by Leonardo da Vinci.

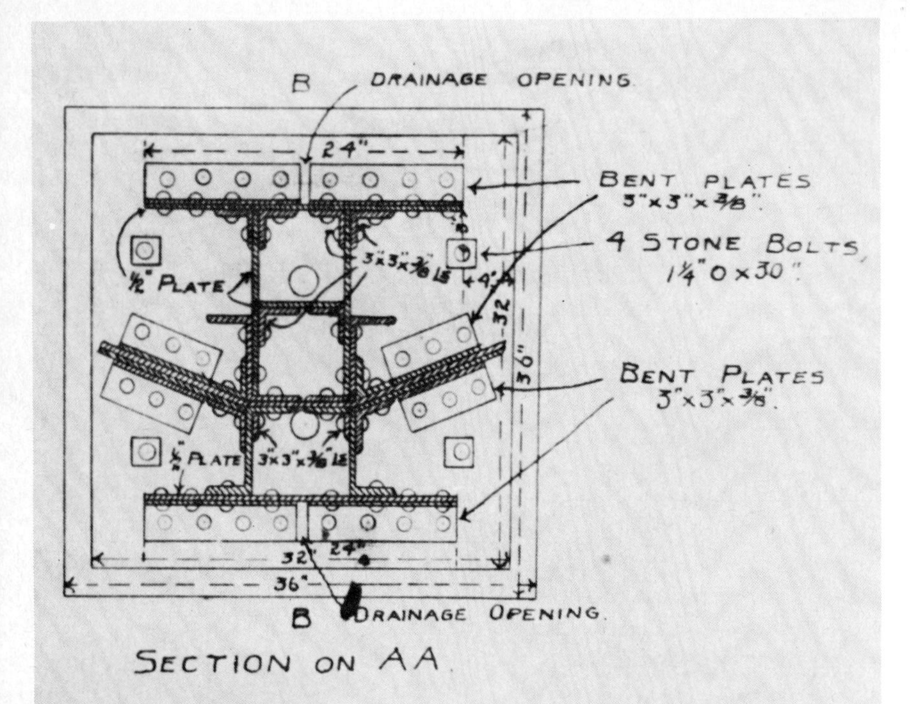

FIG. 1.6 A sectional view of a column of a water tower constructed about 1900. The designer, drafter, and construction supervisor were the same person.

FIG. 1.7 A design team studies a set of drawings. Today, many different types of engineers are involved in the development of a set of design drawings. (*Courtesy of Allen-Bradley Co.*)

instead supervise the technicians and drafters who do the preparation. This trend of the engineer away from the details of graphics is reversing somewhat because of the speed and capability of the interactive computer graphics terminal.

1.4 THE GRAPHIC LANGUAGE

The study of graphics involves three aspects: terminology, skills, and theory. The framework within which you will develop your ability to communicate graphically is briefly outlined here, along with some definitions and the objectives of this textbook.

1.4.1 Definitions

1. *Engineering graphics* is the area of engineering which involves the application of graphic principles and practices to the solution of engineering problems.

2. *Engineering design* is the systematic process by which a solution to a problem is created. Engineering graphics is the common thread that allows convenient communication between the steps in the design process.

3. *Descriptive geometry* is a set of principles which enable the geometry of an object to be identified and delineated by graphic means. It is the theory by which spatial problems involving angles, shapes, sizes, clearances, and intersections are solved.

4. *Computer graphics* involves the application of the capabilities of a computer to the analysis and synthesis of engineering problems and the communication of solutions in a graphic format.

1.4.2 Objectives of the Text

The rapid advancement of electronic technology and its application to the various technologies has signaled the next industrial revolution. The language of graphics will continue to be the basis of communication for engineering and related technical fields. However, the changes we will see in the methods of transferring graphic, written, and spoken material will be astounding. These changes will improve the productivity of industries as well as the quality of products. The requirements for a new engineer will include a better understanding of the fundamentals of engineering graphics and of how these fundamentals are used in the industry. Thus, this textbook has the following objectives:

1. To develop your understanding of the fundamental principles of engineering graphics
2. To develop your ability to apply the fundamental principles to the graphic solution of engineering problems
3. To familiarize you with the various methods of presenting graphic information
4. To develop your ability to visualize in three dimensions
5. To develop your ability to communicate with freehand sketches
6. To develop your ability to communicate with drawings prepared with the aid of common drawing instruments
7. To introduce you to the use of interactive computer graphics in the solution of engineering problems

The study of graphics will provide lessons about the logical approach to a problem, the accuracy of your work, and sensitivity to the correctness of a solution. It will provide an atmosphere of discipline which will carry over into upper-level engineering courses and your work after graduation. Perhaps most important, the study of graphics will enhance your ability to visualize concepts and create unique solutions to problems.

1.5 ENGINEERING GRAPHICS AND THE COMPUTER

The impact of the computer on the engineering community is just beginning to be felt. Of course, hand calculators, desk-top computers, and large computer systems have been in use for many years. However, we are only beginning to see computers used to

FIG. 1.8 A modern engineering accomplishment which was designed using computer graphics. As technology advances, it becomes more important to be able to describe objects in a graphical form for visualization and descriptive purposes. (*Courtesy of Control Data Corporation.*)

FIG. 1.9 An engineer contemplates the results of an engineering analysis on the computer. (*Courtesy of Tektronix, Inc.*)

produce a picture of an object on the screen, allowing the user to view the object from any direction, size the object, change the shape of the object, and obtain a drawing of the object on paper in a fraction of the time required for conventional drawing practices.

It is possible to generate a first design on a computer screen, modify the design, automatically prepare a set of instructions for building the object, and have the object manufactured by a computer-controlled machine, all without human handling of the design or object. While this is not yet common practice, many industries are investing heavily in this computerized design and manufacturing process to increase productivity. You will need to be prepared in both the conventional graphics practices and the computer graphics aspect when you enter the engineering profession.

Most likely you are aware of computer graphics already. It is used on commercial television in advertising, weather forecasting, and commercials. For example, if you live in areas where warnings about severe weather are broadcast, you have seen a band of print move across the screen superimposed over the regular programming. Home video games are another commercial application of computer graphics. Such games as Pac-Man and Space Invaders may have occupied many hours of your leisure time. Other familiar examples of computer graphics are seen in the simulation of flying an airplane or driving a car and the interactive student course scheduling used by many universities, colleges, and secondary schools.

FIG. 1.10 Two boys "hard at work" playing with a home video game.

FIG. 1.11 An artisan operates a numerically-controlled milling machine. The machine is controlled by a computer-prepared tape prepared from design data. (*Courtesy of the U.S. Department of Commerce.*)

You will begin to read about the developments in computer-aided design (CAD), computer-aided engineering (CAE), and computer-aided manufacturing (CAM) as you progress through your program of study. Keep in mind that a common element in all these areas is engineering graphics. Only if you are well prepared in the fundamentals of graphics will you be able to take advantage of the potential of the computer.

chapter 2

The Necessary Skills

2.1 INTRODUCTION

The language of engineering graphics provides methods for the communication of ideas. The means by which the ideas are described are called the tools of engineering graphics. There are literally hundreds of graphics tools and related drawing aids available for communicating ideas in clearly understandable form. In this chapter, we will describe many of the basic instruments used by engineers and discuss the use of these instruments. In work you will encounter in later engineering courses and in engineering practice, you may find the need for tools that are not described here. If you develop the needed skills in using the basic instruments, you will not find it

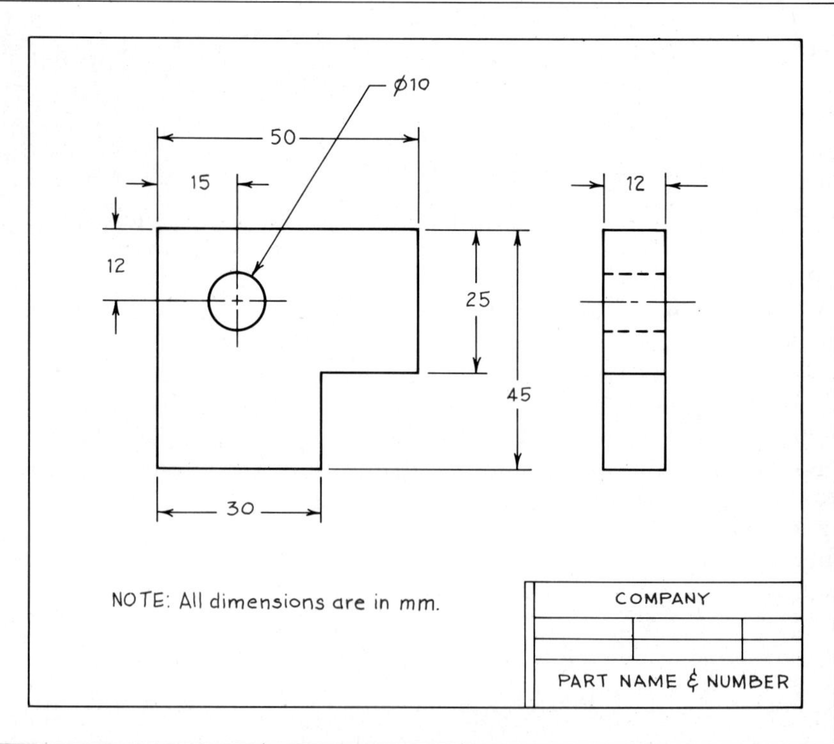

FIG. 2.1 A simple engineering drawing illustrating two views of a part. It demonstrates the use of scales, geometric construction, and lettering. The original drawing by the design engineer would have been a sketch.

difficult to learn how to use the more specialized tools.

We will develop our discussion of the necessary skills in terms of the four tasks required for graphic communication: lettering, geometric construction, measurement with scales, and sketching. The basic instruments required for performing each of these tasks will be introduced along with instructions for their use.

2.2 LETTERING

Figure 2.1 illustrates the use of lettering in graphic communication. The dimensions, titles, and notes are all essential for the correct interpretation of an idea on paper. You will find that good lettering practice is followed on a day-to-day basis in engineering work, whether it is used in the preparation of formal drawings as in Fig. 2.1 or in an informal idea-exchange session.

Lettering is accomplished with freehand techniques or by mechanical means. As shown in Fig. 2.1, freehand lettering is the predominant means by which titles and other features are added to a drawing. The first step in learning to perform good freehand lettering is to learn the correct methods for forming letters, numbers, and symbols and to practice these methods until you can letter correctly and rapidly.

Several styles of lettering, such as Gothic, Old English, and Roman, are used for various applications today. The Roman style, which is shown in Fig. 2.2*a*, may be chosen for its classic lines and decorative serifs. Old English, which is shown in Fig. 2.2*b*, provides flare and embellishment to letters. Engineers, however, prefer the single-stroke Gothic letters and numbers shown in

ABCDEFGHIJKLM
NOPQRSTUVWXYZ
abcdefghijklmnopqrs
tuvwxyz
1234567890

(*a*)

ABCDEFGHIJKLM
NOPQRSTUVWXYZ
abcdefghijklmnopqrs
tuvwxyz
1234567890

(*b*)

FIG. 2.2 Examples of lettering styles generally not used for engineering work but popular in art and advertising.

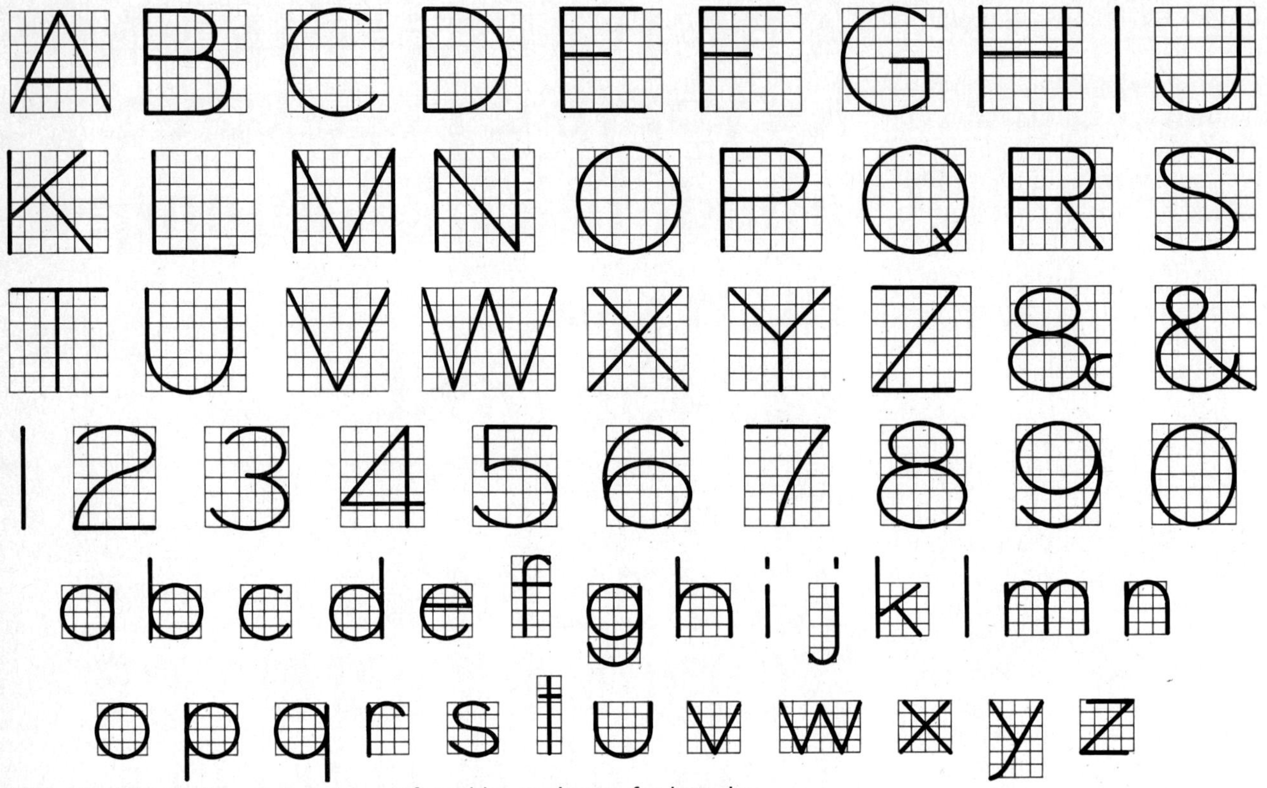

FIG. 2.3 The Gothic alphabet is preferred by engineers for lettering.

Random vertical guidelines

CGOQ
FEHILT
NXYZ
AKMVW4
DUJ
PRB25
S3869

FIG. 2.4 Letters and numbers that require similar strokes are grouped to show similar construction techniques.

Fig. 2.3. The principal reason for using Gothic as the lettering standard is the rapidity with which the letters and numbers can be formed compared with other styles of lettering. The Gothic letters may be uppercase or lowercase. The choice depends on individual or company preference, but once the choice is made, consistency should be maintained.

As you begin to practice lettering, study Fig. 2.3 to learn the correct shape of the letters and numbers. It may be helpful to practice lettering by grouping letters and numbers that require similar strokes. Note the similar strokes required to form each group of letters and numbers shown in Fig. 2.4.

Once you have developed the ability to form individual letters and numbers correctly and rapidly, use the letters to form words and phrases, as illustrated in Fig.

2H

H ENGINEERING LETTERING

FIG. 2.5 Space individual words approximately twice as far apart as the height of the letters.

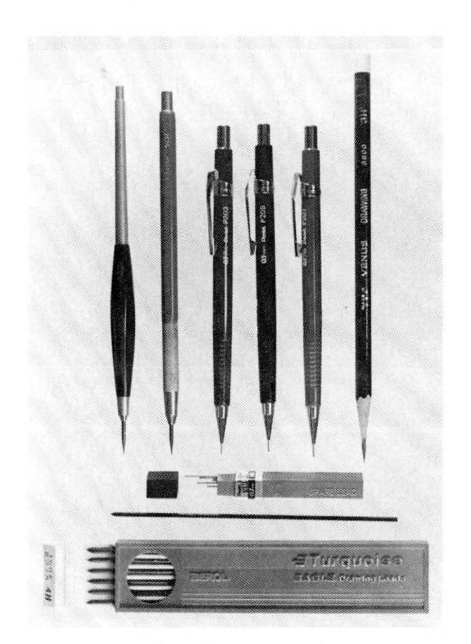

FIG. 2.6 Pencils and erasers are available in a wide range of types and styles.

2.5. Particular note should be given to the spacing between letters and between words. We will illustrate size and spacing of letters and numbers in more detail after describing the equipment required for freehand lettering.

2.2.1 Lettering Equipment

The standard equipment for freehand lettering is a good pencil and eraser. Figure 2.6 illustrates some of the commonly used pencil types and erasers.

Leads for the pencils come in grades ranging from 9H (very hard) to 6B (very soft). For general freehand applications the H, F, or HB grades are the most frequently used. The choice depends on personal preference, the drawing paper, and the intended use of the drawing. Recently, the small-diameter lead (0.5 mm) used in mechanical pencils has become popular. This lead does not need sharpening and provides a uniform line.

If a mechanical or wooden pencil with a larger lead diameter is chosen, a wide selection of points may be achieved through sharpening techniques. Ordinary sandpaper can be used to achieve the desired sharpness. You must exercise care in sharpening the pencil point. A point that is too sharp can break easily or cut into the drawing paper (see Fig. 2.7*a*). Conversely, a point that is allowed to become dull, as shown in Fig. 2.7*c*, will create fuzzy and inconsistent lines.

The best erasers remove marks on the paper quickly and without smudging. The size, shape, and type of the eraser chosen depend on the application. It is a good idea to have several types available for specific erasing needs.

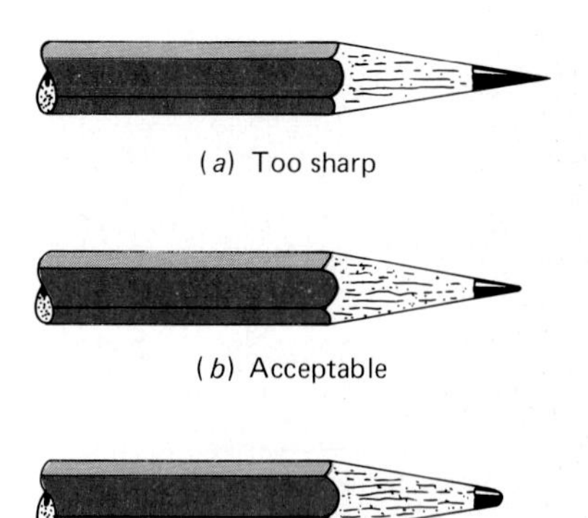

FIG. 2.7 To achieve proper line consistency the lead must be correctly sharpened.

It is helpful to have an erasing shield available similar to the type shown in Fig. 2.8. The shield will allow selective erasing without removing nearby parts of the drawing which are correct. This can save time and enhance the appearance of the finished drawing.

FIG. 2.8 Erasing shield.

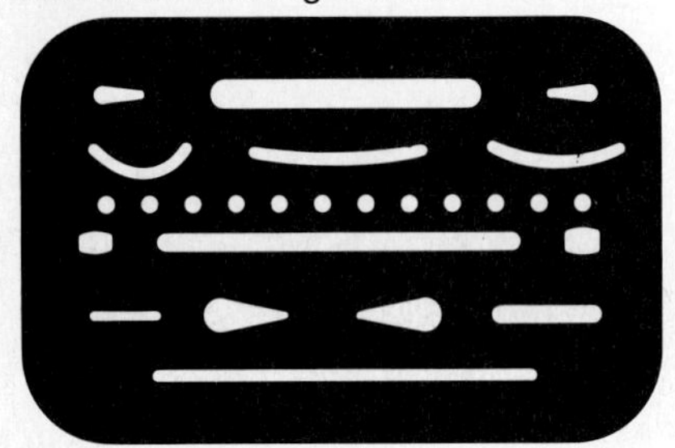

2.2.2 Lettering Techniques

The ability to produce high-quality, consistent freehand letters and numbers is a matter of practice. As is true with any skill, good lettering is developed over time as a result of conscientious effort. You should not become discouraged initially by the amount of time required to form correct letters and numbers. Diligent practice will increase your speed without a sacrifice in quality.

The first step in freehand lettering is to construct guidelines to help you maintain consistent letter size. Figure 2.9*a* is a set of instrument guidelines of light construction set approximately 3 mm apart. Uppercase Gothic letters and numbers can then be formed between the guidelines, with the top and bottom extremities of the letters and numbers just touching the guidelines. When lowercase letters are used, a third guideline is required, as shown in Fig. 2.9*b*. If the Reinhardt system for letter proportions is used, the intermediate guideline is placed two-thirds of the distance from the bottom guideline to the top guideline. For lettering fractions, a set of five guidelines is required, with the fraction constructed twice the height of a single number (Fig. 2.9*c*). Note that neither the numerator nor the denominator of the fraction touches the division bar. The 3-mm size chosen for letter and number height is convenient for most applications

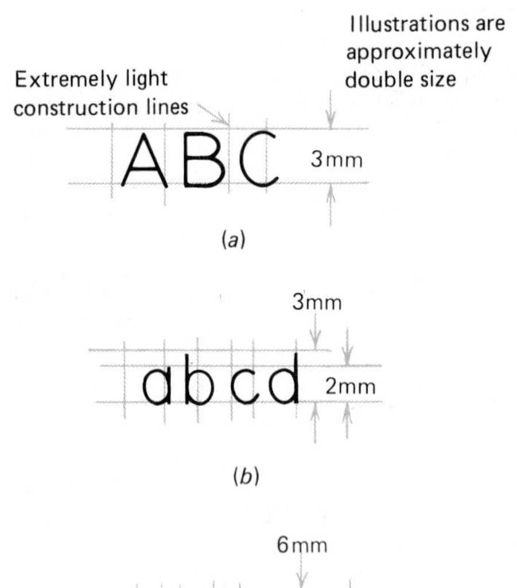

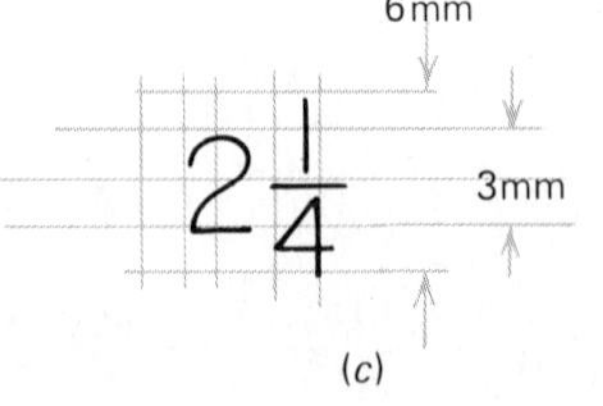

FIG. 2.9 Both horizontal and vertical guidelines should be of very light construction. They should not distract from the lettering.

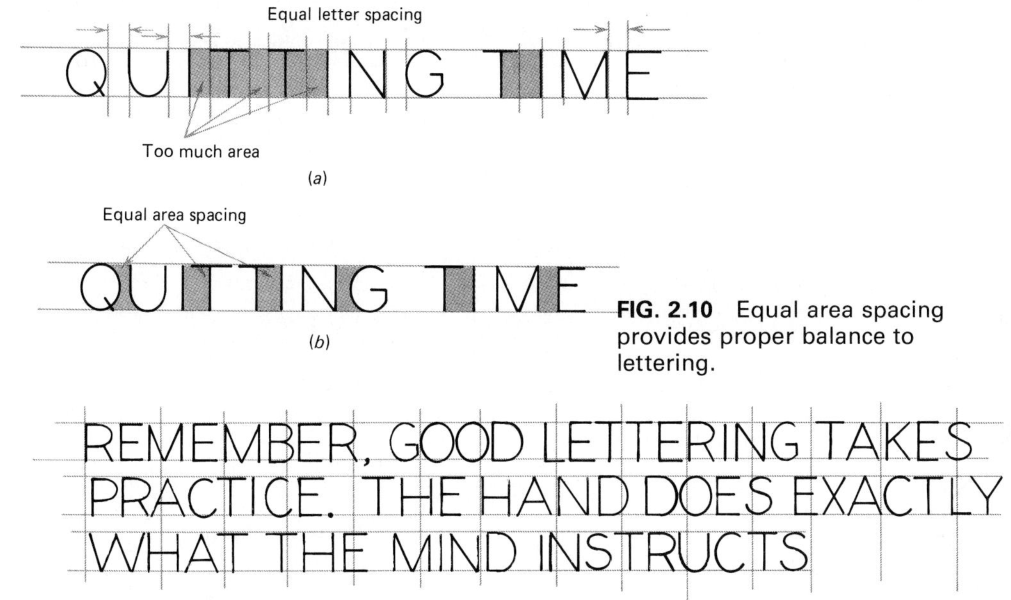

FIG. 2.10 Equal area spacing provides proper balance to lettering.

FIG. 2.11 Freehand lettering.

but is not to be considered a fixed size. You should choose the size of letters which will produce the desired effect in your work. Large drawings or presentation posters require larger lettering than do small drawings.

Maintaining proper spacing between letters is the most difficult part of good lettering. Spacing is a function of which letters or numbers are adjacent. In most cases, you can tell if the spacing is balanced by noting whether the letters or numbers appear compact and whether words and sentences are easily distinguishable. This balance is best achieved by trying to maintain an approximately equal area between letters. Figure 2.10*a* shows a phrase in which the letters are spaced equally. Figure 2.10*b* is the same phrase using the area concept. Practice in lettering will increase the coordination between your hand and eyes, and soon you will be performing good lettering without having to think first about the appropriate sizing, spacing, and shaping of each letter or number (Fig. 2.11).

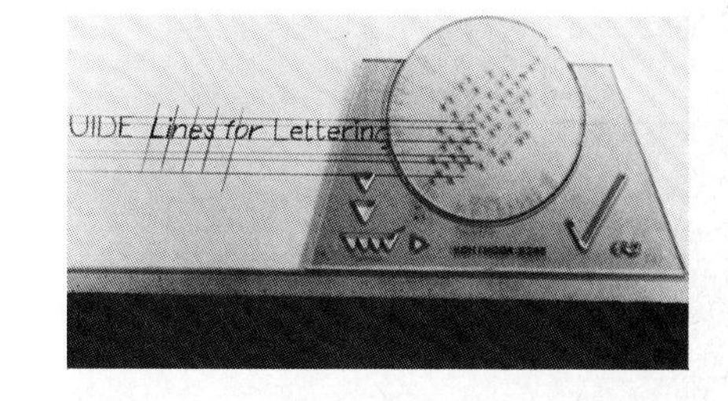

2.2.3 Special-Purpose Equipment

Many mechanical devices are available to help produce high-quality lettering. The Ames lettering guide shown in Fig. 2.12 is used in conjunction with a straightedge to produce uniform guidelines. The actual lettering is performed freehand. Transfer

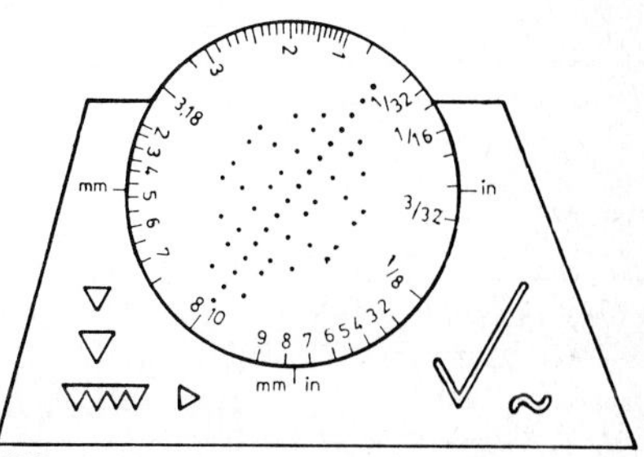

FIG. 2.12 Ames lettering guide.

FIG. 2.13 Transfer letters

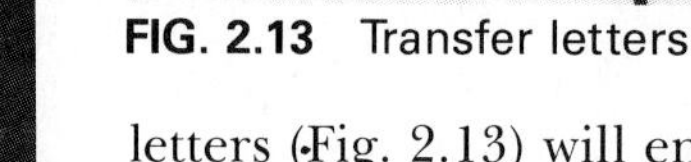

FIG. 2.14 Lettering stencil

letters (Fig. 2.13) will ensure a uniform size and shape of letters, but correct spacing is determined by the user. A lettering stencil, shown in Fig. 2.14, is used when large letters and a formal appearance are required. These devices save time, provide uniformity, and make available a wide range of choices for styles and sizes of letters.

Many pencil drawings are traced over in ink to provide a clear, sharp original drawing from which copies can be made. A mechanical pen, which is shown in Fig. 2.15, is a device that allows for rapid tracing in ink. The ruling pen, which is shown in Fig. 2.16, may be used in freehand inking or with lettering guides and templates. Both styles of inking pens come with various point diameters to accommodate different line widths.

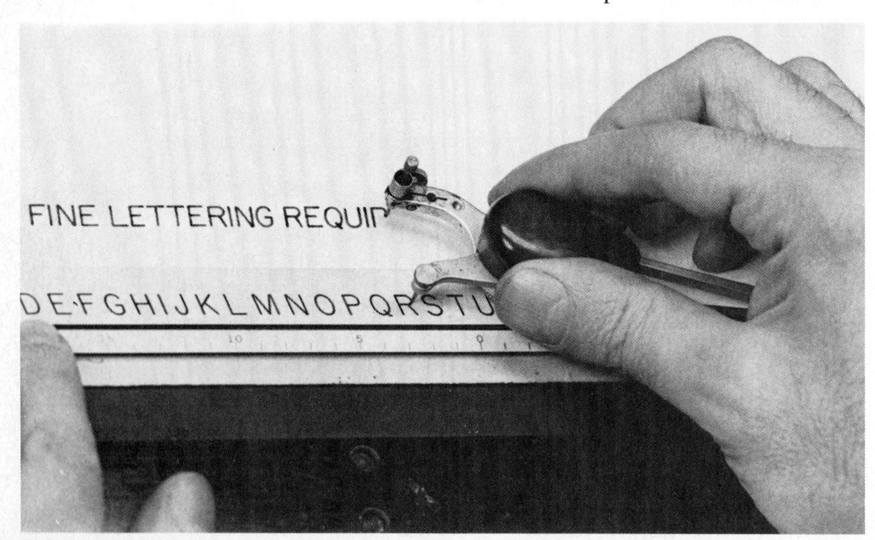

FIG. 2.15 Different types of mechanical systems are available to assist the lettering process.

2.3 DRAFTING EQUIPMENT AND GEOMETRIC CONSTRUCTION

In order to produce quality drawings, proper use of a variety of drafting equipment is necessary. We will illustrate the use of the following essential equipment:

pencil	templates	dividers
eraser	triangles	irregular curve
scales	compass	protractor

This is by no means a complete list of available drafting equipment. Such items as the T square, ruling (ink) pens, drafting machines, reproduction equipment, and mechanical lettering devices will not be discussed here.

FIG. 2.16 Ruling pen that can be used with stencils or freehand letters.

2.3.1 Pencil and Eraser

In the discussion of freehand lettering equipment (Sec. 2.2.1), it was noted that the selection of a pencil and eraser is largely a matter of personal preference. For formal drafting work, the quality of lines becomes important. Figure 2.17 shows a range of line widths (thicknesses) required in formal drafting work. The differences in line characteristics are important in reading or interpreting a drawing. Consistency in line width and intensity must be maintained on an original drawing in order for clear reproductions to be made. Many drawings are produced in ink rather than pencil to achieve the desired reproduction qualities. Ink drawings generally are produced by skilled drafters. The engineer must be able to produce clear pencil drawings with appropriate line widths and contrasts; this can be accomplished through selection of leads with varying hardness and diameters.

2.3.2 Scales

The three-dimensional objects represented on engineering drawings vary greatly in size. For

TYPES OF LINES

Type	Weight	
Object	Thick	
Hidden	Medium	
Center	Thin	
Phantom	Thin	
Extension & Dimension	Thin	
Leader	Thin	
Section	Thin	
Cutting plane	Thick	
Short break	Thick	
Long break	Thin	

FIG. 2.17 All lines should be sufficiently dark for reproduction. Line thickness differentiates between line types.

(a)

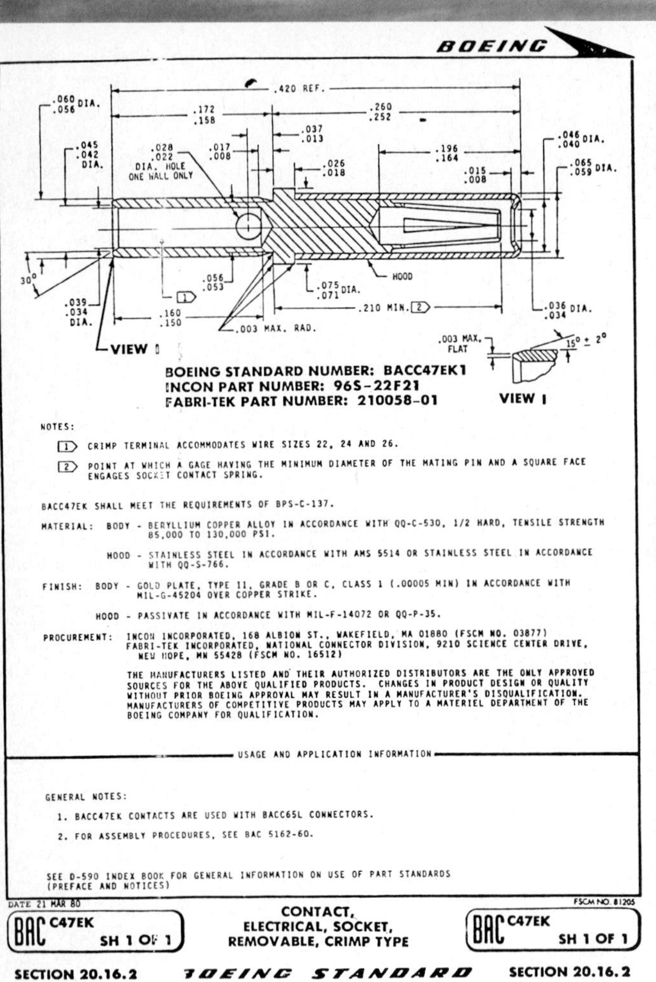
BOEING

VIEW I

BOEING STANDARD NUMBER: BACC47EK1
INCON PART NUMBER: 96S-22F21
FABRI-TEK PART NUMBER: 210058-01

VIEW I

NOTES:

1. CRIMP TERMINAL ACCOMMODATES WIRE SIZES 22, 24 AND 26.
2. POINT AT WHICH A GAGE HAVING THE MINIMUM DIAMETER OF THE MATING PIN AND A SQUARE FACE ENGAGES SOCKET CONTACT SPRING.

BACC47EK SHALL MEET THE REQUIREMENTS OF BPS-C-137.

MATERIAL: BODY - BERYLLIUM COPPER ALLOY IN ACCORDANCE WITH QQ-C-530, 1/2 HARD, TENSILE STRENGTH 85,000 TO 130,000 PSI.

HOOD - STAINLESS STEEL IN ACCORDANCE WITH AMS 5514 OR STAINLESS STEEL IN ACCORDANCE WITH QQ-S-766.

FINISH: BODY - GOLD PLATE, TYPE II, GRADE B OR C, CLASS 1 (.00005 MIN) IN ACCORDANCE WITH MIL-G-45204 OVER COPPER STRIKE.

HOOD - PASSIVATE IN ACCORDANCE WITH MIL-F-14072 OR QQ-P-35.

PROCUREMENT: INCON INCORPORATED, 168 ALBION ST., WAKEFIELD, MA 01880 (FSCM NO. 03877)
FABRI-TEK INCORPORATED, NATIONAL CONNECTOR DIVISION, 9210 SCIENCE CENTER DRIVE, NEW HOPE, MN 55428 (FSCM NO. 16512)

THE MANUFACTURERS LISTED AND THEIR AUTHORIZED DISTRIBUTORS ARE THE ONLY APPROVED SOURCES FOR THE ABOVE QUALIFIED PRODUCTS. CHANGES IN PRODUCT DESIGN OR QUALITY WITHOUT PRIOR BOEING APPROVAL MAY RESULT IN A MANUFACTURER'S DISQUALIFICATION. MANUFACTURERS OF COMPETITIVE PRODUCTS MAY APPLY TO A MATERIEL DEPARTMENT OF THE BOEING COMPANY FOR QUALIFICATION.

USAGE AND APPLICATION INFORMATION

GENERAL NOTES:

1. BACC47EK CONTACTS ARE USED WITH BACC65L CONNECTORS.
2. FOR ASSEMBLY PROCEDURES, SEE BAC 5162-60.

SEE D-590 INDEX BOOK FOR GENERAL INFORMATION ON USE OF PART STANDARDS (PREFACE AND NOTICES)

DATE 21 MAR 80 — FSCM NO. 81205

BAC C47EK SH 1 OF 1 — CONTACT, ELECTRICAL, SOCKET, REMOVABLE, CRIMP TYPE — BAC C47EK SH 1 OF 1

SECTION 20.16.2 — BOEING STANDARD — SECTION 20.16.2

(b)

FIG. 2.18 Many different scales are needed to construct drawings of this modern airliner. The various parts range from the large wing and fuselage surfaces to the small electronics components in the control systems. (*Courtesy of the Boeing Company.*)

example, a set of drawings for a wide-body airliner (Fig. 2.18*a*) will include drawings of tiny electronic components used in the control panel (Fig. 2.18*b*) as well as drawings of the airframe components (Fig. 2.18*c*). These drawings are produced on a two-dimensional surface which may be a computer screen or paper that ranges in size from 8½ × 11 in (216 × 279 mm) to 36 × 48 in (914 × 1219 mm). In order to accomplish this, the drawings must be scaled to the actual object so that correct proportions can be maintained. The scales commonly used for engineering drawings are discussed in Sec. 2.4.

(c)

2.3.3 Dividers

Dividers, like the pair shown in Fig. 2.19, are designed to be operated with one hand and are used for marking distances or transferring measurements. Specified lengths can be obtained from scales or from another drawing and transferred to the drawing being prepared. Figure 2.20 illustrates how the dividers may be used to create a double-sized drawing simply by transferring measurements, thus avoiding the necessity of measuring each length and doubling the measurement.

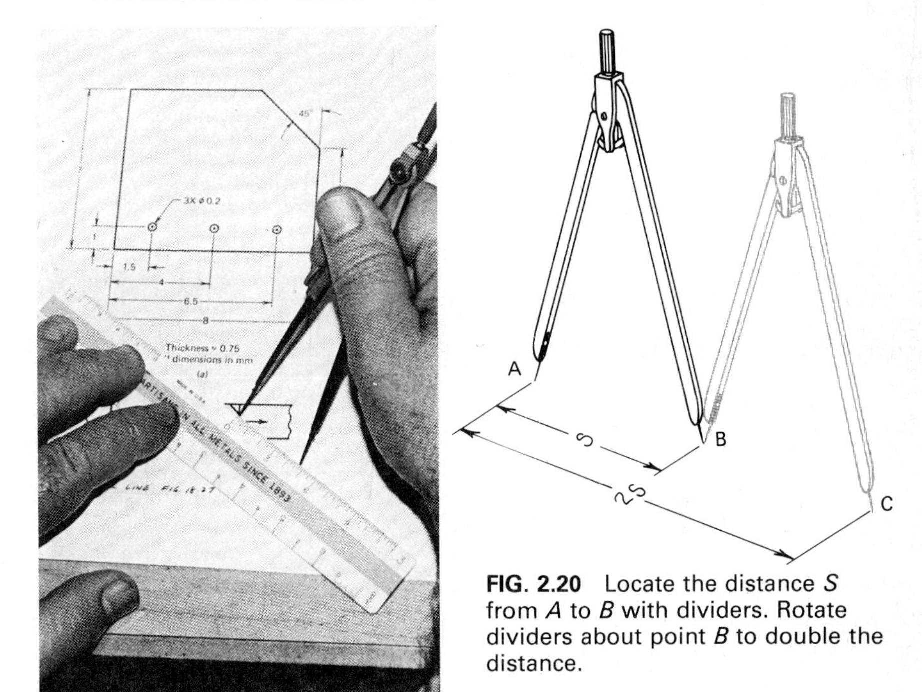

FIG. 2.19 Dividers

FIG. 2.20 Locate the distance S from A to B with dividers. Rotate dividers about point B to double the distance.

2.3.4 Triangles

The capability of rapidly producing straight lines on instrument drawings is provided by the 45° and 30-60° triangles (Fig. 2.21). Using the triangles as a pair, you can generate parallel and perpendicular lines and produce angles of a multiple of 15°.

Parallel lines are produced by establishing one side of a triangle along the given line or line direction. The supporting triangle is then fixed against one of the other sides of the first triangle. The first triangle is slipped along the supporting triangle to any desired position, and the parallel line is drawn (Fig. 2.22).

Perpendicular lines may be produced by either the sliding triangle method or the revolved triangle method. The sliding triangle method is shown in Fig. 2.23. One leg of a triangle is placed along the given line. The supporting triangle is then fixed

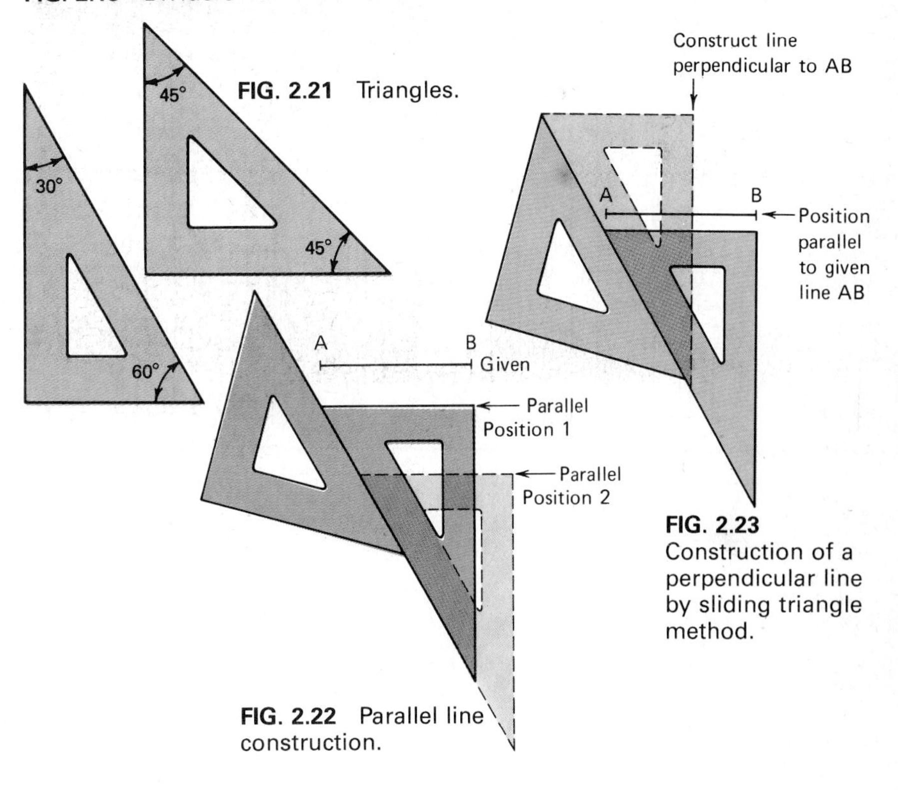

FIG. 2.21 Triangles.

FIG. 2.22 Parallel line construction.

FIG. 2.23 Construction of a perpendicular line by sliding triangle method.

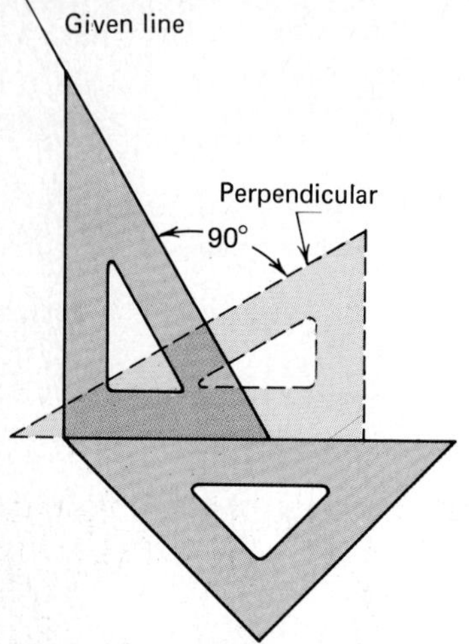

FIG. 2.24 Construction of perpendicular lines by rotating triangle method.

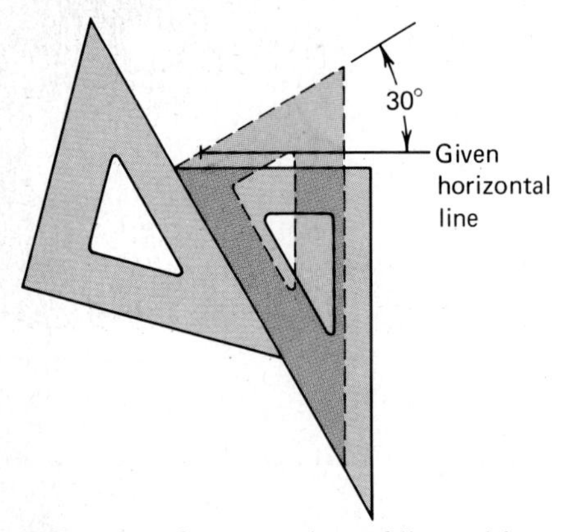

FIG. 2.25 Construction of line with 30° angle from a given line.

against the hypotenuse of the first triangle. Then the opposite leg of the first triangle is positioned by sliding, and the desired perpendicular line is drawn.

The revolved triangle method, which is illustrated in Fig. 2.24, requires fixing the two triangles together so that the given line is along the hypotenuse of the first triangle and the supporting triangle is fixed against one leg of the first triangle. Simply revolve the first triangle until the opposite leg rests against the supporting triangle, and the perpendicular line can be drawn.

If a vertical or horizontal line is established on the paper, any angle which is a multiple of 15° can be constructed relative to the established line. Figure 2.25 shows the construction of a 30° angle relative to a given horizontal line. Many other combinations can be drawn by proper positioning of the two triangles. Once an oblique line is established, the multiples of 15° can be generated by proper placement of the 30-60° and 45° triangles.

A useful application of the triangles may be seen in the division of a given distance into equal parts. In Fig. 2.26*a*, it is desired to divide the distance *AB* into seven equal parts. First, lay off line *AC* of arbitrary length at a convenient angle with *AB*. With a pair of dividers set at a convenient length, mark off points 1 through 7 along *AC*. The length of the spaces 0-1, 1-2, etc., is arbitrary, but they must be identical. Second, connect point 7 with point *B*, using a triangle. Then, using the triangles as in Fig. 2.22, construct lines through points 6, 5, 4, etc. parallel to the line connecting 7 and *B*. This will divide line *AB* into seven equal parts.

A similar procedure can be used to divide a line into equal segments with equally spaced lines perpendicular to the given line (Fig. 2.26*b*). Using triangles, construct a line *EF* perpendicular to the given line *DE*. Then, using a convenient scale, set the initial point of the scale at *D* and rotate the scale until the 7 unit intersects line *EF* (to divide the line into seven equal parts). Mark the points 1-6 and construct lines through these points parallel to *EF*. Each line is perpendicular to *DE*, and they are spaced equally. The division of a line into equal parts is very useful in making charts, tables of data, or special paper for graphing purposes.

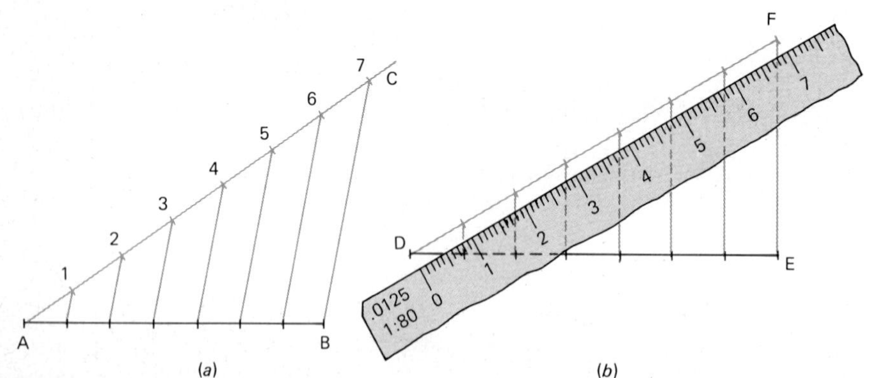

FIG. 2.26 (*a*) Dividing a line into equal segments by parallel line construction. (*b*) Dividing a line into equal segments by perpendicular line construction.

2.3.5 Templates

Many geometric figures and engineering symbols occur so frequently on drawings that special templates are manufactured to reduce the drawing time for these standard figures. Figure 2.27 shows a few of the many templates available to the drafter. Although most templates have a variety of sizes for each symbol or figure, care must be exercised in using these templates on scaled drawings. If it is necessary to show the symbol to the exact scale of the drawing, it may be necessary to draw the symbol instead of using the template.

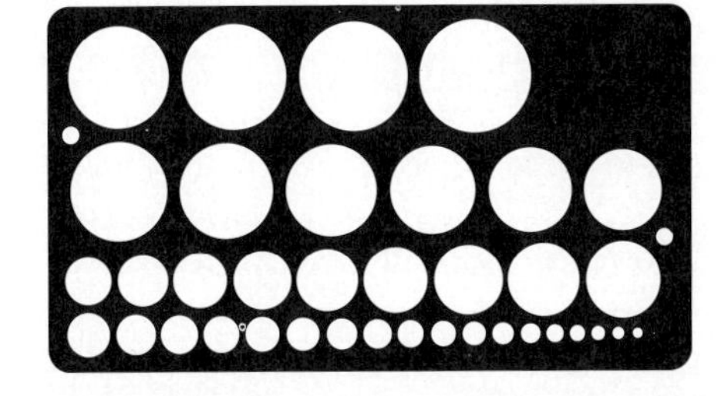

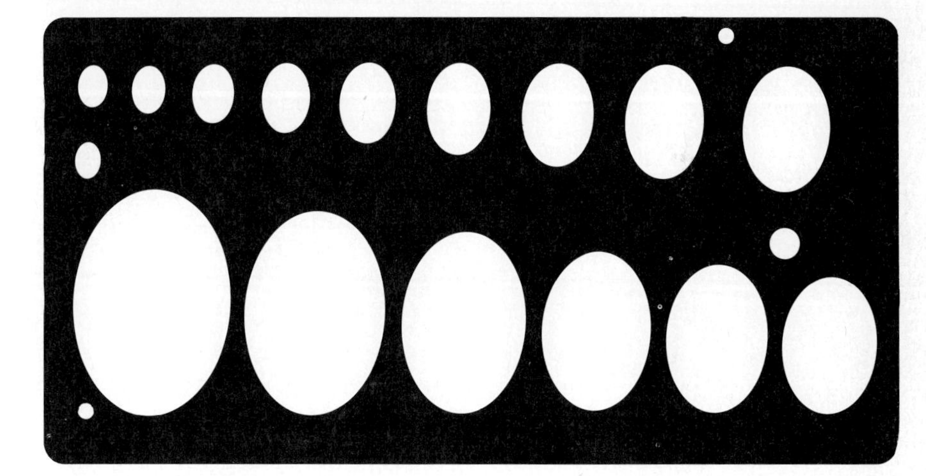

FIG. 2.27 Examples of typical engineering templates.

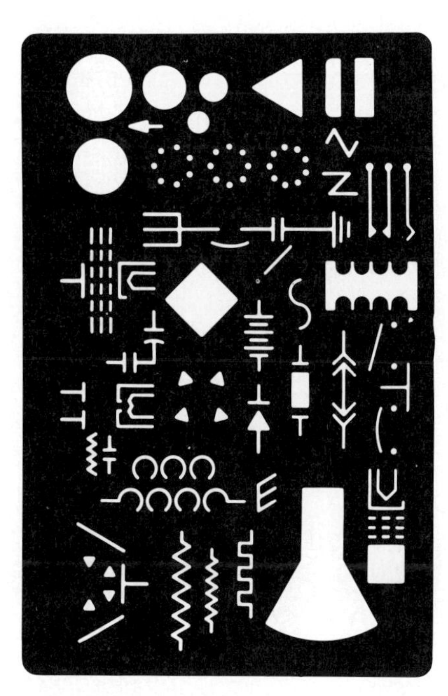

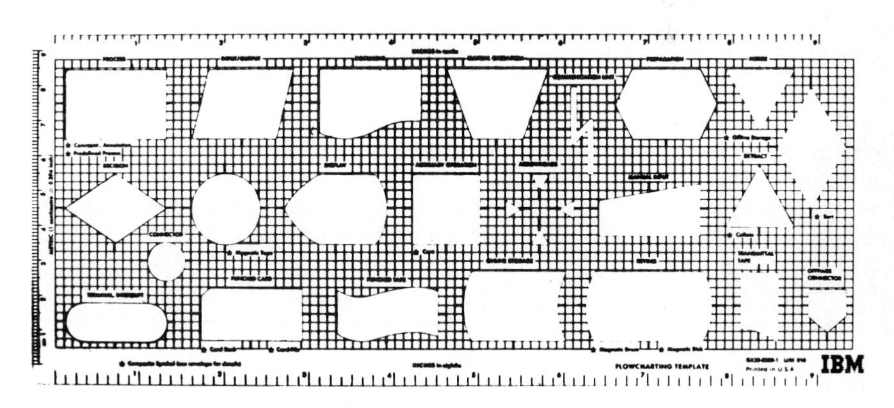

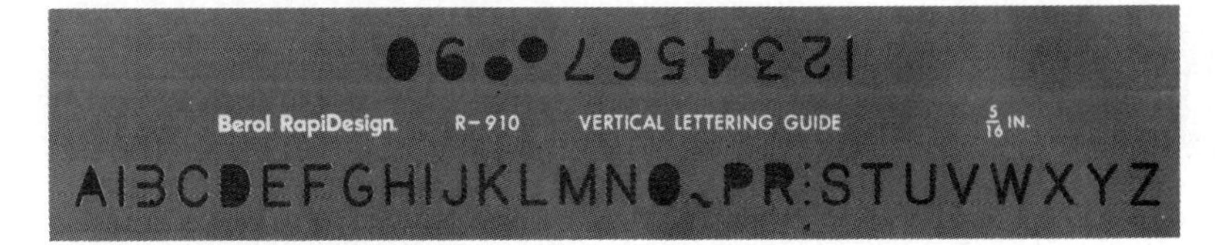

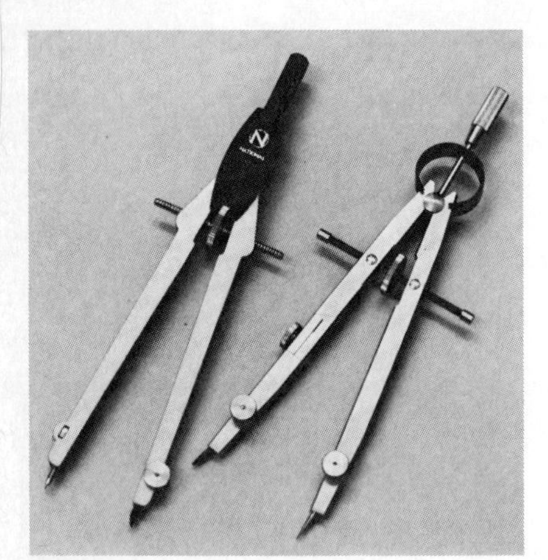

FIG. 2.28 Fixed-bow and rapid-action compasses.

2.3.6 Compass

When a drawing requires a circle or circular arcs drawn to a scale that is not reproducible by any available template, a compass must be used. Figure 2.28 shows the two common types: the fixed bow and the rapid-action bow.

In order to maintain line quality when using a compass, the lead must be properly sharpened and positioned correctly in the compass. A sharpening device such as a metal file or sandpaper file may be used to create the beveled side of the lead, as shown in Fig. 2.29. In addition, the lead used in the compass should be identical in hardness to the lead used for other linework on the drawing. Expert compass work will leave no breaks in the linework. Figure 2.30 shows some acceptable and unacceptable work with a compass.

2.3.7 Irregular Curves

Frequently, a drawing requires curved lines which are not of circular, elliptic, or other common geometric shape. An irregular curve may be used to construct the required curve. Figure 2.31 shows two types of irregular

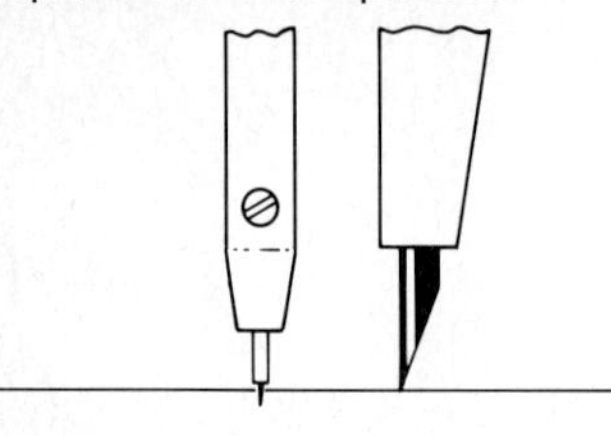

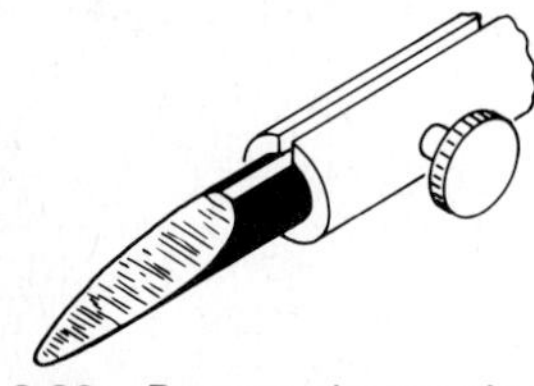

FIG. 2.29 Proper sharpening procedure for compass lead.

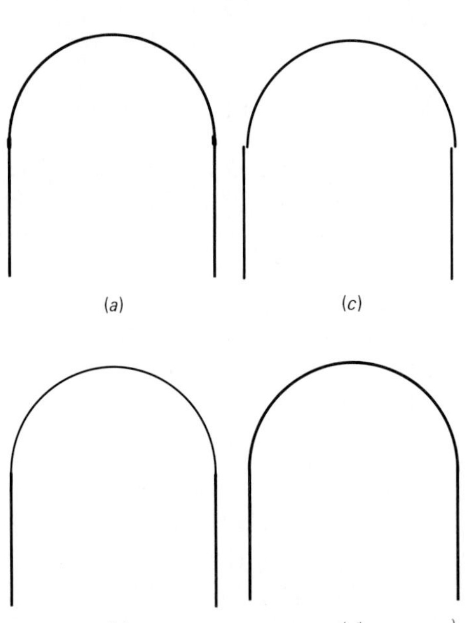

FIG. 2.30 Good compass construction requires practice to avoid unacceptable linework. (*a*) The transition from curve to line is not smooth. (*b*) The line thickness is not uniform. (*c*) The offset is not acceptable. (*d*) An acceptable drawing accomplished by drawing the curve first and then matching the transition points with the lines.

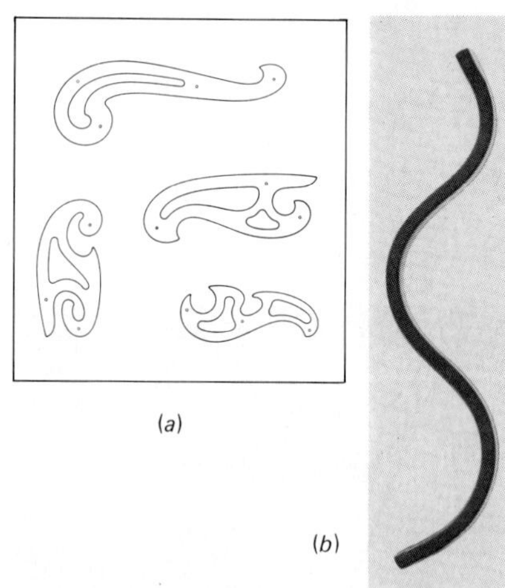

FIG. 2.31 Examples of (*a*) irregular curves and (*b*) a flexible curve.

curves: the fixed curve and the flexible curve. The nature of the curve to be drawn and personal preference dictate the particular type selected.

Perhaps the most frequent application of irregular curves is the graphic presentation of data which requires the connection of a series of points such as those shown in Fig. 2.32. To construct the line properly, begin by sketching a very light construction line approximating the curvature as accurately as possible. Then position the irregular curve, in this case a fixed type, over the first three points. Draw the line connecting points 1 and 2 and then reposition the curve over points 2, 3, and 4, continuing the line from point 2 to point 3. Continue this process through the last point, making sure that the line weight is correct and consistent. If the curvature is not sharp, three or more points may be connected with one setting of the irregular curve. The flexible curve is used similarly; it provides an advantage in that several points may be connected with one setting.

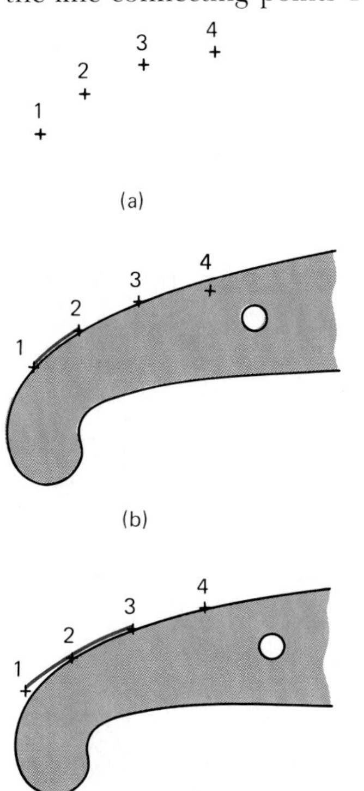

FIG. 2.32 One construction method that can be used to form a smooth curve between points.

2.3.8 Protractor

A typical protractor used for measuring angles is shown in Fig. 2.33. You have most likely used this instrument in a geometry or trigonometry course.

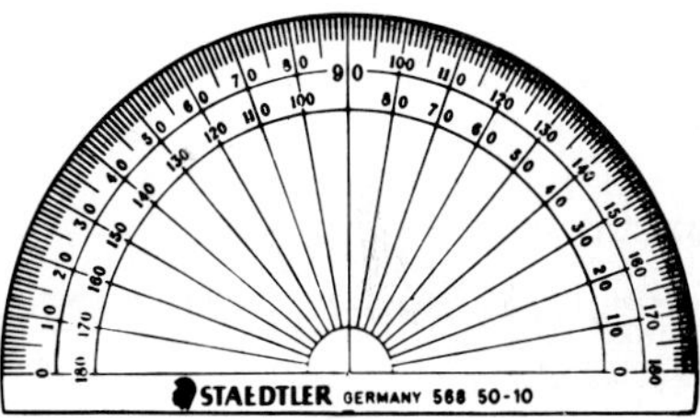

FIG. 2.33 Protractor.

2.3.9 Geometric Construction

Some typical geometric constructions will be described here to illustrate the uses of many of the drafting tools. Additional geometric constuctions will be found in the problems at the end of the chapter.

Example 1.1

In Fig. 2.34, the straight lines AB and CD are given. Connect the lines with a circular arc of radius R tangent to both lines.

Solution

Step 1. Construct lines parallel to AB and CD a distance R from both lines. The intersection of these two lines determines the center O of the circular arc EF. See Fig. 2.34

Step 2. Use a compass to draw arc EF.

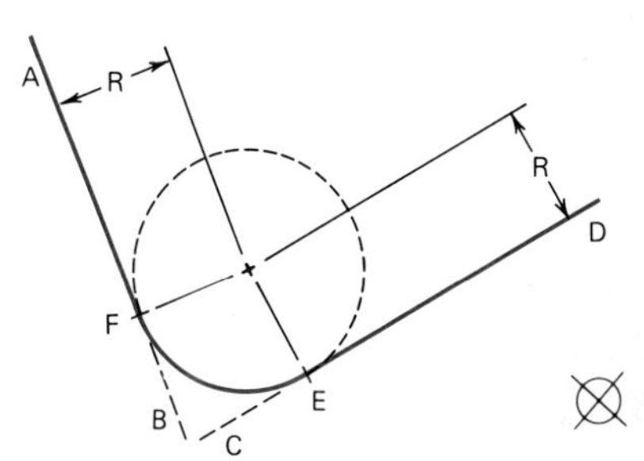

FIG. 2.34 Construction of an arc between two straight lines.

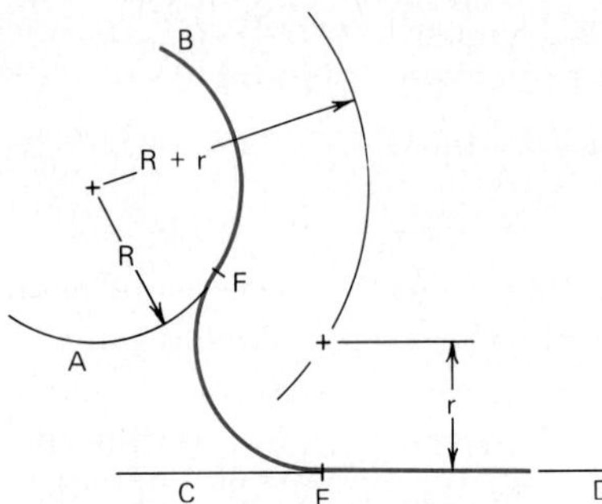

FIG. 2.35 Construction of an arc between a straight line and a circular arc.

Example 1.2
The circular arc *AB* of radius *R* and straight line *CD* are given. Locate the tangent arc *EF* of radius *r*. See Fig. 2.35.

Solution

Step 1. Construct a circular arc of radius *R* + *r* concentric with *AB*.

Step 2. Construct a line parallel to *CD* at a distance *r* from *CD*. The intersection of the lines constructed in steps 1 and 2 is the center *O* of the tangent arc *EF*. See Fig. 2.35.

Step 3. Use a compass to draw arc *EF*.

Example 1.3
The major axis *AB* and minor axis *CD* of an ellipse are given. Construct the ellipse using the approximate four-center method which generates four circular arcs to represent the ellipse. See Fig. 2.36.

Solution

Step 1. Construct line *AC*.

Step 2. With *PC* as radius and *P* as center, locate *E* on *AB*.

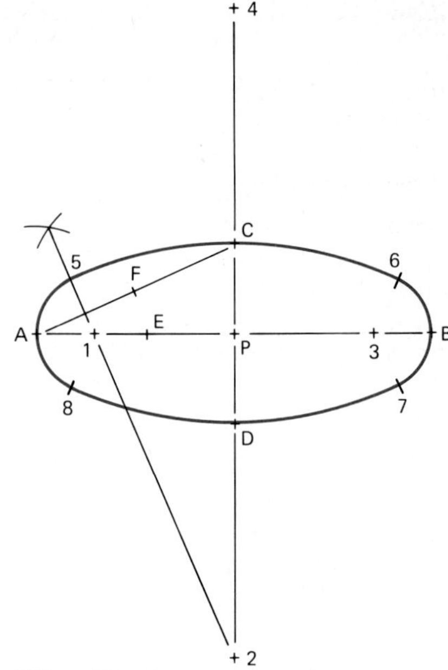

FIG. 2.36 Construction of an approximate ellipse using the four-center method.

Step 3. With *C* as center and *AE* as radius, construct an arc that intersects line *AC* at *F*.

Step 4. Construct the perpendicular bisector of line *AF*. Points 1 and 2, where this bisector intersects the major and minor axes, respectively, are two of the centers.

Step 5. The other two centers 3 and 4 may be located by laying off equal distances *P*-1 and *P*-2 from the center of the ellipse.

Step 6. Locate the terminal points 5, 6, 7, and 8 of each of the circular arcs and draw in each arc, using the appropriate center. See Fig. 2.36.

Example 1.4
The major axis *AB* and minor axis *CD* of an ellipse are given. Construct the ellipse, using the concentric circle method. See Fig. 2.37.

Solution

Step 1. Using a compass with the center at *P*, construct circles of radius *AP* and *CP*.

Step 2. Divide both circles into equal parts (12 would be a minimum).

Step 3. Construct lines such as *P*-2 and *P*-4. Where *P*-2 intersects the inner circle, draw a line parallel to the major axis; where *P*-2 intersects the outer circle, draw a line parallel to the minor axis. The intersection of these lines, points 5, 6, 7, and 8, lies on the ellipse.

Step 4. Construct as many of these points as needed and use an irregular curve to draw the ellipse.

2.4 SCALES

You are familiar with tape measures and rulers, devices which are used to determine the actual length or size of an object. These devices are quite useful when measuring to cut a board or to size a room for carpeting.

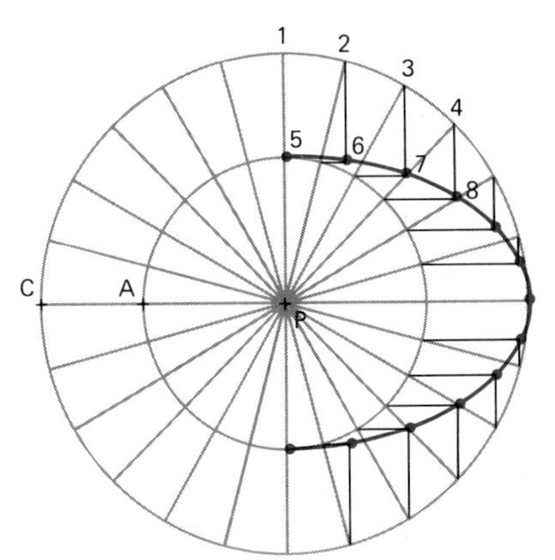

FIG. 2.37 Construction of an ellipse using the concentric circle method.

However, if you wish to represent an object such as a 2 × 12 board 25 ft long on a piece of paper, the common tape measure or ruler is not practical. For example, suppose that you decide to let 3/32 of an inch represent 1 ft of length of the board. A distance of 25 ft would require 25 separate 3/32-in increments to be drawn on the paper. Of course, we could first do a calculation 25 × 3/32 = 75/32 = 2 11/32 in and draw this length on the paper. This procedure would have to be carried out for each line on a drawing.

A scale is an item of drawing equipment that has been carefully graduated (marked) and calibrated (labeled) in convenient increments for the user. Scales enable a user to make size reductions or enlargements rapidly and accurately.

Three scales will be discussed here: metric, engineer's, and architect's. In practice, you may encounter the need for other types of scales, but you will not have difficulty using them if you understand the three scales discussed here.

2.4.1 Metric Scale

The Système International d'Unites (SI) specifies the meter as the base unit of length. Metric scales are convenient to use because adding a prefix to the base unit (meter) has the effect of multiplying the unit by an appropriate power of 10. Common decimal multipliers together with their prefix names and symbols are listed in Fig. 2.38.

MULTIPLIER	PREFIX NAME	SYMBOL
10^3	kilo	k
10^2	hecto	h
10^1	deka	da
10^{-1}	deci	d
10^{-2}	centi	c
10^{-3}	milli	m

FIG. 2.38 Metric prefixes.

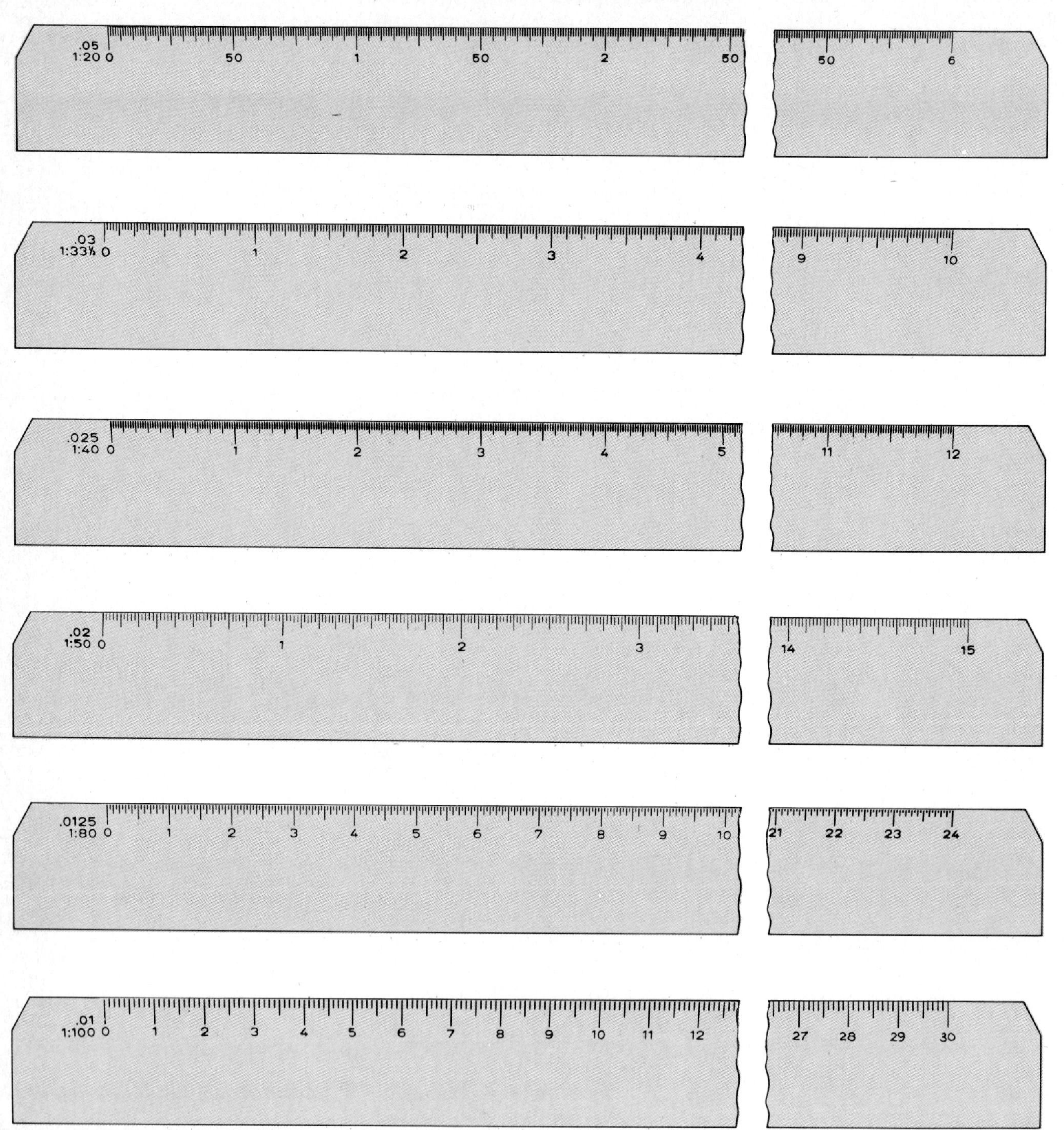

FIG. 2.39 Metric scales.

Metric scales are graduated and calibrated their full length, and each scale is identified by a scale ratio such as 1:100 or 1:50. Six typical metric scales are illustrated in Fig. 2.39.

Since the base unit is the meter, an important characteristic of all metric scales is that the distance from 0 to 1 on the scale represents 1 m. In Fig. 2.40, the ratio 1:100 denotes that the meter has been divided into 100 parts. The distance from 0 to 1 on this scale is actually 0.01 m, but on the 1:100 scale it represents 1 m on the actual object.

On the 1:80 scale, the meter has been divided into 80 parts so that the distance from 0 to 1 is equal to 0.0125 m. When the scale ratio is 1:80, the distance from 0 to 1 measured on a drawing represents 1 m on the actual object.

The flexibility of the metric scale can be demonstrated with another look at the 1:100 scale. Suppose that you wanted a drawing one-tenth the size of the actual object. It would be at a ratio of 1:10. On the 1:100 scale, a scale ratio of 1:10 means that the distance between 0 and 1 on the scale represents 0.1 m. Simply changing the scale ratio by a factor of 10 has produced a new scale. This process allows an unlimited number of scales to be used in which the ratio changes by a power of 10. Table 2.1 illustrates some of the possible scales that can be developed from the 1:100 scale.

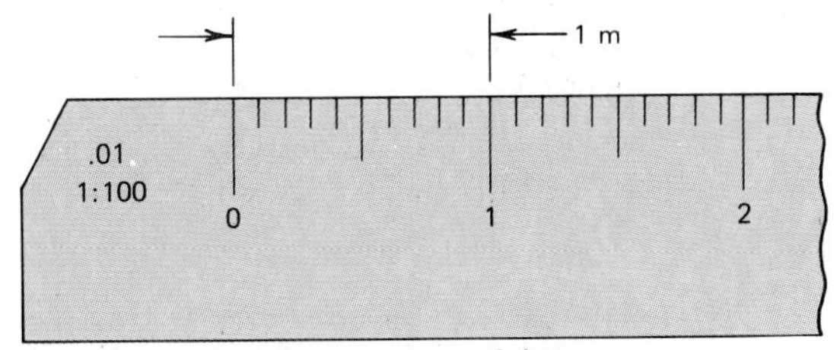

FIG. 2.40 Representation of 1.0 meter.

Many drawings are made full size. On any scale, full size is denoted by a scale ratio of 1:1, indicating that 1 unit on the drawing represents 1 unit on the object. A scale ratio of 1:1 is found by using the 1:100 scale and establishing the distance between 0 and 1 on that scale as 0.01 m.

Verify the readings in Fig. 2.41. Check carefully the prefix given with the units. With SI units, nu-

TABLE 2.1
Scale Ratio Multipliers

1:100 SCALE SCALE RATIO	DISTANCE FROM 0 TO 1 EQUALS			
1:100	1 m =	10 dm =	100 cm =	1000 mm
1:10	0.1 m =	1 dm =	10 cm =	100 mm
1:1	0.01 m =	0.1 dm =	1 cm =	10 mm
1:0.1	0.001 m =	0.01 dm =	0.1 cm =	1 mm

0.9 mm

0 1 2 3 4 5 6 7 8 9 10 11

1:0.01

(*a*)

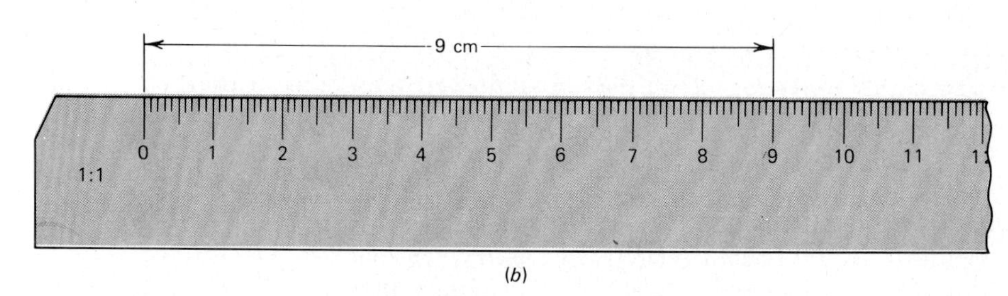

(*b*)

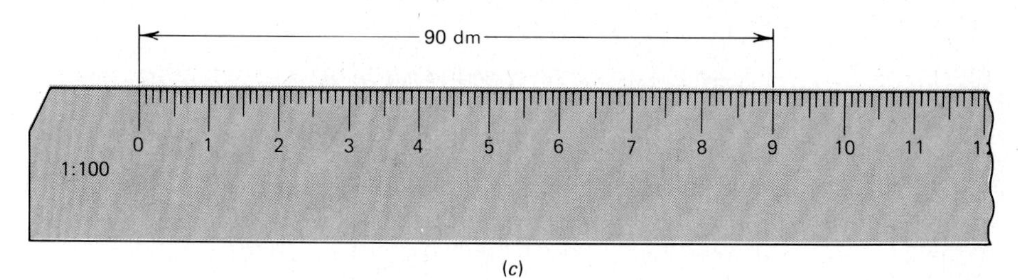

(*c*)

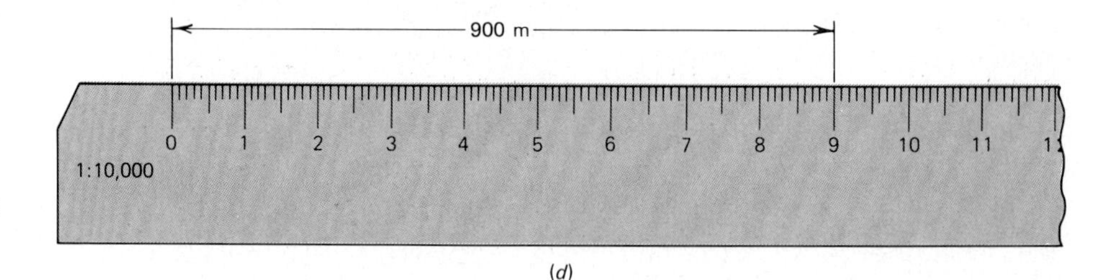

(*d*)

FIG. 2.41 An illustration of the flexibility of the metric scale.

meric values should be written as numbers between 0.1 and 1000, with the decimal point adjusted by the use of an appropriate prefix.

2.4.2 Engineer's Scale

The engineer's scale is based on the inch and, like the metric scale, is fully graduated and calibrated. Six common engineer's scales are shown in Fig. 2.42.

As an example, study the 10 scale in Fig. 2.42. The actual distance from 0 to 1 on this scale is 1 in. Full-scale drawings (scale ratio 1:1) would be constructed with the 10 scale. In this case, a scale equation could be written as 1″ = 1″. The 10 scale can represent a multitude of scales, each one differing by a power of 10. For example, the 10 scale would be used to construct drawings or lines which have scale equations 1″ = 10″, 1″ = 0.01″, 1″ = 1′, 1″ = 100′, and so forth.

The 10 scale in Fig. 2.43 illustrates the versatility of the engineer's scale. For the given measurement, verify the lengths and scale ratios in the table.

The scale ratio is obtained from the scale equation by reducing the scale equation to common units of inches. For example, in Fig. 2.43, the scale equation 1″ = 10′ is equivalent to 1″ = 120″, or a scale ratio of 1:120. This ratio means that the object is 120 times larger than the drawing, a fact that is not immediately obvious when one looks at the scale equation 1″ = 10′.

The scale equation may be calculated if the ratio is known. Suppose the ratio is 1:720. This is equivalent to 1″ = 720″, or if the right-hand side of the equation is written as feet (12″ = 1′), the scale equation becomes 1″ = 60′. The 60 scale now can be used to scale a drawing with numeric values expressed in feet. If the right-hand side of the equa-

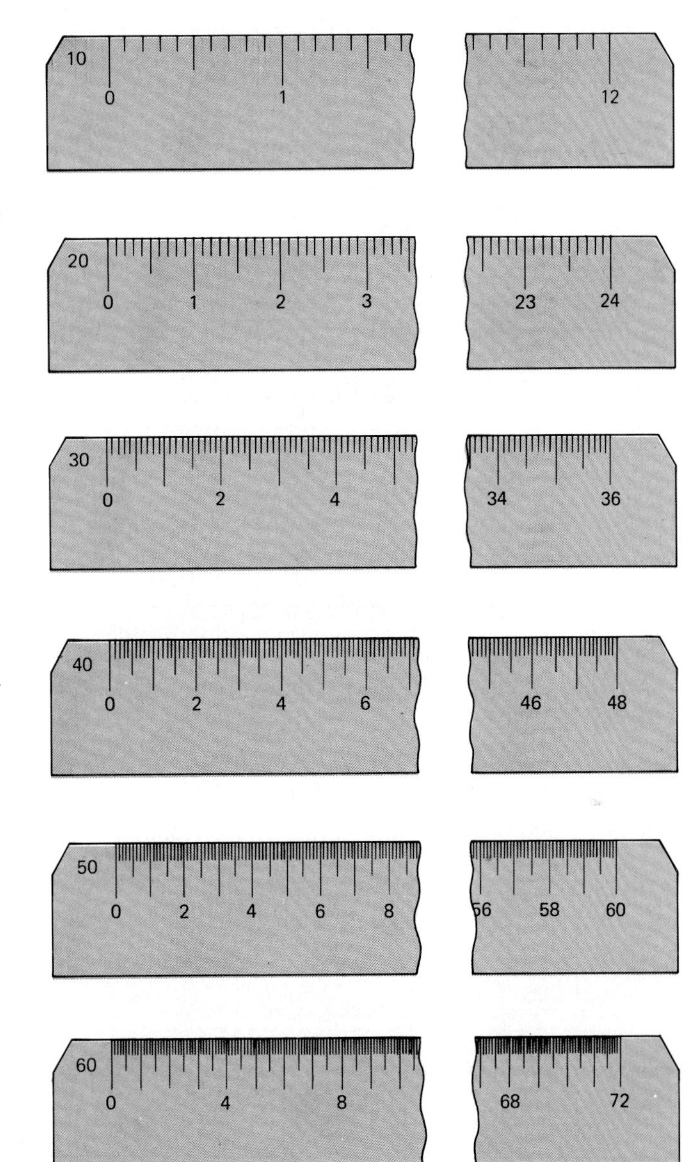

FIG. 2.42 Engineer's scale.

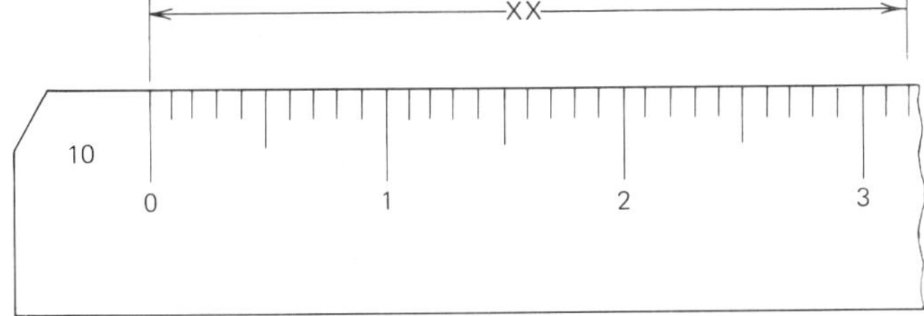

FIG. 2.43 10 Scale

SCALE EQUATION	LENGTH	SCALE RATIO
1″ = 10′	32′	1:120
1″ = 10″	32″	1:10
1″ = 0.1′	0.32′	1:1.2
1″ = 1000′	3200′	1:12000
1″ = 1Yd	3.2Yd	1:36

tion is expressed in yards, the scale equation is written as 1″ = 20 yd, and the 20 scale is used with measurements expressed in yards.

2.4.3 Architect's Scale

A typical architect's scale is triangular-shaped with 11 scales or flat with 7 scales (Fig. 2.44). Each side of an architect's scale contains two scales superimposed in opposite directions, with the exception of the 16 scale.

The 16 scale is the only fully divided scale. It is graduated in sixteenths of an inch, as shown in Fig. 2.45. The distance from 0 to 1 on this scale is equal to 1 in and, as with all other types of scales, may represent various lengths on a drawing. For example, full-scale (12″ = 1′-0), half-scale (6″ = 1′-0), and double-scale (24″ = 1′-0) are possible scale equations for the 16 scale. Ten other arthitect's scales are listed with the corresponding scale ratios in Table 2.2.

TABLE 2.2
Architect's Scales

SCALE EQUATION	SCALE RATIO
3″ = 1′-0	1:4
1½″ = 1′-0	1:8
1″ = 1′-0	1:12
¾″ = 1′-0	1:16
½″ = 1′-0	1:24
⅜″ = 1′-0	1:32
¼″ = 1′-0	1:48
3⁄16″ = 1′-0	1:64
⅛″ = 1′-0	1:96
3⁄32″ = 1′-0	1:128

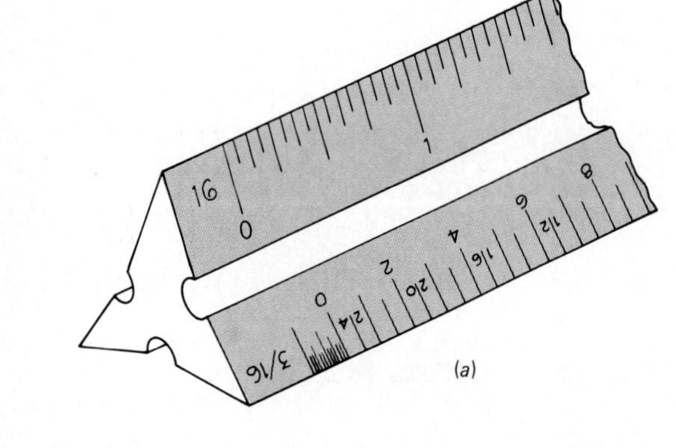

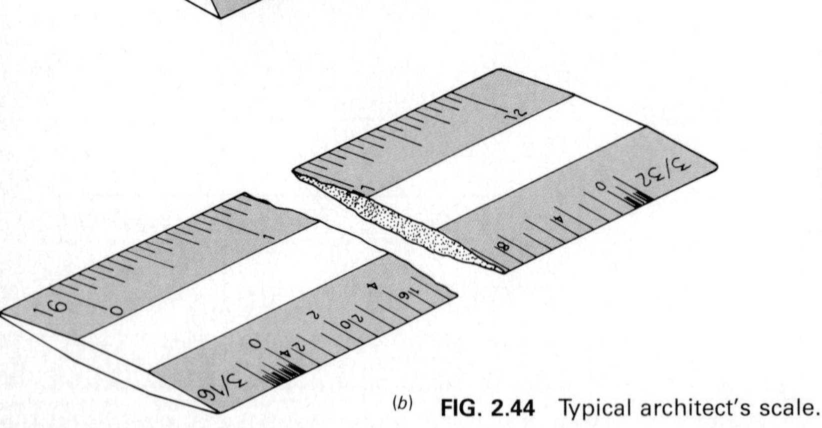

FIG. 2.44 Typical architect's scale.

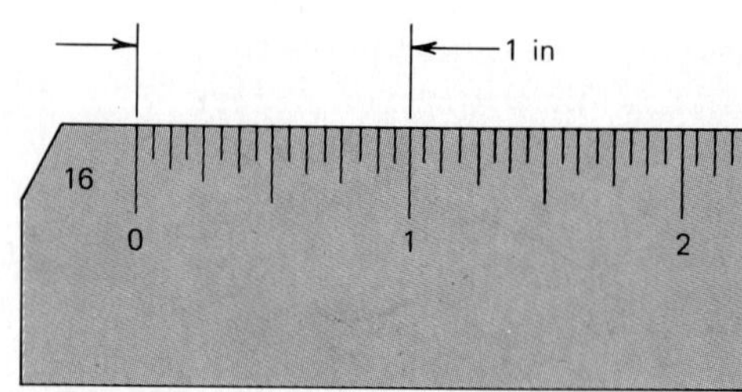

FIG. 2.45 The 16 scale.

Verify the measurement and scale ratios for the lines given in Fig. 2.46.

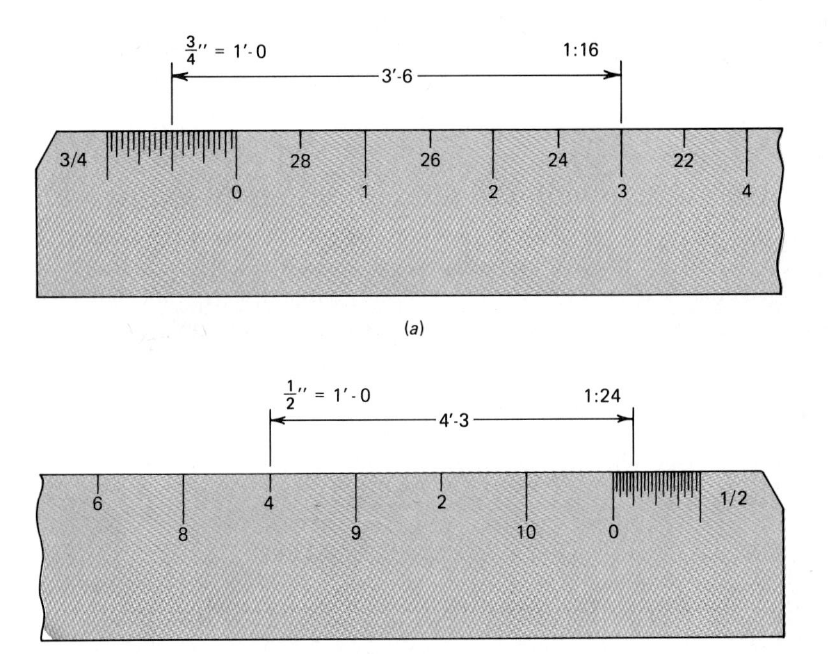

FIG. 2.46 Typical measurements on an architect's scale.

2.5 SKETCHING

Sketching, or freehand drawing, is a quick, convenient method of graphic communication. The form and function of a new idea most likely will start with sketches created by one or more engineers. Creating clear, appropriate sketches is a skill that must be developed through effort and practice.

Figure 2.47 shows common examples of the types of freehand drawings an engineer uses to communicate information. Figure 2.47*a* is a schematic, or block, diagram of a refrigeration process. Blocks are employed to represent the actual devices used in the process. In this case it is more important for the user of the sketch to understand the refrigeration process than to see what each device in the process actually looks like.

The sales forecast chart shown in Fig. 2.47*b* would play an important role in corporate decision making. Overall trends rather than exact times and dollar amounts represent the desired effect of this sketch.

Figure 2.47*c* illustrates a design modification to a gib-head key. The sketch emphasizes the form of the object and illustrates the change from the original design. Each of the sketches in Fig. 2.47 can be prepared in minutes; this

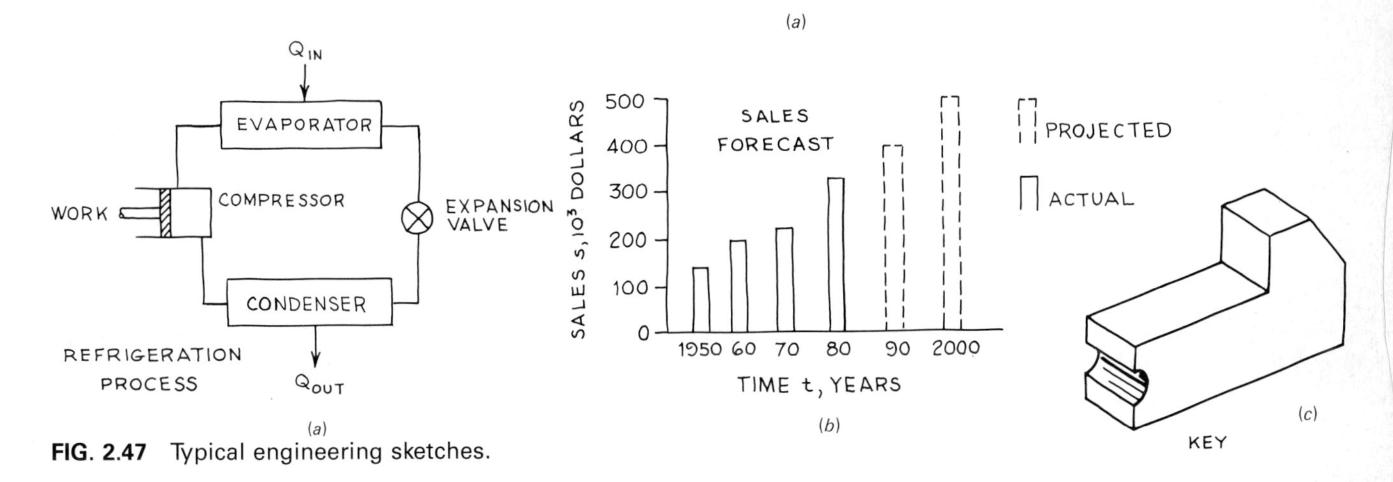

FIG. 2.47 Typical engineering sketches.

represents a powerful means of communicating new ideas and existing conditions.

Sketches may range from those which give quick impressions, as in Figs. 2.47 and 2.48, to those which carefully describe complex objects, as shown in Fig. 2.49. The accuracy and time involved in creating a sketch depend on the ultimate purpose. The engineer will make the decision about the intent of the sketch and the time to be spent in creating the sketch.

2.5.1 Equipment

The equipment for sketching is the same as that used for freehand lettering, as described in Sec. 2.1.1: pencil, paper, and soft eraser. The eventual use of the sketch will dictate the type of paper and the lead hardness to use.

A finished sketch should be clear with bold lines and void of unnecessary lines and smears. A soft eraser should be used to remove construction lines and incorrect markings.

2.5.2 Types of Sketches

We will illustrate three categories of freehand sketches:

1. Sketches which communicate technical data such as charts, graphs, maps, and diagrams. See Fig. 2.50.

FIG. 2.48 Sketch of a carpenter's tape.

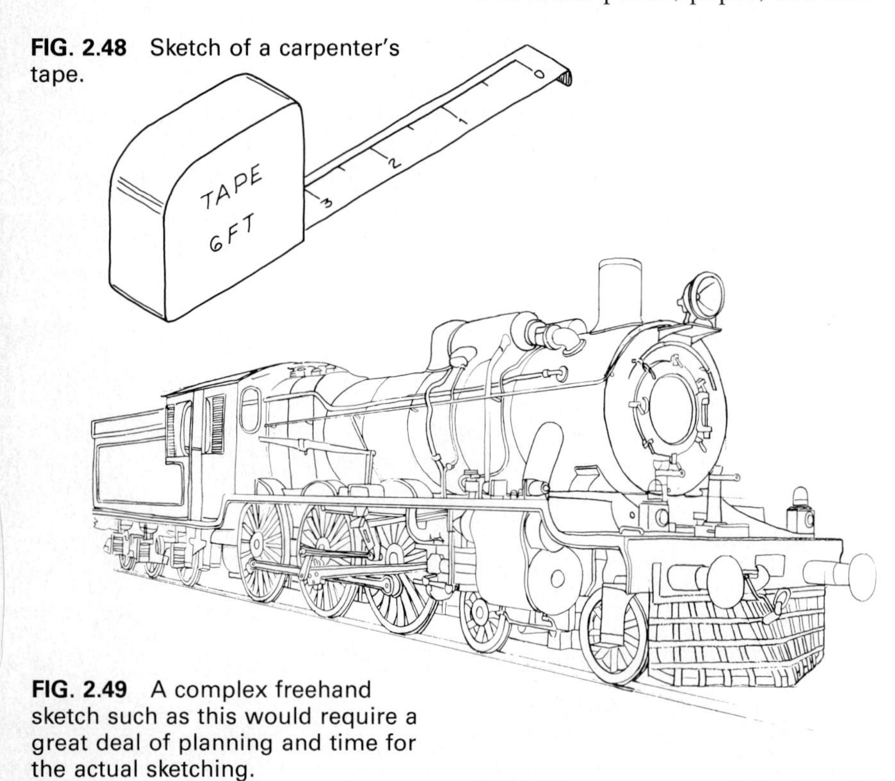

FIG. 2.49 A complex freehand sketch such as this would require a great deal of planning and time for the actual sketching.

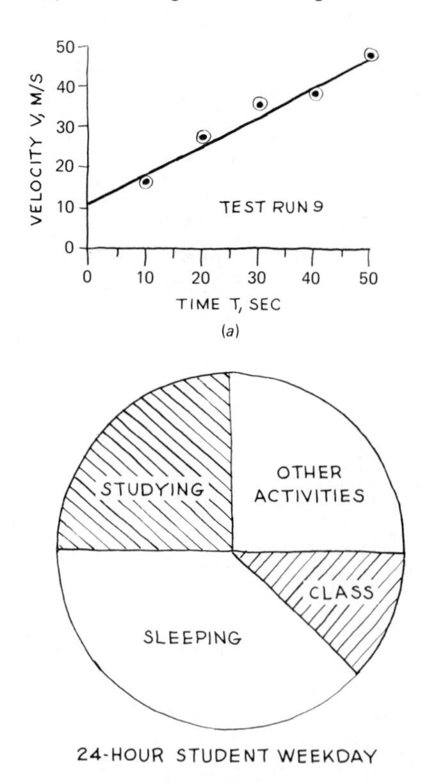

FIG. 2.50 Communication of information using freehand charts and graphs.

2. Sketches which illustrate two dimensions of an object, as shown in Fig. 2.51. Figure 2.51*a* is called a single view and is immediately recognizable as a pair of dividers. In Fig. 2.51*b*, three particular single views of the airplane are sketched, each showing two dimensions. The relative placement of the three single views is important for identification purposes and will be described in Chap. 3.

3. Sketches which are two-dimensional representations of three-dimensional objects, as illustrated in Fig. 2.52. These pictorial sketches represent closely what the observer's eye would see when looking at an object from a point of view that reveals the three general dimensions of length, width, and depth. We will introduce oblique and isometric pictorials in this chapter; a more thorough development of pictorial drawing is presented in Chap. 10.

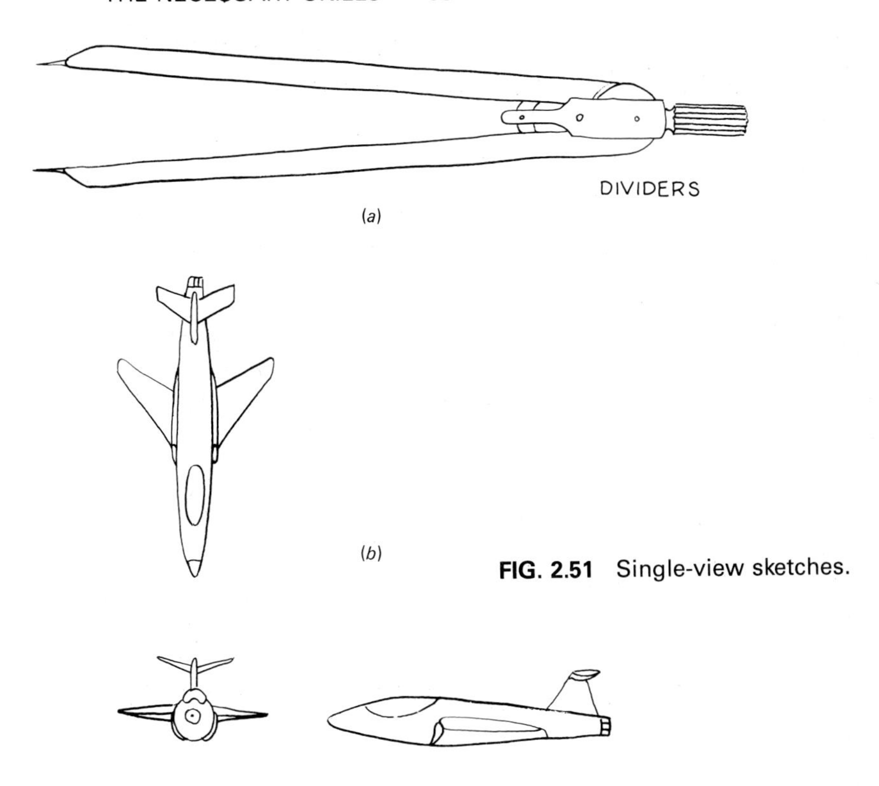

FIG. 2.51 Single-view sketches.

FIG. 2.52 Pictorial sketching. (*a*) The oblique form. (*b*) The isometric form. (*c*) The perspective form, which is used very little by engineers.

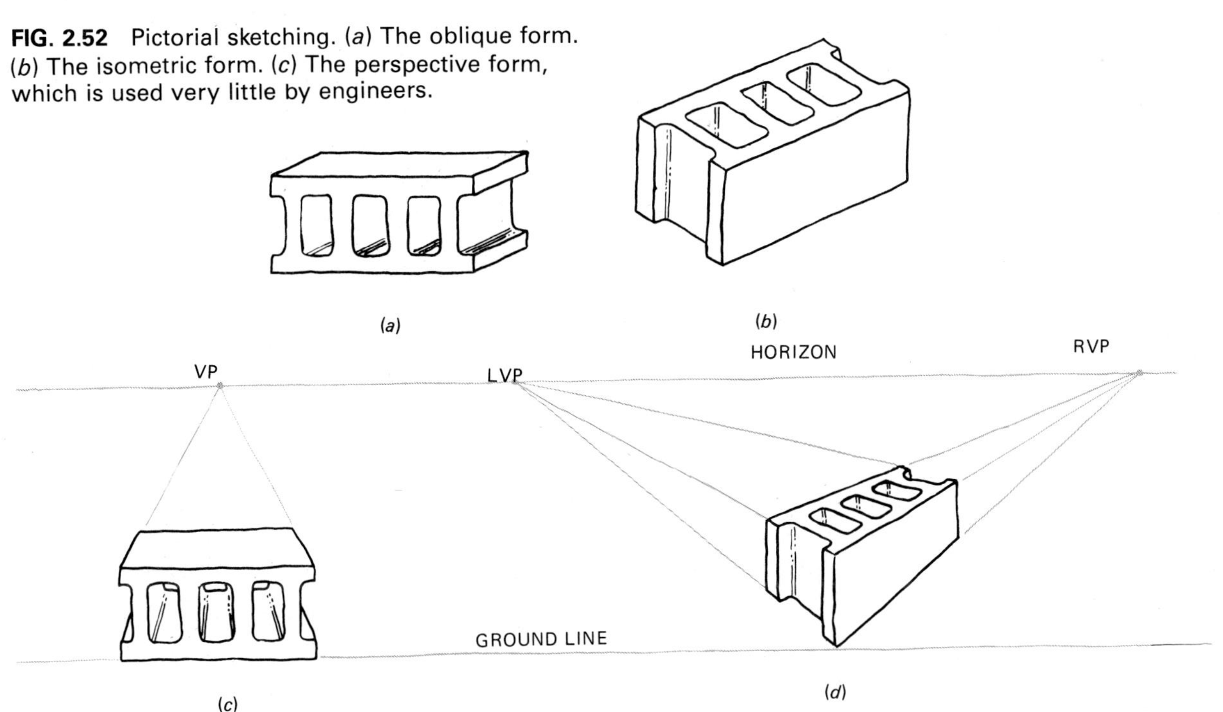

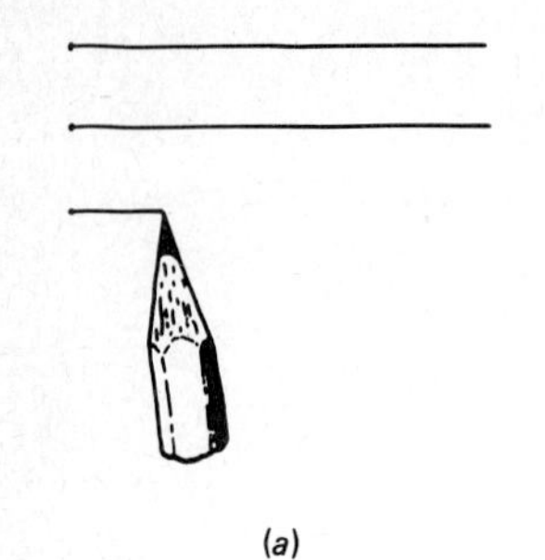

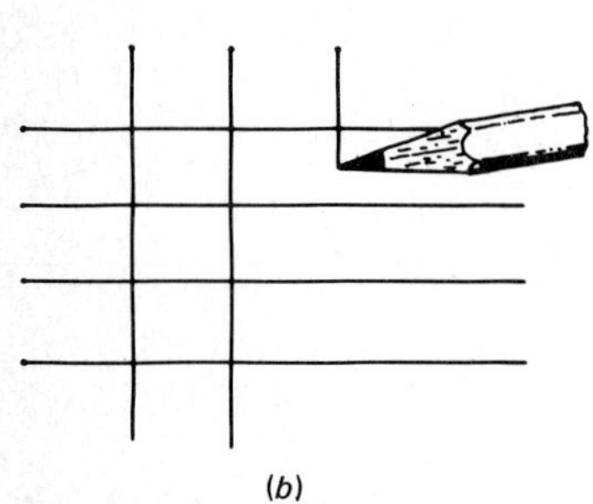

FIG. 2.53 Freehand construction of straight lines.

2.5.3 Steps in Freehand Sketching

We recommend that a specific procedure by followed closely in constructing a quality freehand sketch. These four easily remembered steps apply to all types of freehand drawings.

1. *Plan.* Before putting pencil on paper, visualize the desired sketch in your mind. This would include size of the sketch, intended use of the sketch, orientation of the object being sketched, and amount of details of the object to be included in the sketch.

2. *Outline.* Using very light construction lines, establish the overall proportions, orientation, and major features of the sketch.

3. *Develop.* Check the proportions, make any alterations, and add details.

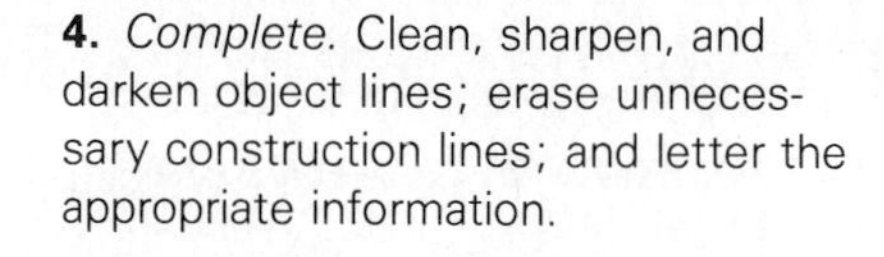

4. *Complete.* Clean, sharpen, and darken object lines; erase unnecessary construction lines; and letter the appropriate information.

2.5.4 Construction Procedure

Good freehand sketches are generated from basic construction techniques and require the ability to put together simple geometric shapes. The first construction technique is the drawing of a straight line between two points. The pencil should be placed on the starting point, and the eyes should be fixed on the terminal point. Then a smooth continuous stroke is made to draw the line between the points.

Once a straight line is mastered, you can practice construction of parallel and perpendicular lines, as illustrated in Fig. 2.53. As you improve in drawing straight lines, practice drawing longer lines and lines that form approximately 30°, 45°, and 60° angles with other lines.

A second technique to master is the construction of a circle, circular arc, or ellipse. Figure 2.54 gives a step-by-step procedure for use in the construction of a circle. The construction procedure for an ellipse is the same except that instead of a radius, the major and minor axes are established.

2.5.5 Pictorial Sketches

The freehand pictorial attempts to show the object the way the eye would view it. A pictorial will show on a two-dimensional surface the three dimensions of the object: length, width, and

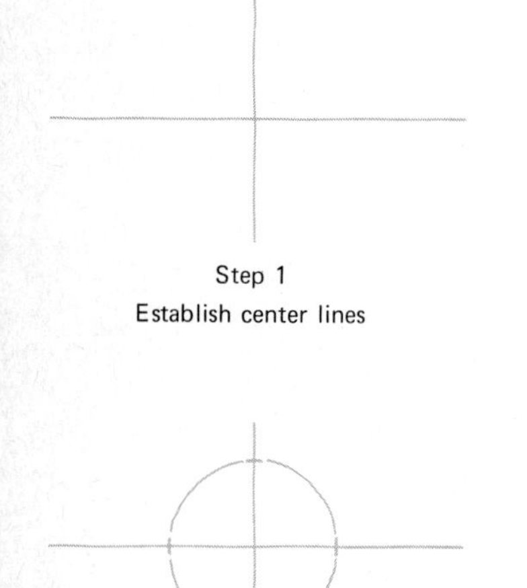

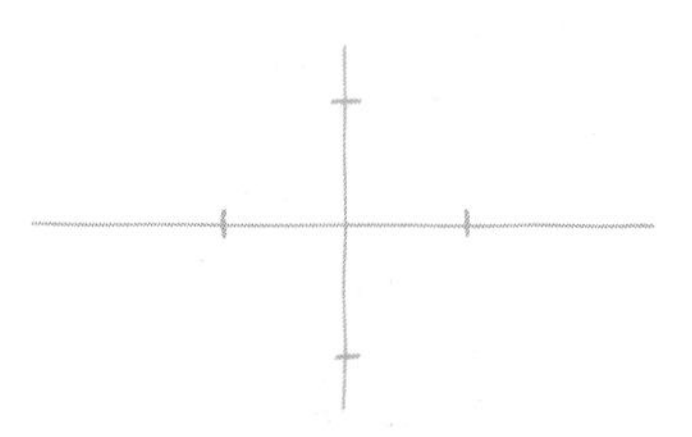

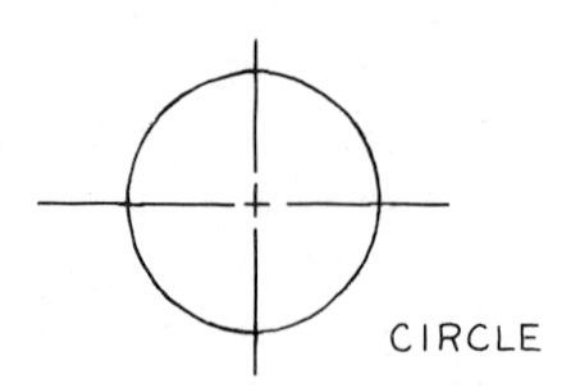

FIG. 2.54 Steps of construction for freehand circle or ellipse.

depth. This representation of three dimensions makes pictorials the most difficult of the freehand drawings.

Most solid objects are constructed as combinations of basic geometric shapes. Figure 2.55 illustrates some of these shapes. Skill in sketching the shapes provides the foundation for drawing more complex objects.

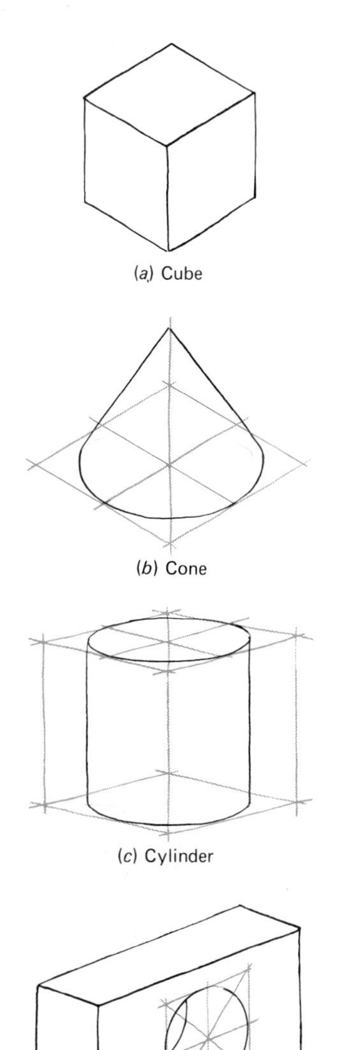

FIG. 2.55 Four simple geometric shapes that are components in most complex objects.

In planning a pictorial sketch, the first decision to be made is choice of viewing direction—the direction in space from which you will "see" the object. This direction is only one of an infinite number of choices. The selection is made to provide the best possible view of the solid object. As an example, take this textbook, close it, and vary its orientation as you look at it. Note the view which shows three dimensions and provides important details of the exterior of the book. This view is the one that you would try to depict in a pictorial.

Two viewing concepts are used frequently in engineering to construct pictorials: oblique and isometric. Both will be shown here for sketching purposes; they are described in more detail in Chap. 10.

Figure 2.56 shows an oblique freehand pictorial. The object is viewed with the front parallel to the paper and the third dimension receding along a variable axis. As a consequence, details in the front face will be in true shape; for example, a circle will be a circle on the sketch. There will be distortion in the surfaces of the object in the direction of the variable axis. Objects which have most of the detail in one face or in parallel faces are quite easy to sketch in oblique.

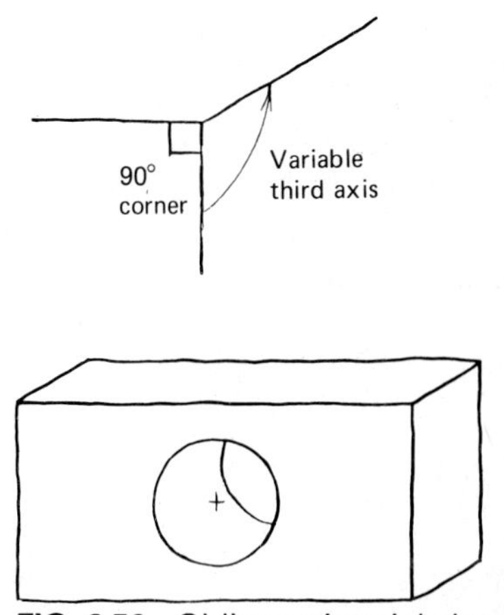

FIG. 2.56 Oblique pictorial sketch.

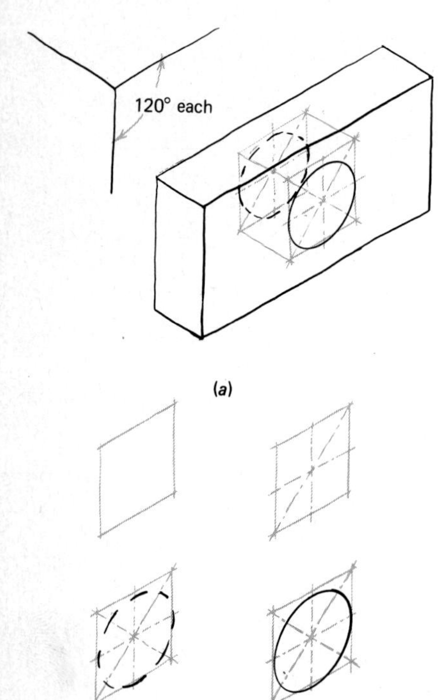

FIG. 2.57 (*a*) Carefully note the construction used to place a circular hole perpendicular to one of the principal axes in isometric. The elliptical construction is on the back face to see if any portion of the hole can be seen. (*b*) 1. Draw the equilateral parallelogram. 2. Sketch the centerlines and long diagonal. 3. Locate the four points of tangency *and* mark the three-quarters point of each half of the long diagonal. 4. Draw the ellipse through all six points.

Figure 2.57*a* is an isometric pictorial. The axes are 120° apart, which means that the three dimensions are seen equally and that details in the three directions are distorted equally when the three-dimensional object is drawn on two-dimensional paper. It is easy to see in Fig. 2.57*a* that the 90° angles on the object are not 90° in the pictorial. Thus, a circle becomes an ellipse when it is drawn in isometric. The step-by-step procedure for freehand construction of an ellipse is demonstrated in Fig. 2.57*b*.

The freehand drawing of a complex object is shown in Fig. 2.58; you should follow the steps as well as the techniques illustrated there. Note how this object is made up of the basic geometric shapes illustrated in Fig. 2.55. A portion of the object has been "cut away" to show some internal detail. This technique and other creative drawing techniques are learned with experience in the use of the graphic language.

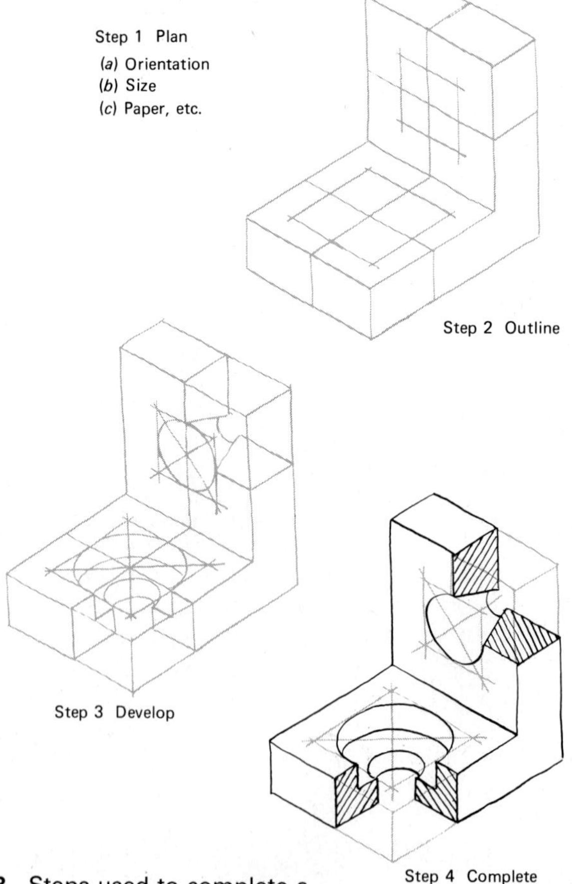

FIG. 2.58 Steps used to complete a sketch.

2.6 METHODS FROM COMPUTER GRAPHICS

The skills discussed in this chapter will enable you to write the language of graphics. Writing the graphic language also can be done with a computer graphics system. For example, Fig. 2.59 represents, with a freehand sketch, an idea for a low-cost sports car. Based on the sketch, permission is given to perform a preliminary design study. A team of design engineers will define the initial geometry of the car components and input these to a computer graphics system. Any component or combination of components can be viewed on the graphics screen or plotted as shown in Fig. 2.60. This definition of geometry is a formidable task and will be discussed in Part Four of this book.

Once the geometry, called a database, is in the computer, design, structural, and power engineers access the data to refine the design. Designers and engineers may analyze, modify, and distribute the design information as the final optimum design is achieved. In a sense, the screen of an interactive computer graphics system serves as a drawing board for designers and engineers. The greatest advantage of the computer graphics system over the traditional drawing board process is speed. It is possible to create and evaluate more potential solutions to a design problem in a much shorter period of time.

Computer graphics does not

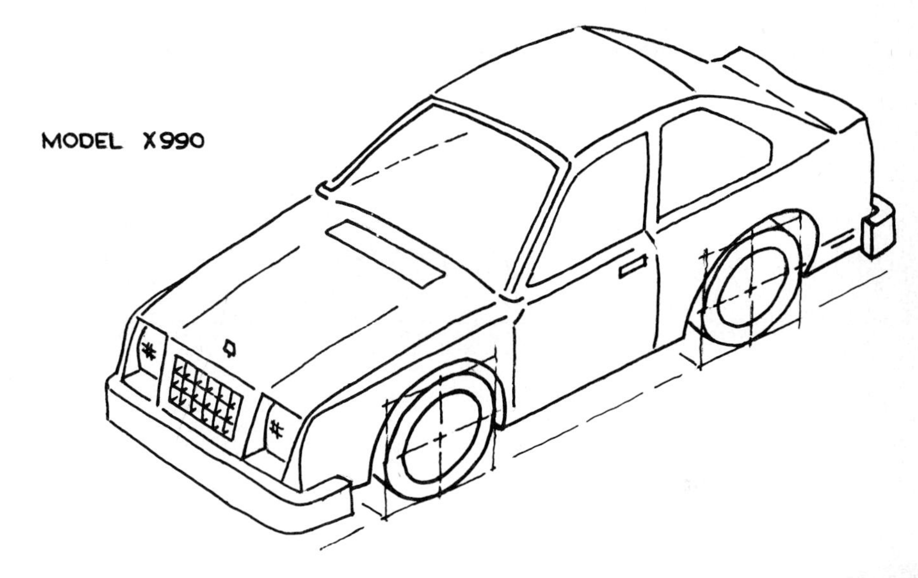

FIG. 2.59 This sketch could be the first step in a model change for a composite for a small car. It would serve as a discussion focus for the design engineering team.

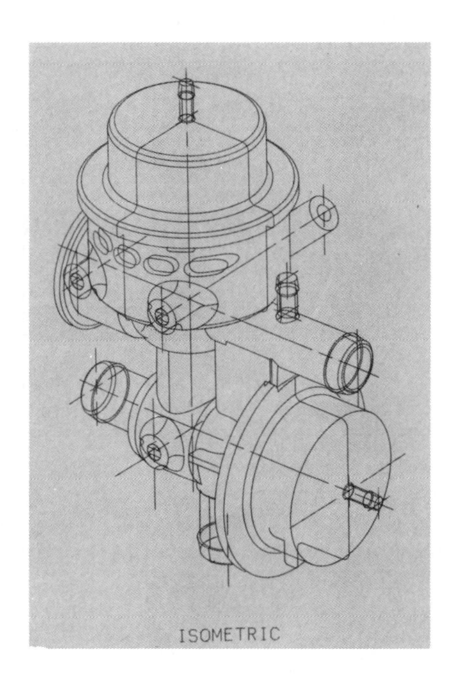

FIG. 2.60 This is a pictorial representation of an antibackfire assembly, one component of a new car model. (*Courtesy of the Ford Motor Company.*)

TEXT COMES IN
MANY SIzES SLOPES
forms
PATTERNS PATTERNS

FIG. 2.61 Computer-generated text.

decrease the need for drafters, designers, or engineers to understand the fundamentals of engineering graphics. It is another tool which gives the technical world a powerful and versatile means with which to communicate information.

Most computer graphics systems have the ability to process text material. Figure 2.61 gives examples of what can be generated with text material. Word

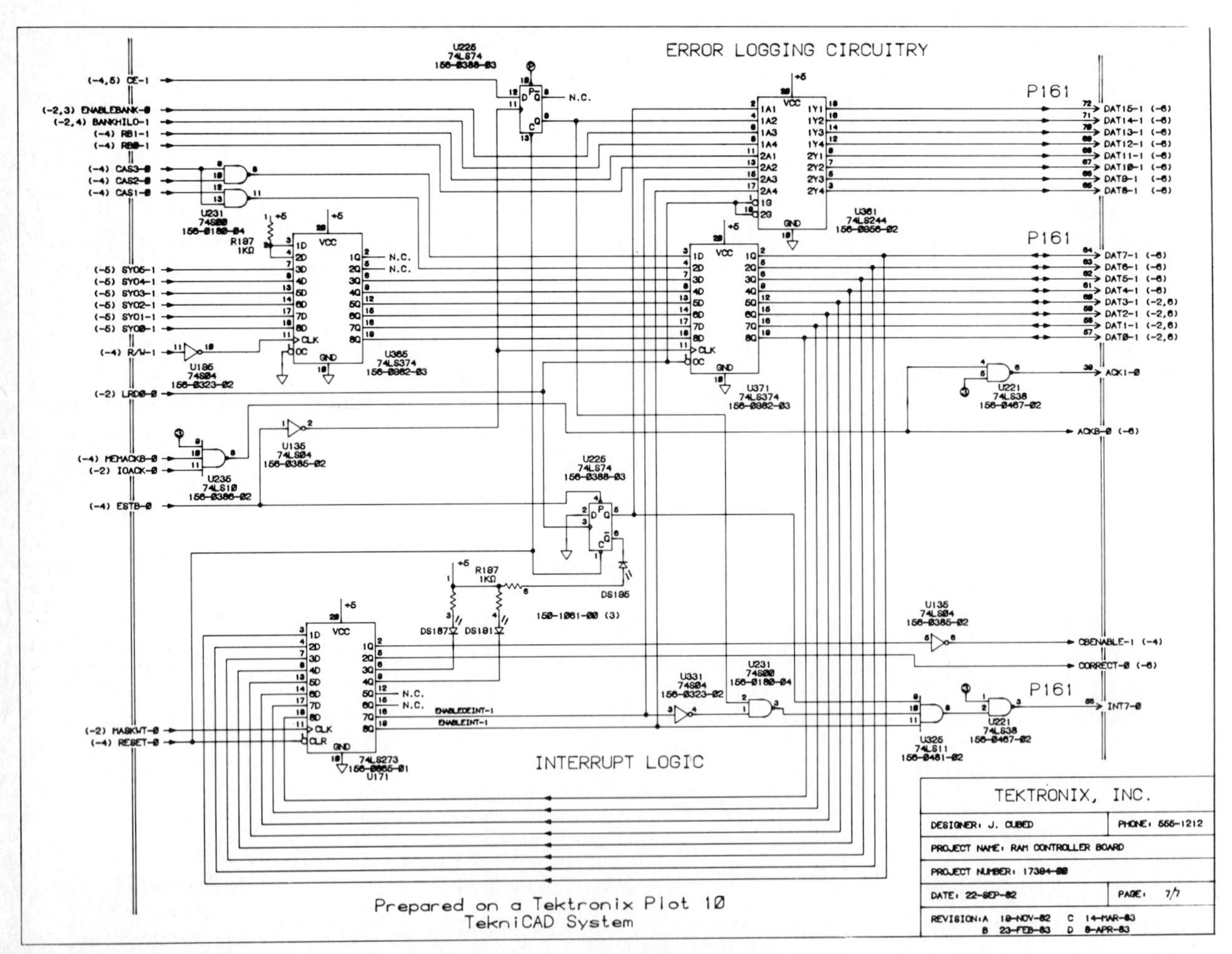

FIG. 2.62 A portion of the circuitry for a computer-controlled board is shown as plotted from a computer database. (*Courtesy of Tektronix, Inc.*)

processors are now commonplace in business. The processors edit and control the printing of letters, manuscripts, product user's manuals, and many other forms of technical information.

Figure 2.62 shows a drawing produced with the aid of a computer. This drawing was plotted from a computer database in less than 5 minutes. All linework and lettering are consistent and can be reproduced exactly. Computer-aided drafting reduces considerably the tedious task of manual drafting and allows the drafter to concentrate on what to do on a drawing rather than how to do the drawing.

Problems

2.1 Construct guidelines approximately 5 mm apart. Using single-stroke Gothic letters and numbers, letter the following:
- (*a*) Alphabet with vertical caps
- (*b*) Alphabet using lowercase letters
- (*c*) Numerals 0 through 9 and different fractions, e.g., $6\frac{1}{2}$

2.2 List five different ways to construct Gothic letters other than freehand.

2.3 Using a set of triangles, lay out a series of parallel and perpendicular lines that are 10 mm apart.

2.4 Construct a vertical line using your triangles. Then draw angles of 15°, 30°, 45°, 90°, 105°, etc., through 360°.

2.5 Construct a line 175 mm long. Then divide the line into 20 equal parts.

2.6 Construct the following geometric shapes:
- (*a*) A rectangle that is 40 × 60 mm
- (*b*) A triangle whose sides measure 35 mm, 50 mm, and 20 mm

2.7 Construct a regular pentagon circumscribing a 60-mm-diameter circle.

2.8 Construct a regular hexagon inscribed in an 80-mm-diameter circle.

2.9 Construct a circular arc of radius 30 mm tangent to lines that intersect at an angle of 75°.

2.10 Given a horizontal line and a circular arc of radius 30 mm with the center 50 mm above line, construct the tangent art.

2.11 Given a major axis of 100 mm and a minor axis of 60 mm, construct an ellipse using the four-center method illustrated in Fig. 2.36.

2.12 Construct the same ellipse given in Prob. 2.11, using the concentric circle method.

2.13 Metric scale exercise. Lay out the values below, using dimension lines that are 10 mm apart, and terminate with an arrowhead at both ends. Start from a vertical line on the left side of the paper.

DISTANCE	SCALE	DISTANCE	SCALE
155 m	1:100	73 cm	1:5
75 mm	1:0.5	1.3 dm	1:1
5.5 dm	1:4	27 cm	1:2
110 m	1:800	100 mm	1:0.8
6 mm	1:0.05	5.8 dm	1:4
12 cm	1:1	0.14 m	1:1

2.14 Engineer's scale exercise. Lay out the following values using dimension lines that are 0.4 in apart and terminate with an arrowhead on each end. Start with a vertical line on the left side of the paper. See page 40.

DISTANCE	SCALE	DISTANCE	SCALE
5.4′	1″ = 1′	2250″	1″ = 400″
175″	1′ = 30″	25″	1″ = 5″
1.15′	1″ = 0.2′	28′	1″ = 5′
27 yds	1″ = 5 yds	0.36″	1″ = 0.06″
3300′	1″ = 600′	185′	1″ = 30′

2.15 Using the four steps of freehand sketching, make a single-view sketch of the following items:

(*a*) Clawhammer (*f*) Automobile
(*b*) Pliers (*g*) Snowmobile
(*c*) Locking plier (*h*) Tennis racket
(*d*) Stapler (*i*) Shoe
(*e*) Tractor (*j*) Wristwatch

2.16 Using the four steps of freehand sketching, make a pictorial sketch of the following objects:

(*a*) Rectangular prism with a drilled hole half through the largest face
(*b*) Coffee cup
(*c*) Portable radio
(*d*) A piece of shop equipment: screwdriver, hammers, pliers, or box wrench
(*e*) Block in Fig. 2.58
(*f*) Automobile that you specify, (e.g., Ford Model A)

chapter 3

Orthographic System Fundamentals

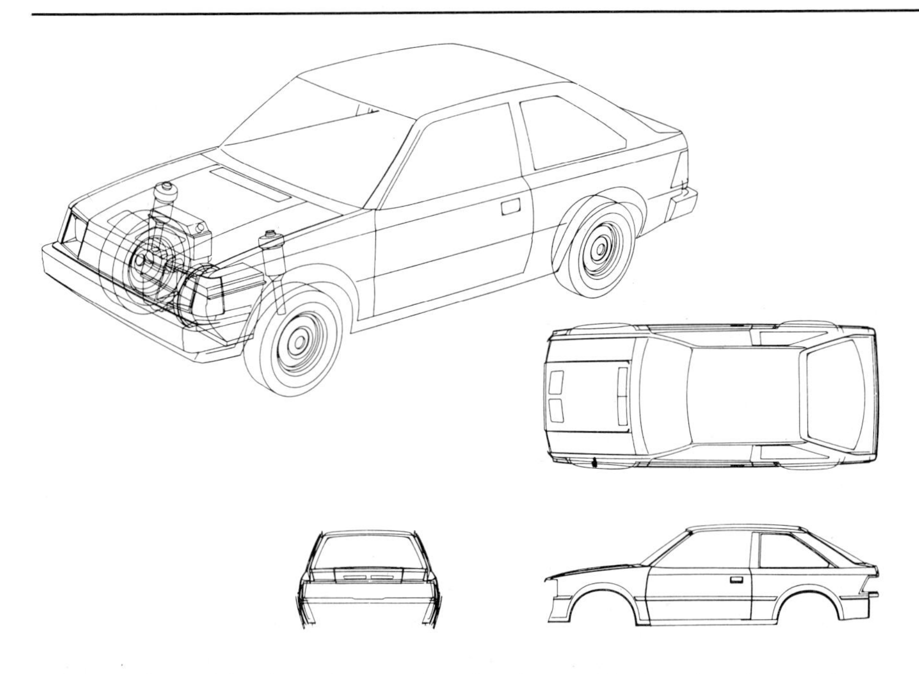

FIG. 3.1 A combination of pictorial and orthographic drawings is used to communicate an automobile body design. (*Courtesy of Ford Motor Company.*)

3.1 INTRODUCTION

The ability to visualize solutions to a problem and communicate these solutions to others is a key element in the problem-solving process. The conceptualization of a problem usually begins with a mental image which must be developed into a tangible form through the use of data (numbers), words, and graphics. Figure 3.1 demonstrates the importance of using graphics to communicate the form and function of the solution to a problem.

The graphic language bridges the gap between a mental picture of a solution and the implementation of the solution in the form of a product, process, or system. Technological advances, particularly in computers, automated

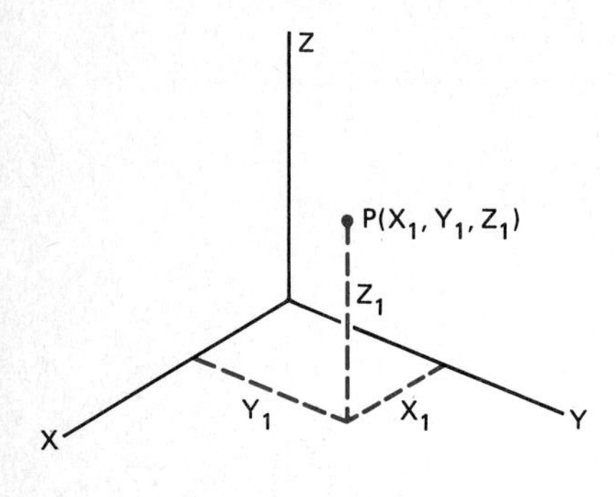

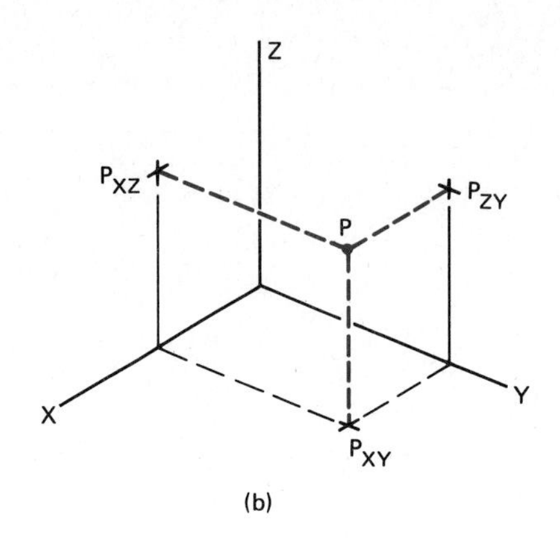

FIG. 3.2 Denoting a point in space.

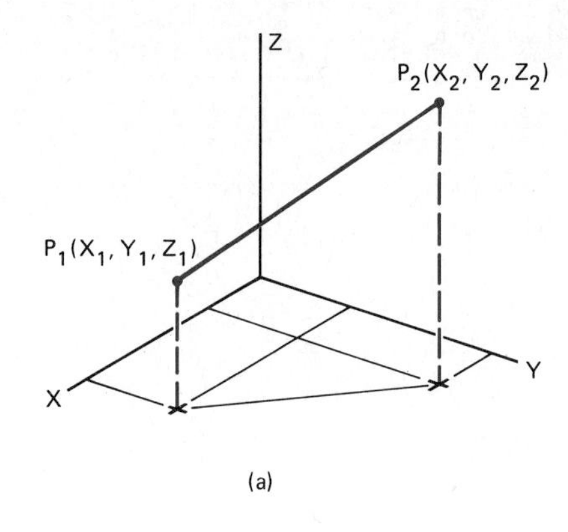

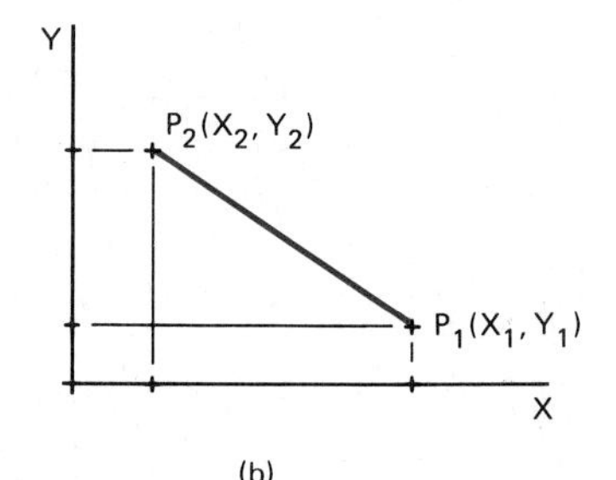

FIG. 3.3 Denoting straight line segments in space.

machines, and mass-production techniques, have accelerated the transformation of ideas into physical reality. In order to understand and use these technological advances in problem solving, you must develop the ability to apply the graphic language effectively.

The most useful area of the graphic language for the engineer is the orthographic system of projection and drawing. This chapter establishes the fundamentals of this system, from which additional theories and applications are developed in the remaining sections of the book. Let us begin with the definitions of the elements of orthographic projection and drawing.

3.2 ELEMENTS OF ORTHOGRAPHIC PROJECTION AND DRAWING

The theory of *orthographic projection,* implemented by a system of *orthographic drawing,* is the basis for producing drawings, as shown in Fig. 3.1. The transformation of an idea into physical reality generally begins with a rough freehand sketch, proceeds to the determination of the geometric relationships of the form and function of the idea, and culminates in a series of drawings which bring the idea into physical reality. The terms orthographic projection and orthographic drawing are often used synonymously, but there is an important distinction. Orthographic drawing involves the practical utilization of the theory of orthographic projection: thus, in order to understand the drawing system, you must understand the theory of projection.

Before a complete definition of orthographic projection can be formulated, certain elements must be explained.

Point An object in space having no dimensions, thus defining a position or location only. A point may also be any position on a line or any position on a two- or three-dimensional object. Figure 3.2*a* represents points in the cartesian coordinate system, and Fig. 3.2*b* depicts the same point as represented in an orthographic projection system. Note that the image of a point is represented in orthographic projection by the plus (+) symbol.

Line The locus (path) of a point moving through space. The path of a line may be straight or curved. A point following a straight-line path moves through space in one direction only. In theory, a straight line is of infinite length and has no thickness. In practice, a straight line is represented by a finite segment between two points and is the shortest distance between these points. Figure 3.3*a* depicts a straight line in a three-dimensional cartesian coordinate system, and Fig. 3.3*b* illustrates a straight line in two dimensions.

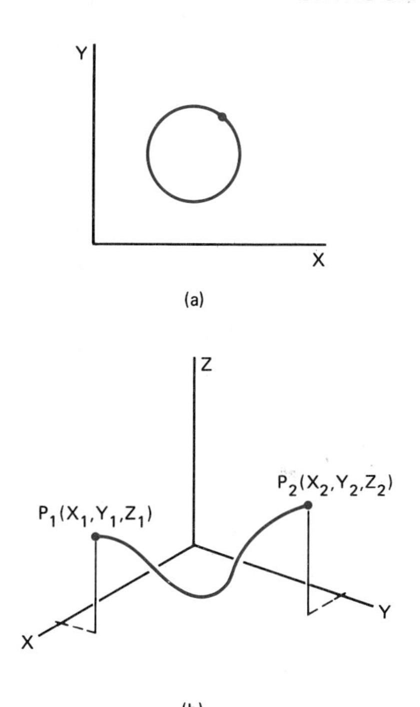

FIG. 3.4 Denoting curved line segments in space.

Curved lines may exist in two-or three-dimensional space. Figure 3.4*a* and *b* illustrate curved lines in two and three dimensions, respectively. A more concise definition of lines and their graphic and mathematical representations is given in Chap. 4.

Plane A two-dimensional area that contains every straight line segment joining any two points on that area. Figure 3.5 shows a plane in space. In theory, the boundaries of a plane are limitless, but for practical application, the dimensions are finite. A plane has no thickness. A more specific definition of planes and their mathematical and graphic representation is provided in Chap. 5.

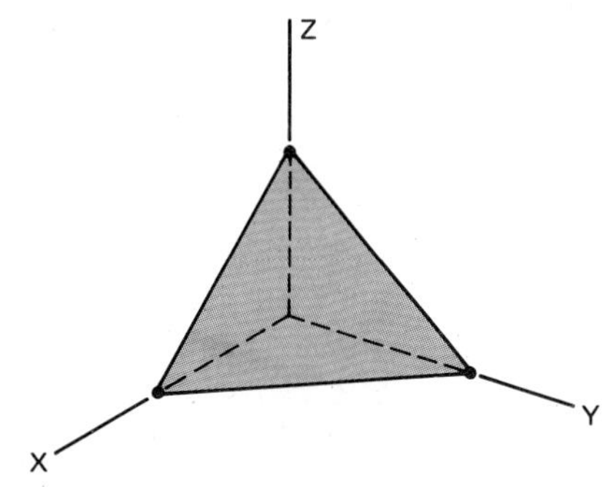

FIG. 3.5 Denoting a plane sector in space.

Object Anything of actual physical substance. An object usually is thought of as a three-dimensional solid or hollow entity having volume and mass, but it may also be a point, line segment, plane, or other surface. The triangular pyramid shown in Fig. 3.6 is an object represented as a combination of points, lines, and planes.

Projectors Rays, or straight lines of sight, emanating from an object into space. The ray emanating from any point on an object in a specified direction will be parallel to the rays emanating in the same direction from all other points on the object. Figure 3.7 illustrates the parallel projectors in one direction from an object.

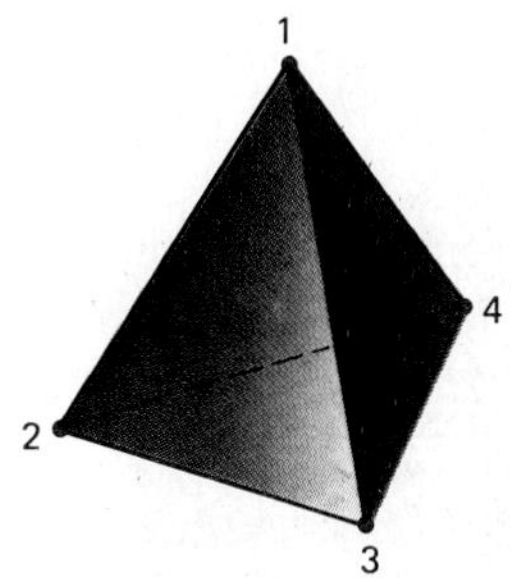

FIG. 3.6 Denoting a solid in space.

Image Plane A plane which is placed perpendicular to projectors emanating in one specified direction from an object. See Fig. 3.7.

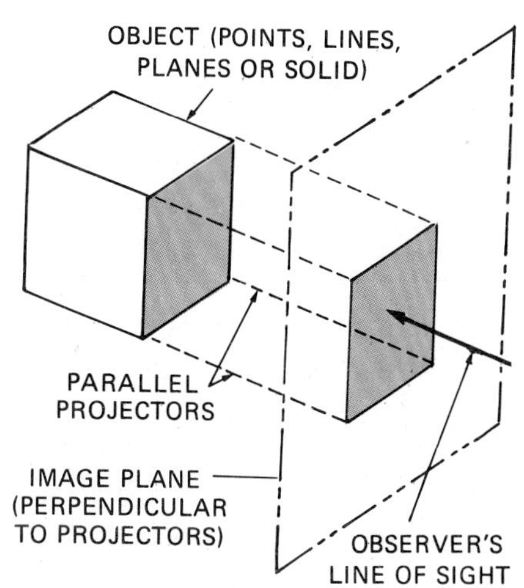

FIG. 3.7 The elements of orthographic projection.

Image The representation of an object formed on the image plane by the intersection of the parallel projectors and the image plane. See Fig. 3.7.

The terms image and view are often used interchangeably. View refers to the orientation or direction of the projectors and the observer's line of sight. This term is usually preceded by a modifier such as a single, multi, front, profile, or elevation.

We are now in a position to define orthographic projection.

Orthographic Projection A procedure by which the image of an object is projected onto an image plane via parallel projectors emanating from the object perpendicular to the image plane. *Ortho* is a Greek term meaning "at right angles."

Study Fig. 3.7, making sure you understand the concept of orthographic projection and the elements which constitute this projection system before continuing to orthographic drawing.

3.3 ORTHOGRAPHIC DRAWING

As stated in the previous section, orthographic drawing is the result of the application of the theory of orthographic projection. It is a practical method of delineating the necessary images of an object in order to illustrate clearly and concisely the form and function of that object. Let us investigate for a moment the pictorial of a three-dimensional object shown in Fig. 3.8. The pictorial is useful for visualization, but it does not completely describe the object. For example, how deep is the hole? Is the inclined face cut through or just partially through the object? As we will see in the next two sections, orthographic views of this object will provide a complete description and eliminate any doubt about its form.

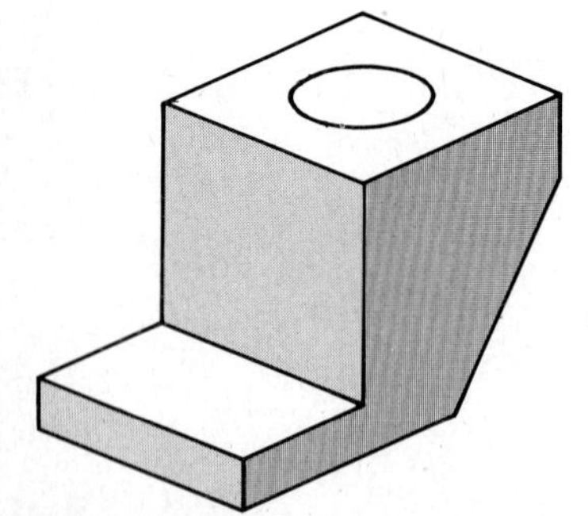

FIG. 3.8 Pictorial representation of a three-dimensional object. Not all of the features can be seen in this presentation.

We classify orthographic drawings as single-view or multiview.

3.3.1 Single-View Drawings

A single-view orthographic drawing illustrates an object from a single direction. The direction is chosen to illustrate two of the three dimensions of height, width, and depth. In Fig. 3.9*a*, a single view of the object from the right side would show the height and depth of the object. Note that the single view (or image) of the object in Fig. 3.9*a* is found on an image plane perpendicular to projectors emanating from the right side of the object. This particular single view is called a profile view (side view). Figure 3.9*b* shows the front view of the object. You must visualize the projectors, the image plane, and your line of sight to understand how this front image is obtained. Also note the features of the object that are more apparent in Fig. 3.9*a* than in Fig. 3.9*b*.

3.3.2 Multiview Drawings

When trying to depict more complex objects, one must construct more than a single orthographic view in order to provide

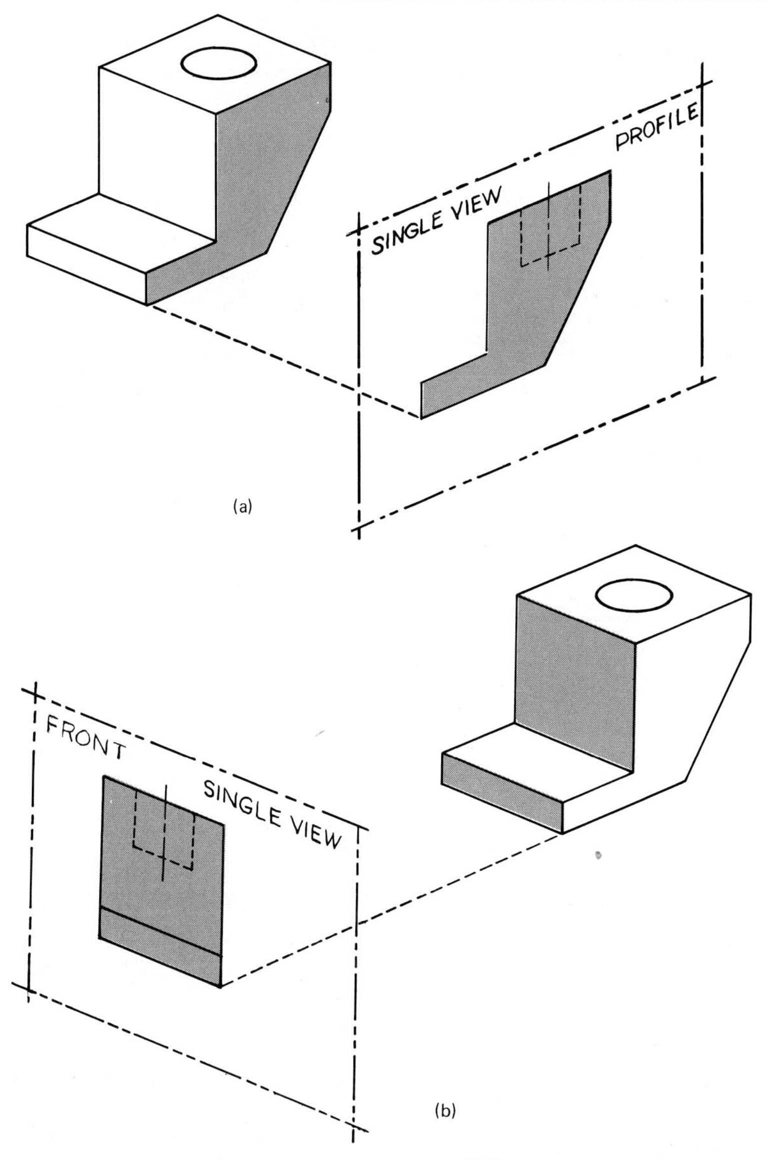

FIG. 3.9 Visualizing a single-view drawing of an object.

a complete description. The system that combines two or more related single views is called a multiview drawing. In a multiview system, the adjacent views are constructed mutually perpendicular and are arranged in a systematic manner so that the object can be observed from any desired position in space.

At this point the concept of a reference line is needed to enable one to construct and visualize multiview drawings. Adjacent orthographic views are identified by appropriately labeled reference

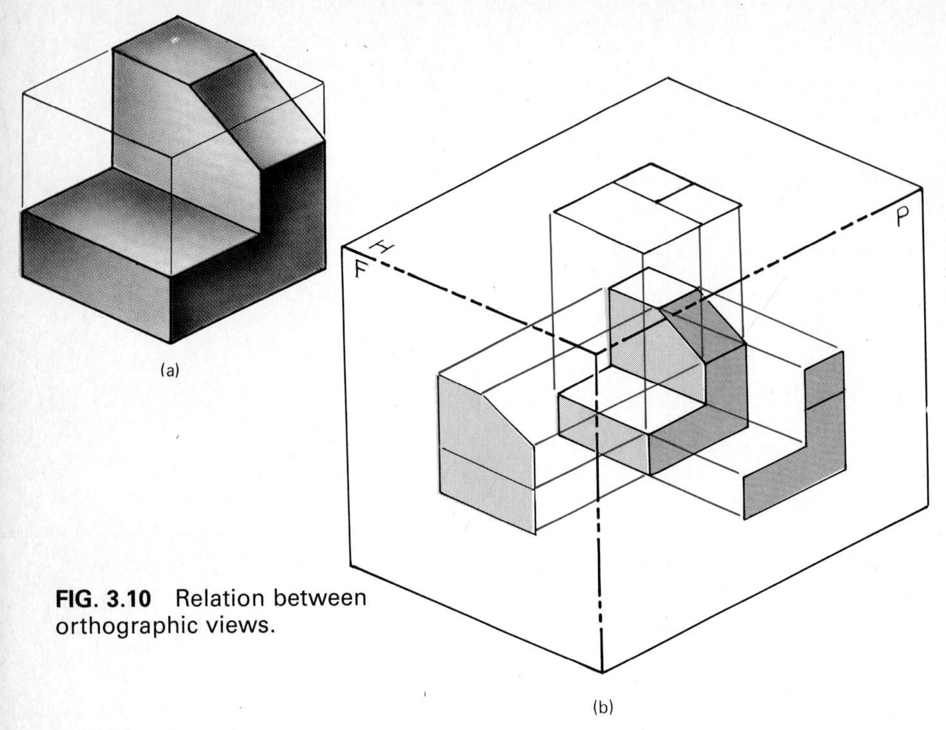

FIG. 3.10 Relation between orthographic views.

lines. Figure 3.10*a* illustrates an object in space. The single views in Fig. 3.10*b* are determined with lines of sight perpendicular to the three image planes H (top), F (front), and P (right side or profile). The image planes H, F, and P are mutually perpendicular, and the lines that identify each of the image planes have significant meaning. For example, the reference line

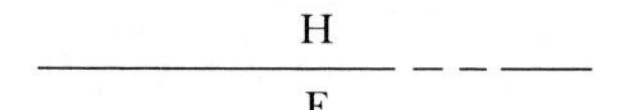

which identifies the top and front image planes, represents

1. The edge of the front image plane when the line of sight is perpendicular to the horizonal image plane
2. The edge of the horizontal image plane when the line of sight is perpendicular to the front image plane
3. The line of intersection of the horizontal and front image planes

Visualize the reference line in Fig. 3.10 which defines the front and profile image planes.

Reference lines are often called hinge lines or fold lines. Multiview drawings not only provide a description of the object depicted, they provide the basis for solving geometry problems involving spatial relationships between objects or parts of objects. The orthographic drawing system requires the correct construction of a multiview drawing before the

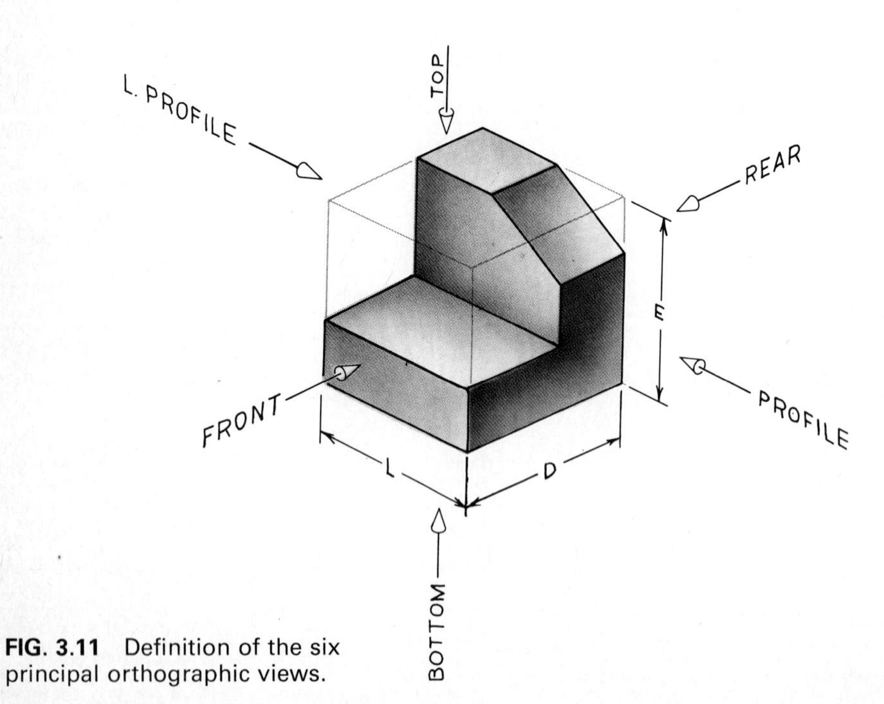

FIG. 3.11 Definition of the six principal orthographic views.

graphic solution to an engineering problem can be performed. This drawing system is based on six principal views and various auxiliary views.

3.3.3 Principal Views

The six principal views illustrated in Fig. 3.11 are front (F), horizontal or top (H), right profile (P), left profile (LP), bottom (B), and rear (R).

When the image planes are transferred from their theoretical positions in space to a plane of paper, they appear as shown in Fig. 3.12. Note again the relative location of each of the views with respect to the reference lines and recall the interpretations of reference lines from Sec. 3.3.2.

The principal image planes may be likened to the six faces of a transparent cube or rectangular prism enclosing the object. One face of the cube is oriented in a horizontal position (the horizontal, or H, view), while another face, adjacent and perpendicular to the horizontal face, becomes the front image plane. The remaining principal planes are visualized similarly. To obtain the layout (multiview drawing) of the six principal views, the transparent cube or prism is "unfolded" to look like Fig. 3.12. To help you visualize the orthographic views in Fig. 3.12, the dimensions L (length), D (depth), and E (elevation or height) are shown. These dimensions are also shown on the pictorial in Fig. 3.11.

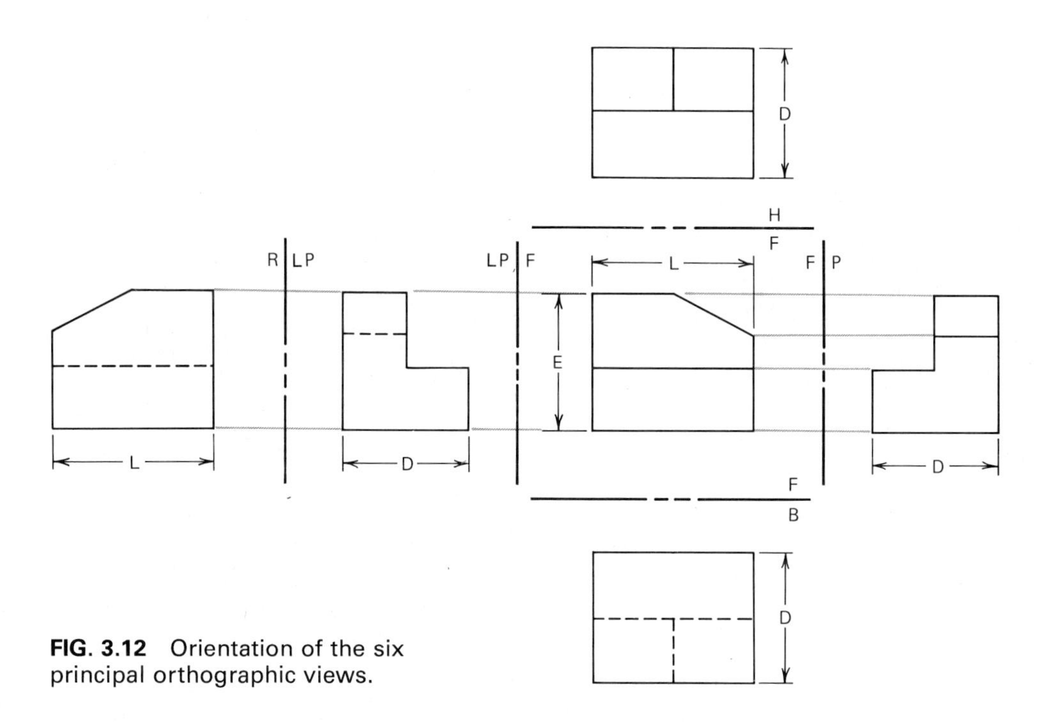

FIG. 3.12 Orientation of the six principal orthographic views.

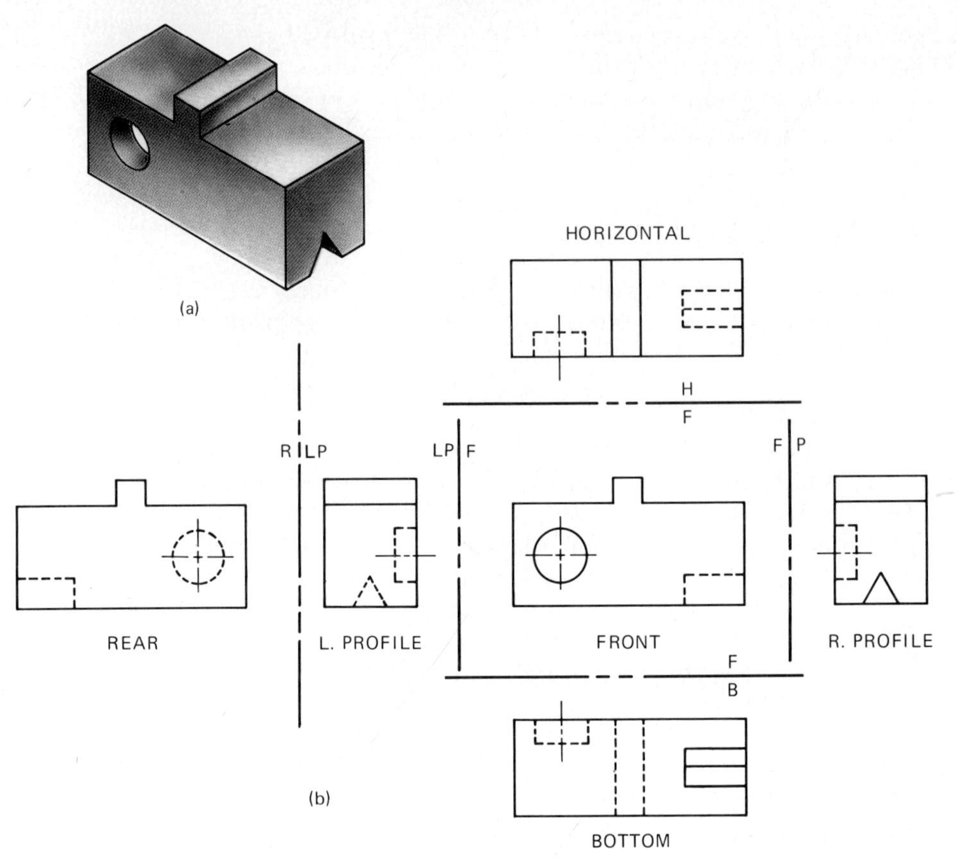

FIG. 3.13 Display of hidden features on a multiview drawing.

During the actual construction of principal orthographic multiview drawings, the H/F (horizontal/front) reference line is generally placed parallel to the bottom edge of the drawing paper. The other image planes are represented by reference lines drawn to achieve predetermined views of an object.

Figure 3.13*a* and *b* depicts a more complex object in a pictorial and in the six principal orthographic views. The dashed lines on the orthographic views indicate feature changes that cannot be seen in the particular view. These are called hidden lines and are described more completely in Sec. 3.6.

3.4 POSITIONING THE OBJECT AND IMAGE PLANES

Positioning the object and image planes in space for the development of orthographic views is usually an arbitrary decision based on a standard or on common practice. If the horizontal and front image planes are constructed as shown in Fig. 3.14, four quadrants 1, 2, 3, and 4 are

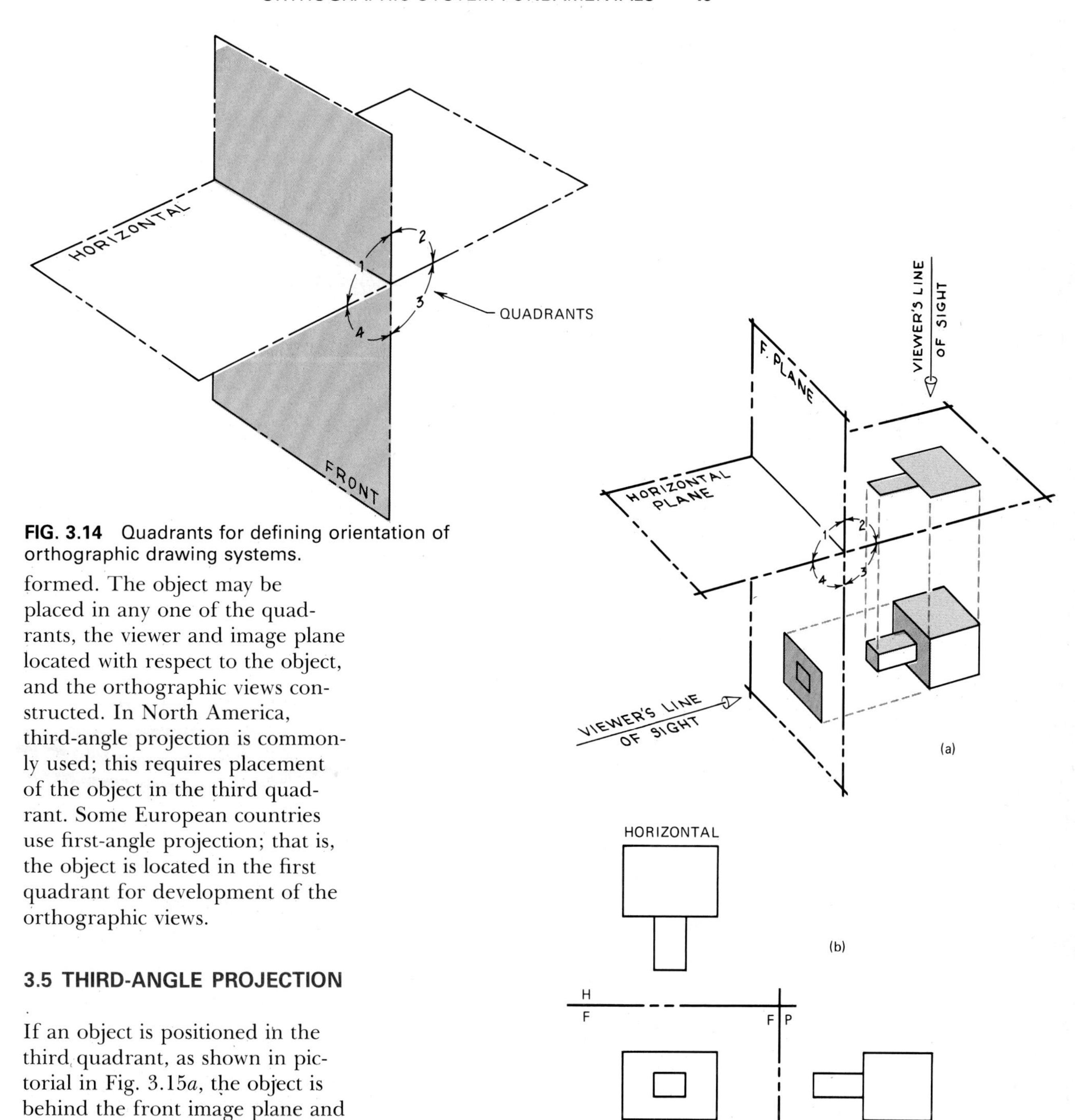

FIG. 3.14 Quadrants for defining orientation of orthographic drawing systems.

formed. The object may be placed in any one of the quadrants, the viewer and image plane located with respect to the object, and the orthographic views constructed. In North America, third-angle projection is commonly used; this requires placement of the object in the third quadrant. Some European countries use first-angle projection; that is, the object is located in the first quadrant for development of the orthographic views.

3.5 THIRD-ANGLE PROJECTION

If an object is positioned in the third quadrant, as shown in pictorial in Fig. 3.15*a*, the object is behind the front image plane and below the horizontal image plane. Figure 3.15*b* shows the resulting horizontal, front, and right pro-

FIG. 3.15 Third-angle projection.

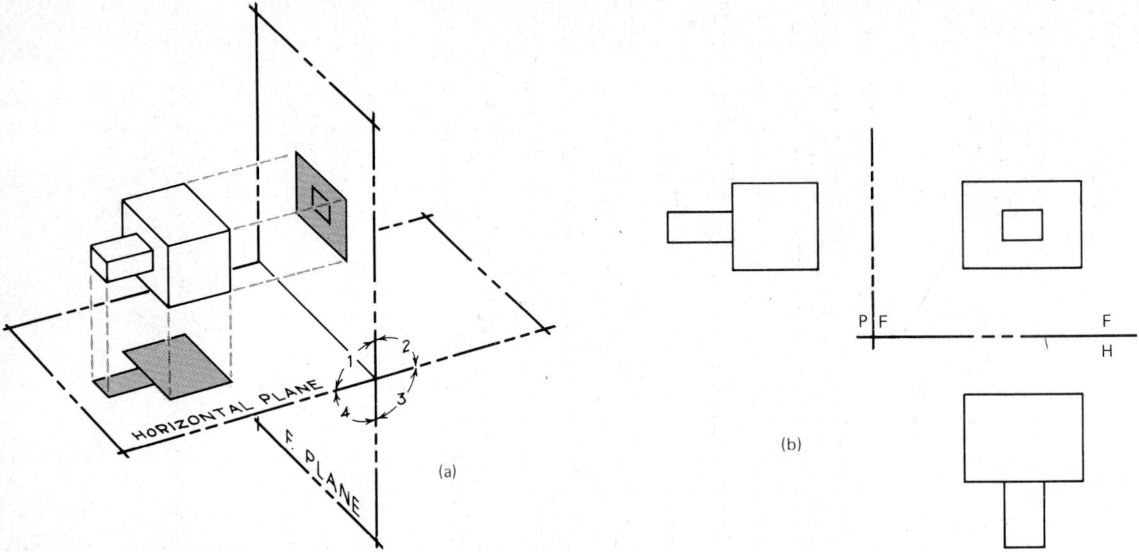

FIG. 3.16 First-angle projection.

file projections of the object. Note that for third-angle projection, the right profile plane is positioned to the right of the object.

As a contrast to third-angle projection, Fig. 3.16*a* shows the object placed in the first quadrant for first-angle projection. In this case, the object is now between the viewer and the image planes. The resulting horizontal, front, and right profile orthographic views are depicted in Fig. 3.16*b*.

3.6 READING AND DELINEATING MULTIVIEW DRAWINGS

To prepare or interpret multiview drawings, you must understand the meaning of the special line symbols used in the views and visualize the object depicted by the related single-view images. The solid or hollow object is assumed to have opaque exterior surfaces.

3.6.1 Line Symbols

Special line symbols are used in orthographic drawing to clarify each view and ensure a single, correct interpretation. Figure 3.17 is a table of the various lines used in orthographic drawing and its applications. Six of these symbols will be discussed here: object lines, hidden lines, center lines, projection lines, construction lines, and reference lines. Other line symbols will be introduced in later chapters as applications require.

Figure 3.18 illustrates the construction procedures for a single-view drawing representing the top view of the object (Fig. 3.18*a*), beginning with a skeleton layout in part b and ending with the finished drawing in part c. The line symbols used are:

Type	Weight	
Object	Thick	
Hidden	Medium	
Center	Thin	
Phantom	Thin	
Extension & Dimension	Thin	
Leader	Thin	
Section	Thin	
Cutting plane	Thick	
Short break	Thick	
Long break	Thin	
Projection & Construction	Thin	

FIG. 3.17 Alphabet of lines for orthographic drawing.

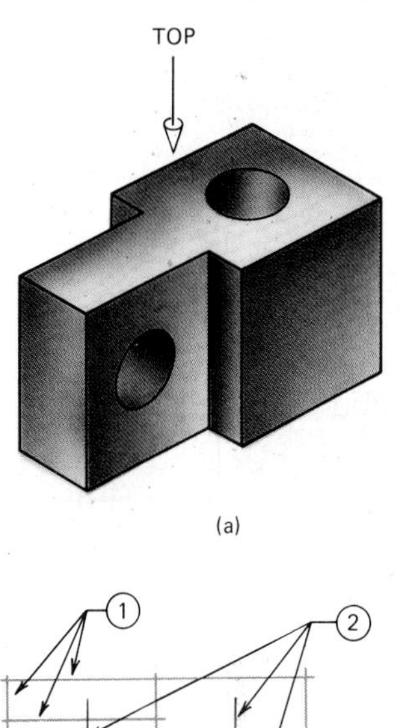

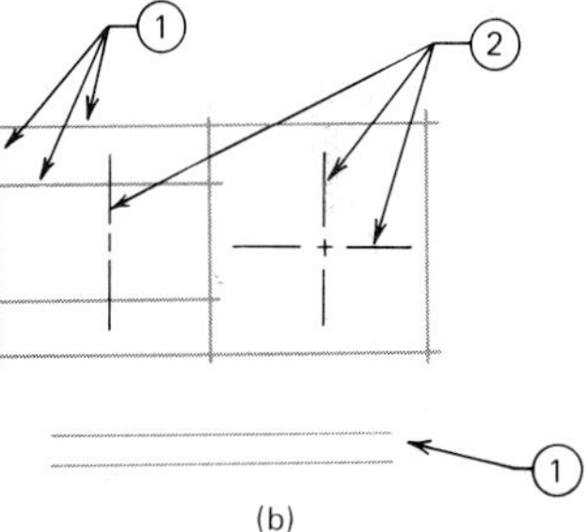

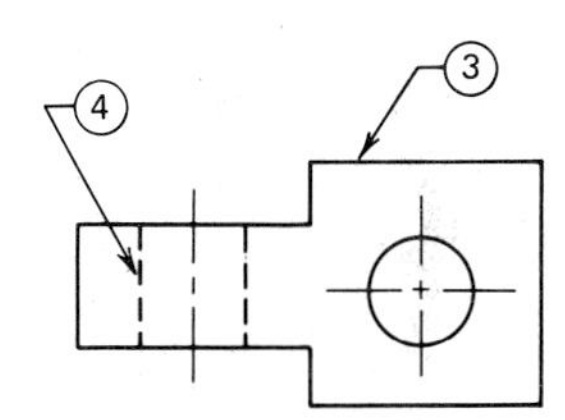

FIG. 3.18 Constructing an orthographic view.

1. Construction lines. Thin, lightweight lines used as guidelines for lettering or for initial layout of views.

2. Center lines. Medium-weight lines constructed as a long dash, short dash, long dash combination. These represent the center of any circle or circular arc and the longitudinal axis of cylinders, cones, spheres, and so forth. Center lines should always be shown on orthographic views.

3. Object lines. Continuous lines of heavy density used to denote the visible lines of an object. They represent the visible surface limits of curved surfaces (e.g., circles), the intersection of planes and other surfaces not meeting in tangency, edge views of planes, and so forth.

4. Hidden lines. Short, intermittent dash lines of medium weight used to indicate object lines which are hidden behind opaque surfaces in an orthographic view.

Figure 3.19 shows the application of two additional line symbols to a multiview drawing.

5. Reference lines. Lines of medium density composed of two long dashes interspaced with two short dashes.

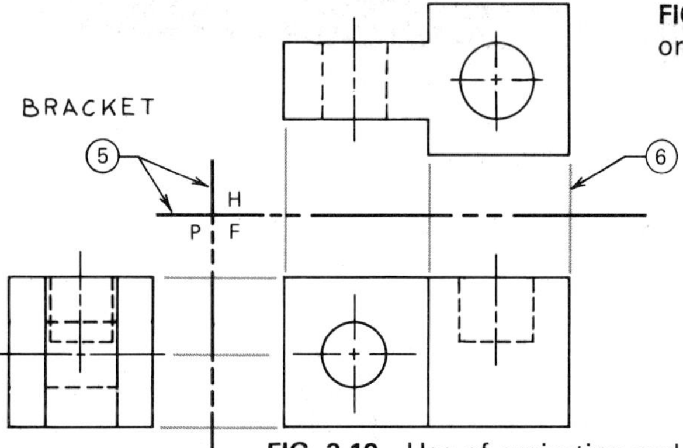

FIG. 3.19 Use of projection and reference lines on an orthographic drawing.

The interpretations of a reference line were given in Sec. 3.3.2. In multiview drawings the reference lines generally are omitted, but specific identification of a view may be made by constructing and labeling a reference line.

6. Projection lines. Thin, lightweight lines that represent the projectors from points on an object perpendicular to an image plane. For many academic exercises, these projectors are left on multiview drawings to show the correct alignment of the views. In industrial applications, projection lines, like reference lines, are implied but do not appear on the final drawings.

3.6.2 Three-Plane Relationship and Common Measurements

A multiview drawing must be visualized as a series (two or more) of single views of an object projected onto image planes. The resulting images are systematically laid out on a plane surface (e.g., the principal views shown in Fig. 3.13*b*) in a manner that defines the three-dimensional space relationships involved. The image planes are positioned with actual or implied reference lines.

When your line of sight is perpendicular to a given image plane, any adjacent image plane will be seen as an edge. The edge is represented by the reference line. Again with respect to Fig. 3.13*b*, if your line of sight is perpendicular to the front image plane, the horizontal, bottom, and right and left profile image planes will appear as edges. Shifting your line of sight perpendicular to the horizontal image plane will reveal the front, rear, and right and left profile image planes as edges.

In order to read or prepare a multiview drawing, it is essential not only to visualize each single view but to understand the relationship betwen the views. This relationship is called the three-plane relationship, and it defines the concept of a common measurement.

In Fig. 3.20*a*, the horizontal (H) and right profile (P) planes are both adjacent and perpendicular to the front (F) plane. This defines a three-plane relationship. The edge view of the F plane is shown in both the H plane and the P plane. Therefore, the perpendicular distance from the H projection of point *A*, labeled A_H, to the edge of F plane is identical to the distance from the P projection of point *A*, labeled A_P, to the edge of the F plane. This is called a common measurement (CM). See Figure 3.20*b*. This concept is valid for any three adjacent views. Its usefulness will be demonstrated more fully when auxiliary views are discussed in Sec. 3.7.

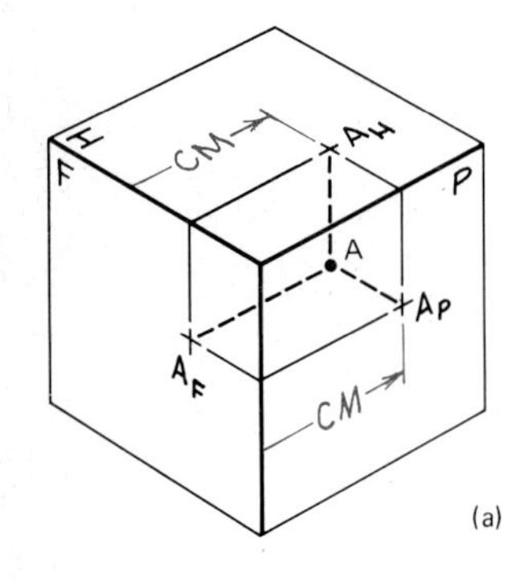

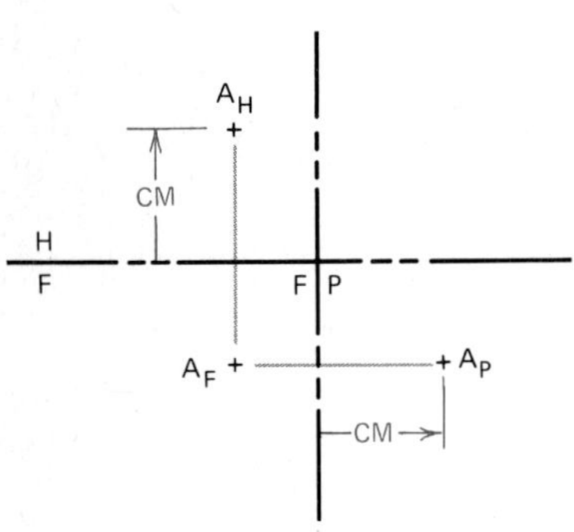

FIG. 3.20 The three-plane relation and common measurements.

3.6.3 Procedures for Delineating a Multiview Drawing

To help you understand the construction of a multiview drawing, a step-by-step outline for a three-view drawing of the triangular pyramid shown in Fig. 3.21*a* will be described.

1. Identify the desired views by drawing appropriate reference lines. Usually two or three adjacent principal views are established. In Fig. 3.21*b*, the H/F and F/P reference lines have been established and positioned on a

PICTORIAL OF A PYRAMID AND DESCRIPTION OF A SIMPLE OBJECT COMPOSED OF POINTS, LINES, AND PLANES

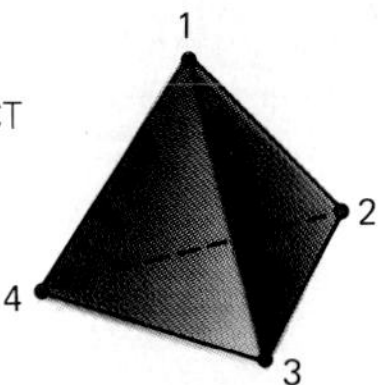

POINT 1 – 5 mm below H, 35 mm left P, 20 mm behind F
2 – 30 mm below H, 15 mm left P, 30 mm behind F
3 – 30 mm below H, 30 mm left P, 5 mm behind F
4 – 30 mm below H, 50 mm left P, 25 mm behind F

(a)

GIVEN INFORMATION FOR HORIZONTAL VIEW:

Point 1 – 35 left P, 20 behind F
2 – 15 left P, 30 behind F
3 – 30 left P, 5 behind F
4 – 50 left P, 25 behind F

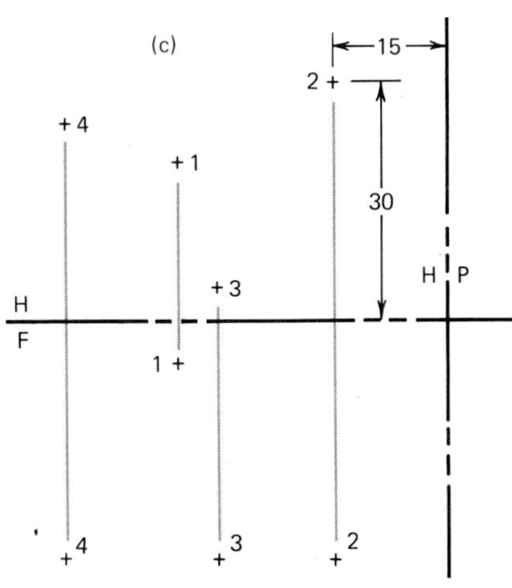

(c)

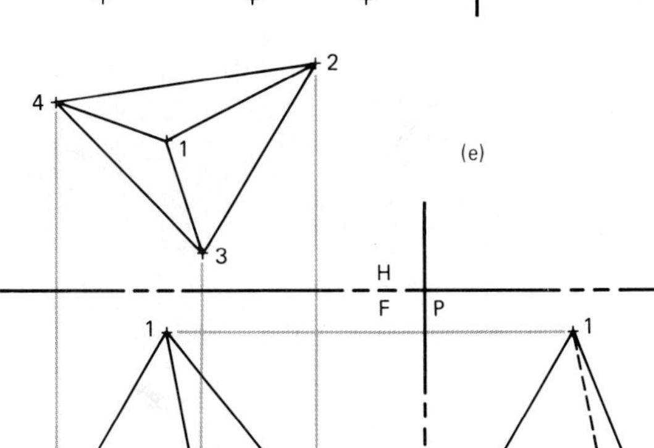

(e)

TO LOCATE FRONT VIEW USE INFORMATION GIVEN:

Point 1 – 5 below H, 35 left P
2 – 30 below H, 15 left P
3 – 30 below H, 30 left P
4 – 30 below H, 50 left P

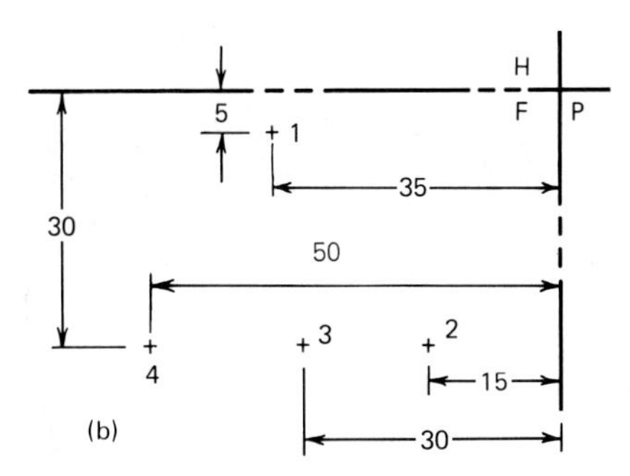

(b)

GIVEN INFORMATION FOR PROFILE VIEW:

Point 1 – 5 below H, 20 behind F
2 – 30 below H, 30 behind F
3 – 30 below H, 5 behind F
4 – 30 below H, 25 behind F

(d)

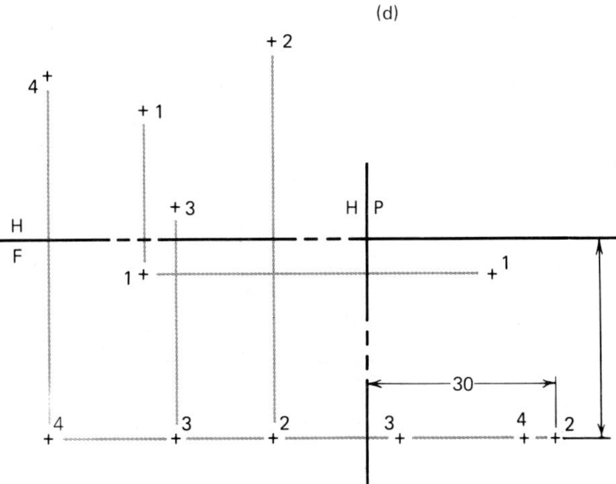

FIG. 3.21 Orthographic projection of a pyramid. The drawing is not full scale.

drawing surface to provide the horizontal, front, and right profile views.

2. Establish the front view of the pyramid by assuming an image plane placed in a vertical position between you and the pyramid, perpendicular to parallel projectors emanating from the object toward the viewer. The image is established where the projectors pierce the image plane. This is accomplished in practice by locating the corner points 1-2-3-4 at their respective distances measured below the horizontal plane and to the left of the right profile plane. Remember that when your line of sight is perpendicular to the front plane, the H/F and F/P reference lines represent, respectively, the edges of the horizontal plane and the right profile plane.

3. Change your line of sight to a vertical position perpendicular to the horizontal plane. Points 1-2-3-4 of the pyramid are projected vertically to the points where the projectors intersect the horizontal plane. This is accomplished on paper by locating the points relative to the front and profile planes. The location of point 2 is labeled in Fig. 3.21*c*. Remember that the H/F reference line now represents the edge view of the front plane and that you see the edge of the right profile plane although you are looking down on it (vertical sight line) instead of in on it (horizontal sight line), as in step 2.

4. Once again, change your line of sight, this time to a position perpendicular to the right profile image plane. Locate corner points 1-2-3-4 in the right profile view by marking the points at their respective distances behind the front plane and below the horizontal plane. Figure 3.21*d* shows the complete three-view projection of points 1-2-3-4, with the location of point 2 in the right profile labeled. It should be clear that the points could have been located in the right profile view by applying the three-plane relationship and marking off the common measurements.

5. Complete the three orthographic views of the triangular pyramid by drawing lines between the points, as shown in Fig. 3.21*e*. Light projection lines are left on the drawing to help you visualize the relationship of the point projections between the three views. The dashed line 1-4 in the right profile view represents a hidden line, meaning that line 1-4 on the pyramid cannot be seen when your line of sight is perpendicular to the right profile plane.

The three-view drawing in Fig. 3.21*e* assumes that the pyramid is solid and opaque. If the pyramid were hollow and opaque, additional hidden lines would be required in each of the three views in order to define the wall thickness. Orthographic projection of more complex objects will be illustrated further in Chap. 9. At this time, it is important to understand that you can delineate orthographic views of an object by projecting appropriate points and lines that define the boundaries of the object.

3.7 EXPANDING THE ORTHOGRAPHIC SYSTEM

Frequently, it is necessary to represent various surfaces of an object with lines in true length and planes in true shape or true size. Also, relationships between objects in space such as clearances between these objects and angles of intersection between planes must be determined. The six principal views may not be sufficient to determine these various relationships. We therefore must expand the orthographic system to include auxiliary views in conjunction with the principal views.

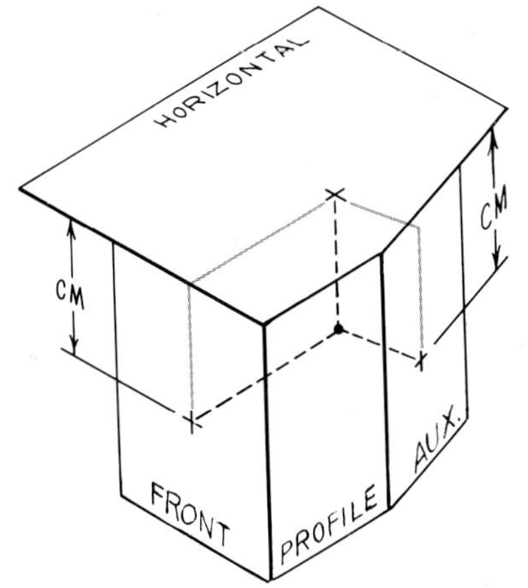

FIG. 3.22 Common measurements in elevation.

Auxiliary Views All orthographic views other than the six principal views which are constructed to determine relationships between objects or parts of objects that otherwise cannot be determined from the principal views. Auxiliary views are classified as elevation, inclined, or oblique.

Auxiliary views enable the viewer to see objects orthographically from any desired position. The three-plane relationships and CM definitions are valid for proper combinations of auxiliary and principal views.

Before describing each class of auxiliary view, we must introduce the term elevation view. Any orthographic view which shows the horizontal (top) view as an edge is referred to as an elevation view. You can see in Fig. 3.12 that of the six principal views, the front, rear, right, and left profile are elevation views showing the CM from any point on the object to the horizontal plane. This CM is a height or elevation. Figure 3.22 illustrates the CM for elevation views.

3.7.1 Auxiliary Elevation Views

Auxiliary Elevation View Any orthographic view other than a principal view which shows the horizontal plane as an edge.

Theoretically, there are an infinite number of possible auxiliary elevation views determined by observing from any position 360° around an object.

Figure 3.23 illustrates several possible auxiliary elevation views in pictorial (Fig. 3.23*a*) and in orthographic projection (Fig. 3.23*b*). The front principal view is included in these figures for reference. Each of the elevation

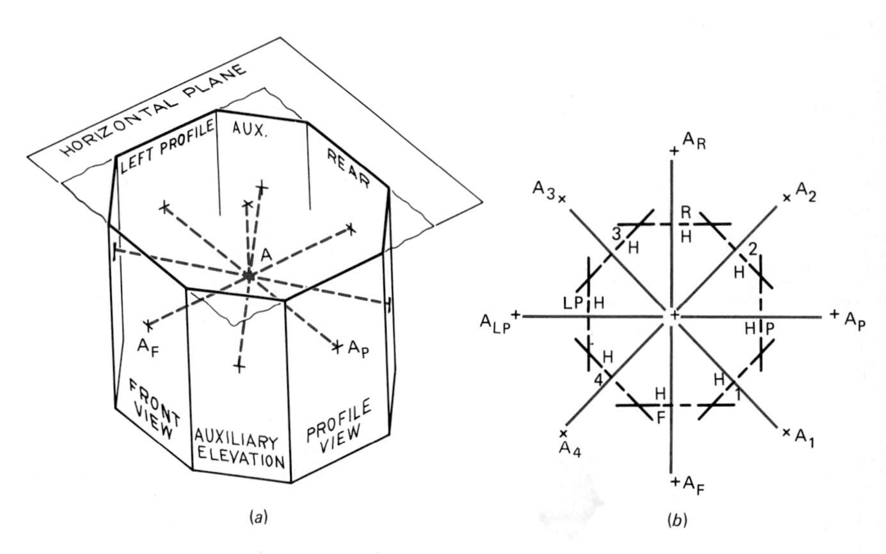

FIG. 3.23 The distance that point A lies below the horizontal plane can be seen in any elevation view.

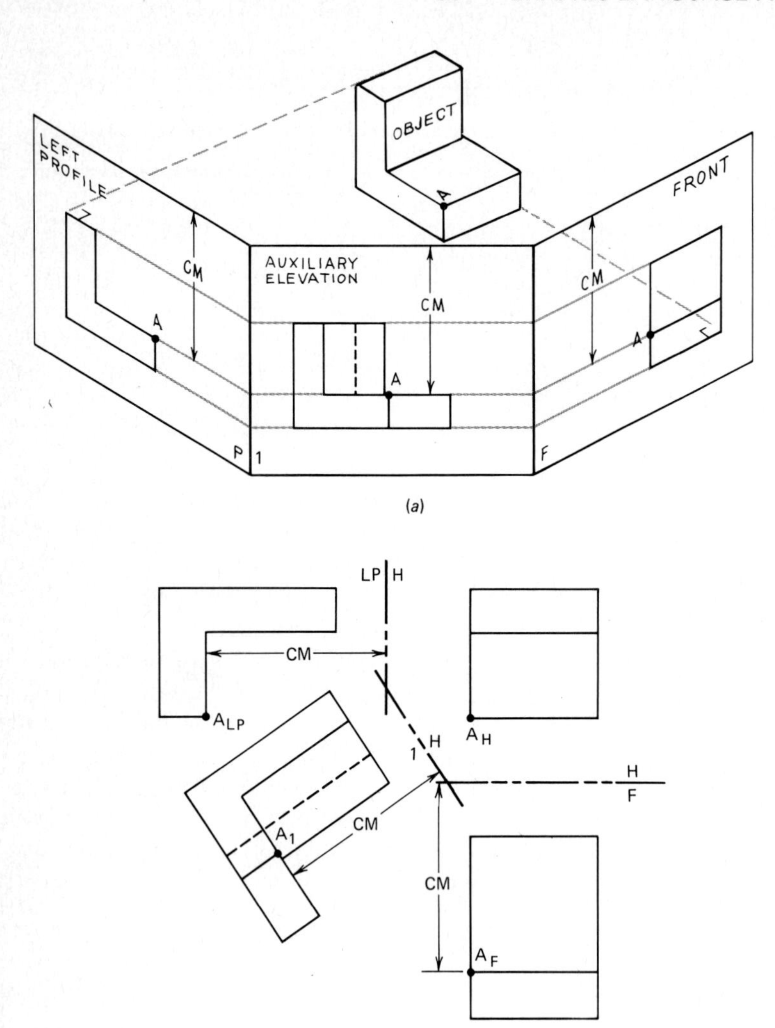

FIG. 3.24 Auxiliary elevation view constructed from given principal views.

views shows the same CM from a point on the object to the horizontal plane. Figure 3.24*a* shows a combination of principal views and one auxiliary elevation view.

The F and LP views are principal elevations, while the 1 view is an auxiliary elevation. The CM from one point, *A*, on the object to the horizontal plane is observable in each elevation view. Note that this CM, height or elevation, is not observed when the line of sight is vertical, that is, perpendicular to the horizontal plane. The corresponding multiview drawing of the object is shown in Fig. 3.24*b*.

3.7.2 Auxiliary Inclined Views

Auxiliary Inclined View Any orthographic view other than a principal view which shows the front, rear, right profile, or left profile plane as an edge.

Figure 3.25*a* shows one of an unlimited number of possibilities for an auxiliary inclined view showing the front plane as an edge. Note that any common measurements for this example would represent a distance behind the front plane and would not be an elevation measurement. Figure 3.25*b* shows how the multiview drawing for this object may appear.

3.7.3 Auxiliary Oblique Views

Auxiliary Oblique View Any orthographic view other than a principal, auxiliary elevation, or auxiliary inclined view. It is projected from an auxiliary inclined or auxiliary elevation view.

Auxiliary oblique views are neither parallel nor perpendicular to any of the principal planes.

Figure 3.26 illustrates an auxiliary oblique view 2 projected from the auxiliary elevation view 1. In order to obtain this projec-

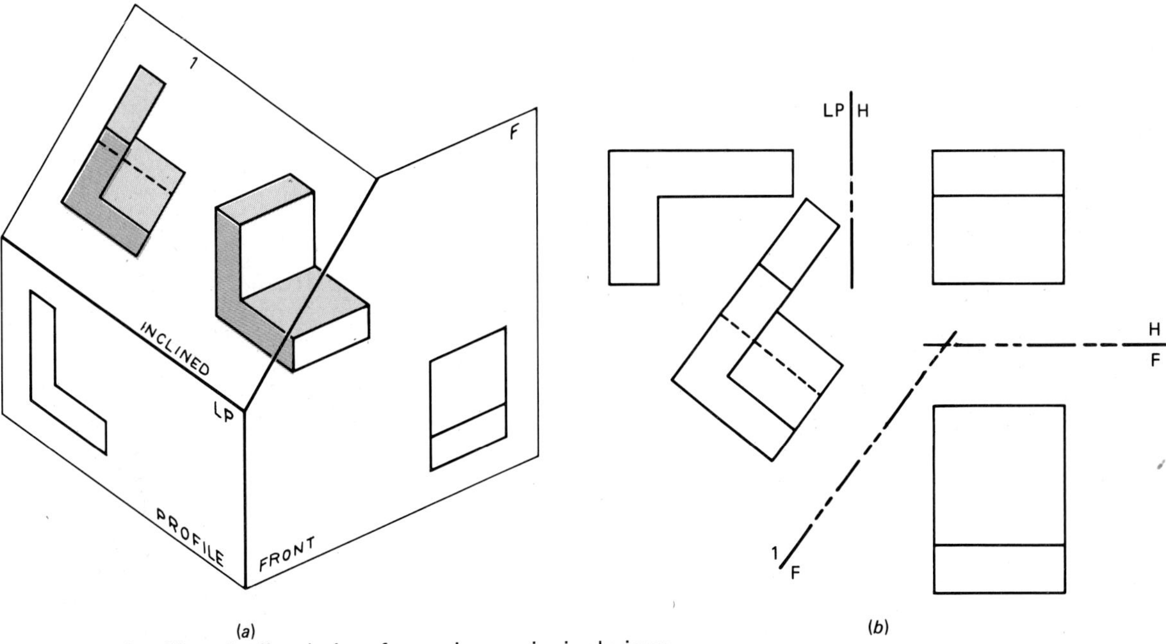

FIG. 3.25 Auxiliary inclined view from given principal views.

tion, the H, 1 and 2 views become the three planes within which common measurements are made. This "second" auxiliary view 2 has an application in the geometry that will be covered in Part Two. The three-plane relationship involving projections into auxiliary oblique views will be illustrated in Sec. 3.8.

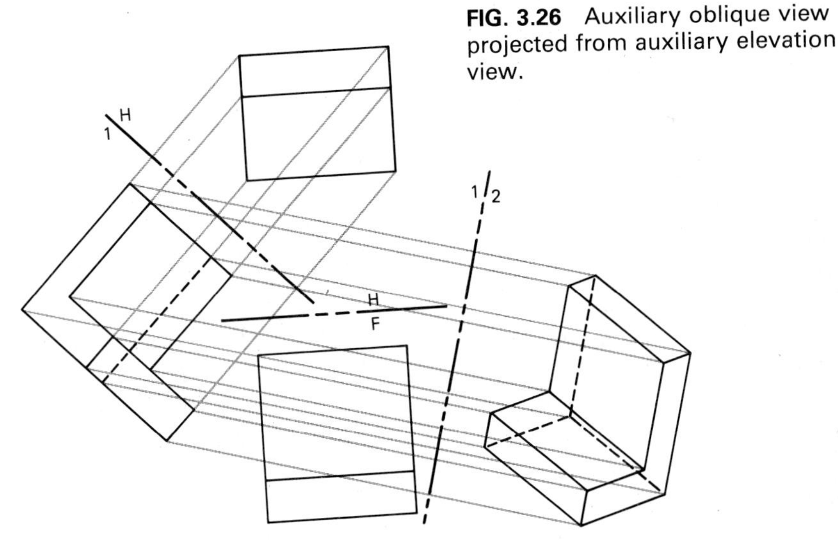

FIG. 3.26 Auxiliary oblique view projected from auxiliary elevation view.

3.8 USING THE ORTHOGRAPHIC DRAWING SYSTEM IN PROBLEM SOLVING

The principles and definitions described thus far in this chapter provide a basis for describing the form and function of three-dimensional objects on a two-dimensional sheet of paper. By preparing a set of orthographic views of an object and arranging the set in an accepted manner, we can show the three dimensions of length, width, and height as well as the external and inter-

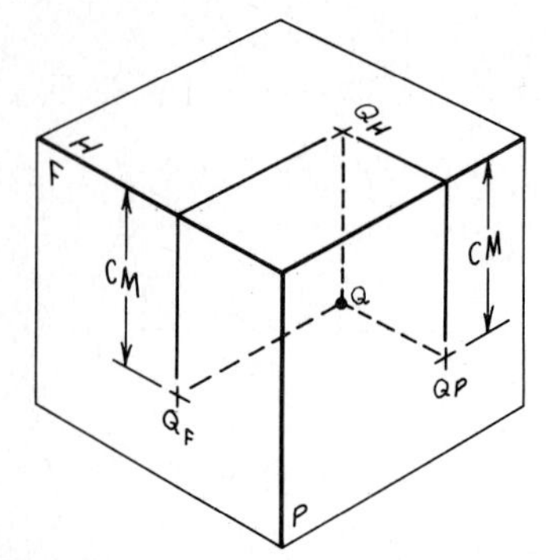

FIG. 3.27 Common measurement.

Specifications for point Q

1 cm behind F
2 cm left P
3 cm below H

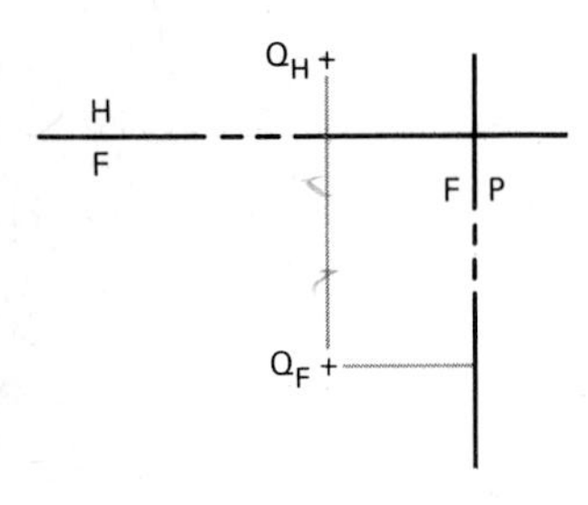

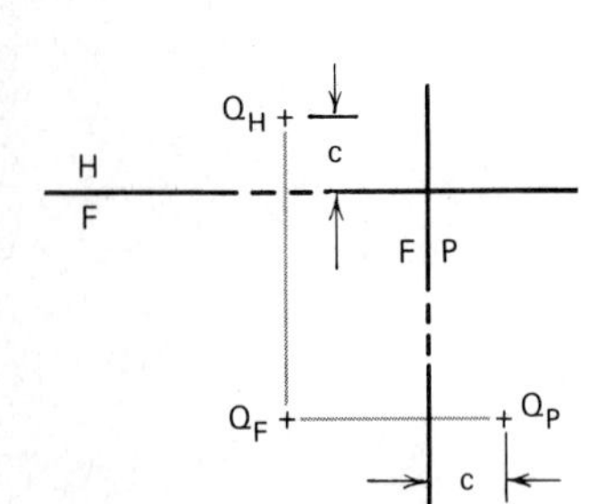

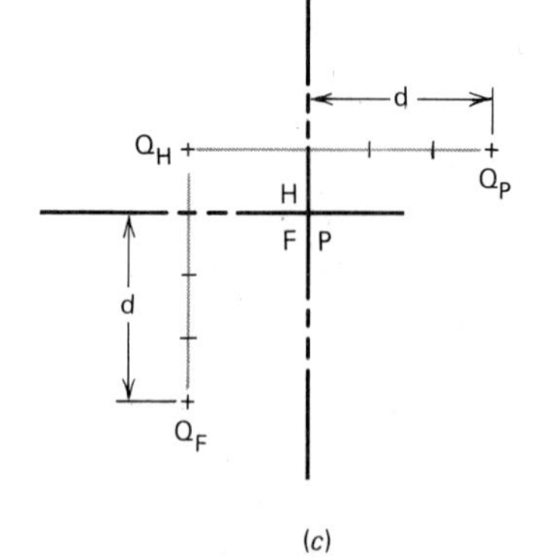

FIG. 3.28 Common measurement—principal views. Drawing not full scale.

nal features of the object. Thus, orthographic drawing is the means by which an engineer can analyze an object by reading the orthographic views or design an object by developing the necessary views to describe the design. Stated another way, orthographic drawing is a tool that an engineer uses to solve problems involving the geometric relationships between objects or between parts of objects.

We will illustrate the practical application of orthographic projection to the description of the geometry of three-dimensional objects. We use a point to demonstrate the procedures, but keep in mind that this single point could represent a specific location on a complex object and that other points on the object would be handled the same way, yielding a multiview drawing of the three-dimensional object.

3.8.1 Projecting an Image through a Series of Views

In Sec. 3.6.2, we introduced the concept of three-plane relationships and common measurements. Figure 3.27 shows this relationship for a point Q in space. Note that the F and P image planes are both perpendicular to the H plane, and thus the perpendicular distance CM from the H plane to the front and profile image of the point is seen in both the F plane and the P plane. This CM can be used to transfer point Q to other views, provided that the three-plane relationship is established.

Figure 3.28*a* establishes the H and F views of point Q from Fig. 3.27. The establishment of the location of Q in two adjacent views is accomplished by the initial specification given in Fig. 3.28*a*. From this specification, we can now project the image of Q into any principal or auxiliary view needed. In Fig. 3.28*b*, the profile image Q_P of point Q is located in two steps. First, a perpendicular projector is drawn from Q_F through the F/P reference line. Second, the CM (c = 1 cm) is then transferred from the H view to the P view as indicated. The H and P planes both show the edge of plane F, and thus the distance c represents the perpendicular distance behind the F plane.

Figure 3.28*c* shows an alternative placement of the right profile view in relation to the H and F views. In this example, the image Q_P is found on a projector through the H/P reference line at a CM of d = 3 cm below the H plane.

The extension of the three-plane relationship to auxiliary views is illustrated in Fig. 3.29. The edge view of plane P is seen in both the F view and the auxiliary inclined view 1 (Fig. 3.29*a*). The CM (e = 2 cm) is the distance from plane P to point Q and is plotted accordingly on the orthographic views given in Fig. 3.29*b*.

Figure 3.30 shows use of a CM to determine the projection of a point Q in an auxiliary elevation view 3. The edge view of plane H is seen in both the F view and the 3 view, and the measurement g = 3 cm is measured from plane H to point Q.

Figure 3.31 shows the extension of the three-plane relationship to auxiliary oblique views. It is assumed that the projection Q_1 (Fig. 3.31*b*) has been found previously and that we wish to find the projection of Q in the 2 plane. A series of three adjacent planes P, 1, and 2 establish the three-plane relationship. The distance f = 3

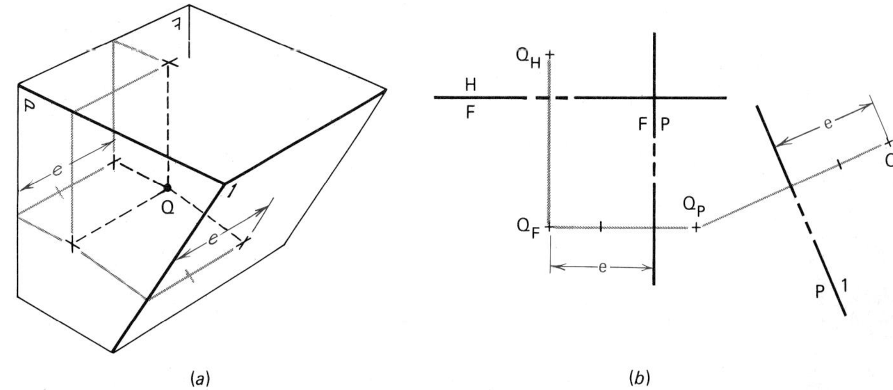

(*a*) (*b*)

FIG. 3.29 Common measurement—auxiliary inclined view.

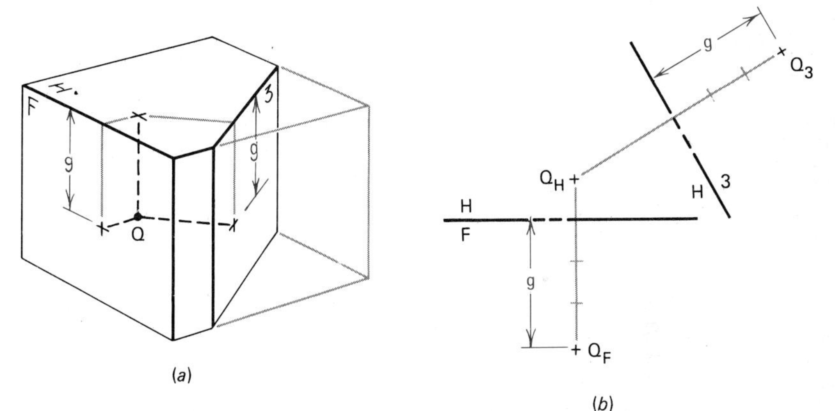

(*a*) (*b*)

FIG. 3.30 Common measurement—auxiliary elevation view.

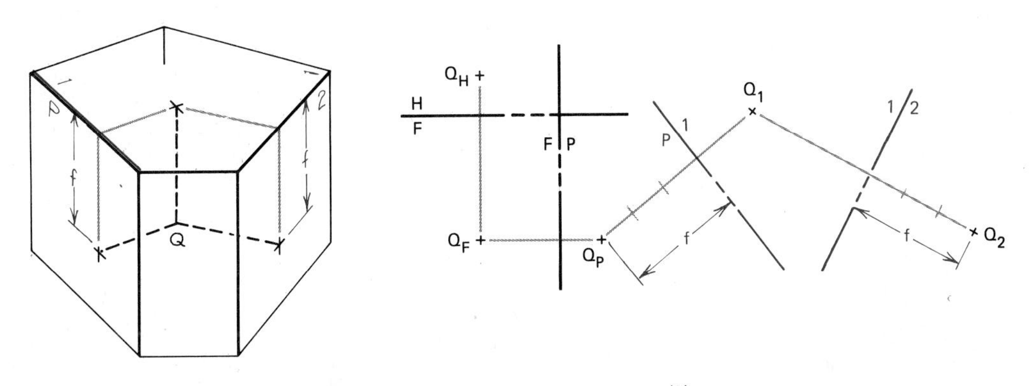

(*a*) (*b*)

FIG. 3.31 Common measurement—auxiliary oblique view.

cm that point Q lies from the edge of plane 1 is measured in the P plane and then transferred to the 2 plane.

If you can visualize the three-plane relationship, locating the common measurement is a straightforward procedure. In most applications, you will be working with the orthographic views and will not have pictorial drawings like Figs. 3.29*a*, 3.30*a*, and 3.31*a* available.

3.8.2 Solution Views and Solution Procedures

As noted previously, orthographic drawing is used to delineate particular features of objects and to show the true sizes and shapes, true distances, and true angles. As in any problem-solving procedure, you must carefully follow a planned sequence that will lead to the correct solution. To use orthographic drawing in solving geometry problems, you need first a thorough knowledge of the orthographic fundamentals that have been described in this chapter; second, the final orthographic view, called the solution view, which shows the desired result; and third, the sequence of steps necessary to achieve the solution view, the solution procedure.

Solution View
Any orthographic view showing the correct graphic solution to a problem in spatial relationships.

Solution Procedure
The steps or action needed to attain the solution to a problem.

This involves constructing the appropriate orthographic views in proper sequence, leading to the solution view.

Solving problems using the orthographic drawing system may be likened to planning and taking an automobile trip. First, you recognize the need for a trip (establish and understand the problem). Next, you pinpoint the destination (visualize the solution view). Third, you proceed on the trip over a planned route of roads and stops (following the solution procedure). In addition, you must be familiar with the operating characteristics of the automobile, know how to read maps and road signs, and have the desire to complete the trip. This may be compared with knowing the graphics tools available to you, having the skills to use the tools, understanding the orthographic system, and having the confidence to determine the problem solution successfully.

At this point, descriptions of solution views and solution procedures for finding the true length of a line, the point view of a line, the edge view of a plane, and the true size of a plane surface will be presented. Further work in problems in spatial relationship will be undertaken in later chapters.

3.8.3 True Length of a Line

Solution View
Any orthographic view of a line that has been projected onto an image plane parallel to the line.

Solution Procedure

1. Draw a reference line (hinge line) parallel to any given projection of the line.

2. Project the line into this view, using the endpoints.

3. Label the projection TL for true length.

You can see the true length of a line by looking perpendicular to it (verify this by moving your pencil about in front of you). Note that according to orthographic principles, your line of sight will be perpendicular to a line if you look through an image plane parallel to the line. The true-length image will be the longest image of the line that you can project. Figure 3.32 shows the H and F projections of line *AB*. Note that the true length of the line is seen in the front view since the front plane is parallel to the line.

The H and F projections of line *XY* are shown in Fig. 3.33*a*. To find the true length of this line, a reference line F/l is constructed parallel to the F projection of the line. The line *XY* is projected onto the auxiliary inclined view 1 (Fig. 3.33*b*), which is the solution view for true length. The problem is completed by labeling the endpoints of the line and designating it as TL. You should note that the correct true length may also be found by constructing a reference line parallel to the H projection of *XY* and projecting the line into the resulting auxiliary elevation view.

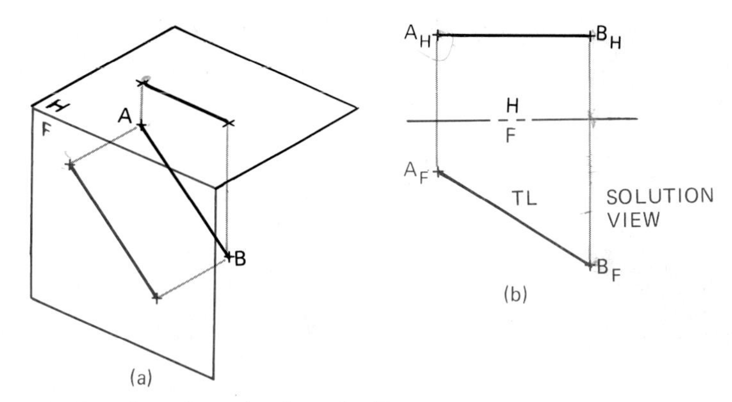

FIG. 3.32 True-length projection of a line.

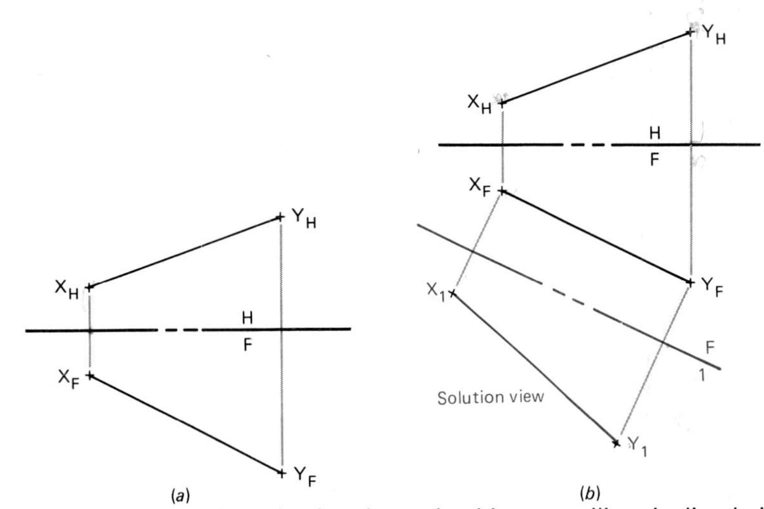

FIG. 3.33 True-length projection determined in an auxiliary inclined view.

3.8.4 Point View of a Line

Solution View

Any view projected onto an image plane perpendicular to a true-length view of the line.

Solution Procedure

1. Find a true-length view of the line.

2. Draw a reference line perpendicular to the true-length projection of the line.

3. Project the endpoints of the line into this solution view. The endpoints are superimposed.

You can visualize the point-view solution by moving your pencil in front of you. Look perpendicular to the pencil to see true length and then imagine viewing the pencil through a plane perpendicular to either end of the pencil.

Figure 3.34*a* shows the H and F projections of line *UV*. True length of the line is determined in the auxiliary elevation view 2 in Fig. 3.34*b*. Finally, the point view is shown and labeled on the auxiliary oblique view 3 (Fig. 3.34*c*). Note that the reference line 2/3 is perpendicular to the TL view of the line.

3.8.5 Edge View of a Plane

Solution View

Any view on an image plane perpendicular to the true-length projection of any line in the plane.

Solution Procedure

1. Find any true-length line in the plane. For example, any line in the plane that is parallel to the horizontal plane will appear in true length in the horizontal view. In the same manner, any line parallel to the front plane will appear in true length in the front view. Figure 3.35*a* illustrates these particular lines.

2. Construct a reference line perpendicular to the true-length projection of the line found in step 1.

3. Project points defining the plane sector into the view defined in step 2.

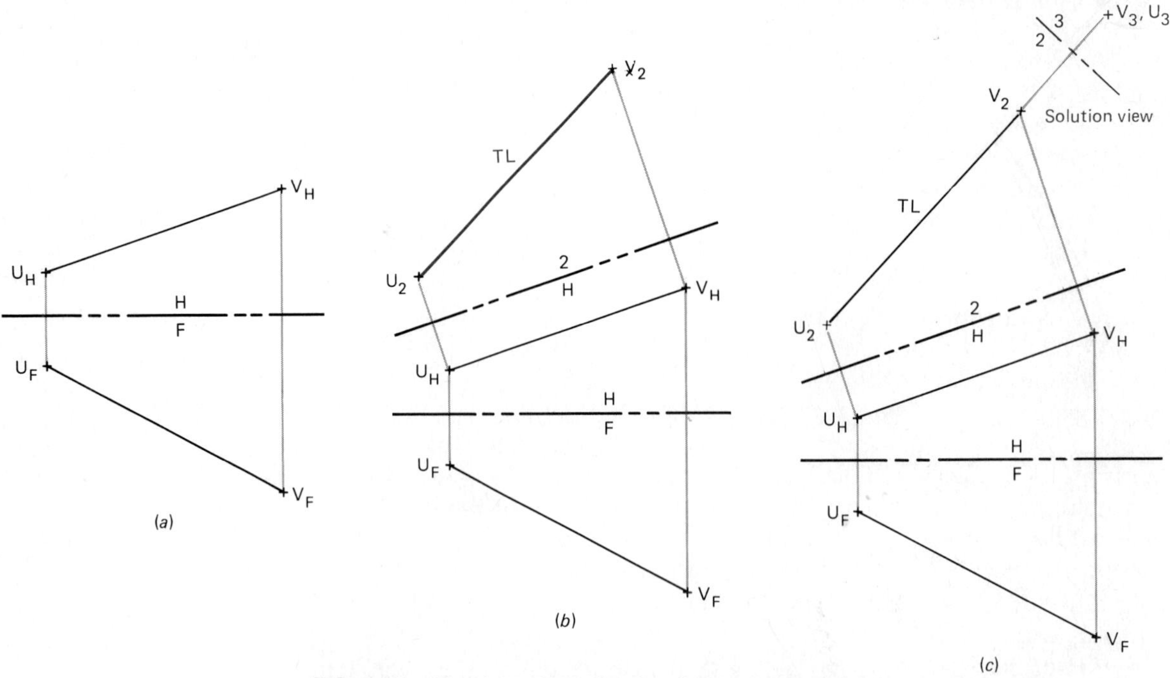

FIG. 3.34 True-length projection in an auxiliary elevation view. Point view is found in an adjacent oblique view perpendicular to the true-length projection.

4. All points in the plane sector will lie on a straight line in the solution view representing an edge view of the plane. Label EV.

Figure 3.35*a* shows the H and F projections of a triangle-shaped plane sector bounded by the three lines *XY, XZ,* and *YZ*. Note that line *XY* is parallel to the horizontal plane and thus appears in true length in the H view. We use this true-length projection to find the edge view of plane *XYZ* by constructing the H/1 reference line perpendicular to this true-length projection. In Fig. 3.35*b*, the three points, *X, Y,* and *Z,* are projected into the 1 plane. The three points appear as a straight line, which is an edge view of plane *XYZ* as seen in an auxiliary elevation view.

The given projections of a plane sometimes do not reveal a true-length projection of a line on the plane. To find an edge view in this case, we first must locate a true-length line by using the properties of lines shown in Fig. 3.34. Consider the plane sector *RST* specified in Fig. 3.36*a*. None of the lines defining the boundary of the plane appears in true length in the given views. In Fig. 3.36*b*, the H projection of a line *TU* is constructed parallel to the F plane. The F projection of this line is in true length. Continuing with the solution procedure, the edge view of the plane sector *RST* is found in the 2 view shown in Fig. 3.36*c*.

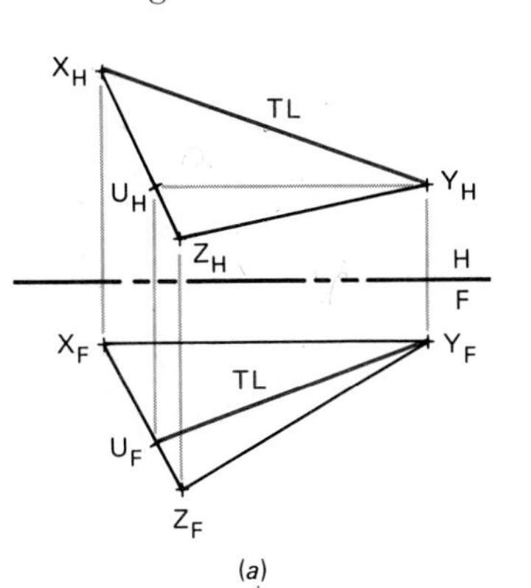

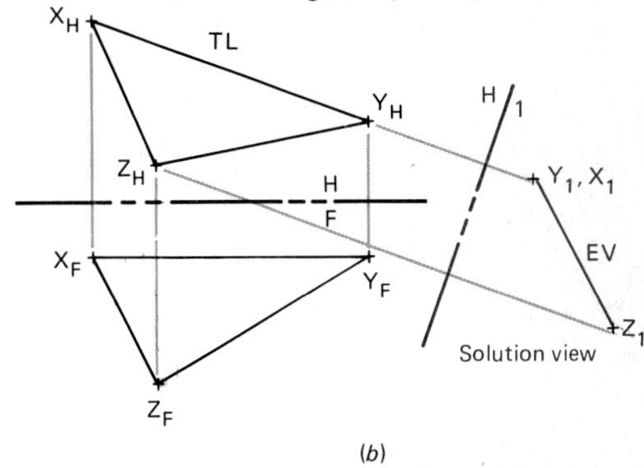

FIG. 3.35 Edge view of a plane sector from given principal views.

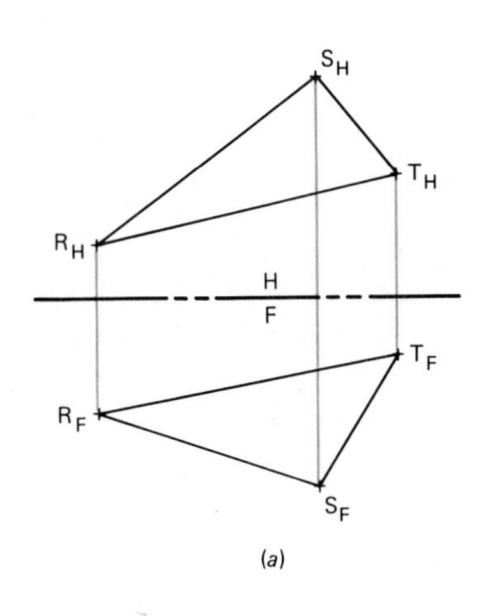

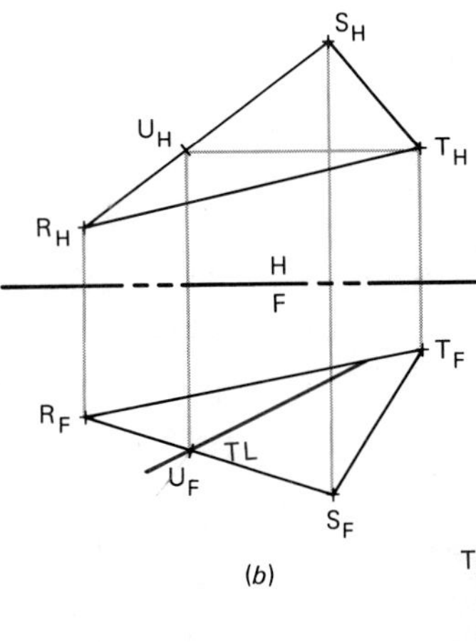

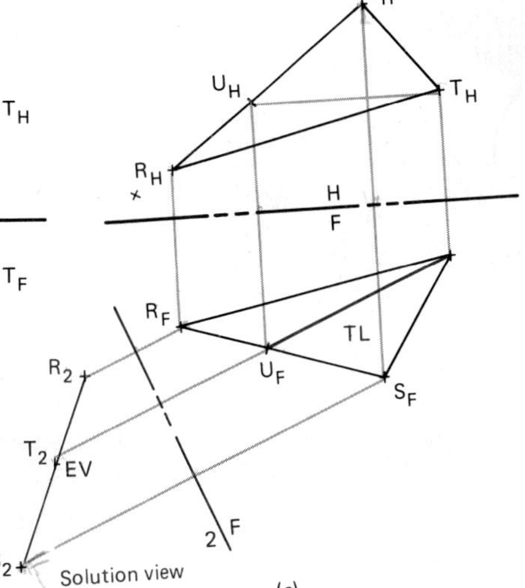

FIG. 3.36 Edge view of a plane sector from given principal views.

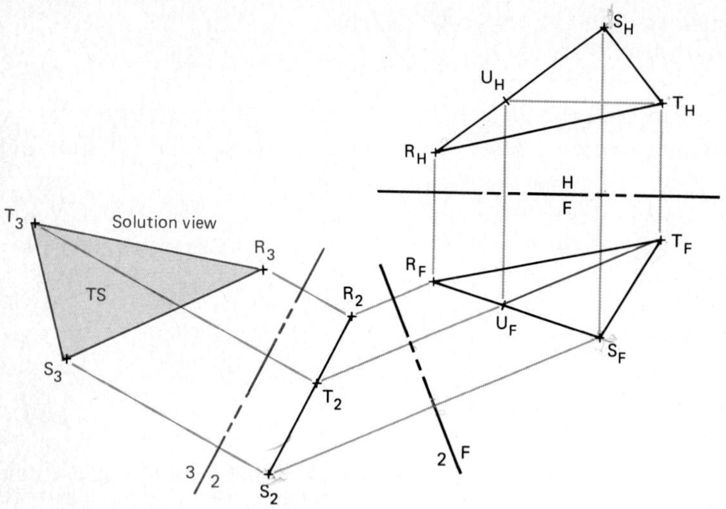

FIG. 3.37 True shape of a plane sector.

3.8.6 True Shape of a Plane Sector

Solution View
Any view projected onto an image plane parallel to the edge view of the given plane sector.

Solution Procedure

1. Obtain an edge view of the plane sector.

2. Construct a reference line parallel to the edge-view projection.

3. Project points defining the plane sector into the view defined in step 2.

4. Draw the true shape of the plane in the solution view and label as TS.

As an example problem to find true shape, consider again the plane sector *RST* as given in Fig. 3.36*a*. To find the true shape, we would begin by determining the edge view shown in Fig. 3.36*c*. Figure 3.37 illustrates the additional step needed to obtain the true shape once the edge view is known. The reference line 2/3 is constructed parallel to the edge-view projection; points *R*, *S*, and *T* are located in the solution view 3 and are connected; and the resulting projection is labeled TS. If the drawing is constructed to a specified scale, the true-shape solution view can be used to determine the dimensions of the plane boundaries, the angles between the boundaries and the area of the plane sector. If this is the case, the true size of the plane sector may be computed.

3.8.7 Problem-Solving Considerations

In the preceding discussion of solution views and solution procedures, the problem to be solved required the construction of additional views to complete the solution. In many cases, the correct solution can be determined in more than one way. For example, to find the point view of line *UV* in Fig. 3.34, we could have begun the solution by constructing a reference line parallel to the front projection of the line, determined the true length in an auxiliary inclined view, and then found the point view. If no other conditions were placed on the solution, either procedure would be correct. We will see in later chapters that certain additional specifications involving the solutions to geometry problems using the orthographic drawing system will limit the choices of solution views and procedures.

When given a problem to solve and at least two orthographic

views describing the spatial relationships, you should first inspect the given views carefully to see if they meet the specifications for a solution view. If so, the solution needs only to be acknowledged with appropriate labeling. Figure 3.38 specifies two views of a plane sector *ABC* for which the true shape is desired. Inspection reveals that an edge view of the plane is shown parallel to the F plane. This satisfies the solution view requirement for true shape; thus, the front projection of *ABC* is a true-shape projection and simply needs to be identified by labeling with a TS, as indicated on the figure. There is no need to find the H or any other view of plane *ABC* in order to satisfy the solution criteria.

Problems involving complex three-dimensional objects may require more careful analysis before you begin a solution procedure. Figure 3.39*a* presents two views of a solid pyramid for which the true shape of face *BCD* is desired. Concentrating on *BCD* as a plane rather than on the entire solid pyramid *ABCD*, we see that line *CD* is in true length in the F plane (Fig. 3.39*b*). By constructing reference line F/1 perpendicular to this true-length projection, we can find an edge view of plane *BCD*. Continuing with construction of reference line 1/2 and projecting only points *BCD* into the 2 plane, we determine the true shape. This solution view must be labeled carefully to indicate that only face *BCD* is being considered and that this is not a view of the entire pyramid. The projection of point *A* is shown in planes 1 and 2 for reference only.

As in any problem-solving procedure, it is first necessary to understand the problem and then to carry out the solution procedure logically and efficiently. The orthographic drawing system provides the means to the graphic solution of geometry problems if you adhere to the correct procedures. Attempts to use shortcuts most likely will lead to incorrect results.

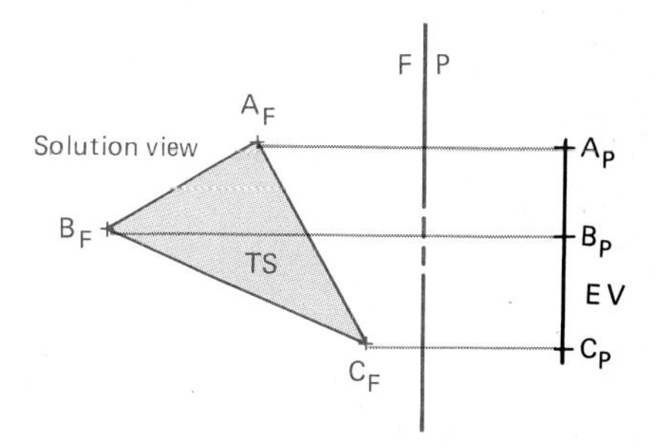

FIG. 3.38 Recognizing edge view and true-shape views in a multiview drawing.

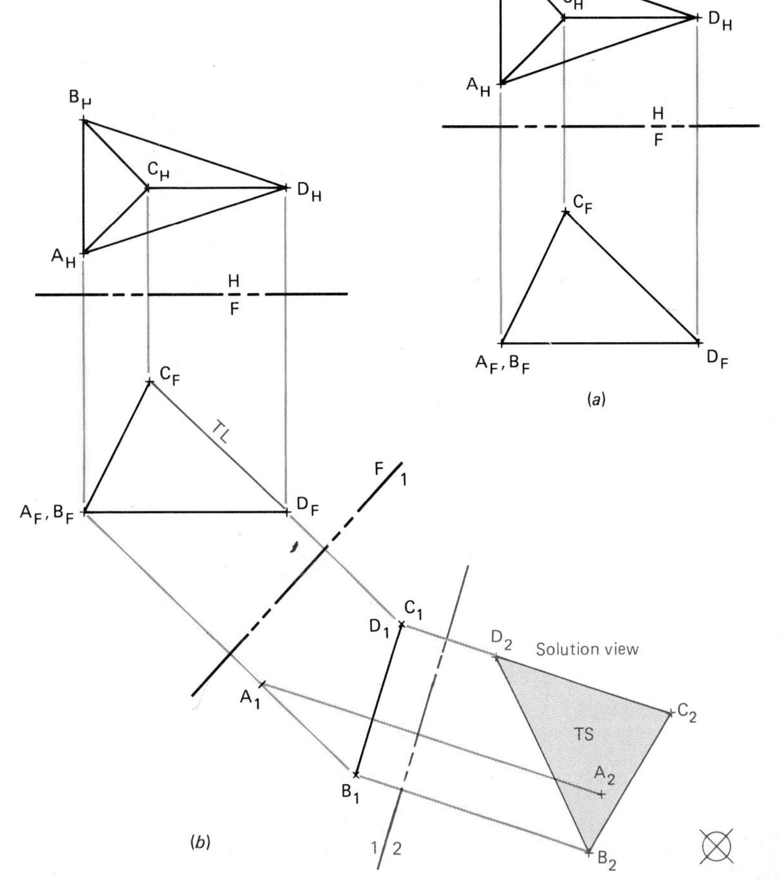

FIG. 3.39 Using orthographic techniques to determine true shape of a plane which is part of the surface of a solid.

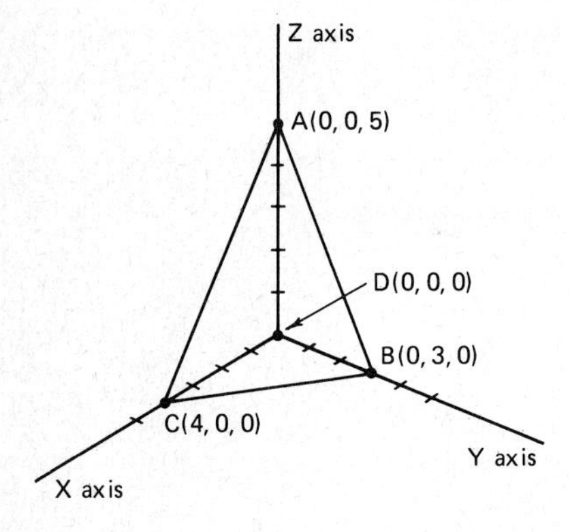

Points: A, B, C, D
Lines: AB, AC, AD, BC, BD, CD
Planes: ABC, ABD, ACD, BCD

FIG. 3.40 Designating points of a solid for computer representation.

3.9 THE ORTHOGRAPHIC DRAWING SYSTEM WITH COMPUTER GRAPHICS

The orthographic drawing system allows us to describe without doubt the shape and characteristics of a three-dimensional object. To effect the orthographic drawing system, we can use any or all of the graphics tools available, such as scales, triangles, and T squares. The important fact is that we use two-dimensional paper to represent three-dimensional geometry and that the orthographic drawing system provides the means to represent the geometry completely in a two-dimensional format.

A computer graphics system can provide a representation of a three-dimensional object on a two-dimensional surface (cathode-ray tube, or CRT) exactly as the orthographic drawing system can, but the computer must operate from a numeric database (describing the solid) rather than from a pictorial drawing or other graphic or written description of an object.

Let us see how an object of known geometry can be set up on a computer graphics system and then see how the computer can be used to generate orthographic images that will allow you to determine the geometric properties of the object. Figure 3.40 shows a triangular pyramid set in a cartesian coordinate system. The space coordinates (X, Y, Z) of corner points *ABCD* are specified along with a listing of the lines and planes that form the shape.

The database for the pyramid can now be developed for storage in the computer. One procedure for doing this is as follows:

1. Describe one surface (plane) point by point in a specified order. For example, plane *ABC* could be generated by *A* (0, 0, 5), *C* (4, 0, 0), *B* (0, 3, 0), and *A* (0, 0, 5). Note that the direction of points inputted for plane *ABC* was counterclockwise as you look normal to *ABC* from outside the pyramid. Each of the other planes must have its corner points inputted in the same consistent fashion or you will not be able to use the geometry correctly.

2. Describe a second surface, say, *ACD*. The data points would be *A* (0, 0, 5), *D* (0, 0, 0), *C* (4, 0, 0), and *A* (0, 0, 5), again following the set procedures.

3. Describe the remaining two surfaces.

The nature of the database generation for this simple object was described in detail to point out that a knowledge of geometry and the principles of engineering graphics is a necessary prerequisite to making the computer graphics system function in a manner that will be beneficial to you.

Industrial and educational computer graphics user facilities usually have a software package which will accept data and produce screen drawings of the geometry from any desired position. At this point in your study of graphics, you need not know what the software is, but only how to use it to produce what you want.

Assume that you have followed

the instructions and have inputted the geometry into a computer system. You would now like to see some different views of the pyramid on the screen. The CRT presents a two-dimensional picture. If, for example, the *xyz* coordinate system were oriented on the screen so that the *z* axis was pointing out of the screen, you would effectively have an orthographic view showing the *xy* dimensions of the object (Fig. 3.41). The software can accomplish this coordinate rotation if you specify a line of sight from where you would like to be to the origin. One way to specify this line of sight is to choose a viewpoint. The computer software can then generate the line of sight from the viewpoint to the origin and "rotate" the object on the screen for you to view.

For ease of visualization we will define the principal orthographic views in terms of the *xyz* coordinate system. If we look down the z axis, we will define the resulting image on the xy plane as the front view. Then looking down the x axis will correspond to the right profile view and looking down the y axis will correspond to the horizontal view. The other principal views are similarly established.

As an example, suppose you chose (0, 0, 10) as a viewpoint for the pyramid (refer to Fig. 3.42*a*). You would then be looking down the *z* axis toward the origin and would see the front view of the pyramid as the right triangle shown in Fig. 3.42*b*. If you chose (10, 0, 0) as your viewpoint, the *x*

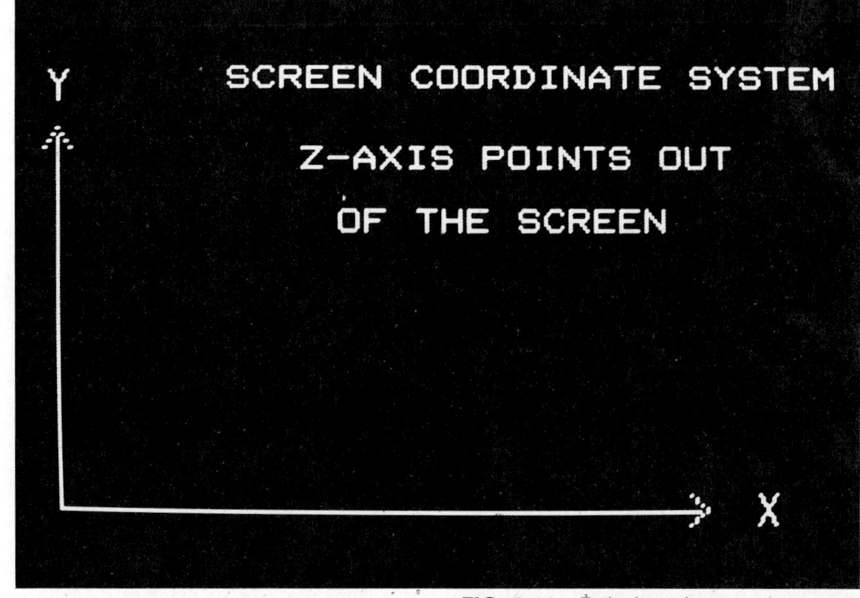

FIG. 3.41 Relating the *xyz* coordinate system to the computer screen.

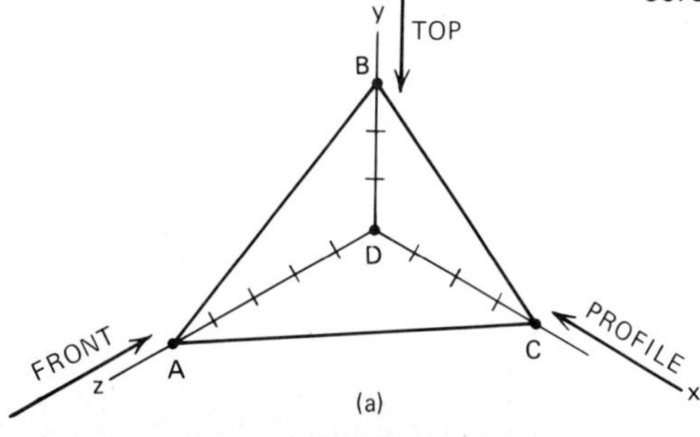

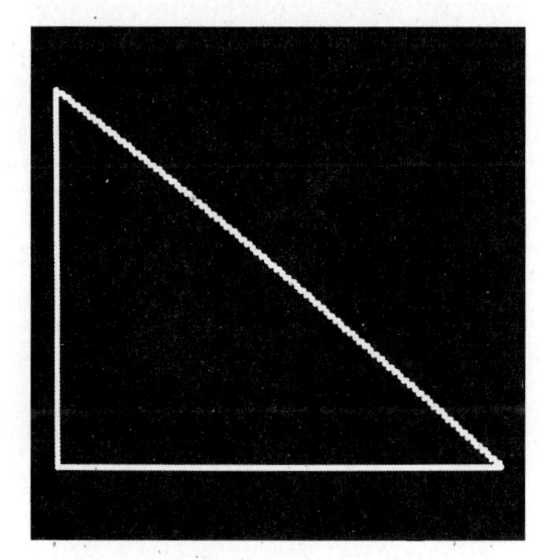

FIG. 3.42(a) A pyramid shown with a coordinate system that is aligned with three of the principal orthographic views. (*b*) The front view of the pyramid can be seen on the computer screen by choosing a viewpoint to look down the *z* axis.

axis would be the line of sight (pointing out of the CRT), and you would see the right profile view shown in Fig. 3.43.

If you choose (10, 10, 10) as a viewpoint, you get the picture that is shown in Fig. 3.44. Note that you now see three dimensions to the pyramid, which means that you are now looking at a pictorial.

We will list a few questions about the pyramid and how you would view it on a CRT in order to see certain geometric properties. These questions can be answered without access to a computer system. If you have access

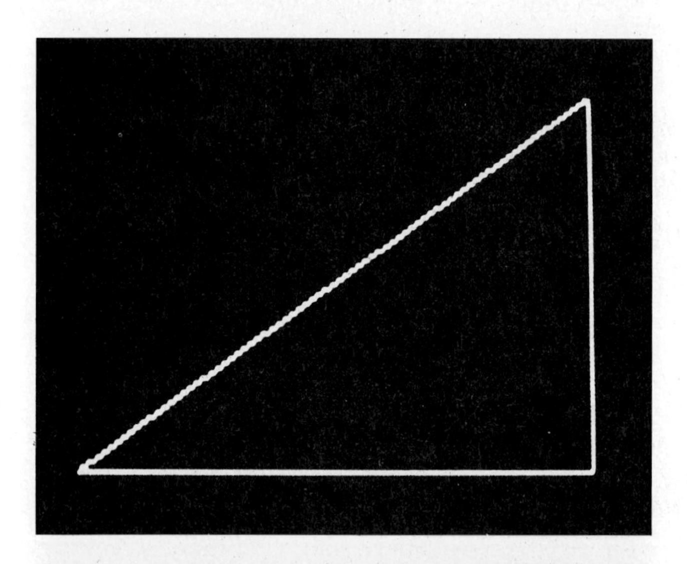

FIG. 3.43 Right profile view of the pyramid seen by looking down the *x* axis.

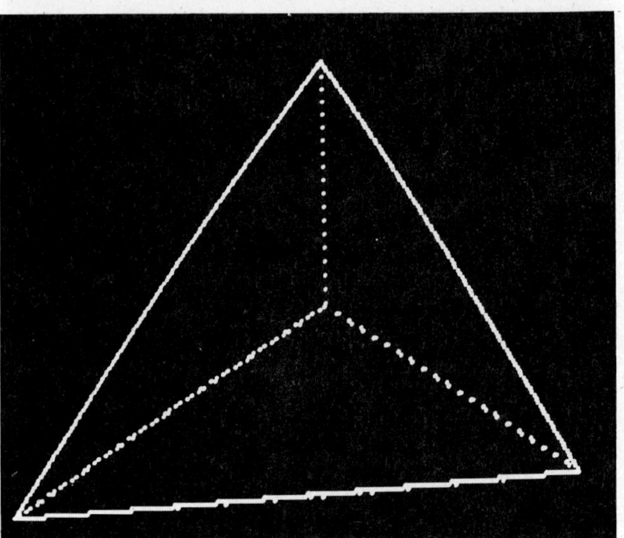

FIG. 3.44 Pictorial of the pyramid.

to a system with the appropriate software and have the time available, you may wish to generate some orthographic views of the pyramid (or any other solid object) and verify the answers to these and similar questions.

1. What lines are shown in true length in the views depicted in Figs. 3.42 and 3.43?

2. What is one viewpoint from which you can see the true length of line *AC*?

3. What is one viewpoint from which you can see the true shape of *ABD*? *BCD*? *ACD*?

4. What is one viewpoint from which you can see the true shape of *ABC*? This requires some analytic geometry calculations or graphic analysis from Part Two in order for you to get a correct answer. You can come very close, however, by inspection.

A computer graphics system thus can produce orthographic views in the same manner as you can by drawing on paper. The advantage of the computer is the speed at which the views are generated, allowing the user to view many in a short amount of time, in contrast to drawing a few views on paper in a large amount of time. As you saw in describing the geometry of the pyramid in Fig. 3.40 and in answering the questions above, computer graphics can assist you only if you have a thorough understanding of the fundamentals of engineering graphics.

Problems

3.1 Without reference to notes or text material, list the four essential features of orthographic projection. Check your answers after completing the assignment.

3.2 Incorporating the four essential features of orthographic projection, write a formal definition of orthographic projection. Check your answer.

3.3 Describe explicitly three different possible interpretations of what the F/P reference line represents; classify completely the planes involved. Do the same for any H/1 reference line, any F/2 reference line, and any 2/3 reference line. Check your answers.

3.4 Without reference to your notes or text, define:
(a) A point *(b)* A line *(c)* A plane *(d)* A solid

3.5 Sketch and label the six principal views of the orthographic drawing system.

3.6 Illustrate with sketches the following orthographic views. Label carefully and state classifications:
(a) Auxiliary inclined *(b)* Auxiliary elevation *(c)* Auxiliary oblique

3.7 Number the points in the F projection of the pyramid shown in Fig. 3.45 that correspond to those shown in the H view.

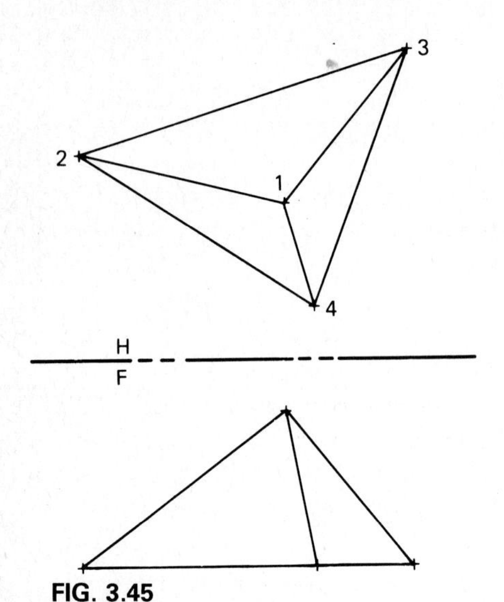

FIG. 3.45

3.8 Number all corner points of the cube in the three related orthographic views shown in Fig. 3.46.

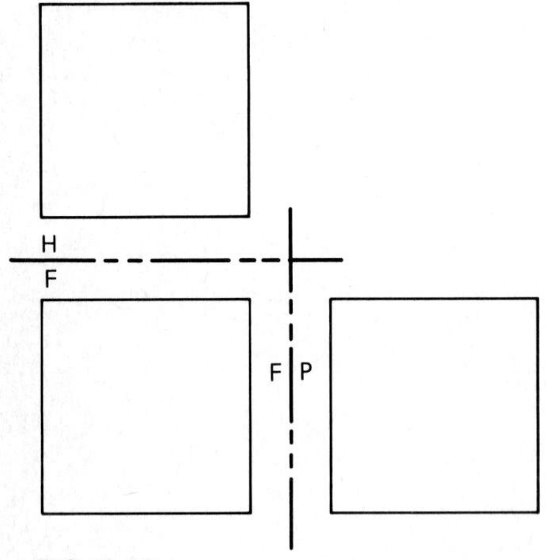

FIG. 3.46

3.9 Redraw in correct orthographic projection the given views of the object represented in Fig. 3.47. Carefully locate the CM and number the points of all intersecting lines in each view.

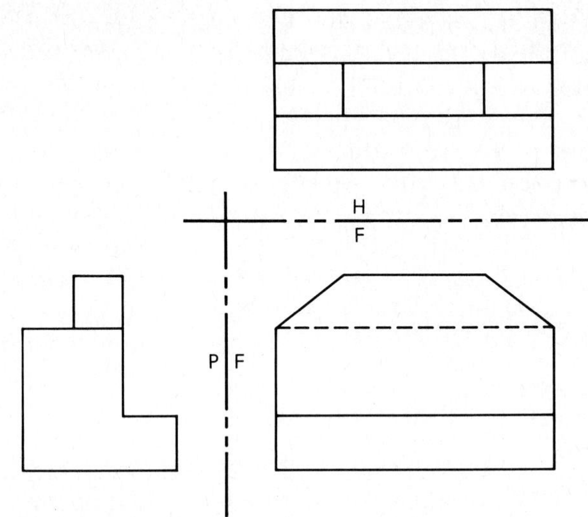

FIG. 3.47

3.10 Construct six principal views, one auxiliary inclined view, one auxiliary elevation view, and one auxiliary oblique view of a cube. Use valid orthographic projection in relation to established reference lines, indicate all common measurements, and number corner points correctly in all views.

3.11 Complete the front view and construct the left and right profile views of the object shown in Fig. 3.48. Use correct line precedence for solid, hidden, and center lines.

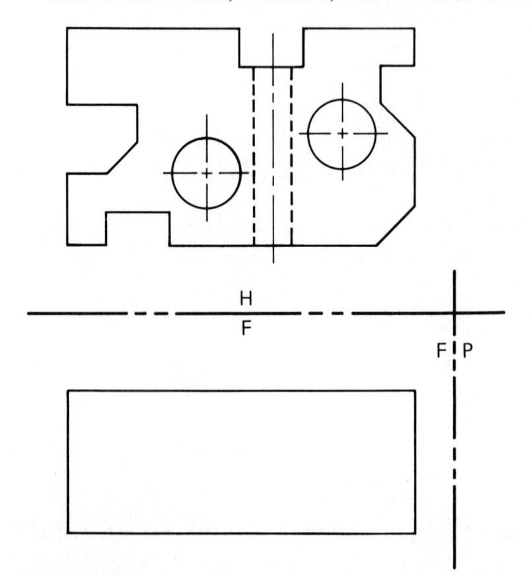

FIG. 3.48

3.12 Construct a set of principal-view orthographic reference lines. Locate point *A* (the apex of a triangular pyramid) 40 mm left of P, 25 mm behind F, and 5 mm below H; point 1, 55 mm left of P, 30 mm behind F, and 30 mm below H; point 2, 40 mm left of P, 5 mm behind F, and 30 mm below H; point 3, 10 mm left of P, 40 mm behind F, and 30 mm below H. Project all points in the horizontal, front, and right profile views. Connect the points with appropriate line symbols to represent an opaque pyramid in multiview orthographic projection. Label the points in all views.

3.13 Using the points from Problem 3.12, expand the multiview drawing of the pyramid by additional adjacent views positioned similarly to those indicated by the reference lines given in Fig. 3.49. Number the points of all intersecting lines in all views and indicate common measurements.

3.14 Redraw Fig. 3.21 and then find and label the true lengths of the specified lines, using solution views indicated as follows:

TL of line 1-4 using an auxiliary elevation view

TL of line 1-3 using an auxiliary inclined view

TL of line 1-2 using an image plane that is perpendicular to the P plane. Classify this view.

Show the point view of each of the lines and classify the image planes used for each.

3.15 Redraw the pyramid shown in Fig. 3.21. With the necessary solution views, obtain the edge view of plane 1-3-4 and the true-shape view of plane 1-2-3. Number the points in all views.

3.16 Draw the truncated prism shown in Fig. 3.50 and find the true size of the surface *ABC*.

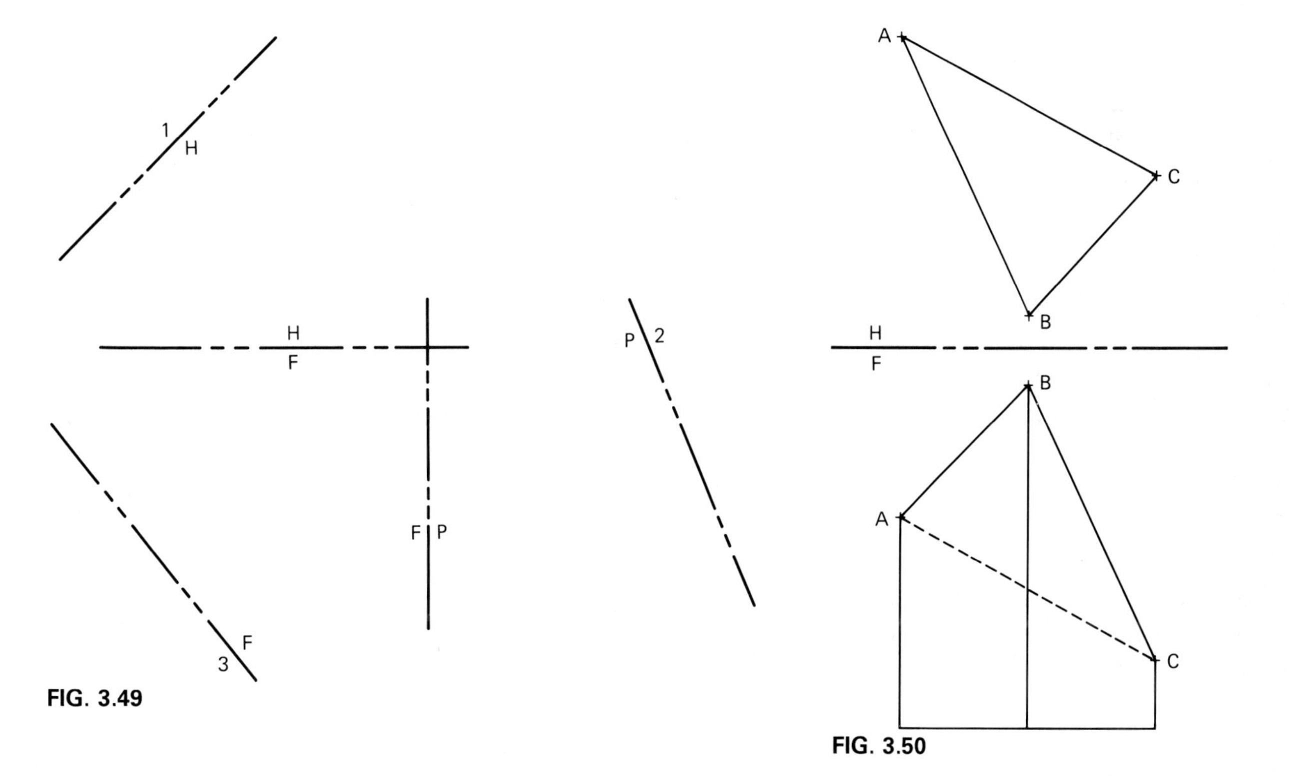

FIG. 3.49

FIG. 3.50

3.17 Draw the orthographic views necessary to show the true shape of all features of the objects shown in Fig. 3.51.

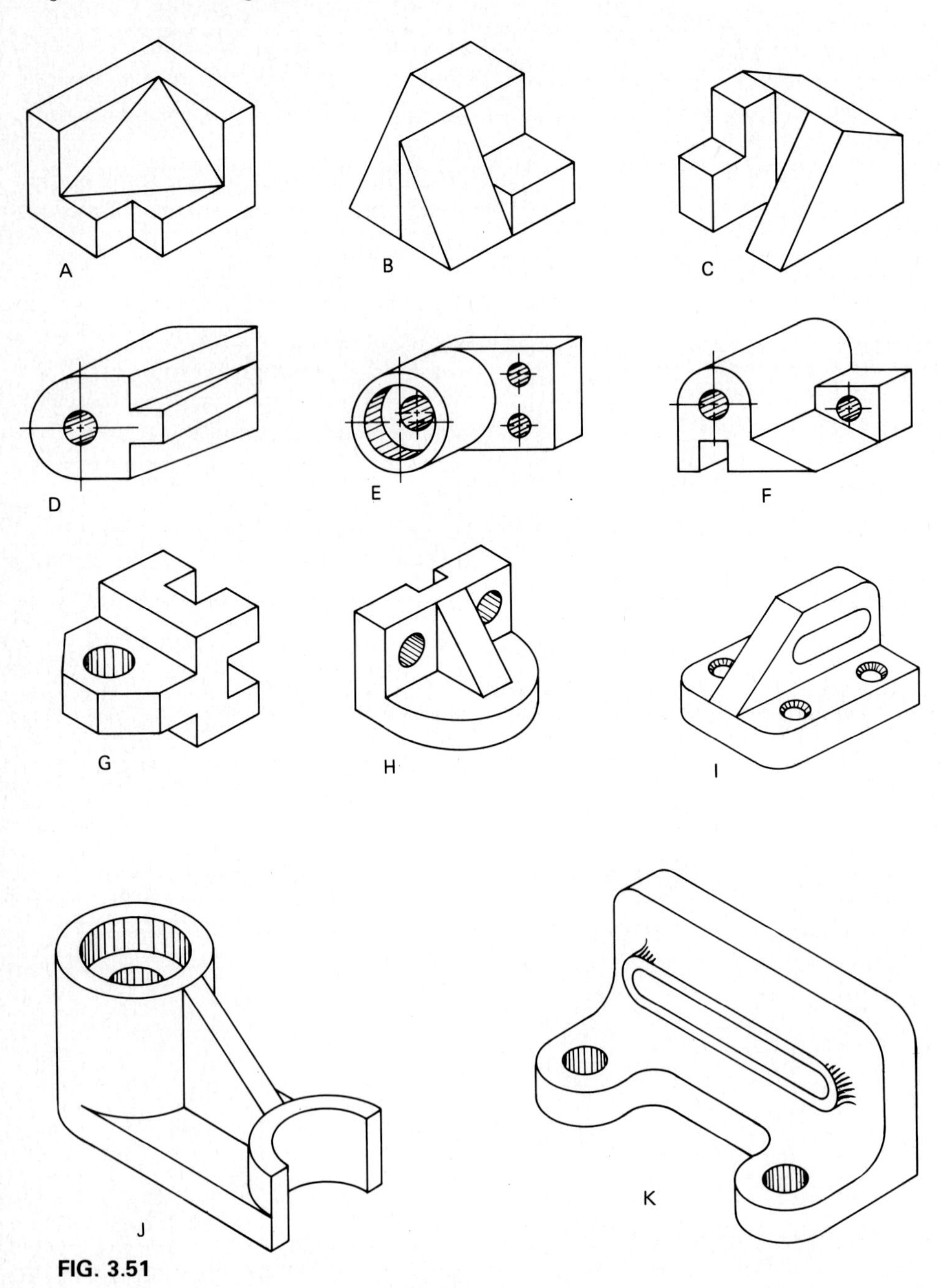

FIG. 3.51

PART TWO

GRAPHICAL THEORY

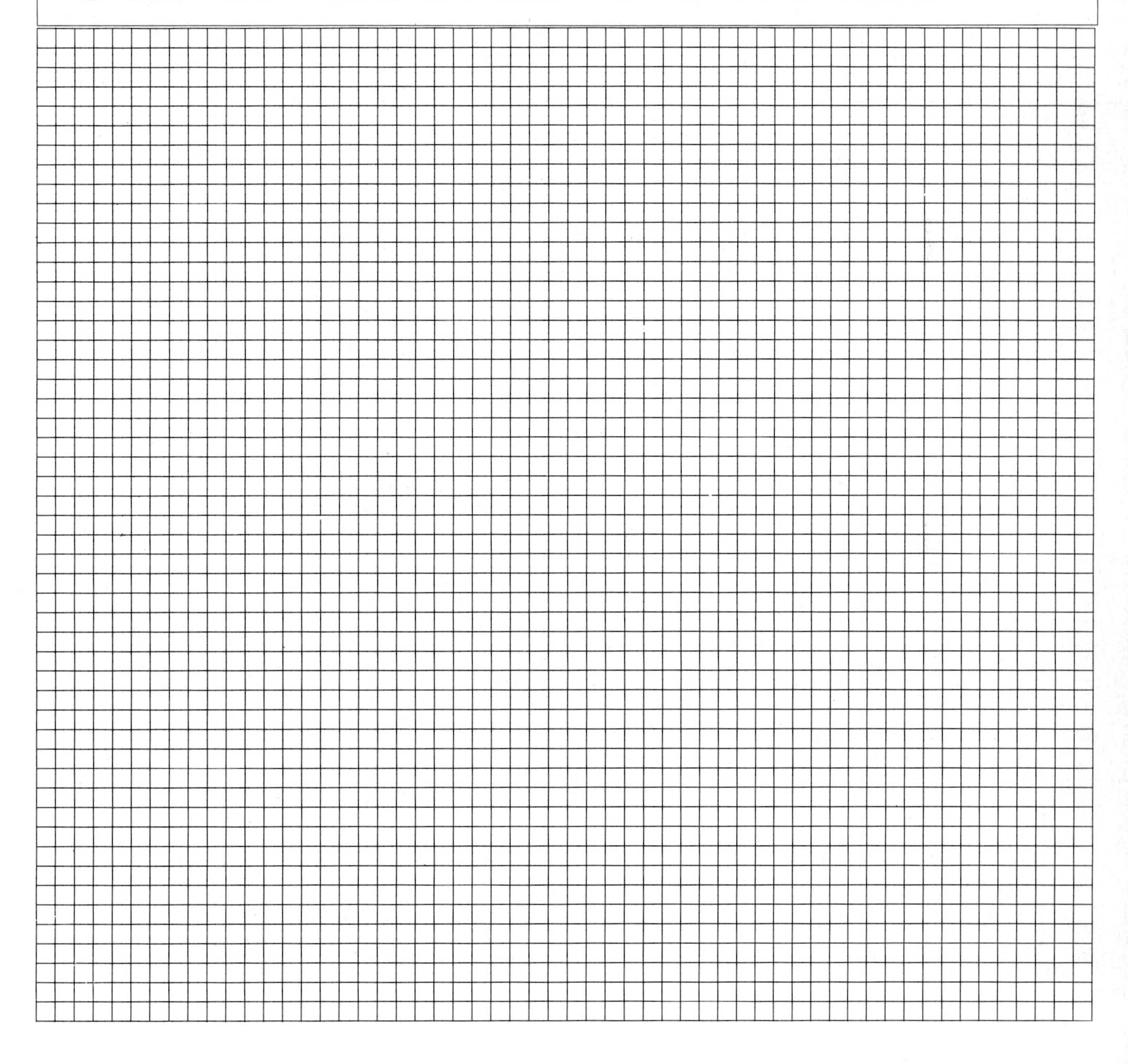

chapter 4

Geometry of Straight Lines

4.1 INTRODUCTION

Lines are essential components in the representation and geometry of two- and three-dimensional surfaces. Thus, knowledge of the classification of lines and their orthographic projection characteristics is critical in solving problems that involve surfaces. Understanding how to find the true length and the point view of a line is essential before more significant geometry problems can be attempted. For example, the theory of lines is important in the layout of pipeline routes.

This chapter will be restricted to the explanation of straight lines. Theoretically, a straight line is of infinite length. In practical usage, however, a straight line is represented with fixed endpoints (that is, a line segment). Be aware that a given line segment may be extended or shortened to facilitate solving various graphics problems. See Fig. 4.1.

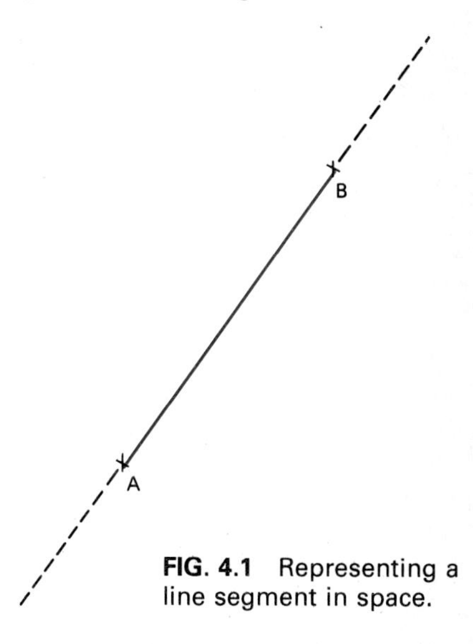

FIG. 4.1 Representing a line segment in space.

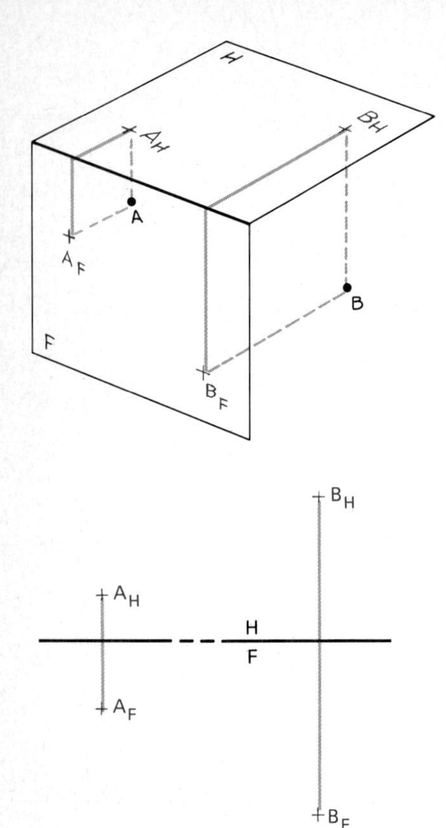

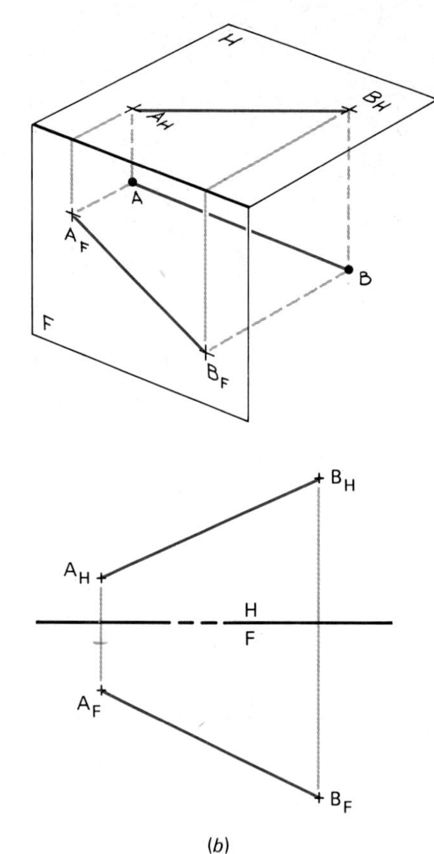

FIG. 4.2 (*a*) Projecting two points to determine a line orthographically. (*b*) Representing a straight-line segment orthographically.

4.2 PROJECTION OF A STRAIGHT LINE

The fundamentals of orthographic projection of points, lines, planes, and three-dimensional objects were presented in Chap. 3. The study of straight-line projection was introduced in Chap. 3, and the projection principles are reviewed here.

As can be seen in Fig. 4.2*a*, two points can be represented independently in adjacent orthographic views. If these points are connected with a straight line, a line segment is defined as in Fig. 4.2*b*. The line is identified by labeling endpoints with a letter and a subscript that identifies that projection. The line may be projected onto other planes by use of the three-plane relation and common measurements.

4.3 CLASSIFICATION OF STRAIGHT LINES

Straight lines can occupy an infinite variety of positions in space. For convenience in describing these various positions for use in the orthographic system, they can be grouped into the seven basic positions shown in Fig. 4.3.

As you study Fig. 4.3, note that the classification of a line is defined by the plane or planes to which it is parallel. Therefore, a line that is parallel to the horizontal plane of projection is classified as horizontal, a line parallel to the frontal plane of projection is classified as front, and so forth. A line parallel to two planes of projection is classified according to both, such as horizontal-profile. The special case of a front-profile line is more com-

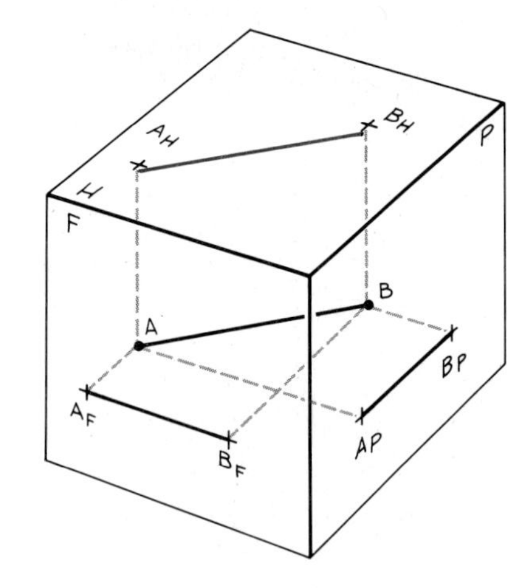

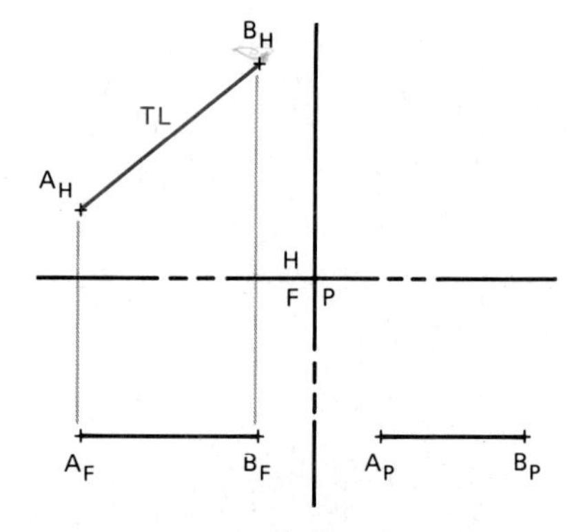

(*a*) Horizontal line

FIG. 4.3 Line classification.

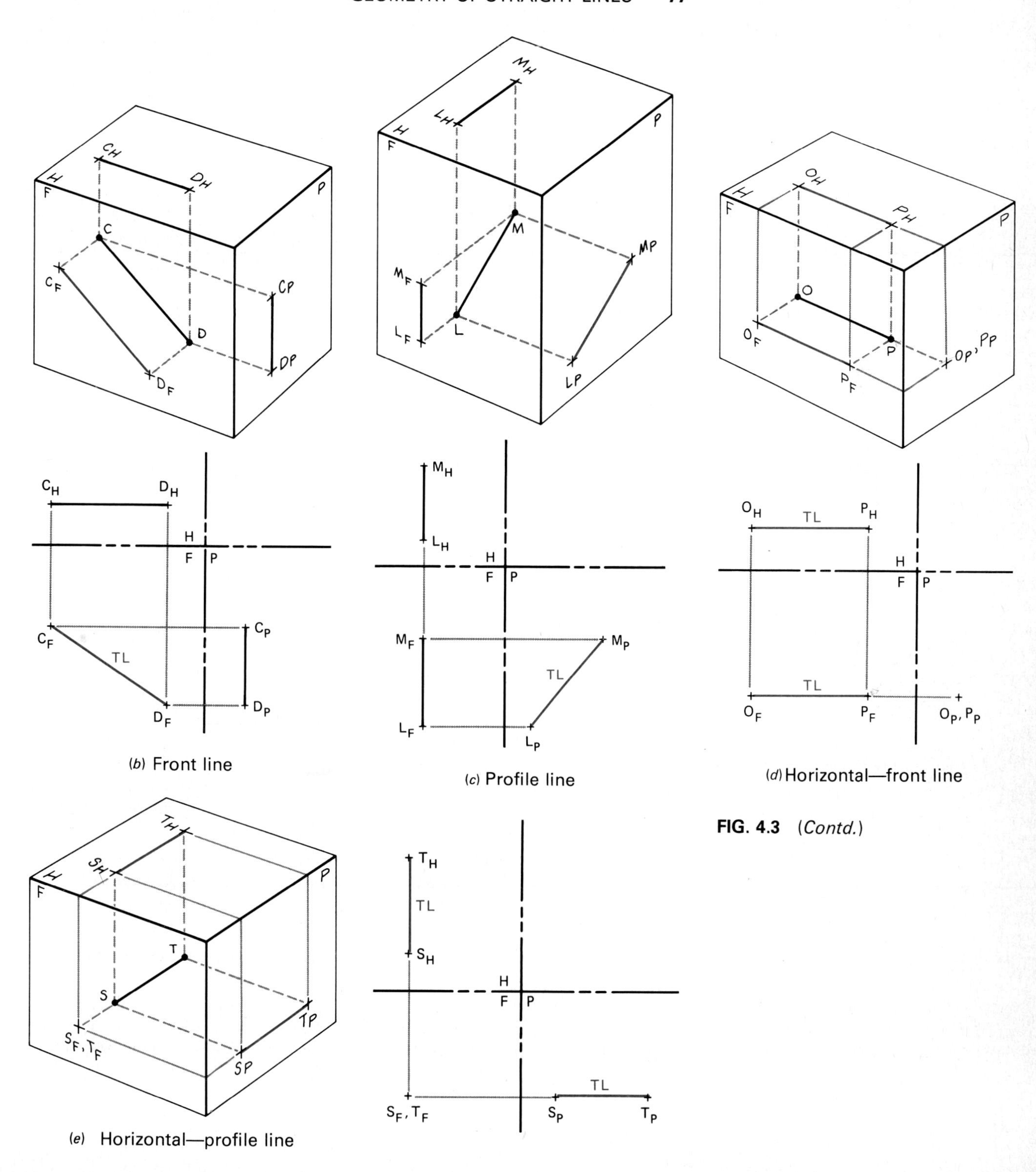

(*b*) Front line

(*c*) Profile line

(*d*) Horizontal—front line

(*e*) Horizontal—profile line

FIG. 4.3 (*Contd.*)

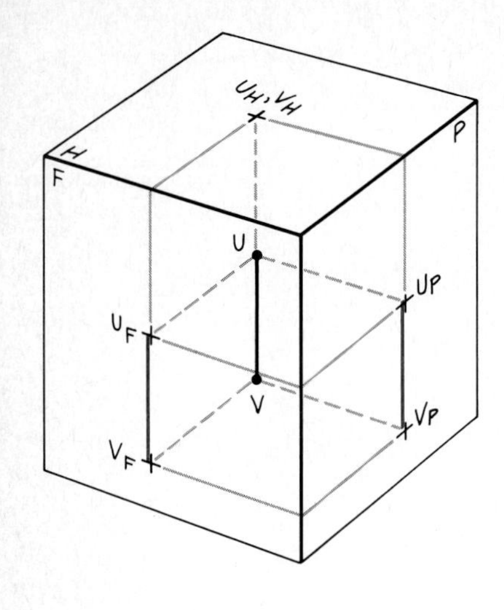

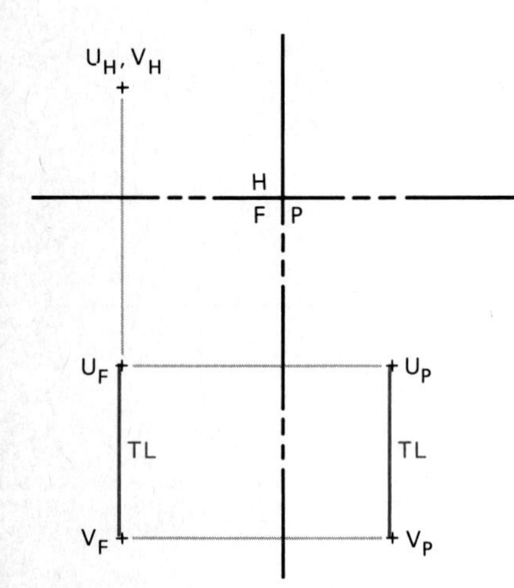

(f) Front—profile (vertical) line.

FIG. 4.3 (*Contd.*)

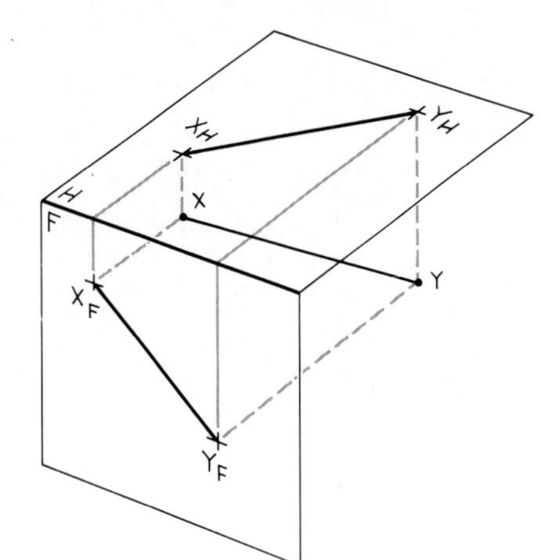

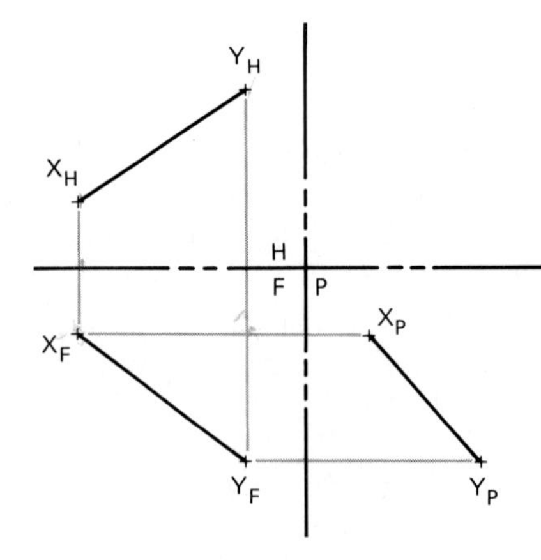

(g) Oblique line

monly identified as a vertical line. A line that is not parallel to any of the principal planes of projection is termed oblique.

4.4 SOLUTION VIEW AND SOLUTION PROCEDURE

In Chap. 3, it was emphasized that the method of using the orthographic system for solving geometry problems is first to visualize the solution view and then to determine and complete the appropriate solution procedure; that is, establish your objective and proceed with direct action to achieve that objective.

You will remember that the *solution view* is the ultimate orthographic view that provides visual proof that the problem is solved, whereas the *solution procedure* is the step-by-step sequence of action required to arrive at the solution view.

It is important to recognize that depending on the problem involved and its orientation in space, the solution view may be any principal or auxiliary orthographic view.

4.5 BEARING

The angular orientation of a line measured from a north-south direction is called the bearing. In the orthographic system of drawing, it is always measured in the horizontal plane of projection (sometimes called the plan view or map view).

To assist you in visualizing the bearing of a line, consider the map of the United States in Fig. 4.4, with Kansas City as the origin point of a line and other major cities as the terminal point. Note the approximate bearings of these lines: Minneapolis, Minnesota, is due north; Houston, Texas, is due south; Washington, D.C., is due east; and San Francisco, California, is due west. By drawing lines to other major cities, you see that the bearing of the line to Phoenix, Arizona,

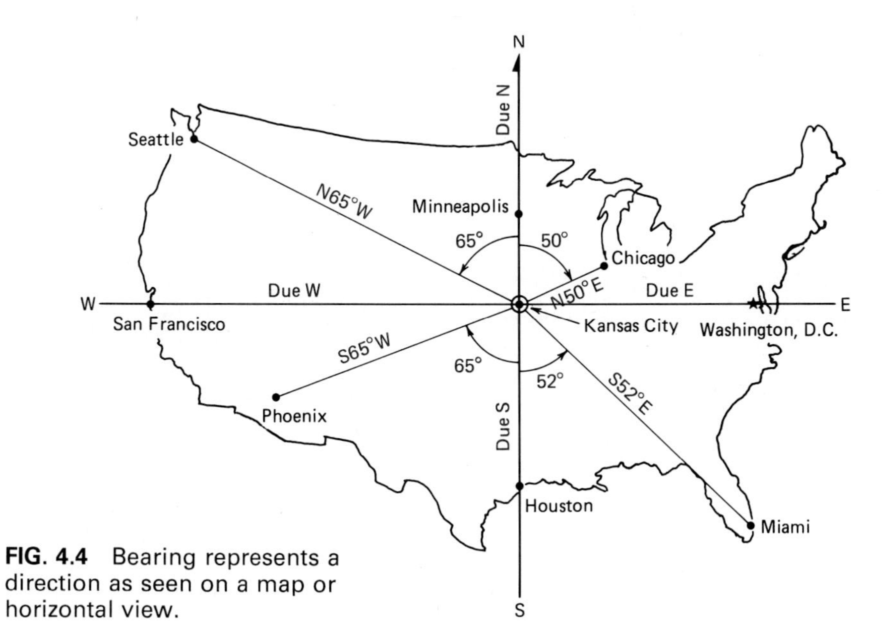

FIG. 4.4 Bearing represents a direction as seen on a map or horizontal view.

measures 65° west of south. Similarly, the bearing of the line to Seattle, Washington, is 65° west of north; to Chicago, Illinois, it is 50° east of north; and to Miami, Florida, it is 52° east of south. Note that bearing has nothing to do with how far it may be from the origin to the end of a line or whether the line goes up or down but deals only with the direction traveled in moving from the origin to the end of the line.

4.5.1 Solution View: Bearing of a Line

The solution view for the bearing of a line is the horizontal (map) view of the line. The horizontal view is used to find the acute angle between the horizontal projection of the line and a north-south line drawn through the origin. Unless otherwise specified, the origin is taken to be the first character in a two character sequence denoting the line. Thus, for lines $\overline{PQ}$ and $\overline{23}$, the origins are P and 2 respectively. For lines $\overline{QP}$ and $\overline{32}$ the origins are Q and 3.

4.5.2 Solution Procedure: Bearing of a Line

1. Find the horizontal view of the line if it is not given.

2. Draw a north-south line through the origin point of the horizontal projection of the given line. A north-south line is perpendicular to the H/F reference line.

3. Use a protractor to measure the acute angle between the north-south line and given lines.

4. Record the bearing as illustrated in Figs. 4.4 and 4.5. Abbreviations for north, south, east, and west are satisfactory in most cases. The proper format for labeling bearing is: Nβ°E (or W), Sβ°E (or W), DUE-N, DUE-E, DUE-S, or DUE-W.

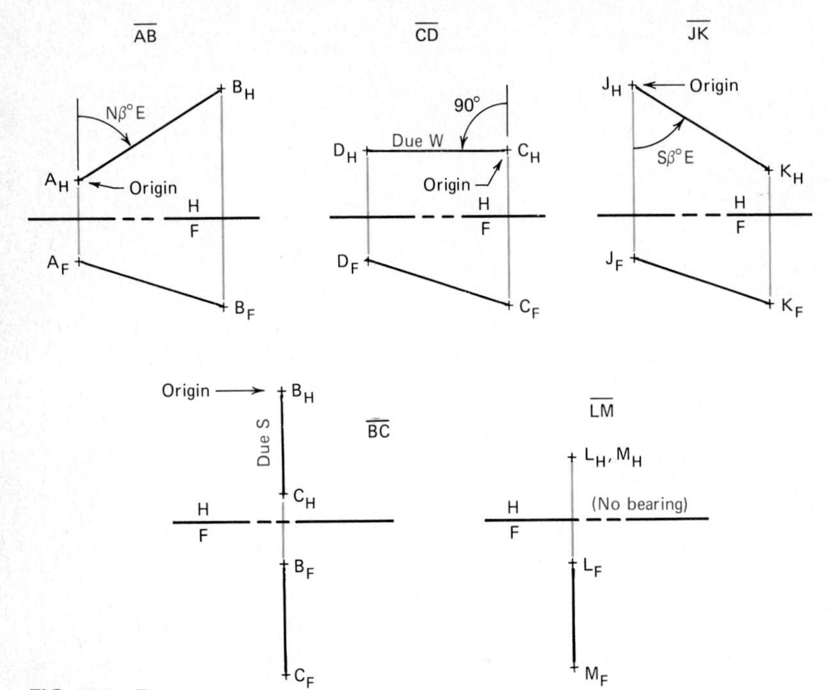

FIG. 4.5 Example bearing designations.

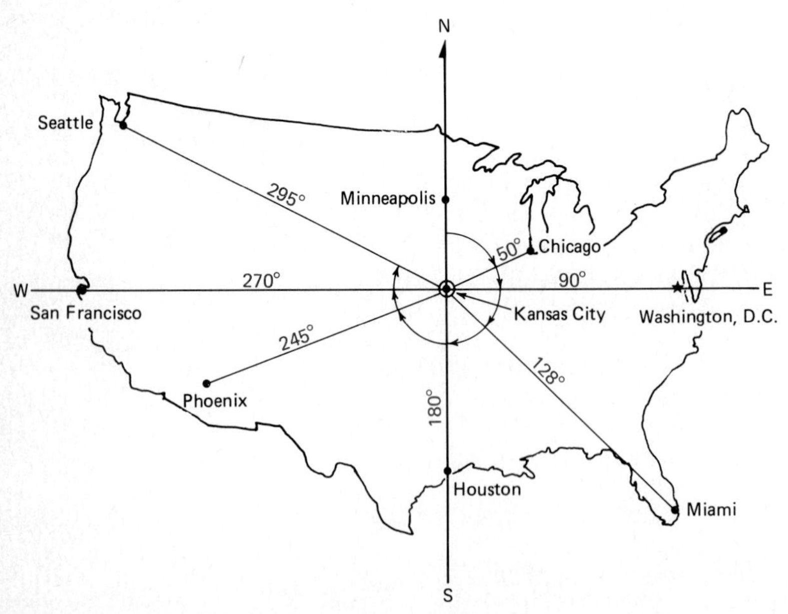

FIG. 4.6 Azimuth measured clockwise from north.

Note True length of a line, as will be discussed more fully in Sec. 4.6, is not a factor in the bearing of a line. Bearing is associated with the acute angle between the projected view in the horizontal plane and the north-south line. The horizontal view of the line will generally be foreshortened because the origin and terminal points are not at the same distance from the horizontal projection plane. This causes no concern because you are only interested in the direction you would go to get from the origin to the terminal end of the given line, not whether you rise or fall.

4.5.3 Azimuth

A related measure of direction of a line in the horizontal view is the azimuth. This is defined as the angle in degrees measured clockwise from a reference direction. In most applications, such as surveying, the reference direction is due north. Thus, the azimuth can have values from 0° up to 360°. Values of azimuth to the cities cited in Sec. 4.5 are shown in Fig. 4.6.

4.6 TRUE LENGTH OF A LINE

The graphic solution to numerous problems is dependent on the use of the true length and point view of any straight lines involved. A line in any given projection may appear as a point foreshortened to some extent or in its true length. Thus, it is essential to understand how the relative positions of a line and a projection plane affect the projection of the line on the plane.

4.6.1 Solution View: True Length of a Line

The solution view for true length is any orthographic view of the line that has been projected onto an image plane parallel to the line. Figure 4.7*a* illustrates a line with its true length visible on the front plane. In general, an additional view to those given will be needed to find the true length.

4.6.2 Solution Procedure: True Length of a Line

Referring to Fig. 4.7*b*,

1. Draw a solution view reference line parallel to any established projection of the line segment.

2. Project the endpoints of the line segment to the solution view and connect them with a straight line.

3. Label the line segment TL (true length).

Note Vertical, horizontal, front, and profile lines are each parallel to one or more of the principal planes of projection. Therefore, the true length of these lines will be seen on at least one of the principal planes. Review the classification of lines and note how this principle holds true in Fig. 4.3, parts a through f. (The true-length projections are labeled in this figure.)

4.7 POINT VIEW OF A LINE

As in the case of true length, knowing how to project a line to a plane of projection showing the point view of the line segment will be useful for solving more meaningful graphics problems. The procedure will be used extensively in later applications.

4.7.1 Solution View: Point View of a Line

The point view of a line is any view projected onto an image plane that is perpendicular to a true-length projection of the line.

4.7.2 Solution Procedure: Point View of a Line

Referring to Fig. 4.8,

1. Obtain a true-length projection of the line segment as outlined in Sec. 4.6.2.

2. Draw a reference line perpendicular to the true-length projection of the line segment.

3. Project the endpoints to the solution view so that one endpoint will be superimposed over the other. Label the coincident points to designate the line as a point view.

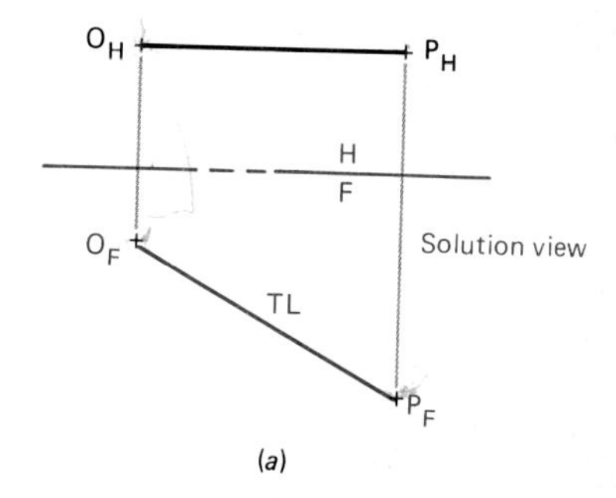

(*a*)

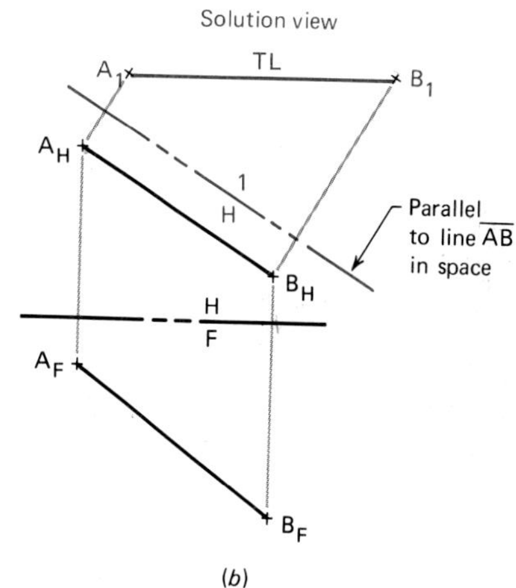

(*b*)

FIG. 4.7 Solution views for true length.

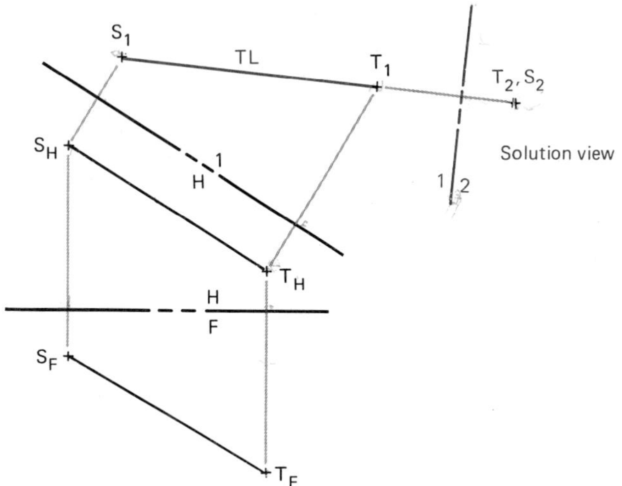

FIG. 4.8 Solution view for point view.

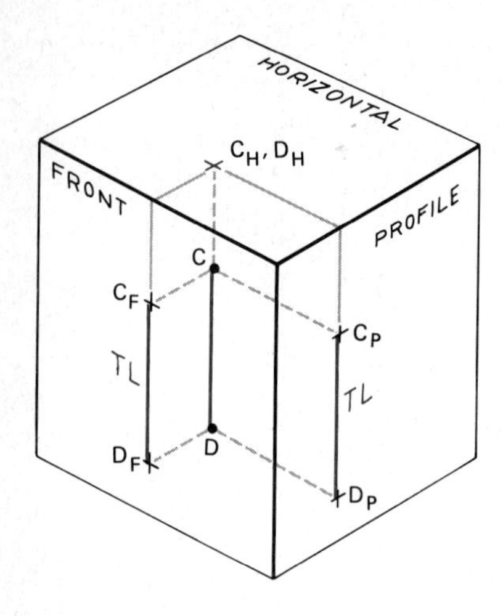

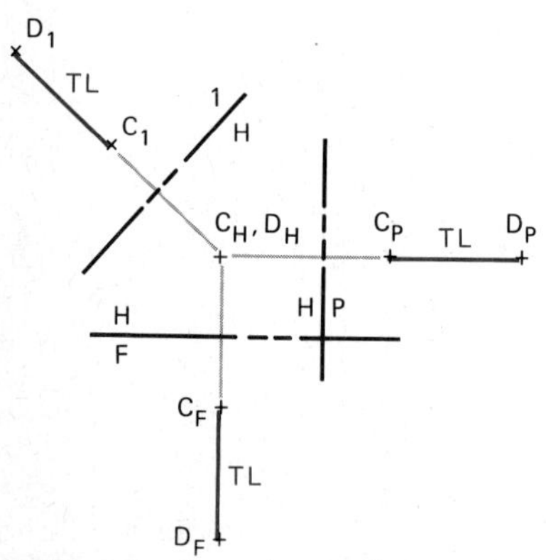

FIG. 4.9 Any view adjacent to a point view will show true length.

Figure 4.9 illustrates the relationship between the true-length view and the point view of a line. Notice that *any* view adjacent to the point view of a line will show the line in true length. This fact will be important later as more complicated graphics solutions are introduced.

4.8 INCLINATION OF A LINE

Inclination describes the deviation of a line from the horizontal plane. From its origin, a line may be horizontal (zero inclination), may rise (positive inclination), or may fall (negative inclination).

Inclination may be expressed in terms of slope angle, slope, or grade, which will be explained in subsequent paragraphs. However, the solution view for graphically showing the inclination of a line is the same whether the inclination is expressed as slope angle, slope, or grade.

4.8.1 Solution View: Inclination of a Line

The inclination of a line can be measured only in an elevation view showing the true length of the line. See Fig. 4.10.

4.8.2 Solution Procedure: Inclination of a Line

Referring to Figs. 4.10 and 4.11,

1. Construct a reference line parallel to the horizontal projection of the given line segment.

2. Project the endpoints of the line to the solution view, which is an elevation view showing the line in true length. The line is in true length be-

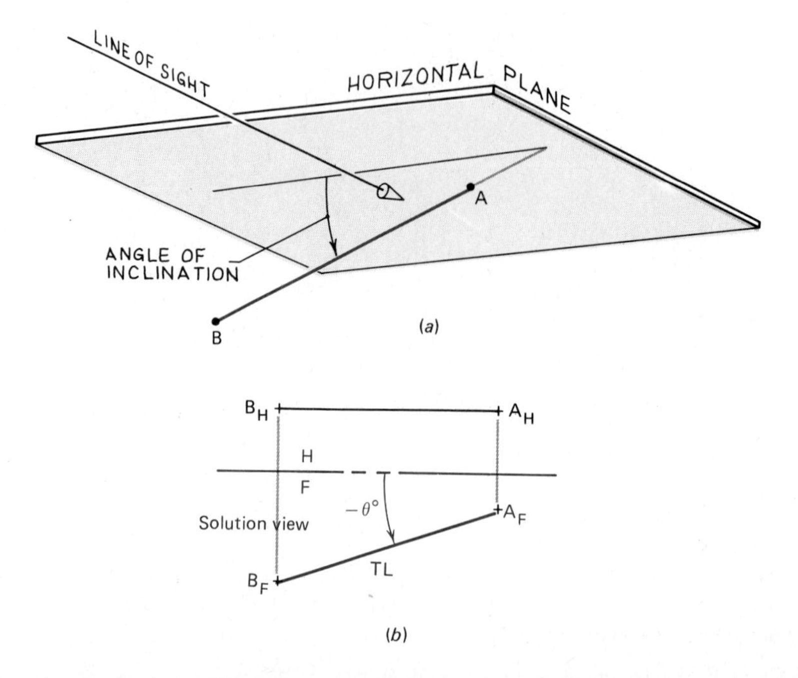

FIG. 4.10 Viewing the true angle of inclination. Note that the inclination of the line is negative if the line falls away from the horizontal plane.

cause the reference line was constructed parallel to a line projection. The view is an elevation view because it is perpendicular to the horizontal plane of projection.

3. Measure the deviation of the true-length projection of the line from the horizontal plane. See Sec. 4.8.3 for the method of finding slope angle, Sec. 4.8.4 for slope, and Sec. 4.8.5 for grade.

4.8.3 Slope Angle

The slope angle is the true angle a line makes with the horizontal plane expressed in degrees. Once the solution view is found, the slope angle is the acute angle between the line segment and a line parallel to the horizontal plane drawn through the origin of the line. If the line falls (origin to terminus) from the horizontal plane, the slope angle is negative; if it rises, the angle is positive. See Fig. 4.11 for examples of finding the slope angle.

4.8.4 Slope

Slope is defined as the tangent of the slope angle; stated another way, it is the ratio of the vertical rise or fall over the horizontal run of a line.

If the slope angle is known, the slope may be determined quickly on a scientific calculator. Graphically, the slope may be determined in the solution view by using the line segment as the hypotenuse of a right triangle. Construct a horizontal (side adjacent) line from the origin point 100 units long (use any convenient scale) and a vertical (side opposite) line to intersect the hypotenuse. The value of the vertical leg (using the same scale) divided by that of the horizontal leg (100 units) equals the slope. This technique is illustrated in Fig. 4.12. As in the case of the slope angle, slope has an attached sign: positive if the line rises, negative if it falls.

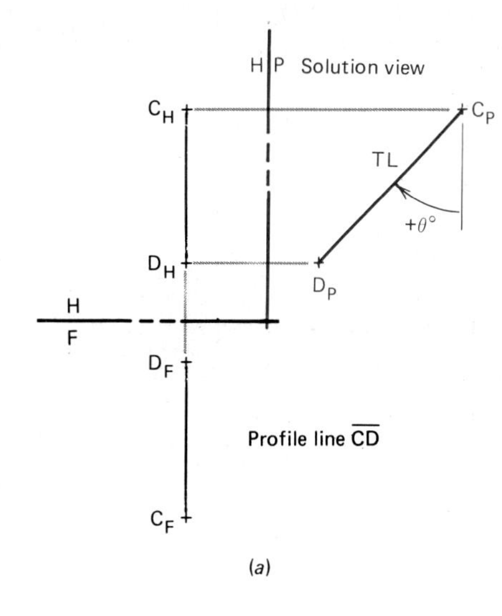

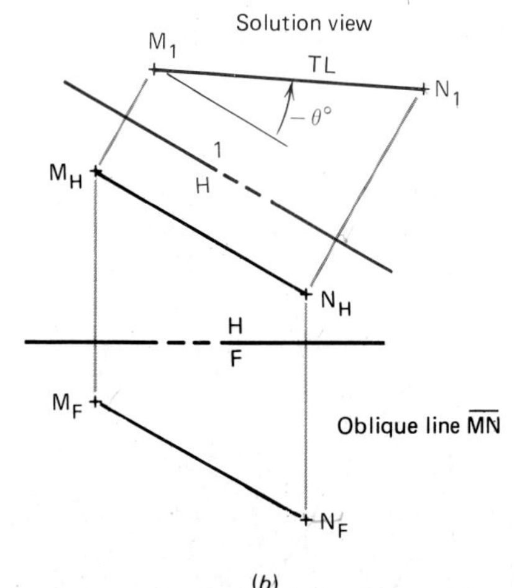

FIG. 4.11 Solution views for angle of inclination.

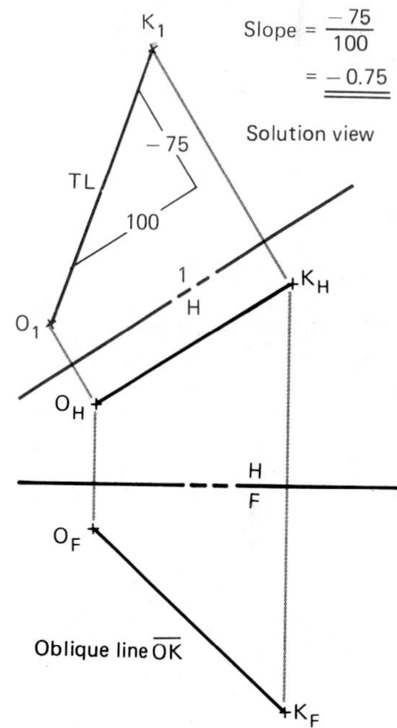

FIG. 4.12 Specification of slope.

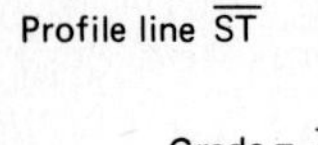

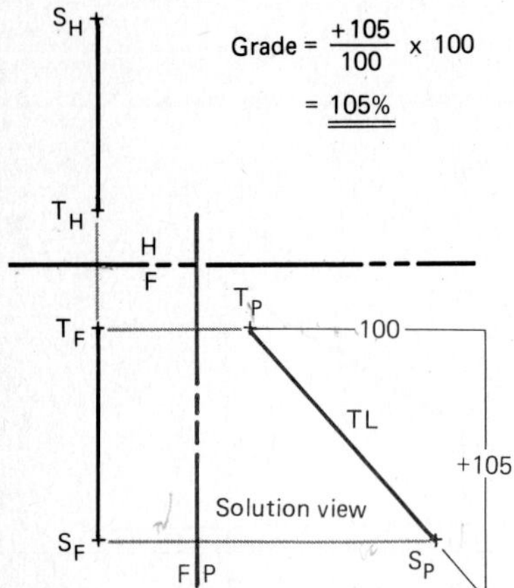

FIG. 4.13 Specification of grade.

4.8.5 Grade

The grade (or percent grade) is obtained by multiplying the slope by 100 percent. Therefore, the graphic procedure for finding the grade is identical to that for finding the slope, except for the final multiplication. The sign of the grade follows from the slope value. Figure 4.13 is an example of a grade calculation.

4.9 GRAPHIC SOLUTION OF INCLINATION PROBLEMS GIVEN LINE SPECIFICATIONS

In the practical work of engineering design, problems involving lines may be described in terms of line specifications. Length, bearing, grade, and the coordinates of one point on the line describe a line as adequately as three orthographic views of the line. However, in this case you must create the views from the specifications.

An example of a problem involving inclination of a line given by specifications only follows. The analysis of the data and a step-by-step graphic procedure are shown in Fig. 4.14, leading to the final solution.

4.9.1 Problem Description

Draw the front, horizontal, and other necessary views of line *AB* that has the following specifications:

True length: 30 mm
Bearing: S30°E
Grade: +140 percent

4.9.2 Problem Analysis and Observations

1. All views must be in correct orthographic projection.

2. You can recognize that the line is oblique from the inclination and bearing specifications. Therefore, the true inclination will not be shown in the front view, and this view cannot be drawn initially.

3. An auxiliary elevation view showing the true length of the line will be needed to construct the inclination.

4. The required auxiliary elevation plane must be parallel to the horizontal projection of line *AB* so that *AB* will project in true length. Therefore, you must have a fixed position of the line in the top view before you can orient the auxiliary plane properly.

5. You can establish only the direction (bearing) of *AB* in the top view. The projected length is not known; that is, the location of B_H cannot be placed intially in relation to A_H.

6. In this problem, the location of point *A* in the horizontal view is not specified. Thus, you can locate it arbitrarily, although you should use some care in doing so to avoid difficulties later. Notice that the bearing from point *A* is S30°E. This means that point *B* will be closer to the H/F reference line than *A* is. If the bearing were NE or NW, the opposite would be true. You should locate *A* far enough behind the front plane so that point *B* will not overlap the H/F reference line. Such an overlap causes visualization problems.

4.9.3 Solution Procedure

Refer to Fig. 4.14 for the steps described below.

Step 1. Anticipate the need for and the direction of expansion of the or-

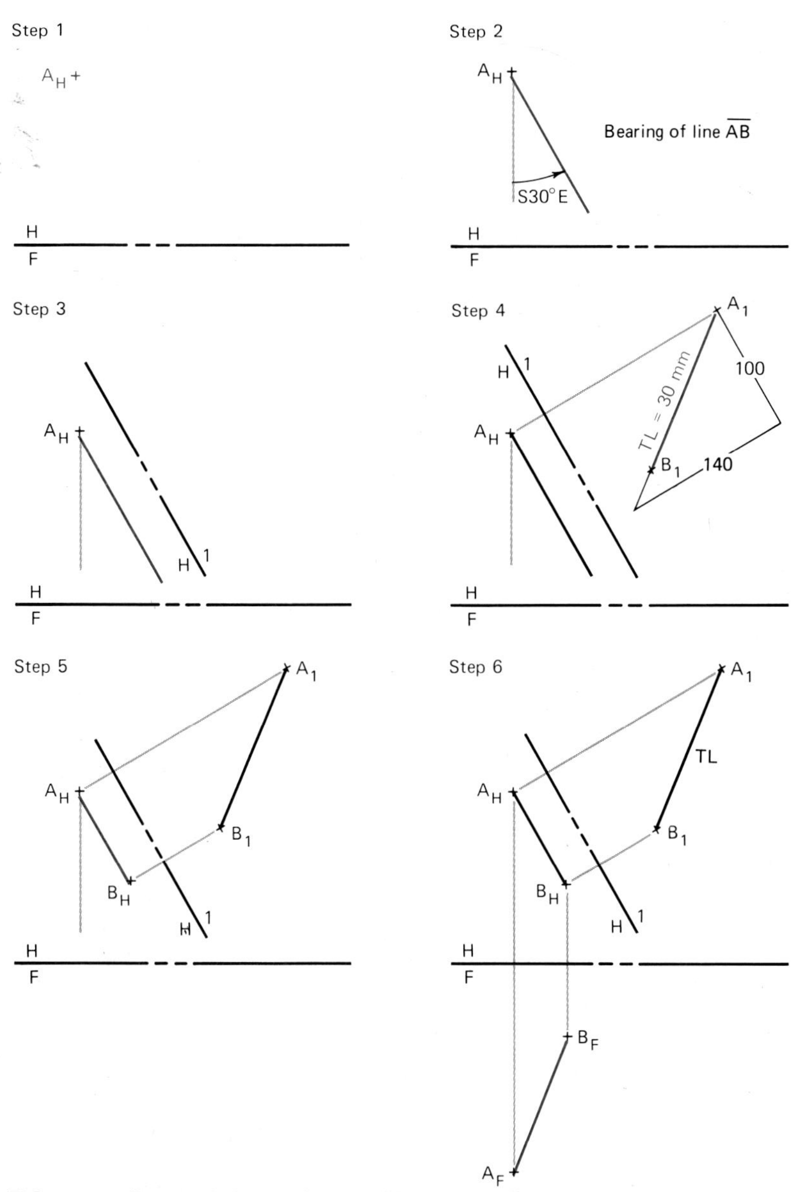

FIG. 4.14 Determining orthographic views of a line from given specifications. Drawing not full scale.

thographic system for required auxiliary views and locate the H/F reference line. Fix the position of point A_H approximately 35 mm behind the front plane (recall that a line will never project longer than its true length).

Step 2. Draw a line in the direction of the given bearing (S30°E) from the origin point A_H. The length of this line is not known yet, and so you cannot locate point B_H.

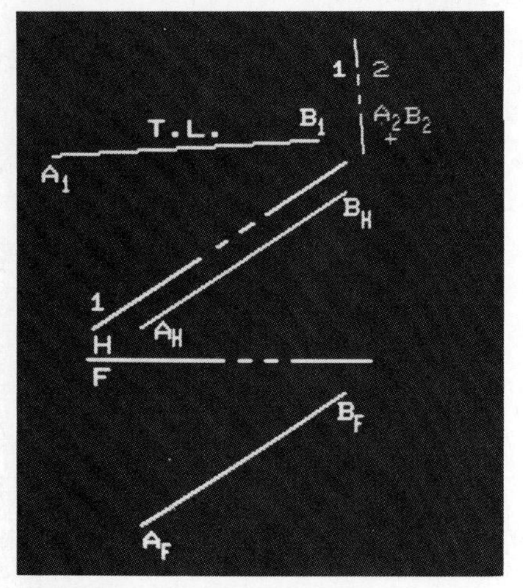

FIG. 4.15 Representation of true length and point view on a computer graphics screen.

Step 3. Draw a reference line H/1 parallel to the incomplete horizontal projection of *AB*.

Step 4. Fix the position of point A_1 at some convenient height, anticipating the space needed to draw line *AB* with a positive inclination. The position of A_1 is arbitrary because the elevation of neither *A* nor *B* was specified. Be sure point A_1 is in correct orthographic alignment with A_H. Again, try to avoid having the elevation view of *AB* overlap the horizontal plane.

Use A_1 as the origin. With any convenient scale, lightly construct a horizontal line 100 units long to produce the adjacent side of a right triangle. Construct a vertical line (side opposite) 140 units long, using the same scale.

Draw the hypotenuse of the right triangle from A_1 through the vertex. This establishes a line along which *AB* will be found. Because the inclination is positive, be sure the line is inclined upward toward the H/1 reference line from point A_1. Measure off 30 mm on this line from point A_1, thus fixing the position of B_1.

Step 5. Project from B_1 orthographically to B_H, which completes the top view of *AB*.

Step 6. Project the completed top view of *AB* to the front view, using standard orthographic procedures.

4.10 METHODS FROM COMPUTER GRAPHICS

The practical applications of computer graphics to line problems are manifold because lines occur so frequently in solids, where they may represent intersections of plane surfaces or edge views of planes. Further discussion of such applications will occur as topics on pictorials and solids are introduced.

Lines are handled in a fundamentally different way on the computer from the way they are in orthographic projection, although the result is the same. Computer software requires the user to define a line segment by specifying the three-dimensional coordinates of the segment's endpoints. These points are then used to establish an equation for the line. Once that is done, you may define a viewing plane by describing its position and orientation. The software will allow you to request a plot of the line projected onto the plane. By using three planes—horizontal, front, and right profile—you can create a three-view orthographic picture of the line that will appear exactly as if you had drawn it with instruments on paper.

The bearing can be seen in the top view as usual, but the true length and inclination are generally not shown in the principal views. You could define a new plane (auxiliary elevation) parallel to the line and project the segment onto it to find the true length and inclination.

Alternatively, you might ask that the line be rotated in three-dimensional space so that it becomes parallel to the front plane; thus, its projection would appear in true length. If the rotation is done about a vertical axis through the origin of the line, the front view will also show the line's true inclination. Note, however, that the rotation will change the bearing of the line since the line's projection will occupy a different position in the horizontal view. Further rotation about a horizontal axis could provide a point view of the line in the top view.

Problems

4.1 State the classification and possible bearing(s) of a line that projects as a point in:
(a) The front view
(b) The profile view
(c) The horizontal view
(d) An auxiliary elevation view
(e) An inclined view perpendicular to the F plane
(f) An inclined view perpendicular to the P plane

4.2 Write a concise description of the solution view for:
(a) True length of a line
(b) Bearing of a line
(c) Inclination of a line
(d) Point view of a line

4.3 A vertical line will always appear true length in any ______ view.

4.4 The projection of a line on a plane parallel to the line will always project in ______.

4.5 What is the slope angle of a profile line having an angle of +30° with the front plane?

4.6 A profile line having a −20° slope angle will have what angle with the front plane?

4.7 Draw the front, horizontal, and necessary auxiliary views to show the following lines:
(a) *AB:* 3 cm TL, S30°W, −0.7 slope
(b) *CD:* 35 mm TL, S30°E, +30° slope angle
(c) *JK:* 3.5 cm TL, N30°W, +120 percent grade

4.8 Point *M* of line *MN* is 10 mm behind the front plane and 40 mm below the horizontal plane. Draw the necessary views of line *MN*, which is TL 38 mm, N45°W, +90 percent grade.

4.9 Point *R* of line *RS* is 2 cm behind the F plane and 2.5 cm below the H plane. Point *S* is 2 cm west and 3 cm north of *R* at an elevation 4 cm below point *R*. Determine the bearing, slope angle, and point view of line *RS*.

4.10 Profile line *TV* has a slope of +1.2 and a TL of 30 mm. Show all necessary views and a point view of line *TV*.

4.11 Point *A* is 10 mm behind the front plane and 15 mm below the horizontal plane. Point *B* is north of *A* and 20 mm west of *A*, at an elevation 30 mm lower than *A*. Line *AB* makes an angle of 60° with the front plane of projection. Determine the bearing and grade of line *AB*.

Note In Problems 4.12 through 4.17, redraw the lines freehand and label (TL) for true length, (B) for bearing, and (Θ) for slope angle.

4.12 Find and label the true length, bearing, slope angle, and classification of the lines shown in Fig. 4.16.

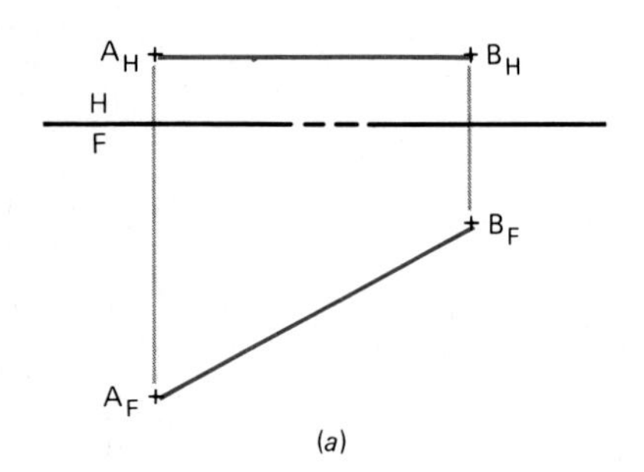

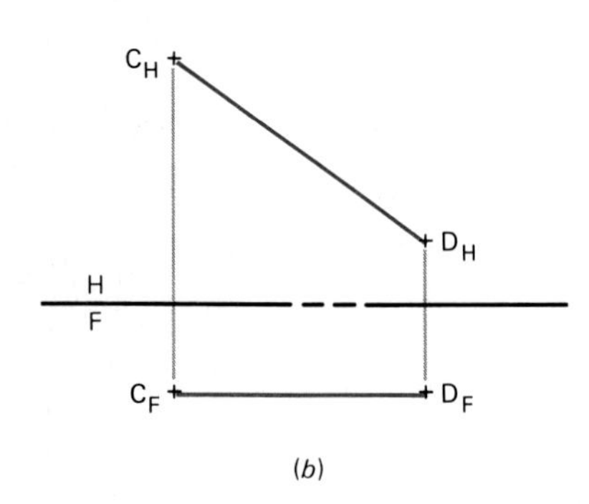

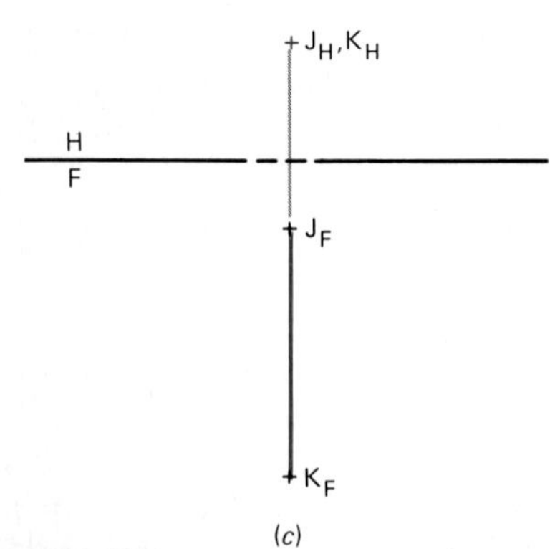

FIG. 4.16

4.13 Find and label the true length, bearing, slope, and classification of the lines shown in Fig. 4.17.

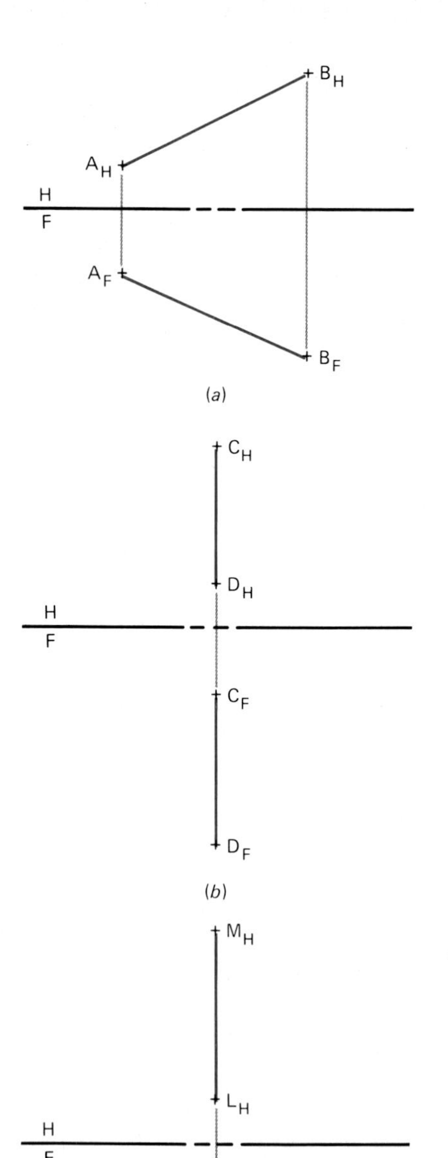

FIG. 4.17

4.14 Find and label the true length, bearing, grade, and classification of the lines shown in Fig. 4.18. Complete the given views and any additional views necessary.

4.15 Find and label the true length, bearing, and slope of the lines shown in Fig. 4.19.

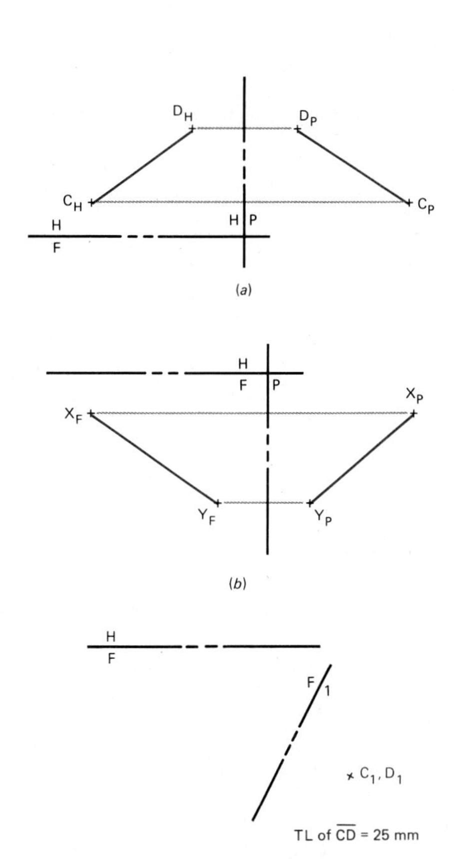

FIG. 4.18

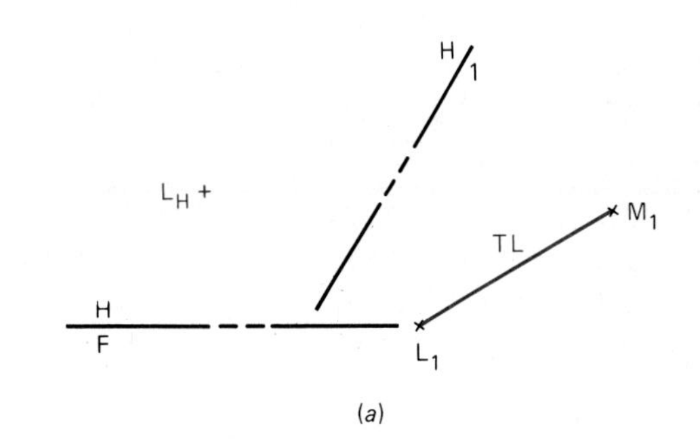

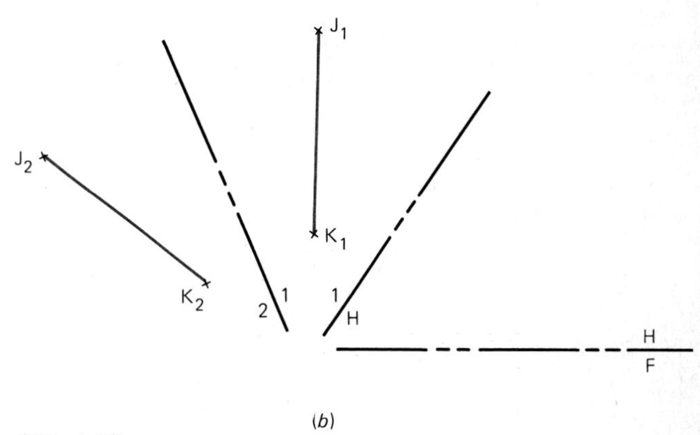

FIG. 4.19

4.16 Find and label the true length, bearing, and slope angle of the lines shown in Fig. 4.20.

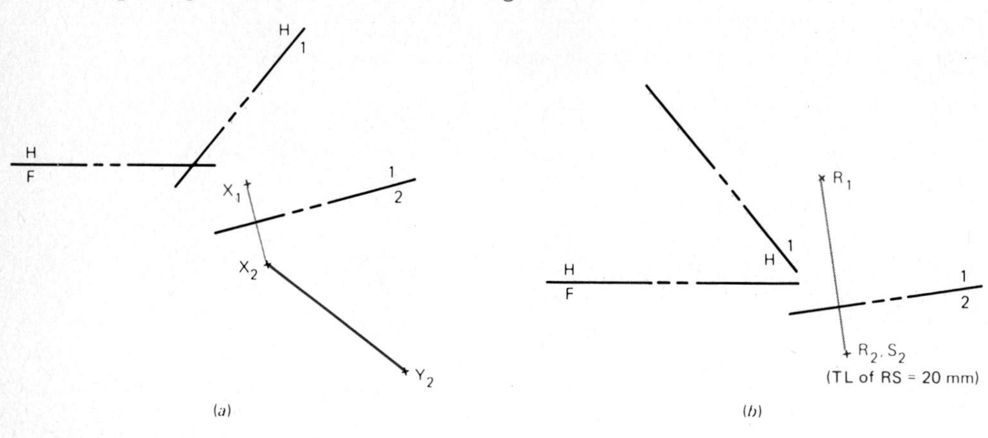

FIG. 4.20

4.17 Find and label the bearing and grade of the lines shown in Fig. 4.21.

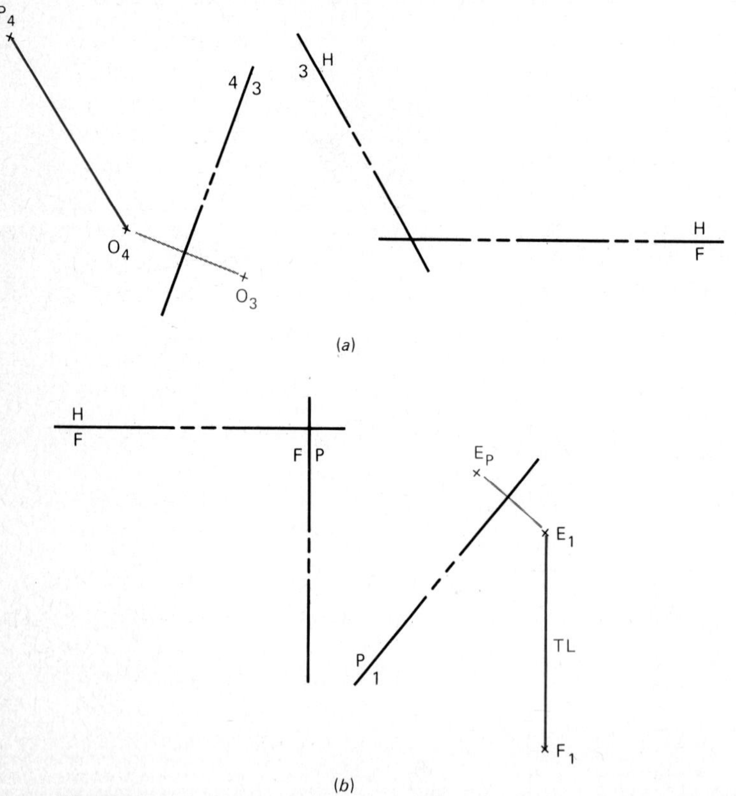

FIG. 4.21

chapter 5

Geometry of Planes

5.1 INTRODUCTION

Planes or plane surfaces can be seen in almost any three-dimensional object. These surfaces must be described carefully in a design so that position, size, and intersections can be found for construction or manufacturing purposes.

A plane is defined as a flat area or space with two dimensions. A straight line touching any two points in the area will lie wholly within the area. Theoretically, a plane is infinite in area, incorporates an infinite number of points and lines, and has zero thickness. In the orthographic drawing system, a plane is represented and positioned in space by a plane sector (a finite part of its infinite area).

5.2 REPRESENTATION OF A PLANE

Planes can be represented orthographically in four ways. In each case only a sector of the entire plane will be drawn. First, a plane can be represented by three points that are not in a straight line, as shown in Fig. 5.1.

FIG. 5.1 Defining a plane with three points.

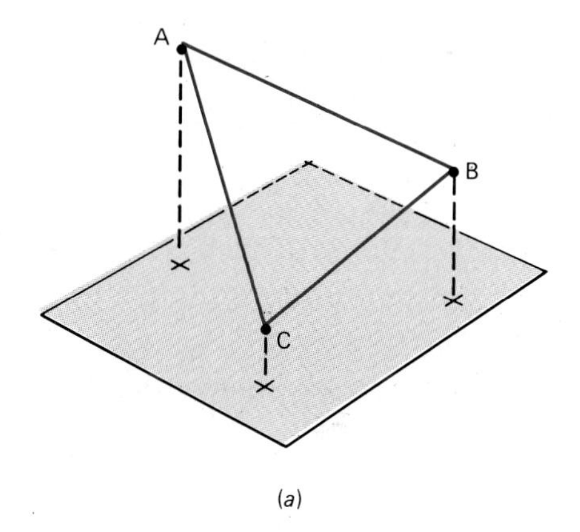

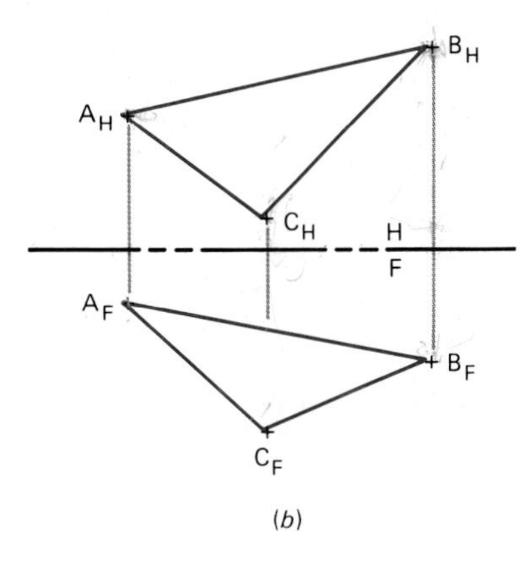

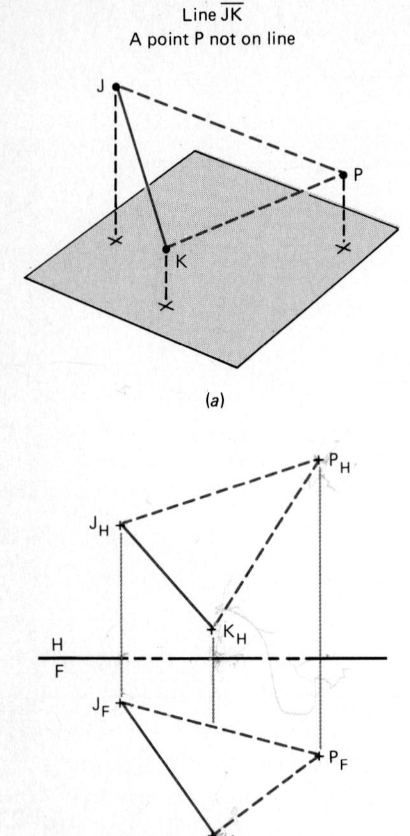

FIG. 5.2 Defining a plane with a line and a point.

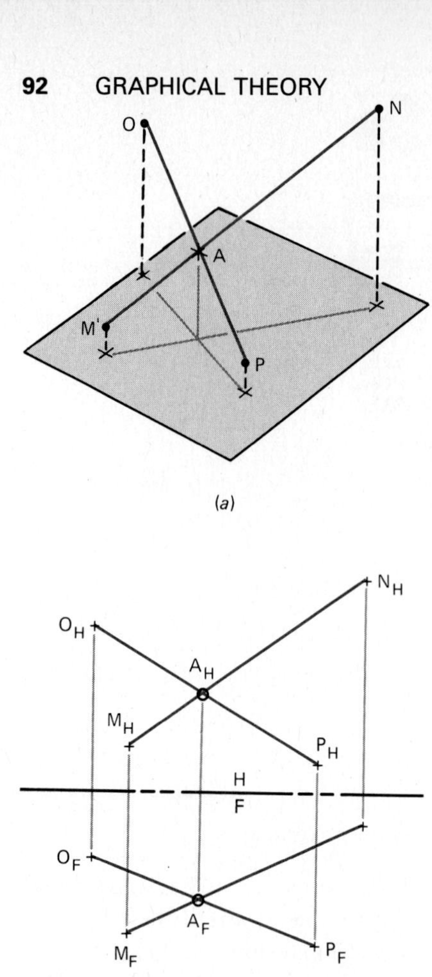

FIG. 5.3 Defining a plane with two intersecting lines.

Visualization of the plane sector is enhanced by connecting the three points with straight lines.

Second, a straight line and a point not on the straight line can represent a plane, as shown in Fig. 5.2.

Figure 5.3 represents a plane surface in space by means of two intersecting straight lines. Notice that the point of intersection, *A*, projects orthographically between the adjacent views in Fig. 5.3*b*.

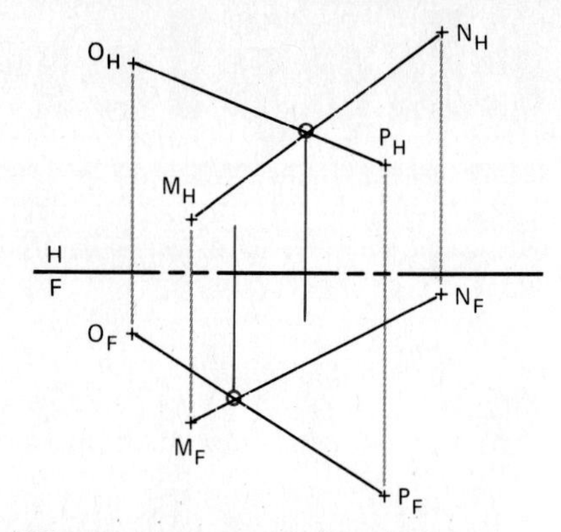

FIG. 5.4 Nonintersecting lines do not form a plane. The apparent points of intersection do not align orthographically.

Figure 5.4 illustrates adjacent views of two lines that are nonintersecting. The apparent points of intersection seen in both views do not align orthographically. Therefore, these lines do not form a plane.

Occasionally, three views are necessary to determine whether two lines intersect. Figure 5.5*a* shows a pair of lines that do not intersect, and Fig. 5.5*b* shows a similar pair that do intersect.

The fourth way to define a plane in space is to use two parallel straight lines, as illustrated in Fig. 5.6. As with nonintersecting lines, you must be careful not to make wrong interpretations as a result of lines appearing parallel when in fact they are not.

Note that if the front and horizontal projections of oblique lines each show the lines to be parallel, there can be no doubt that they are actually parallel. See Fig. 5.6*b*. However, if the H and F projections of profile lines appear parallel, they may not be parallel (Fig. 5.7*a)* or may be parallel (Fig. 5.7*b*).

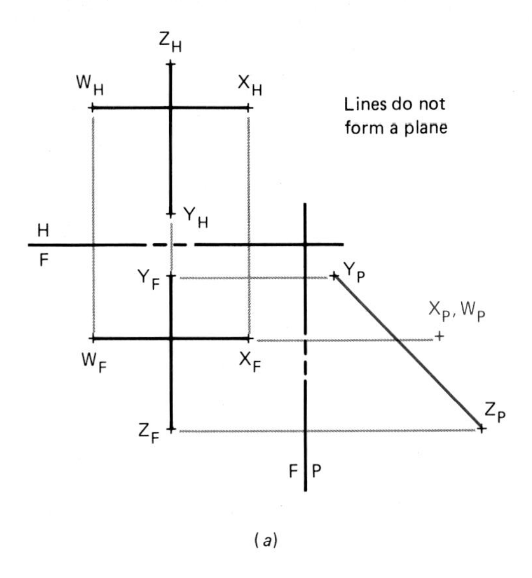

(a)

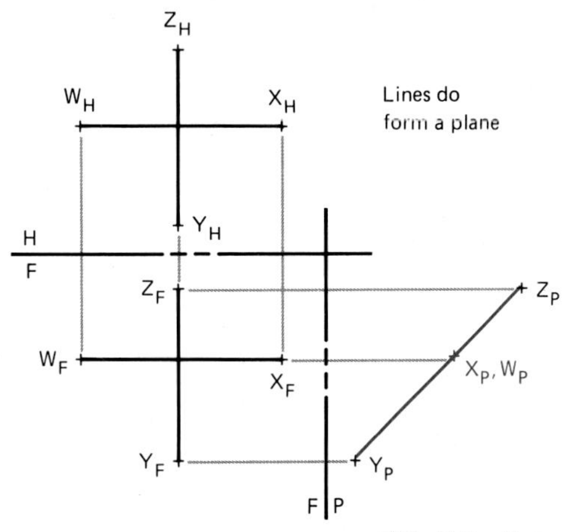

(b)

FIG. 5.5 Determining if two lines intersect in space.

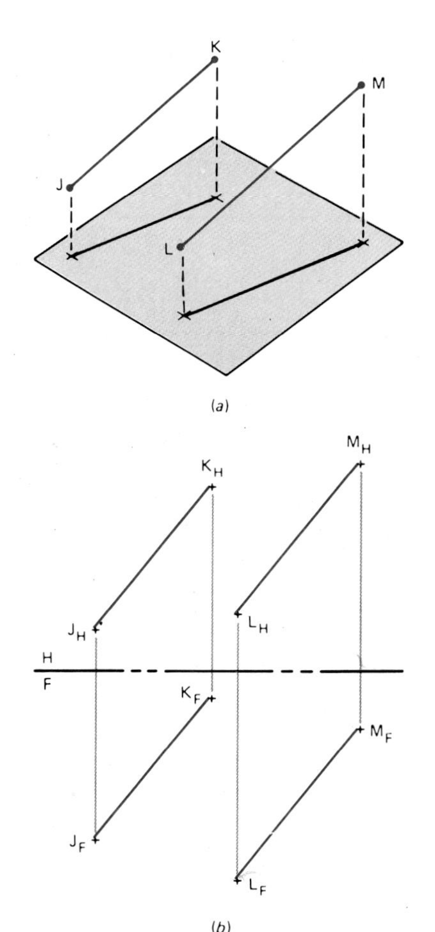

FIG. 5.6 Any projection of oblique parallel lines will show parallelism.

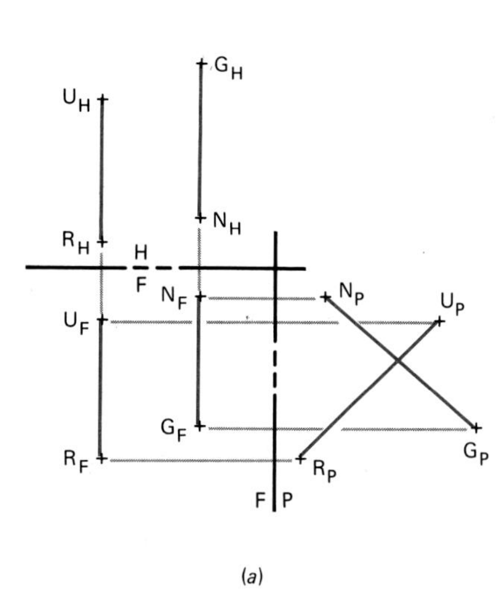

(a)

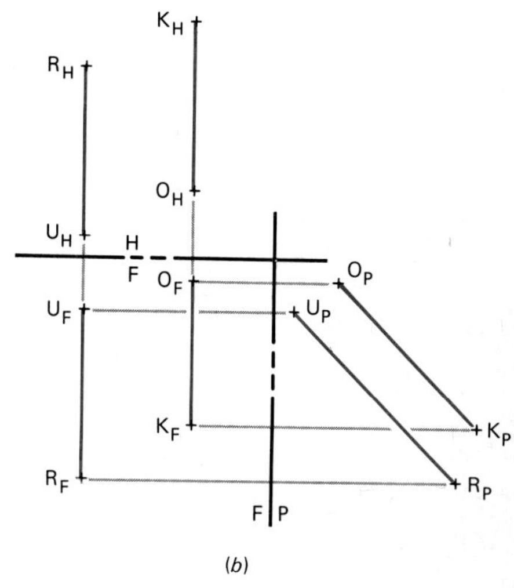

(b)

FIG. 5.7 H-F projections are not sufficient to show parallelism of profile lines.

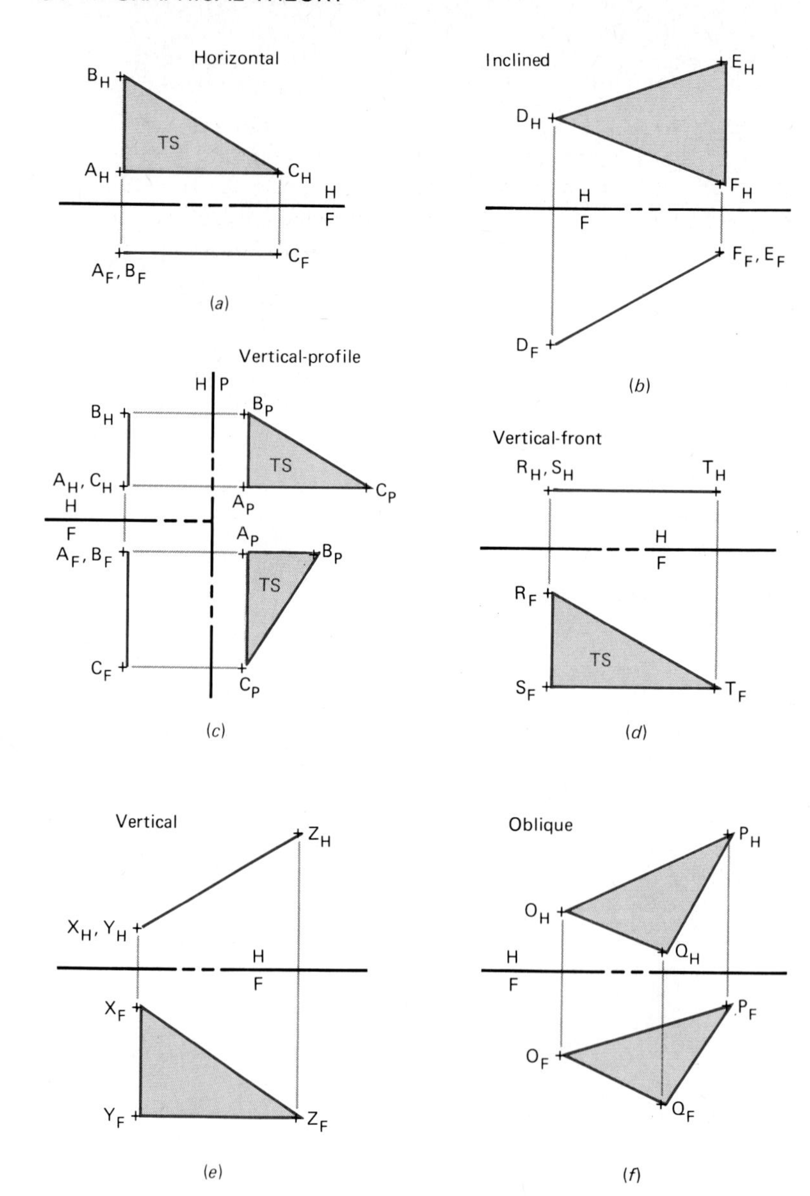

FIG. 5.8 Classification of planes.

5.3 POSITIONS OF PLANES (CLASSIFICATION)

The six basic positions of planes are shown in Fig. 5.8; the positions are horizontal, inclined, vertical-profile, vertical-front, vertical, and oblique. There are, of course, an infinite number of positions included among the six classified positions. For example, an inclined plane (perpendicular to the front principal plane) may vary from nearly horizontal to nearly vertical, and a vertical plane (perpendicular to the horizontal principal plane) may be located anywhere between a vertical-front position and a vertical-profile position.

As with lines, the classification of planes is based on their relationship with the principal planes of the orthographic drawing system. That is, a horizontal plane is parallel to the principal horizontal plane as represented by the H/F reference line, a front plane is parallel to the principal front plane, and so forth.

To avoid confusion between the *image planes* of the orthographic system (front, horizontal, and so forth) and the various other planes that may be described in orthographic drawing, we will refer to these other planes as *object planes.*

5.4 EXTENDING A GIVEN PLANE SECTOR

In solving spatial geometry problems involving plane surfaces, it is frequently necessary to extend or change the representation of a given plane sector. Such extensions can be easily accomplished without a change in the original position of the sector. Often you can simply extend given boundary lines of the sector; in other cases you may need to locate specific lines on the given plane sector. Keep in mind the definition and nature of a plane as you study the illustrations of extending plane sectors in Fig. 5.9.

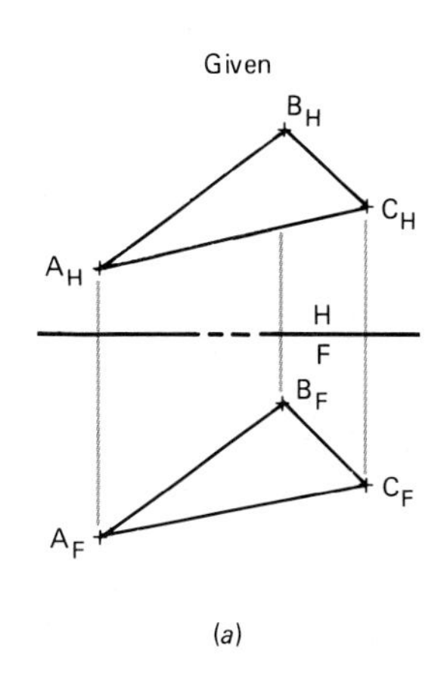

FIG. 5.9 Extending a plane sector. Orthographic rules must be followed. (*b*) Line $\overline{BC}$ of the triangular sector extended to point D. (*c*) Oblique line $\overline{BE}$ constructed parallel to $\overline{AD}$. (*e*) A parallelogram formed as the plane sector. (*f*) $\overline{JLO}$ formed and then $\overline{MP}$ constructed parallel to $\overline{JLO}$.

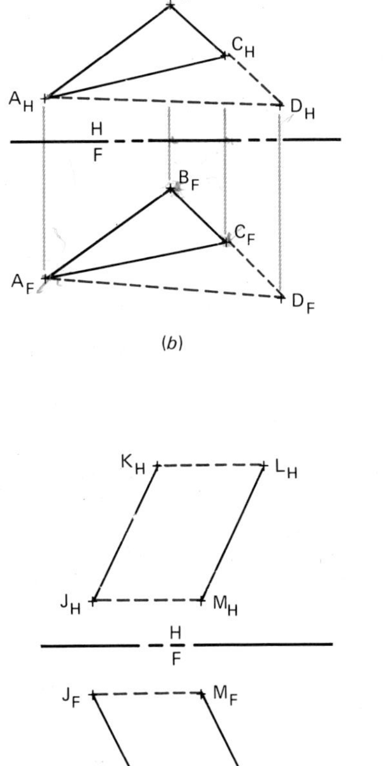

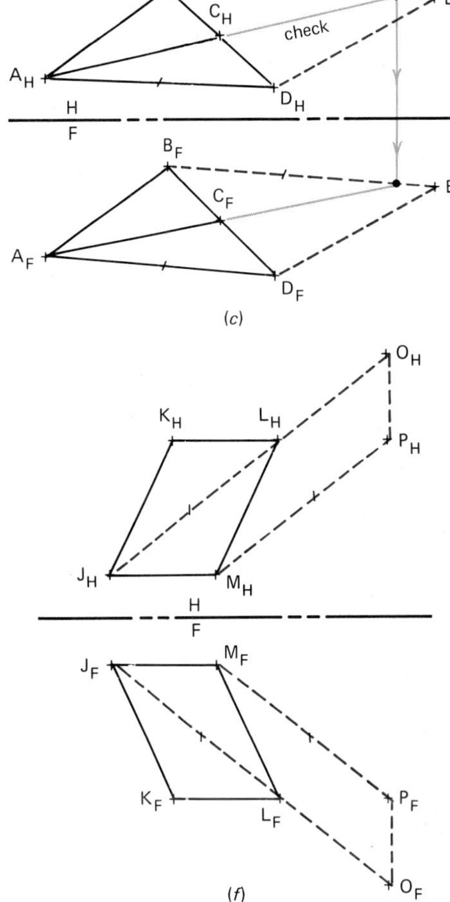

Given

K_H L_H

J_H M_H

H/F

J_F M_F

K_F L_F

(*d*)

Of course, in orthographic drawing, line extensions or additions must be in correct projection; that is, if you extend a line in one view, the same line should be properly located and extended in adjacent views. Note that you must intersect at least two points in a given sector in order to define any new line you wish to locate in that sector. Adjacent views must show these intersections in correct projection.

5.5 LOCATING SPECIFIC LINES ON PLANES

In subsequent sections, you will see the practical applications of locating specific lines such as horizontal and front lines on plane sectors. Figure 5.10 shows how to locate a horizontal line and a front line on a given plane. Figure 5.11 uses the properties of parallel lines to locate a line *OR* in true length when one leg of the plane sector is a profile line.

5.6 LOCATING SPECIFIC POINTS ON PLANES

The principles of orthographic projection, parallel lines, and intersecting lines may be used to locate specific points on a plane sector or on a plane sector extended. Remember that if a point lies on a line and the line lies on a plane, the point lies on the plane.

Figure 5.12 shows the procedure for locating in all given

Given

S_H R_H T_H H F S_F T_F R_F

(*a*)

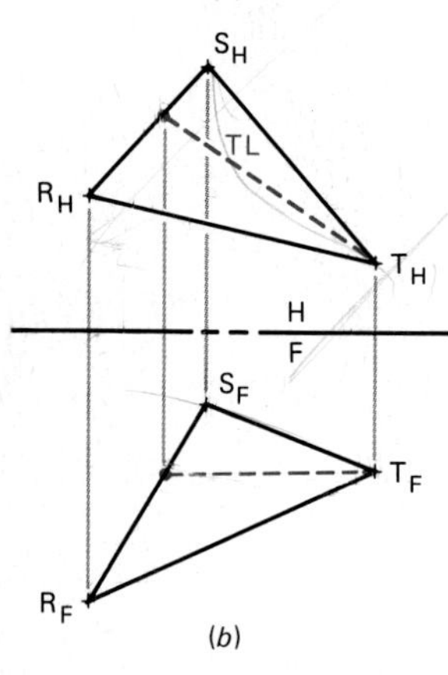

(*b*)

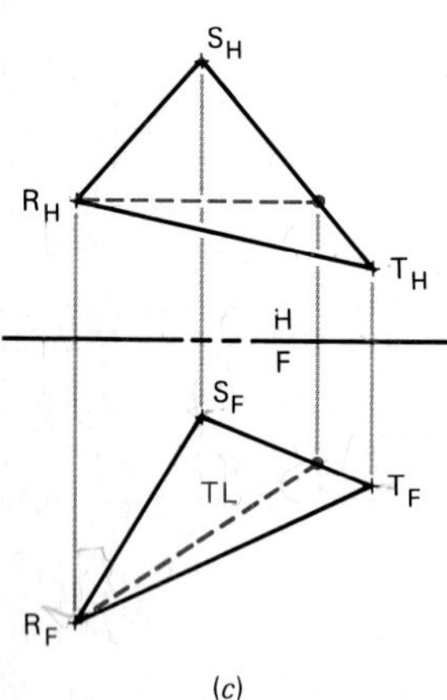

(*c*)

FIG. 5.10 A horizontal line is located on plane RST in *b*. In *c*, a front line is located.

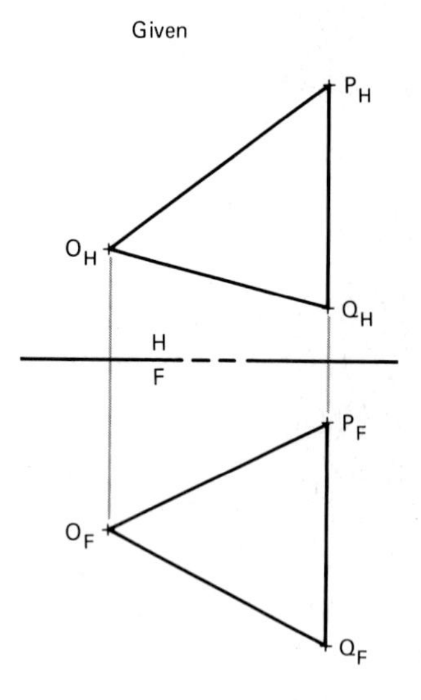

(*a*)

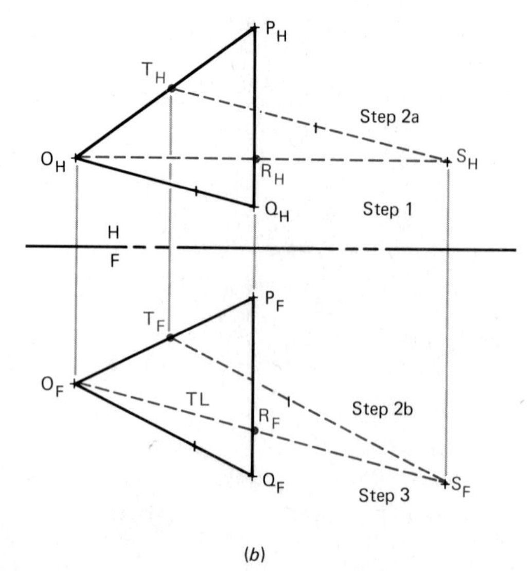

(*b*)

FIG. 5.11 $\overline{ORS}$ is drawn in top view parallel to H/F reference line. Line $\overline{ST}$ is constructed parallel to $\overline{OQ}$ in top view and then projected by using parallelism to front view. R_F is now located.

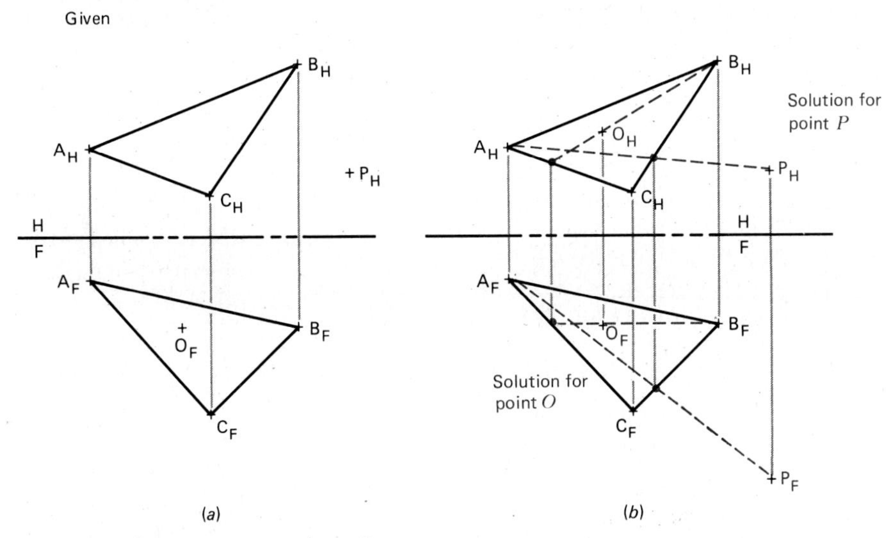

FIG. 5.12 The principle of intersecting lines is used to locate the projections O_H and P_F of points O and P which lie on the plane.

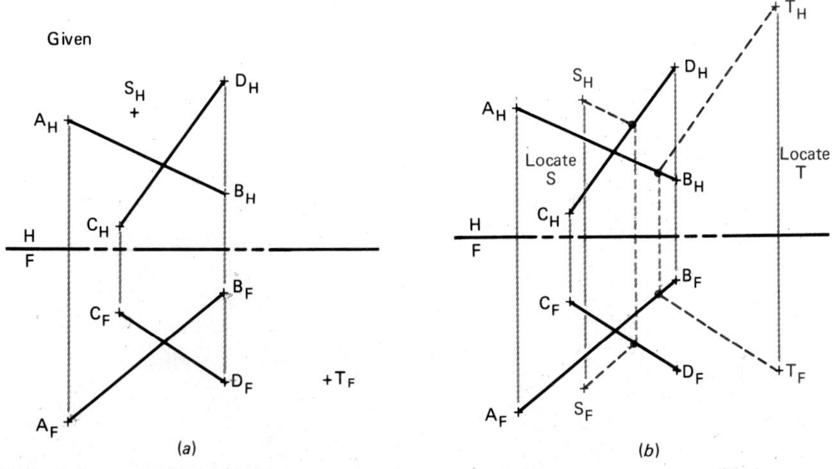

FIG. 5.13 Parallel and intersecting lines are used to locate the projections S_F and T_H of points S and T which lie on the plane.

views the points *O* and *P* which lie on the plane sector (or extended sector) *ABC*. In this case intersecting lines and the rules of projection determine the location of the points in the view adjacent to the given view.

Figure 5.13 locates points *S* and *T* by applying the principles of parallel and intersecting lines in conjunction with the rules of projection. Both Fig. 5.12 and Fig. 5.13 should be studied carefully before you read further.

5.7 EDGE VIEW OF A PLANE

Spacial geometry solution methods may require that a view of an object plane as an edge be found.

5.7.1 Solution View: Edge View of a Plane

The solution view showing an edge view of a plane is any orthographic view showing any line in the plane as a point.

5.7.2 Solution Procedure: Edge View of a Plane

Refer to Fig. 5.14.

1. Identify or locate a true-length line on the plane sector.
2. Position a reference line perpendicular to the true-length view of the line.
3. Project the true-length line as a point along with other points defining the plane sector into the solution view.

In certain cases a given plane may appear as an edge in one of the principal planes (review Fig. 5.8). However, the solution view principle is still valid in such cases.

It may be helpful to place a pencil (simulating a line) against a drafting triangle or other plane surface. Now position this physical representation of a line on a plane so that you see the end view of the pencil. You will also see the edge view of the plane surface. This demonstrates the principle involved in the solution view for graphically showing the edge view of an object plane.

5.8 TRUE SHAPE OF A PLANE

Figure 5.15 illustrates the sequence of graphical procedures required to determine the true

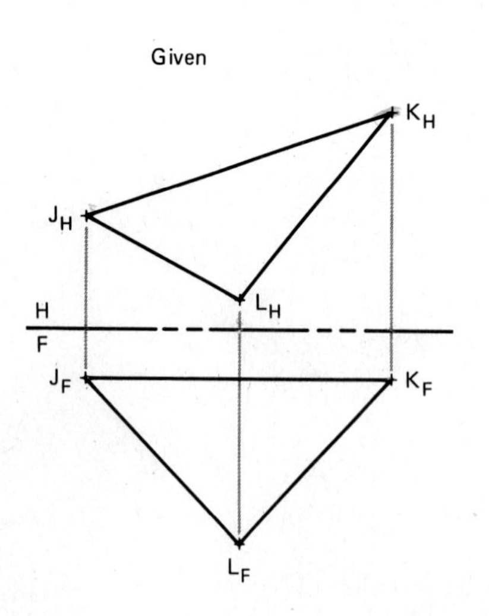

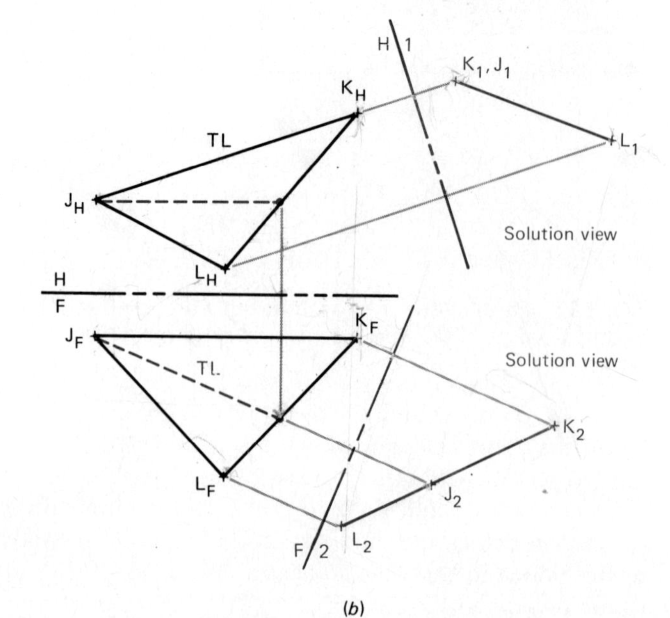

FIG. 5.14 Two possibilities for the edge view of plane JKL.

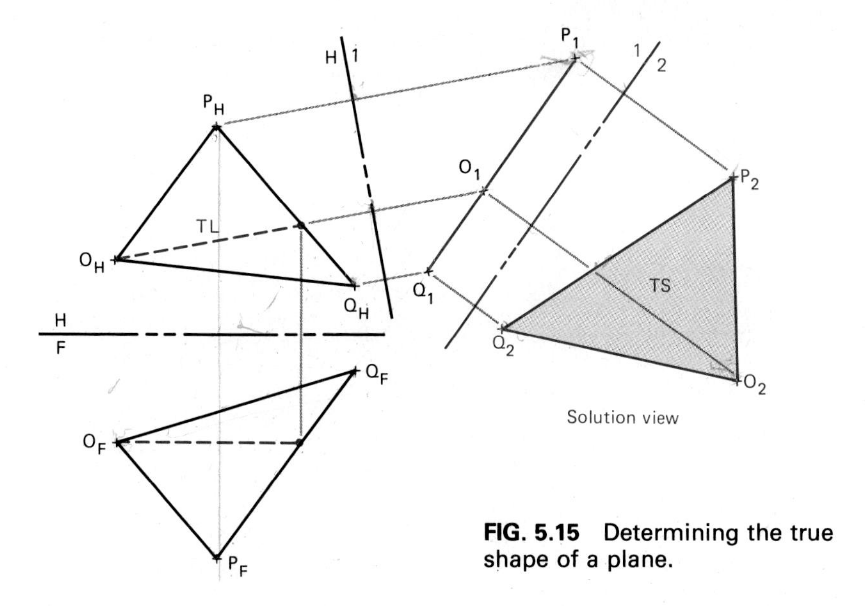

FIG. 5.15 Determining the true shape of a plane.

shape (TS) of a plane sector. First visualize the solution view by viewing the position relative to the object plane in order to see its true shape. If you select a drafting triangle to simulate an object plane and position it so that your line of sight is perpendicular to the surface of the object plane, you will see its surface in true shape.

5.8.1 Solution View: True Shape of a Plane

The solution view for the true shape of a plane is any orthographic view obtained on an image plane positioned parallel to the object plane.

5.8.2 Solution Procedure: True Shape of a Plane

1. Obtain the edge view of the object plane as described in Sec. 5.7 and shown in Fig. 5.14.

2. Position a reference line parallel to the edge view of the object plane.

3. Project all points and lines of the object plane to the image plane indicated by the reference line.

It may again prove helpful to use a transparent triangle to represent the object plane and another to represent the image plane. With the image plane (triangle) held perpendicular to your line of sight, rotate the object plane until you see it in true shape. Note that the two triangles are then parallel; this illustrates the solution view principle for the true shape of a plane. Front and horizontal object planes appear in true shape in the front and horizontal views, respectively, as can be seen in Fig. 5.8. If the drawing scale is known, the true shape view can be measured and the true size determined.

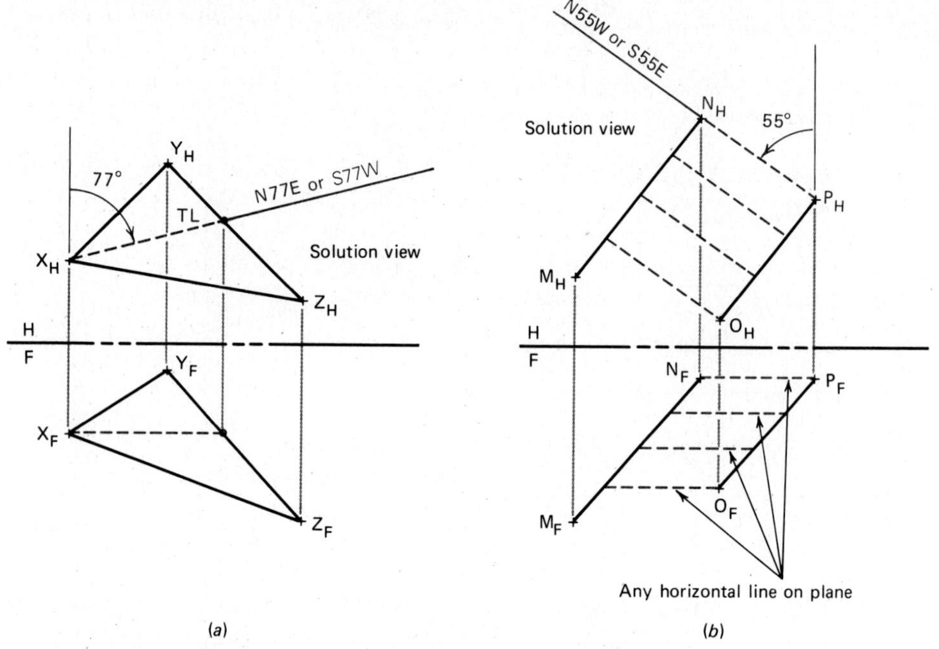

FIG. 5.16 Determining the bearing (strike) of a plane.

5.9 BEARING, OR STRIKE, OF A PLANE

The bearing of a plane is the compass direction of a horizontal line on the object plane. The bearing is measured in the top view, as is shown in Fig. 5.16. Because there is no fixed origin point for a plane, a horizontal line on a plane actually has two directions. For example, a due north line is also due south, and a line N30°E is also S30°W. It is customary procedure to register only one of these two directions.

Strike is a technical term for bearing that is used by mining engineers. For all practical purposes, the terms are synonymous.

5.10 INCLINATION, OR DIP, OF A PLANE

Fundamentally, the inclination of an object plane describes its deviation from a horizontal position. The amount of deviation can be expressed alternatively as slope (inclination) angle, slope, grade, or dip, a term used by mining engineers. The solution view for inclination is the same no matter which term is used for expressing magnitude.

5.10.1 Solution View: Inclination, or Dip, of a Plane

The solution view for the inclination, or dip, of a plane is any elevation view showing the edge view of the object plane.

5.10.2 Solution Procedure: Inclination, or Dip, of a Plane

Refer to Fig. 5.17b for the following steps.

1. Locate a horizontal line in the object plane.

2. Position a reference line H/1 (auxiliary elevation) perpendicular to the true length (top view) of the horizontal line.

3. Project the edge view of the object plane into the elevation view 1.

4. Determine the deviation of the object plane from a horizontal position. (Reference line H/1 provides the horizontal reference.)

The angle between the H/1 reference line and the edge view can be measured in the 1 view; it is the slope angle or dip angle of the plane. If the slope or grade of the plane is required, either can be obtained mathematically from the slope angle or can be graphically found by the use of a slope triangle similar to that shown in Figs. 4.12 and 4.13.

The direction of slope or the dip of an object plane may also be required. A line on the plane that is perpendicular to the bearing, or strike, line and is directed toward the side of lower elevation is called the slope, or dip, line. Because this line has the maximum declination (negative inclination) of any line on the object plane, it therefore has the same declination as the plane. The proper direction of the slope line is visible only in the top view and points toward the compass quadrant in which the plane declines. Dip is expressed as 40°DIP NE, 25°DIP SE, and so forth. See Fig. 5.17.

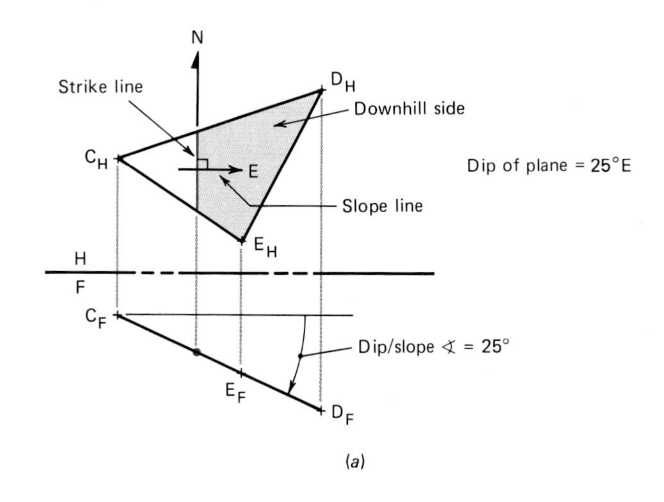

(a)

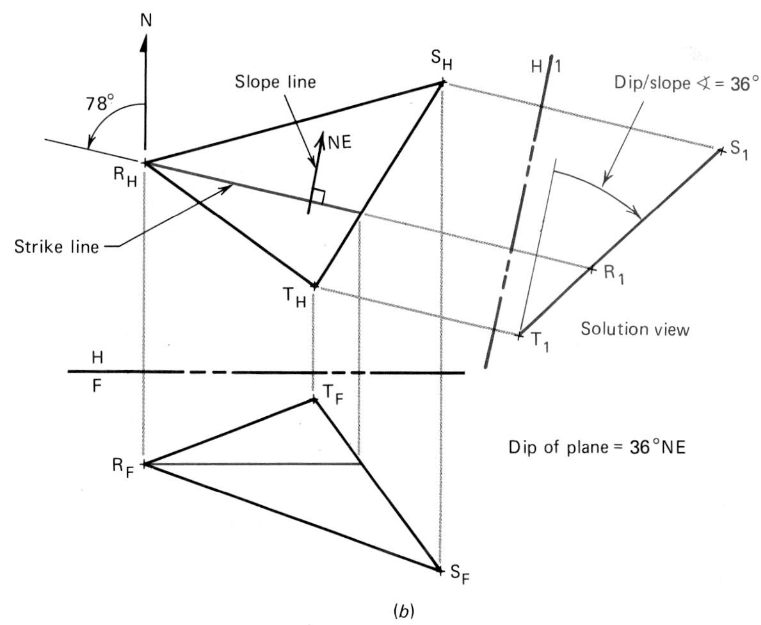

(b)

FIG. 5.17 Determining inclination (dip) of a plane.

5.11 GRAPHIC DELINEATION OF A PLANE SECTOR FROM WORD SPECIFICATIONS

The preceding sections have shown that a plane can be represented by a sector and can be positioned in space by means of its bearing (strike) and inclination (dip). Since the sector shown is only a part of the infinite plane, its shape is immaterial. It may be square, rectangular, triangular, circular, and so forth.

5.11.1 Problem Description

The problems examined so far began with a given plane sector, and we described such a sector by finding specifications such as true shape, strike, and dip. Now we wish to investigate the reverse problem, in which the specifications are given and the plane is to be constructed. This type of problem is more difficult and requires that each of the principles involved be thoroughly understood.

Draw the front, horizontal, and any other necessary views of a plane *ABC* that has the following specifications:

Equilateral triangle, 30-mm sides
AB is strike line, bearing N 70°E
Slope (dip): 45°NW

Figure 5.18 demonstrates the step-by-step procedure for the graphic solution. We will trace the solution procedure by analyzing the given specifications carefully.

5.11.2 Problem Analysis and Observation

1. The given bearing and inclination indicate that plane *ABC* is an oblique plane. Thus, the true size will not be observed in the front or horizontal views.

2. A true-size view (parallel to an edge view of *ABC*) must be obtained.

3. An edge view of the sector can be obtained on an image plane positioned perpendicular to a true-length line on the object plane. (Figure 5.15 is an example.)

4. Since inclination of a specified amount is required, only an elevation view showing the object plane as an edge will be acceptable.

5. The image plane needed to obtain an elevation view of the object plane as an edge must be perpendicular to the true-length projection of a horizontal line on the object plane.

6. The bearing, or strike, line of a plane projects in true length in the top view, so that the strike line must be found first.

7. Consideration of the dip of the object plane is necessary to maintain clarity in the adjacent views (this will avoid overlapping projections later). By using side *AB* as the strike line and drawing a short line perpendicular to it with an arrow in the specified direction for dip, you will get some idea (but not a precise one) where C_H eventually will be located in the top view.

5.11.3 Solution Procedures

Refer to Fig. 5.18 for the steps described below.

Step 1. Position line $A_H B_H$ to accommodate the eventual location of C_H.

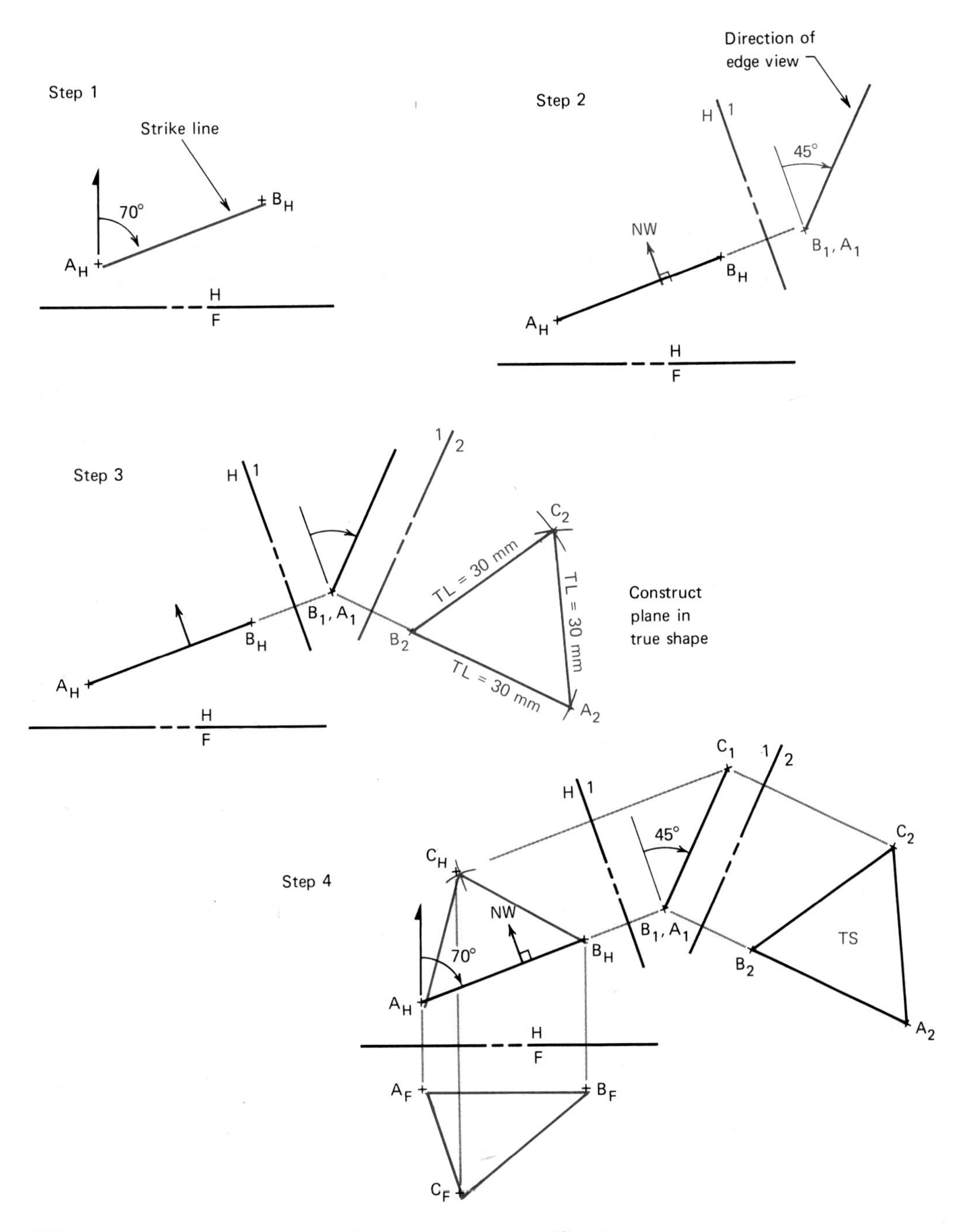

FIG. 5.18 Determining a plane from geometry specifications.

Step 2. Place the H/1 reference line perpendicular to the strike line A_HB_H. When you project line *AB* as a point into elevation view 1, consideration should be given to its distance below the horizontal plane (H/1 reference line). As the object plane dips downward from the strike line, A_1B_1 can be located a short distance below the horizontal plane. An indefinite edge view of *ABC* can now be drawn. C_1 cannot be located until after the true-size view has been completed.

Step 3. Locate an image plane (1/2) parallel to the edge view of *ABC* according to the solution view principle for true shape. Project points *A* and *B* to view 2. Because this is a true-shape view, it is now valid to construct the entire equilateral triangle *ABC*.

Step 4. Project point *C* back through view 1 to view H. Then locate the entire triangle *ABC* in view F, thereby solving the problem.

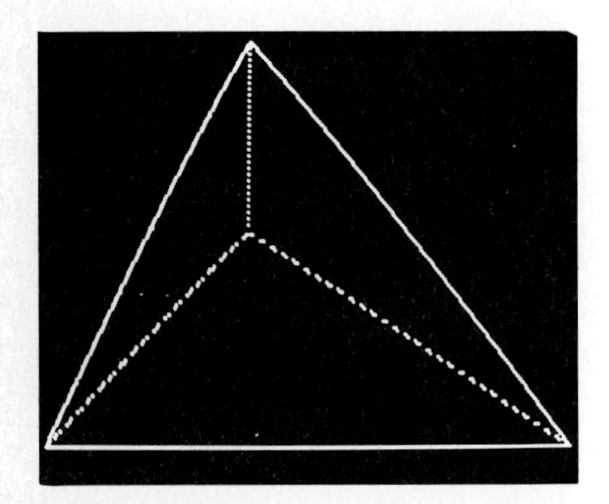

FIG. 5.19 The pyramid described in Fig. 3.42 is viewed from a direction which shows the oblique surface in true shape.

5.12 METHODS FROM COMPUTER GRAPHICS

The description of a plane on the computer can be accomplished in any of the four ways discussed in Sec. 5.2. To simplify the discussion, let us limit ourselves to a three-point description. Specification of three coordinate points is sufficient to determine the equation of a unique plane. If these three points can be thought of as establishing a triangular segment of the plane, we can project this segment onto any image plane that we can describe.

By using the software capability of rotation, we can rotate the plane about appropriate axes in order to position the plane in any of the six classifications discussed in Sec. 5.3.

We can visualize this rotation process as physically reorienting the plane in a fixed-coordinate system (or set of orthographic image planes, if you will), or we can visualize ourselves moving around a fixed plane and stopping at any point to "view" the plane sector. If we consider ourselves to be movable, in effect we are placing an image plane perpendicular to the line of sight between ourselves and the plane. What we see through this image plane is equivalent to the views we create on paper.

Consider the necessity of establishing the slope angle and true shape of the triangular segment. We know that to see true shape, we must move to a position where the line of sight is normal to the segment. To find the slope angle, we must first have a horizontal line of sight and then move around the plane until we see an edge view through the image plane.

Practical applications of true-shape views of planes are numerous in detailed drawings of parts or assemblies where detail of interest exists on a plane which is not seen in true shape in any of the principal views. If the drawing is available in a numeric database, a simple rotation will allow an individual to draw the plane in true shape on a cathode-ray tube and then produce hard copy if required.

Problems

5.1 Concisely describe the solution view for:
(*a*) Plane as an edge
(*b*) Inclination of a plane
(*c*) True shape of a plane

5.2 Classify the following planes:
(*a*) A plane that projects as an edge parallel to the H/F reference line in the front view
(*b*) A plane that projects in true shape in the front view
(*c*) A plane that projects as an edge parallel to the H/P reference line in the horizontal view
(*d*) A plane that is not perpendicular to any of the principal planes of projection
(*e*) A plane that projects in true shape in an auxiliary elevation view
(*f*) A plane having a strike of N30°W and a dip angle of 60°

5.3 In your own words, define an infinite plane.

5.4 Using a plastic triangle to represent various positions of a "plane sector" in space, illustrate such classifications as horizontal, vertical, front, profile, inclined, and oblique. Sketch related orthographic views to represent each classification.

5.5 Intersecting lines *WX* and *YZ* form a plane with a strike of N60°E and a NW dip of 60°. Show the true-shape view and the front view of plane *WXYZ*.

5.6 The true shape of plane sector *ABC* is shown as an equilateral triangle (3 cm each side) in an auxiliary elevation view. Line *AB* has a strike of N30°W. What is the dip and classification of plane *ABC*? Show the front view of the plane.

5.7 A 30-mm-square plane sector *MNOP* dips NW at 45°. The diagonal line *MO* of the square sector represents the strike (N45°E). Show the front and all necessary views of plane *MNOP*.

Note: In Probs. 5.8 through 5.11, redraw the given information freehand and label strike (B) and dip (Θ).

5.8 Find and label the strike, dip, and true shape of the planes shown in Fig. 5.20.

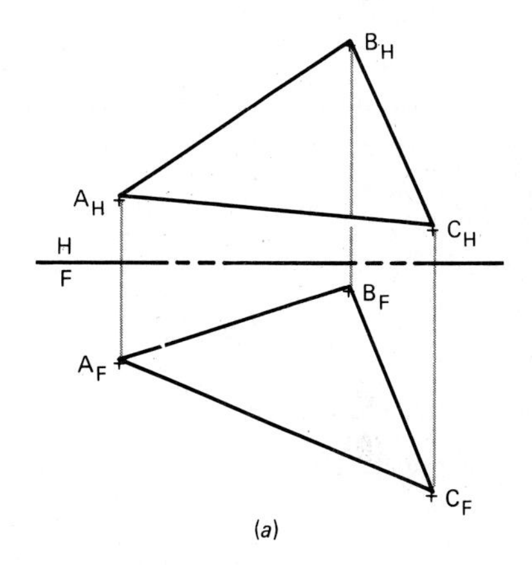

(*a*)

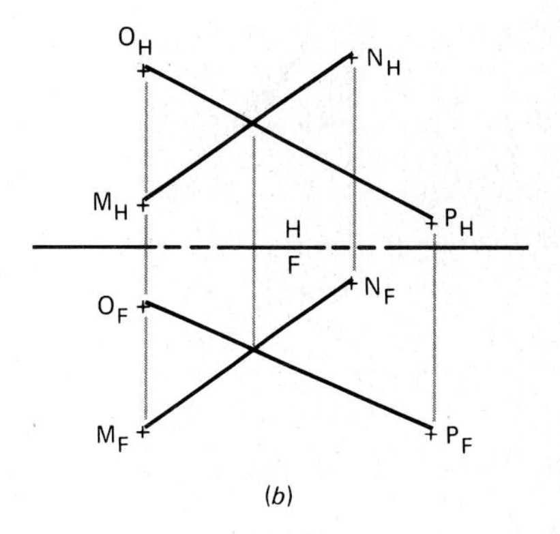

(*b*)

FIG. 5.20

5.9 Find and label the strike and dip of the planes shown in Fig. 5.21.

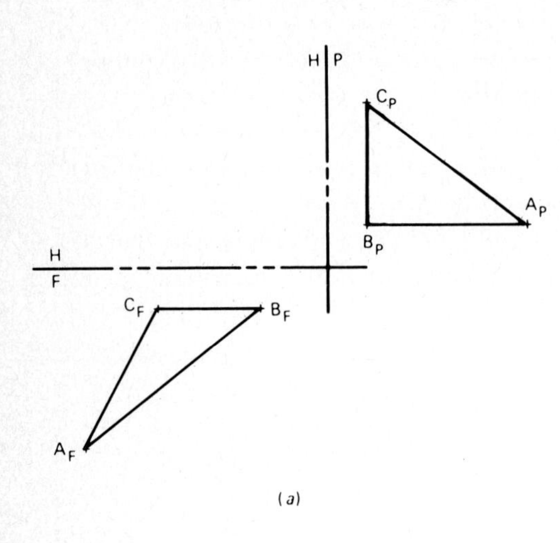

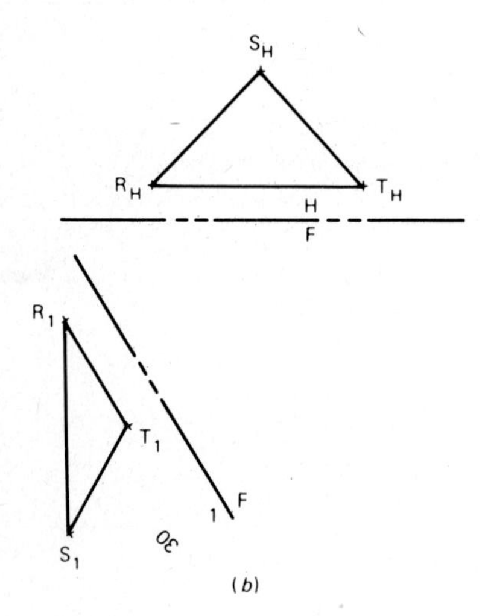

FIG. 5.21

5.10 Determine the slope angle and true shape of planes *VAB, VAD,* and *ABCD,* as shown in Fig. 5.22.

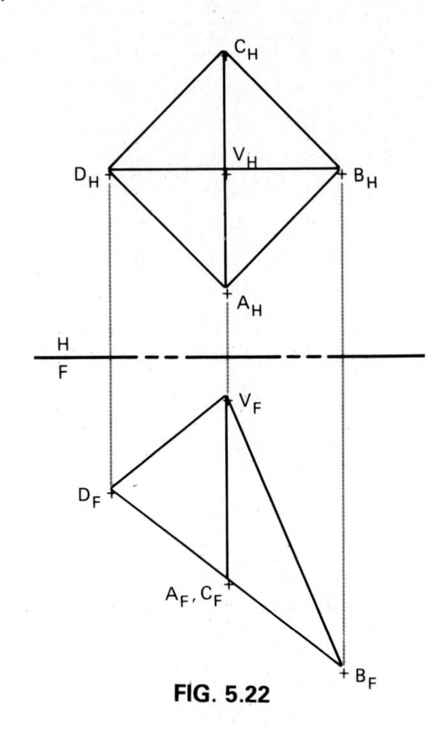

FIG. 5.22

5.11 Determine the slope angle and true shape of each of the planes in the prism shown in Fig. 5.23.

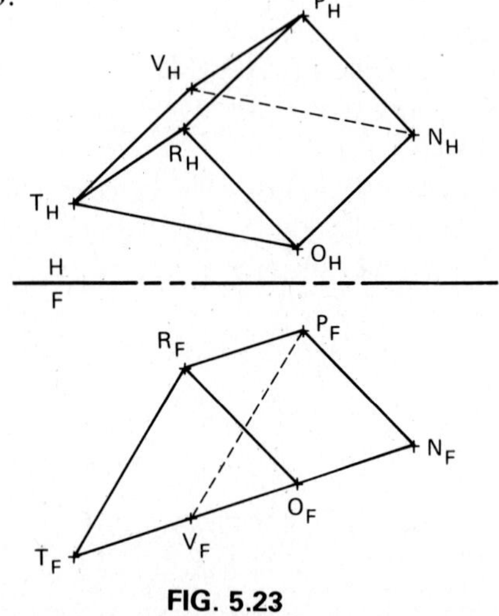

FIG. 5.23

chapter 6

Point, Line, and Plane Relationships

6.1 INTRODUCTION

The geometry of relations between points, lines, and planes is essential to the design and fabrication of industrial products. For example, parallel and perpendicular lines involving clearances or connections between related components are frequently encountered in the development of these products. The principles, concepts, and procedures outlined in this chapter are fundamental for an understanding of the geometry problems encountered in engineering.

Graphic accuracy is always important in the solution of geometry problems, but accuracy deserves special attention with respect to clearance and connector problems. The accuracy is critical for graphically verifying the validity of complex solutions. This will become apparent in the example problems that will be discussed and illustrated.

6.2 PARALLEL AND PERPENDICULAR LINES

The pictorial and orthographic drawings of the cube shown in Fig. 6.1 illustrate the fact that if two lines are parallel, they will always appear parallel in any projection. On the other hand, lines perpendicular to each other will project perpendicular only in views where either or both lines project in true length.

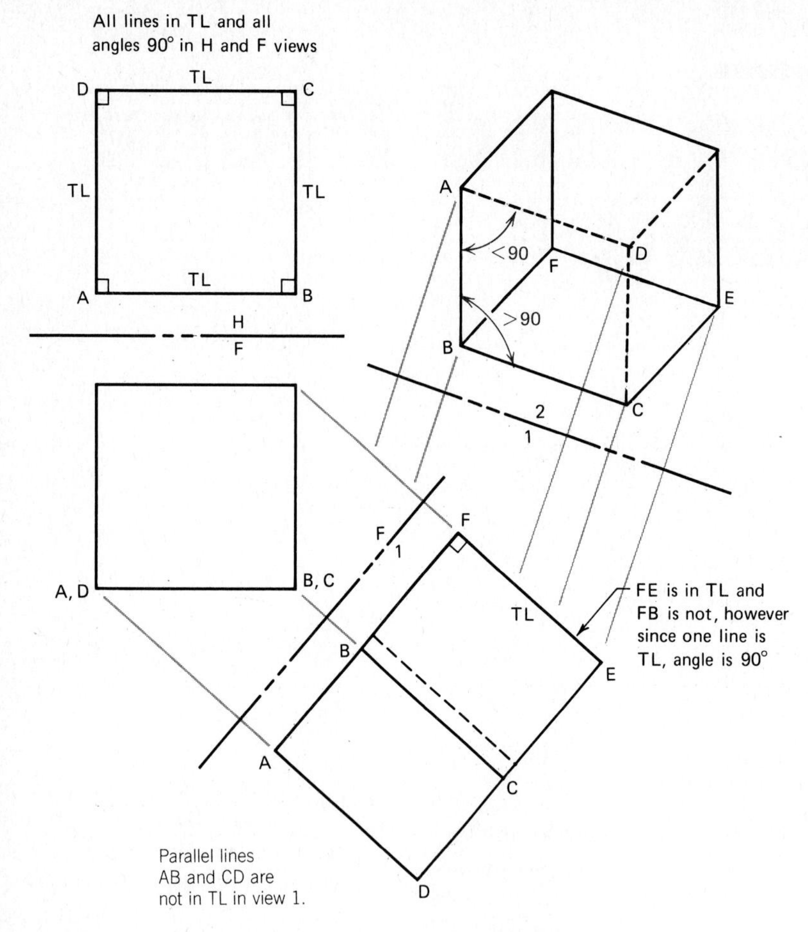

FIG. 6.1 Parallel and perpendicular lines in orthographic projection.

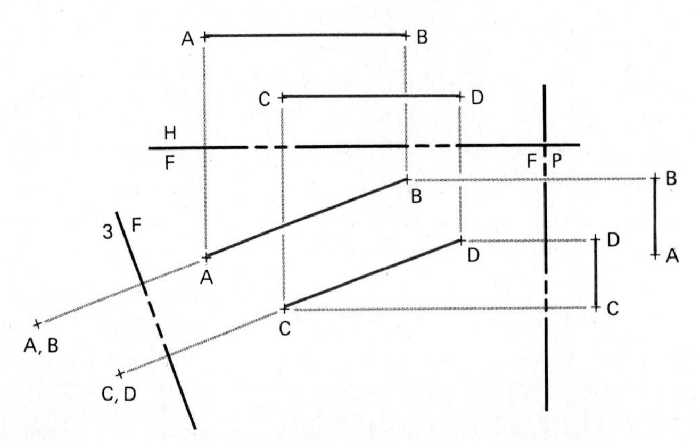

FIG. 6.2 Orthographic projections of parallel lines.

Identify several sets of lines that are mutually parallel and others that are mutually perpendicular in the pictorial view (view 2) of the cube in Fig. 6.1. Note that lines *AB* and *CD* appear parallel in all orthographic views, even when they are foreshortened. The projected angle between perpendicular lines *AB* and *BC* is greater than 90° in view 2, while the angle between perpendicular lines *AB* and *AD* is less than 90° in the same view. This occurs because none of the lines *AB, BC,* and *AD* are in true length in this view.

Figure 6.2 illustrates the principle of parallelism with two line segments. View 3 of this figure shows the lines as points; this is only possible if the lines are parallel.

Figure 6.3 clearly shows the principle of perpendicularity. In all views where either or both of the lines project in true length (views F, 1, 2, and 3), the known right angles project at 90°. On the other hand, in views H and P, where neither line projects in true length, the projected angles are not 90°, even though the lines are known to be perpendicular.

View subscripts have not been used in Figs. 6.1, 6.2, and 6.3. Your instructor may suggest that you continue to use subscripts, but we will not use them in the remainder of the material on graphic theory in order to keep the figures as simple as possible. In cases where views overlap, we may use subscripts to prevent confusion.

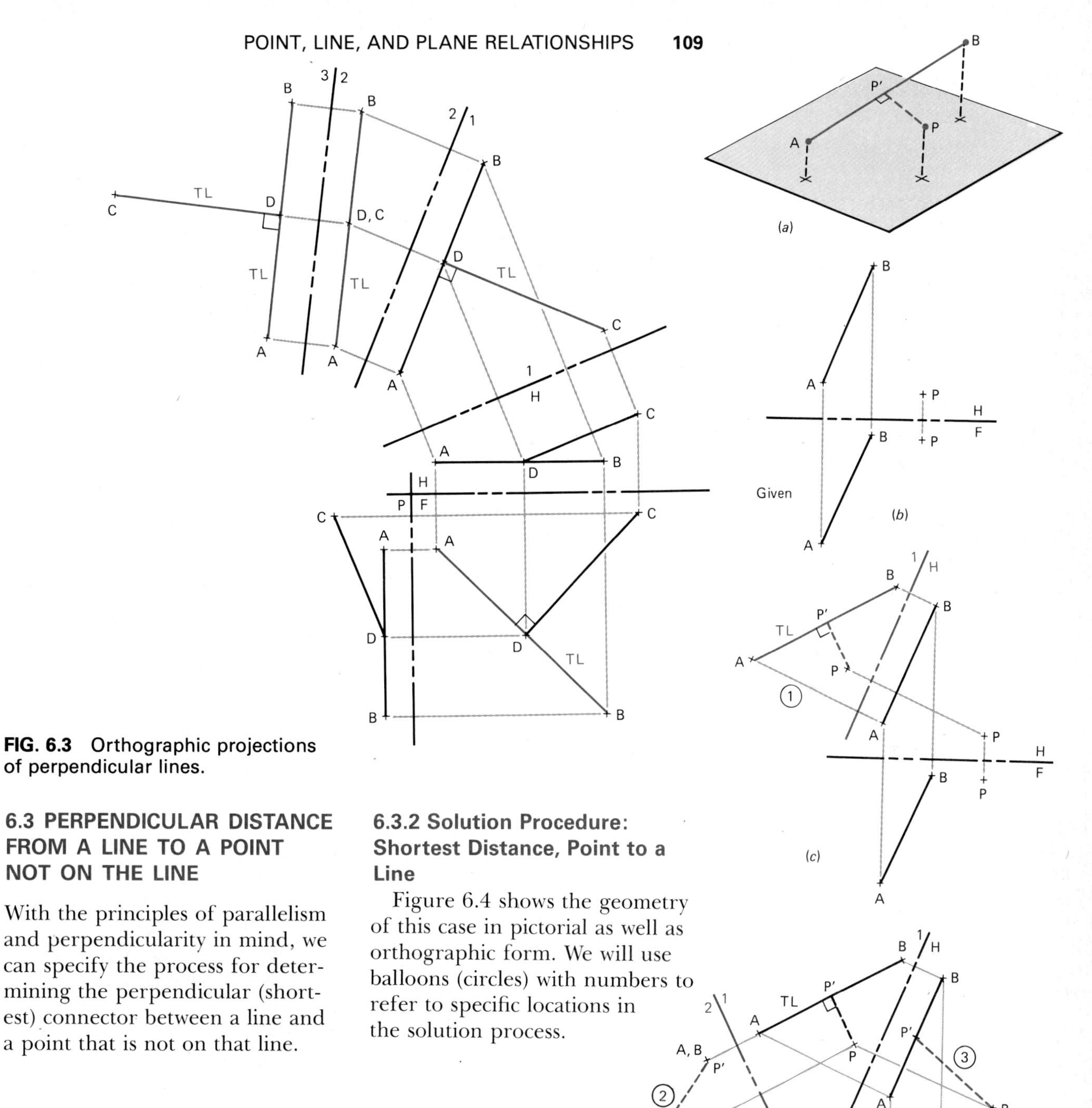

FIG. 6.3 Orthographic projections of perpendicular lines.

6.3 PERPENDICULAR DISTANCE FROM A LINE TO A POINT NOT ON THE LINE

With the principles of parallelism and perpendicularity in mind, we can specify the process for determining the perpendicular (shortest) connector between a line and a point that is not on that line.

6.3.1 Solution View: Shortest Distance, Point to a Line

The solution is found in any orthographic view showing a point view of the line together with the projection of the point.

6.3.2 Solution Procedure: Shortest Distance, Point to a Line

Figure 6.4 shows the geometry of this case in pictorial as well as orthographic form. We will use balloons (circles) with numbers to refer to specific locations in the solution process.

FIG. 6.4 Shortest (perpendicular) distance from a point to a line.

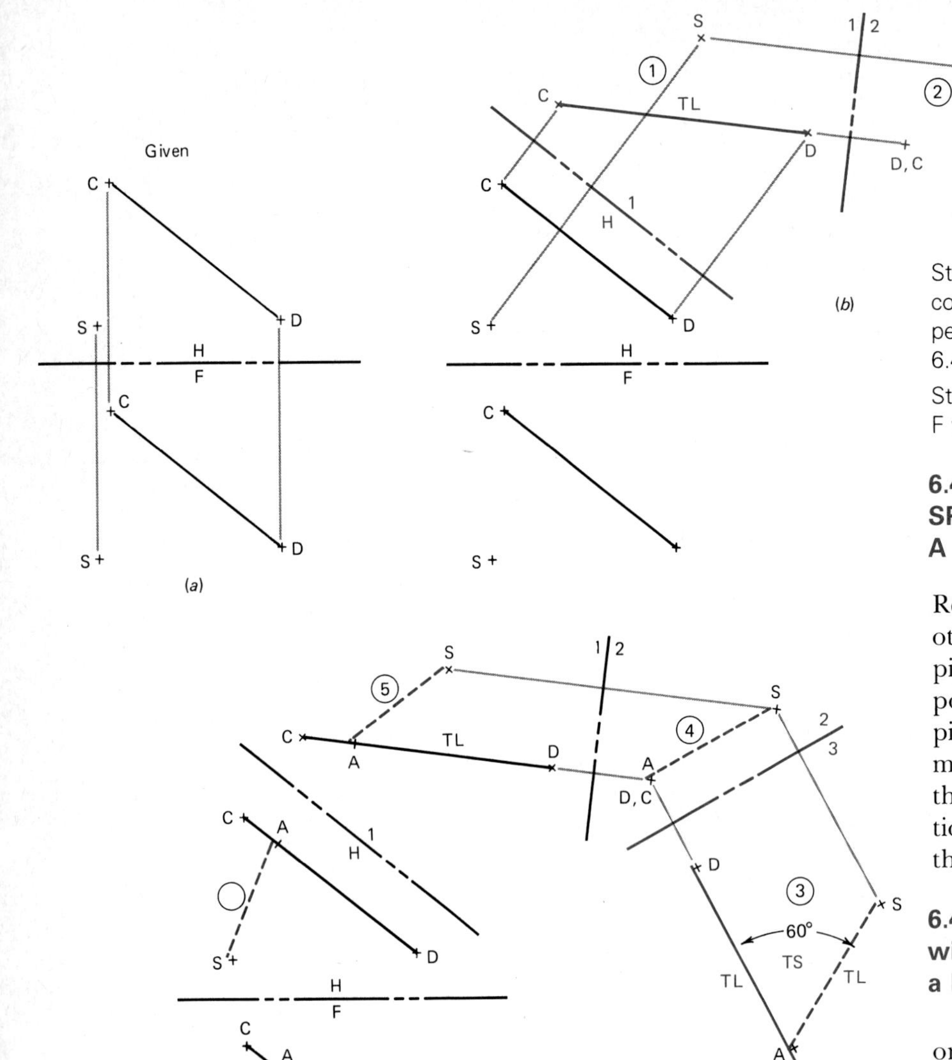

FIG. 6.5 Connector at a specified angle from a point to a line (line method.)

Step 1. Find the true length of line *AB*. See Fig. 6.4*c*, ①. From the concept of perpendicularity, the connector *PP′* will appear at right angles to *AB* in view 1 and therefore can be constructed. View 1 locates the connector but does not show it in true length.

Step 2. Locate the true length of the connector *PP′*, as in ②. Line *AB* appears as a point in view 2. See Fig. 6.4*d*.

Step 3. Project point *P′* into the H and F views. See ③ and ④.

6.4 CONNECTOR WITH A SPECIFIED ANGLE: POINT TO A LINE

Refineries, chemical plants, and other industries may require a piping connector from a specified point to an established straight pipe. Finding a connector that meets the straight pipe at other than 90° represents an application of the material presented in this section.

6.4.1 Solution View: Connector with a Specified Angle, Point to a Line

The solution is found in any orthographic view showing the true size of a plane containing the line and the designated point. Two procedures are available to accomplish this.

6.4.2 Solution Procedure: Connector with a Specified Angle, Point to a Line (Line Method)

Step 1. Obtain the true-length view of line *CD* described in Fig. 6.5*b*, along with the projection of point *S*; see ①.

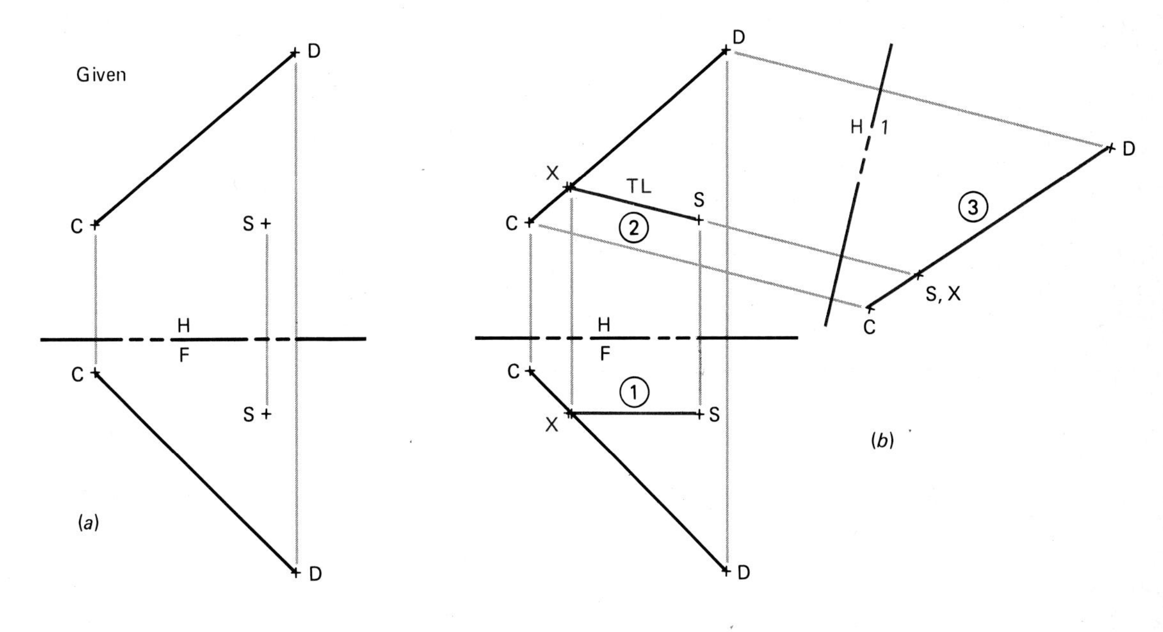

Step 2. Construct a point view of *CD* and also project point *S*; see ②.

Step 3. Place a reference line 2/3 parallel to the plane containing line *CD* (point of view) and point *S*. Project *CD* and *S* into view 3; see ③. See Fig. 6.5*c*.

Step 4. Draw the connector *SA* at the desired angle with line *CD*; see ③. The line was drawn at 60° in this example.

Step 5. Project the connector *SA* into all remaining views; see ④, ⑤, ⑥, and ⑦.

6.4.3 Solution Procedure: Connector with a Specified Angle, Point to a Line (Plane Method)

One less orthographic view is required if some preliminary construction is done in the given H and F views. Figure 6.6 shows the procedure using the plane method.

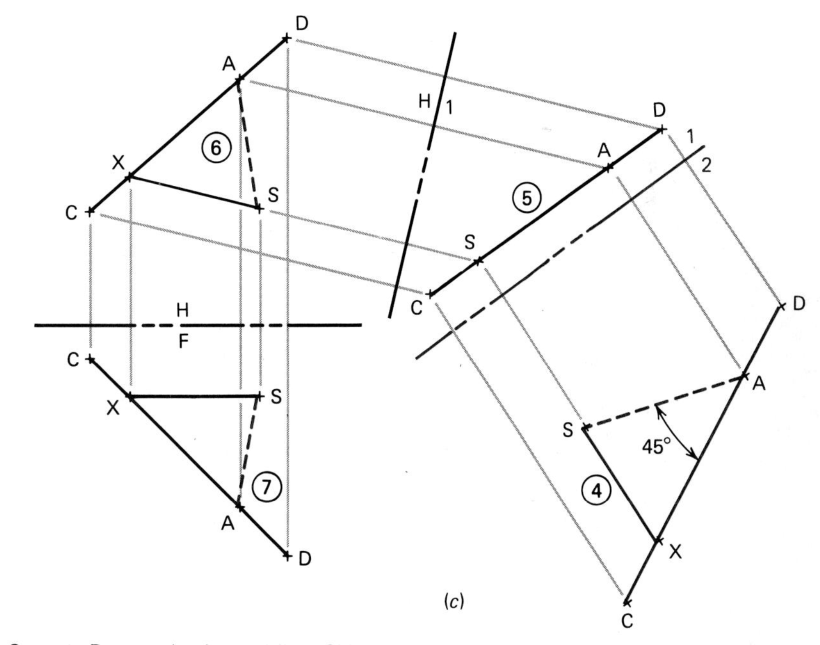

Step 1. Draw a horizontal line *SX*, which identifies a plane *CDS*; see ① and ②.

Step 2. Project the edge view of plane *CDS*; see ③.

Step 3. Position reference line 1/2 par-

FIG. 6.6 Connector at a specified angle from a point to a line (plane method).

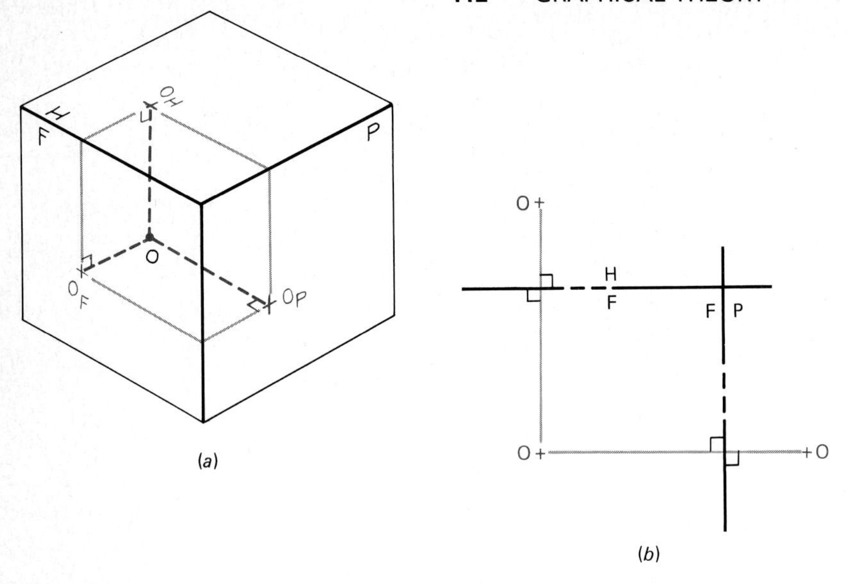

FIG. 6.7 An orthographic projector is the shortest distance from the point to the image plane.

allel to the edge view of *CDS* to obtain its true-shape view. Draw a line from *S* to intersect *CD* at point *A* (the angle was chosen to be 45° in this example); see ④.

Step 4. Project *SA* into the remaining views; see ⑤, ⑥, and ⑦.

6.5 PERPENDICULAR DISTANCE: POINT TO A PLANE

Briefly review the fundamental concepts of orthographic projection related to the principal projection planes H, F, and P shown in Fig. 6.7. Note that point *O* is

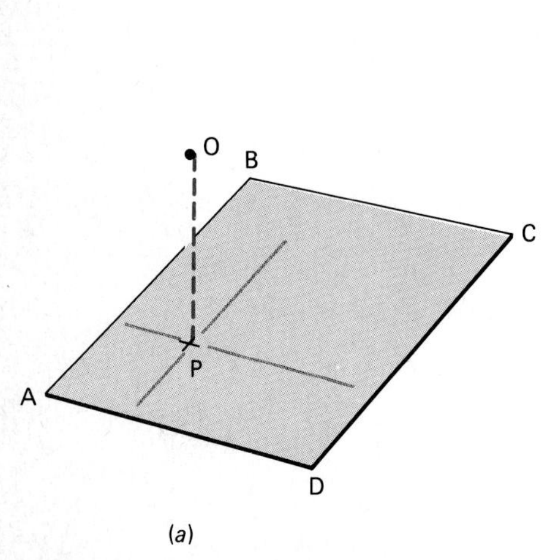

FIG. 6.8 Shortest connector (perpendicular distance) from a point to a plane. Application of the principles of perpendicular and parallel lines is the key element in the solution procedure.

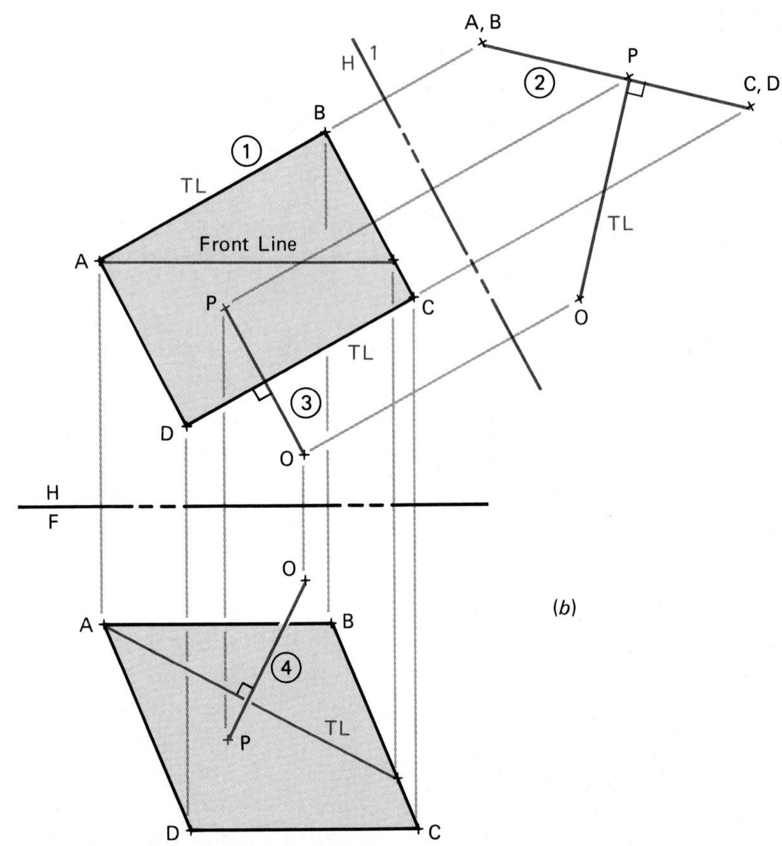

projected 90° to the edge view of the principal plane in each case (the shortest distance from the point to the plane). The same idea is used to determine the shortest distance (perpendicular connector) from a point to any designated oblique plane.

6.5.1 Solution View: Shortest Connector, Point to a Plane

Any orthographic view showing the edge view of the designated oblique plane and the projection of the specified point will define the shortest connector.

6.5.2 Solution Procedure: Shortest Connector, Point to a Plane

Step 1. Locate a horizontal line *AB* or *CD* on plane *ABCD* of Fig. 6.8. The line will appear in true length in the top view; see ①.

Step 2. Place the H/1 reference line perpendicular to the true-length view of *AB* to find the edge view of *ABCD*; see ②.

Step 3. Draw the connector from point *O* so that it meets the edge view of *ABCD* at 90°. The connector *OP* appears in true length in view 1; see ②.

Step 4. Project the connector to the H and F views; see ③ and ④. Since view 1 of line *OP* is in true length, the H projection must be parallel to reference line H/1. This fact allows you to locate point *P* in the top view. If the connector is parallel to reference line H/1, it intersects *CD* at 90°, again illustrating the relationship of perpendicular lines. This relationship is again seen in the F view with the projection of the connector and a front line. The connector is drawn as a solid line so that correct visibility with the plane sector can be represented.

An additional example is offered in Fig. 6.9. The balloon numbers match the steps above. Note the dashed-line portion of the connector in the F view of Fig. 6.9*c*. This indicates that the

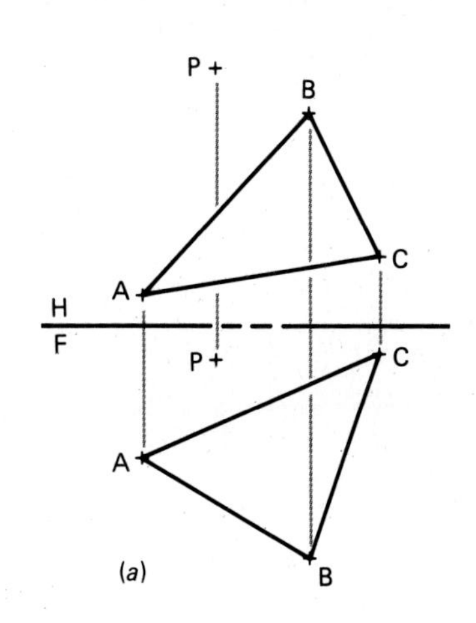

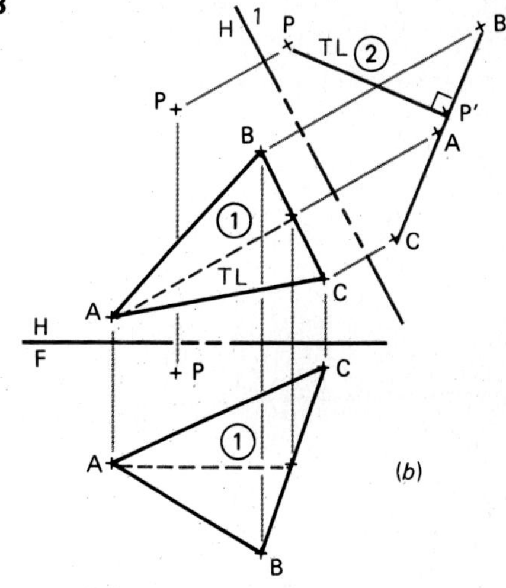

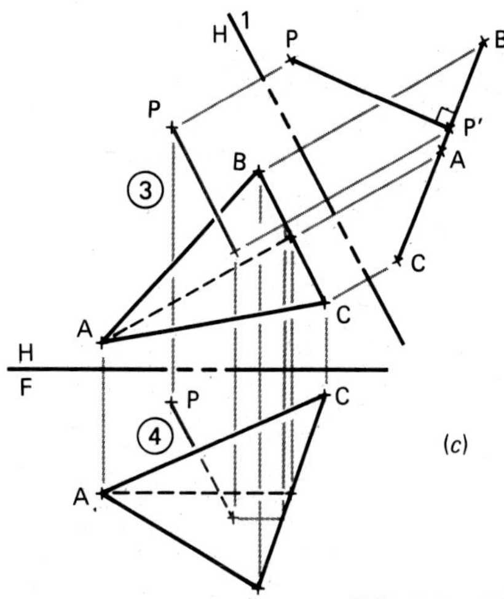

FIG. 6.9 Shortest connector from a point to a plane.

line is behind the plane when looked at from the front view; it is not visible if the plane is opaque.

6.6 VERTICAL CONNECTOR BETWEEN SKEW LINES

Of the various connectors between nonintersecting, nonparallel (skew) lines, the vertical connector is one of the most useful. It is also the simplest to locate in most orthographic problems.

6.6.1 Solution View: Vertical Connector between Skew Lines

The horizontal (top) orthographic view showing the projections of the two lines reveals the vertical connector.

6.6.2 Solution Procedure: Vertical Connector between Skew Lines

Step 1. Examine the horizontal view of the lines and observe the point where the lines appear to cross, as in Fig. 6.10*b*. (Extend them if necessary, as in Fig. 6.10*c*.) This is the point view of the vertical connector; see ①.

Step 2. Project to the front view to see the true length of the vertical connector; see ②.

Theoretically, there is a vertical connector between any two skew lines unless they lie in parallel vertical planes. If the apparent

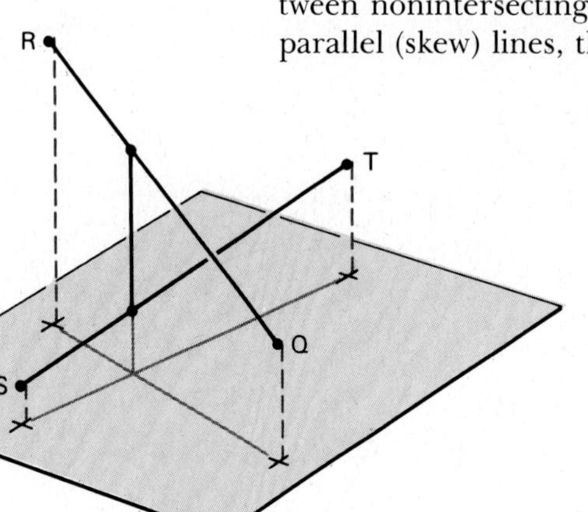

(*a*)

(*b*)

(*c*)

FIG. 6.10 Vertical connector between skew lines.

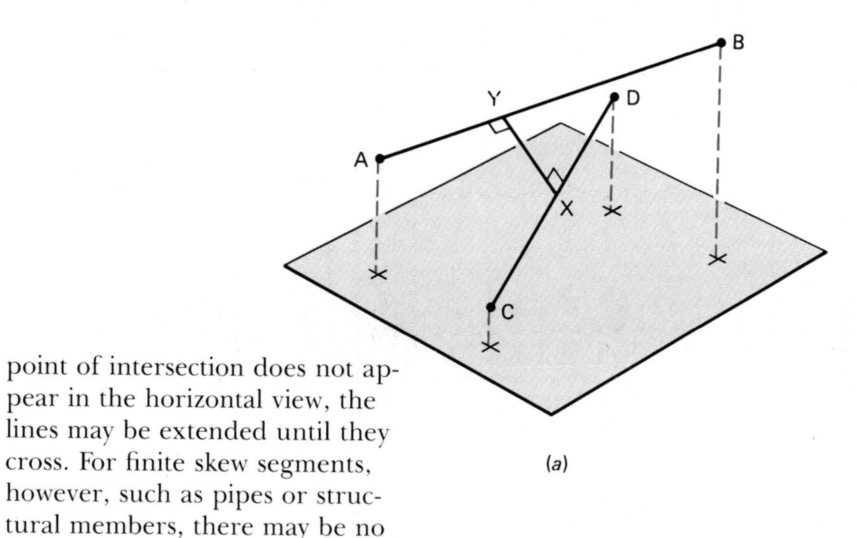

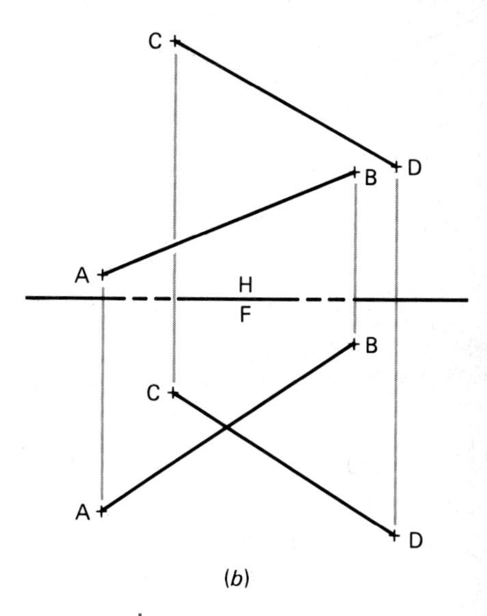

point of intersection does not appear in the horizontal view, the lines may be extended until they cross. For finite skew segments, however, such as pipes or structural members, there may be no cross points, so that no vertical connector exists.

6.7 PERPENDICULAR CONNECTOR BETWEEN SKEW LINES

There are two methods of determining the perpendicular connector between skew lines. The line method will be explained here, and a second method will be introduced later.

6.7.1 Solution View: Perpendicular Connector between Skew Lines (Line Method)

The perpendicular connector is found in any orthographic view showing the point view of one line and the projection of the other.

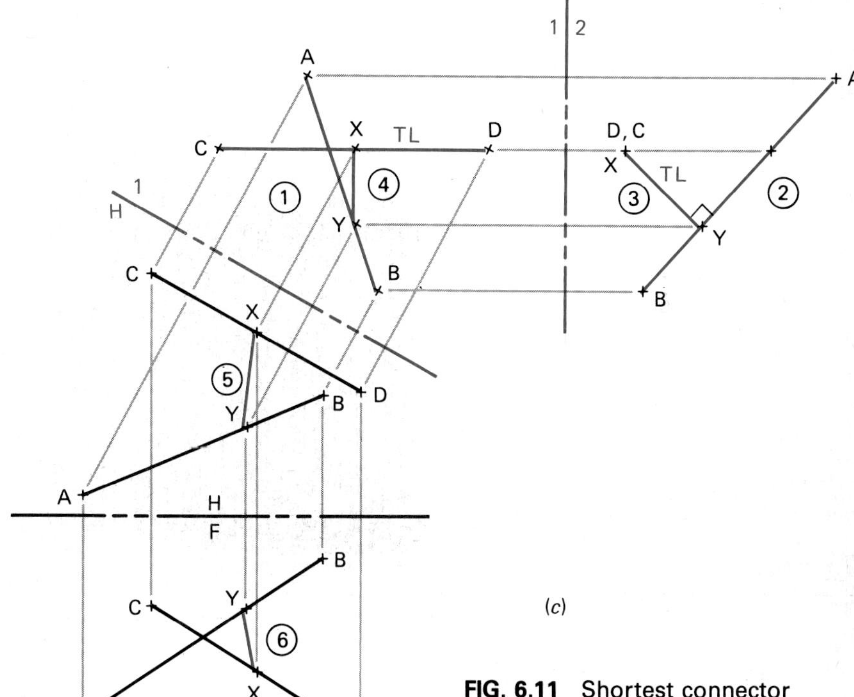

FIG. 6.11 Shortest connector (perpendicular distance) between skew lines.

6.7.2 Solution Procedure: Perpendicular Connector between Skew Lines (Line Method)

Step 1. Project the true-length view of one of the skew lines (line *CD* in Fig. 6.11*c*) along with the projection of the other (*AB*); see ①.

Step 2. Place reference line 1/2 perpendicular to the true-length view of *CD* and project both lines into view 2. Line *CD* will appear as a point; see ②.

Step 3. Draw the connector *XY* from the point view of *CD* to meet line AB at 90°; see ③. This results in *XY* being perpendicular to both lines and appearing in true length in view 2.

Step 4. Project *XY* into view 1, where it intersects *AB* at *Y* and is parallel to reference line 1/2; see ④.

Step 5. Complete the H and F views; see ⑤ and ⑥.

6.8 CONSTRUCTION OF A PLANE WHICH CONTAINS ONE SKEW LINE AND IS PARALLEL TO A SECOND SKEW LINE

In order to solve certain connector problems involving skew lines, it is necessary to construct a plane which contains one of the skew lines and is parallel to the second. A brief example of the method will be illustrated here and then used in subsequent sections.

In Fig. 6.12, we wish to construct a plane which contains line *CD*, with the plane being parallel to line *AB*. Begin by constructing a line from *D* parallel to *AB* in both the H view and the F view; see ① and ③. Locate point *X* by constructing a horizontal line *CX*; see ② and ④. Thus, line *CX* in the horizontal view is in true length, and a point view can be easily found by constructing reference line H/1 and projecting into the 1 view. The 1 view illustrates the parallelism of line *AB* and plane *CDX*.

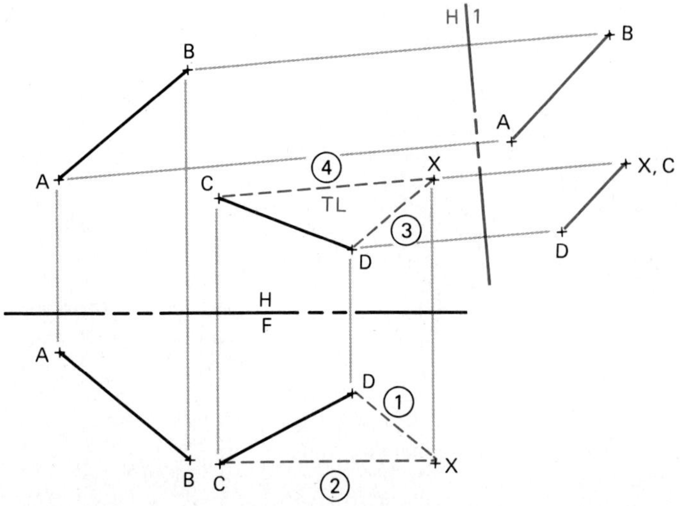

FIG. 6.12 Construction of a plane containing one line and being parallel to a second line.

6.9 SHORTEST HORIZONTAL CONNECTOR BETWEEN SKEW LINES

All methods of finding the shortest horizontal connector between skew lines are based on the principle that the true length of their connector is the same as the shortest horizontal distance between parallel planes that contain the skew lines.

6.9.1 Solution View: Shortest Horizontal Connector between Skew Lines

The solution appears in a view perpendicular to a preceding elevation view that shows the edge views of parallel planes containing the skew lines.

6.9.2 Solution Procedure: Shortest Horizontal Connector between Skew Lines

Refer to Fig. 6.13 for the following steps.

Step 1. Construct a "positioning" plane containing one of the lines so that the plane is parallel to the other

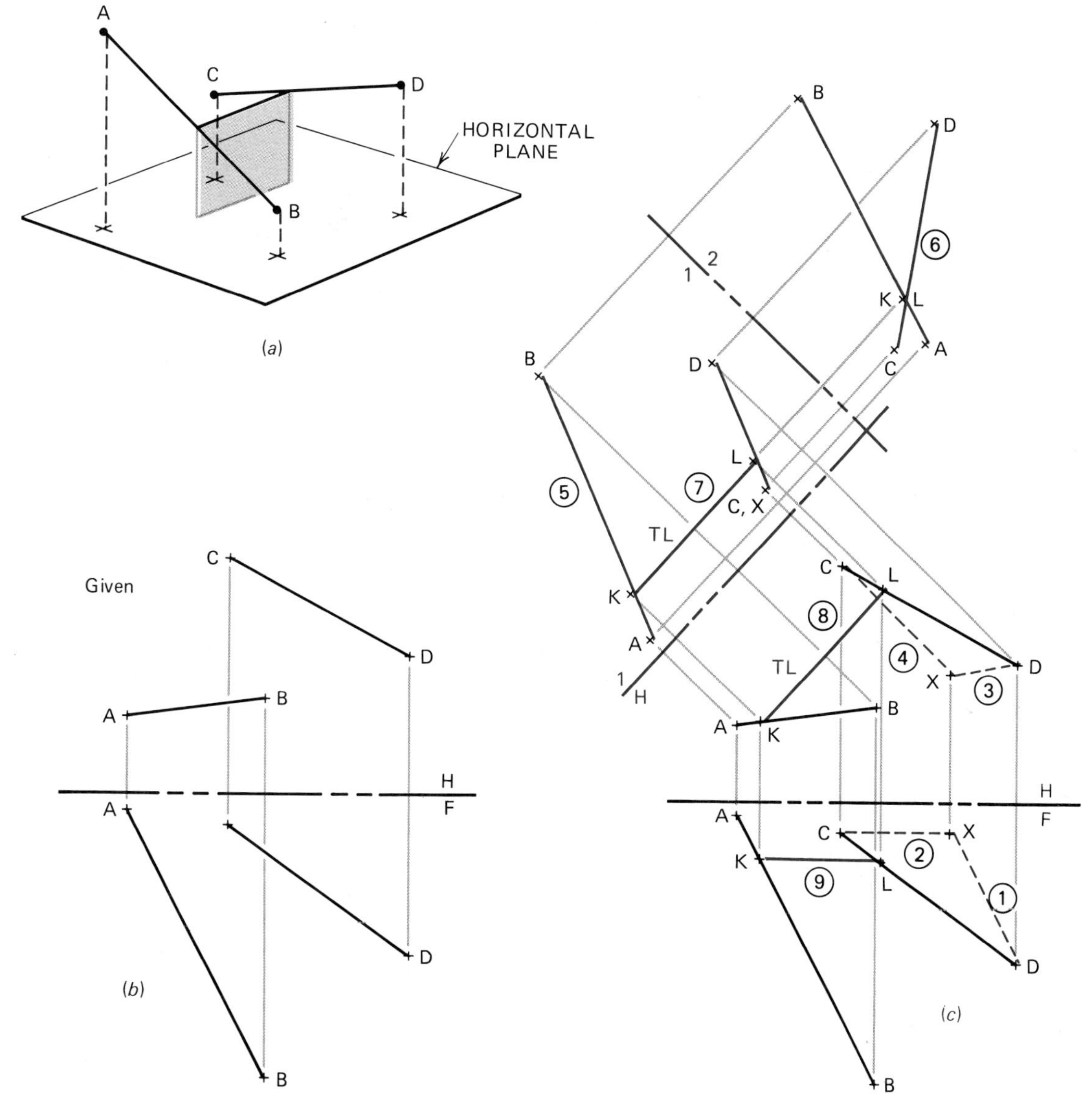

FIG. 6.13 Shortest horizontal connector between skew lines (plane method).

skew line (Fig. 6.12); see ①. This plane is used to locate the view in which the skew lines will appear parallel. Accuracy is critical. It should be noted that it is necessary to construct only a single plane containing one skew line that is parallel to the other skew line. A similarly constructed plane containing the other skew line will automatically be parallel to the first plane.

Step 2. Project the horizontal line in the plane just constructed into the H view; see ③ and ④.

Step 3. Project the edge view of plane *CDX* and line *AB* to view 1, where the two lines *AB* and *CD* appear parallel; see ⑤.

Step 4. Draw reference line 1/2 perpendicular to reference line H/1 and project lines *AB* and *CD* into the 2 view; see ⑥. The point where lines

AB and *CD* appear to cross in view 2 is the location of the point view of the shortest horizontal connector (labeled *KL*). Sometimes one or both of the skew lines will have to be extended until they cross. If so, the lines will have to be extended in all other views to allow the projection of the connector.

Step 5. Project *KL* back to views 1, H, and F; see ⑦, ⑧, and ⑨. The 1 and H views of *KL* must both be parallel to reference line H/1 and show *KL* in true length. Also, the front view of *KL* must be parallel to the H/F reference line since it is a horizontal connector.

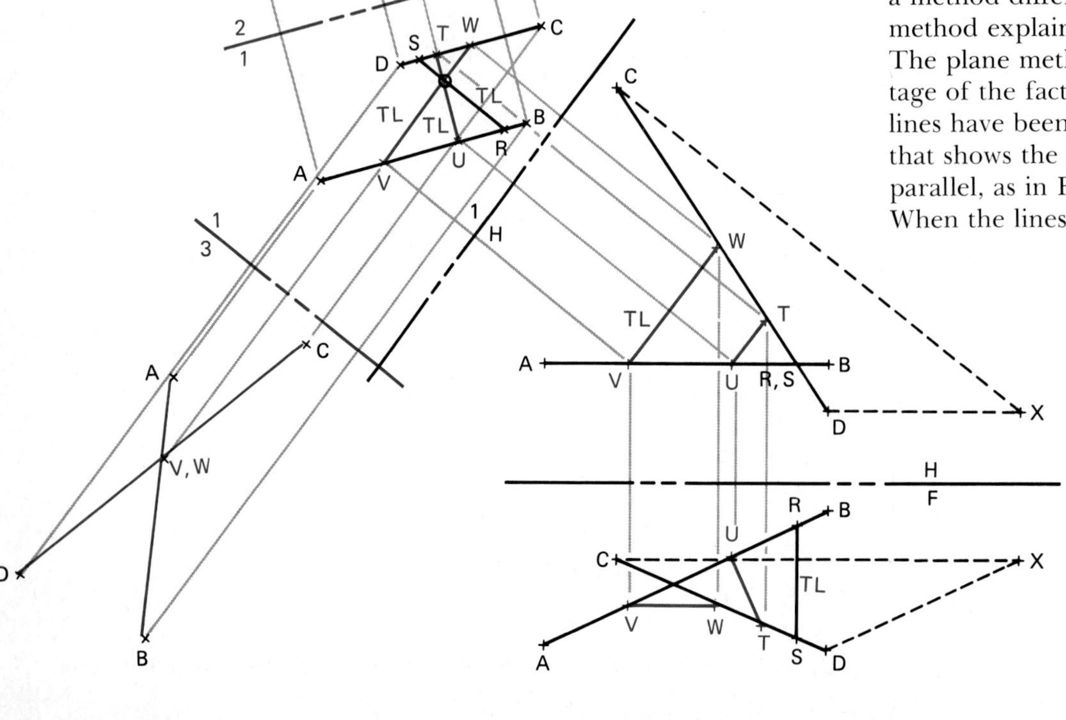

FIG. 6.14 **Projection of vertical, horizontal, and perpendicular connectors.**

Accuracy in constructing the positioning plane is necessary to ensure success.

6.10 VERTICAL, PERPENDICULAR, AND HORIZONTAL CONNECTORS BETWEEN SKEW LINES

A combination of vertical (*RS*), perpendicular (*TU*), and shortest horizontal connector (*VW*) is illustrated in Fig. 6.14. After elevation view 1 has been accurately positioned by means of plane *CDX* (parallel to *AB*), any desired straight-line connector can be located with only one additional view for each connector. Taken independently, the procedure for each connector is similar to that outlined in Sec. 6.9.

The shortest perpendicular connector *TU* has been located by a method different from the method explained in Sec. 6.7. The plane method takes advantage of the fact that the two skew lines have been located in a view that shows the lines appearing parallel, as in Fig. 6.14, view 1. When the lines appear parallel, an infinite number of choices appears to exist for the common perpendicular. The correct location of the shortest perpendicular connector is found in a view where the lines appear in true length (view 2 in this example).

Again, good accuracy is required in constructing the "positioning" plane so that the solutions for the connectors will be correct.

6.11 SHORTEST GRADE CONNECTOR: PLANE METHOD

As illustrated in Sec. 6.13, any desired straight-line connector can be located in a single additional view that is properly aligned to the view showing the skew lines apparently parallel. The perpendicular connector (*TU*) in Fig. 6.14 has a slope of about 1.3 (130 percent grade) as seen in view 1.

6.11.1 Solution View: Shortest Grade Connector

The shortest grade connector is found in an orthographic view perpendicular to a line with the same grade as the connector located in an elevation view where the skew lines appear parallel.

6.11.2 Solution Procedure: Shortest Grade Connector

Step 1. Construct a plane containing line *AB* that is parallel to line *CD*, as in Fig. 6.15. Locate a horizontal line in plane *OAB* so that *OB* appears in true length in the H view; see ①.

Step 2. Find the edge view of *OAB* which then gives an elevation view where the skew lines appear parallel; see ②.

Step 3. Construct a grade triangle in view 1 corresponding to the desired grade of the connector; see ③. Here, for a grade of 70 percent, the horizontal leg is established at 100 units and the vertical leg is drawn at 70 units. The required connector is then constructed parallel to the hypotenuse of this triangle.

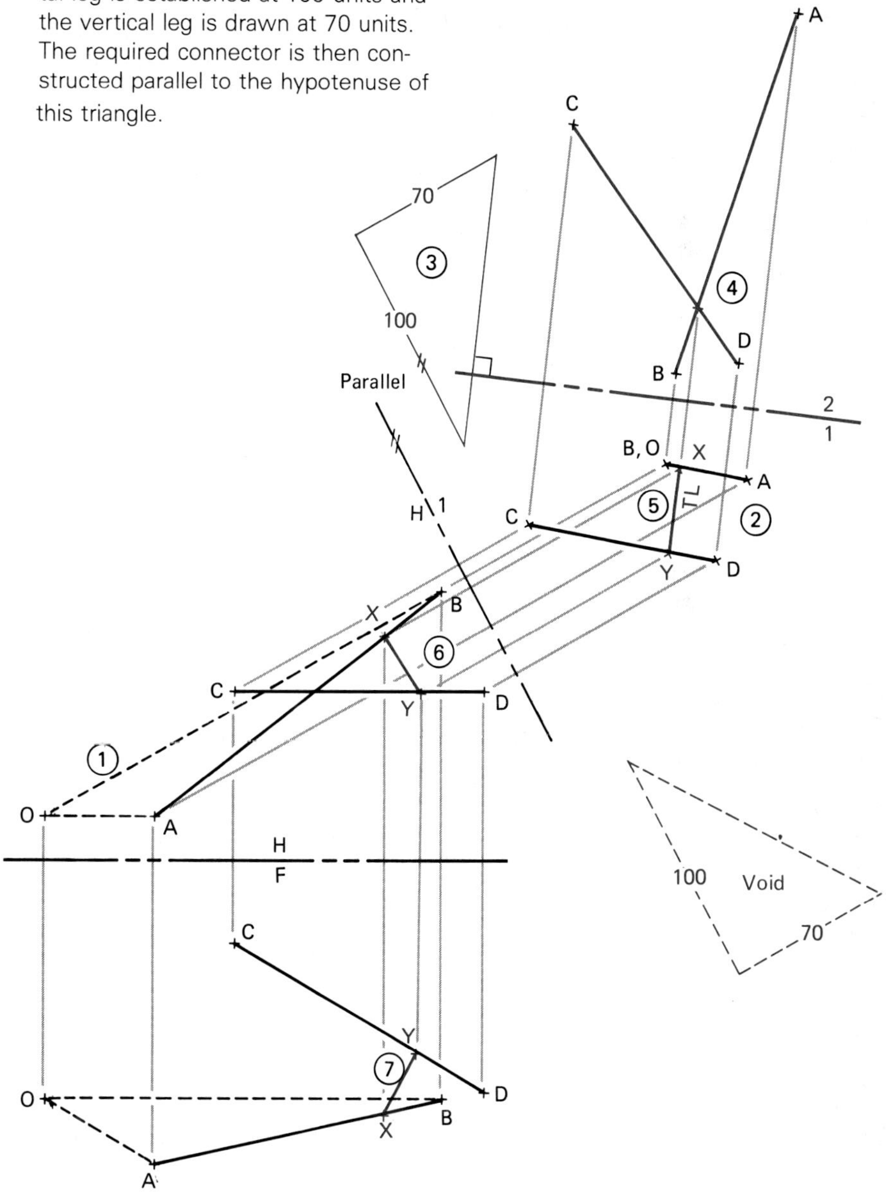

FIG. 6.15 Shortest grade connector between two lines (plane method).

Step 4. Place reference line 1/2 perpendicular to the hypotenuse of the grade triangle, thus creating a view which shows a point view of the grade connector *XY*; see ④.

Step 5. Project connector *XY* into all views; see ⑤, ⑥, and ⑦.

In the general case, there are two possible connectors having the same grade, but one will usually be shorter than the other. By inspection of the position of the skew lines at the point where they appear parallel, you should be able to quickly tell which will be shorter. In the particular case depicted in Fig. 6.15, the voided grade triangle has its hypotenuse nearly parallel to view 1 of the skew lines, and thus there is no second connector with a 70 percent grade.

In cases where skew lines lie in parallel vertical planes, there are two possible connectors of equal length for a given grade. When perpendicular skew lines lie in parallel horizontal planes, there are an infinite number of possible connectors of equal length for each specified grade. Examples of these special cases will appear in the problems at the end of the chapter.

6.12 SHORTEST HORIZONTAL CONNECTOR: SHORTCUT METHOD

As stated previously, all methods of determining the shortest horizontal connector between two skew lines are based on the principle that such a connector will be the same length as the shortest horizontal distance between parallel planes. A simpler and more-accurate method than that given in Sec. 6.11.2 involves projecting onto a projection plane positioned parallel to the true-length horizontal line in the plane that contains one of the lines and is parallel to the other skew line. This elevation view will directly show the point view of the desired shortest horizontal connector at the place where the skew lines cross. The skew lines may have to be extended to find the crossing point.

Figure 6.16 shows this shortcut method along with the conventional method (dashed lines).

6.13 SHORTEST CONNECTORS: VERTICAL, HORIZONTAL, GRADE, AND PERPENDICULAR: SHORTCUT METHOD

A useful phenomenon in the geometry of connectors between skew lines, including vertical, horizontal, grade, and perpendicular connectors, is that they intersect a common horizontal axis. This axis is parallel to the parallel planes that contain the skew lines involved. The horizontal axis will project as a point in the elevation view where the skew lines appear parallel. All these connectors intersect the axis in this view. Therefore, any two shortest connectors projected onto this elevation view will accurately locate the axis, simplifying the location of the other desired connectors.

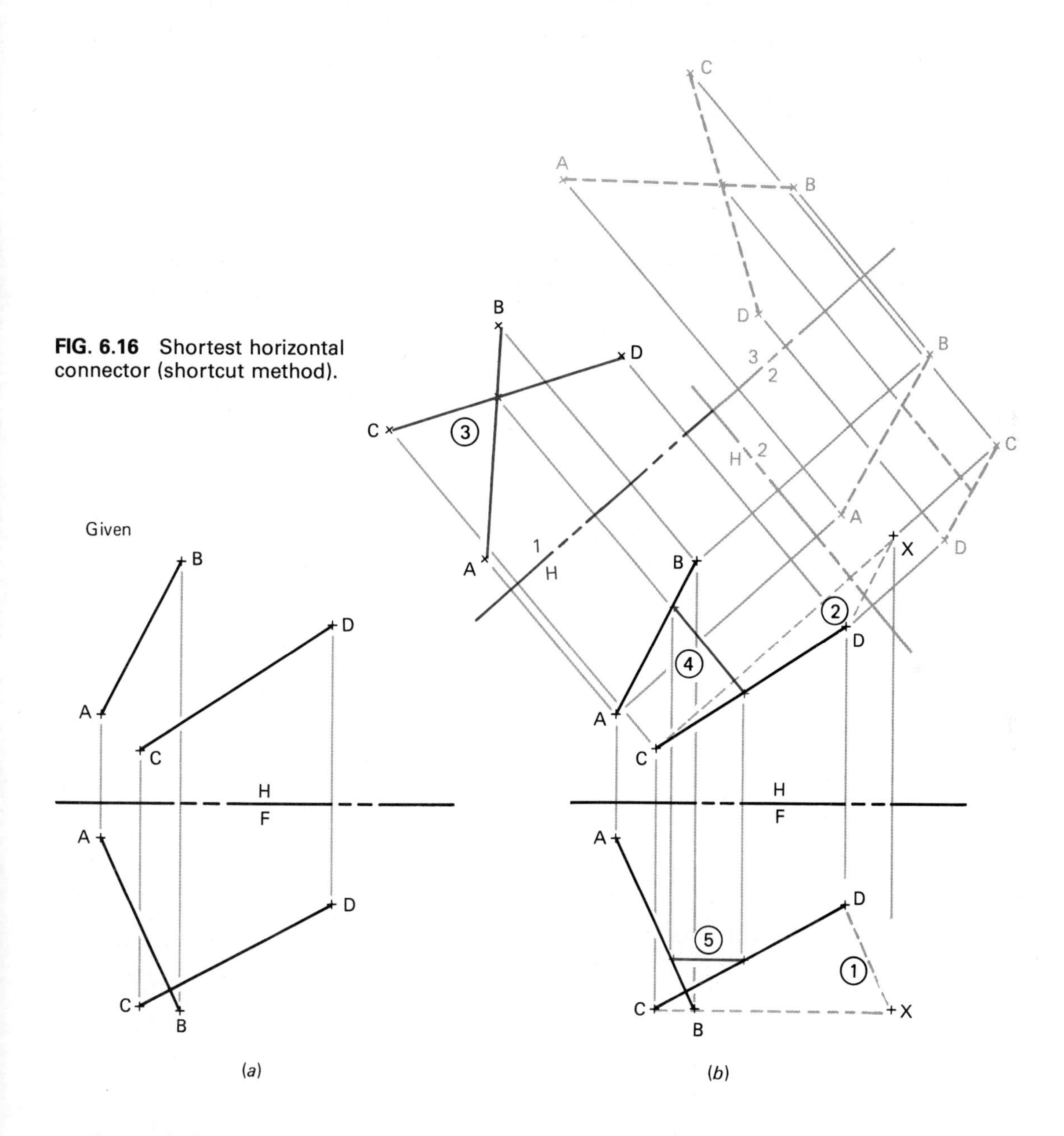

FIG. 6.16 Shortest horizontal connector (shortcut method).

The vertical connector is the easiest to locate accurately. This connector combined with any of the other connectors in this elevation view will locate the desired axis. The shortest horizontal connector found by the shortcut method is a good choice. No additional views are required to locate and determine the true length of any of the other connectors.

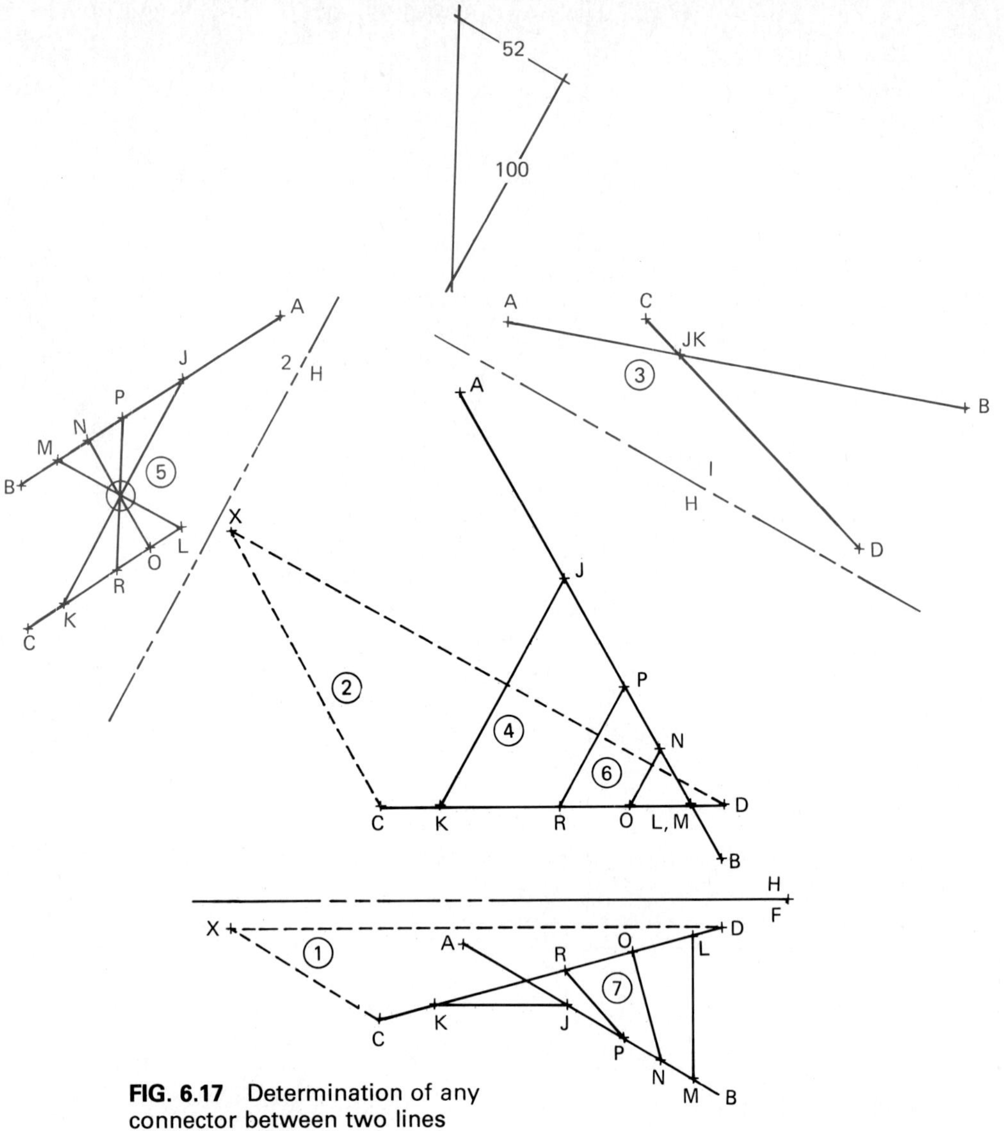

FIG. 6.17 Determination of any connector between two lines (shortcut method).

6.13.1 Solution Procedure: Shortest Connectors, Shortcut Method

Figure 6.17 shows the steps described below.

Step 1. Construct a plane containing line *CD* that is parallel to line *AB*. Then locate the F and H views of horizontal line *DX*; see ① and ②.

Step 2. Establish elevation view 1 with reference line H/1 parallel to the true-length projection of *DX*. Then locate the point view of the shortest horizontal connector *JK*; see ③.

Step 3. Project *JK* to the horizontal view; see ④. Accuracy is imperative here.

Step 4. Find skew lines *AB* and *CD* in view 2 (the elevation view in which they appear parallel). Project *JK* and *LM* (vertical connector) to view 2, thereby locating the point view of the axis through which all the connectors will pass. Construct the perpendicular connector *NO* at 90° to the projections of *AB* and *CD*. Draw the grade connector *PR* (52 percent grade in this example) through the point view of the axis; see ⑤.

Step 5. Project *PR* and *NO* to the H view; see ⑥. All connectors should appear parallel in the H view.

Step 6. Locate all the connectors in the F view; see ⑦. Line *JK* must project parallel to the H/F reference line in the F view.

6.14 SHORTEST FRONT CONNECTOR: SHORTCUT METHOD

Figure 6.18 illustrates how the fundamental principles of geometry can be adapted to locate the shortest front connector. The procedure is similar in all respects to that for locating other connectors, except that a front line is located in the plane containing one of the skew lines instead of a horizontal line.

6.14.1 Solution Procedure: Shortest Front Connector, Shortcut Method

Step 1. Construct plane *CDX* that contains line *CD* and is parallel to line *AB*, as seen in Fig. 6.18; see ① and ②. Line *XD* is a front line.

Step 2. Place reference line F/1 parallel to the true-length projection of *XD* and project lines *AB* and *CD* to view 1; see ③. This locates the point view of the shortest front connector *YZ*.

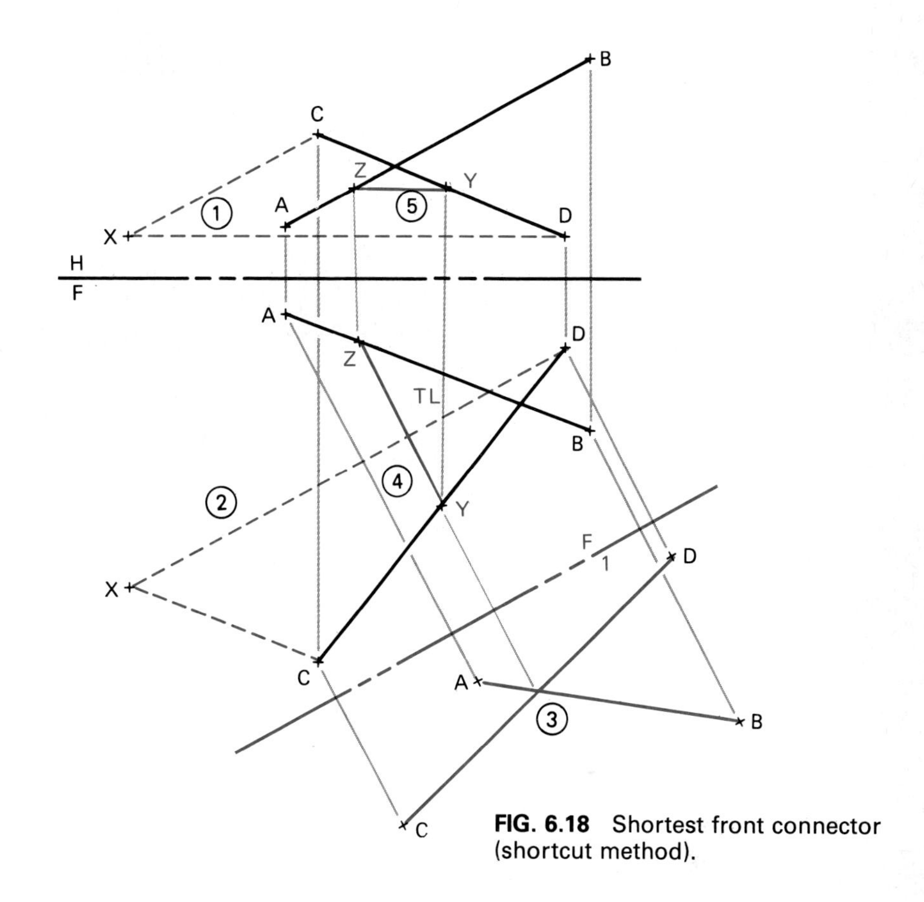

FIG. 6.18 Shortest front connector (shortcut method).

Step 3. Project *YZ* to the F and H views; see ④ and ⑤. Line *YZ* must project parallel to the H/F reference line in the H view.

6.15 METHODS FROM COMPUTER GRAPHICS

In Chaps. 4 and 5, we discussed the use of the computer to help solve some geometry problems that involve lines and planes. The ideas presented there apply equally well for dealing with connectors. A desired connector can be found if one can orient the viewing position properly. Much

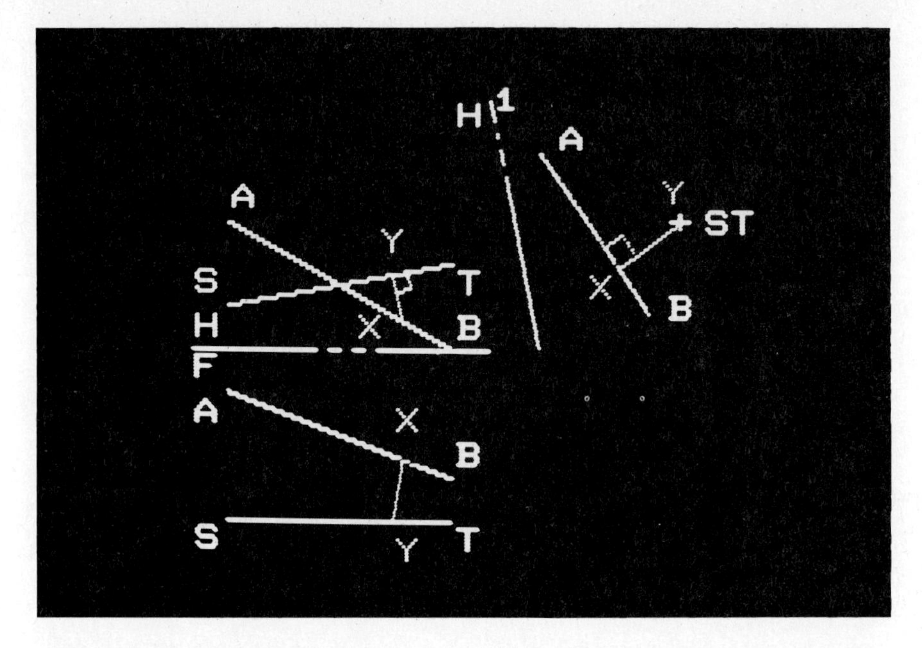

FIG. 6.19 Computer graphics representation of the perpendicular connector between two lines.

of the material presented in Chap. 6 has been directed toward finding such a view. When using graphics methods, we may need one or more intermediate views in order to place a reference line at the proper orientation. In computer graphics, we should be able to ask that the previously defined skew lines be oriented in any position we desire. The connector of interest can then be seen on the screen and its length can be calculated from the equations of the lines together with conditions specified by the connector, horizontal or vertical, for example. Figure 6.19 illustrates a perpendicular connector and indicates how it would appear if generated by computer graphics.

Problems

6.1 By example problem, construct a situation where two lines are drawn parallel in adjacent views but where the lines are not parallel in space.

6.2 Two perpendicular lines will appear at 90° in any view that shows one or both lines in true length. What situation must be excluded?

6.3 It can be stated that a line that is perpendicular to a plane is perpendicular to all lines in that plane. What equivalent principle can be applied to the edge view of the plane?

6.4 For two skew lines in space, explain how to determine the perpendicular connector by:
  (a) Line method
  (b) Plane method

6.5 Explain the usefulness of knowing that the vertical, horizontal, grade, and perpendicular connectors between skew lines intersect a common horizontal axis.

6.6 For Fig. 6.20:

(*a*) Draw a line parallel to line 12 that passes through point *A*.

(*b*) Find the shortest connector from point *A* to line 12.

(*c*) If there exists a vertical connector from point *A* to line 12, find it; if not, where would A_H have to be located so that such a connector would exist?

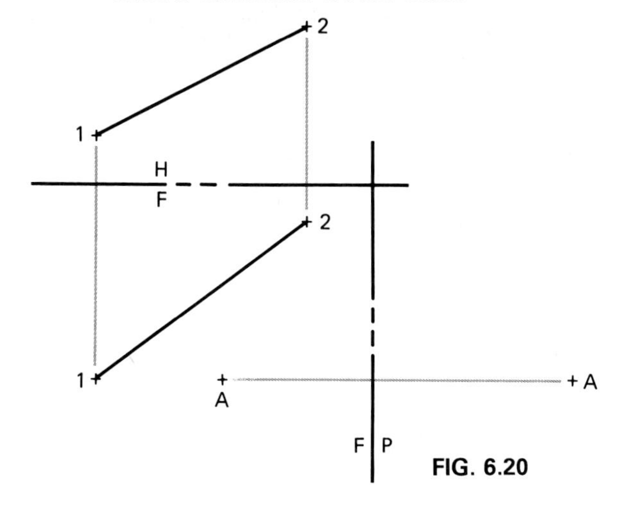

FIG. 6.20

6.7 For line 12 and point *A* given in Fig. 6.20, find a connector from *A* to 12 that makes a 45° angle with 12.

6.8 In Fig. 6.21, find the shortest connector from point *V* to plane *XYZ*. Show the horizontal and front views.

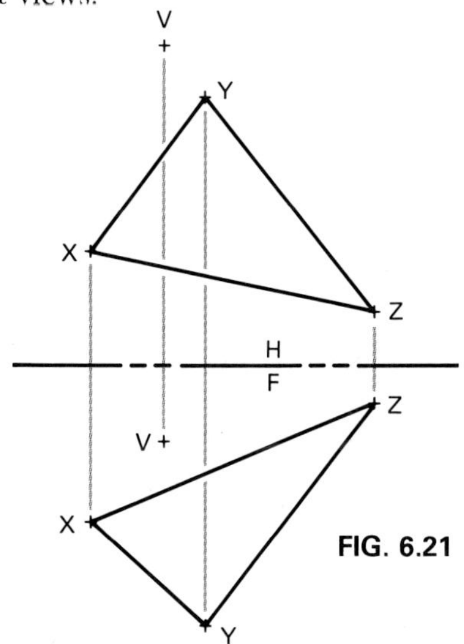

FIG. 6.21

6.9 Locate the vertical connector between the lines in parts *a* and *b* of Fig. 6.22.

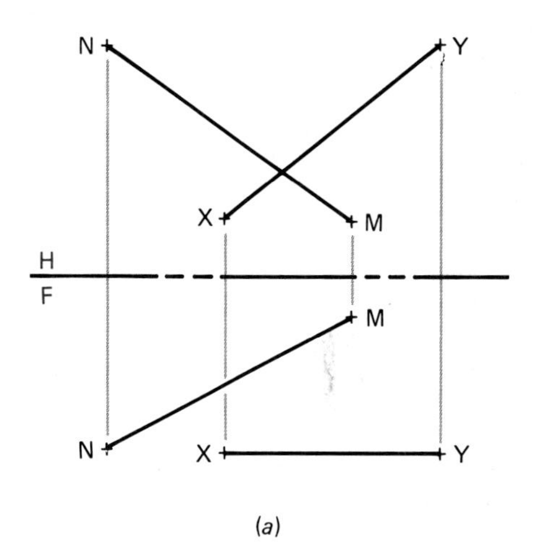

(*a*)

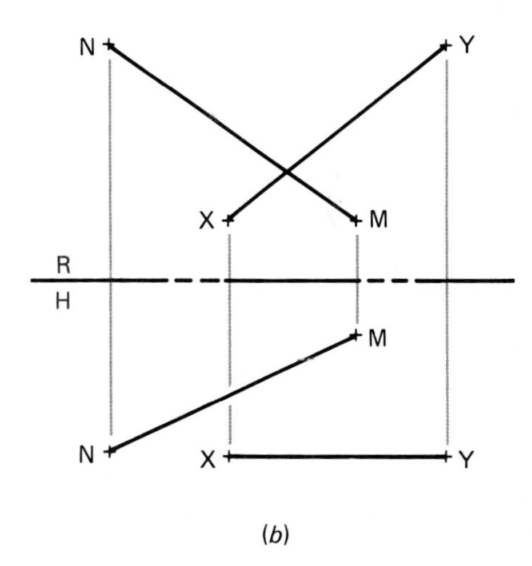

(*b*)

FIG. 6.22

6.10 For the lines in parts *a* and *b* of Fig. 6.22, construct the shortest connector in each case. Show the front view of each.

6.11 Graphically determine the shortest horizontal, the shortest perpendicular, and the vertical connectors for lines *RS* and *OP* as shown in Fig. 6.23.

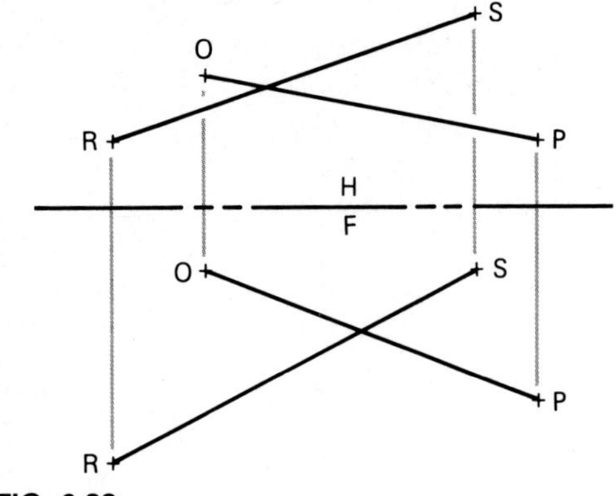

FIG. 6.23

6.12 In preparation for constructing two tunnels, you imagine their plan view as seen through an *xy* grid and note that one tunnel extends from $x = 0$ ft. $y = 100$ ft at a depth of 20 ft to a point $x = 50$ ft, $y = 0$ ft at a depth of 60 ft. The other tunnel extends from $x = 0$ ft, $y = 0$ ft to $x = 50$ ft, $y = 100$ ft at a constant depth of 10 ft. If you treat the data given as being applied to the tunnel center lines, what are the largest equal diameters that the tunnels may have and have a clearance of 10 ft? Determine graphically.

chapter 7

Surfaces: Generation and Development

7.1 INTRODUCTION

A common manufacturing technique is to cut a pattern from flat metal stock and then bend, fold, and form the stock into the desired product. The well-known "tin can" container for soup, vegetables, and fruits found on the grocery shelf is made from three flat pieces of metal. One piece is formed into a hollow cylinder, and the other two pieces are attached as ends, one before filling and the second after filling the can. This chapter describes the methods of designing the patterns for various surfaces. Figure 7.1 shows some of the many surfaces that exist.

We will approach the discussion of surfaces by first defining several terms that are common to all surfaces. We will then indicate how some specific surfaces are generated and developed. Chapter 8, in turn, will describe how some of these various surfaces may intersect.

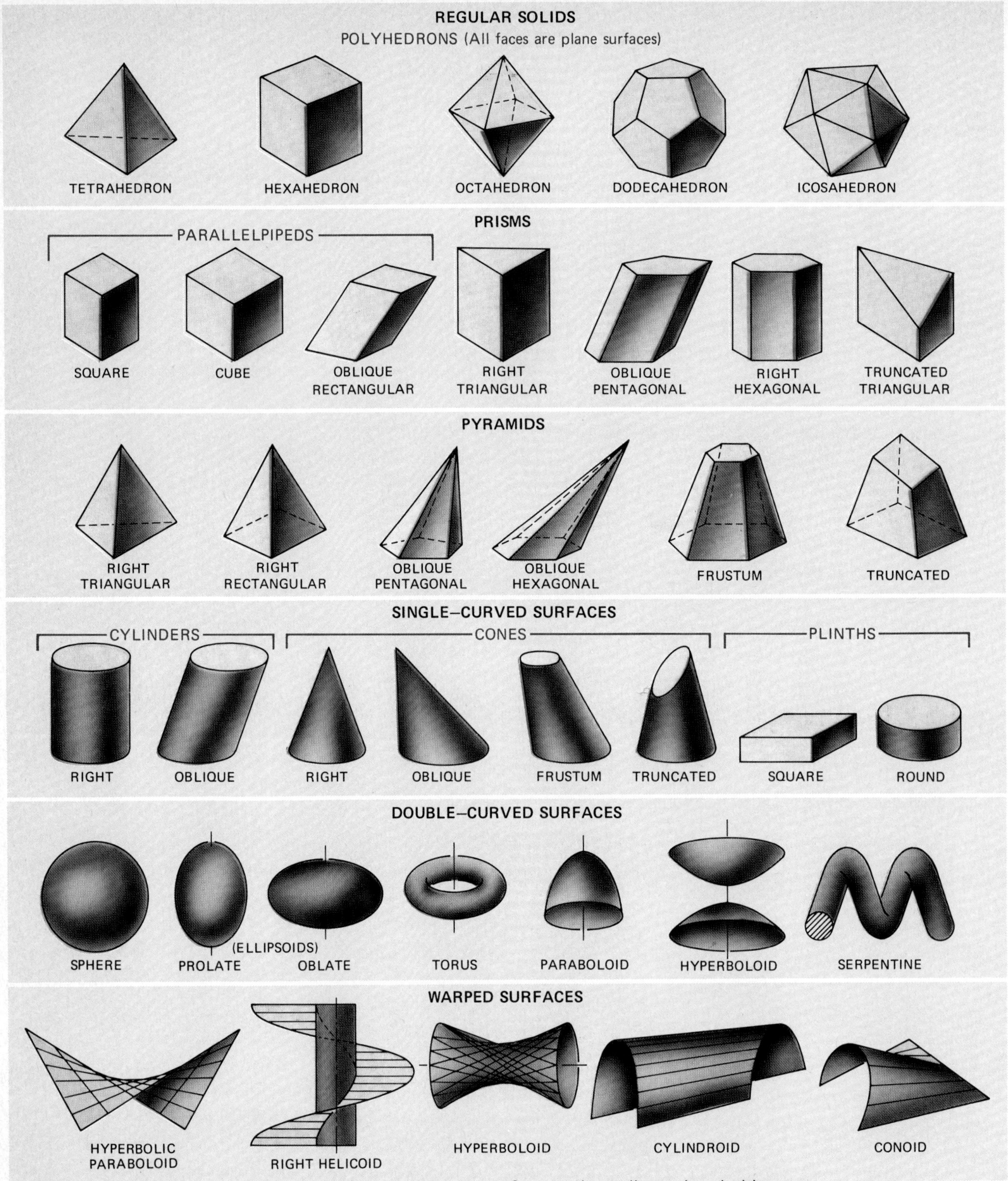

REGULAR SOLIDS

POLYHEDRONS (All faces are plane surfaces)

TETRAHEDRON HEXAHEDRON OCTAHEDRON DODECAHEDRON ICOSAHEDRON

PRISMS

PARALLELPIPEDS

SQUARE CUBE OBLIQUE RECTANGULAR RIGHT TRIANGULAR OBLIQUE PENTAGONAL RIGHT HEXAGONAL TRUNCATED TRIANGULAR

PYRAMIDS

RIGHT TRIANGULAR RIGHT RECTANGULAR OBLIQUE PENTAGONAL OBLIQUE HEXAGONAL FRUSTUM TRUNCATED

SINGLE–CURVED SURFACES

CYLINDERS

RIGHT OBLIQUE

CONES

RIGHT OBLIQUE FRUSTUM TRUNCATED

PLINTHS

SQUARE ROUND

DOUBLE–CURVED SURFACES

SPHERE PROLATE (ELLIPSOIDS) OBLATE TORUS PARABOLOID HYPERBOLOID SERPENTINE

WARPED SURFACES

HYPERBOLIC PARABOLOID RIGHT HELICOID HYPERBOLOID CYLINDROID CONOID

FIG. 7.1 Solids and surfaces which define the geometry of many three-dimensional objects.

7.2 DEFINITIONS: GENERATION AND DEVELOPMENT OF SURFACES

A surface is the cover, envelope, or exterior face of an object. Theoretically, a surface is an area having length and width but no thickness. A surface may also be an area generated by a moving line (straight or curved) generatrix guided by a line (straight or curved) directrix or plane director.

Although a theoretical surface has no thickness and therefore has no volume, actual surfaces (of metal or plastic, for example) must have some measurable thickness. An auto body, refrigerator case, and aircraft skin are some examples of practical surfaces.

A *generatrix* is a straight or curved line the movement of which describes a surface.

An *element line* is any fixed position of the generatrix.

A *directrix* is a point or line (which can be straight or curved) that guides the movement of the generatrix.

A *director* is a plane to which a moving generatrix remains parallel.

A *developable surface* is a surface that can be folded or rolled out to form a plane without distortion. This plane is called a development or pattern.

An *approximately developable surface* is not accurately developable but can be assembled to become the desired contour by joining a series of small sectors that have been distorted by stretching, compressing, or forming. A sphere is an example of a surface that can be approximately developed.

A *nondevelopable surface* is one that must be formed, molded, or shaped. It is a surface that cannot be laid out in a plane as a true development or pattern.

A *formed or molded surface* is a surface that is produced by pouring liquefied material such as metal, plastic, or concrete into a form or mold so that when the material solidifies, it takes the desired shape. Some warped and double-curved surfaces can also be formed by punching, pressing, stretching, twisting, or cutting.

A *constructed surface* is a surface such as a hyperbolic paraboloid, cylindroid, or other warped surface that can be constructed by offsetting adjacent elements such as rafters or other structural members in a fixed warped position and then completing the surface with flexible materials such as concrete or rubber.

A *ruled surface* is an accurately developable surface generated by a straight-line generatrix. Planes and single-curved and warped surfaces are ruled surfaces.

A *double-ruled surface* is a nondevelopable surface generated by two different straight-line generatrices. Two intersecting straight lines can be drawn through any point in the surface. The hyperbolic paraboloid is an example.

Figures 7.2 and 7.3 summarize some developable and non-

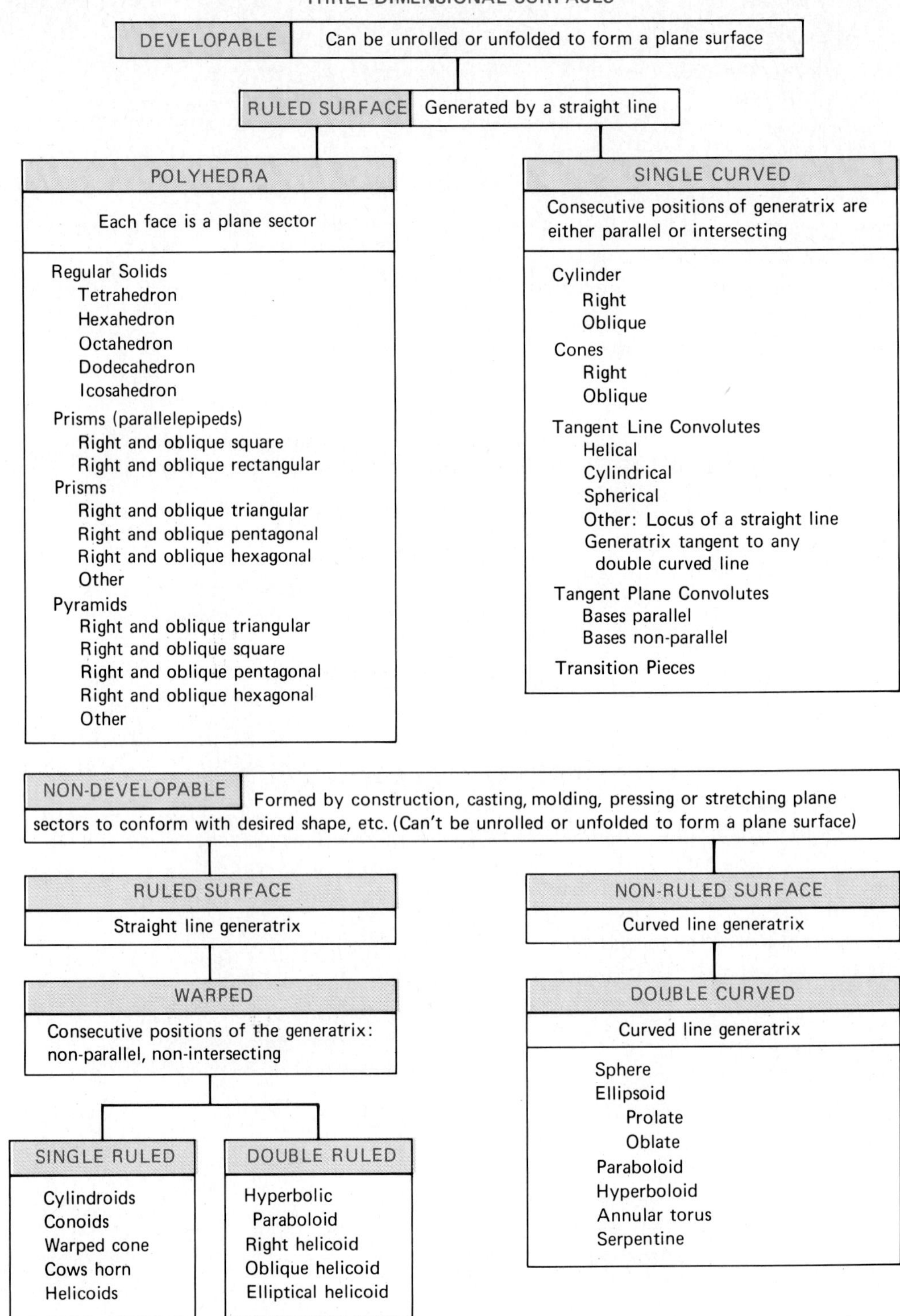

FIG. 7.2 Categories of surfaces according to development characteristics.

FIGURE 7.3
Nature and Use of Surfaces

NAME OF SURFACE	GENER-ATRIX	DIRECTRICES	PLANE DIRECTOR	ACCUR-ATELY DEVELOP-ABLE	POSSIBLE USE
Polyhedra (*all*)	Straight line	Straight line	No	Yes	Boxes; buildings
Cylinder	Straight line	Axis and circle	No	Yes	Pipes; tanks
Cone	Straight line	Fixed point; straight and curved lines	No	Yes	Funnel; transition
Tangent line convolute	Straight line	Cylindrical or spherical helix	No	Yes	Auger; conveyor
Tangent plane convolute	Plane	Two dissimilar straight and curved lines	No	Yes	Shelter
Transition piece	Straight line	Straight line and straight and curved line	No	Yes	Furnace plenum
Cylindroid	Straight line	Two nonparallel straight and curved lines	Yes	No	Subway entrance
Conoid	Straight line	One straight line; one straight and curved line	Yes	No	Subway entrance
Hyperbolic paraboloid	Straight line	Two nonintersecting nonparallel straight lines	Yes	No	Roofs; structures
Right helicoid	Straight line	Straight-line axis and helix	No	No	Stairway; conveyor
Oblique helicoid	Straight line	Straight-line axis and helix	No	No	Auger; screw thread
Sphere	Circle	Axis-diameter of circle	No	No	Balls; tanks
Ellipsoid-prolate	Ellipse	Axis-major diameter	No	No	Fuel tank; spotlight
Ellipsoid-oblate	Ellipse	Axis-minor diameter	No	No	Water tank
Paraboloid	Parabola	Axis of symmetry	No	No	Reflector
Annular torus	Circle	External axis	No	No	Life preserver
Serpentine	Circle	Helical axis	No	No	Spring; corkscrew

FIG. 7.3 Nature and use of surfaces.

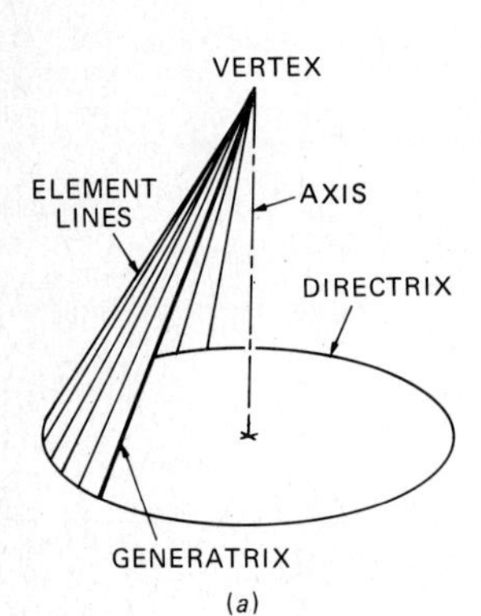

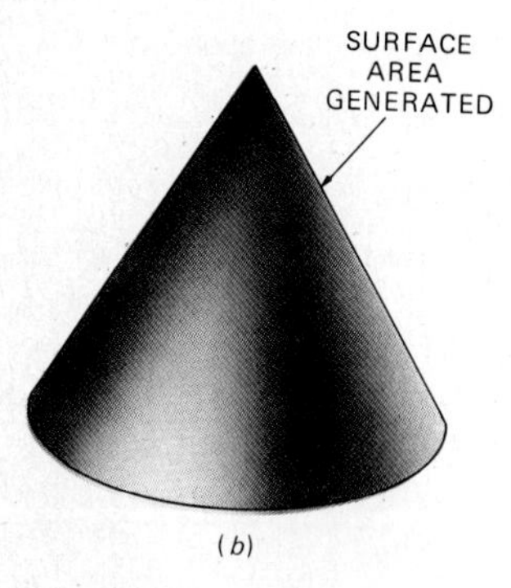

FIG. 7.4 The right cone is an example of a ruled surface.

developable surfaces that are used in manufacturing and construction.

Figure 7.4 illustrates pictorially the generation of a ruled surface, specifically a right circular cone.

7.3 GENERATION OF SURFACES: GENERAL RULES

Although pictorial drawings often are helpful in visualizing the three-dimensional characteristics of geometric surfaces, accurate developments and intersections require orthographic views of the surface(s) involved. The following rules apply when generating a surface:

1. A minimum of two adjacent orthographic views generally are required.

2. Principal bases, vertexes, contour, and overall sizes of the three-dimensional object are drawn.

3. The appropriate positions of line directrices or the plane director are established.

4. An optimum number (frequently 12 to 16) of judiciously spaced element lines are drawn to help describe the visual appearance and expedite formation or development.

5. Element lines are usually numbered in a logical sequence beginning with the shortest and proceeding in a clockwise rotation.

6. Procedures for determining the true length of the various element lines in order to aid development or construction of the surface are implemented.

7. Generally, surfaces are considered to be opaque. When you darken element lines, the appropriate line symbol for visible or hidden lines should be used.

7.4 DEVELOPMENTS: GENERAL RULES

1. Developments are usually laid out inside up to facilitate bending or rolling into shape. See Fig. 7.5.

2. Generally the shortest element line is used as a starting and ending place for the pattern.

3. If a clockwise sequence has been used on the generation drawing with number 1 as the shortest element, the pattern can be laid out in a corresponding sequence which ensures that it will be inside up. Special attention must be paid to nonsymmetric patterns to ensure correct results.

4. Bend or fold lines (usually coincident with element lines) that guide rolling or forming of the final product and trim or cut lines should be clearly marked on the pattern.

5. Weld or rivet laps are frequently added to the pattern to provide extra strength at the splice (location where the end element lines of a pattern are placed together and soldered, welded, riveted, or otherwise fastened).

7.5 PLANE SURFACES: DEFINITION

A plane surface is a ruled surface generated by a straight-line generatrix moving parallel to its original position while contacting a straight-line directrix that is not parallel to the generatrix.

Regular and irregular polyhedrons are bounded by plane surfaces and are exactly developable. Because all such polyhedrons are relatively simple and are generated and developed in a similar way, only a few, such as right and oblique prisms and pyramids, will be discussed here. Right refers to all objects with axes perpendicular to their bases; oblique refers to objects with axes that are not perpendicular to their bases.

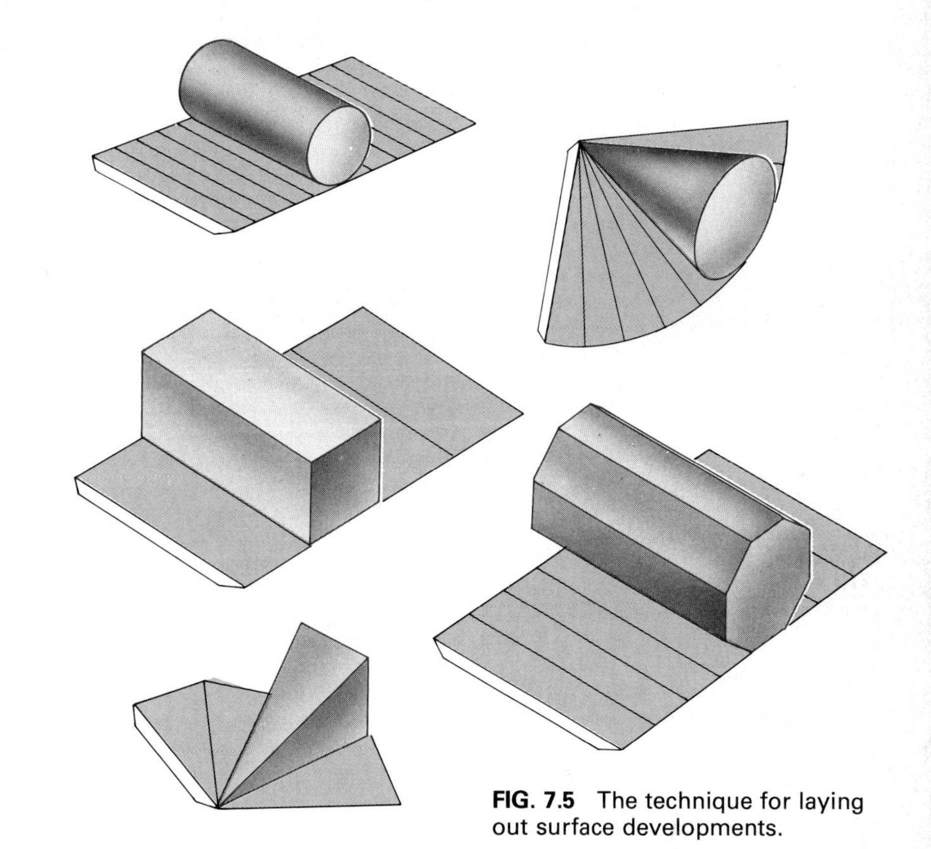

FIG. 7.5 The technique for laying out surface developments.

7.6 RIGHT HEXAGONAL PRISM

Refer to Fig. 7.6 for an illustration of the generation and development of a right hexagonal prism. Up to this point in the text, solutions to example problems have been outlined in detail in order to provide you with a quick reference capability as you carried out the steps. Hereafter, only the steps themselves will be given.

7.6.1 Generation

Draw adjacent orthographic views. One view should show the edge view of all six sides of the prism, and the other should show the elements in correct projection.

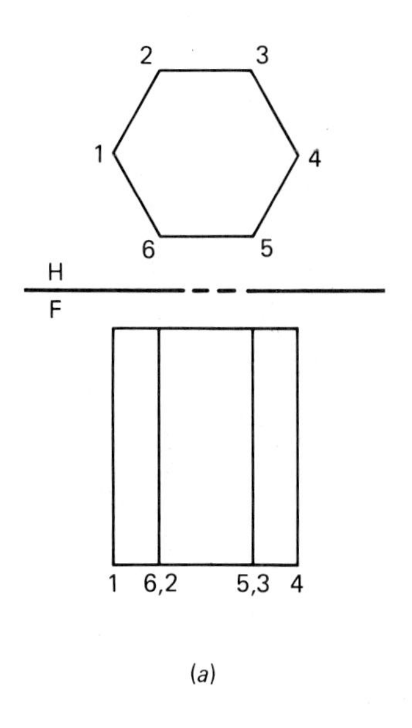

(a)

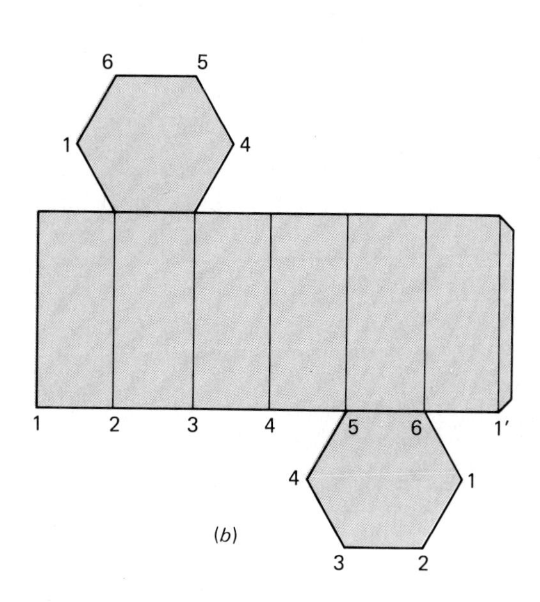

(b)

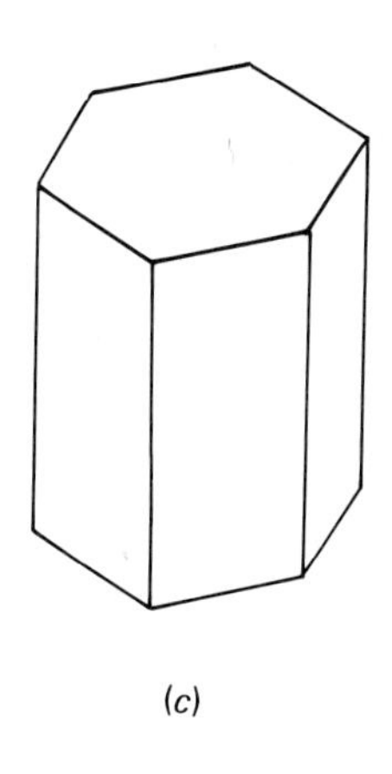

(c)

FIG. 7.6 Generation and development of a hexagonal prism.

Number the element lines in a clockwise manner, as indicated in Fig. 7.6*a*. Note that the front view shows the true length of all the element lines (they are vertical lines in this case). Also note that the true distances between the element lines are shown in the top view.

7.6.2 Development

Lay out two parallel lines (top and base of the prism) spaced at the true height of the prism.

Establish element line number 1 perpendicular to the top and base lines of the layout on the left side of the drawing surface.

Step off the respective true distances between element lines. True distances are seen in the horizontal view.

Number the element lines 1 through 6 back to 1′ from left to right as shown in Fig. 7.6*b*.

Attach the top and bottom and the lap segment if required. The top and bottom of the prism are shown in true size in the top projection of the generation drawing.

Show bend lines on the lateral surface of the inside-up layout.

7.7 OBLIQUE TRUNCATED HEXAGONAL PRISM

An oblique truncated hexagonal prism can be generated and developed by means of the same technique described in Sec. 7.6. However, two adjacent views often fail to show all the necessary information, such as true lengths of elements or the true distance between elements. In Fig. 7.7, the true distances are not shown in the H and F views. To find the true distance between elements 2–3, 3–4, 5–6, and 6–1, an auxiliary inclined view is drawn perpendicular to the element lines. If the object was positioned so that the element lines were vertical (top view identical to the inclined view in Fig. 7.7), an auxiliary view would be unnecessary.

It is important to remember that the true lengths of elements, the true perpendicular distances between them, and the true position relationship of all element lines must be determined before you lay out a development of the object.

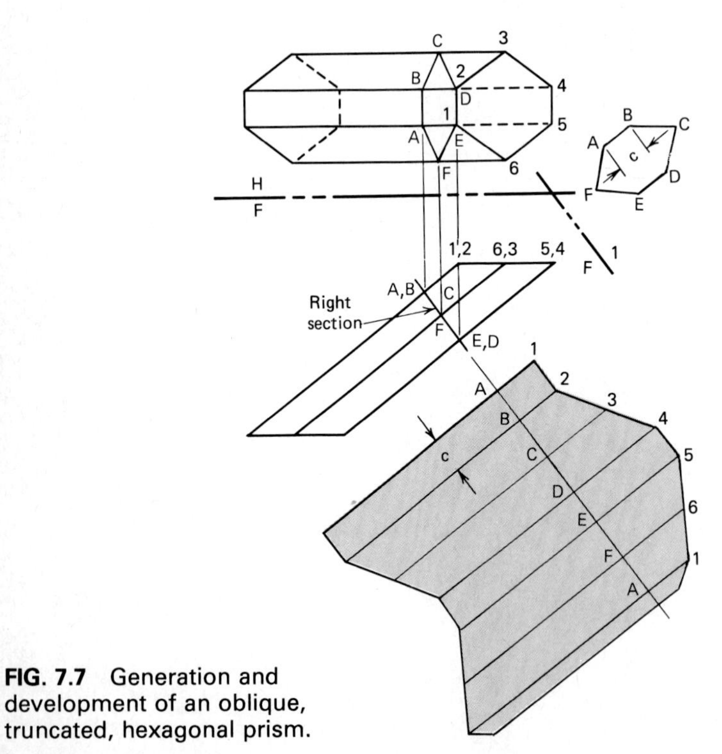

FIG. 7.7 Generation and development of an oblique, truncated, hexagonal prism.

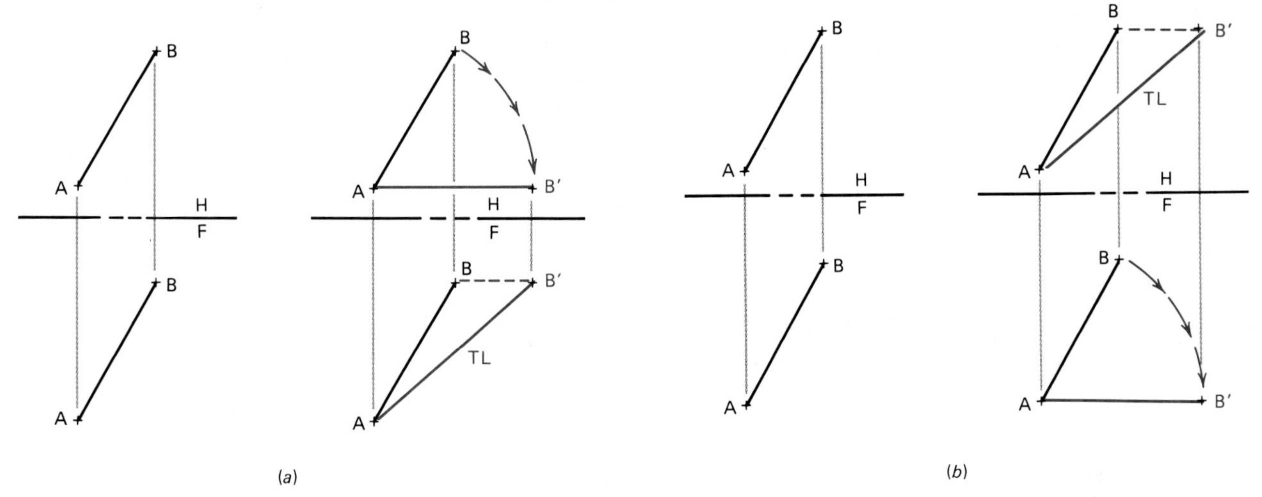

FIG. 7.8 Finding true lengths by rotation.

7.8 TRUE LENGTHS AND TRUE SHAPES BY ROTATION

The use of rotation to find true lengths of element lines or true shapes of surfaces is a convenient method that is particularly appropriate for use in developing surfaces. This method is fast and accurate and requires less room on a drawing than does the auxiliary-view method.

7.8.1 Procedure: Rotation of Lines for True Lengths

Refer to Fig. 7.8 and follow the procedure below to determine the true length of a line by rotation.

Assume a point view of the axis of rotation in any orthographic view of the given line (usually at one of the endpoints of the line). The axis of rotation is at point *A* in Fig. 7.8*a*.

Rotate the line segment to a position parallel to a selected adjacent view (that is, parallel to the selected reference line). *AB* is rotated to *AB′* in the horizontal view.

Project the new (rotated) position of the line to the adjacent view. Only the dimensions parallel to the reference line will change in this adjacent view. The front view shows the true length *AB′*. By beginning with the axis of rotation in the front view, you will derive the same true length in the horizontal view. See Fig. 7.8*b*.

7.8.2 True Shape of a Plane Surface by Rotation

The procedure for finding the true shape of a plane sector by rotation is similar to the method for finding the true length of a line segment; it is illustrated in Fig. 7.9.

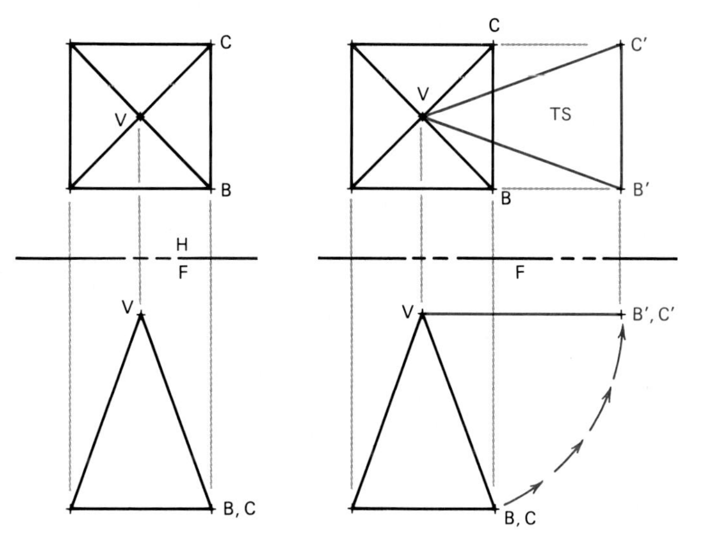

FIG. 7.9 Finding true shape by rotation.

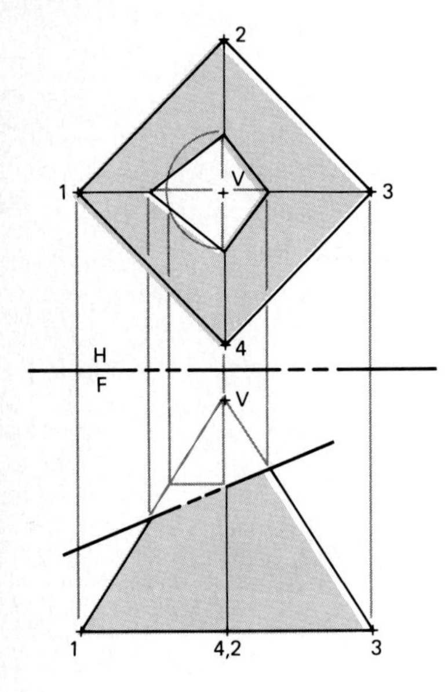

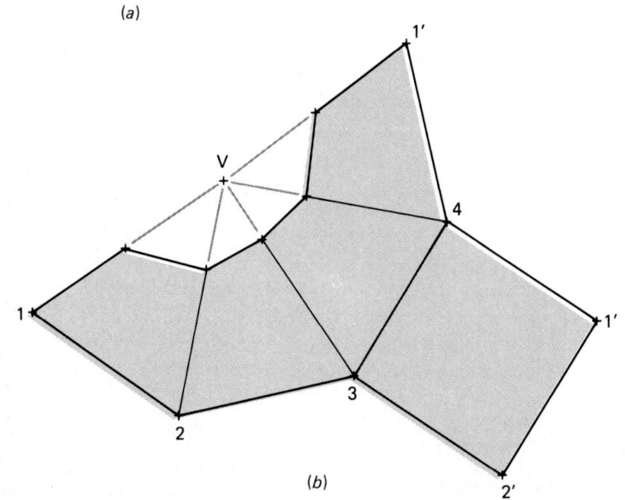

FIG. 7.10 Generation and development of a right square pyramid.

Establish the point view of an axis in any view that shows the plane surface as an edge (usually coincident with some point or point view of a line on the plane surface). In Fig. 7.9, plane *BCV* is an edge in the front view, and so *V* can be the point view of the axis.

Rotate the edge view of the plane to a position parallel to a selected adjacent projection plane.

Project the new position of the plane sector to the adjacent projection plane. Again, only the dimensions parallel to the reference line will change in the true-shape view.

7.9 TRUNCATED RIGHT SQUARE PYRAMID

A right square truncated pyramid can be formed by first constructing four equal triangular plane surfaces that have a common vertex and a common square base. The truncation can be accomplished later in the procedure.

7.9.1 Generation

Draw two adjacent views positioned as shown in Fig. 7.10*a*. Number the element lines clockwise. The true lengths of the sides of the triangles are found in the front view, and those for the square base are shown in the top view.

Draw the edge view of the cutting plane representing the truncation at the desired position in the front view.

Project points on the element lines cut by the cutting plane to the top view. Profile lines *V*–2 and *V*–4 must be rotated to a front position to find the true distance to the point of cut.

Connect the points just found in the top view to define the truncated surface.

7.9.2 Development

Scribe an arc with a radius equal to the true length of *V*–1 or *V*–3. See Fig. 7.10*b*.

Starting at point 1 of the arc, measure off four chords, each equal in length to one side of the base.

Draw light (construction) element lines from the arc center *V* to the chord-arc intersection points and identify as *V*–1, *V*–2, *V*–3, *V*–4, and *V*–1′.

Locate true-length distances from *V* to the point where each element is cut by the cutting plane; connect these points to identify the truncated top edge of the development. Darken the outline of development.

Add the true-shape square area of the base connected to the base edge of one of the triangular planes.

7.10 TRUNCATED RIGHT TRIANGULAR PYRAMID

The generation and development of a truncated right triangular pyramid requires procedures that are typical for all three-dimensional objects made up of plane surfaces.

7.10.1 Generation

Draw adjacent orthographic views showing the projection of the cut surface and theoretical vertex, with light-weight lines defining its height and position relative to the base plane. See Fig. 7.11*a*.

Number the corner points clockwise, starting with the shortest element.

Use procedures previously described to determine the true shapes of the surfaces.

7.10.2 Development

Lay out the development in the manner prescribed for the square pyramid, starting with the shortest element line and continuing in the sequence established on the generation views.

Draw the top and bottom surfaces if required. See Fig. 7.11*b*. Note that in contrast to Fig. 7.10*b*, a triangular "cap" is included in the development of Fig. 7.11*b*. The true shape of the cap is determined by rotating the plane in Fig. 7.11*a* to a horizontal position (F view) and marking the outline with dashed lines in the H view.

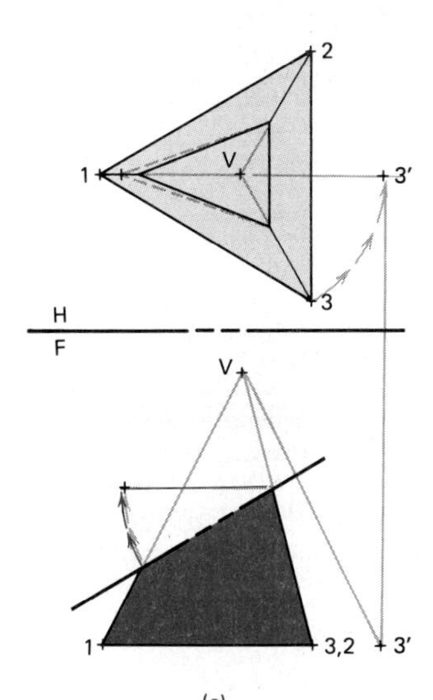

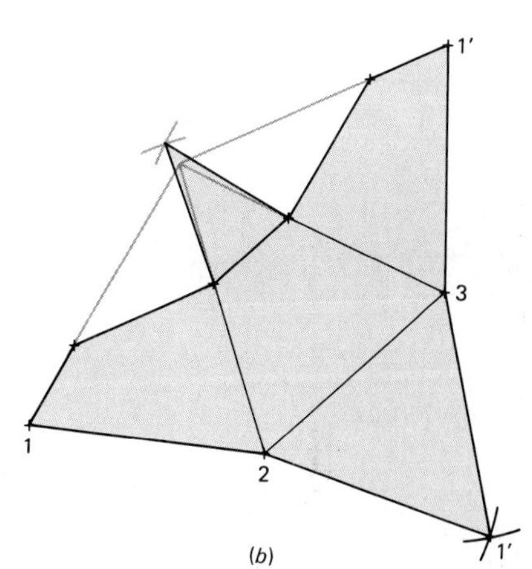

FIG. 7.11 Generation and development of a right triangular pyramid.

7.11 TRIANGULATION

Triangulation is a method of constructing the true shape of a triangular plane by scribing three intersecting arcs, equal respectively to the true lengths of the sides of the triangle. This technique can be used in laying out the true-shape developments of various pyramids.

The procedure for triangulation given the three sides (*A*, *B*, and *C*) of a triangle is as follows (refer to Fig. 7.12*a*):

Draw side *A* in any desired position.

With the compass point at either end of side *A*, scribe an arc of radius *B*.

With the compass point at the opposite end of side *A*, draw an arc of radius *C* that intersects the first arc.

Connect the point of the intersecting arcs to the ends of side *A*, forming triangle *ABC*.

This technique is also effective in developing a trapezium (a quadrilateral having no two sides parallel). This is accomplished by dividing the given trapezium into

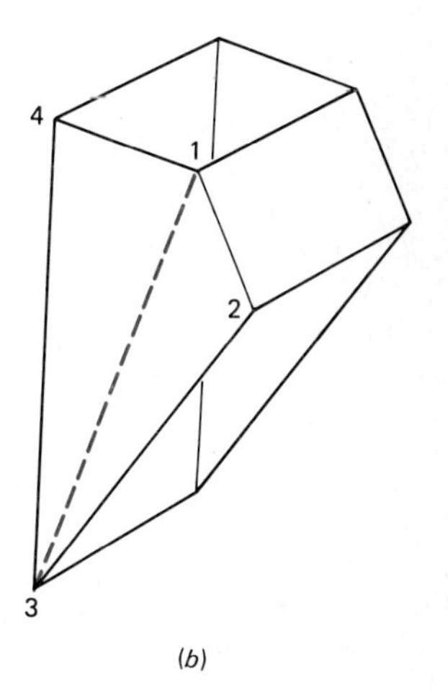

FIG. 7.12 Use of triangulation in development work.

two adjacent connected triangular surfaces. Then triangulation is used to develop these adjacent triangles. An object for which this technique is appropriate is shown in Fig. 7.12*b*, where side 1–2–3–4 is divided into two triangles by dashed line 1–3.

7.12 TRUNCATED OBLIQUE TRIANGULAR PYRAMID

The technique of triangulation just described can be used effectively in developing a truncated oblique triangular pyramid.

7.12.1 Generation

Draw adjacent orthographic views, as shown in Fig. 7.13*a*.

Label vertex *V* and number the elements.

Establish the true length of each oblique element line.

7.12.2 Development

Draw element line *V*–1 as shown in Fig. 7.13*b*.

From point 1, scribe an arc of radius 1–2.

From point *V*, scribe an arc of radius *V*–2 to intersect the first arc.

Continue the process of triangulation in sequence until all triangular surfaces are defined, including the base.

Darken the perimeter of the development and show the bend (element) lines lightly.

7.13 SINGLE-CURVED SURFACES

Developable ruled surfaces having either parallel or intersecting elements are called single-curved surfaces. They are generated with straight-line generatrices that are guided by various curved-line directrices. Figure 7.2 lists surfaces classified as single-curved. These include cylinders, cones, tangent line and tangent plane convolutes, and various

FIG. 7.13 Rotation and triangulation are used in the generation and development of a truncated oblique triangular pyramid.

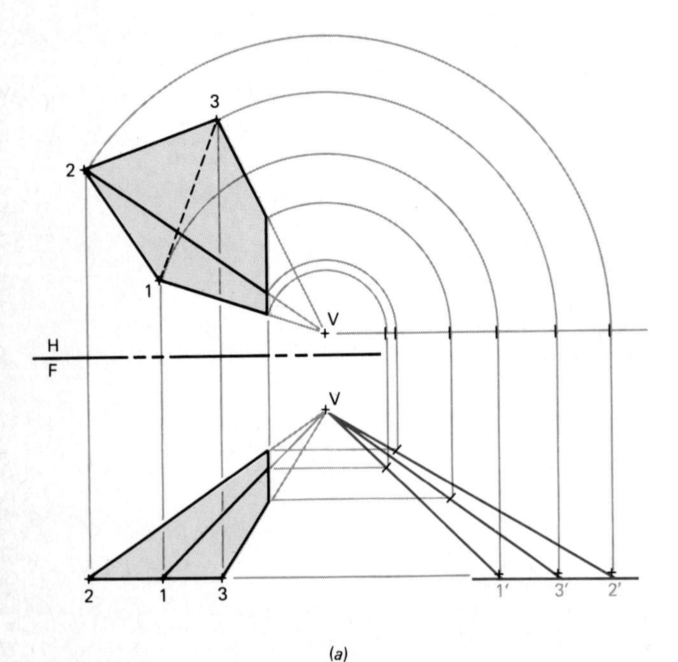

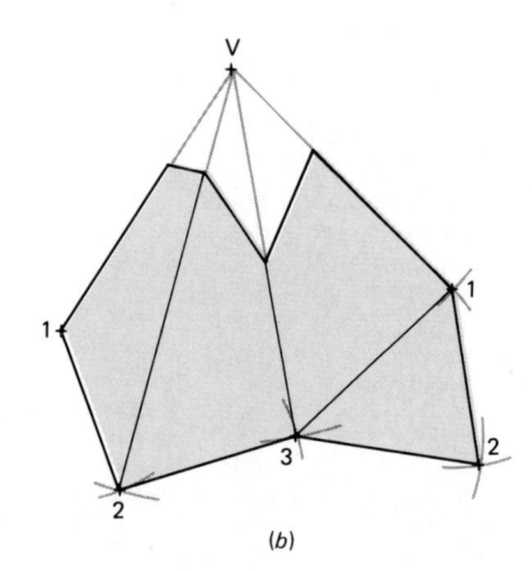

transition pieces. Sections 7.14 through 7.20 describe the procedures for generating and developing several single-curved surfaces.

7.14 TRUNCATED RIGHT CIRCULAR CYLINDER

A right circular cylinder is generated by a straight-line generatrix revolved concentrically around and parallel to a straight-line axis.

7.14.1 Generation

Draw the circular and rectangular views of the cylinder as shown in Fig. 7.14, with the truncated plane shown as an edge in the rectangular view.

Add any special details such as the small hole illustrated in Fig. 7.14.

Divide the circular view into an appropriate number of equally spaced elements (point view of the element lines). Twelve element lines is a reasonable number for a cylinder truncated as in Fig. 7.14. If small details such as the hole shown in the front view are involved, additional elements should be spaced in the area to ensure an accurate development.

Number the elements in a clockwise manner, starting with the smallest element. All elements of a right cylinder appear in true length in the rectangular view.

7.14.2 Development

Lightly lay out a rectangle equal in height to the longest element and equal in length to the circumference of the cylinder (π times the diameter).

Beginning with element 1, lay out true heights and relative positions of the element lines by projection (Fig. 7.14) or by a transfer of the dimensions with dividers or a scale.

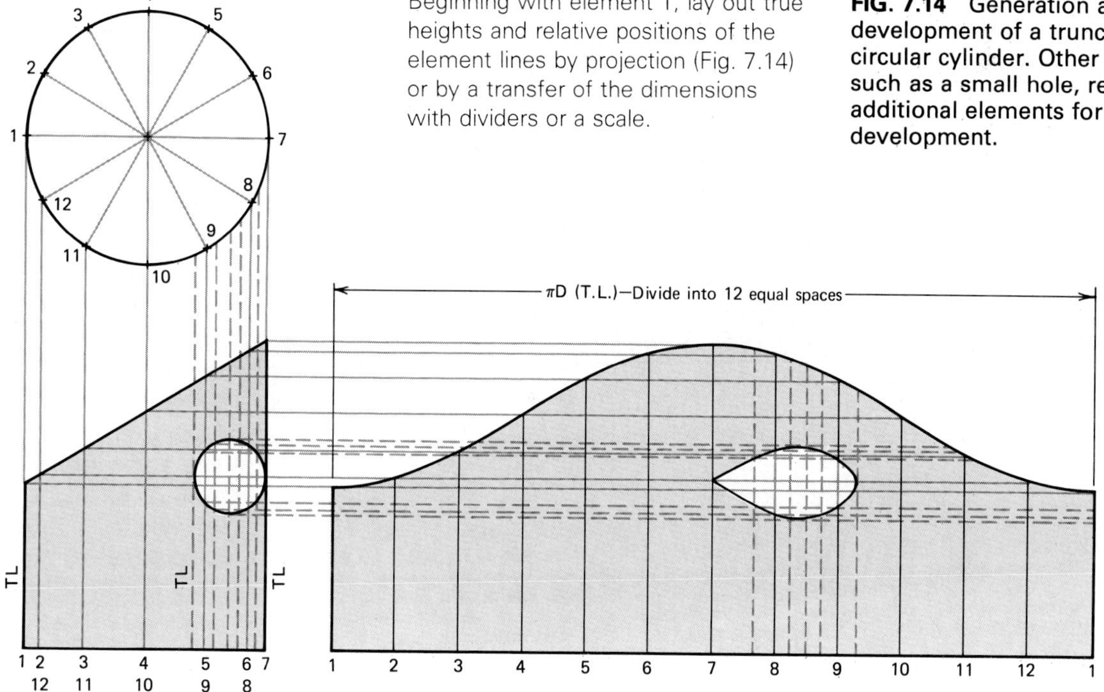

FIG. 7.14 Generation and development of a truncated right circular cylinder. Other features, such as a small hole, require additional elements for accurate development.

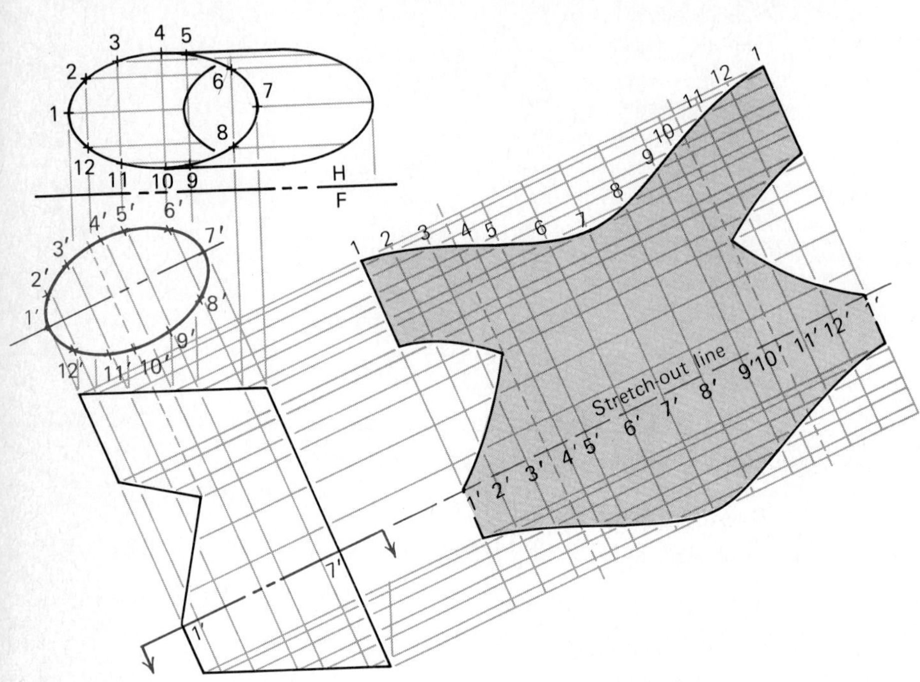

FIG. 7.15 An auxiliary view perpendicular to the axis of the cylinder provides true arc distances for the development of an oblique elliptical cylinder.

Add supplementary element lines for special features such as the small hole. True spacing of these supplementary lines can be seen in the circular view of the cylinder, where the element lines appear as points.

Darken the perimeter of the development and the outline of special features.

Prisms and cylinders have parallel elements that remain parallel when unfolded or rolled out into a plane surface. A cross section perpendicular to the axis of such an object will roll out as a straight line, with the element lines perpendicular to this straight line called a stretchout line. The stretchout line is convenient as a base line to use for positioning element lines and special features. An example is shown in Fig. 7.15.

7.15 OBLIQUE ELLIPTIC CYLINDER

The cross section perpendicular to the axis of some cylindrical pipes or ducts is elliptic in shape. These ducts may have elliptic bases (Fig. 7.15) or circular bases (Fig. 7.16).

An elliptic cylinder is generated by a straight-line generatrix revolved in an elliptic path around and parallel to a straight-line axis.

7.15.1 Generation

Draw the desired size and shape of the opening, either elliptic (top view of Fig. 7.15) or circular (top view of Fig. 7.16).

Locate and number the elements in the adjacent views (not necessarily equally spaced).

Project an auxiliary view perpendicular to the true-length view of the axis of the cylinder. The resulting point view of the elements will show the true arc distance between them.

Because of the difficulty involved in determining the exact mathematical circumference of an ellipse and in dividing the elliptic view of this cylinder equally, a more practical approach is generally followed. Elements are placed close together along the shorter radii of the ellipse so that the chord distance between the elements nearly equals the arc distance between them. The chord distances are then stepped off with dividers on a stretchout line to provide a satisfactory approximation of the true spacing

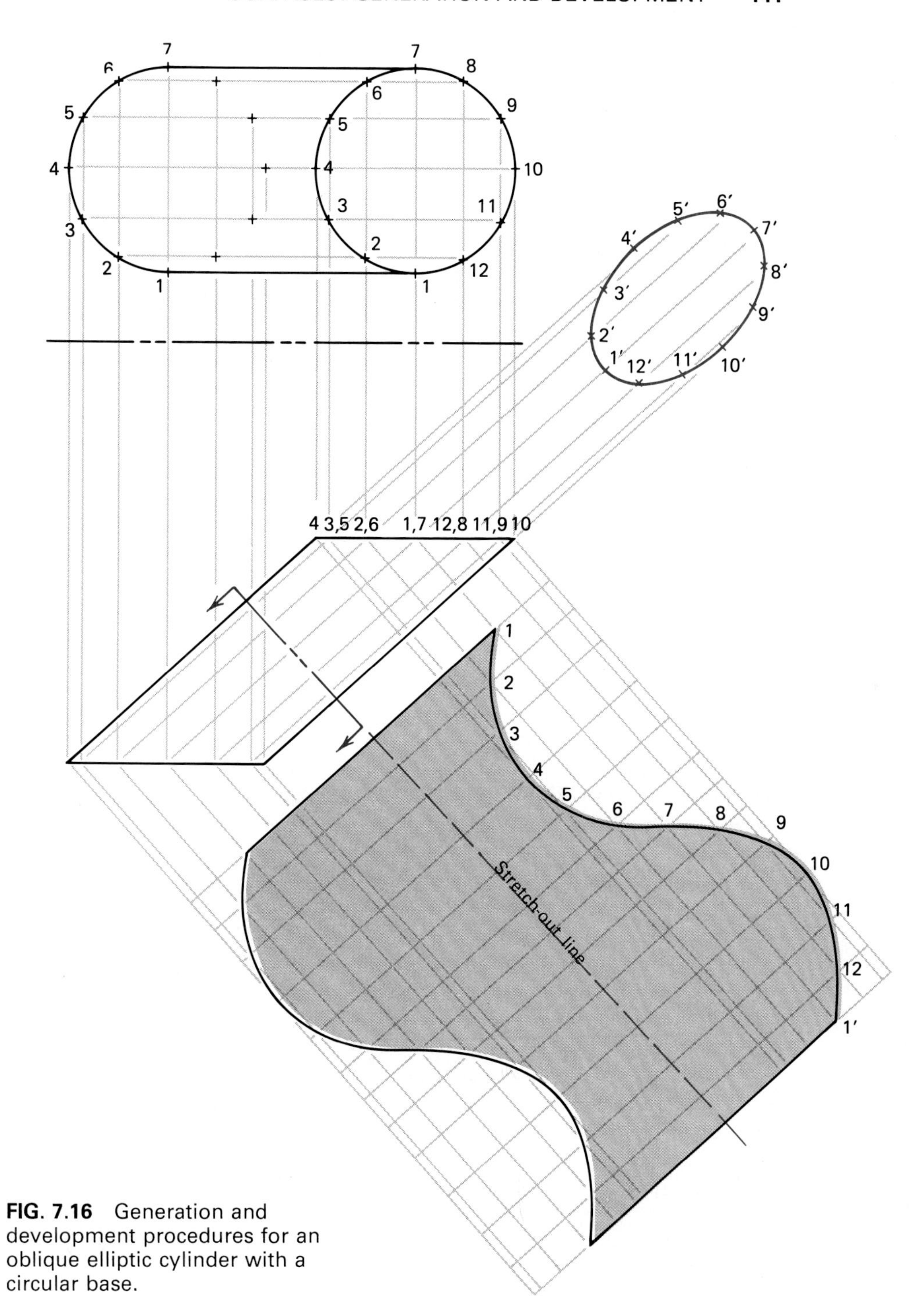

FIG. 7.16 Generation and development procedures for an oblique elliptic cylinder with a circular base.

between element lines on the development. This procedure is illustrated in Figs. 7.15 and 7.16.

7.15.2 Development

Mark off and number element lines along the stretchout line.

Project or transfer the true length of each element line.

Locate points to define any special cuts or openings.

Darken the outline of the development.

7.16 TRUNCATED RIGHT CIRCULAR CONE

A right circular cone is generated by a straight-line generatrix fixed at one end (vertex) while the other end revolves in a circle concentric to an axis through the vertex. A cross section perpendicular to the axis through any part of the cone is a circle. Cones are single-ruled surfaces.

7.16.1 Generation

Draw adjacent circular and triangular views of the cone with the desired base diameter and height. See Fig. 7.17.

Divide the circular base view into equally spaced and numbered elements that converge at the vertex of the cone.

Project the intersection of each element and the base to the front view and then draw the elements converging at the vertex.

Show the cutting plane for the truncation as an edge in the triangular view and project points cut by the plane to the top view to describe the shape of the truncated surface.

Determine the true length of each element by rotation, using the top view of the vertex as the rotation axis.

7.16.2 Development

Scribe an arc with a radius equal to the slant height of the cone.

Calculate the included angle Θ of the development from the formula $\Theta = {}^{r}/_{s}$ (360°), where r is the radius of the base and s is the slant height of the cone, base to vertex. The calculation can be simplified by measuring one unit along the slant surface of the cone from the vertex with any convenient scale and then measuring the base radius at that height with the same scale. Since the value of s is then equal to 1, the formula reduces to $\Theta = (r)(360°)$. This is illustrated in Fig. 7.17.

Divide the development into the same number of equally spaced parts as in the generation drawing, number them in the same sequence, and lay out the true element lengths to define the truncation curve. If the elements are closely spaced, the chord lengths may be used to approximate the arc lengths for element spacing.

Darken the perimeter of the completed pattern.

7.17 TRUNCATED OBLIQUE CONE

An oblique cone is generated with a straight-line generatrix revolving in an elliptic path around a straight-line axis. It is a single-ruled surface and is truly developable. A cross section perpendicular to the axis is an ellipse.

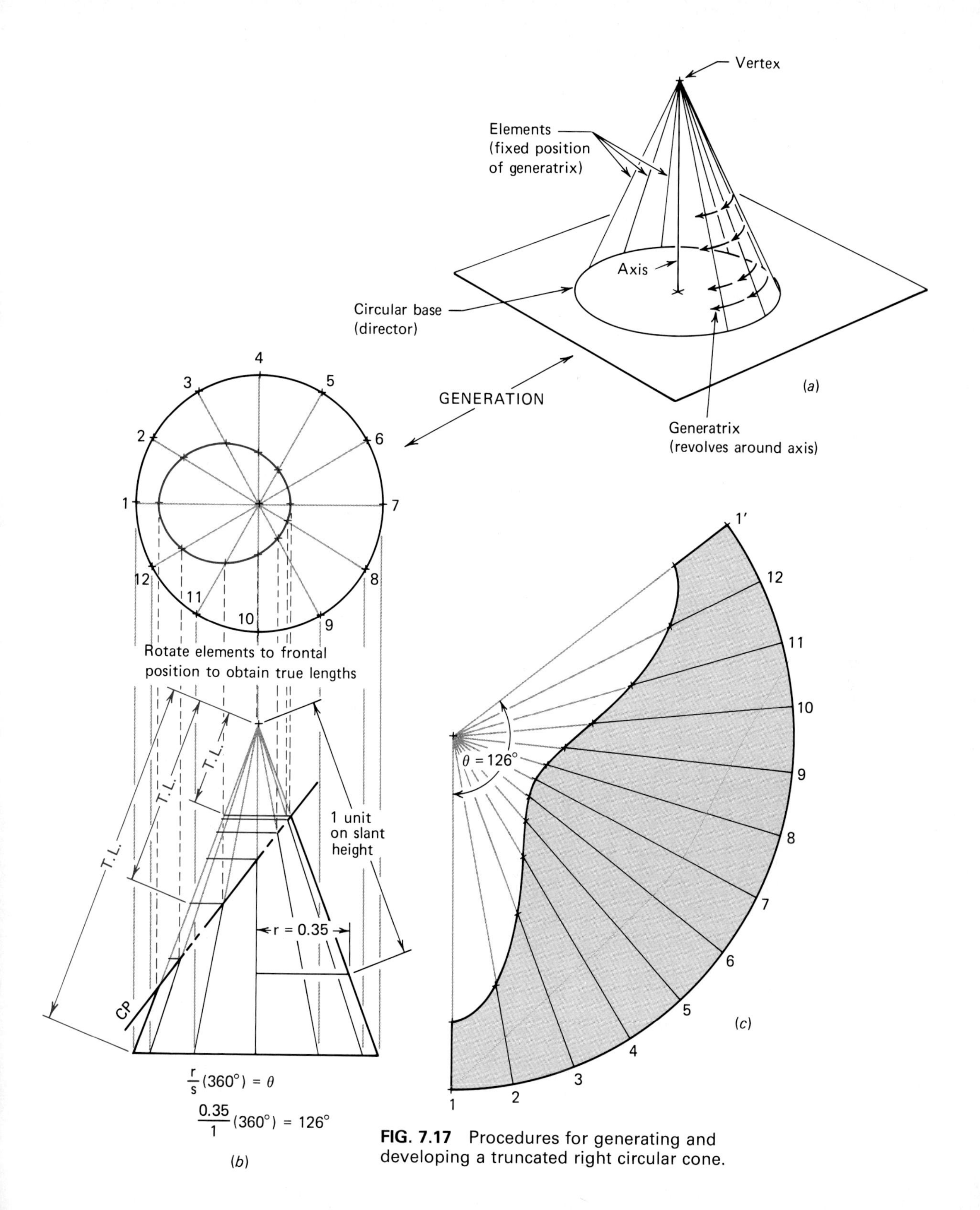

FIG. 7.17 Procedures for generating and developing a truncated right circular cone.

Frequently, truncated oblique cones form the transition between large cylindrical ducts and smaller cylindrical or elliptic ducts.

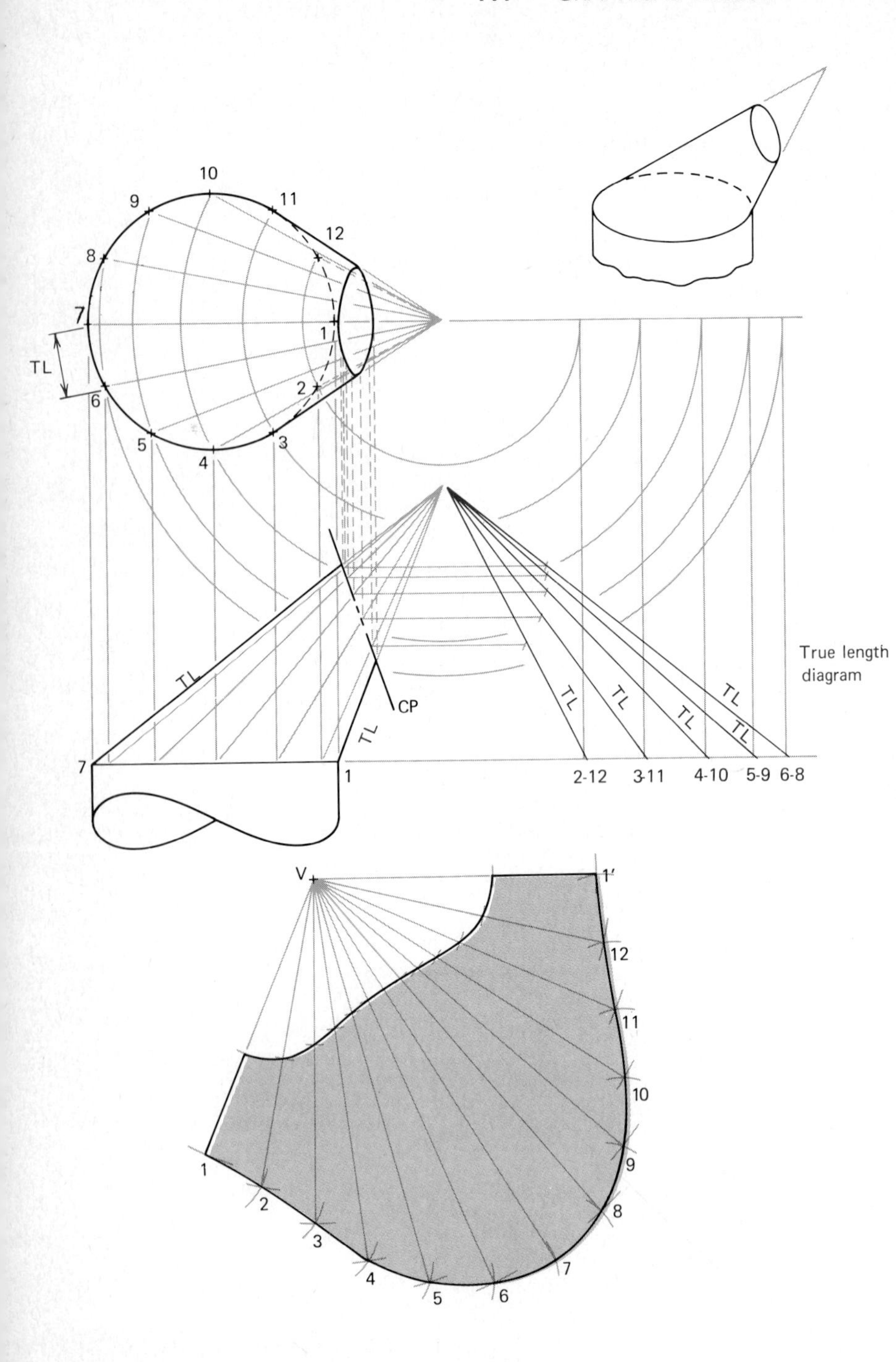

FIG. 7.18 Development of a truncated oblique cone, a surface often used as a transition piece in ductwork.

7.17.1 Generation

Draw adjacent views showing the base plane (circular or elliptic) in its true shape in one view and the triangular shape in the other, as in Fig. 7.18. The vertex is not centered over the base for an oblique cone.

Draw and number the appropriate element lines.

Show the edge view of the cutting plane where the truncation is to be made.

Construct a true-length diagram of all oblique elements.

Note the true arc length along the base from element to element in the top view. The circular base is horizontal.

7.17.2 Development

Starting with the shortest element, lay out consecutive sectors of the development by the triangulation technique introduced in Sec. 7.11. In cases where the base of the oblique cone projects as a circle, the exact length of the base arc between elements can be calculated mathematically. However, from a practical standpoint it is appropriate to use the chord distance between elements in the development. When you smooth the perimeter of the pattern, most of the loss that results from using the chord is regained.

Darken the perimeter of the development.

7.18 HELICAL TANGENT LINE CONVOLUTE

A helical tangent line convolute is generated by a straight-line generatrix always moving tangent to a cylindrical helix. Either a right-hand or a left-hand surface is possible.

7.18.1 Generation

Draw H and F views (circular and rectangular projections, respectively) of a cylinder of the desired diameter and height. See Fig. 7.19*a*.

Create an appropriate number—12 in this example—of equally spaced elements in both views.

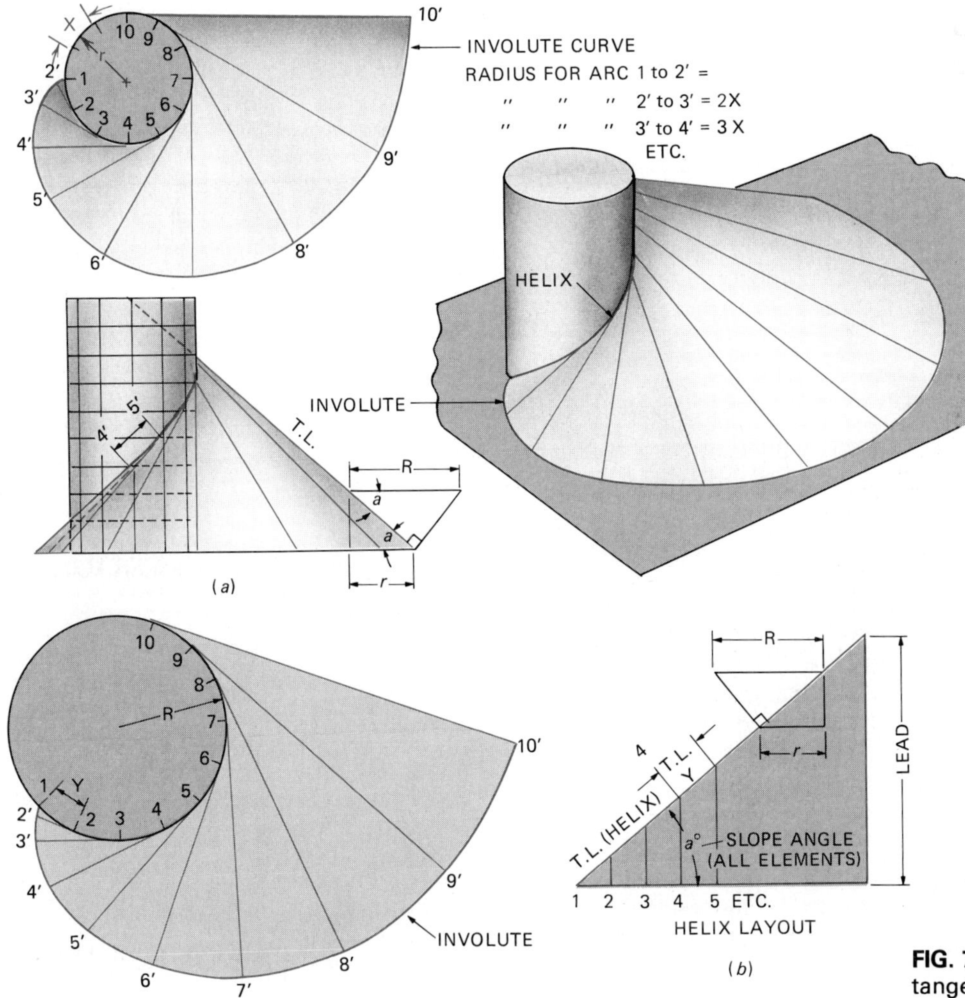

FIG. 7.19 Development of a helical tangent line convolute requires accuracy and a carefully executed step-by-step procedure.

Divide the rectangular view into the same number of equal divisions, bottom to top. Only three fourths of the cylinder height is shown in Fig. 7.19.

Draw a right-hand helix at the required lead (the amount the helix advances parallel to the cylinder axis while progressing around the cylinder a complete 360° turn). Only a three-fourths turn is shown in Fig. 7.19, and so the lead is figured on a proportional basis; that is, it advances 9 units vertically while progressing 9 units around the diameter of the cylinder.

In the top view, construct an involute curve, which is defined as the locus of a point on the end of a taut cord unwinding on a plane from around a circle or polygon. A satisfactory approximate involute can be drawn with a compass by taking centers at successive tangent points and scribing a series of connected arcs between successive tangent lines. The radius of each arc is increased by the chord distance between the respective tangent points each time the point of the compass is advanced. This procedure is illustrated in Fig. 7.19*a*.

Lay out the development of the helix, which is simply the stretchout of the circumference of the cylinder (three fourths of the circumference is illustrated in Fig. 7.19*b* and the proportional lead height. The hypotenuse of the right triangle formed is the true length of the helix winding three fourths of the way around and up the cylinder surface. The distance Y shown in Fig. 7.19*c* is the true distance between the points where the numbered element lines are tangent to the helix.

Determine the radius R shown in Fig. 7.19*c* either mathematically from $R = r/\cos a$, where r is the radius of the helix cylinder and a is the helix angle, or graphically. The graphic technique is achieved by the triangular construction drawn on the hypotenuse of the helix development triangle which represents the true slope angle of all elements in the convolute surface.

7.18.2 Development

Scribe a circle with radius R, as in Fig. 7.19*c*.

Mark equally spaced points a distance Y apart along the perimeter of the circle.

Draw tangent lines to the left of each point.

Construct successive connected arcs between the tangent lines to form the developed involute curve. The radius of each successive arc should be Y longer than the preceding one.

Darken the perimeter of the entire convolute development. Although the helical convolute surface illustrated is limited by a plane surface at the base of the cylinder, the surface could be extended indefinitely or could be limited by a larger cylinder concentric to the first.

7.19 TANGENT PLANE CONVOLUTE WITH PARALLEL BASES

A tangent plane convolute is a surface generated by a plane moving tangent to two dissimilar curved directrices lying in different planes.

7.19.1 Generation

Draw related views to the desired size, as shown in Fig. 7.20*a*. In this case the upper base is circular and the lower is elliptic.

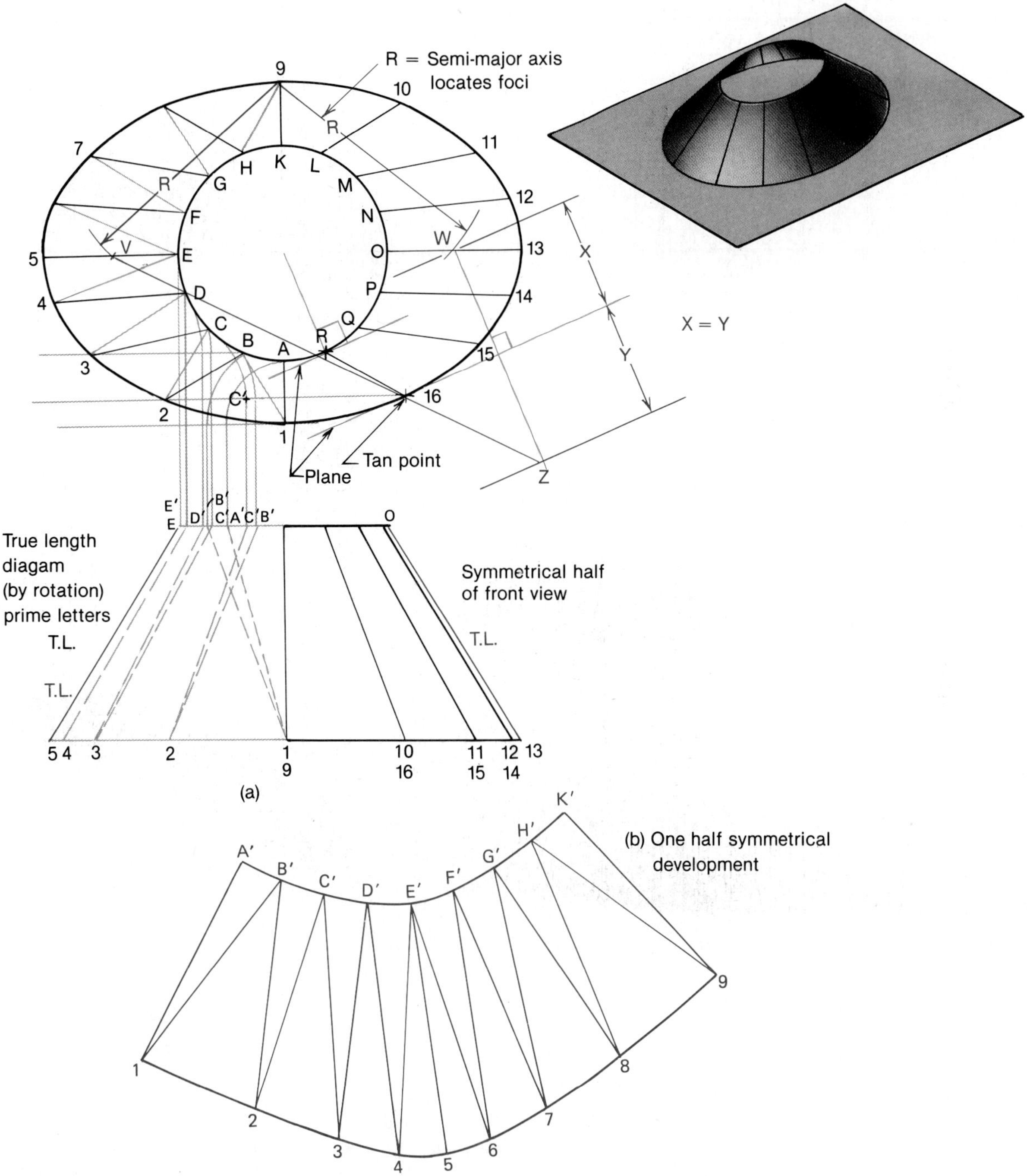
R = Semi-major axis
locates foci
X = Y
Tan point
Plane
True length
diagam
(by rotation)
prime letters
T.L.
T.L.
Symmetrical half
of front view
T.L.
(a)
(b) One half symmetrical
development

FIG. 7.20 Generation and development of a tangent plane convolute with parallel bases.

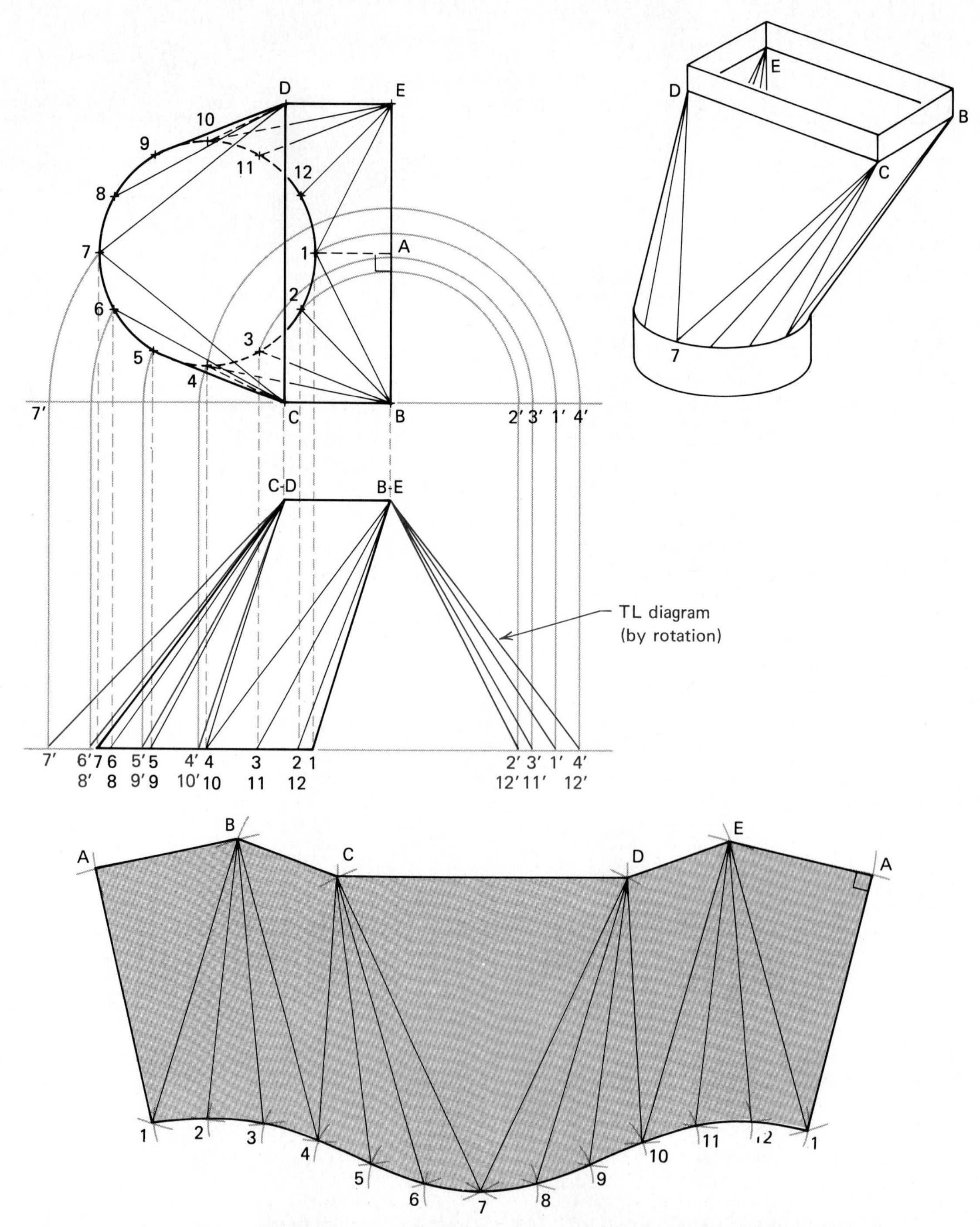

FIG. 7.21 Transition pieces are generated and developed using a combination of procedures for several geometric shapes.

Divide the circular base into equally spaced points, 16 in this example. An even number is appropriate here because of the symmetry of the object.

Draw in elements *A*–1, *E*–5, *K*–9, and *O*–13. This can be done because the object is symmetric. The left half of the front view in Fig. 7.20*a* has been omitted so that the space can be used for the true-length constructions.

Construct a radial line from the center of the circle to point *R* (lower right of top view in Fig. 7.20*a*) and draw a line perpendicular to it at *R* which provides a tangent to the circle.

Draw a second line parallel to the tangent line just drawn. Make it tangent to the perimeter of the ellipse.

To locate the exact point of tangency, construct a line from focus *W* of the ellipse perpendicular to the tangent to the ellipse. Then extend the line so that the *X* and *Y* distances are equal, thus locating point *Z*. From *Z*, draw a line to focus *V*. Where the line intersects the ellipse, you will find the point of tangency labeled 16 on the drawing.

Repeat the process to locate tangency points for elements *P*–14 and *Q*–15.

Transfer tangent points to the remaining quadrants with dividers or a compass. This is possible because of symmetry.

Draw diagonal lines across each trapezium sector formed by the elements, creating a series of connected triangular sectors.

Rotate all nonfront element lines in one quadrant to front positions, as shown in the true-length diagram. It is necessary to handle only one quadrant because there are three other elements, one in each quadrant, identical in length to each element considered.

7.19.2 Development

Lay out a symmetric half development by the triangulation technique that was described previously. See Fig. 7.20*b*.

7.20 TRANSITION PIECES

Combinations of various surfaces which form a continuous surface connecting dissimilar openings are called transition pieces. Plenums for furnaces requiring a connection between square or rectangular openings and round or other openings is an example. Three or more geometric shapes are often combined. Figure 7.21 shows a transition piece from a rectangular to a circular opening.

7.20.1 Generation

Draw adjacent orthographic views showing the combined surfaces in correct projection.

Construct and number element lines for each part of the combined form as you would for the individual shapes.

7.20.2 Development

Use any convenient method or combination of methods to find the true length of each element.

By triangulation, lay out the plane sectors in proper sequence.

7.21 DOUBLE-CURVED SURFACES

Surfaces generated with a single-curved line generatrix and guided by an axis or a second

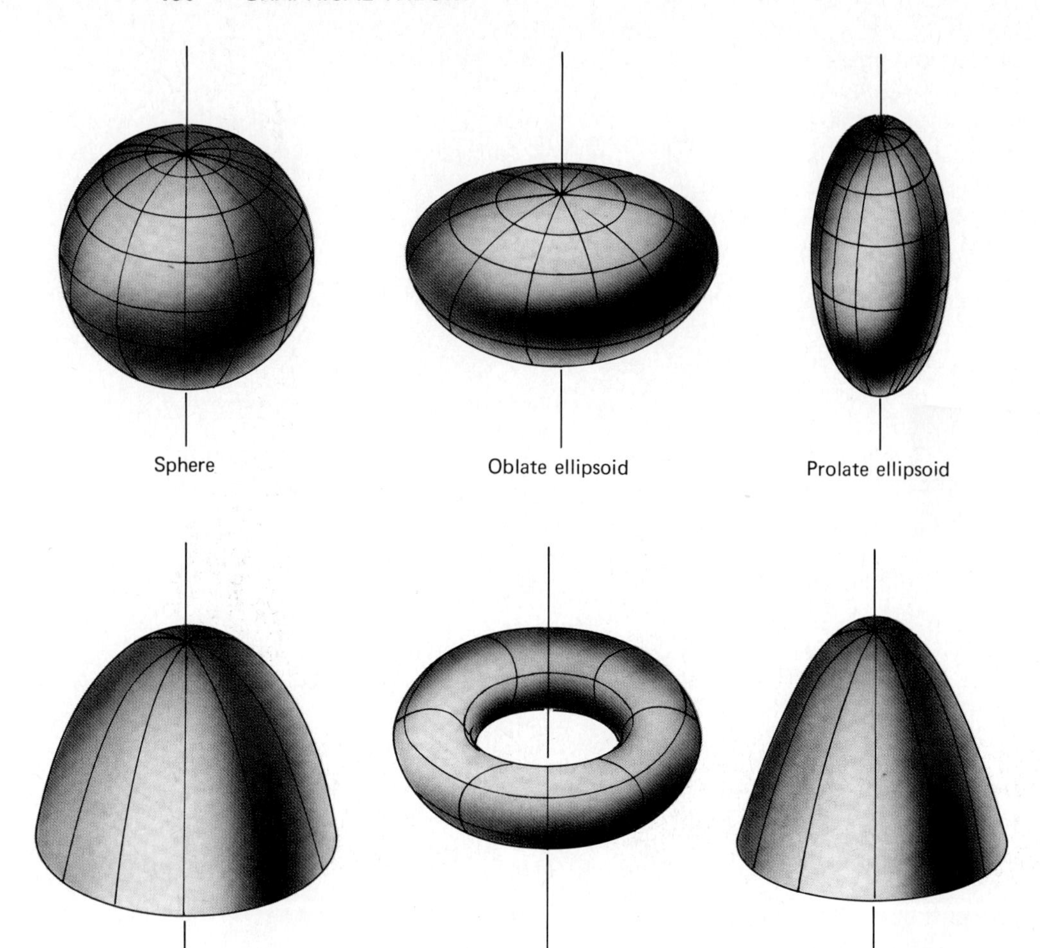

FIG. 7.22 Examples of double-curved surfaces.

single-curved line directrix are known as double-curved surfaces. Such surfaces contain no straight-line elements.

Common double-curved surfaces include the sphere, oblate or prolate ellipsoid, paraboloid, hyperboloid, and annular torus. These are illustrated in Fig. 7.22.

7.22 SPHERE

Spheres are perhaps the most common and useful double-curved surfaces. The perfectly symmetric nature of the surface is aesthetically pleasing and functional for a wide variety of uses. Balls of all types are commonly used in the recreational area. Ball bearings, bulbs, balloons, storage tanks, and undersea vessels are among the hundreds of industrial uses.

7.22.1 Generation

A sphere is generated by a semicircular generatrix rotated

around its diameter. It is a double-curved surface, and it cannot be developed exactly. However, an accurate approximate development can be made by dividing the sphere into small plane sectors and pressing or forming them into the spherical contour.

7.22.2 Representation

The contour of a sphere viewed from any position is a great circle (a circle formed on the surface of a sphere by the intersection of a plane that passes through the center of the sphere, showing its maximum circumference). Because of this unique attribute, adjacent views of a sphere fail to differentiate between the three basic dimensions that are rather easily seen in most other geometric shapes. One tends to believe that the circle seen in the top view of a sphere is the same circle seen in the front or profile views, but this is not true (Fig. 7.23). A point may appear to be in the same relative position on a sphere in adjacent views when in reality it is not on the surface at all (see Fig. 7.24).

Study Figs. 7.23 and 7.24 for a better understanding of the nature of a sphere before you try to do approximate developments or intersection problems involving spherical surfaces.

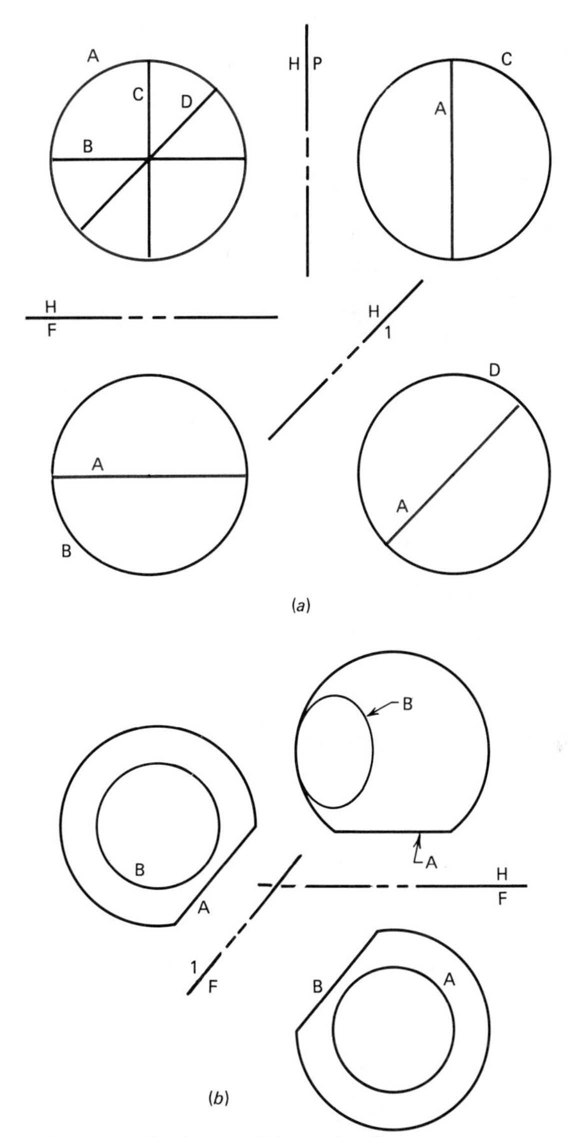

FIG. 7.23 Orthographic projection characteristics of spheres.

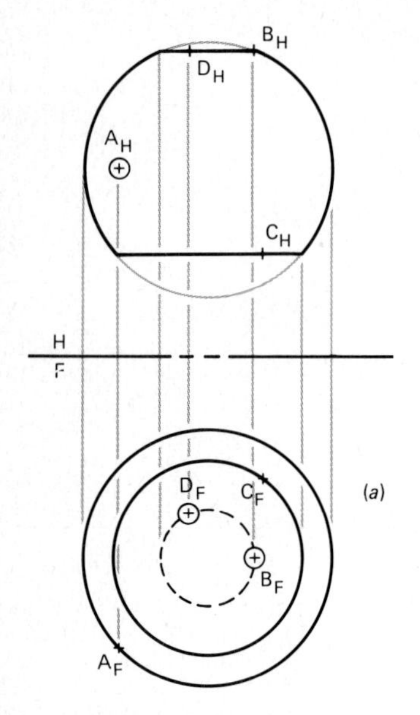

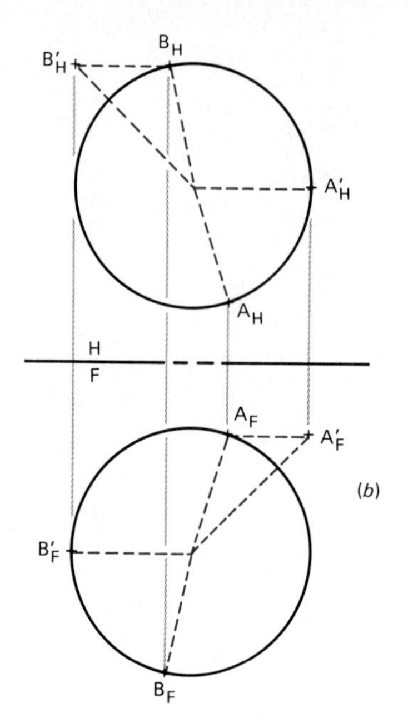

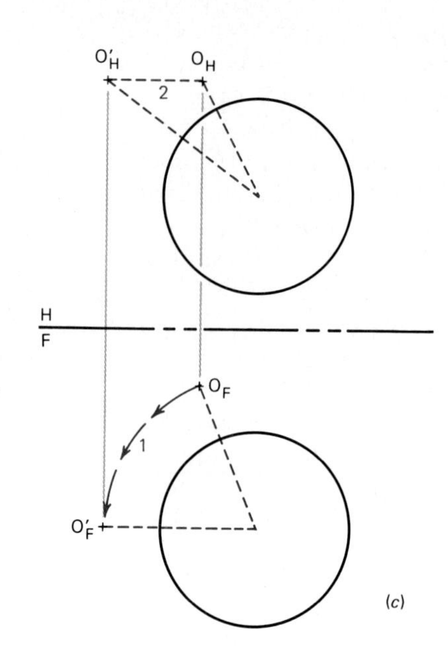

FIG. 7.24 Procedures for determining whether a point lies on or off a spherical surface.

7.23 LINES AND POINTS ON A SPHERE: POINTS ADJACENT TO A SPHERE

Figure 7.23*a* shows four related views of a sphere. The great circle in one view projects as a diameter in an adjacent view. In Fig. 7.23*b*, a front cutting plane *A* has truncated the sphere, leaving a circular plane exposed in the front view. An inclined cutting plane *B* shown as an edge in the front view appears as a circle in the inclined view but as an ellipse in the top view.

Figure 7.24*a* shows points *A, B, C,* and *D* on the surface of the sphere. Point *A* lies on the lower hemisphere of the sphere on the great circle that appears as an edge in the top view and a circle in the front view. The small circle around point A_H in the top view is the hidden point symbol. This symbol should be used in cases where the sphere is opaque and the point is on the far side of the sphere. When the point is visible, it may be designated by a small *v* under the notation B_H or may be left unmarked.

Point *B* lies on the far side of the great circle that is seen as a circle in the top view and at the diameter of the sphere in the front view. Therefore, *B* is hidden in the front view.

Points *C* and *D* each lie on the surface of the sphere. They can be located by projecting them to circular lines cut by truncating planes passed through the sphere.

Figure 7.24*b* shows points *A* and *B*, which appear at first glance to lie on the surface of the sphere but actually do not.

To obtain the true position of a point relative to a sphere, draw a line from the point to the center of the sphere in two related views. If the true length of the line appears in either view, the position of the point can be confirmed. If not, you must get the true length. If the rotation method is used, the center of rotation must be the center of the sphere, as shown in Fig. 7.24*b* and *c*. In these illustrations, none of the points *A*, *B*, and *O* lie on the surface of the sphere.

7.24 APPROXIMATE DEVELOPMENT OF A SPHERE

If you develop small plane sectors and form them before joining, a sphere can be accurately constructed. Two mehods for the approximate development of a sphere are the *zone,* or *polyconic,* method and the *sector,* or *polycylindric,* method. The zone method is described in Fig. 7.25 and the sector method, in Fig. 7.26.

Adjacent views of the sphere are drawn, and the desired number and type of divisions are detailed. Although there is a practical limit, the smaller the plane sectors, the more accurately they will conform to a spherical shape.

True lengths and widths are determined by methods indicated in Figs. 7.25 and 7.26.

Note that the "W" distances are equal segments of the great circle in the front view (total of 16 in this example). Each of the horizontal planes in the front view is a boundary of a zone of the sphere. The other bounda-

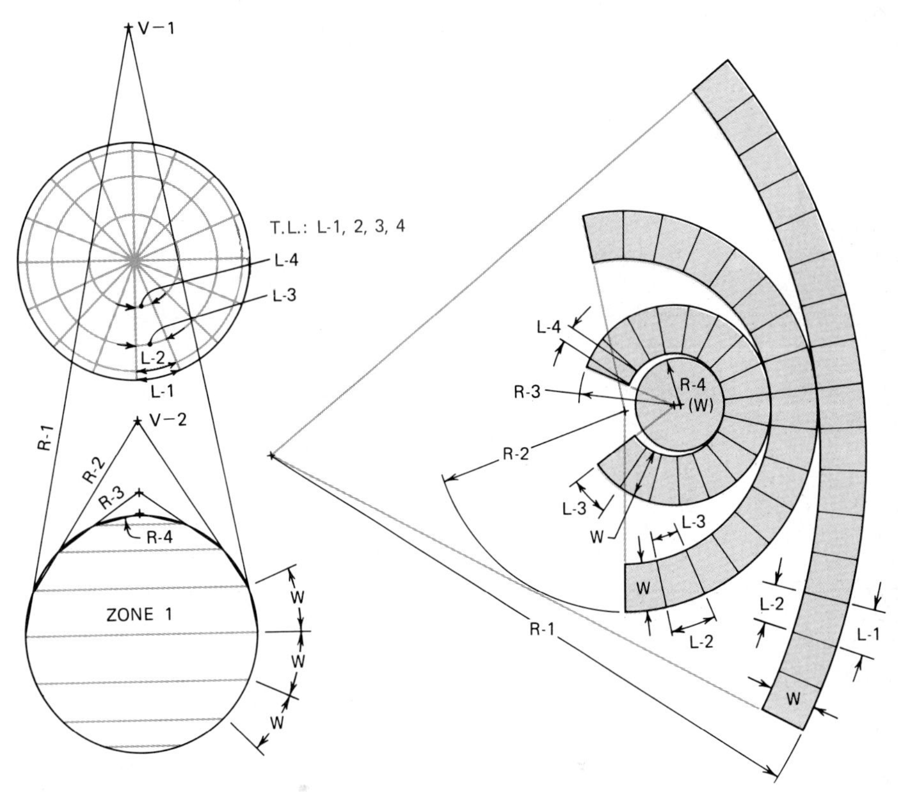

FIG. 7.25 Approximate development of a sphere using the zone method.

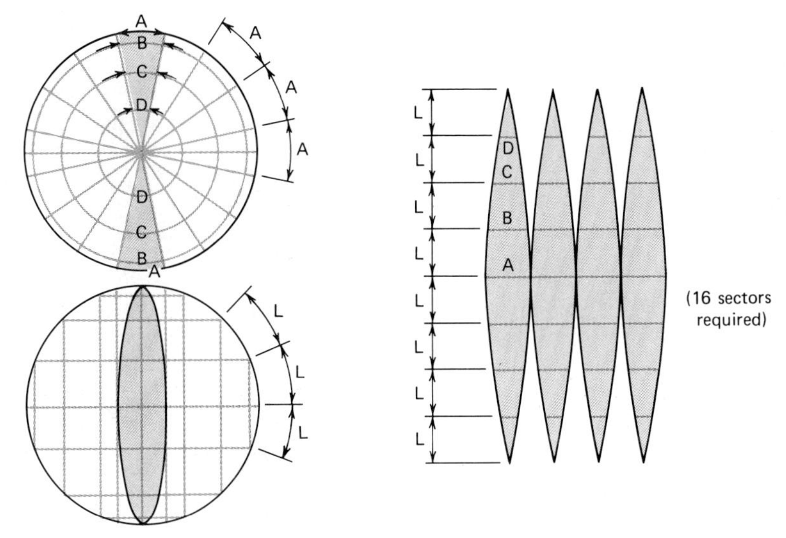

FIG. 7.26 Approximate development of a sphere using the sector method.

ry of the zone is the arc distance connecting the endpoints of successive horizontal planes. Each zone is then developed with the cone frustrum technique. For example, the slant height R-1 of the cone for zone 1 is found by extending the chord lines to the intersection V-1. Other slant heights are found in a similar manner, as shown in Fig. 7.25.

A half development is required for a pattern when one uses the zone method.

One true-shape sector can serve as a pattern for all the sectors required for the sector method. See Fig. 7.26 for the procedures.

Placements inside up or outside up may be important if the surface treatment differs on either side.

7.25 OBLATE AND PROLATE ELLIPSOIDS

The oblate ellipsoid (Fig. 7.22) is generated by an elliptic generatrix rotated around its minor axis. Similarly, the prolate ellipsoid is generated by rotating the elliptic generatrix around its major axis. Each results in a double-curved surface that can be approximately developed.

The development is achieved by use of the same method that was applied to the sphere in Fig. 7.26. There are many industrial items, such as liquid storage tanks, pontoons, and dirigibles, which are examples of items for which the ellipsoid can be used.

7.26 WARPED SURFACES

Warped surfaces are ruled surfaces defined by a generatrix moving so that no two adjacent positions are parallel or intersecting. Warped surfaces are not developable but can be constructed by stretching, pressing, or fairing techniques with concrete, plastic, or other pliable materials.

Included in the general classification of warped surfaces are hyperbolic paraboloids, hyperboloids of revolution, right and oblique helicoids, conoids, cylindroids, warped cones, and a unique surface called the cow's horn.

7.27 HYPERBOLIC PARABOLOID

The hyperbolic paraboloid is generated with a straight-line generatrix moving parallel to a plane director while contacting two nonparallel, nonintersecting straight-line directrices. It can be ruled in two directions (double-ruled). This surface is found as a part of bridge abutments, roof or wall structures, bows of ships, and storage structures.

Figure 7.27 shows the construction of a hyperbolic paraboloid. The procedure is as follows:

Draw adjacent views of the directrices. They must be nonintersecting and nonparallel.

Locate the edge view of the plane director.

Establish element lines parallel to the edge view of the plane director.

Project the elements to the adjacent view.

Experimentation with various positions of the directrices and plane director may be necessary to get the right degree of warp. In engineering applications, this warp may be fixed by related features of a structure.

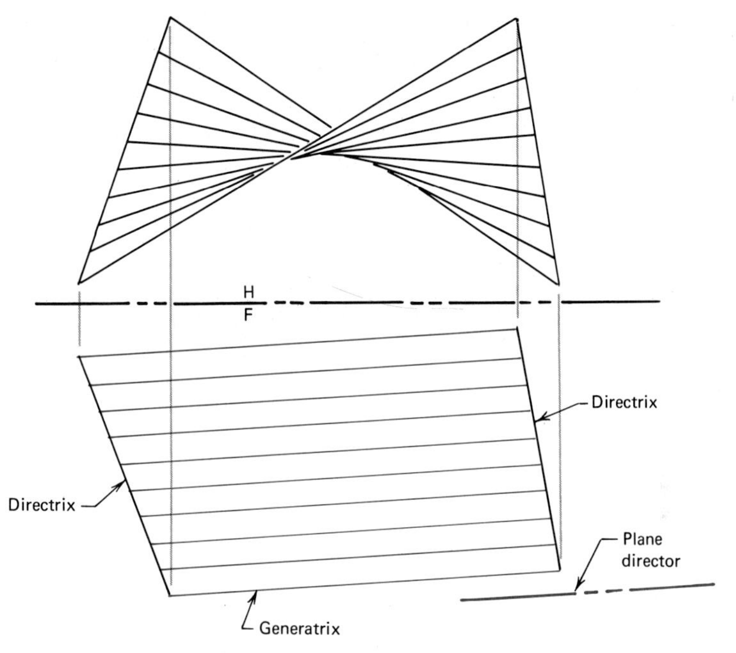

FIG. 7.27 Procedures for generating a hyperbolic paraboloid; one of the warped surfaces.

7.28 RIGHT CONOID

The right conoid is generated by a straight-line generatrix moving parallel to a plane director while contacting one straight-line directrix perpendicular to the plane director and a circular directrix in a plane parallel to the line directrix. Conoids can be used in architectural structures and tents and, when truncated, as transition pieces.

A right conoid may be represented as follows:

Draw adjacent views of two directrices to give the desired height and diameter. For a vertical right conoid, the line directrix must be horizontal and perpendicular to the plane director (the front plane can serve as the plane director). See Fig. 7.28.

Construct sufficient element lines to define the surface (equally spaced is desirable).

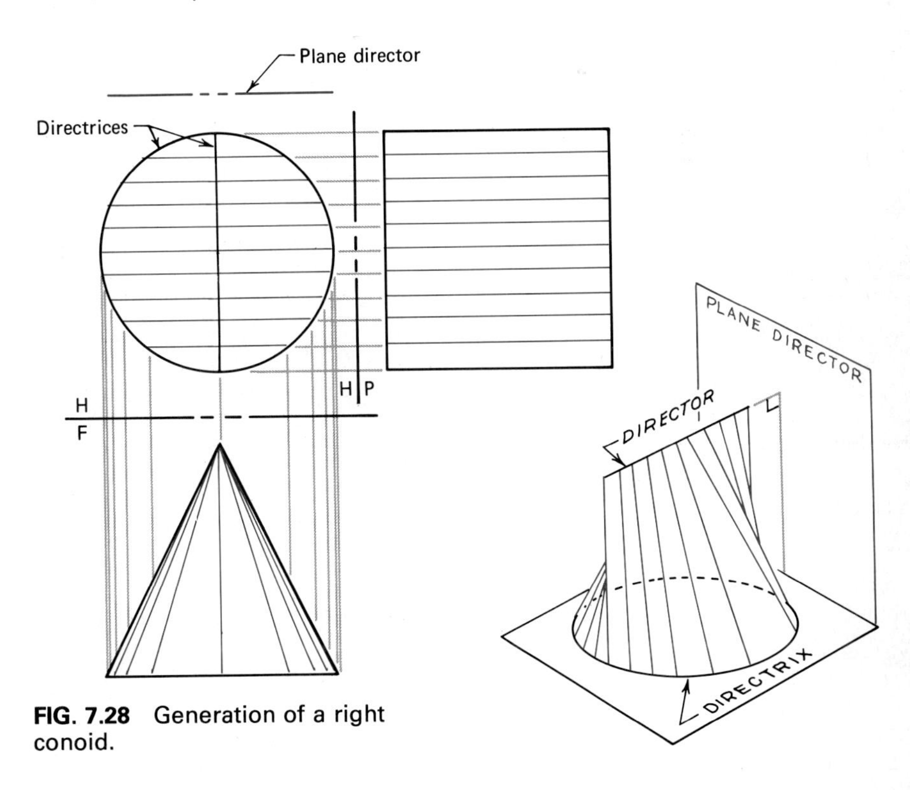

FIG. 7.28 Generation of a right conoid.

7.29 OBLIQUE CONOID

The oblique conoid is single-ruled and is generated by a straight-line generatrix moving parallel to a plane director while contacting a circular base circle (directrix) and a straight-line directrix that is not perpendicular to the plane director. The

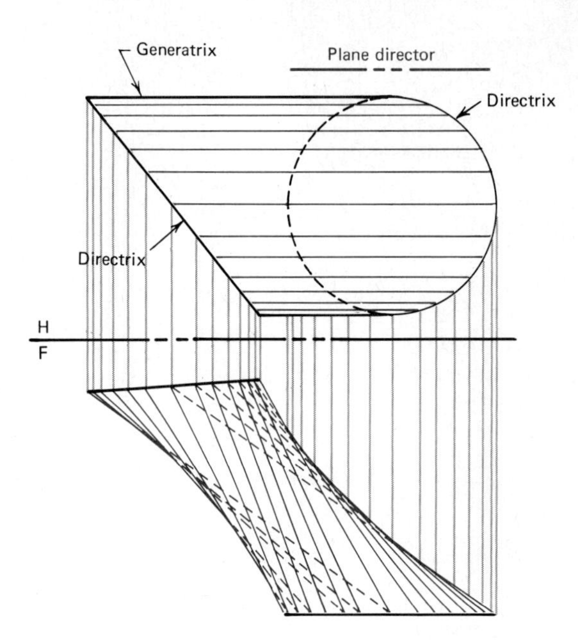

FIG. 7.29 Generation of an oblique conoid.

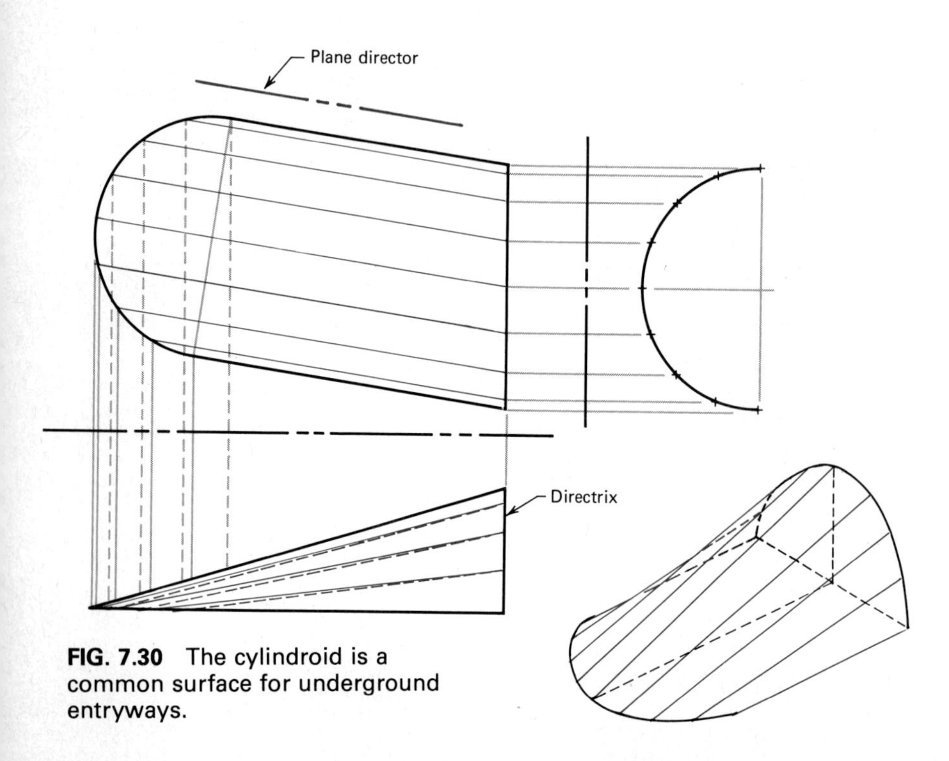

FIG. 7.30 The cylindroid is a common surface for underground entryways.

oblique conoid can be drawn in accordance with the method for the right conoid, except that the line directrix must be neither parallel nor perpendicular to the plane director. See Fig. 7.29.

7.30 CYLINDROID

A cylindroid can be generated by a straight-line generatrix moving parallel to a plane director while contacting two single-curved directrices that are not in parallel planes. See Fig. 7.30. The result is a single-ruled nondevelopable surface.

The procedure for constructing a cylindroid is similar to that for conoids and hyperbolic paraboloids. The cylindroid can be used for subway and underground entryways, tents, and some unique business structures.

7.31 RIGHT HELICOID

The right helicoid can be applied to spiral stairways, slides for children, and fire escapes. It is generated by a straight-line generatrix moving in contact with a cylindrical helix and perpendicular to the axis of the helix. The surface is single-ruled.

A right helicoid is drawn by constructing adjacent views of concentric cylinders, drawing a helix as illustrated in Fig. 7.19, and completing it as shown in Fig. 7.31.

7.32 CIRCULAR HYPERBOLOID

The generation of a circular hyperboloid is achieved by a

straight-line generatrix that is at an oblique angle to and is rotated around an axis but that does not contact the axis. The generatrix is guided by circular directrices concentric to the axis. This surface is used in the design of storage bins, pulleys, and cooling towers, for example.

The circular hyperboloid is drawn by the following procedure:

Construct parallel circular directrices concentric to a common axis.

Draw a gorge circle concentric with the larger circles. This defines the minimum diameter of the surface.

Construct the first element of the surface so that it contacts the circular directrices and is tangent to the gorge circle.

Decide on the number of elements required and draw them, as illustrated in Fig. 7.32.

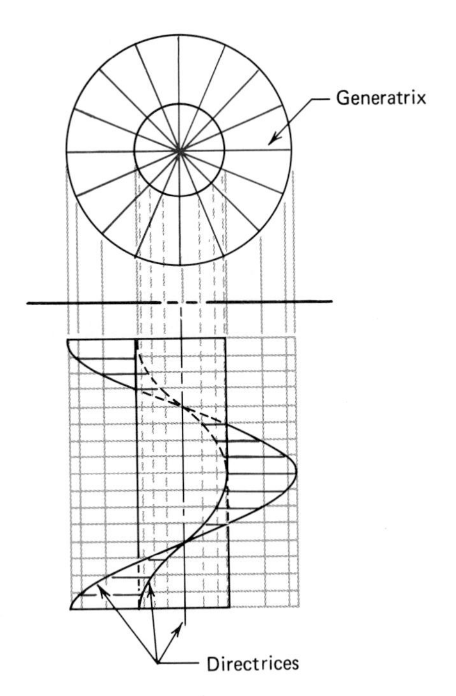

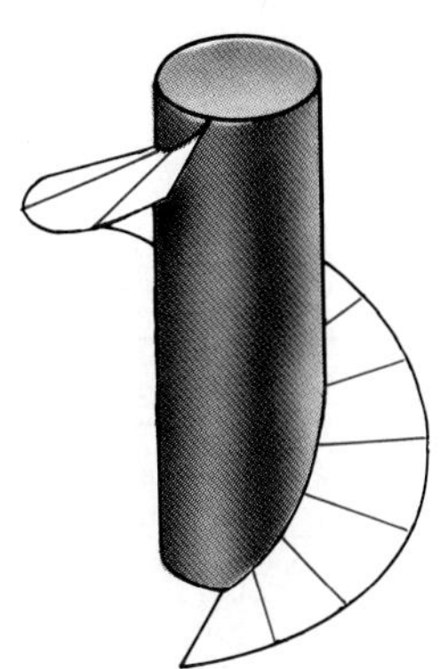

FIG. 7.31 Generation of a right helicoid.

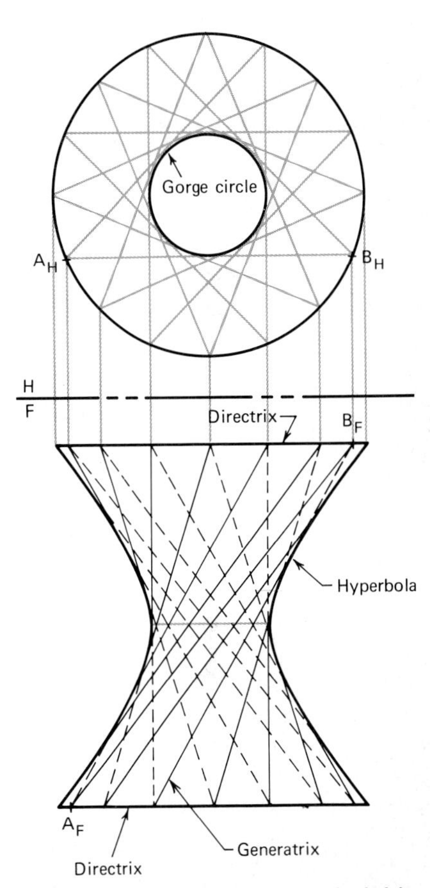

FIG. 7.32 The circular hyperboloid surface is often the basis for design of cooling towers for power plants.

7.33 METHODS FROM COMPUTER GRAPHICS

Let us consider the truncated cone shown in Fig. 7.17 as an example to show how a computer may be used in preparing a pattern. It is possible to use a method similar to that given in Sec. 7.16, with the computations being done by the computer.

The circular base can be described by an equation, and the vertex is a point in space. Once that is done, you can select points, say, 100 of them, equally spaced around the base circle and determine the equation of the line (element) between each successive base circle point and the vertex. Because of the rapid computations, it is possible to take 100 or more points around the base rather than the 16 points used in Fig. 7.17.

The truncation of the cone is produced by a plane which can be defined by an equation. With the equation of the plane and the equation of the first element, you can compute the point of intersection between the two and therefore can find the distance along the element from the base circle to the intersection point.

The distance from the base circle to the vertex provides the radius of the 2-*D* pattern, and the circumference of the base circle allows us to calculate the included angle (126° in this example). Combining this information with the distances from the base circle to the truncation plane along each element previously calculated, we can then lay out the problem.

With 100 element lines, it may be sufficient to connect the points along the irregular side of the pattern with straight-line segments. If greater accuracy is required, a curve-fitting technique can be employed to provide a smooth curve connecting all points on the irregular side.

Your first reaction may be that it seems like a lot of work to do for a simple truncated cone. You are, of course, correct. The advantage of doing the programming work for a truncated cone comes with future cases, when it will be necessary only to provide the base radius, vertex location, and truncation plane orientation to produce a high-quality pattern in minutes.

Think about what you might do to use the computer to produce a pattern for each skin panel in the aircraft shown in Fig. 7.33, bearing in mind that the geometry of the surface already has been defined and stored in the computer.

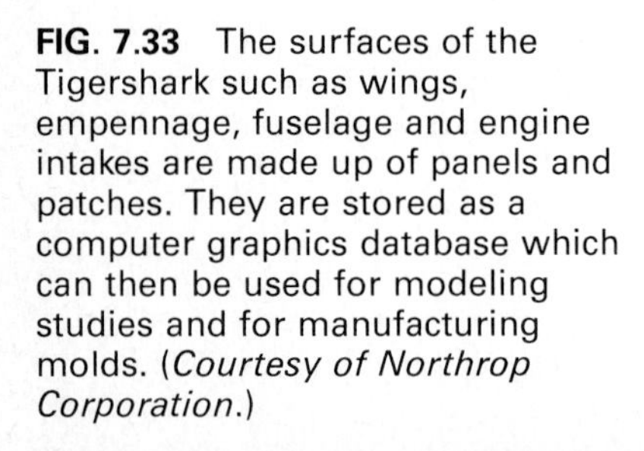

FIG. 7.33 The surfaces of the Tigershark such as wings, empennage, fuselage and engine intakes are made up of panels and patches. They are stored as a computer graphics database which can then be used for modeling studies and for manufacturing molds. (*Courtesy of Northrop Corporation.*)

Problems

7.1 Write definitions for the following terms: *(a)* surface, *(b)* generatrix, *(c)* element line, *(d)* directrix, *(e)* plane director, *(f)* developable surface, *(g)* approximately developable surface, *(h)* nondevelopable surface, *(i)* constructed surface, *(j)* ruled surface, *(k)* double-ruled surface, *(l)* plane surface, *(m)* single-curved surface, *(n)* double-curved surface, *(o)* warped surface.

7.2 List as many general rules for the generation of surfaces as you can. Then verify your list from the text and add any that you forgot.

7.3 List as many general rules for the development of surfaces as you can. Then verify your list from the text and add any rules that you omitted. Repeat this exercise until you can list them all without referring to the text.

7.4 After constructing top and front views, draw the development for each of the following:

(a) A right pyramid 8 cm high with a 6-cm-square base

(b) A right cone 8 cm high with a 6-cm-diameter base

(c) A right cylinder 8 cm high with a 6-cm-diameter and a truncated plane surface at a 45° angle to the axis from the top extreme right element, cutting completely through the cylinder

7.5 After constructing the top and front views, draw the development of a 6-cm-diameter right circular cylinder 8 cm high with a 4-cm-diameter hole perpendicularly through the cylinder at its midheight and tangent to its extreme right element.

7.6 After constructing the top and front views, draw the development of a 8-cm-high right cone with a 8-cm-diameter base. The center of a 4-cm-diameter hole through the cone (perpendicular to its axis) is located 3 cm above the base of the cone. The perimeter of the hole is to be tangent to the extreme right element of the cone.

7.7 Make a pattern (partial) development of the tangent plane convolute shown in Fig. 7.34.

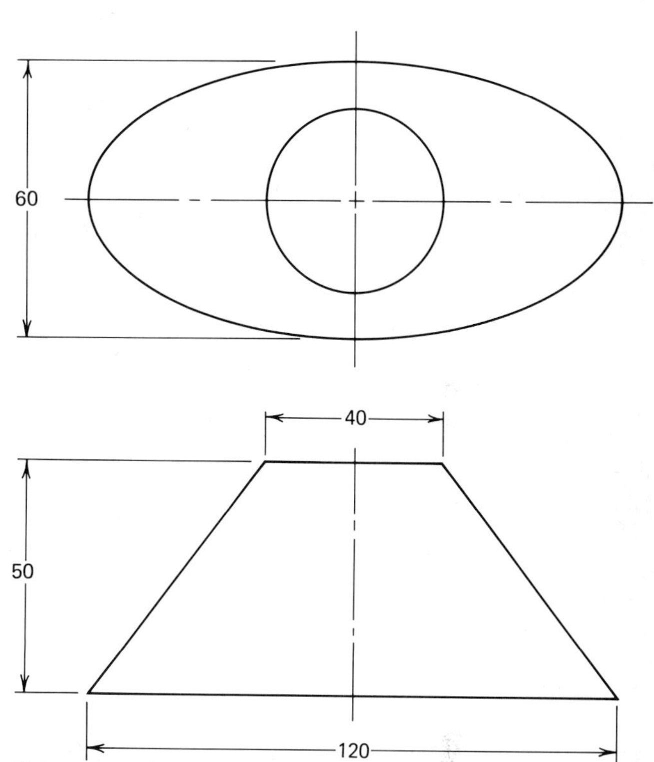

FIG. 7.34

7.8 Draw the top and front views and the approximate (partial) development of a 70-mm-diameter sphere by each of the following:

(a) Sector method

(b) Zone method

7.9 Draw the top and front views of a 50-mm-diameter sphere.

(a) Show the H and F projections of point *A*, which is centered (front or back) on the perimeter of the horizontal hemisphere (great circle). Assuming opaqueness of the surface, indicate if the point is visible (A) or hidden Ⓐ.

(b) Show in H and F planes the only position(s) possible in space for point *B*, which would be precisely 25 mm (TL) from both point *A* and the surface of the sphere.

(*c*) Show in H and F planes any three points *C*, *D*, and *E*, each of which would be precisely 25 mm from point *B* and 15 mm from the surface of the sphere.

7.10 Show the H and F projections of a hyperbolic paraboloid with eight elements, using a front plane director and the given directrices. See Fig. 7.35.

7.11 Using directrices as shown in Fig. 7.36 and a front plane director, complete the H and F views of the conoid, using elements intersecting 16 equally spaced points on the circular base *C*.

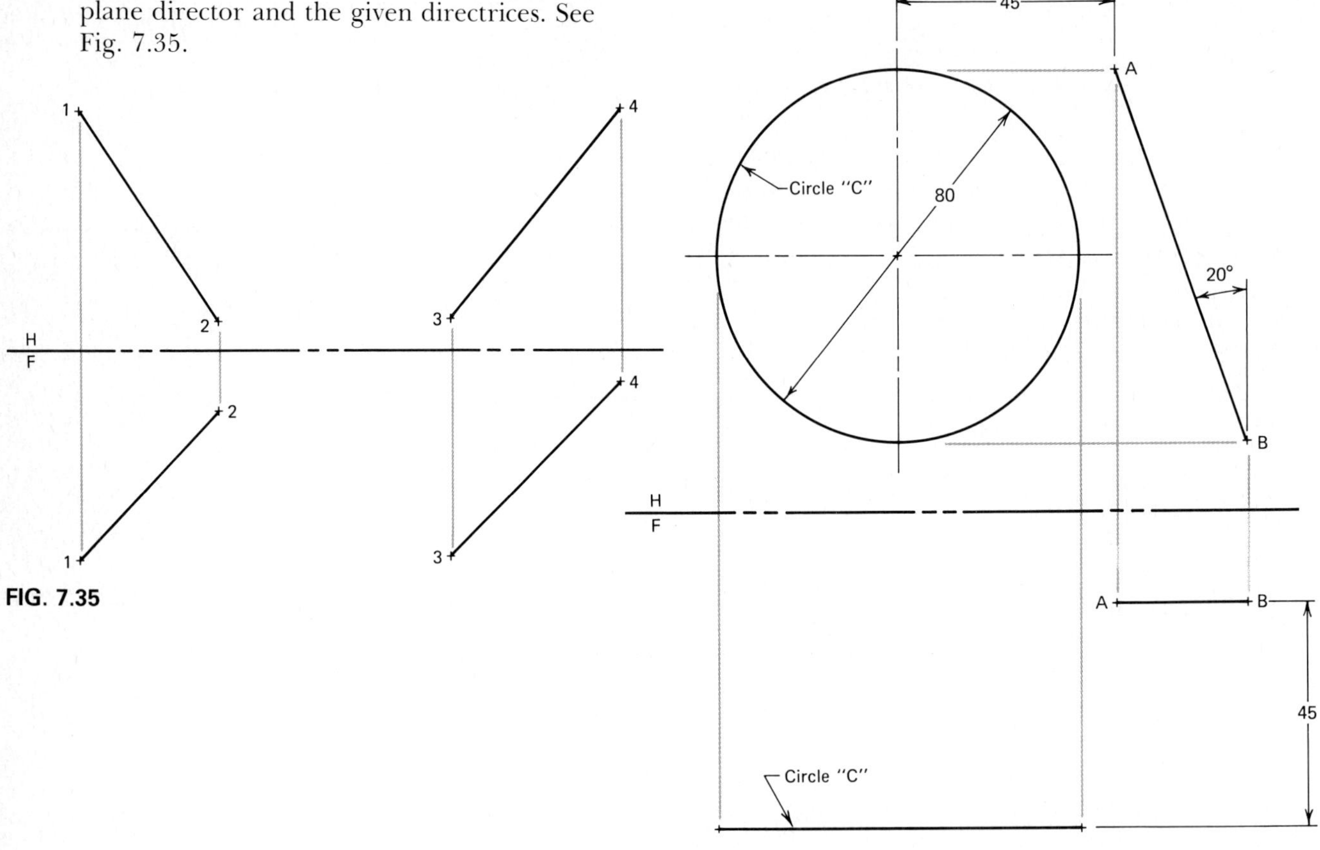

FIG. 7.35

FIG. 7.36

7.12 Draw the top and front views of the cylindroid indicated in Fig. 7.37. Assume that the surface is opaque and use 12 or more judiciously spaced elements, with solid or hidden line symbols to indicate visibility.

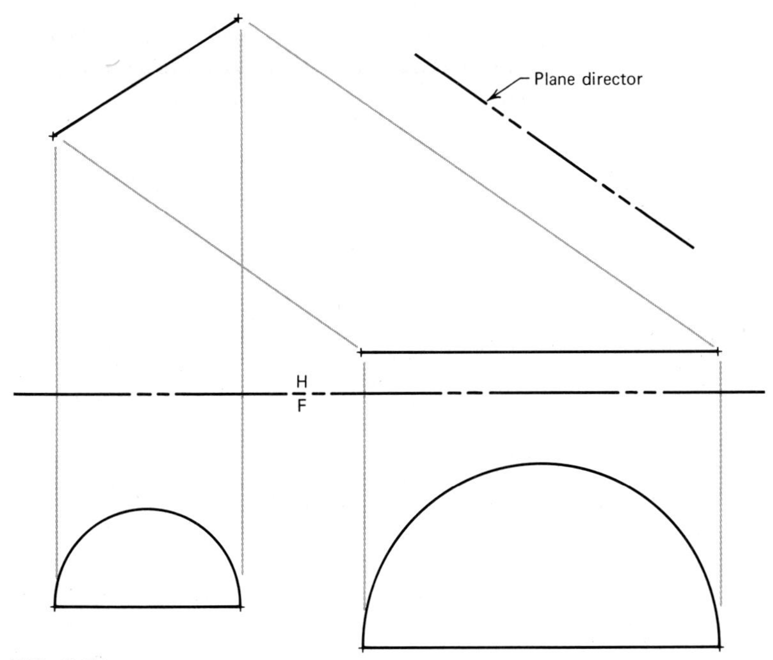

FIG. 7.37

7.13 Complete the H and F views of the transition piece shown in Fig. 7.38, incorporating a cylindroid with plane surfaces. Assume opaqueness. Choose judiciously spaced elements using solid and hidden line symbols to describe the entire object.

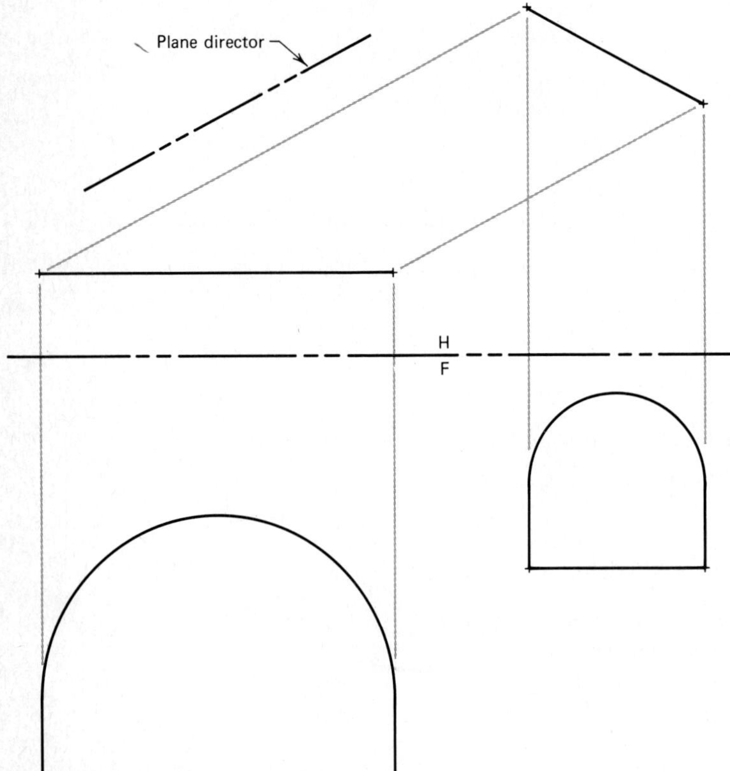

FIG. 7.38

7.14 Draw the top and front views of a cylindrical right helicoid similar to that shown in Fig. 7.31. The inner cylinder diameter is 50 mm, the right-hand helix lead is 60 mm, and the outer diameter of the helicoid is 80 mm.

7.15 Draw the top and front views of a circular hyperboloid similar to the one shown in Fig. 7.32. The bottom circle (directrix) is 80 mm, the top circle is 60 mm, the height is 80 mm, and the slope angle of 16 equally spaced elements is 47°.

chapter 8

Intersections

8.1 INTRODUCTION

An observing eye will discover innumerable examples of surface and solid intersections that occur in nature or are created by humans. In nature, trees intersect the surface of the earth, and limbs intersect tree trunks. Likewise, engineers design the intersection of wings with the fuselage of an airplane and the intersection of a windshield with the body of an automobile.

Most solids are composed of surfaces which intersect or form a junction between two or more surfaces. Your knowledge of how surfaces are generated will be supplemented by a description of intersecting surfaces.

8.2 LINE OF INTERSECTION

The line of intersection between surfaces or solids is the visible or smoothly blended joint where these surfaces or solids are connected. It is the line of demarkation where there is a sudden change in the direction of plane surfaces or where the juncture of two dissimilar surfaces takes place. An intersection consisting of a single point is called a piercing point.

In the case of intersecting plane surfaces, the resulting line of intersection is a straight line or a series of connected straight lines. When curved or warped surfaces intersect, the line of intersection is usually curved and irregular. Delineating curved lines of intersection is more complicated than is the case when the line is straight. Such problems are solved by locating several points along the line of intersection by means of the methods outlined in this chapter.

There are several important rules and considerations pertaining to most problems involving intersecting surfaces. These rules should be studied before you investigate specific situations.

8.3 GENERAL RULES AND CONSIDERATIONS

1. Determine whether the surfaces to be connected are basically two-dimensional or three-dimensional.

2. If the surfaces are three-dimensional, ascertain whether the objects are solid or hollow. Frequently they are hollow—for example, sheet-metal heating and cooling ducts.

3. Select the method of solution—either direct edge view or cutting-plane method.

4. Determine the necessary orthographic views, size, position of the objects, and so on.

5. Draw with light construction lines until you are certain of achieving the correct results.

6. Label critical features by using a logical sequence of numbers or letters to identify key element lines and points on the line of intersection.

7. When using the cutting-plane method, place planes through the intersecting objects that will provide simple element lines on both objects; that is, try to create straight lines or circles rather than ellipses, parabolas, or other irregular curves.

8. Use an optimum number of cutting planes when employing the auxiliary plane method. Five to eight will suffice for most problems.

9. Remember that there is more than one way to solve most intersection problems. Choose the most ap-propriate one when you have the option.

10. Draw accurately when delineating problems involving lines of intersection to ensure a high-quality result.

8.4 LINE INTERSECTING A PLANE

A fundamental technique used in solving intersection problems is to find the point of intersection and visibility of a line and a plane surface.

You will note that lines parallel to a plane or to certain single-curved surfaces will not intersect the plane or surfaces. An example is a line parallel to the axis of a cylinder. Also, although a theoretical piercing point between an infinite plane and a nonparallel line can be found, it is not always practical to do this when dealing with finite surfaces. Therefore, if a given element line on one object does not pierce the finite part of a second object, the surfaces involved do not intersect along that particular line.

8.4.1 Edge-View Method

Although Fig. 8.1 involves a two-dimensional (plane) surface, the solution view coincides in principle with that associated with Fig. 8.5*a* and *b*, which shows a three-dimensional surface. In Fig. 8.5*a*, a single-curved surface is

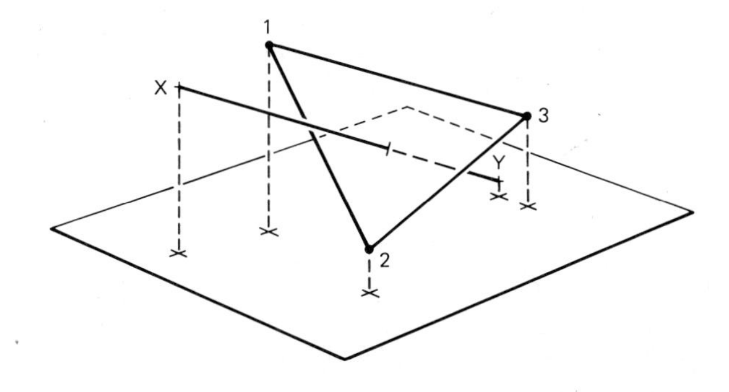

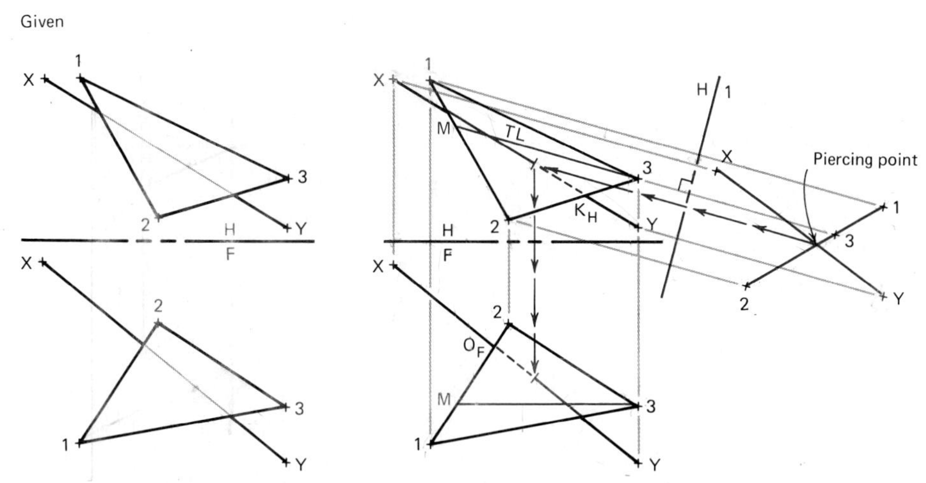

FIG. 8.1 Edge-view method for determining the piercing point of a line and plane.

viewed as an edge, in contrast to Fig. 8.1, in which a plane is viewed as an edge.

8.4.1.1 Solution View. The solution view is any orthographic view that shows the edge view of the plane and the projection of the line.

8.4.1.2 Solution Procedure. Given plane 123 and line *XY* as shown in Fig. 8.1, to find the piercing point, proceed as follows:

Draw a horizontal line *M*3 in plane 123.

Construct reference line H/1 perpendicular to the true-length view of *M*3. Project the edge view of plane 123 and also project line *XY* into the same view.

Project the intersection (piercing point) back into the H and F views.

Determine visibility in the F view by projecting point O_F (the apparent point of intersection of *XY* and 12) to the H view. Observe that line 12 is in front of *XY* at this point. This means that if the plane sector is opaque, it will hide the segment of the line from point O_F to the piercing point in the front view. Therefore, the remaining segment of line *XY* will be visible from the piercing point to *Y* in the F view.

Proceed in a similar manner to establish visibility in the H view. Select an apparent intersection (K_H) of *XY* and line 23. Project point *K* to the F view. Note that line 23 is closer to the H plane at this point than is the corresponding point on *XY*. Thus, the segment of *XY* from K_H to the piercing point is hidden, and the remainder of the line is visible.

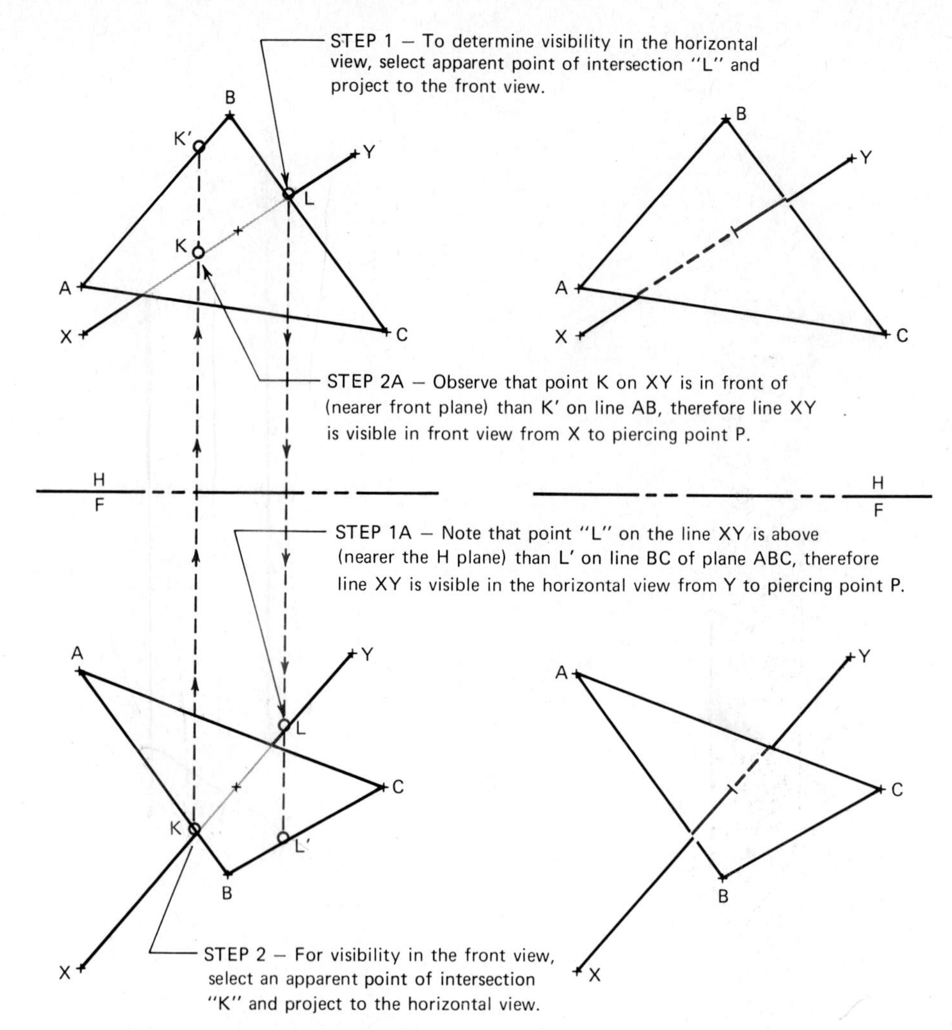

FIG. 8.2 Determining visibility in orthographic projection.

Figure 8.2 will help you understand visibility. Thorough understanding of these concepts will be helpful when you want to determine the visibility of more complex objects in later problems.

8.4.2 Auxiliary-Plane Method

An alternative to the edge-view method is the auxiliary-plane method for finding the point of intersection of a line and a plane.

8.4.2.1 Solution View. The solution view is a view that is adjacent to a projection which shows the edge view of an auxiliary plane that contains the given line. This solution view indicates the common point that exists between the line of cut of the auxiliary plane on the given plane and the given line. This common point is called the piercing point.

1130

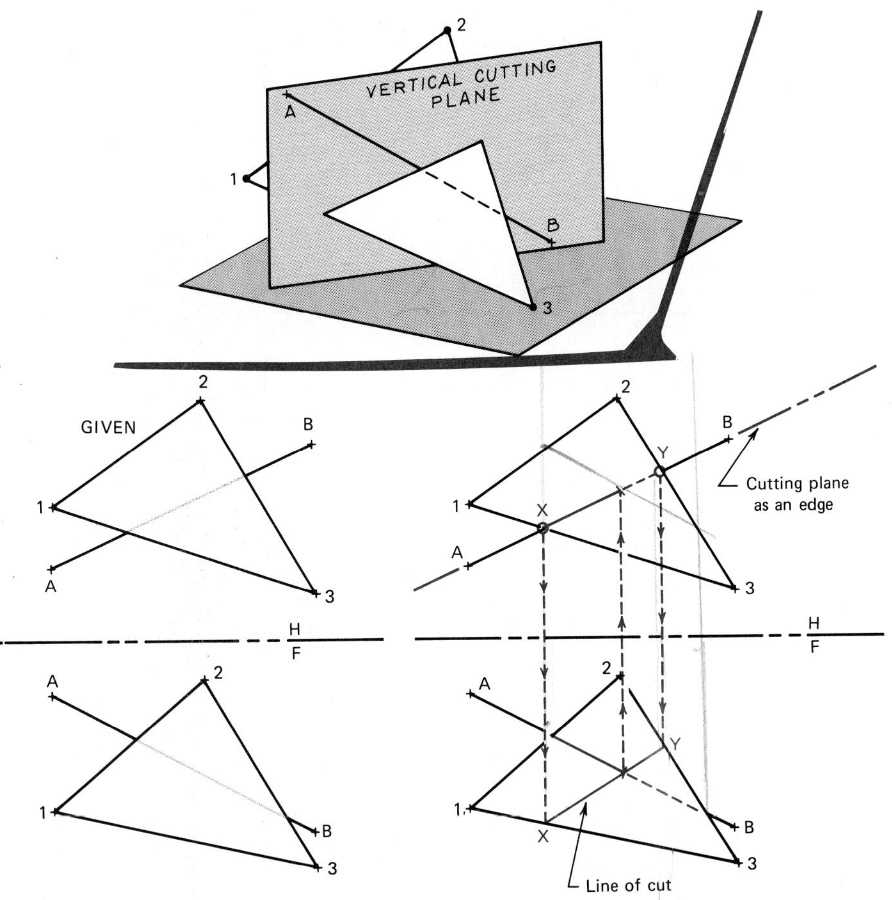

FIG. 8.3 Auxiliary-plane method for determining piercing point.

8.4.2.2 Solution Procedure

Place a cutting plane (in this case vertical) coincident with line $A_H B_H$ that intersects the boundaries of plane 123 at X_H and Y_H. Refer to Fig. 8.3.

Project line XY to the front view, thereby establishing the piercing point as the common point to AB and XY.

Project the piercing point to the horizontal view.

Determine visibility by procedures described previously.

A second example of finding the piercing point between a line and a plane is presented in Fig. 8.4. In this case, the cutting plane is placed in the front view.

8.5 LINE INTERSECTING A CYLINDER OR PRISM

The auxiliary-plane method of determining the piercing point of a line with a plane surface can be used effectively to determine

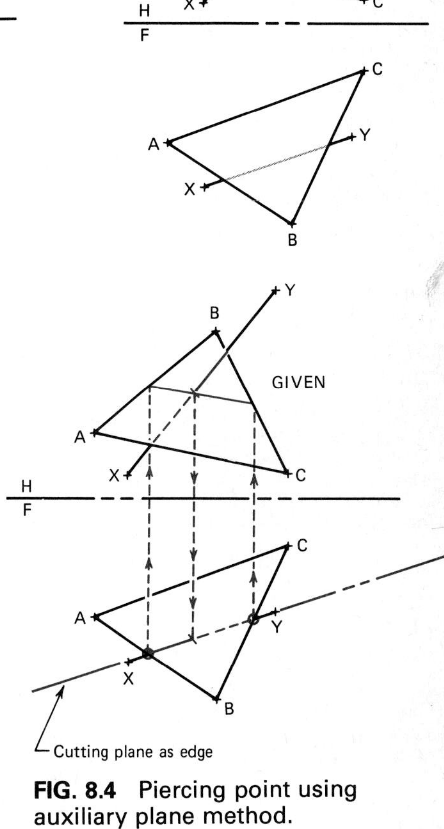

FIG. 8.4 Piercing point using auxiliary plane method.

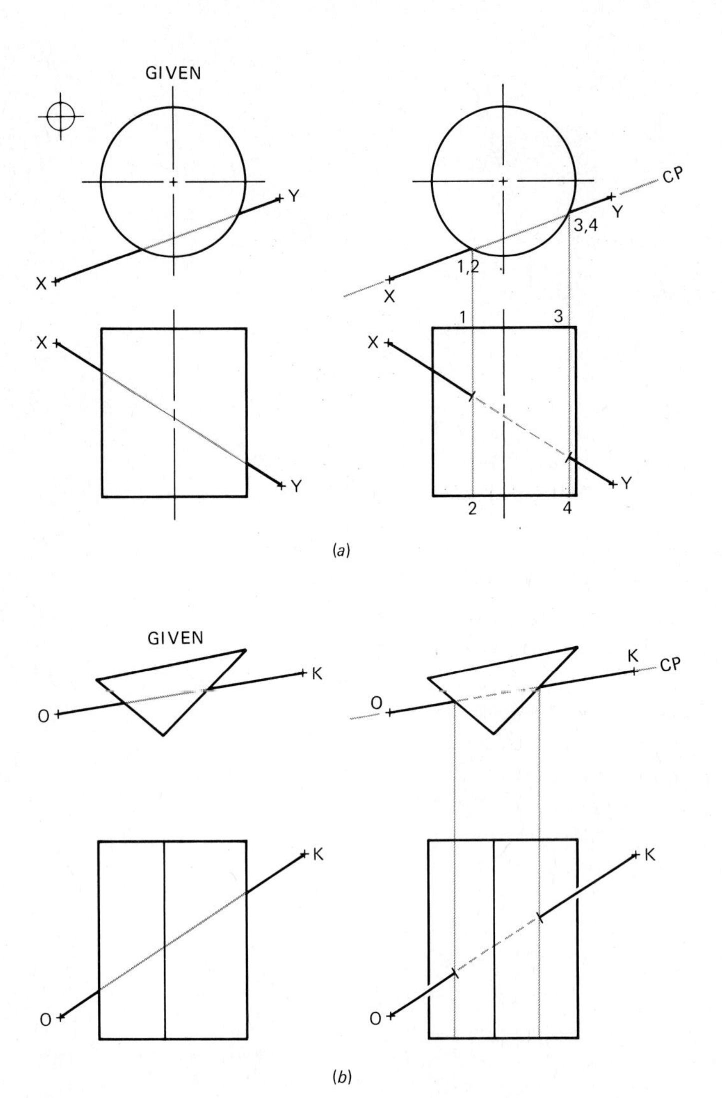

FIG. 8.5 Piercing points of a line and a cylinder or prism.

where a line intersects either a cylinder or a prism. See Fig. 8.5 for illustrations.

8.5.1 Solution View

The solution view is any orthographic view that shows the intersection points between a given line and the line of cut created by a cutting plane containing the given line as it passes through the surface of the object.

8.5.2 Solution Procedure

Pass a cutting plane (vertical in this case) including line *XY* through the cylinder in the horizontal view. See Fig. 8.5*a*.

Project the lines of cut made on the cylinder by the cutting plane (12 and 34) to the front view.

Establish the two piercing points where line *XY* and lines 12 and 34 intersect in the front view. Note that an inclined cutting plane seen as an edge in the front view could have been used. It would cut an ellipse on the cylinder that would project as a circle in the top view. The points where this circle intersects line *XY* are the piercing points.

Complete visibility.

An example of a line piercing a prism is given in Fig. 8.5*b*. The method here is similar to that for the cylinder. The details of the solution are left to the reader.

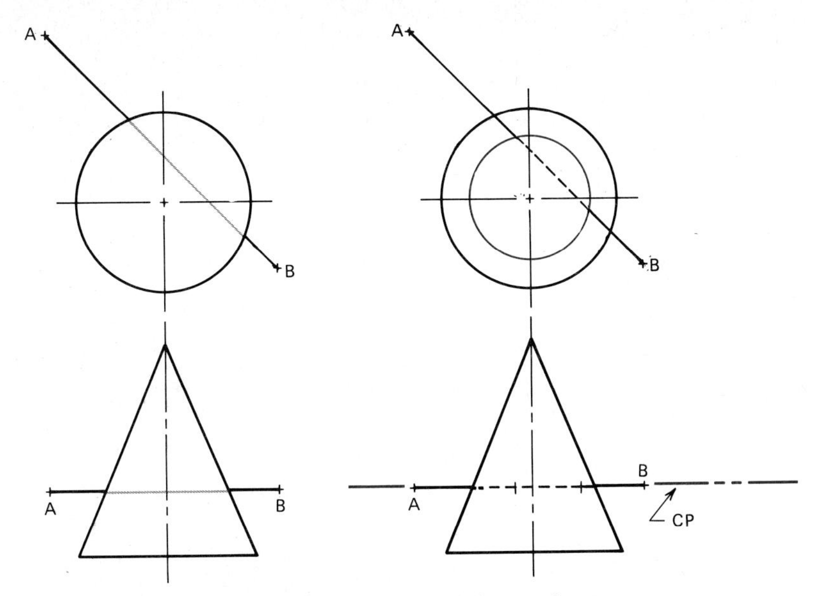

FIG. 8.6 Piercing points of a horizontal line and a cone.

8.6 HORIZONTAL LINE INTERSECTING A RIGHT CONE

The cutting-plane method can again be used in finding the two intersections of a line and a cone. In this case you must be careful to select a cutting plane that intersects the surface of the cone with a simple line of cut such as a straight line or a circle. In Fig. 8.6, it is clear that if a vertical cutting plane were used, the line of cut would be a hyperbolic curve. Therefore, a better selection would be a horizontal cutting plane, as shown. The line of cut is then circular.

The solution view and solution procedure given in Sec. 8.5 apply here as well.

8.7 OBLIQUE LINE INTERSECTING A RIGHT CONE

With an oblique line is involved, as shown in Fig. 8.7, it is not advantageous to show an edge view of a cutting plane in either the horizontal view or the front view. You might be required to plot either an ellipse or a hyperbola as the line of cut between the auxiliary plane and the cone. But a more convenient cutting plane can be chosen, as described below. The solution view remains the same as the one given in Sec. 8.5.1.

8.7.1 Solution Procedure

Construct plane *ABV* in the H and F views that contains the given line *AB* and the cone vertex *V*.

Place a horizontal line *BD* in plane *ABV*.

Project plane *ABV* and the cone into view 1 in order to see plane *ABV* as an edge.

Extend plane *ABV* so that it passes through the base of the cone, resulting in cutting lines *V*1 and *V*2 on the cone.

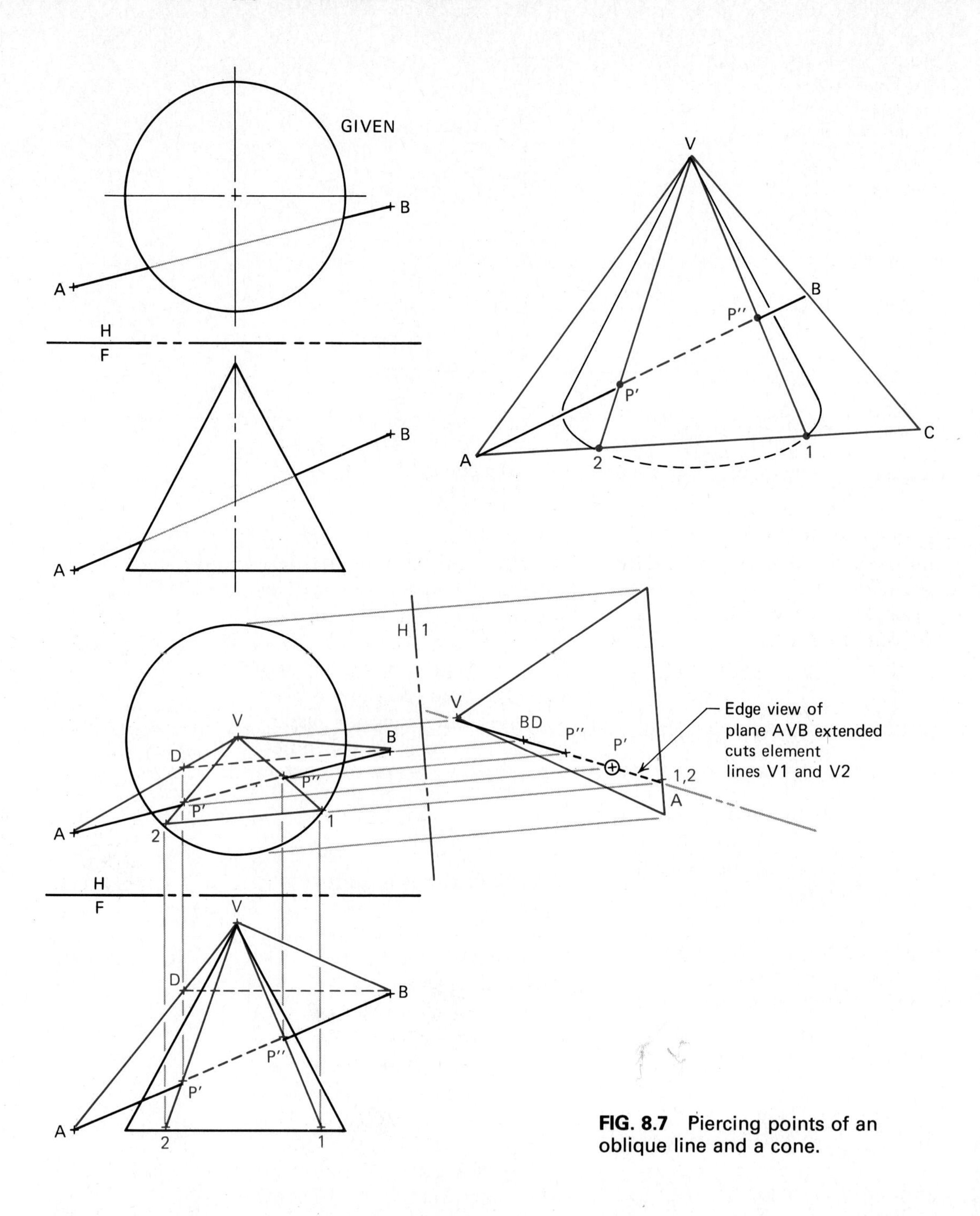

FIG. 8.7 Piercing points of an oblique line and a cone.

Project points 1 and 2 back to the H and F views. Note that the piercing points are where cutting lines *V*1 and *V*2 intersect the given line *AB*.

Determine visibility. Because both points are on the front (visible) surface of the cone, the line segments from *A* to *P′* and from *P″* to *B* are visible in both the H and the F projections. Only *P″B* is visible in elevation view 1.

8.8 LINE INTERSECTING A SPHERE

The auxiliary-cutting-plane method is also useful in finding the piercing points of a straight line with double-curved surfaces. Remember to choose a cutting plane that produces a simple line of cut on the object. The solution view remains the same as in previous sections. Two solution procedures are given. The first applies to line *AB* in Fig. 8.8, and the second to line *XY*.

8.8.1 Solution Procedure 1 (Line *AB*)

Pass a vertical cutting plane through line *AB* and the given sphere. See Fig. 8.8*b*.

Project the front view of the circle cut by the cutting plane on the sphere. The circle intersects line *AB* at *P* and *P′*, the piercing points in the front view.

Project *P* and *P′* to the horizontal view.

Show visibility. The line and the piercing points are in front of the great circle (major diameter) of the sphere in the H view. Point *P* is in the lower hemisphere and is therefore hidden in the top view.

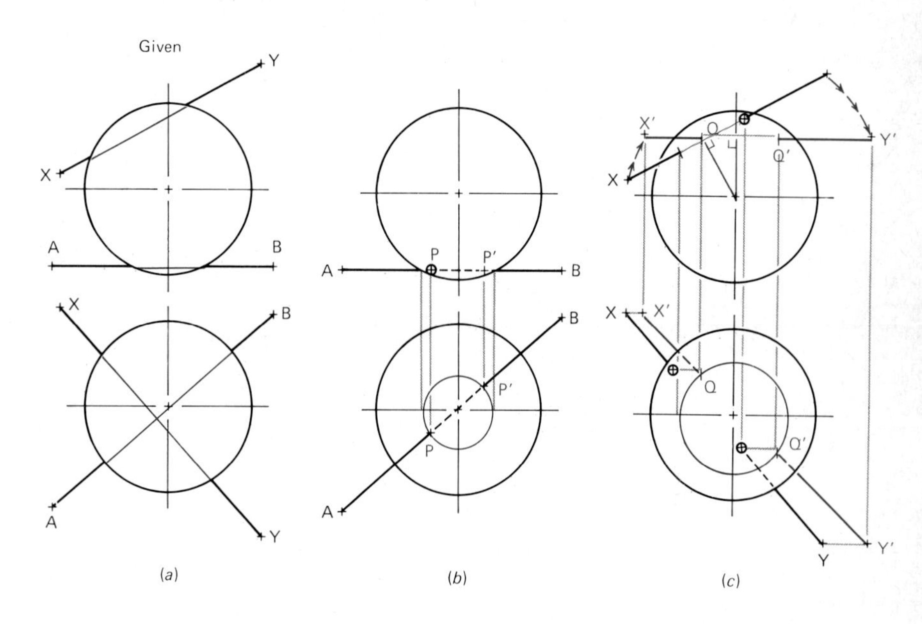

FIG. 8.8 Piercing points of a line and a sphere.

8.8.2 Solution Procedure 2 (Line *XY*)

Revolve line *XY* to the front position *X′Y′*, using a vertical axis through the center of the sphere as shown in Fig. 8.8*c*. As the sphere revolves, any new position is coincident with the original position.

Project the new position of *X′Y′* to the front view.

Pass a vertical-front cutting plane through *X′Y′* and the sphere in the horizontal view.

Draw the circular line of cut in the front view. The intersection of this circle and *X′Y′* provides the piercing points *Q* and *Q′* in the revolved position.

Revolve the piercing points to the original position of line *XY* in the H view, using the same vertical axis as before.

Project the piercing points to the F view.

Show visibility. Both piercing points lie behind the great circle of the sphere and thus are not visible in the front view. One point is in the upper hemisphere and the other in the lower; therefore, one is visible and the other is hidden in the H view. The opaque surface of the sphere hides all the lines within the sphere.

8.9 LINE INTERSECTING A WARPED SURFACE

Although the procedure becomes somewhat tedious when warped surfaces are involved, the basic principle used in determining piercing points remains the same. An auxiliary plane cutting a warped surface results in a line of intersection that is an irregular curved line. Such a line cannot be

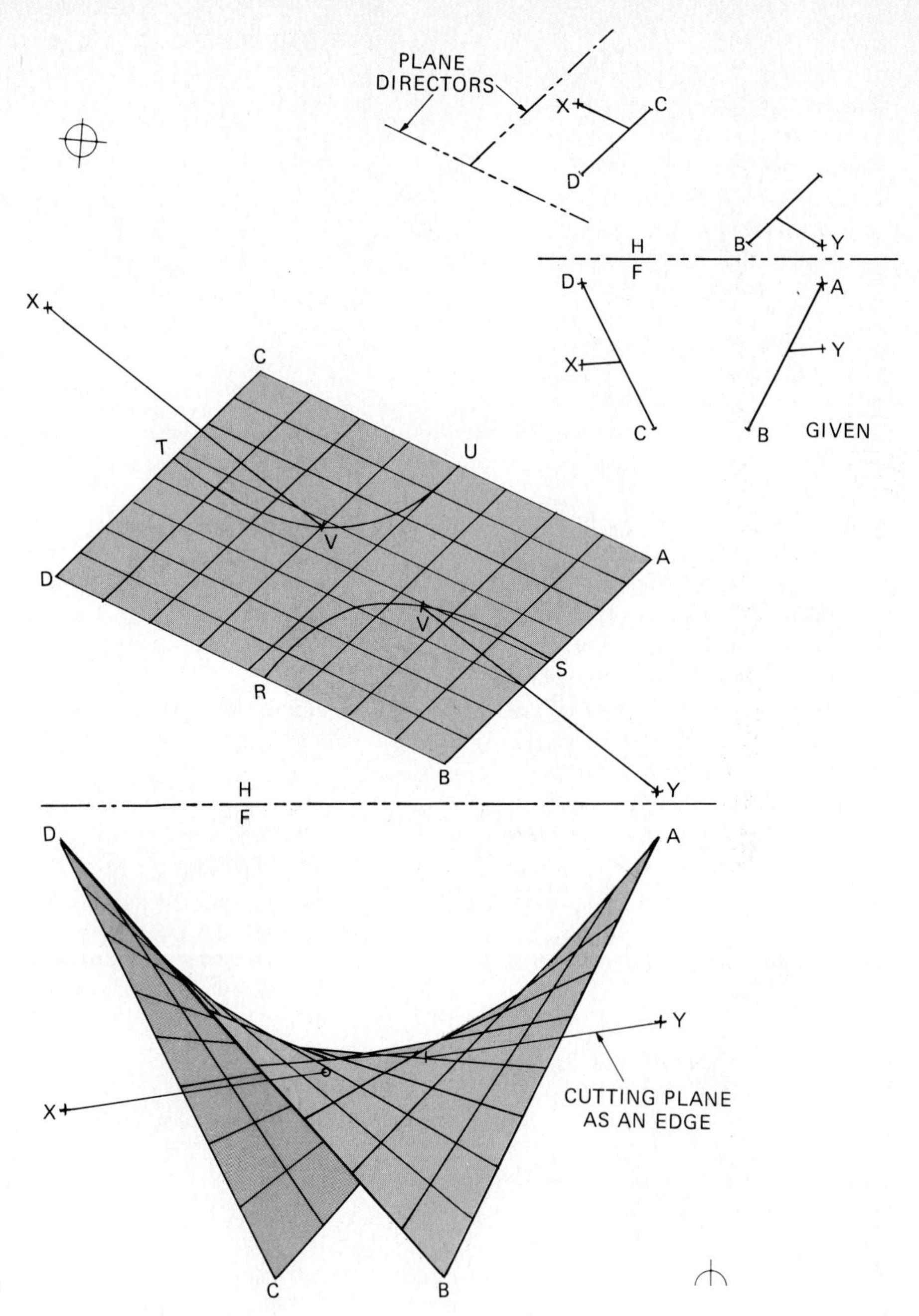

FIG. 8.9 Piercing points of a line and a warped surface, a hyperbolic paraboloid in this example.

drawn with a straightedge or compass. Therefore, several points must be located along the line and smoothly connected with a french curve (irregular curve).

The solution view remains the same as before. Because the procedures for various warped surfaces are similar, only the hyperbolic paraboloid is discussed here.

8.9.1 Solution Procedure

Delineate the surface of the hyperbolic paraboloid, as shown in Fig. 8.9.

Show the position of line *XY* lightly.

Pass a cutting plane through the given line *XY* and the surface of the hyperbolic paraboloid.

Project the points where each of the elements on the surface is intersected by the cutting plane to the horizontal view.

Draw the curved lines *RS* and *TU* in the top view with a french curve.

Locate the piercing points where *RS* and *TU* intersect line *XY*.

Determine visibility.

8.10 PLANE INTERSECTING A PLANE; DIHEDRAL ANGLE: EDGE-VIEW METHOD

Two planes will intersect in a straight line. This line of intersection may be located by two methods: the edge-view method and the auxiliary plane method. We will outline the edge-view method first. Viewing the line of intersection as a point will show the dihedral angle between the planes.

8.10.1 Solution View

The solution view is any view showing the edge of either of the two given planes and the projection of the other.

8.10.2 Solution Procedure

Locate a horizontal line *AK* in plane sector *ABC*, as shown in Fig. 8.10.

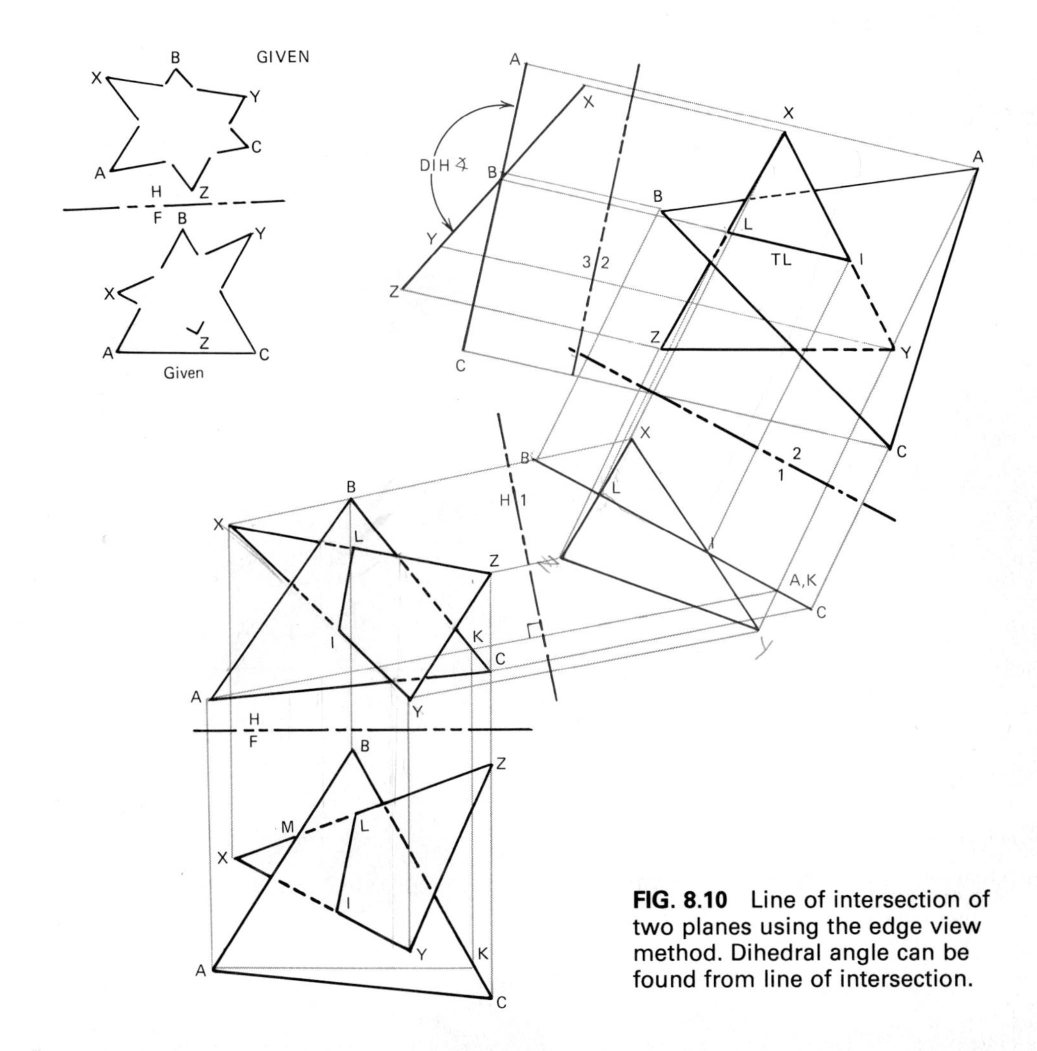

FIG. 8.10 Line of intersection of two planes using the edge view method. Dihedral angle can be found from line of intersection.

Construct reference line H/1 perpendicular to the true-length projection of *AK* and project both planes into elevation view 1. Note that you could have constructed a front line that would appear in true length in the F view and then used an inclined view for the solution view.

Determine the indefinite line of intersection *LI* where the edge view of plane *ABC* apparently cuts plane *XYZ*.

Project points *L* and *I* back to the adjacent H view. The definite line of intersection is now found as the points on line *LI* which are common to both planes *ABC* and *XYZ* (in this example all points on *LI* found in the 1 view are common to both planes).

Establish visibility by checking where boundary elements of the two plane sectors appear to intersect. The plane sector boundary closer to the viewer in a given projection will be visible; the other will be hidden. For instance, visibility in the F view can be determined by studying the apparent intersection of the plane boundaries *AB* and *XZ* at point *M*. Projecting this point *M* to the H view shows that *AB* is closer to the viewer than is *XZ*. Therefore, plane *ABC* is visible in the F view. Complete the visibility in the F view, remembering that visibility reverses when the line of intersection is crossed. You may wish to review Sec. 8.4.1.2 on visibility.

The dihedral angle can be found from any view showing the line of intersection as a point.

8.10.3 Solution View

Any orthographic view showing the point view of the line of intersection.

8.10.4 Solution Procedure

Construct a reference line parallel to any given view of the line of intersection. Reference line 1/2 is shown in Fig. 8.10.

Project the planes and the line of intersection, now in true length, into this view (2 view in Fig. 8.10).

Construct a reference line perpendicular to the true length view of the line of intersection. See reference line 2/3 in Fig. 8.10.

Project the line of intersection as a point and the planes as edges into this view (3 view in Fig. 8.10). The dihedral angle can now be labeled and measured.

8.11 PLANE INTERSECTING A PLANE; DIHEDRAL ANGLE: AUXILIARY CUTTING-PLANE METHOD

The cutting-plane method can be used for finding the line of intersection of surfaces as well as finding the piercing point between a line and a surface, as described previously.

8.11.1 Solution View

The solution view is any view showing two points on the line of intersection between plane sectors. These points are found at the intersection of common lines cut by auxiliary planes passed through the given plane sectors.

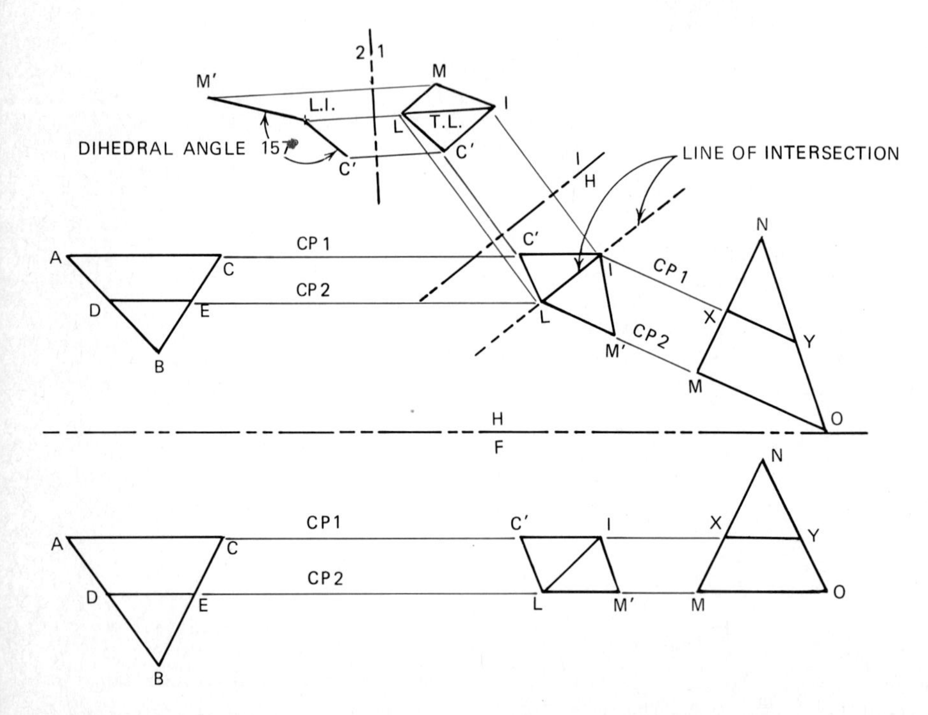

FIG. 8.11 Line of intersection and dihedral angle using the auxiliary plane method.

8.11.2 Solution Procedure

Pass two parallel horizontal cutting planes CP1 and CP2 through plane sectors *ABC* and *MNO,* as illustrated in Fig. 8.11. The cutting planes need not be parallel but should be positioned well apart on the plane sectors to ensure better accuracy. Figure 8.12 shows the effect of cutting-plane position on the accuracy of the result. The use of parallel cutting planes helps improve the accuracy of projections to adjacent views.

Project the lines of intersection *DE, AC, XY,* and *MO* to the horizontal view.

Locate points *L* and *I* where the 1 and 2 lines, respectively, intersect in the horizontal view. Points *L* and *I* lie on the theoretically infinite line of intersection between the two planes represented by sectors *ABC* and *MNO.*

Consider visibility. The entire plane sectors *ABC* and *MNO* are visible since the plane sectors do not extend to the line of intersection in the views drawn.

Figure 8.11 shows how the dihedral angle is found for plane sectors that are separated. The plane sector *ABC* is extended to include *LIC′* and sector

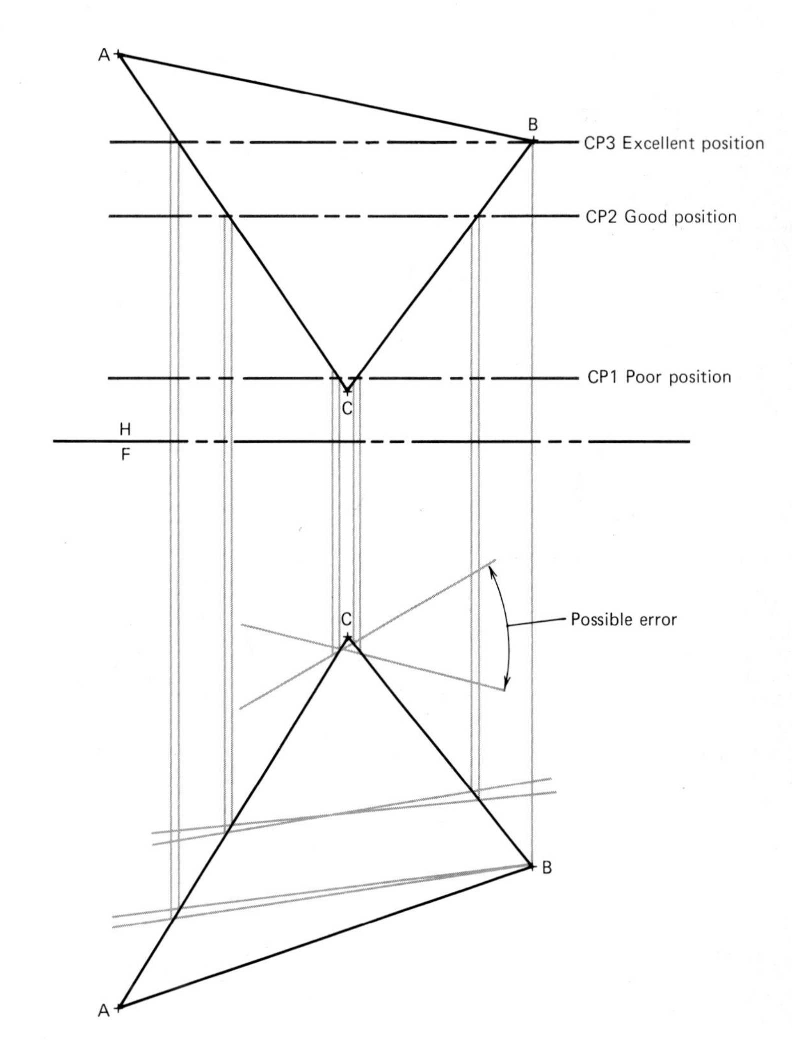

FIG. 8.12 Positioning of the cutting planes is very important in determining lines of intersection.

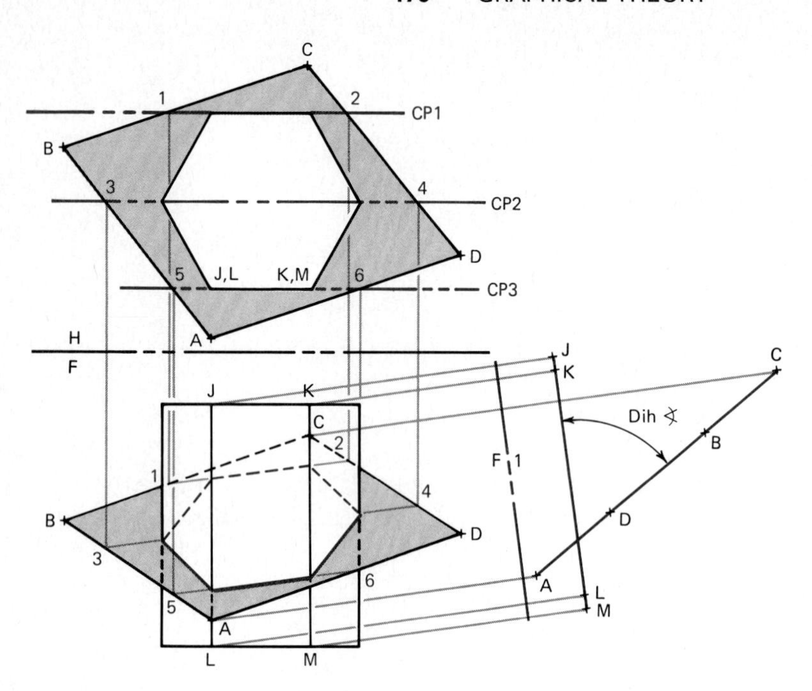

FIG. 8.13 Intersection of an oblique plane and a prism using the auxiliary-plane method.

MNO is extended to include *LIM'*. The dihedral angle is then found by projecting the point view of *LI*, as indicated in Fig. 8.11.

8.12 OBLIQUE PLANE INTERSECTING A HEXAGONAL PRISM: AUXILIARY PLANE METHOD

The method that involves finding the given plane as an edge, as illustrated in Fig. 8.10, can be used to determine the line of intersection between a plane and a prism. In many cases, however, it is faster and equally accurate to use a series of auxiliary cutting planes, as has been done in Fig. 8.13. The solution view is the same as the one defined in Sec. 8.5.

8.12.1 Solution Procedure

Establish vertical cutting planes CP1, CP2, and CP3 through corner points (actually element endpoints) of the prism that also intersect the boundary lines of the given plane sector *ABCD*.

Lightly draw element lines on the prism in the front view (all are shown, but only *JL* and *KM* are labeled).

Draw in lightly the lines of intersection in the horizontal view 12, 34, and 56 resulting from the respective cutting planes 1, 2, and 3 intersecting plane *ABCD*.

Locate the points of intersection between the vertical element lines and the lines cut by the auxiliary planes. Note how elements *JL* and *KM* intersect the lines of cut 56. These three lines are common to (lie in) cutting plane CP3.

Show visibility by checking each segment of the line of intersection.

Darken the element lines to show correct visibility (consider both the prism and the plane *ABCD* to be opaque).

Other design problems such as determining the dihedral angle between plane *ABCD* and plane *JLMK* (front face of the prism) can now be solved, as shown in inclined view 1.

Two new terms must be introduced for use in subsequent work: critical elements and turnaround points.

Critical elements are those elements which are located where abrupt changes in the direction of surface occur. For example, the six corner elements of the prism in Fig. 8.13 and the edge of the pyramid in Fig. 8.14 are critical elements.

Turnaround points are points on the line of intersection between objects in partial intersection in which the line of intersection reverses its direction and goes back to its starting point by a different route. Point 7 in Fig. 8.15 is a turnaround point. However, it should be noted that the line of intersection between two three-dimensional objects, whether they are partial or full intersections, will result in a single continuous line.

8.13 OBLIQUE PLANE INTERSECTING A TRIANGULAR PYRAMID: GIVEN PLANE AS AN EDGE METHOD

The line of intersection between the triangular prism and plane *XYZ* shown in Fig. 8.14 can be determined by the auxiliary plane method, but some difficulty is encountered with profile line *AV*. The point on line *AV* that is intersected by a cutting plane must be rotated with the line to a front position in order to project the front view to the horizontal projection. The line must be rotated back to its original position to complete the line of intersection between the plane and the pyramid.

Thus, it is simpler to project the edge view of plane *XYZ* along with the entire pyramid into an auxiliary view (see Fig. 8.14).

8.13.1 Solution View

The solution view is any view that shows the given plane as an edge and the projection of the complete pyramid.

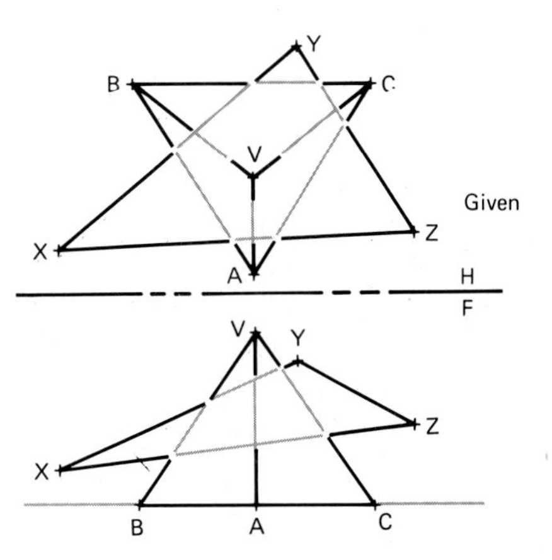

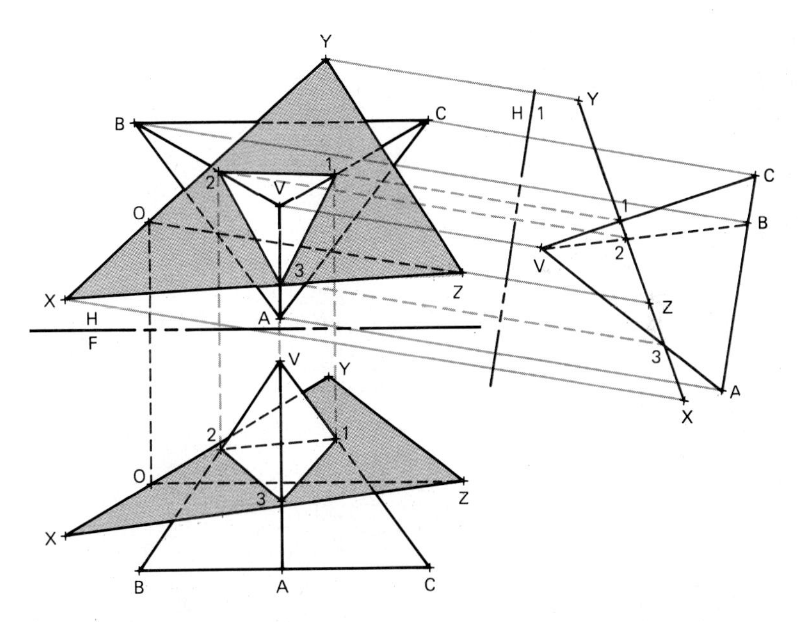

FIG. 8.14 Intersection of an oblique plane and a triangular prism using the edge-view method.

8.13.2 Solution Procedure

Draw horizontal line *OZ* in plane *XYZ* and project plane *XYZ* and the pyramid onto an image plane perpendicular to the true-length projection of *OZ*.

Find points 1, 2, and 3, where *VC*, *VB*, and *VA*, respectively, are cut by *XYZ*.

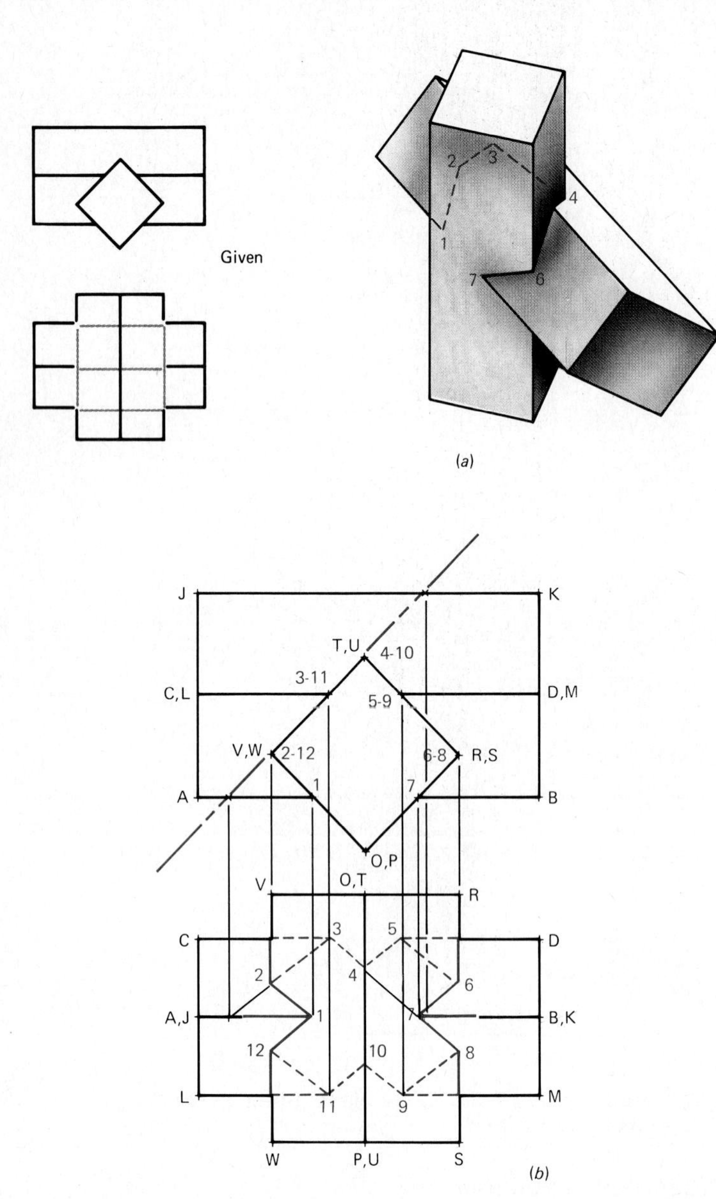

FIG. 8.15 Intersection of two prisms using auxiliary plane method.

Project points 1, 2, and 3 into the H and F views.

Draw the complete line of intersection in both the H view and the F view by connecting the points with straight-line segments. Check visibility.

Darken the boundary lines of *XYZ* and the edge elements of the pyramid, using correct line symbols.

8.14 PRISM INTERSECTING A PRISM: AUXILIARY-PLANE METHOD

The principles for determining the line of intersection between two solids by the auxiliary plane method are the same as those for the intersection between a plane sector and a three-dimensional object. The problem is more complex only because there are more surfaces involved; therefore, more construction is required to draw the line of intersection.

In the case of intersecting prisms (Fig. 8.15), the cutting planes are simply extensions of the plane surfaces that make up one of the prisms. They are used to find the piercing points of corner elements of the other prism.

8.14.1 Solution View

The solution view is any view showing the intersection of corners or other critical elements of one object with surfaces of the other object.

8.14.2 Solution Procedure

Extend the edge view of plane *VWTU* of the vertical prism across the top view of the horizontal prism. Use this edge-view plane to find the piercing points of the corner elements. Work with construction-weight lines.

Project points on *AB, CD, JK,* and *LM where extended plane VWTU* intersects these critical elements of the horizontal prism to the F view.

Draw in lightly the lines of intersection formed on each of the four planes of the horizontal prism by connecting points that lie on the same plane.

Construct the F projection of the critical elements of the vertical prism and locate the piercing points.

Connect the piercing points that lie on the same plane to form the reversed but continuous loop line of intersection shown in Fig. 8.15.

Determine visibility and darken lines, considering the prisms to be opaque.

8.15 CYLINDER INTERSECTING A CYLINDER: AUXILIARY-PLANE METHOD

Again, auxiliary cutting planes can be effectively used to determine the line of intersection between two cylinders.

8.15.1 Solution View

The solution view is any view showing the intersection of element lines on each cylinder common respectively to a series of cutting planes passed through both objects.

8.15.2 Solution Procedure

Draw construction-weight adjacent views of the cylinder in the desired positions. See Fig. 8.16.

Construct a profile view of the horizontal cylinder.

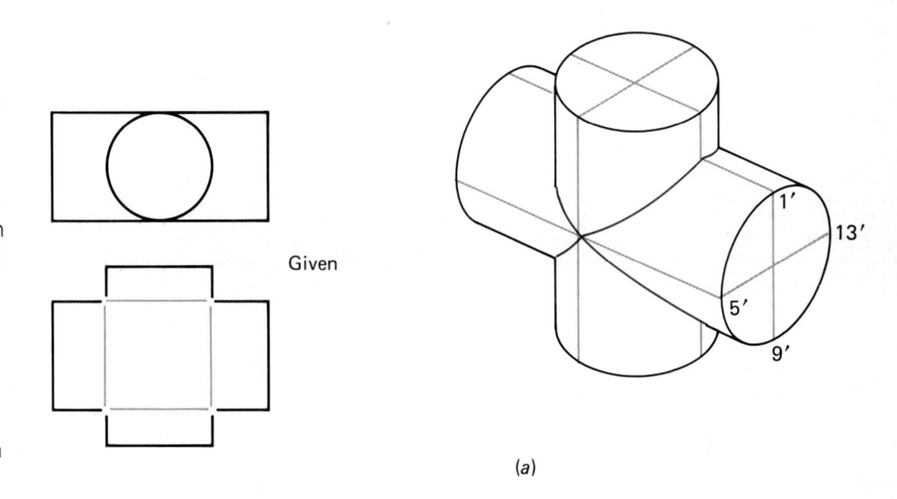

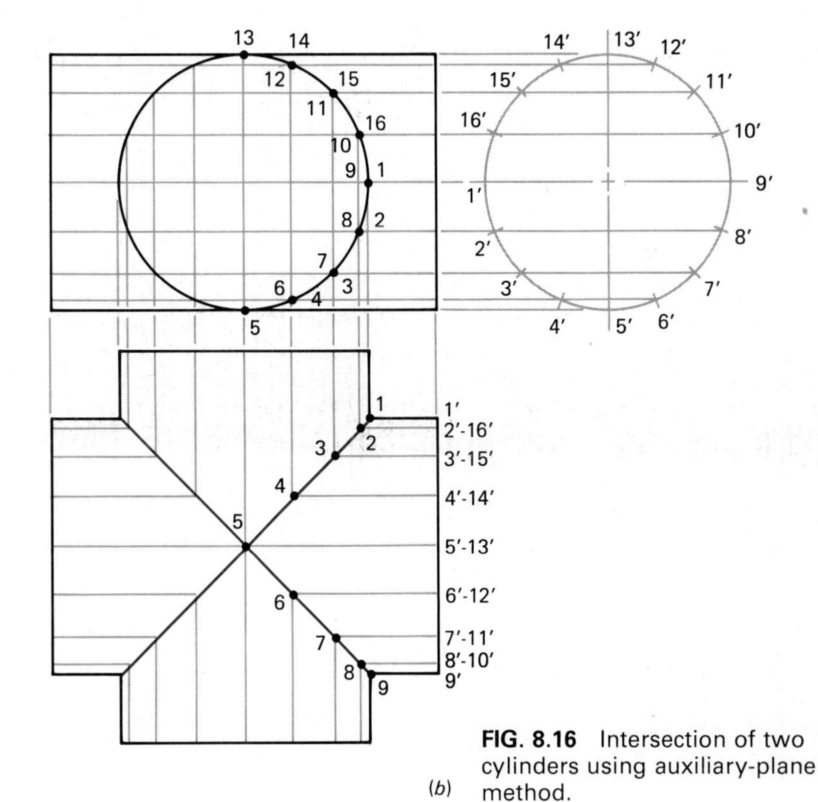

FIG. 8.16 Intersection of two cylinders using auxiliary-plane method.

Pass a series of vertical cutting planes through the H and P views of the cylinder, showing common elements intersected by the cutting planes on each cylinder.

Identify the points where common elements intersect in the front view and then draw the lines of intersection. There appear to be two lines of intersection that continue on both sides of the intersection when cylinders of equal radius intersect fully. In reality, it is one continuous line.

Darken the element lines to terminate at the line of intersection if desired.

8.16 CYLINDER INTERSECTING A CYLINDER: AUXILIARY-PLANE METHOD SIMPLIFIED

Figure 8.17 shows a convenient use of auxiliary planes to determine the intersection of two cylinders. One of the cylinders has been cut at 45° with its axis to facilitate projection from the top to the front view of elements cut by common cutting planes. The solution view and solution procedure are the same as the ones given in Sec. 8.15.

8.17 PRISM INTERSECTING A PYRAMID: AUXILIARY-PLANE METHOD

When a prism intersects a pyramid, the auxiliary plane method of determining the line of intersection is again a logical choice.

8.17.1 Solution View

The solution view is any view showing the piercing points of critical elements of one object with the surfaces of the other object.

8.17.2 Solution Procedure

Pass vertical planes through the critical element lines of the prism and an inclined cutting plane through *VN* in the front view of the pyramid. See Fig. 8.18.

Identify points 1 through 6 where the critical elements intersect the lines created by the cutting planes. (These are the piercing points of each of the elements.)

Project lines cut on planes *VPN, VNM,* and *BC*234 by the auxiliary cutting planes.

Draw straight-line segments between the piercing points lying on a given plane, as shown in the numbered sequence 1 through 6 in Fig. 8.18. Show visibility.

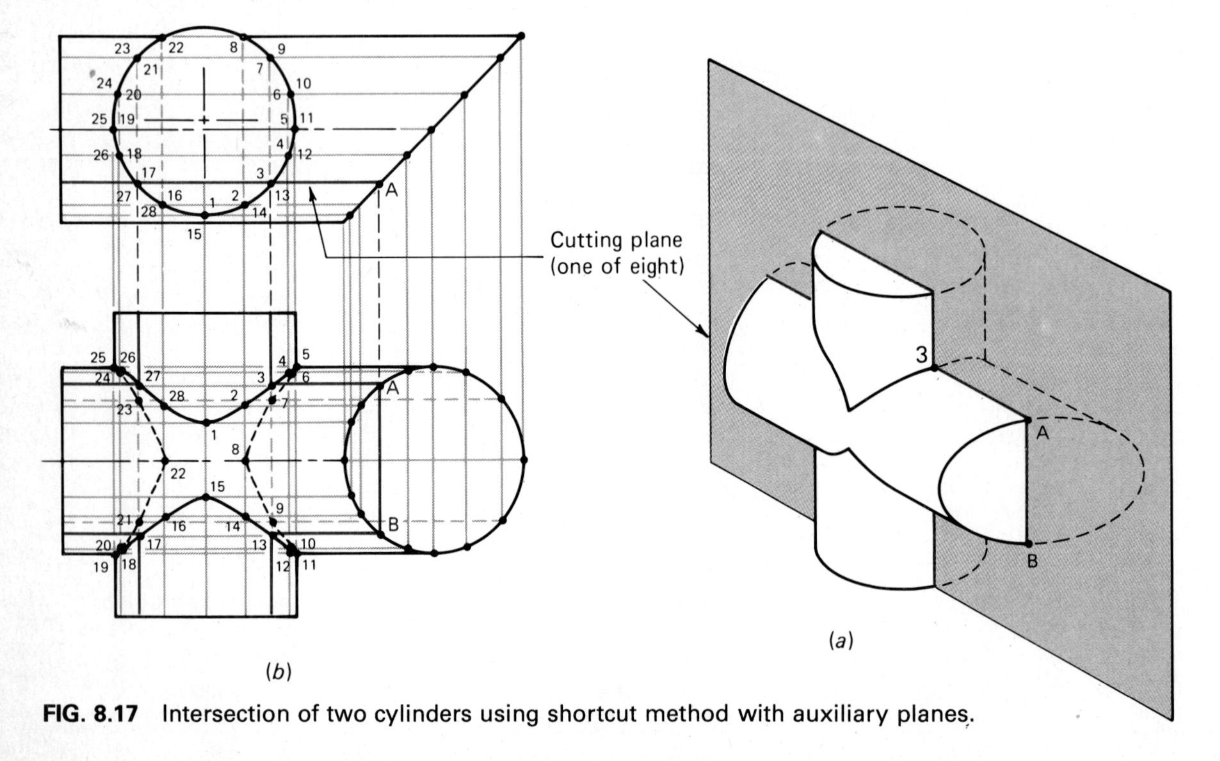

FIG. 8.17 Intersection of two cylinders using shortcut method with auxiliary planes.

8.18 PRISM INTERSECTING A CYLINDER: AUXILIARY-PLANE METHOD

It is clear that the auxiliary-plane method is used extensively to solve many problems involving intersecting surfaces. Where curved surfaces are involved, the line of intersection is at least partially curved. Because of this, more cutting planes are required to accurately locate more points on the line of intersection so that a smooth curve can be drawn between the points. The solution view is the same as the one given in Sec. 8.15.

8.18.1 Solution Procedure

Draw lightly the front, top, and profile views of the cylinder and prism in the desired position. See Fig. 8.19.

Pass a series of vertical cutting planes through the two objects (front planes will be satisfactory), creating a series of common intersecting element lines. Be sure to pass the cutting planes through critical elements so that the turnaround points can be accurately located.

Identify intersections between common element lines (those cut by the same auxiliary plane) with a sequential numbering system, as shown in Fig. 8.19.

Draw in a smooth curve (or a straight-line segment where appropriate) between points. Show proper visibility.

Darken the critical edge lines of both objects until they intercept the line of intersection or in some cases pass by it. The surfaces of each object are terminated at the line of intersection; that is, they do not pass through each other.

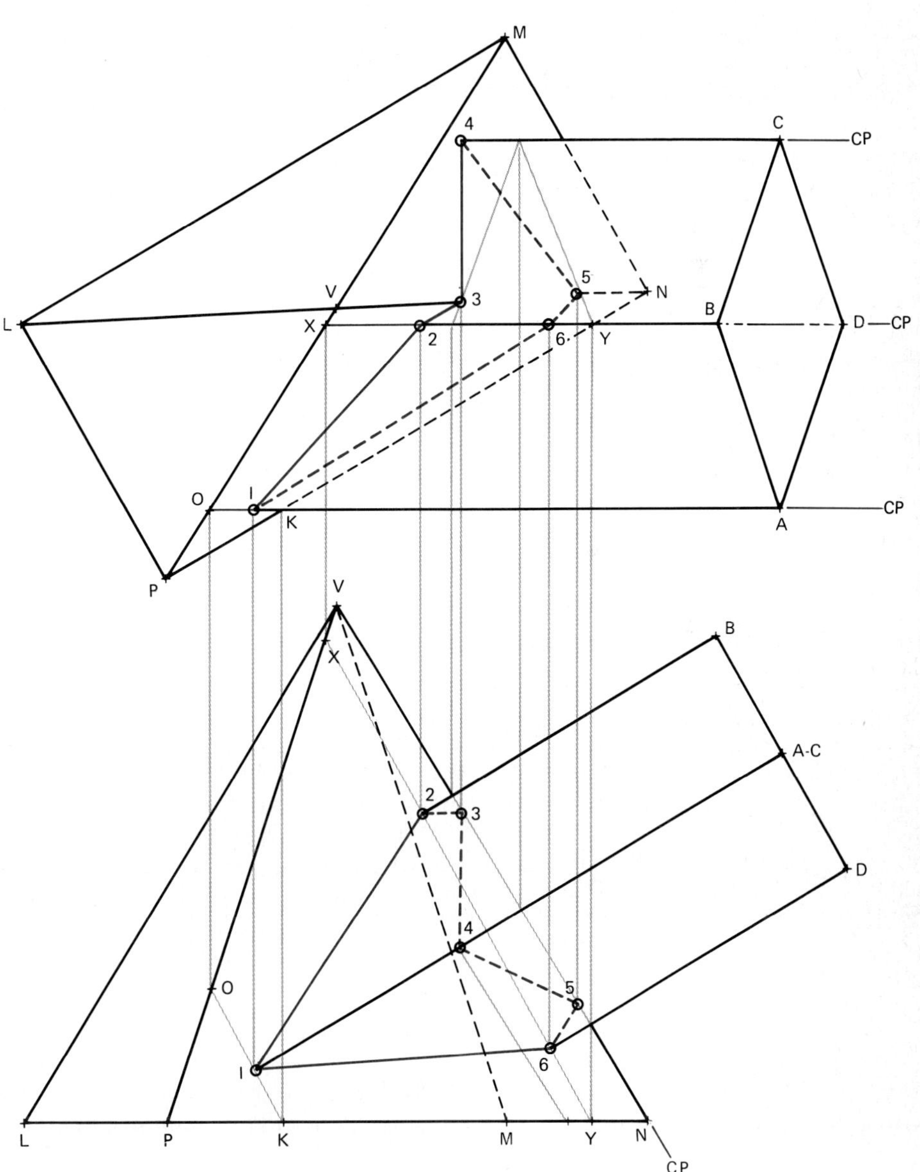

FIG. 8.18 Intersection of a prism and a pyramid using auxiliary-plane method.

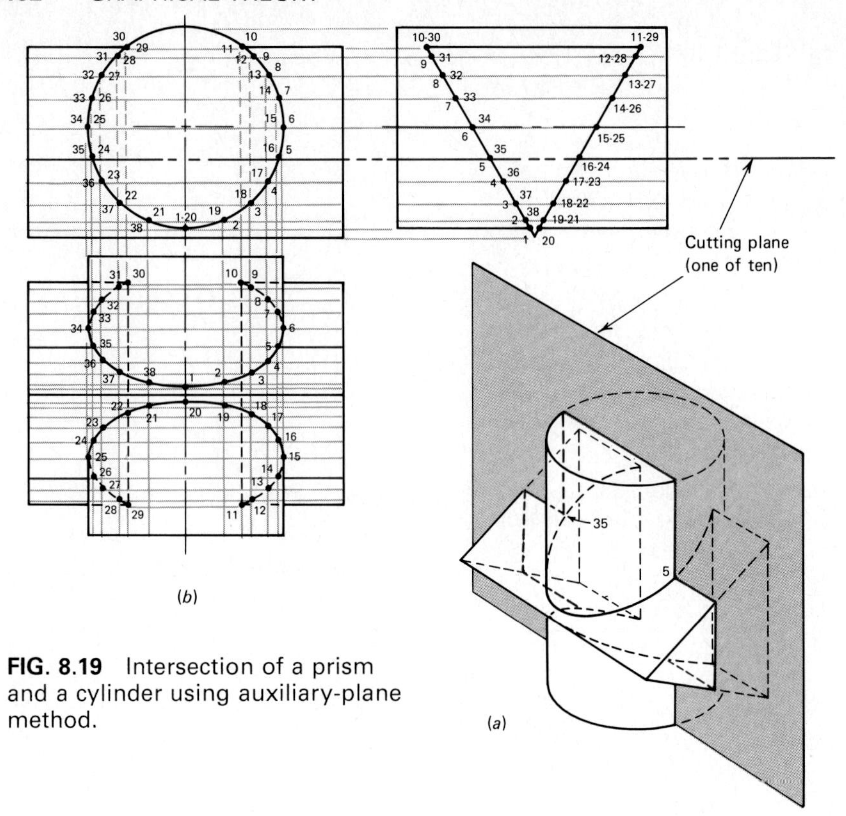

FIG. 8.19 Intersection of a prism and a cylinder using auxiliary-plane method.

8.19 CYLINDER INTERSECTING A CONE: AUXILIARY-PLANE METHOD

As surfaces become more complex, greater care must be used in the application of the auxiliary-plane method. Remember that the general rule states that auxiliary cutting planes should be positioned relative to both objects so that simple lines of cut (straight or circular) will appear on both objects.

In Fig. 8.20, if vertical cutting planes are used, the lines of cut on the cone will be hyperbolic. If horizontal cutting planes are used, a series of concentric circles will be generated on the cone in the top view while straight lines appear on the cylinder. The solution view is the same as the one described in Sec. 8.16.

8.19.1 Solution Procedure

Cut the cylinder in the top view at an angle that will produce a circular front view, thereby avoiding the need for an extra view. Draw the circular front view first and project it to the top view. The projected diameter of a cylinder remains the same in all views.

Pass cutting planes parallel to the axis of the cylinder to produce common element lines on both objects: straight lines on the cylinder, circular lines on the cone.

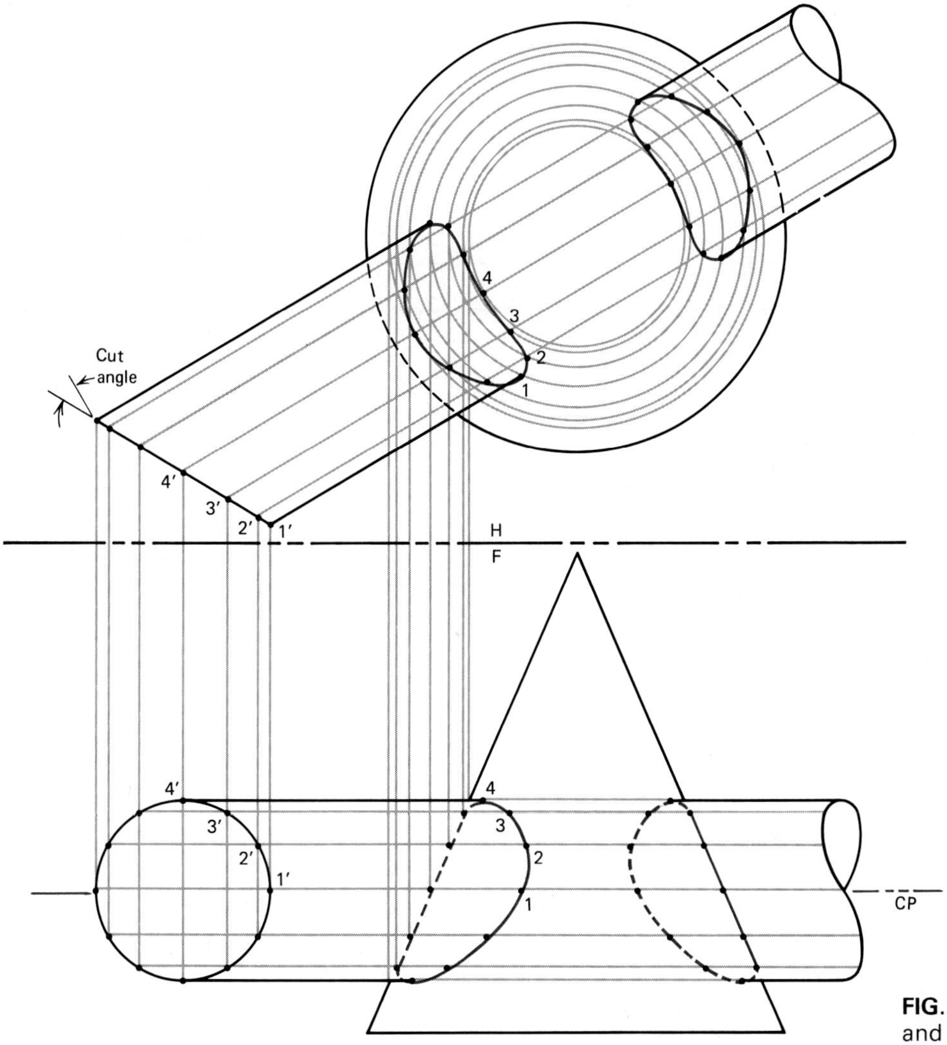

Horizontal cutting plane parallel to cylinder axis and perpendicular to cone axis

FIG. 8.20 Intersection of a cone and cylinder using auxiliary-plane method.

Identify intersection points of common elements and proceed as in previous examples.

8.20 CYLINDER INTERSECTING A SPHERE: AUXILIARY-PLANE METHOD

Careful placement of the cutting planes is again the major consideration. Front cutting planes create ellipses on the cylinder shown in Fig. 8.21. Through the use of horizontal cutting planes, simple lines are cut on both objects. The solution view from Sec. 8.16 applies here as well.

8.20.1 Solution Procedure

Position the H and F views of the objects as desired.

Cut the nonintersecting end of the cylinder so that it projects as a circle in the F view.

Pass a series of horizontal cutting planes to intersect critical elements and points as shown.

Determine intersection points between common elements and complete the line of intersection in a manner similar to previous examples.

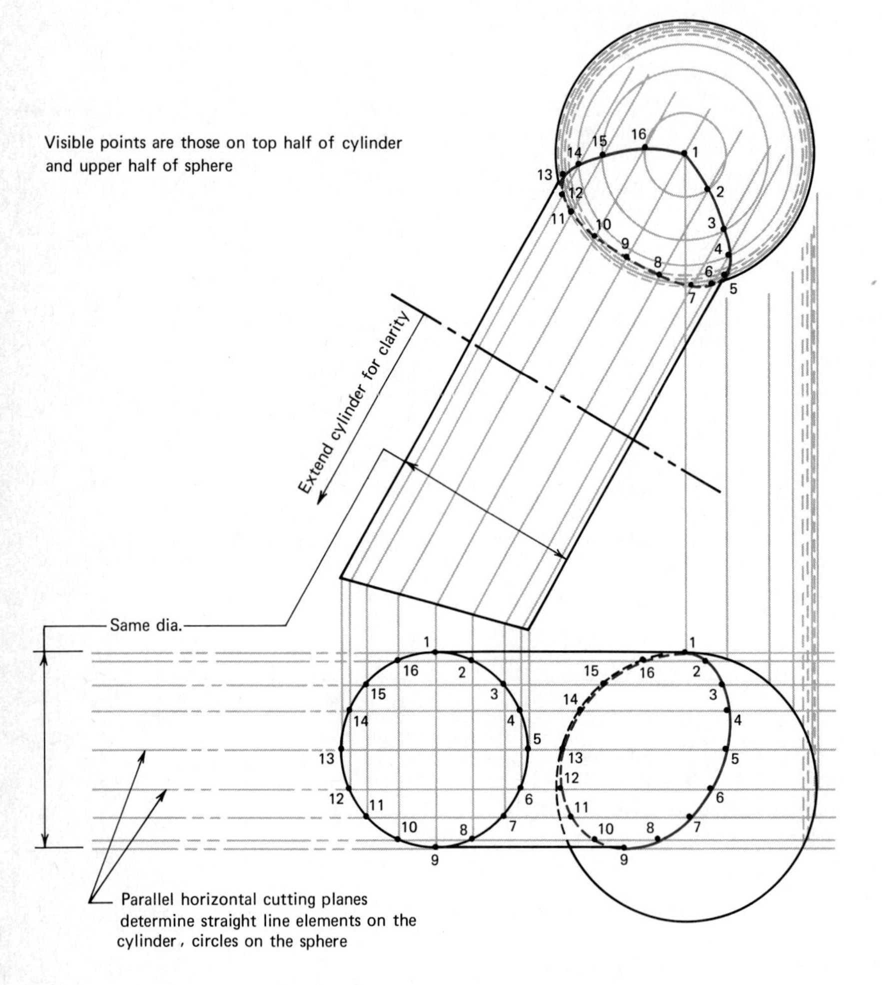

FIG. 8.21 Intersection of a cylinder and sphere using auxiliary-plane method.

8.21 INTERSECTION OF TWO OBLIQUE CYLINDERS: TRACE OF OBLIQUE CUTTING-PLANE METHOD

When two related views of intersecting cylinders that are oblique to the principal planes of projection are given, traces of a series of cutting planes parallel to the axes of the cylinder are used to locate the line of intersection. This eliminates the need for the auxiliary views that would be required if cutting planes were used as before.

A trace of a cutting plane is the line intersecting the base of a cylinder, cone, prism, or pyramid that determines the endpoints of element lines. The element lines are known to be cut parallel to the axes of cylinders or prisms and to the vertex of cones or pyramids.

8.21.1 Solution View

The solution view is any view showing the intersection of element lines on the surfaces that are common to cutting planes parallel to the axes of the cylinders.

8.21.2 Solution Procedure

Draw in lightly the H view of the intersecting cylinders by aligning the axes of the cylinders in the top view with guiding-plane director *ACB*, as shown in Fig. 8.22.

Establish the circular view of each cylinder base in the top view. The cylinders shown in Fig. 8.22 slant upward and are oblique to both the H and the F planes of projection.

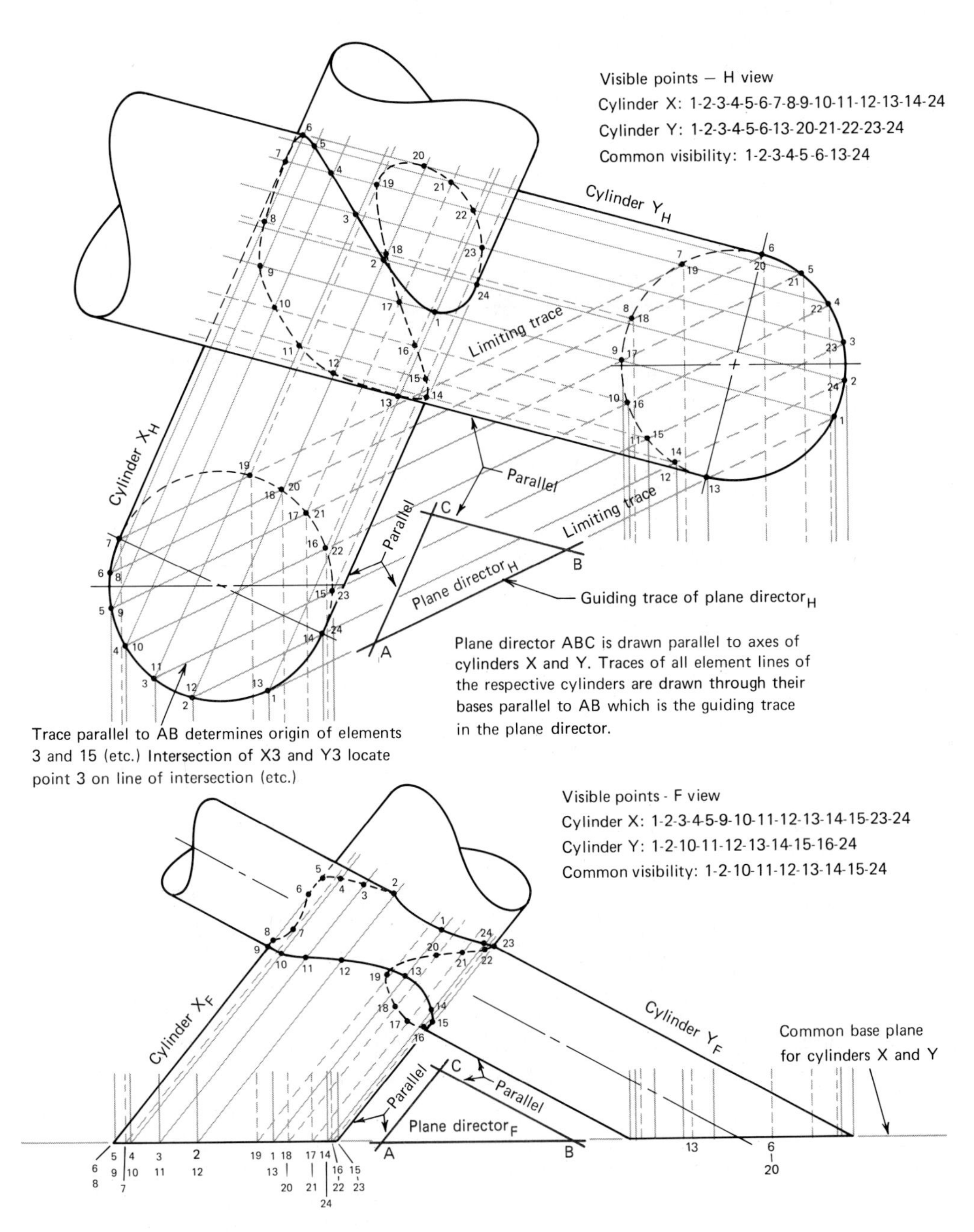

FIG. 8.22 Intersection of oblique cylinders using traces of oblique cutting planes.

Control the amount of intersection of the cylinders by the position of the traces. Those in Fig. 8.22 are offset only slightly. To intersect completely, the limiting traces (outside traces) can be positioned so that at least one of them is tangent to the base circle.

Draw the guiding trace *AB* of plane director *ACB* parallel to the limiting traces in the H view.

Project the bases of the cylinder to a common horizontal base plane. Refer to notes on Fig. 8.22.

Construct the front view of the plane director. Points *A, B,* and *C* must project orthographically. *AC* and *BC* may be positioned at the desired angle within practical limits.

Draw the critical elements of the F view of the cylinders parallel to *AB* and *CB* of the plane director.

Place intermediate traces in the H view parallel to the cylinder axes, one or two of which should be positioned through or near critical points on the base planes, such as 5 and 9, 14 and 27, 17, 3, and 23. These are end-points of elements that lead to turn-around points on the line of intersection. Care in the selection of intermediate traces will ensure the accuracy of the line of intersection.

Draw elements from the base trace points parallel to the axes of the cylinders in both views.

Determine the intersecting points of common elements and number them in a logical sequence.

Use a french curve to connect the intersection points for a smooth line of intersection.

Darken appropriate lines, considering visibility as in previous examples.

8.22 OBLIQUE CYLINDER INTERSECTING A CONE: TRACE OF A RADIAL CUTTING-PLANE METHOD

Although fundamentally the same method as that described in Sec. 8.21, the use of cutting-plane traces for problems involving geometric shapes with vertexes involves one major difference. The planes cut through the objects, leaving radial rather than parallel traces on the bases of the objects. However, the cutting planes still cut straight-line elements that are parallel on cylinders or prisms and converge to the vertex on cones and pyramids.

8.22.1 Solution View

The solution view is any view showing the intersection of common element lines cut by a series of planes through the vertex of the cone and parallel to the axis of the cylinder.

8.22.2 Solution Procedure

Draw the desired position in the H view of the bases and outside elements of both objects. See Fig. 8.23.

Draw the line of intersection of all cutting planes used. This line is called the cutting-plane hinge line. Experiment with various limiting traces on the bases of the cone and cylinder. These limiting traces control the amount and position of the intersection. The limiting traces and eventually all intermediate traces must converge and intersect the cutting-plane hinge line at its piercing point with the base plane of both objects. See Fig. 8.23 for this construction.

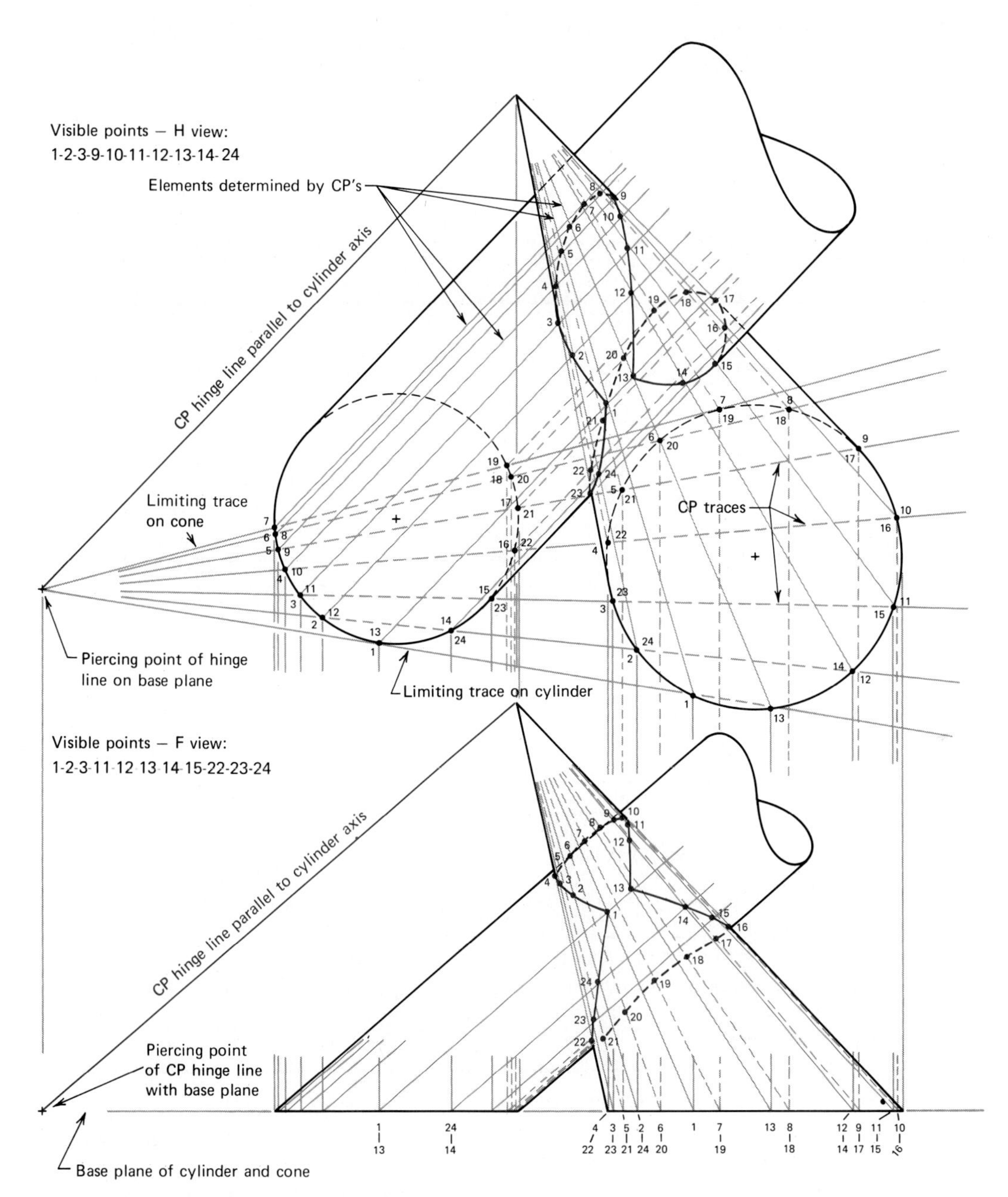

FIG. 8.23 Intersection of a cone and oblique cylinder using traces of radial cutting planes.

Project the circular bases of both objects and the piercing points of the cutting-plane hinge lines to the edge view of the base plane on the F view.

Project the vertex of the cone to the F view, select its elevation, and draw the critical elements of the cone from the vertex to the base plane in projection with the H view.

Draw the F view of the cutting-plane hinge line from the cone vertex to its piercing point with the base plane.

Construct the F view of the cylinder by drawing the axis and critical elements from the base parallel to the cutting-plane hinge line.

Position intermediate traces in the H view radiating from the piercing points of the cutting-plane hinge line across the bases of the cylinder and cone at or near critical points such as 7, 21, and 23 on the cylinder and 1, 3, 7, 9, and 13 on the cone.

Number the intersections of the traces and the two bases, paying particular attention to turnaround points. Remember that elements cut by a common plane should have matched numbers to facilitate identification on the line of intersection.

Draw in the element lines on both objects in both views.

Identify points on the line of intersection by finding where common elements intersect. Number these points in sequence.

Darken the line of intersection and element lines, showing proper visibility.

8.23 METHODS FROM COMPUTER GRAPHICS

Consider for a moment how you might use a computer to approach the problem of intersection of two planes as illustrated in Fig. 8.10. You could write an equation to describe each plane in the cartesian coordinate system. The equations represent planes of infinite extent rather than sectors, as would be necessary with a paper and pencil solution.

The line of intersection from analytic geometry is the line made up of common points on the two planes. Mathematically, this is given by a single equation for a line.

To show plane sectors on a cathode-ray tube, think of each plane boundary as a line, which it is, and program the computer to draw lines between the corner points in three-dimensional space.

The triangular sectors of the planes in Fig. 8.10 could then be readily plotted in one view and by the use of a rotation program, they could be rotated to provide any desired view. One such view would be view 1, which shows one plane as an edge. The line of intersection is clearly seen in view 1.

At this point, you could compute the coordinates labeled L and I, providing a way of plotting the finite lines of intersection in the other views.

The question of visibility still remains. A variety of methods for removal of hidden lines can be incorporated into the program.

Problems

8.1 (A) Without reference, describe in writing what is meant by a line of intersection between surfaces.

(B) Study the general rules relating to intersection of surfaces. Without reference, list as many of them as possible. Check against the text, add rules that you omitted, and repeat the exercise until you know the principles involved in these rules.

8.2 Without reference to text or notes, write a description of the procedure(s) for solving the following problems:

(a) Line intersecting a plane
(b) Line intersecting a cylinder or prism
(c) Horizontal line intersecting a right cone
(d) Oblique line intersecting a right cone
(e) Line intersecting a sphere
(f) Line intersecting any warped surface
(g) Line of intersection between two planes
(h) Line of intersection between an oblique plane and a hexagonal prism
(i) Identification of critical elements
(j) Identification of turnaround points
(k) Line of intersection between two prisms
(l) Line of intersection between two cylinders

8.3 Show the H and F projections of the point of intersection (piercing point) between the line and plane shown in Fig. 8.24. Assume the plane to be opaque and show correct visibility of line XY. Use the plane-as-an-edge method and then check the solution with the auxiliary-plane method.

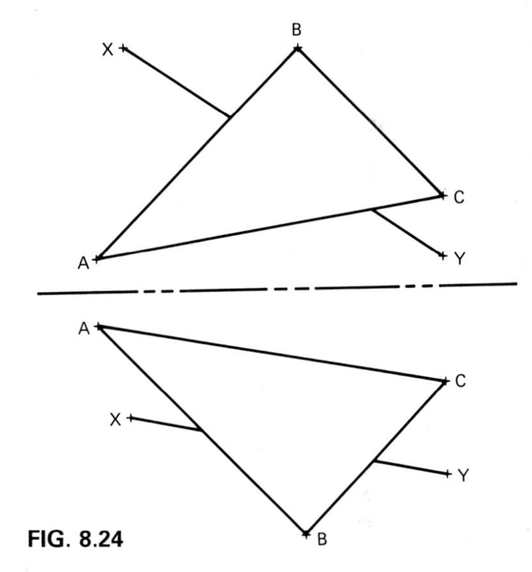

FIG. 8.24

8.4 A television antenna is to be anchored to the roof of a structure, as indicated in Fig. 8.25. Show the H and F projections of the anchor points on the roof. When stretched tightly, the cables are each to have a slope angle of 50°. The cables are all attached to the antenna at point *A*.

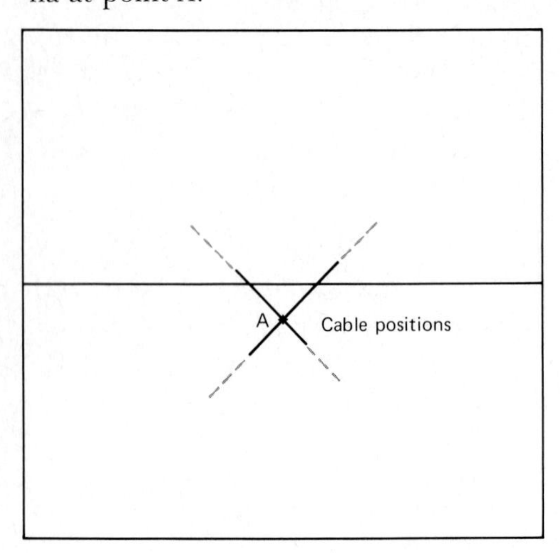

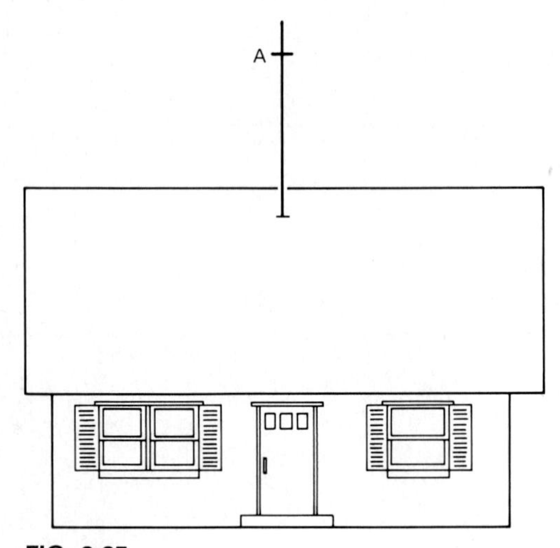

FIG. 8.25

8.5 Show the H and F projections of the piercing points of lines *AB* and *XY* with the cylinder represented in Fig. 8.26. Using P for visible and Ⓟ for hidden piercing points, show correct visibility for both piercing points and line segments.

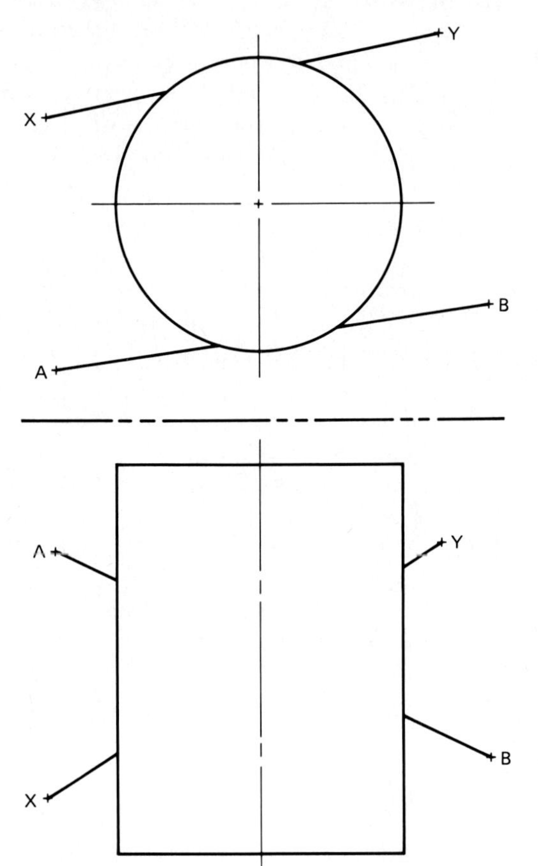

FIG. 8.26

8.6 Lines AB and CD intersect the sphere shown in Fig. 8.27. Show the piercing points, indicating visibility in the H and F views.

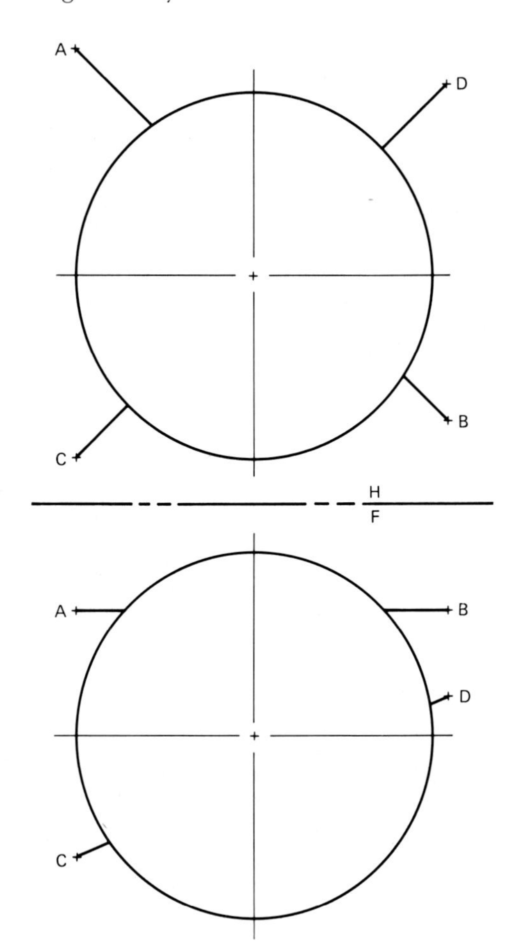

FIG. 8.27

8.7 Complete the H and F projections of the plane sectors ABC and XYZ in Fig. 8.28. Assume that both sectors are opaque. Show the finite portion of the line of intersection and proper visibility of the plane sector in both views.

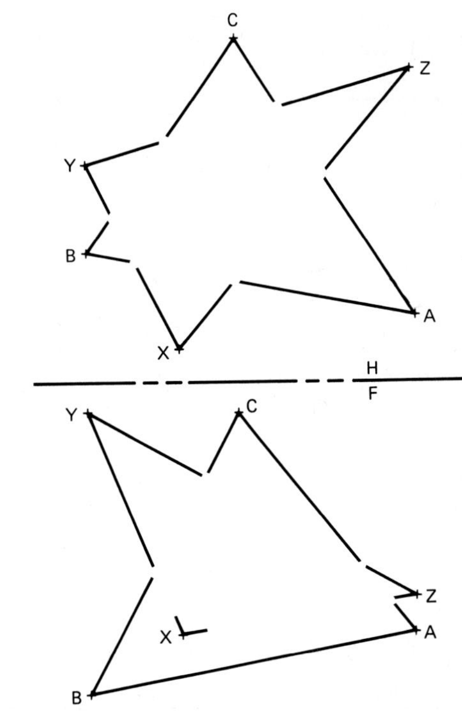

FIG. 8.28

8.8 Determine the line of intersection between plane *ABC* and the pyramid shown in Fig. 8.29. Show the H and F projections.

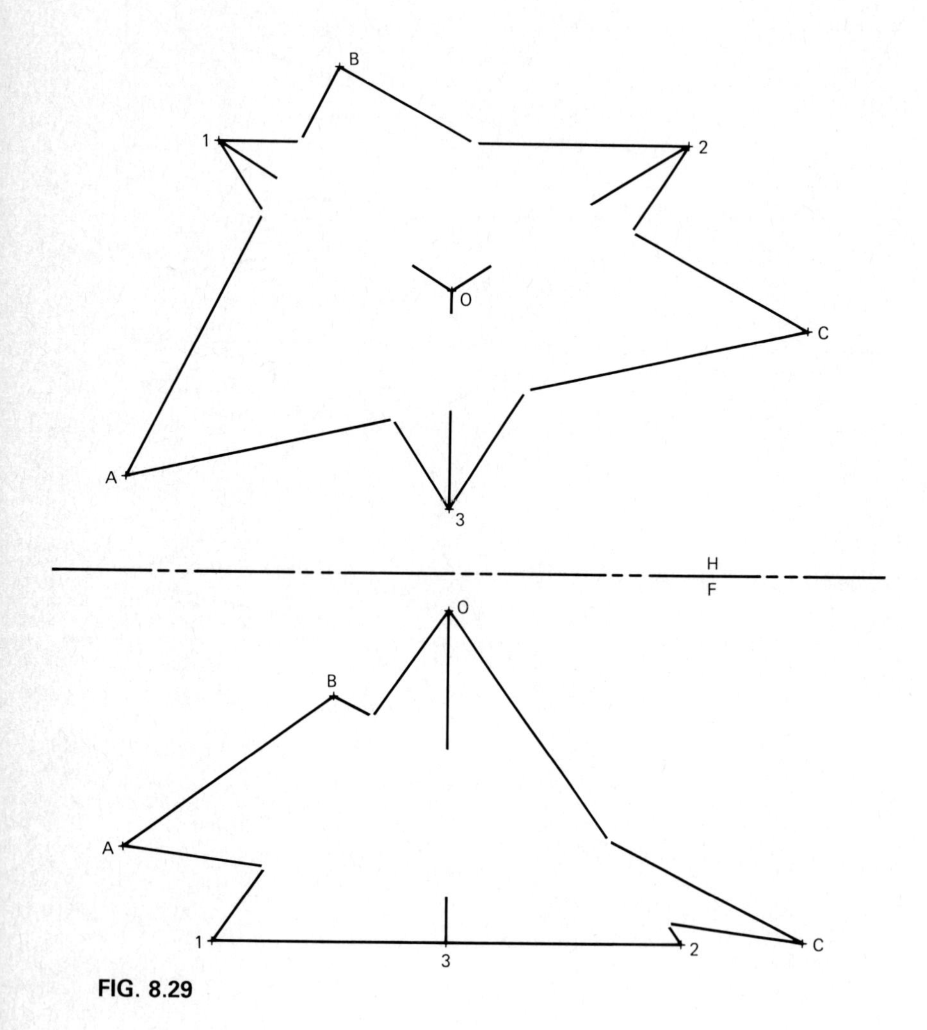

FIG. 8.29

8.9 Determine the line of intersection of the two prisms shown in Fig. 8.30.

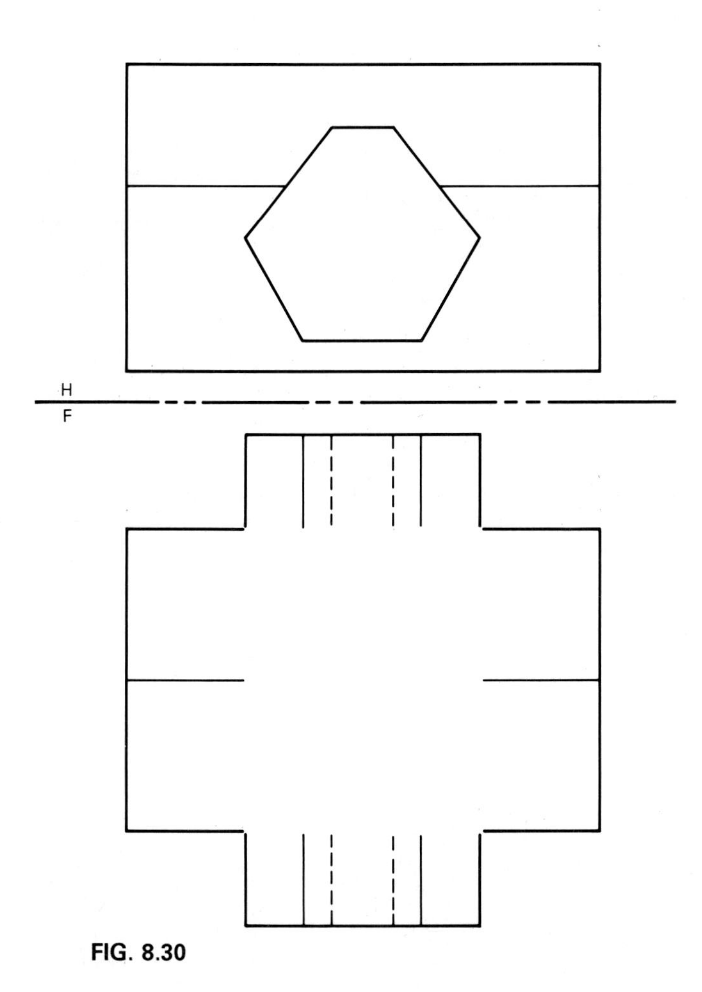

FIG. 8.30

8.10 Determine the complete line of intersection between the prism and the pyramid shown in Fig. 8.31.

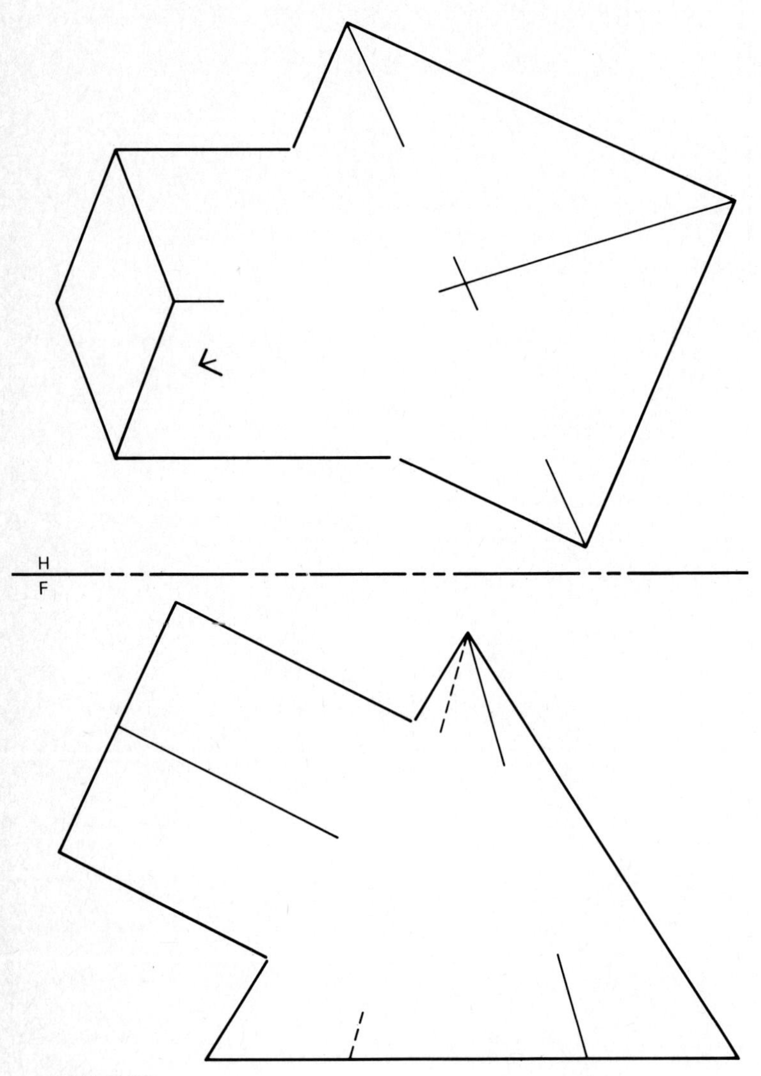

FIG. 8.31

8.11 Show the complete line of intersection between the cone and the right cylinder shown in Fig. 8.32.

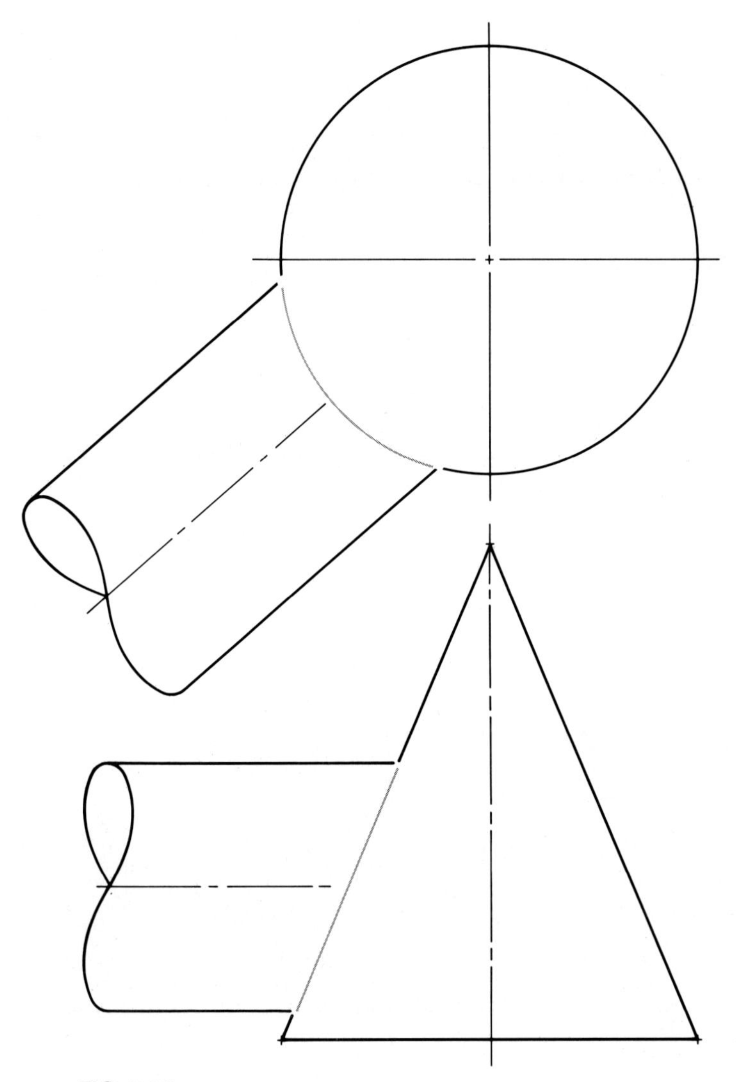

FIG. 8.32

8.12 Show the H and F projections of the complete line of intersection between the cone and the cylinder shown in Fig. 8.33.

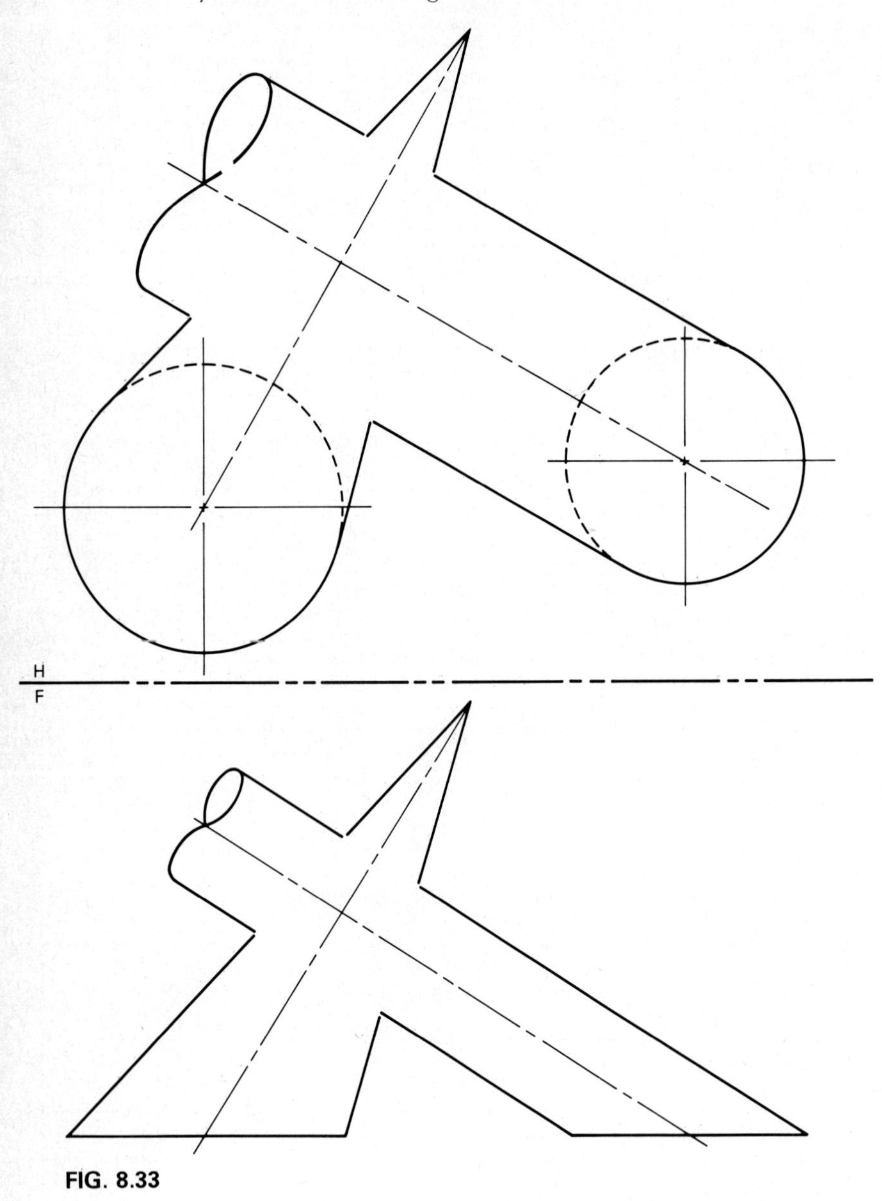

FIG. 8.33

PART THREE

GRAPHICAL APPLICATIONS

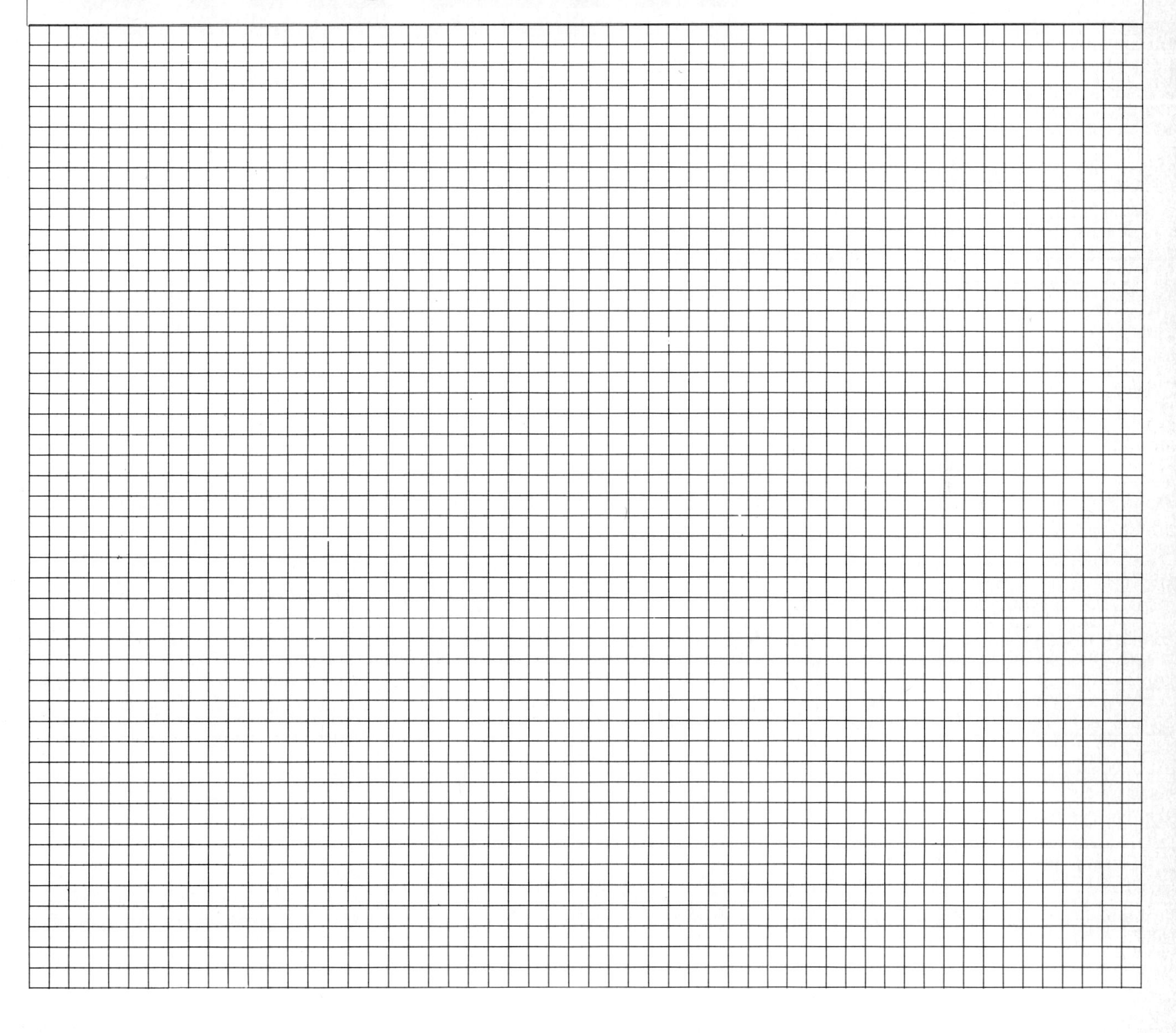

chapter 9

Solids

9.1 INTRODUCTION

As a practicing engineer, you have a responsibility to communicate graphically with a wide range of people. On the one hand, you will communicate with peers in your area of speciality who understand the graphic language. On the other hand, you will spend time explaining concepts and three-dimensional objects to people who understand little of this language. The purpose of this chapter and the other chapters in Part Three is to help you learn to communicate with those who understand the language. You will use the orthographic projection techniques discussed in Chap. 3 to produce multiview drawings that describe three-dimensional objects precisely. Your goal will be complete communication of the geometry of an object so that only one interpretation is possible.

To accomplish this goal, first you must learn standards and methods which have been developed and universally adopted in different areas of technology. These accepted standards and methods of producing complete drawings will enhance the understanding of your work by others.

Second, you must be able to interpret correctly and use work others have performed; that is, you must develop an ability to "read" drawings. Techniques for reading drawings will be discussed in this chapter. Practice in using these techniques gradually will improve your ability to read a drawing. For instance, there are many similarities between reading a drawing and reading a musical composition. Obviously, it takes considerable practice to read and interpret even the simplest representation of a musical piece, and the same is true for reading and interpreting drawings. Skill in reading drawings improves with practice, but the ability to interpret complex drawings correctly requires study. A superficial study probably will lead to errors and limit your capabilities in graphic communication.

This chapter concentrates on methods for defining three-dimensional geometry uniquely; Chap. 10 describes how pictorial representation enhances the viewer's understanding of objects. The internal details of a solid object can be described by means of the section views illustrated in Chap. 11. The size of an object is specified with the dimensioning techniques outlined in Chap. 12. Chapter 13 describes the numerous fasteners used to secure objects. Part Three concludes with Chap. 14, which discusses various types of drawings that must be produced to turn an idea in the mind of an engineer into a physical reality.

FIG. 9.1 A model of a proposed short-range passenger aircraft that exhibits complicated three-dimensional geometries which must be communicated between designer and fabricator.

9.2 LOCATION AND CONSTRUCTION OF VIEWS

Techniques from orthographic projection, which were presented in Chap. 3, are used here to produce multiview drawings of objects. Six views (top, front, right profile, left profile, rear, and bottom) are defined as principal views. These views, or as many of them as are needed, are always drawn in orthographic alignment and are located in specific positions relative to each other so that the drawing will be readily understandable.

Once the selection of the front view has been made (the criteria for selection will be discussed later in this chapter), the top view is placed directly above it in orthograhic alignment, the right profile view is placed immediately to the right of the front view, and so forth, as shown in Fig. 9.2. The alternative position of the right profile view is acceptable when better use of the available space on the paper can be made. The standard position is more common, but you will often see drawings that use the alternative approach. You will note that the orientation of the alternative right profile view is changed so that it is in correct orthograhic alignment.

You must always ensure that views are aligned orthographically. This is particularly important in reading and interpreting drawings. The space between views is sometimes chosen so that the drawing will fit the paper used; but more important, when dimensions are to be added to the drawings, the majority of them will be placed between views. If the purpose of the multiview drawing is to sketch the shape of an object only in order to aid verbal communication, little concern need be given to spacing views, although alignment must be maintained. But if the drawing is to be dimensioned, adequate space should be left to prevent crowding of that information.

The most common multiview arrangement consists of three views: top (horizontal), front, and right profile. The simple object in

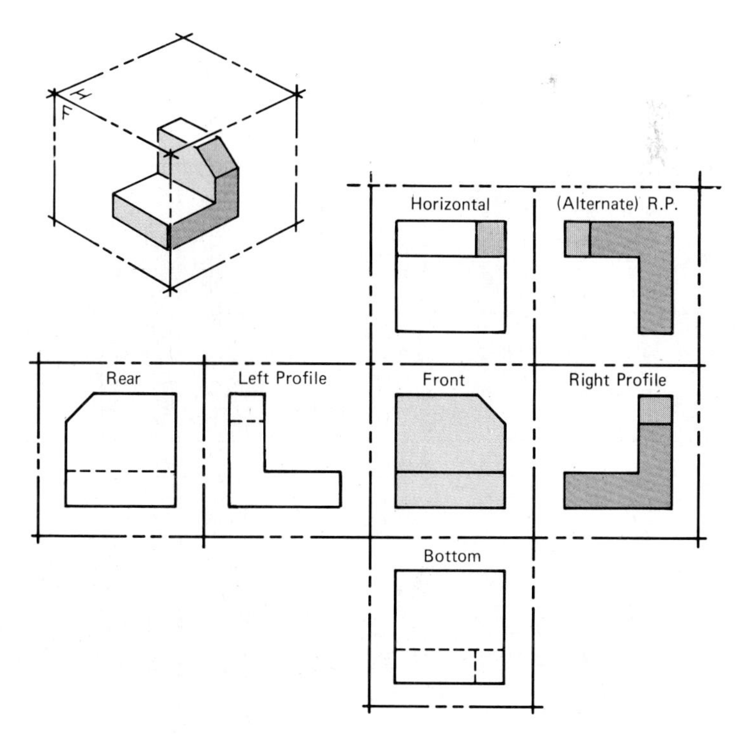

FIG. 9.2 The correct placement of the six principal views.

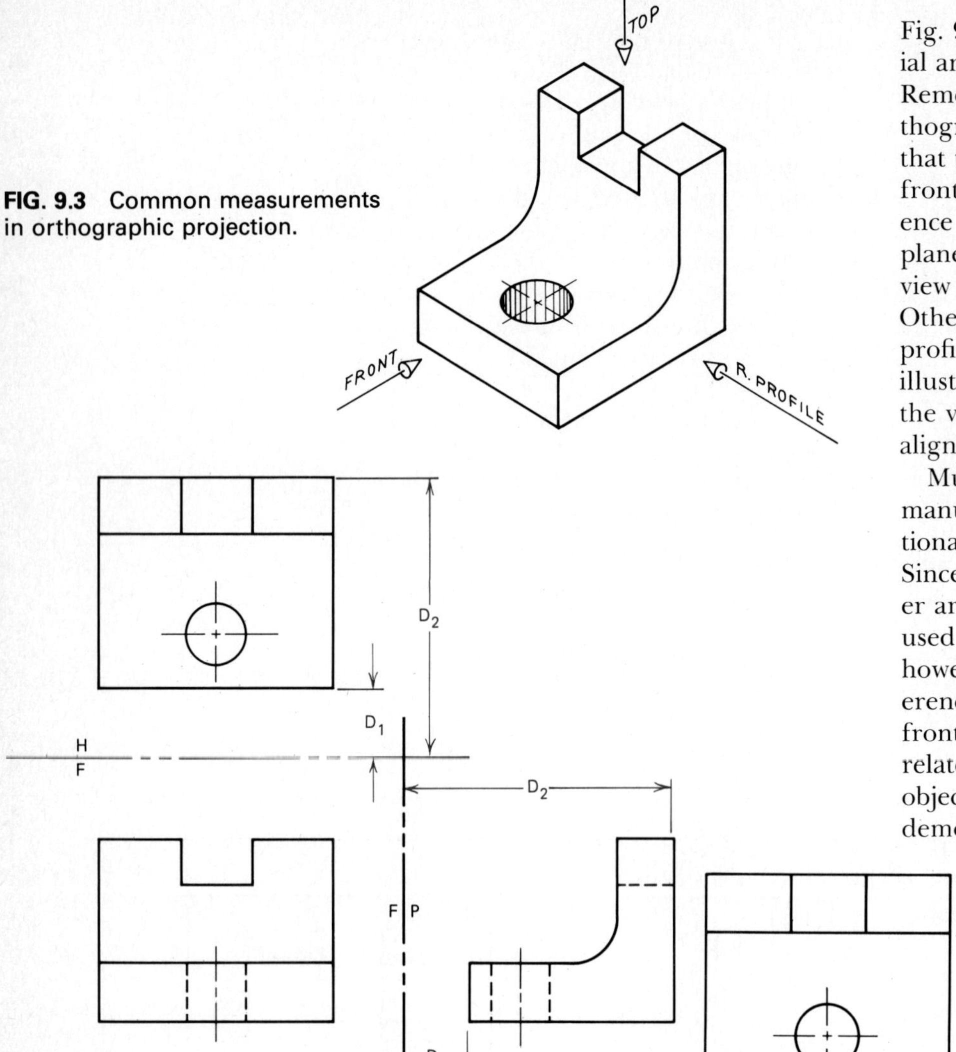

FIG. 9.3 Common measurements in orthographic projection.

Fig. 9.3 is shown both as a pictorial and as a multiview drawing. Remember that the theory of orthographic projection prescribes that the distance D_1 from the front of the object to the reference line (edge view of the front plane) will be seen in the top view and the right profile view. Other points in the top and right profile views are also related, as illustrated by D_2. Note again that the views are in orthographic alignment.

Multiview drawings made for manufacturing purposes conventionally omit the reference lines. Since the reference lines no longer are shown, they cannot be used to lay off distances. We can, however, choose some other reference that is present, such as the front surface of the object, and relate the remaining parts of the object to it. This technique is demonstrated in Fig. 9.4.

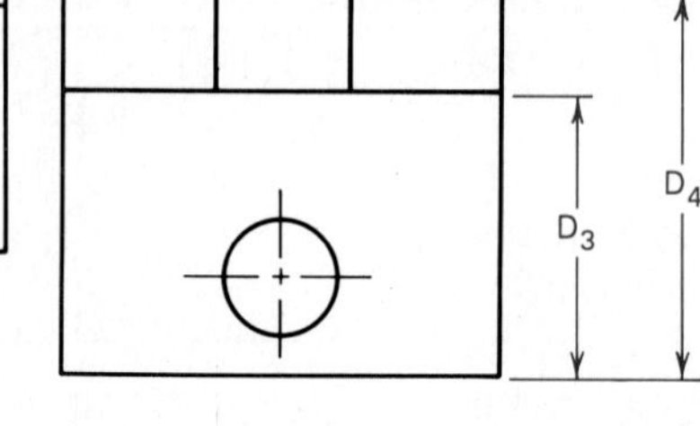

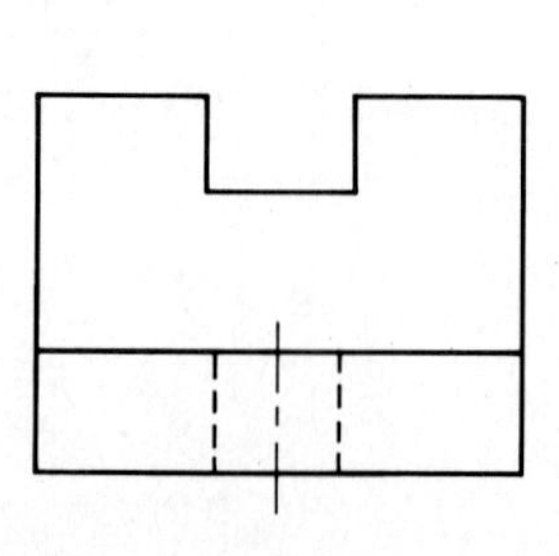

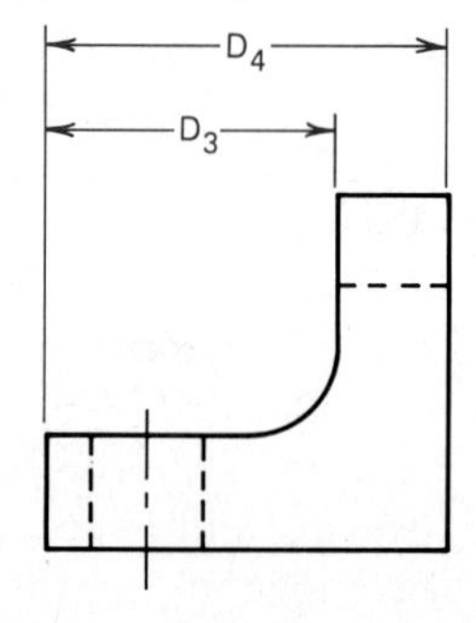

FIG. 9.4 Correct alignment of the top, front, and right profile of the object in Fig. 9.3, omitting the reference lines. The common measurements are taken from the object dimensions rather than the reference lines.

As a designer, you will be producing either multiview drawings or sketches from an image of the object that you have in your mind. Thus, you will be making drawings that will allow others to share your thoughts or actually build the object you have in mind. With practice, this new object can be rotated so that it can be viewed from any direction, all within your imagination. You then will be able to describe on paper what you "see" in your mind. To begin with, it may be easier to learn to draw objects that exist physically. When techniques for drawing multiviews have been mastered, you will be able to concentrate on the creative aspects of design.

Let us consider how we can take an object, look at it from various viewpoints, and create a multiview drawing that will describe its shape or form precisely. For this purpose, look again at the object shown in Fig. 9.3 in pictorial form. We have seen the multiview drawing of this object, but let us now go through the steps necessary to produce the drawings and then discuss these steps in some detail.

In Fig. 9.5, we note that the object will fit inside a rectangular box with dimensions of height, width, and depth. We can then

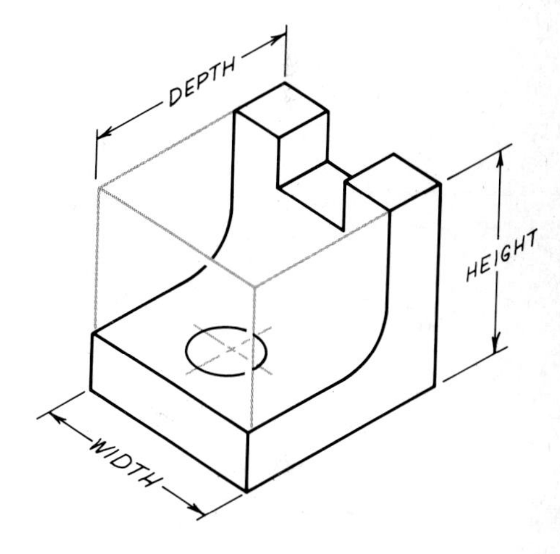

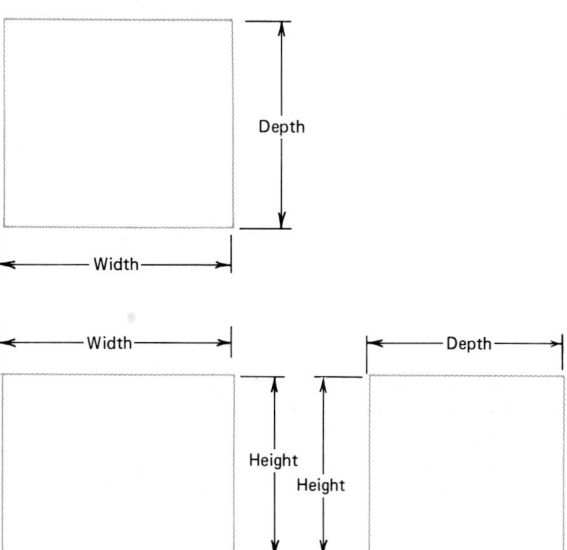

FIG. 9.5 To begin layout of the orthographic views, the overall height, width, and depth of the object are determined and shown as rectangles in the orthographic alignment.

construct the outlines of the box in each view as shown. We do this, of course, by looking at the object from the top, front, and side to obtain the necessary dimensions. The space left between the rectangular outlines depends on the space requirements needed later for dimensioning. Allow plenty of space to avoid crowding the dimensions. The space between the top and front views may be greater or smaller than that between the front and side views. The outlines are aligned vertically and horizontally according to the rules of orthographic projection.

With the position of each view defined, we can begin to locate features. The circular hole will show as a circle in the top view. We can fix its center by means of center lines that are located from the edges of the outline. Refer to Fig. 9.6. Note that the center lines are shown in all three views even though the circular contour will appear only in the top view. In the front and side views, we must imagine where the center of the hole is since we cannot actually see the hole from these viewing positions.

In Fig. 9.7, we add the remaining details. The top and right profile views show the depth of the vertical back of the object D in true length. The diameter of the hole E determines the size of the circle in the top view and the distance between the dashed lines in the other two views. The dimension C is seen in the front and right profile views. The radius R of the fillet is used to draw the arc in the right-side view.

The final feature to be added in Fig. 9.7 is the slot. Dimensions F and G define its depth and position. Although the slot cannot be seen in the profile view, a horizontal dashed line (hidden line) is placed to locate the bottom of the slot within the solid.

Let us review what we have done to create a multiview drawing of the object.

1. We have considered the overall size of the object viewed from the top, front, and right side and have placed rectangles in orthographic alignment in the proper positions.

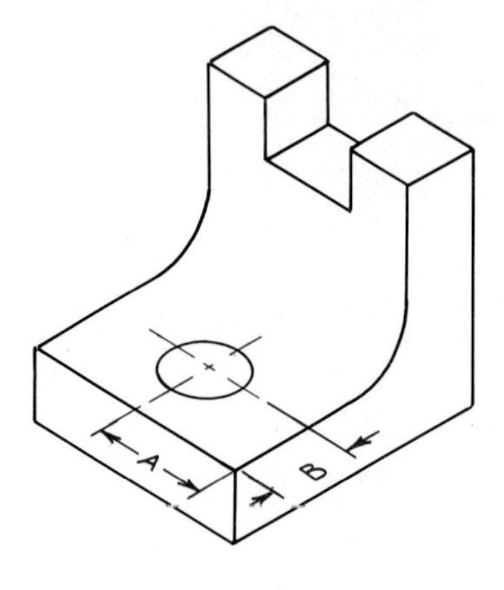

FIG. 9.6 Centers are located and centerlines are constructed on the orthographic views.

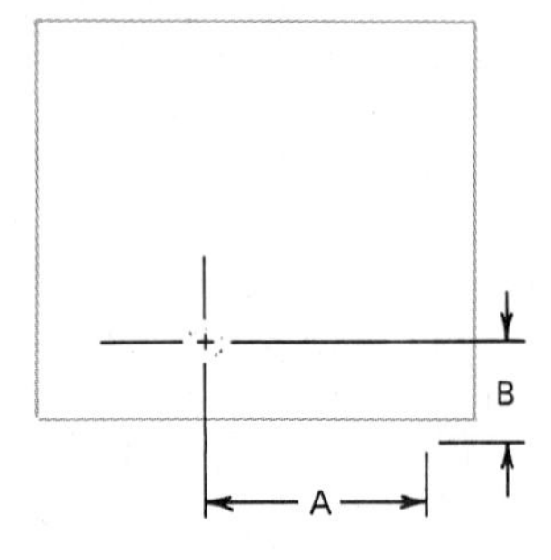

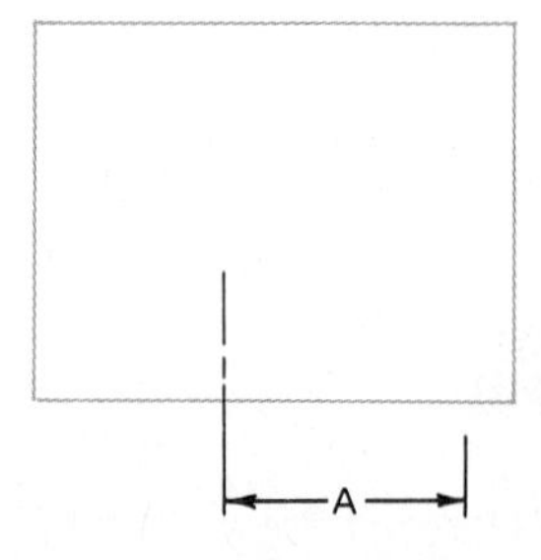

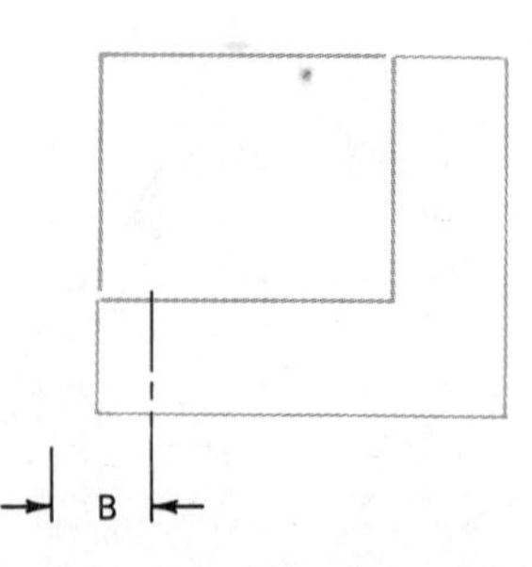

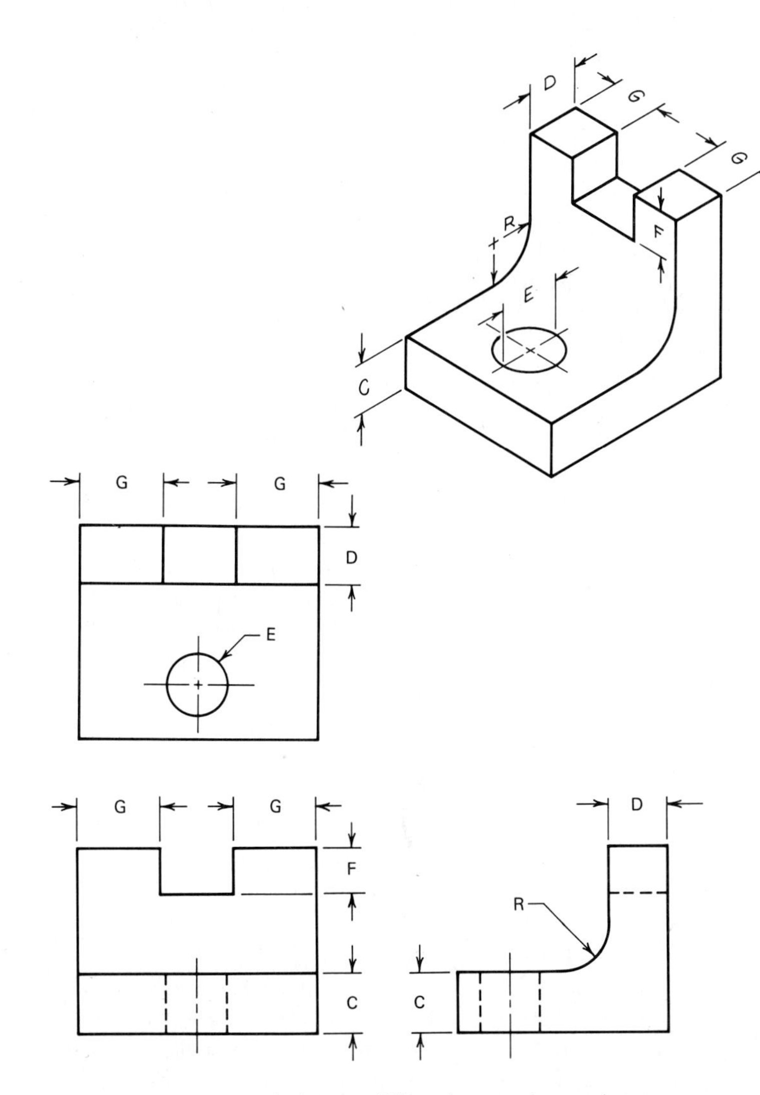

Note: These dimensions are for layout and illustration – not for manufacture.

FIG. 9.7 Features are added and construction lines are erased.

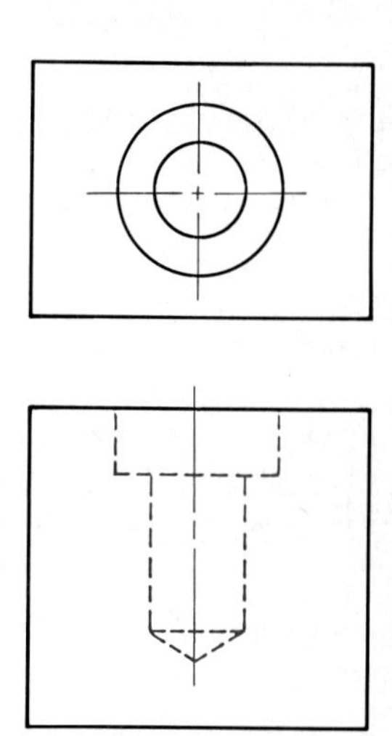

FIG. 9.8 Use of dashed lines to show the shape of a drilled and counterbored hole.

2. We have located center lines of circular features in each view.

3. We have added features one at a time in all views with appropriate solid or dashed lines.

Note that we worked on all views rather than completing one view entirely before moving to the next. As a final check, we should look at the physical object from each direction to be certain that all features have been shown in the proper view or views. Size description (dimensions) must still be added for a complete drawing, as will be discussed in Chap. 12.

9.3 LINE STANDARDS

The alphabet of lines was introduced in Chap. 2. Now we wish to examine some questions that may arise in using the alphabet of lines. What should be done when different types of lines coincide? If a dashed line meets another line, should the dashed line begin with a dash or a space? How should center lines be used for circular features in multiview drawings?

Dashed lines are used to represent surface edges or intersections that are not visible from the viewing position. The feature may be on the opposite side of the solid, such as a slot or a notch, or inside the solid, such as a hole. Figure 9.8 shows a counterbored hole which must be represented on the drawing with dashed lines in the front view.

With respect to dashed lines, we will consider methods of handling the following situations:

1. Intersection of a dashed line and a solid line

2. Extension of a solid line into a dashed line

3. Intersection of a dashed line with a dashed line

4. Closely spaced parallel dashed lines

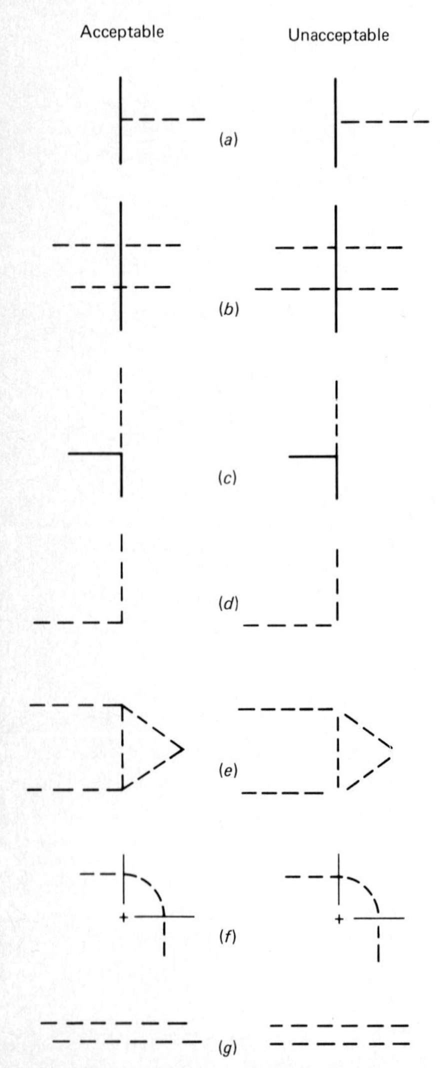

FIG. 9.9 Hidden-line conventions.

Figure 9.9 gives some guidelines to follow when dashed lines are used. When a dashed line terminates at a solid line (Fig. 9.9*a*) at 90° or at some other angle greater than zero, the dashed line should begin with a dash. If a dashed line continues beyond a solid line, the intersection treatment should be symmetric, with the space preferred to the dash for crossing the solid line, although either is acceptable (Fig. 9.9*b*).

In cases where a dashed line extends from a solid line (Fig. 9.9*c*), leave a space so that it is readily apparent where the solid line ends. Dashed lines may meet normally, as in Fig. 9.9*d*, or at some other angle, as occurs at several locations in Fig. 9.9*e*. The dashed lines should meet with solid segments at the junction, which clearly establishes the point of intersection.

The special case of a dashed line representing a circular feature is noted in Fig. 9.9*f*. At the point where the circular arc becomes tangent to the straight-line segment, the dash should meet the center line on the arc side. This leaves a space next to the center line on the straight-line segment.

Closely spaced parallel dashed lines, as shown in Fig. 9.9*g*, should have the spaces staggered slightly. The term closely spaced will apply to parallel lines spaced about 1 cm or less apart.

The actual sizes of the dashes and spaces must be adapted somewhat to the drawing size and the space available for showing a particular feature. Reasonable values are 3-mm dashes and spaces slightly smaller than 1 mm. All the acceptable methods we have discussed here may be violated occasionally in the interest of clarity.

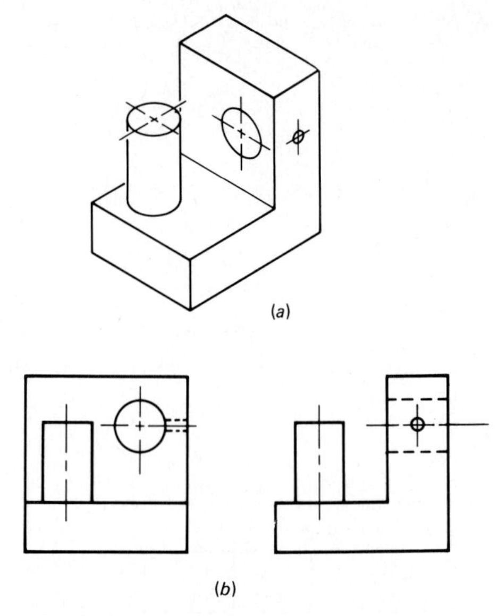

FIG. 9.10 Centerlines are drawn in all views of features with circular characteristics.

Center lines in multiview drawings are used to show the position of the theoretical center of circular features of an object. Center lines are shown in both the circular (contour) view and the rectangular view of cylinders and holes (sometimes called negative cylinders). In pictorials, center lines may be used to indicate that features are actually circular even though they do not appear so. See Fig. 9.10*a*.

Center lines consist of alternating long dashes and short dashes separated by small spaces. The long dashes may extend up to 40 mm long, and the short dashes should be about 3 mm long. The spaces will normally be less than 1 mm long, again subject to drawing size.

In the circular view, the short dashes of two perpendicular center lines should form a small cross as they intersect at the center of the feature. If the circle is so small that spaces are impractical, the center lines may be drawn solid, as was done for the small hole in Fig. 9.10. Note that the center lines extend beyond the circular view by approximately 3 mm.

The rectangular view of a cylinder or a hole should also contain center lines to help identify the feature as circular. Here too the center line extends about 3 mm beyond the ends. Center lines should be drawn so that they appear thinner than the object lines in the view. They should, however, remain crisp and easily visible.

When different types of lines are superimposed in a view, the following order of precedence should be used:

1. Object lines
2. Dashed lines
3. Center lines

Thus, an object (solid) line is drawn if it is superimposed over either a dashed line or a center line. Dashed lines are shown if they coincide with center lines. Superposition of different line types occurs at three places, marked as *A*, *B*, and *C* in Fig. 9.11. At *A* in the top view, the center line of the semicircular top of the object and the dashed line representing the edge of the hole coincide. The guidelines state

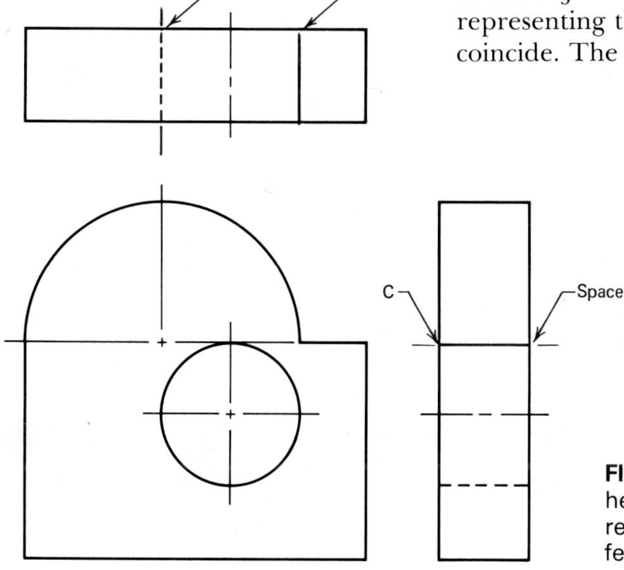

FIG. 9.11 Line precedence helps to establish unique representation of certain features.

that the dashed line should be drawn. You will note, however, that because the center lines extend a few millimeters beyond the feature outline, they are still seen in the top view. A small space must be left between the object line and the center line extension in this case.

An object line and a dashed line occupy the same position at *B*. The solid line has been drawn in preference to the dashed line. Position *C* in the right profile view shows three types of lines superimposed. You can see that the center line of the semicircular top, a dashed line representing the limit of the hole, and an object line resulting from the intersection of surfaces all occupy the same space. The guidelines for line precedence dictate that we draw the solid line, use the center line extensions, and not show the dashed line.

9.4 SELECTION OF VIEWS

The ultimate purpose of a multiview drawing is to provide accurate and complete communication. We will offer some suggestions that you can use in deciding how many and which views to draw. Keep in mind that the final purpose is communication of your work.

The number of views that are necessary to define different objects is certainly not fixed. For example, a standard circular dowel 2 cm in diameter and 1 m long requires no views at all; a verbal description is adequate and represents a much faster and cheaper way of specifying the dowel than does a drawing. On the other hand, the engine for an Indianapolis 500 race car will require hundreds of principal views plus numerous auxiliary views to describe features that do not lie in the principal planes.

Flat objects such as the automotive thermostat gasket shown in Fig. 9.12 can be described by a single view together with a note that specifies the constant thickness of the part. Here only one view contains significant contour. When the part is dimensioned, the hole dimensions can be used to make it clear that the features are holes, not protrusions.

Long cylindrical objects that contain relatively few details can often be represented by two views. The form of the shoulder bolt with an internal lubrication passage shown in Fig. 9.13 can be understood from only two views.

In most applications, three appropriately chosen views will be sufficient to describe objects with contours in the three dimensions of height, width, and depth. You normally would use the top, front, and right profile views. You can help ensure that these

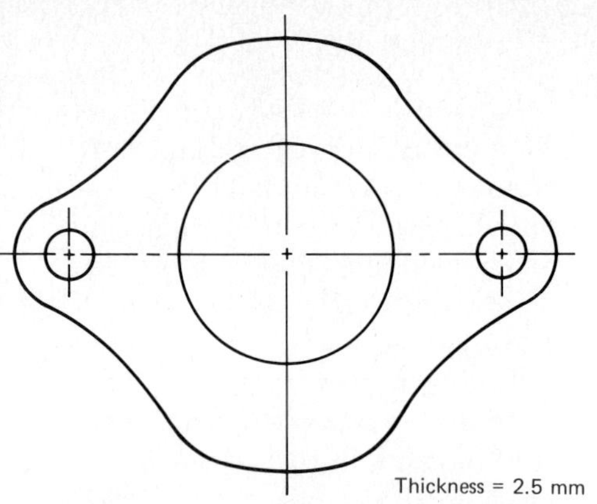

FIG. 9.12 Very thin, flat objects can be uniquely described with a single view plus a note for thickness.

three views are appropriate by selecting the side of the object with the most contour to be the front view. Place the object so that this side is parallel to the front plane, with the remaining sides parallel to the other principal planes. If you orient the object so that the most irregular sides are on the top and right sides, these contours will appear in the top and right profile views. This procedure may place the object so that the front view is not what is normally thought of as the front. Consider, for example, the aircraft depicted in Fig. 9.14. The nose (front) of the aircraft should

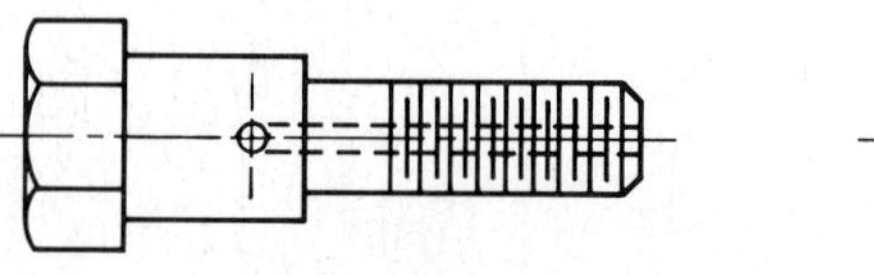

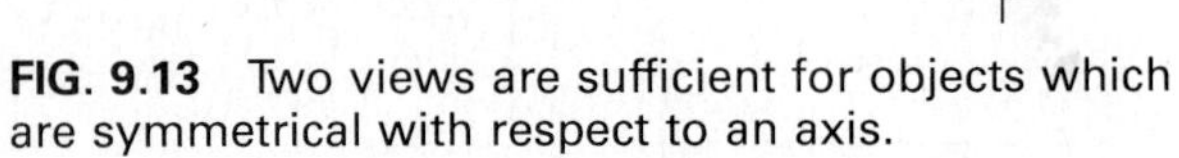

FIG. 9.13 Two views are sufficient for objects which are symmetrical with respect to an axis.

be shown in the right profle view, and you should choose as the front view the side of the vehicle which shows the characteristic contours. Note that dashed lines have been omitted in the interest of clarity. Including all the dashed lines would create a totally unintelligible drawing for an object of this complexity.

Some objects can be described equally well by a left-side view or a right-side view or by the bottom rather than the top view. In such situations (all other considerations being equal), follow these guidelines:

1. Choose the right profile over the left profile.

2. Choose the top over the bottom.

3. Choose the view which minimizes the number of dashed lines.

Much of the discussion in this section really deals with the concept of necessary or essential views. Any object has a minimum number of views which must be drawn (necessary views) to describe it completely and accurately. When deciding which views are necessary and which are not, you must not allow the reader any freedom of assumption. The object must be understood in one way and one way only. We sometimes have a tendency to take it for granted that the reader will assume that a feature has rectangular contour in a view we haven't drawn, when in fact some other interpretation is also possible. In multiview drawing we must avoid the tendency, ex-

FIG. 9.14 The front view is chosen to represent the most or largest contours even though this view may not be commonly called the front. An automobile would be another example. The front of the automobile is not the best choice for the orthographic front view.

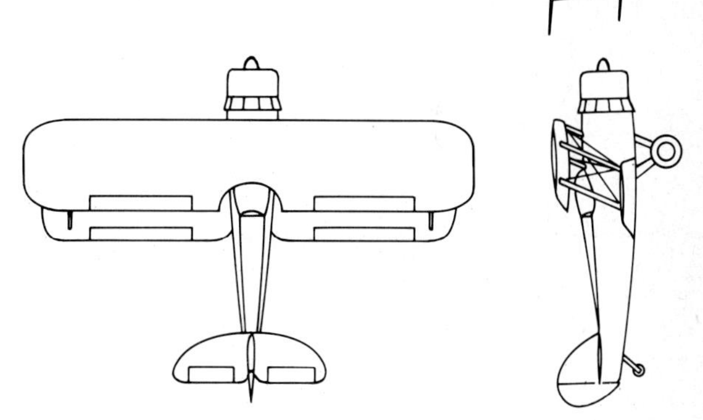

FIG. 9.15 Which view of this computer graphics terminal would you choose for the front orthographic view?

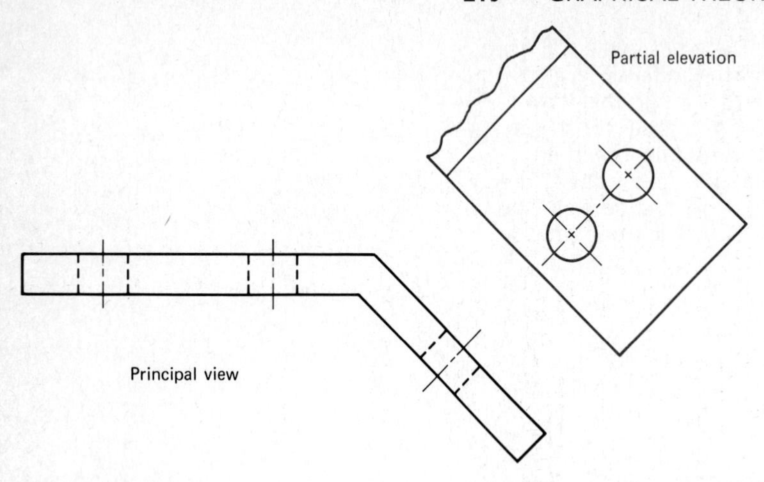

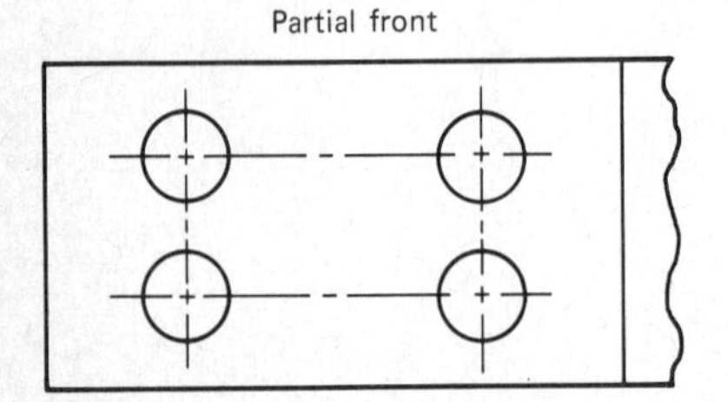

FIG. 9.16 Partial views can be used to advantage for illustrating features.

amine the final drawings with an eye toward any possible misinterpretation, and correct the drawings so that no ambiguity exists. If an additional view is needed, we must provide it. Inclusion of an additional view or partial view that is technically not essential but significantly improves the readability of the work can be a good investment of time.

Parts that contain important contours on nonprincipal planes as well as on the principal planes create a special problem of description. An acceptable alternative to a complete view is a partial principal or auxiliary view that shows a contour in true shape.

You may or may not be able to eliminate one of the principal views; the choice depends on the object. The angle bracket drawn in Fig. 9.16 can be adequately described by one principal view and two partial views.

9.5 CONVENTIONAL PRACTICES RELATING TO MULTIVIEW DRAWINGS

We have emphasized the importance of maintaining correct orthographic alignment for multiview drawings. The purpose is to maintain a universal language of graphics so that each reader of a given drawing will interpret the graphics in exactly the same way. There are instances, however, where the true orthographic projection may lead to incorrect interpretation. To avoid these instances, certain violations of orthographic rules are allowed. The exceptions, which are called conventional practices, have been adopted universally. Some will be introduced here, and others will be introduced in Chap. 11.

The conventional practice illustrated in Fig. 9.17 applies to radially arranged features. The right profile view B shows the three small holes 1, 2, 3 of the coupler in correct orthograhic projection. If you study the hidden lines in view B, you may conclude that holes 1 and 2 are closer to the axis of the coupler than hole 3 is, when in fact all three holes are

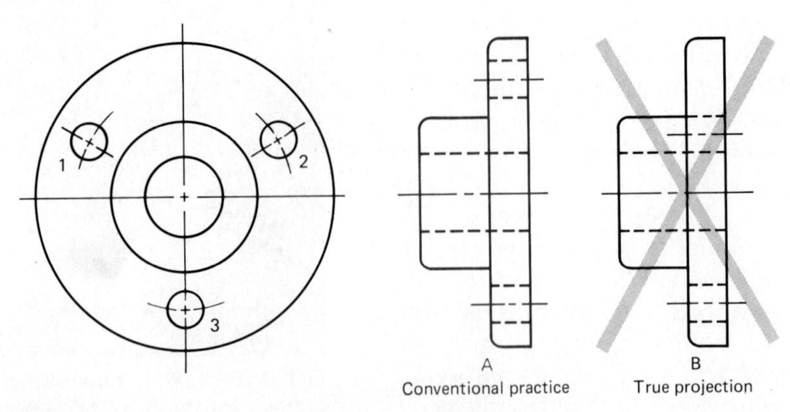

FIG. 9.17 Conventional practice for radial features.

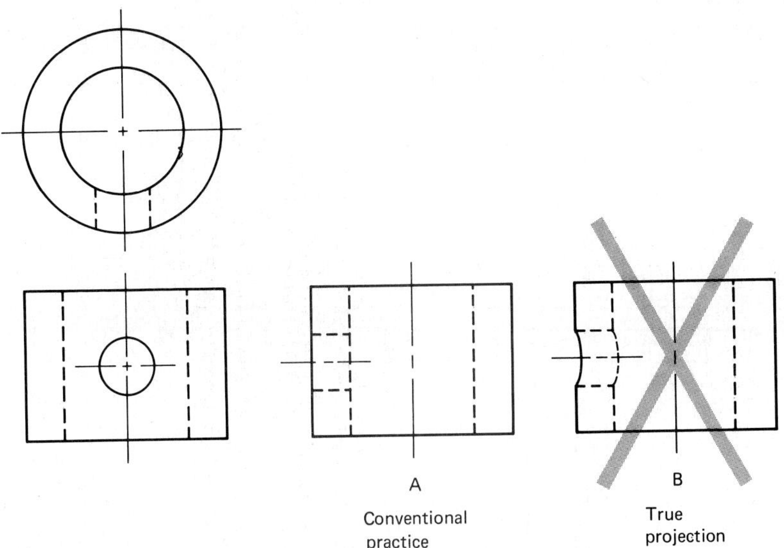

FIG. 9.18 A simplified drawing of a complex intersection.

the same distance from the axis. To avoid this potential error in interpretation, the conventional practice calls for the profile view A to be drawn. View A, although not in orthographic alignment with the front view, does provide a more realistic representation of the location of the three small holes. You will note that the correct positions for the holes can be ascertained from the front view.

A second conventional practice is illustrated in Fig. 9.18. When a hole is drilled perpendicular to the axis of a cylinder, a complex intersection is created. If the true projection is desired, as shown in view B, the line of intersection must be found and drawn. Conventional practice allows the true intersection to be disregarded in favor of showing a simplified arrangement as long as the true intersection is not important to the function of the object. If the intersection is critical, it must be drawn in true projection. This conventional practice applies both to holes (negative cylinders), as illustrated in Fig. 9.18, and to protrusions (positive cylinders or prisms) which are appended to the surface of a cylinder. You may use the conventional practice in either case.

In Fig. 9.17, we imagined the holes 1 and 2 to be revolved in the front view to a position where the profile projection would realistically represent the radial distance of the holes from the axis. This conventional practice also applies to ribs, spokes, tabs, key-

Conventional practice

True projection

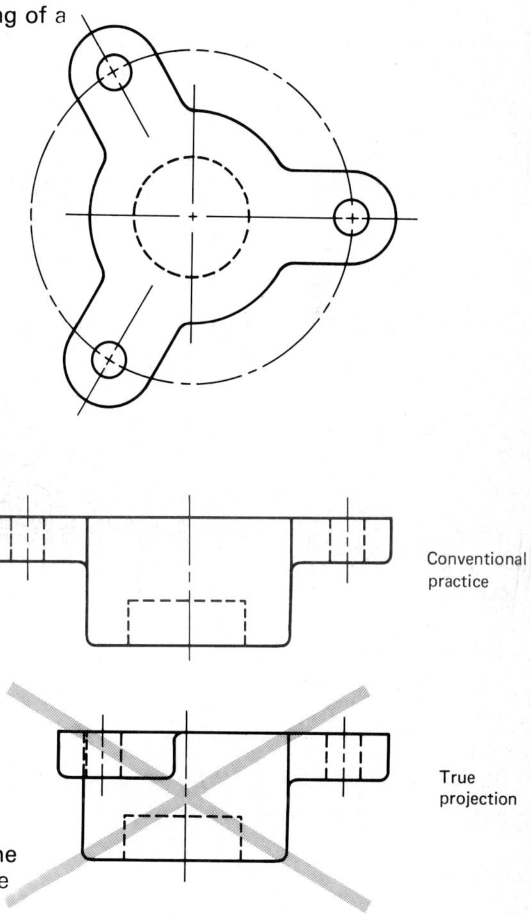

FIG. 9.19 The symmetry of the object is better shown with the conventional practice.

ways, and so on. In addition to providing a more realistic view, this practice reduces the difficulty of drawing compared with the true projection. The two front views shown in Fig. 9.19 contrast the true projection of the tabs and the conventional practice. In fact, the true projection gives the impression of a nonsymmetric object, which is not the case.

Figure 9.20 illustrates the conventional practice of representing ribs. Note again that conventional practice violates true orthographic projection but presents a clearer view of the object.

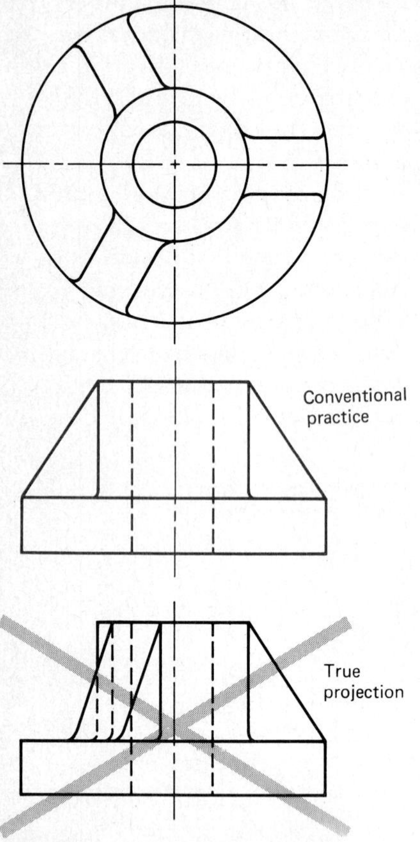

FIG. 9.20 Conventional practice for ribs.

9.6 READING A DRAWING

In elementary school you probably learned to read by associating words with objects. Later you

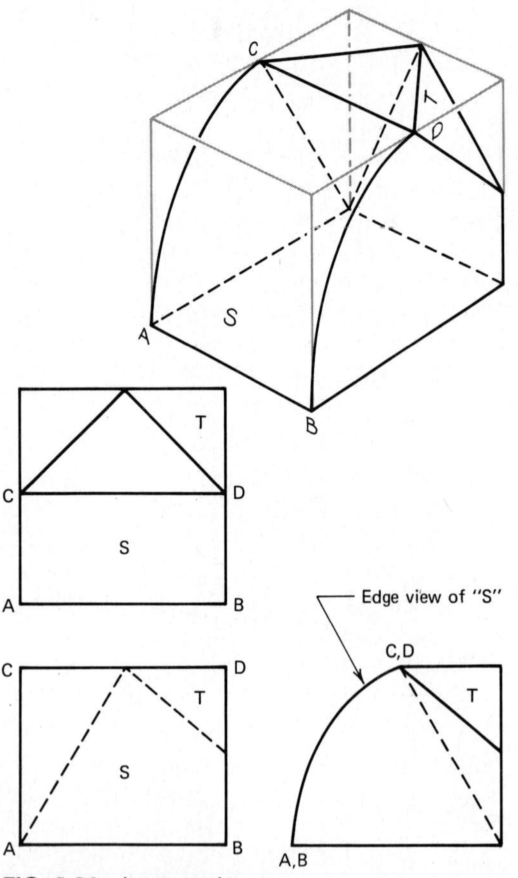

FIG. 9.21 Interpreting an orthographic drawing.

found that some words do not describe objects, and so you had to expand your learning methods. The dictionary could be relied on if you could not recall the meaning of a familiar word or if you encountered a new word. Even then, you found that many words had several meanings and that the correct meaning could be discovered only from the context in which a word was used.

Reading a drawing is much like dealing with a word that has several possible meanings. You can determine the correct meaning of a drawing only by studying all available views presented in that drawing. You interpret the information on a drawing by using your "graphics" dictionary: the sum of all you have learned about the principles and practices of drawing. We will add to your dictionary by providing some tips on what a feature on a drawing can represent. You must make the final determination by accounting for all information contained in a drawing.

Any line on a drawing, solid or dashed, may represent the limiting element of a surface, the edge view of a surface, or the nontangent intersection of two surfaces. A line may appear in true length, foreshortened, or as a point. You can determine which of these conditions exists by studying an adjacent orthographic view. In Fig. 9.21, line *AB* is the limiting element of the curved surface *ABDC,* as seen in the horizontal view. Line *AB* is also the line of intersection between the curved surface and the bottom plane of the object as seen in the profile view.

To understand the concept of a limiting element better, find a cylindrical object such as a pen or pencil and view the cylinder as a rectangle. Two of the lines you see in the rectangular view represent the extremities of the cylinder surface but do not represent a line of intersection, since the cylinder surface is continuous. Thus, the orthographic view of a cylinder shows the limiting elements of the surfaces as well as the intersection lines between the cylinder surface and the circular ends.

Line *CD* represents the nontangent intersection of the curved surface with the top planar surface. This intersection is seen as a point in the profile view. The presence of line *CD* tells us that the curved surface is not tangent to the top planar surface; if the two surfaces were tangent, no line would exist. Lines *AC* and *BD* represent the edge view of the curved surface which is seen in the profile view.

Another observation which will help you interpret drawings is that a plane surface will have the same basic shape in all views where it does not appear as an edge. Thus, rectangular, triangular, T-shaped, and U-shaped areas will maintain these shapes in any view which does not show an edge of the shape. In addition, curved surfaces, such as *ABDC* in Fig. 9.21, will appear in the same shape in any view which does not show the edge of the surface.

Surface *T* in Fig. 9.21 appears as a triangle in all three views because it is not perpendicular to any principal planes and therefore does not appear as an edge. Surface *T* is hidden in the front view, but the dashed line still allows recognition of the triangular shape.

Curved surface *ABDC* is labeled as area *S* and is seen as a rectangle in the horizontal and front views and as an edge in the profile view. A keen imagination is necessary in interpreting drawings, but it must be used cautiously. For example, the top and front views of *S* may lead you to conclude that it is a sloped plane surface, whereas further study of the profile view will reveal its curved nature. It is important to go from view to view and eliminate possible interpretations you have imagined until you have the correct interpretation.

For simple objects you can use the technique of constructing a small pictorial sketch to help interpret the drawing. You start with a basic prism shape and add or subtract features such as cylinders, cones, and prisms as they are recognized from the orthograhic views. Once the sketch is complete, a final check can be performed to convince yourself that all lines and surfaces on the drawing have been interpreted properly.

To illustrate further the need to interpret lines and surfaces on orthograhic drawings carefully,

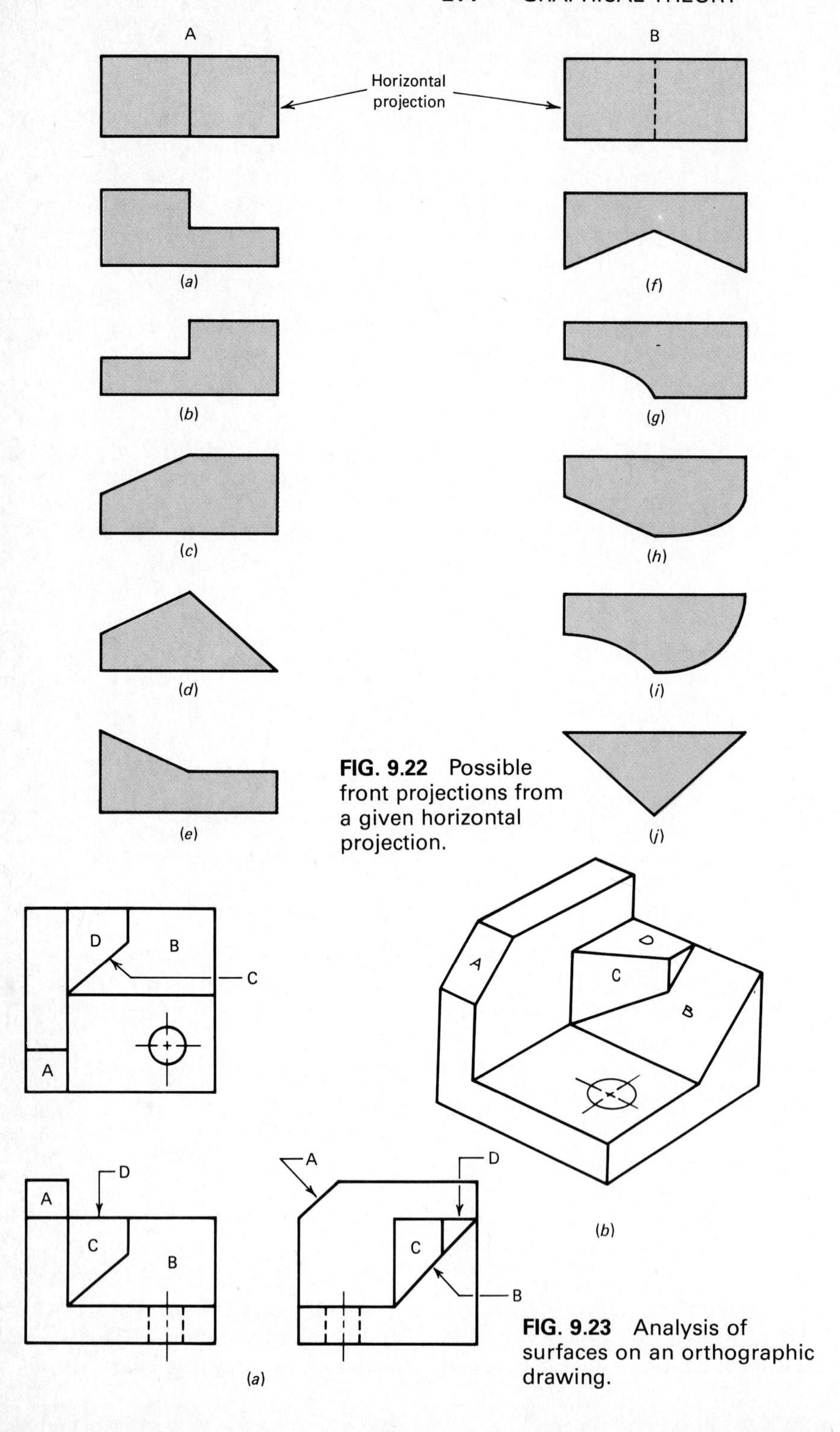

FIG. 9.22 Possible front projections from a given horizontal projection.

FIG. 9.23 Analysis of surfaces on an orthographic drawing.

look at the horizontal views A and B shown in Fig. 9.22. Then look at the possible front views that are drawn from the horizontal view. You probably can add several other consistent front views. This variety of things a single line can represent on a drawing means that you must carefully examine each feature and be sure you have the correct interpretation. If you can find more than one correct interpretation, an additional view or a note on the drawing is needed.

Let us analyze the object in Fig. 9.23*a*, using the concepts we have discussed. The first observation might be that the object is basically a rectangular prism, with the only curved portion being the circular hole noted in all views. The front view shows an L-shaped component, which information in the top view confirms. Plane surface *A* is seen in the top and front views as a foreshortened projection; in the profile view it is seen as an edge.

The same characteristic shape identified as *B* in the top and front views suggests that these are two views of the same surface. No similar shape appears in the side view; therefore, the edge view of the surface is found there. The conclusion is that area *B* is a slanted plane surface with a portion either cut away or added on.

The front and side view of plane *C* can be readily seen. There is an area in the top view with a similar shape, but the process of elimination shows that surface *C* must be vertical; thus, it shows as an edge in the top view. The side view confirms that there is an add-on to surface *C*.

All of this analysis results in the pictorial shown in Fig. 9.23*b*. A mental image of the top, front, and side confirms that the multiview drawing in Fig. 9.23*a* does represent this object.

9.7 APPLICATINS FROM COMPUTER GRAPHICS

Multiview drawings are primarily made up of straight lines and circular arcs which can be readily drawn on a computer graphics terminal. Typical graphics programs allow you to draw a line between any two points on the screen. The line may often be specified as solid or dashed at any angle. In very simple graphics systems a dashed line may have to be generated by a series of short, solid lines separated by spaces. Circles are made up of a sufficiently large number of straight-line segments which the eye will interpret as a smooth curve. In simple systems the segments and their locations must be defined. More sophisticated packages allow you to specify the radius and center position of a circle; then the required circle can be drawn automatically. A circular arc can be created by limiting the included angle of the arc to a value lower than 360°. Specific routines are often available to draw center lines or reference lines and to control the line weight (thickness). Characters (al-

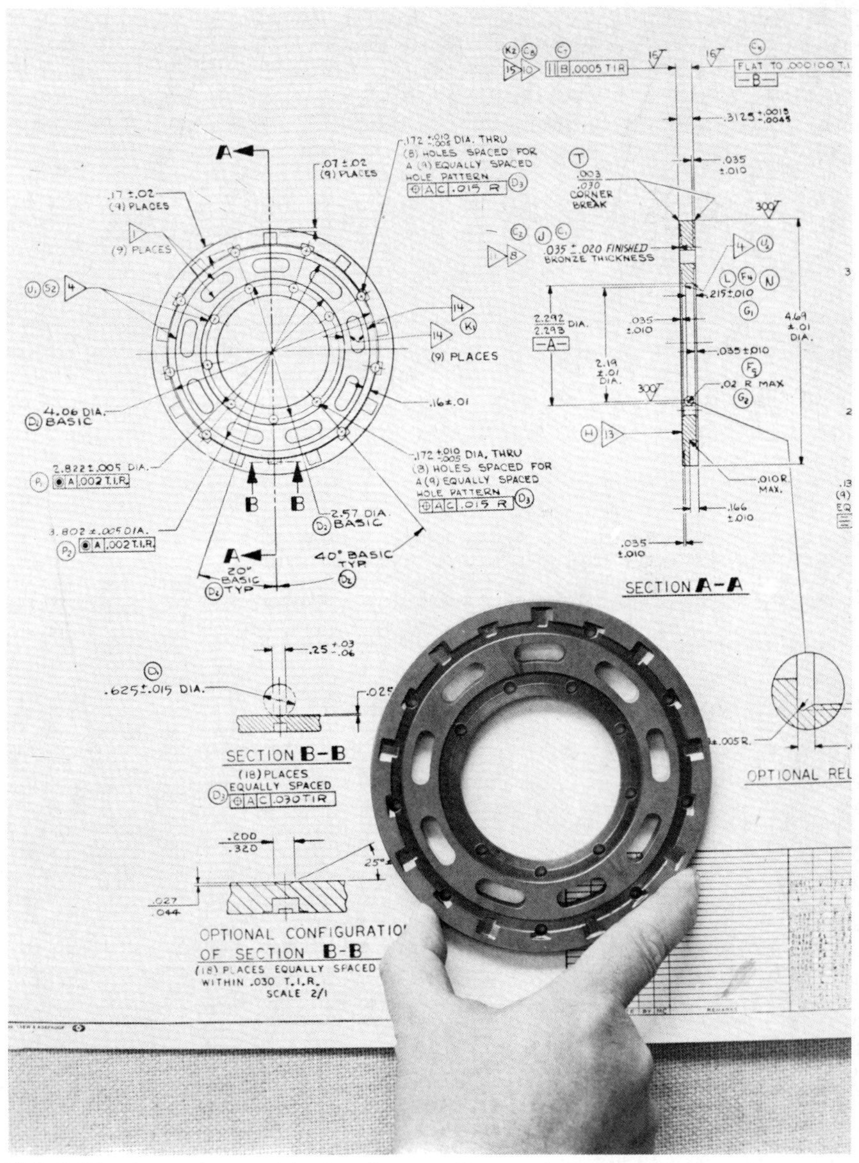

FIG. 9.24 An orthographic drawing used to produce the actual part.

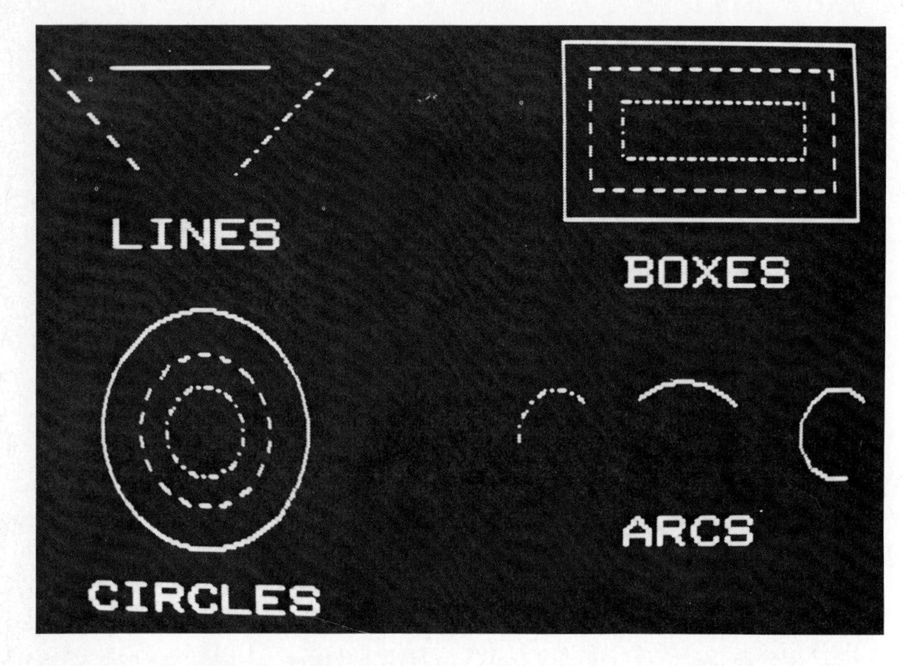

FIG. 9.25 Representation of line characteristics on a computer graphics terminal. Hidden lines can be represented; but unless special considerations are made in the software, the line conventions shown in Fig. 9.9 will not necessarily be followed.

FIG. 9.26 An engineer using a digitizing pen to locate features quickly on a computer graphics screen. *(Courtesy of Control Data Corporation.)*

phabetic and numeric) of varying types are generated to your size and location specifications.

The most advanced graphics systems allow you through the use of a light pen or a joy stick and control keys essentially to point to the spot where you want a feature. Then, if you describe it as a circle, rectangle, etc., with specified size, the computer will respond by placing the feature on the screen. If you have erred on the placement, you can "drag" the feature across the screen to the proper location. You also can change size or rotate a feature to adjust it to your design concept. Once you are satisfied with the drawing, you can obtain a copy on paper by using a peripheral plotting device.

A significant advantage of such a system for doing industrial drawings is that the drawings are stored in some computer medium (tape, disk, and so on). The information can be quickly returned to the screen, where necessary changes (redesign, perhaps) can be done and a new drawing produced without changing the majority of the former drawing information. If a company wishes, all segments of the company, from design to purchasing to manufacturing, will have access to the exact same drawing of each part. This is particularly advantageous when engineering changes become necessary or when time delays in drafting and distribution of new drawings causes unnecessary expense and waste.

With all the advantages of computer-aided design and drawing, there are still cases where handwork is essential. For example, simple one-time design drawings can be done more easily and cheaply directly on paper. Small companies may be unable to afford the cost of a computer-grahics system, although the cost is dropping to a level that is within the reach of more and more companies.

Computer graphics can be very helpful, but it does not replace your knowledge of orthographic projection or conventional practice for drawings. It simply provides you with an excellent analytic and drafting tool which expands the amount and complexity of work you are capable of performing. The creative component of design must still be provided by the human being at the controls.

FIG. 9.27 A step-by-step process for developing the database for an object. Although shown here in pictorial, the database can be manipulated to present orthographic views of the object. *(Courtesy of Tektronix.)*

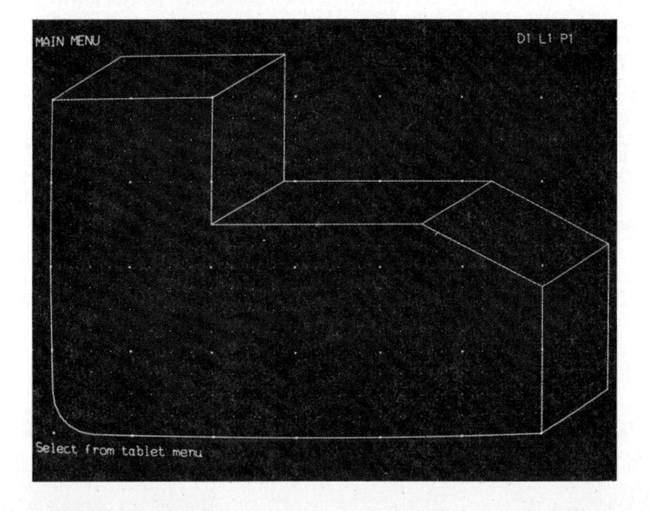

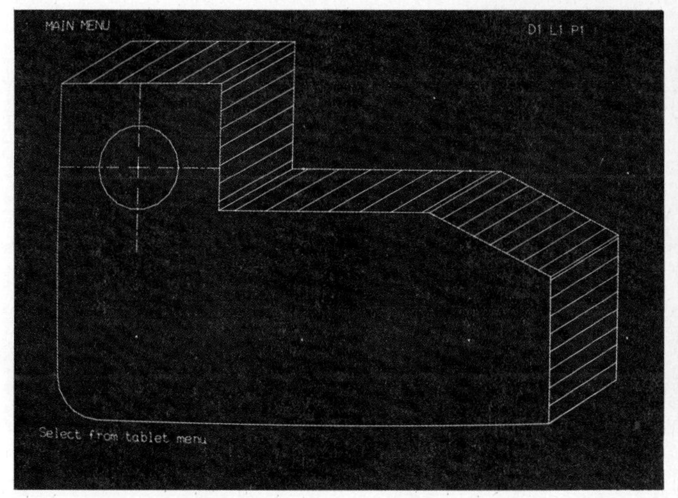

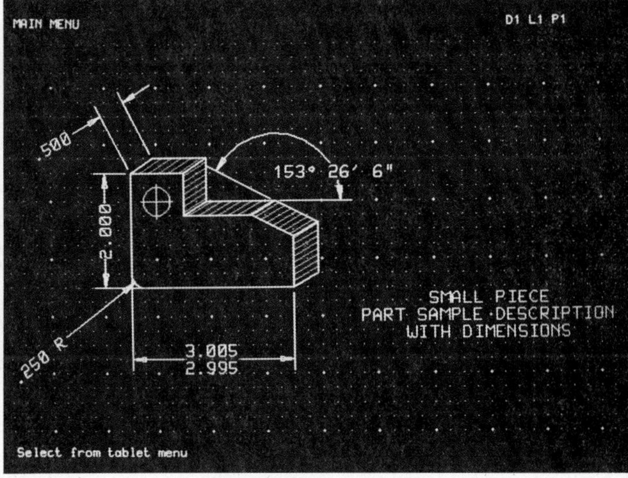

Problems

9.1 Draw or sketch the orthographic views shown in Fig. 9.28*b* and number each visible surface (not the edge view of planes) to correspond to the numbers shown on the surfaces in the pictorial in Fig. 9.28*a*.

9.2 For Fig. 9.29, follow the procedures specified in Problem 9.1.

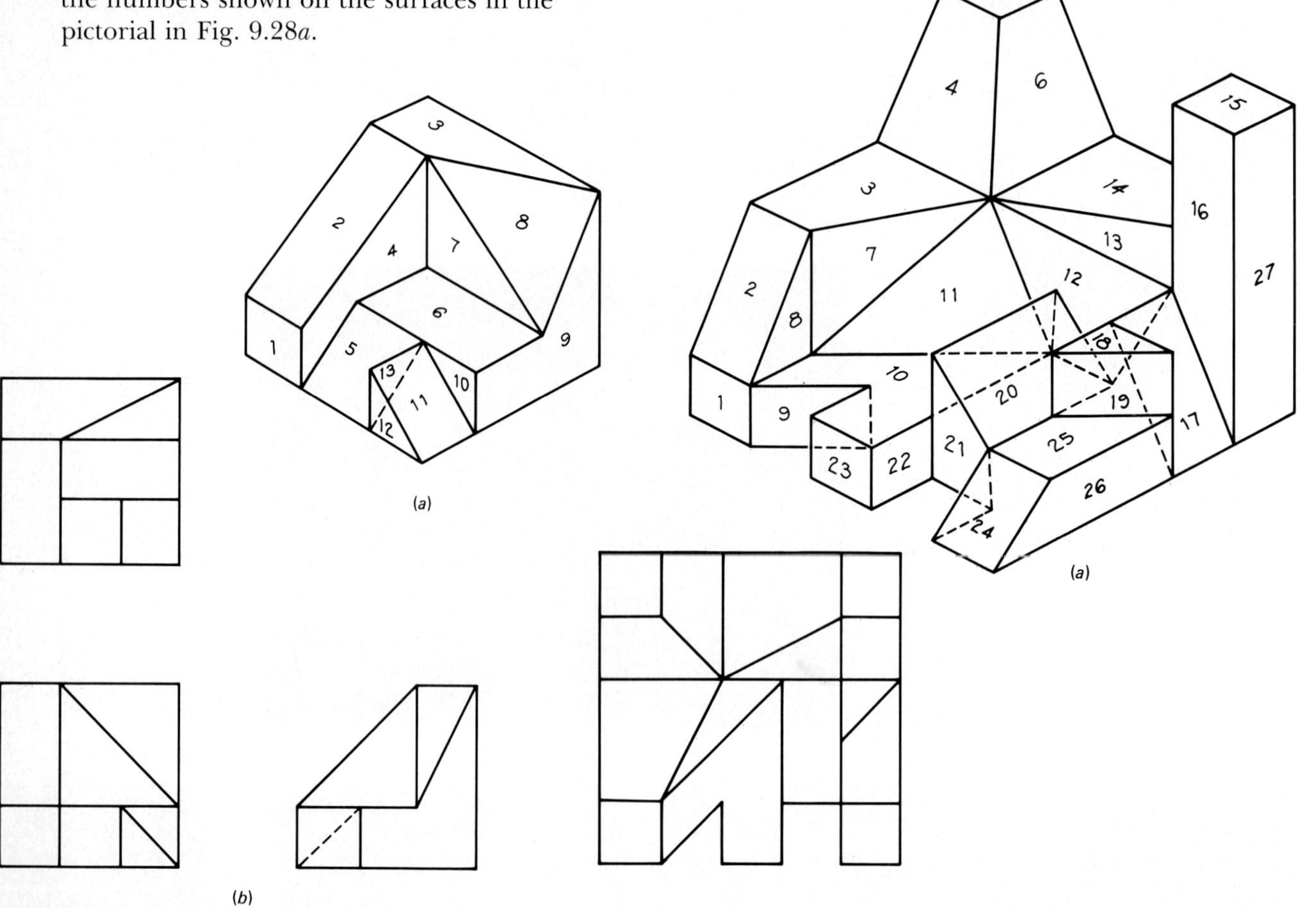

(b)

FIG. 9.28

(b)

FIG. 9.29

9.3 Sketch or draw the correct orthographic views for the objects assigned from Fig. 9.30.

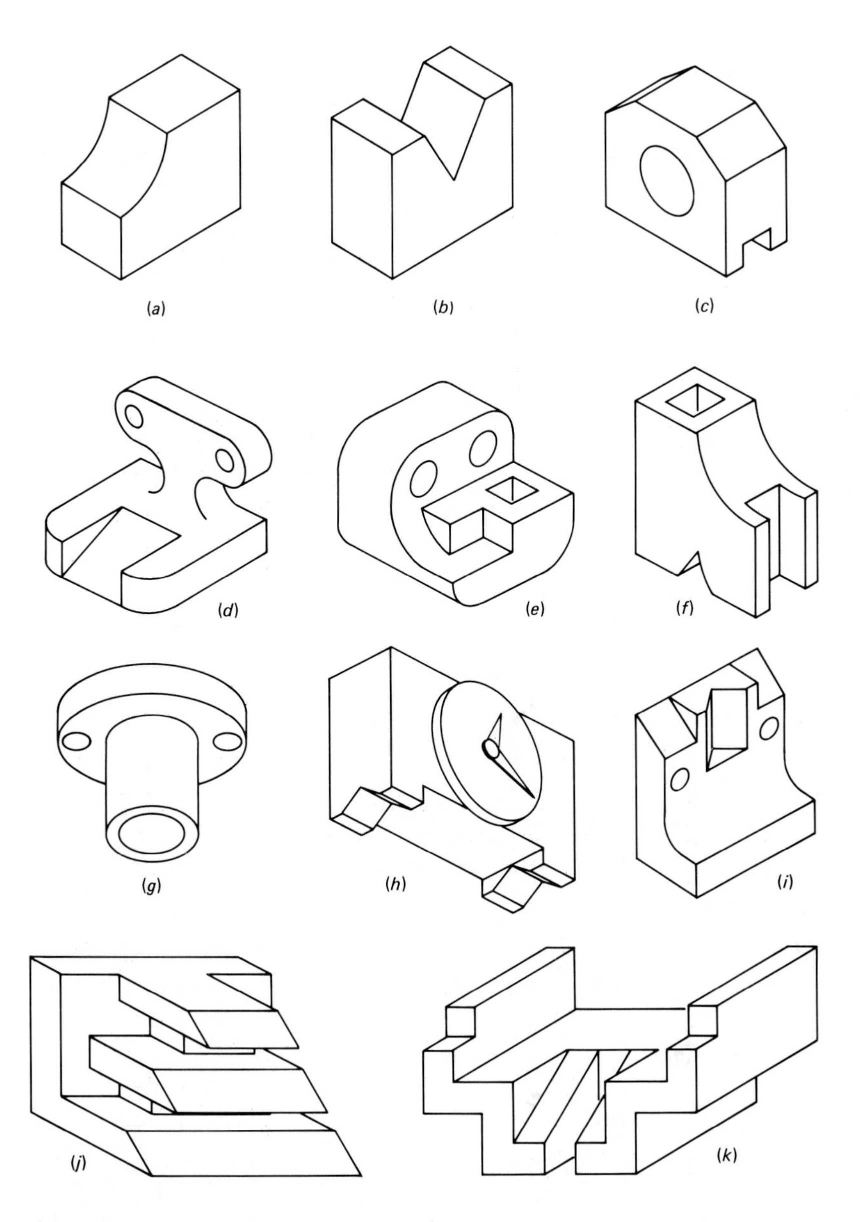

FIG. 9.30

9.4 Sketch or draw the orthographic views shown in Fig. 9.31. Complete the missing views and sketch a small pictorial to verify the orthograhic interpretation in each case.

9.5 Sketch or draw related orthographic views for the object shown in Fig. 9.32. Arrange the views as suggested by the arrows shown.

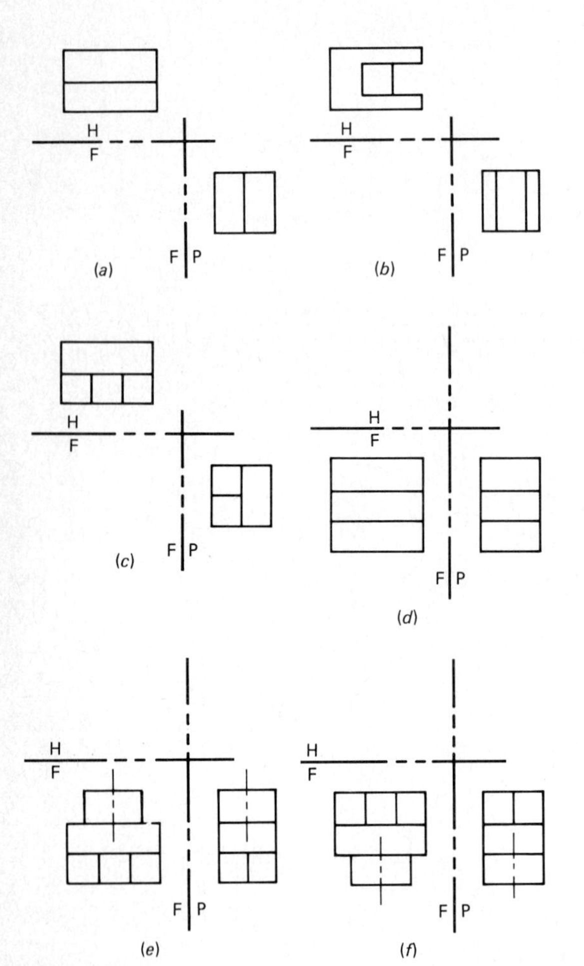

FIG. 9.31

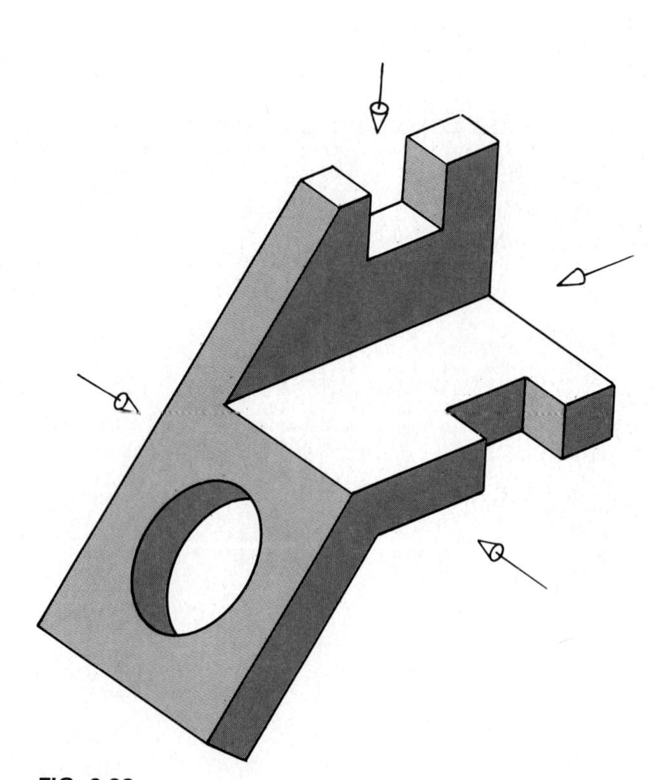

FIG. 9.32

9.6 Sketch or draw the given H projection of Fig. 9.33 and substitute a "conventional practice" front view for the one shown.

9.7 Sketch or draw the given F projection of Fig. 9.34 and substitute a "conventional practice" profile view for the one shown.

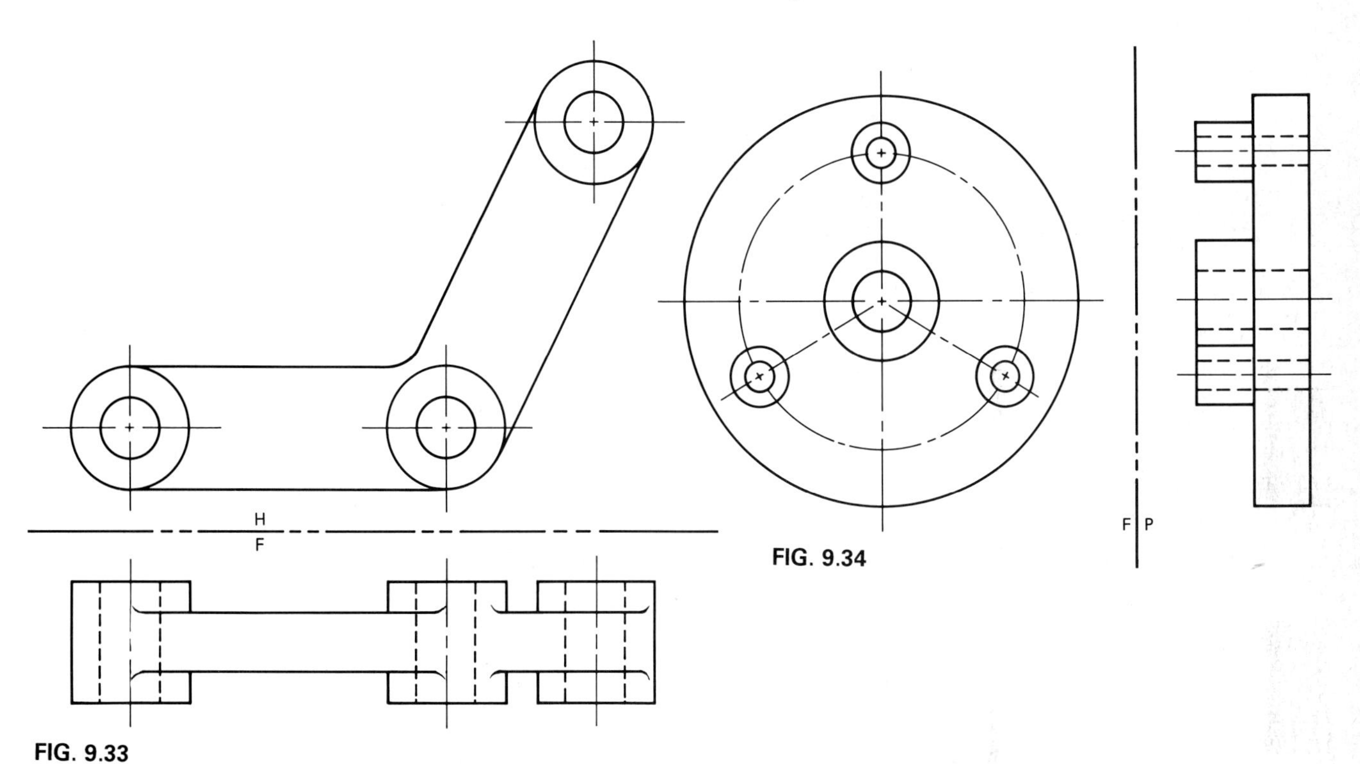

FIG. 9.33

FIG. 9.34

chapter 10

Pictorials

10.I INTRODUCTION

Pictorials are a popular form of graphic communication because they attempt to convey what the eye would see when viewing an object. Figure 10.1 shows a pictorial which portrays the form of the object much more clearly to a person inexperienced in reading drawings than a set of orthographic drawings would. The pictorial conveys size, shape, and the geometric relations between surfaces on an object just as orthographic views do. However, pictorials are difficult to construct accurately and do not represent the object as completely as do orthographic views.

A pictorial shows all three dimensions of an object in a single view. The orientation of the object in a pictorial permits emphasis on one or more of the object's surfaces. For example, in Fig. 10.1, the three visible surfaces have equal emphasis in the pictorial.

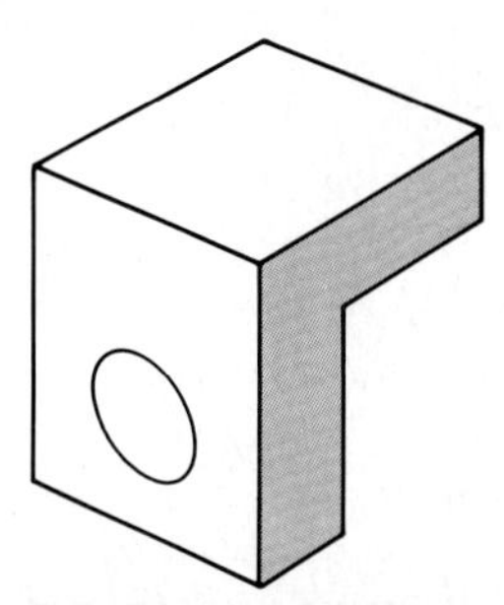

FIG. 10.1 A pictorial attempt to represent what the eye would see.

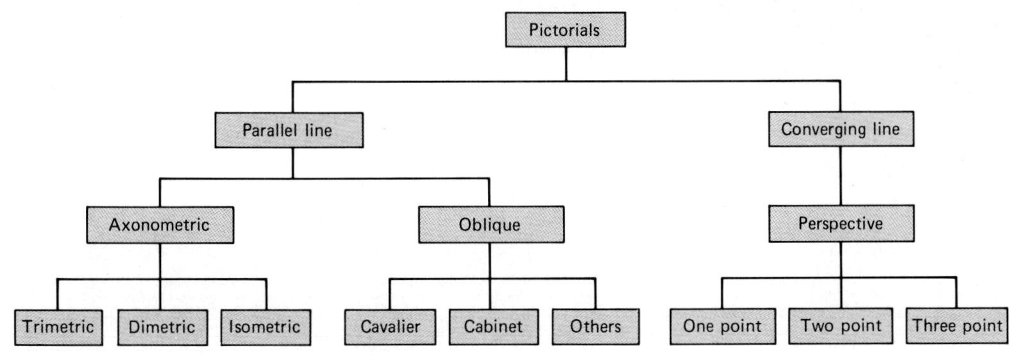

FIG. 10.2 The classification of pictorials.

The chart in Fig. 10.2 classifies pictorials into the categories of parallel-line or converging-line drawings. Each of these forms will be discussed.

10.2 THEORY OF PICTORIALS

A pictorial is the representation of a three-dimensional object on a two-dimensional surface (plane). The most popular means of accomplishing this are axonometric, oblique, and perspective drawings. We will define each of these briefly and illustrate the theoretical bases.

10.2.1 Axonometric Projection

Figure 10.3 shows the elements of orthographic projection in which a two-dimensional face *ABCD* of the object is placed parallel to the projection plane and the image of this face appears on the projection plane. This process allows us to view other faces of an object, such as the top and the right side. If you imagine the object in Fig. 10.3 as being rotated 45° about the vertical line *AD* and the orthographic view as being constructed as shown in Fig. 10.4, you will see two faces of the object. Note that lines on the object that are no longer parallel to the projection plane are foreshortened.

Now visualize the object in Fig. 10.4 as being rotated forward; that is, fix point *D* and move point *A* closer to the projection plane. See Fig. 10.5. In the orthographic image three surfaces, and thus the three dimensions of length, width, and height of the object, will be visible, producing the axonometric form of parallel-line pictorials.

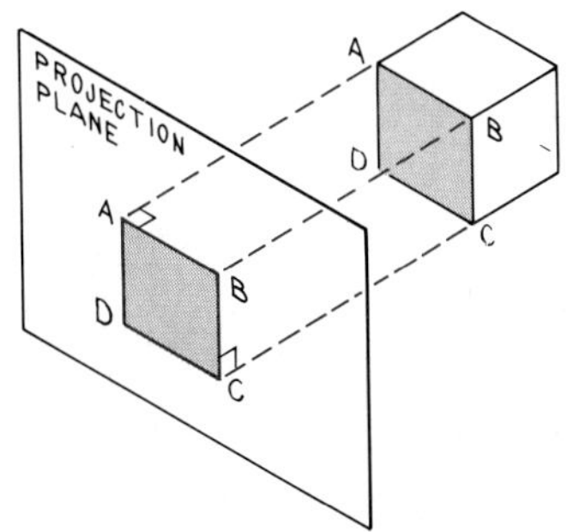

FIG. 10.3 Orthographic projection.

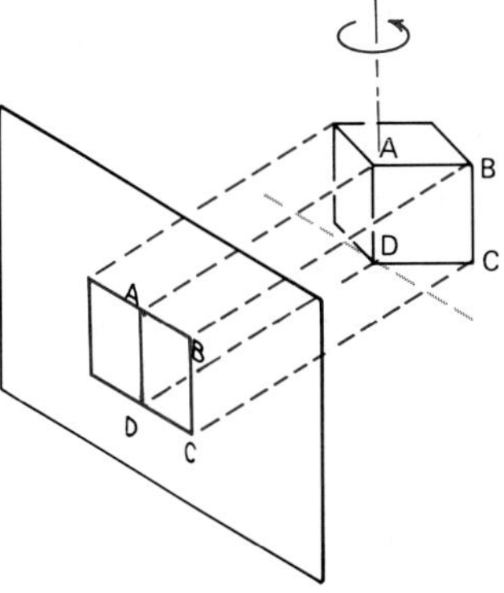

FIG. 10.4 If the object is rotated or if the projection plane is placed properly, more than one surface of an object can be viewed. Two faces are visible in this view.

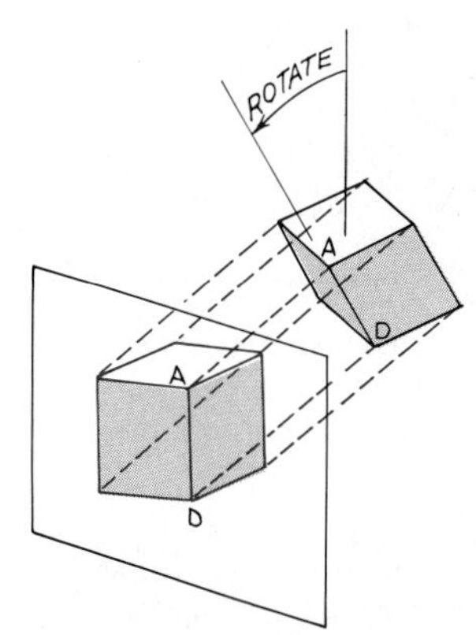

FIG. 10.5 Three surfaces of an object can be viewed with proper positioning of the projection plane.

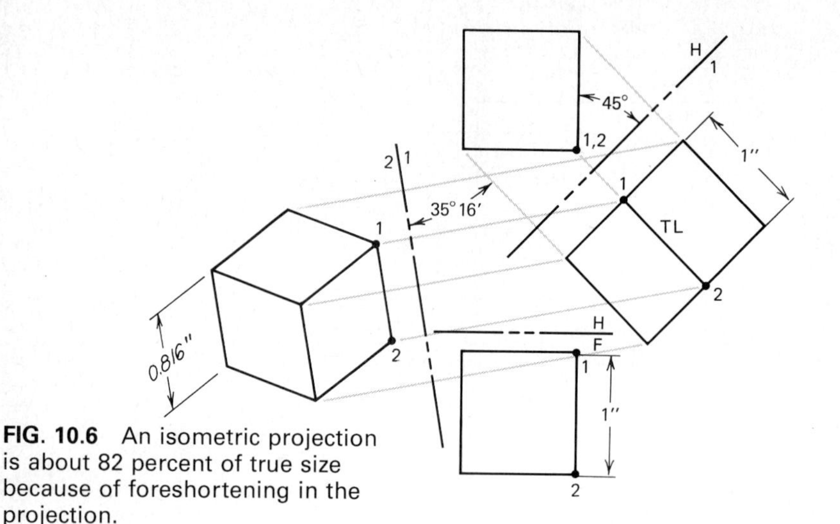

FIG. 10.6 An isometric projection is about 82 percent of true size because of foreshortening in the projection.

FIG. 10.7 Once an angle of intersection other than 90° has been established between the projection plane and the projectors, all other projectors are parallel to this one. The angle of intersection here is 45°.

Thus, the orthographic projection principles we have learned enable us to produce a pictorial by positioning the object so that three surfaces will be visible on the projection plane. The actual orientation of the object with respect to the projection plane determines which type of axonometric projection is depicted.

Perhaps the most common axonometric projection is isometric. This projection requires that the three axes representing the dimensions of length, width, and height each make the same angle with the projection plane. In other words, each of the three dimensions receives equal emphasis in the pictorial.

Figure 10.6 shows the isometric projection of a cube 1 inch on a side. From the given H and F views, the H/1 reference line is established at an angle of 45° and the cube is projected into the 1 view. This is identical to the rotation of the object shown in Fig. 10.5. Then a reference line 1/2 is drawn at an angle of 35°16′ with a perpendicular to the H/1 reference line. The resulting 2 view is the isometric pictorial of the cube. As you can see in Fig. 10.6, the isometric projection is foreshortened to approximately 82 percent of its original value in the projection process. Note that the procedure shown in Fig. 10.6 is a specific application of the orthographic projection illustrated in Fig. 10.5.

10.2.2 Oblique Projection

The concept of oblique projection is illustrated in Fig. 10.7. An object is placed with one face parallel to the projection plane, and parallel rays are projected to intersect the projection plane at an angle other than 90°. The resulting image is a pictorial with one face in true shape. When the ray intersects the plane of projection at 45°, a cavalier oblique projection is generated.

The cavalier projection is illustrated in Fig. 10.8. If a ray from point A intersects the projection plane at 45°, this ray may lie anywhere on a cone which is partially shown in the F projection plane. By selecting different locations of point A on the cone such as B, B', and B'' and by projecting the remainder of the object, we can generate three oblique pictorials, as shown in Fig. 10.8. Note that the face originally parallel to the projection plane appears exactly the same in all views. The front (orthographic) view is included in Fig. 10.8 to locate point A as the cone vertex.

Rather than choose a point on the cone, it is more convenient to select an angle which defines the receding axis of the oblique pictorial. The choice of angle allows different faces of the object to be seen with the true-shape surface. For example, the angle of receding axis determined from B'' shown in Fig. 10.8 allows us to see the left side and the top of the object. Figure 10.9 illustrates some possible views with different receding axis angles. Certain views, such as the one labeled "Avoid," should not be used because of their confusing appearance.

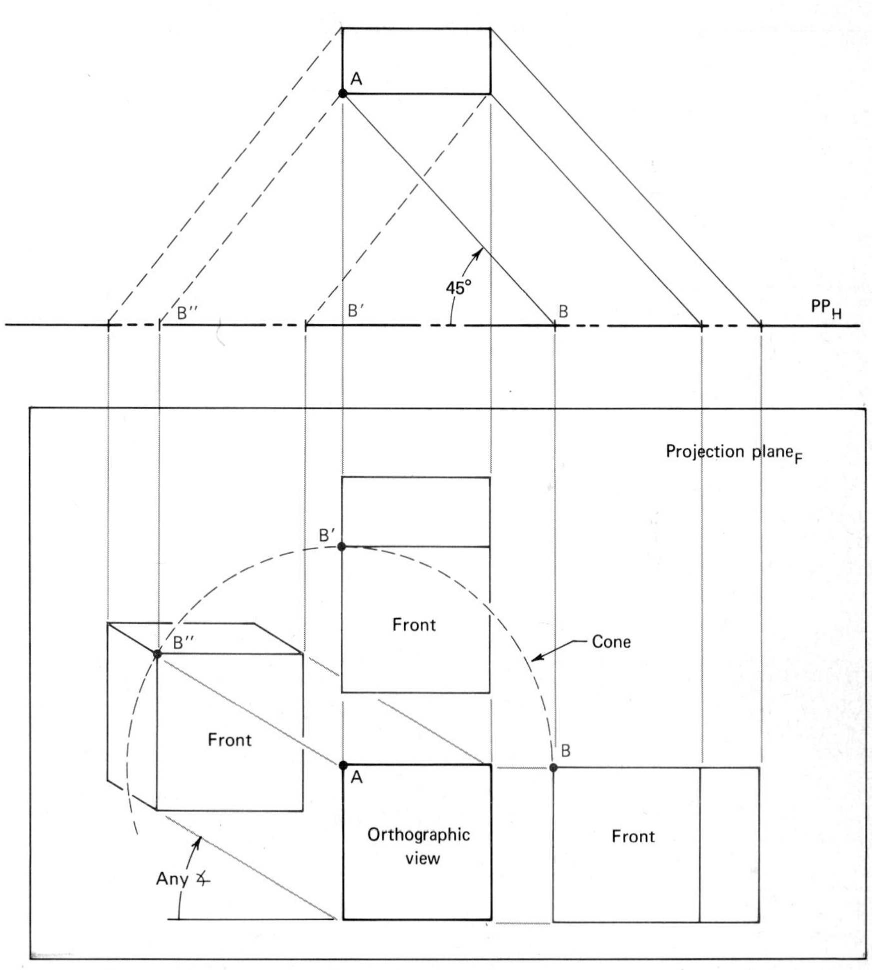

FIG. 10.8 A 45° angle of intersection of the projectors and the projection plane, called cavalier projection, allows for several views of the object depending on where the projector intersects the image plane. The position determined from the point B″ provides a satisfactory view of the object.

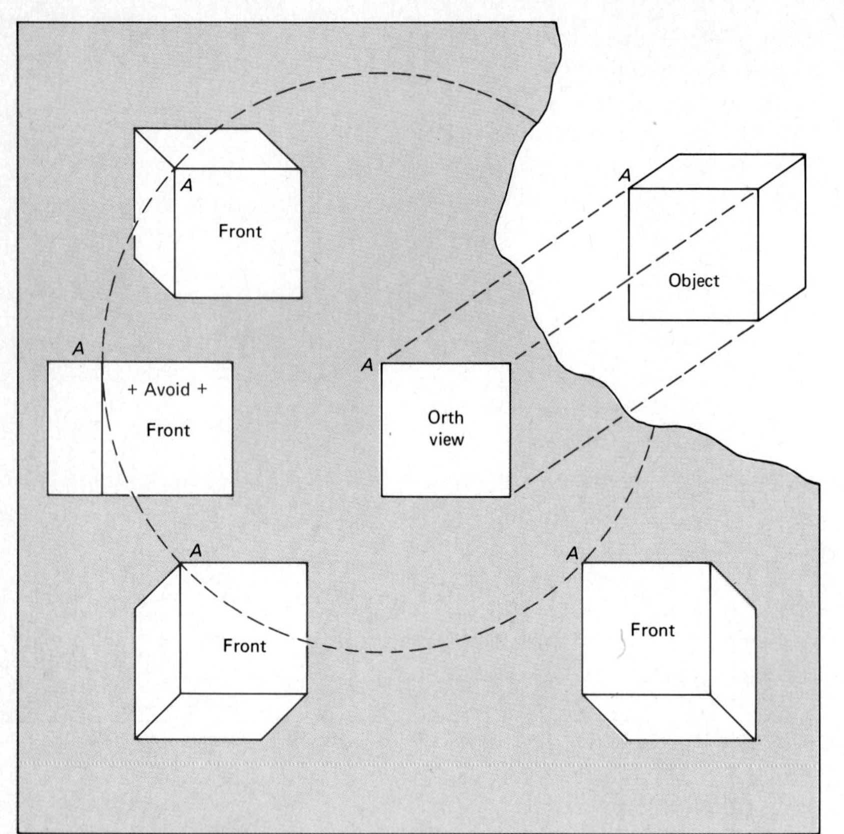

FIG. 10.9 The receding axis determines which surfaces will be visible in an oblique projection.

10.2.3 Perspective Projection

Perhaps the most realistic pictorial representation is perspective projection. In perspective theory, the rays from points on an object converge to a point, producing a similar effect to that of a camera lens or a human eye. Because of this phenomenon, parallel lines on an object receding from the viewer converge to a single point at infinity.

There are three classes of perspective. One-point, or parallel, perspective, as illustrated in Fig. 10.10, places one surface parallel to the projection plane; the receding parallel lines converge to a single point, called a vanishing point (VP), on the horizon. Two-point, or angular, perspective places the object at an angle to the projection plane so that two sets of parallel lines converge to two different vanishing points on the horizon. See Fig. 10.11. In three-point perspective, receding lines parallel to the x, y, and z axes of the cartesian system converge at three vanishing points.

Although perspective provides a very realistic representation, it

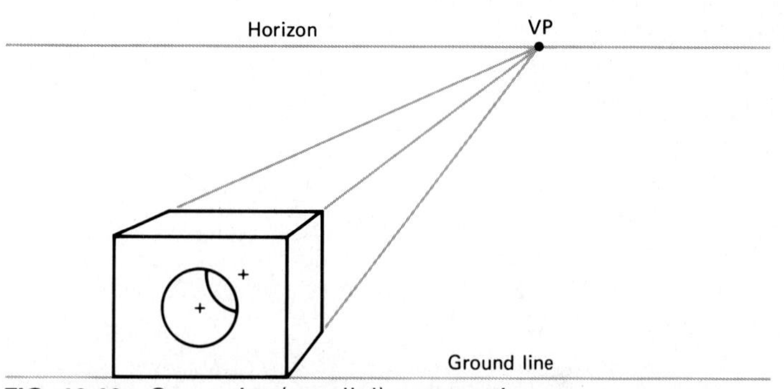

FIG. 10.10 One-point (parallel) perspective.

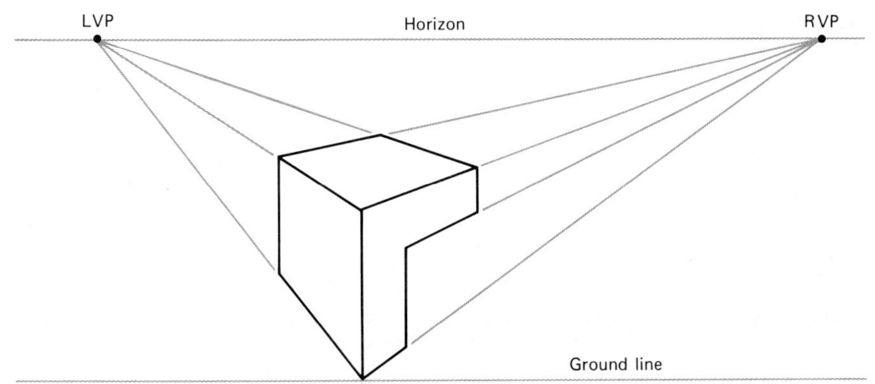

FIG. 10.11 Two-point (angular) perspective.

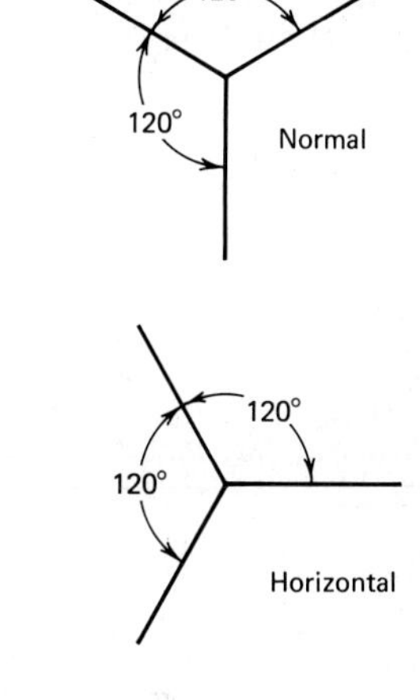

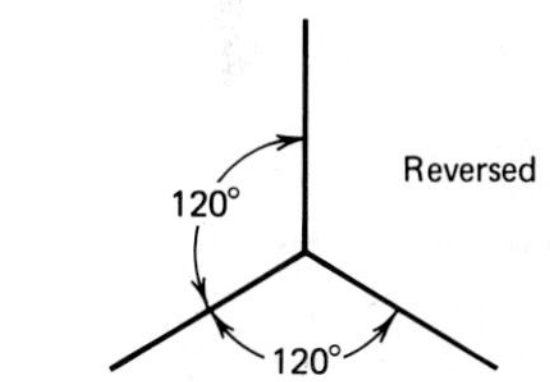

FIG. 10.12 The standard positions of the axes for isometric drawings.

is not extensively used in engineering practice because of the drawing time required. Therefore, we will not cover perspective in greater detail.

10.3 PICTORIAL DRAWINGS

In practice, most theories of pictorial projection are modified to simplify the drawing task and provide a clearer representation. When incorporated in construction, these modifications lead to pictorial *drawings* as contrasted to pictorial *projections.*

10.3.1 Axonometric Drawings

From the definition of axonometric projection, you can see that it is possible to place the object in an infinite number of positions to obtain an orthographic projection showing three surfaces. Axonometric drawings are parallel-line drawings which orient the three surfaces of an object at specified angles to the projection plane and allow certain modifications in projection theory in order to simplify construction. Axonometric drawings are classified as isometric, dimetric, and trimetric.

Isometric drawings position the three principal axes which define the three dimensions of an object at angles of 120°. There are also three standard positions of the isometric axes—normal, horizonal, and reversed—as indicated in Fig. 10.12. The isometric drawing shows true lengths along and parallel to the principal axes, as shown in Fig. 10.13. In effect,

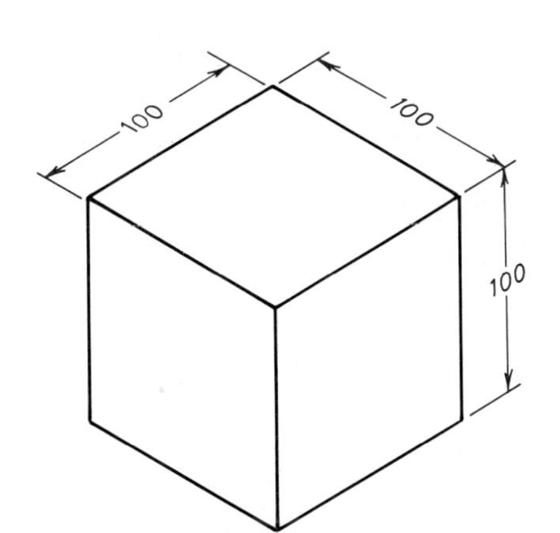

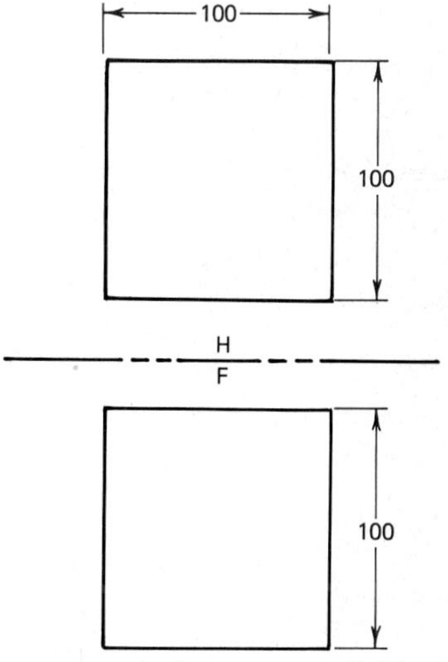

FIG. 10.13 True-length distances can be measured along or parallel to the axes of an isometric drawing.

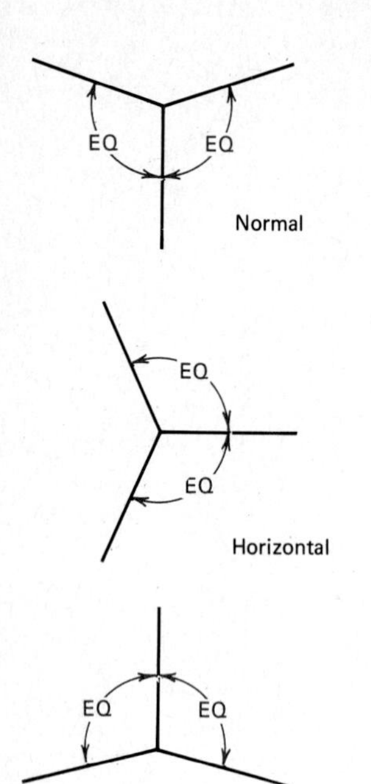

FIG. 10.14 Positions of the axes for dimetric drawings.

foreshortening from the isometric projection is ignored. This results in more rapid drawing since isometric scales need not be used.

Dimetric drawings require that two of the three angles of the principal axes be equal whereas the third angle can vary. By appropriate selection of the third angle in dimetric, we can emphasize or suppress one surface in the drawing. Figure 10.14 illustrates dimetric positions of the axes and the required two equal angles.

Trimetric drawings allow the axes to form any three unequal angles. This allows one to select surfaces for particular emphasis on the basis of the amount of contour or the desired effect on the viewer. Figure 10.15 shows three axonometric drawings of objects with different contour. If an object has contour on three surfaces, isometric is a logical choice. If there is contour on only one or two surfaces, trimetric or dimetric may be chosen to highlight the contour.

Additional flexibility is possible when the type of axonometric drawing is combined with the different positions of the axes. You should experiment with combinations to come up with the most realistic pictorial. Figure 10.16 shows three choices of combinations for three different objects.

Construction of an axonometric pictorial begins with the selection of the type (isometric, dimetric, or trimetric) and the position of the axes (normal, horizontal, or

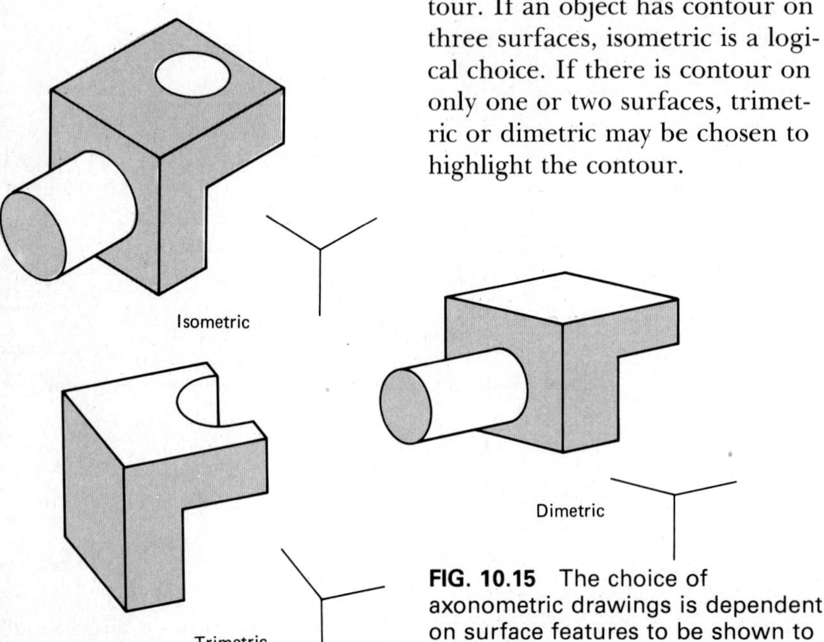

FIG. 10.15 The choice of axonometric drawings is dependent on surface features to be shown to advantage.

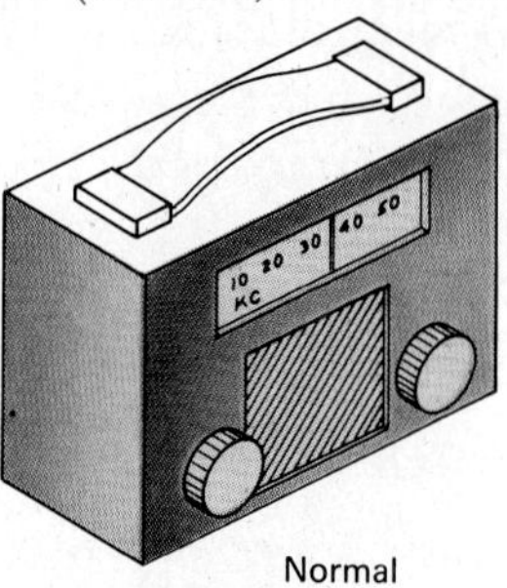

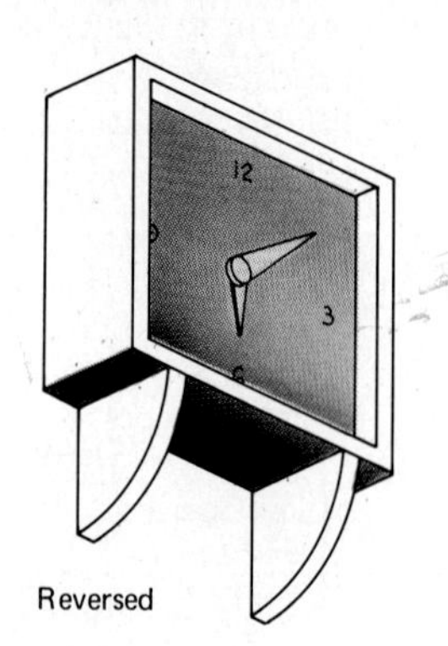

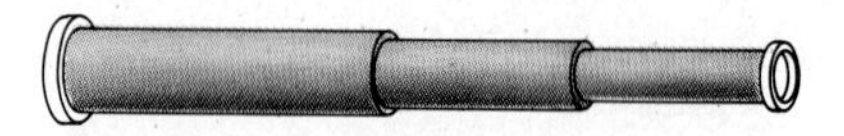

FIG. 10.16 An object represented in pictorial can be enhanced by an appropriate combination of an axis position and type of axonometric drawing.

reversed). As previously stated, trimetric affords the most latitude by enabling you to vary the orientation of the axes until you show the object to its best advantage. Dimetric restricts the choices somewhat but still allows a great deal of freedom. Isometric is the most restrictive format, but it is the most frequently used axonometric pictorial because it can be constructed quickly with a parallel rule and a standard set of triangles.

An isometric drawing has some disadvantages, which are noted in Fig. 10.17. A 45° corner cut, as shown in Fig. 10.17*a* , will appear as an edge. The pyramid shape in Fig. 10.17*b* may be confused with an open rectangular box if isometric representation is used.

In axonometric drawings, lines which are not parallel with the principal axes must be constructed by first locating the endpoints. Remember that lines along or parallel to the principal axes are drawn in true length. Figure 10.18 illustrates construction techniques that must be used to produce an axonometric pictorial.

Consider lines 1–4 and 2–3 in Fig. 10.18, which are not parallel to any of the principal axes. To construct these lines in pictorial, first locate endpoints 1 and 2 by constructing the true length line, X_1, on the pictorial. Apply the same process to locate endpoints 3 and 4 by constructing the true length line, X_2. Note that angles cannot be measured on an orthographic view and laid out on the pictorial. In axonometric drawing

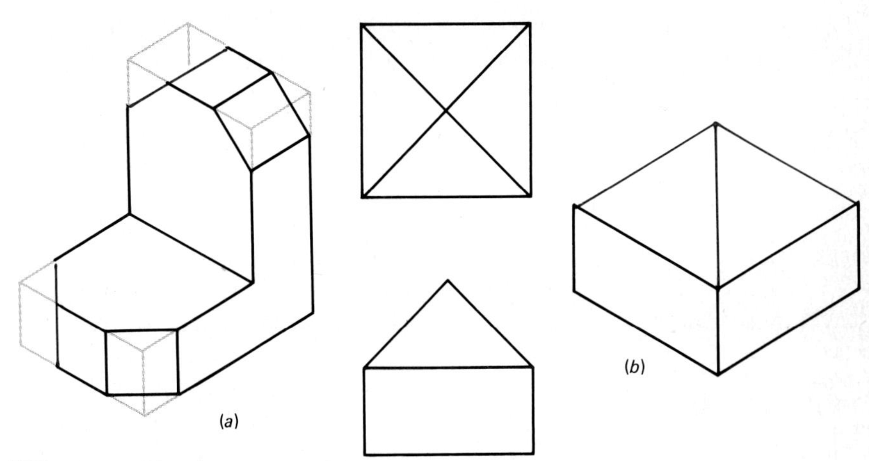

FIG. 10.17 Disadvantages of isometric drawings.

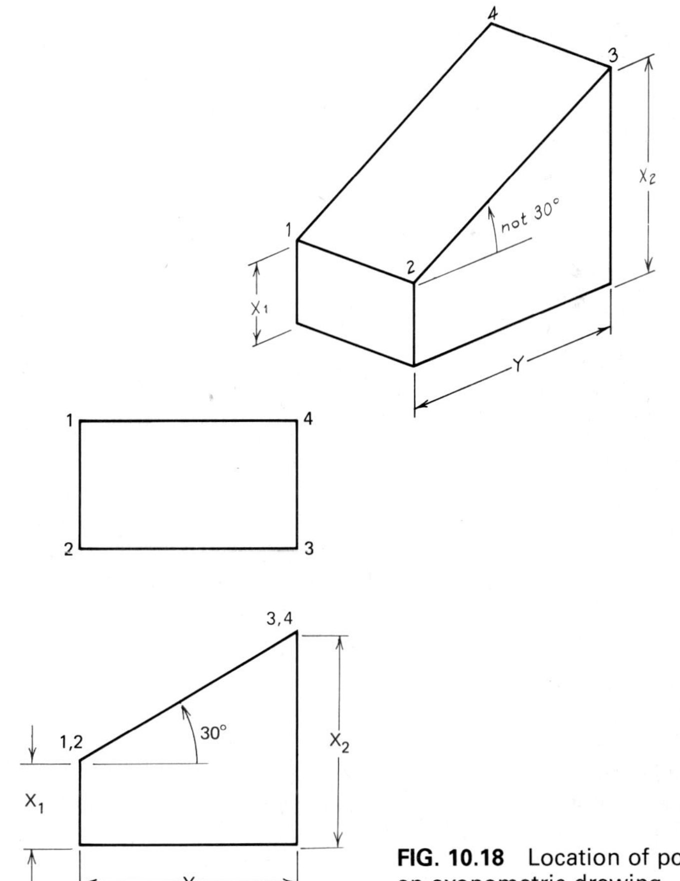

FIG. 10.18 Location of points on an axonometric drawing.

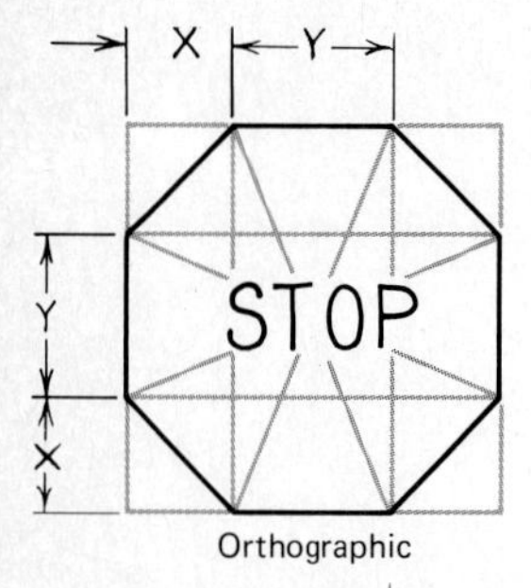

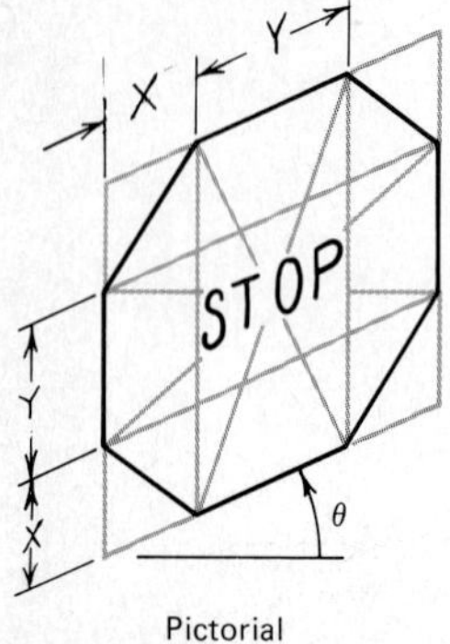

FIG. 10.19 Axonometric construction using distances along the axes.

there are no true-size angles on the pictorial. For example, a 90° angle will appear as 60° or 120° if it is formed by lines that are parallel to the principal axes.

Figure 10.19 shows that any number of straight lines that are not parallel to the principal axes can be constructed by locating the endpoints.

Curved surfaces can be constructed conveniently in orthographic projection by means of techniques that were discussed in Chap. 2. For example, circle construction can be done freehand, with a compass or a circle template. See Fig. 10.20. Converting this circle to its equivalent in an axonometric pictorial requires a careful construction procedure.

A circle in axonometric becomes an ellipse. The construction is accomplished as follows:

1. Construct an equilateral parallelogram with sides equal to the diameter of the circle. The parallelogram must be oriented in exactly the same way as the surface on which the circle is to be drawn. Locate the center as shown in Fig. 10.21.

2. If you have access to an ellipse template, it may be possible to construct the ellipse rapidly. For example, if the circle is to be placed on an isometric pictorial parallel to any of the principal surfaces, the 35° template will provide the correct elliptic shape. Ellipse templates come in a variety of sizes and angle ranges. A complete set will have an angle range of 15° to 60° in increments of 5°, and each angle will be associated with a number of diameters. Many times, however, the template size will not correspond to the desired construction. Other methods exist to approximate an elliptic shape. One of these, the four-center method, requires a compass.

3. The four-center construction is shown in Fig. 10.22. Begin by constructing perpendiculars to the sides of the parallelogram at the tangent points. The circular arc *AB,* approximating a segment of the desired ellipse, has its center at *C,* the intersection of the perpendiculars from *A* and *B*. The compass is used to draw in this segment of the ellipse. Similarly, arc *BD* has its center at *E,* and this segment can be drawn. Continue until the four circular arcs are drawn, forming an approximate ellipse.

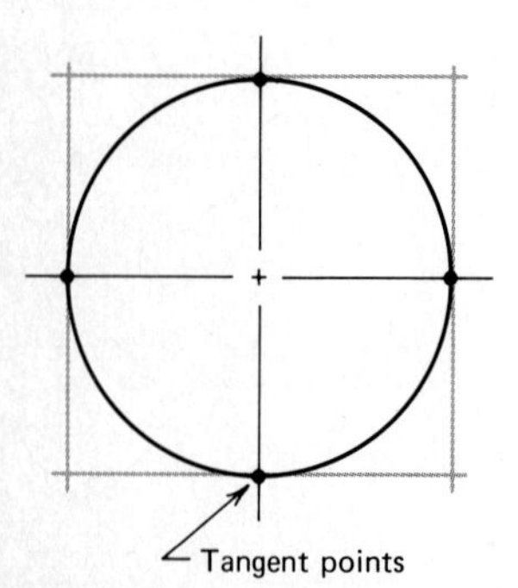

FIG. 10.20 Tangent points of a circle and circumscribed square form the basis for circle construction in axonometric drawings.

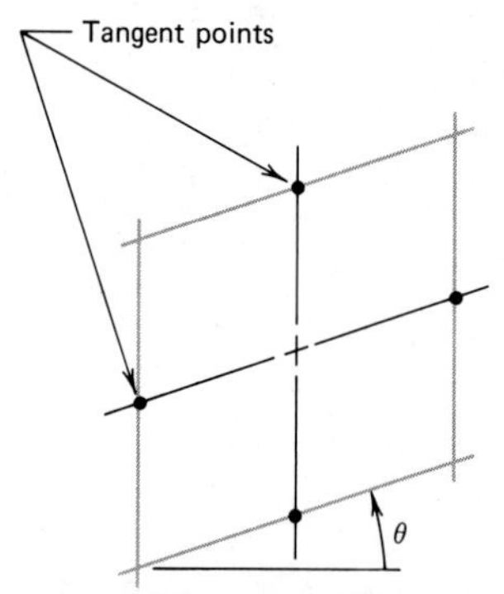

FIG. 10.21 Establishing the tangent points for representing a circle in axonometric drawing.

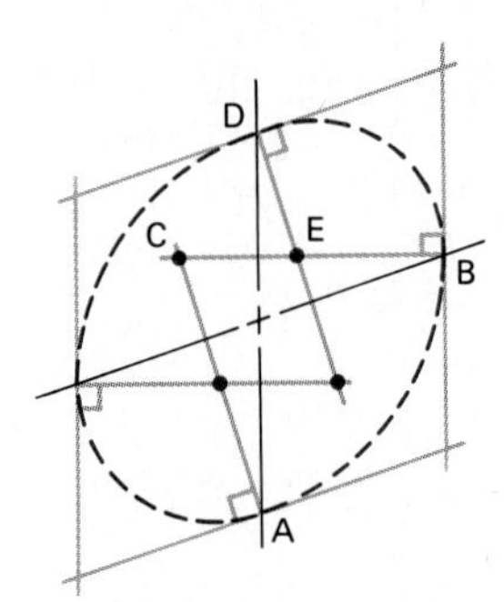

The construction of a curved line in axonometric can be accomplished by plotting from the orthographic view a series of points on the pictorial. These points are often called offsets. The procedure is illustrated in Fig. 10.23. Begin by forming a grid pattern on the orthographic view. The density of the pattern depends on the accuracy of the line desired in the pictorial. Construct the same grid pattern on the pictorial. This procedure lends itself best to freehand construction in cases where the points can be located approximately within the grid. You can see that for instrument drawings, points on the curve can be selected, and very accurate offsets can then be transferred to the pictorial. Finally, an irregular curve can be used to draw the line in pictorial.

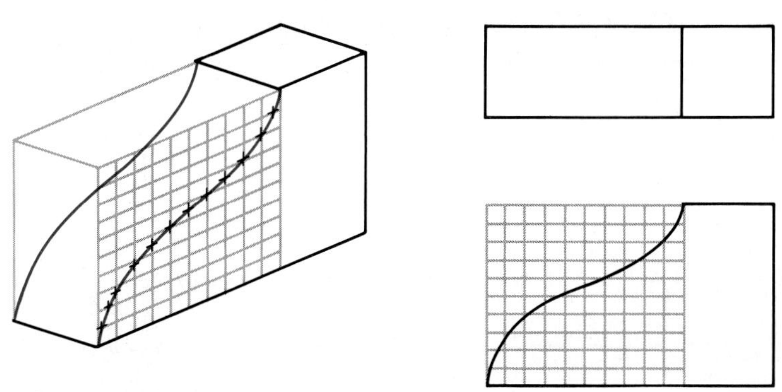

FIG. 10.23 Offset method for construction of irregular curves in axonometric.

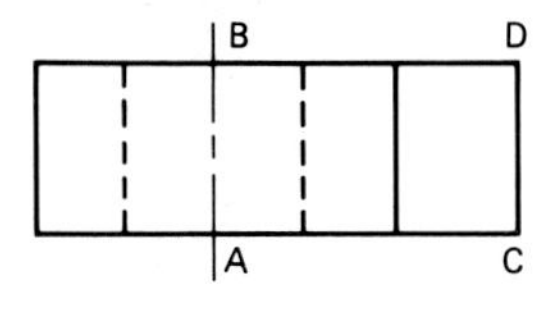

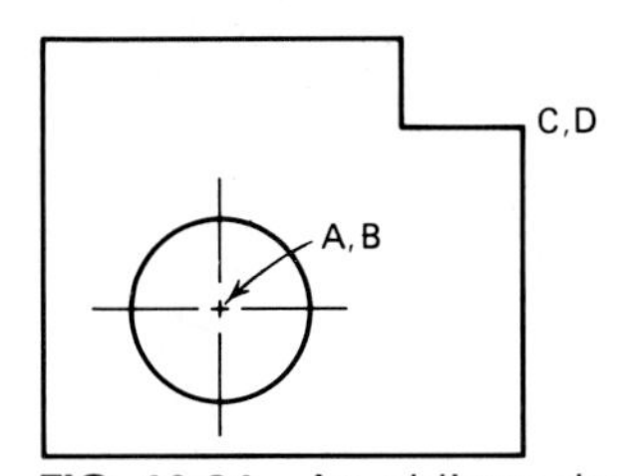

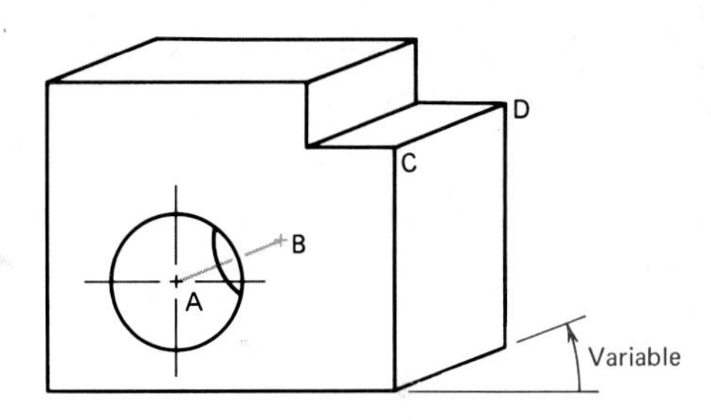

FIG. 10.24 An oblique drawing.

10.3.2 Oblique Drawing

Oblique drawings require one surface of the object to be parallel with the projection plane. Thus, features on this surface and on any parallel surface are constructed in true shape. Figure 10.24 shows an oblique pictorial. Note that circles remain circles and that angles and lengths are in true size in the front face of the pictorial. The receding axis angle can be varied to best illustrate the object.

Figure 10.25 shows that once the surface to be positioned parallel to the projection plane has been selected, the receding axis can be drawn to feature the object from the top, bottom, right side, or left side. In this case, the best view is the one that shows the notch to best advantage.

Oblique drawings are classified as cavalier, cabinet, or general. The classification is determined from the ratio of the actual scaled receding dimension in orthographic to the receding dimension constructed in pictorial. When the ratio is 1:1, the oblique drawing is classified as cavalier. When the ratio is 1:0.5, that is, when the receding dimension on the pictorial is one-half that of

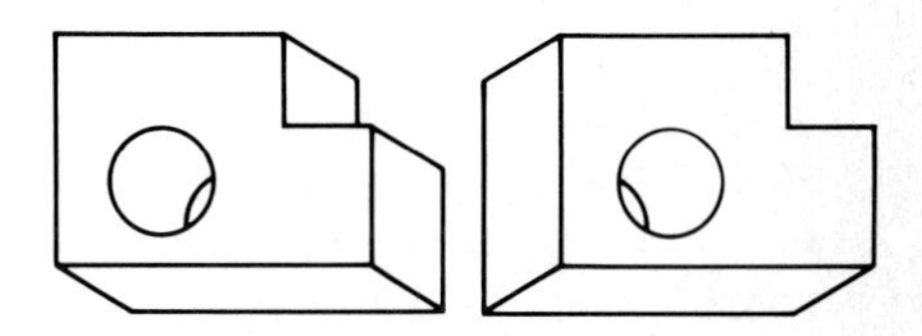

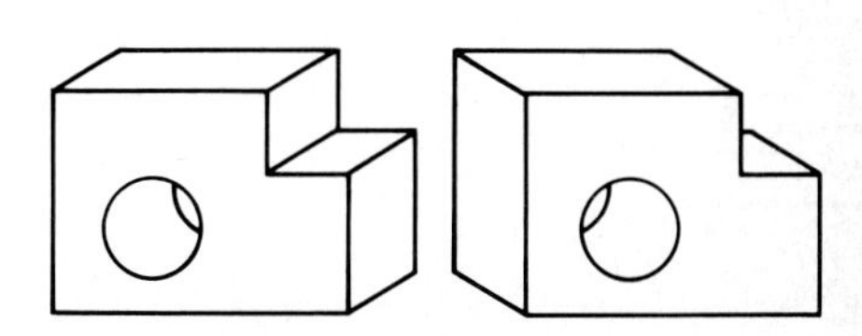

FIG. 10.25 Selection of the most appropriate direction for the receding axis.

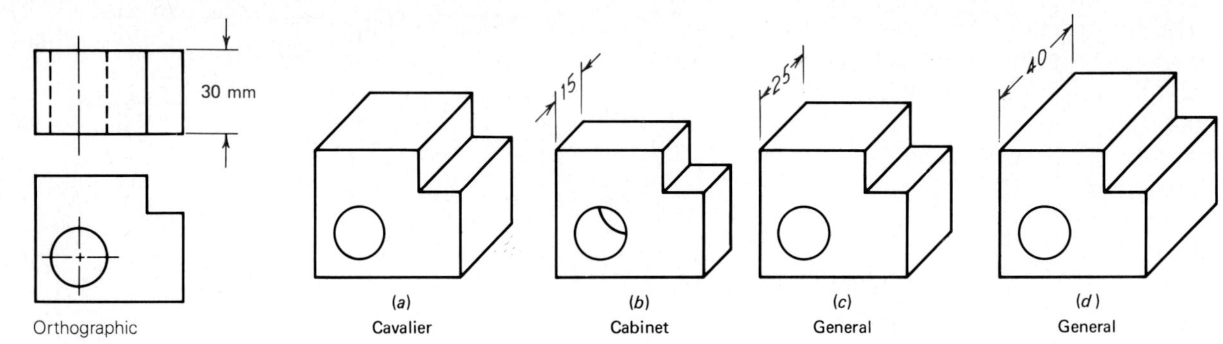

FIG. 10.26 Classification of oblique drawings.

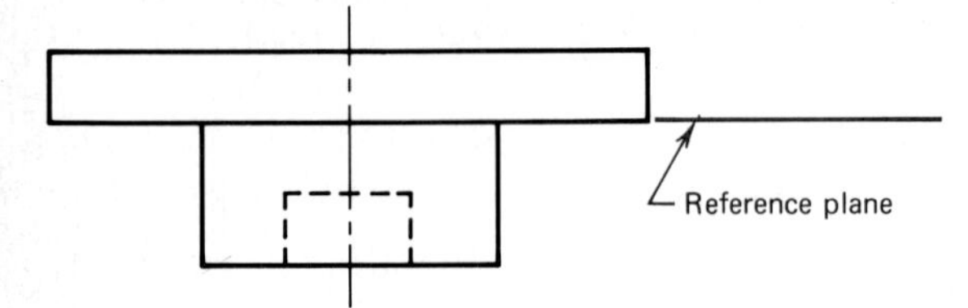

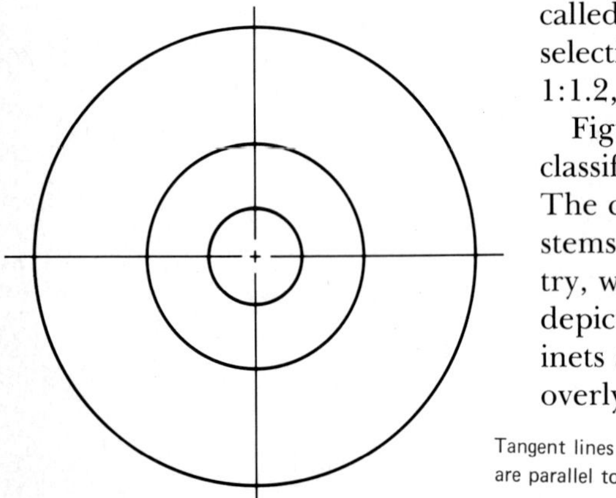

FIG. 10.27 Planning the construction of an oblique pictorial.

the actual object, the drawing is called cabinet oblique. Any other selection of ratio, such as 1:0.3 or 1:1.2, is classified as general.

Figure 10.26 illustrates the classification of oblique drawings. The cabinet drawing (l:0.5 ratio) stems from the furniture industry, where the cavalier drawing depicted the receding axis of cabinets and chests of drawers as overly long and distorted. The 1:0.5 ratio makes furniture drawings much more realistic.

Care must be exercised in using oblique pictorials. The object under consideration should have contour, particularly circular contour, in one plane or in parallel planes. The receding dimension should be relatively short and, if possible, the long dimension of the object should be parallel with the projection plane. When a choice must be made between the best receding axis and the long dimension parallel to the projection plane, place the most irregular contour parallel to the projection plane. If the long dimension must be placed in the receding direction, a cabinet drawing will enhance the object by reducing the distortion.

Careful planning is required before you draw an oblique pictorial of an object such as the one in Fig. 10.27. First, choose the scale. Next, choose a reference plane as a starting point. For example, if you choose the reference plane labeled in Fig. 10.27, you can draw a circle of correct diameter with a compass or template and then construct a line in

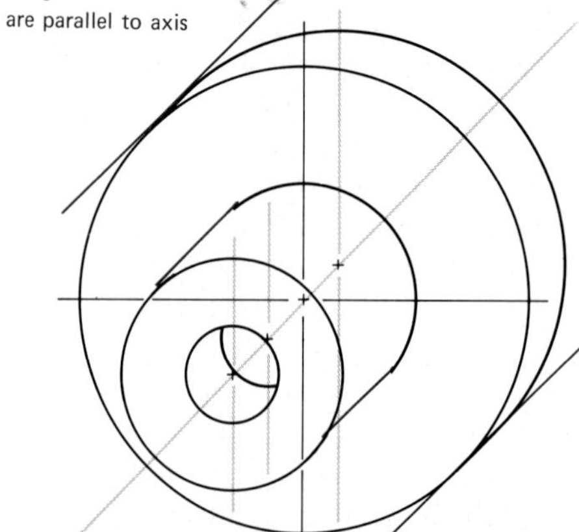

FIG. 10.28 Completed oblique drawing. Construction lines may be erased.

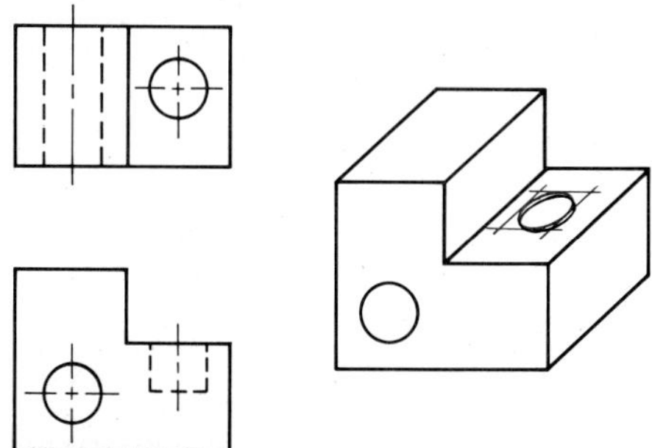

FIG. 10.29 Construction on a receding surface in oblique.

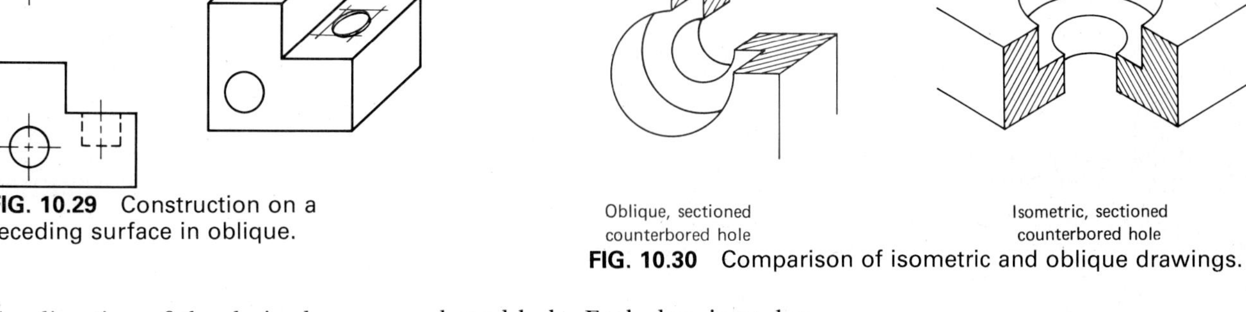

FIG. 10.30 Comparison of isometric and oblique drawings.

the direction of the desired receding axis. See Fig. 10.28. Other planes of the object can now be drawn by moving along the receding axis the appropriate distance as computed from the ratio (cavalier, cabinet, or general). After you draw the correct tangent lines parallel to the receding axis, the pictorial is complete.

If contours exist in one or more receding planes, great care must be taken in selecting the type of oblique drawing to represent the object. For example, if a drilled hole is to be shown in a receding plane, a cavalier drawing (1:1 ratio) will permit use of a four-center ellipse or ellipse template. See Fig. 10.29. If any other ratio is chosen, the circle will have to be plotted by offset methods. In this case, an axonometric drawing may be a better choice.

Oblique drawings allow for rapid instrument construction, particularly if the contours are circular and are in parallel planes. Figure 10.30 compares an oblique drawing with an isometric drawing of a sectioned, counterbored hole. Each drawing takes about the same amount of time if templates are used. However, if a compass is used, the oblique requires less time.

10.4 SHADING

Shading can be used effectively on pictorials to provide realism and accentuate features or details. The countersink in Fig. 10.31 is easier to visualize when straight-line shading is used for elements of the cone.

Normally, shading is used for purposes of clarifying surfaces. The conventional position of the light source is to the front and upper left of the object. Surfaces perpendicular to the light direction will receive little if any tone. Surfaces that are not illuminated will be darkest or in shade.

Shading can be provided on pencil drawings by either lines or continuous tone. Line shading requires that tone be produced by line weight and spacing. Drilled holes are a good example; heavy lines with close spacing produce

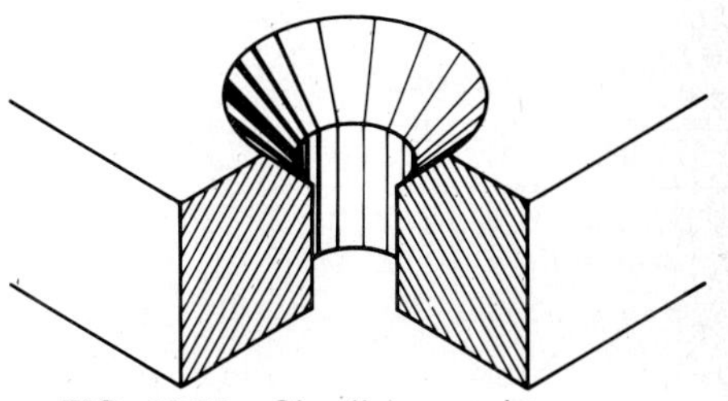

FIG. 10.31 Shading used to enhance pictorial drawings.

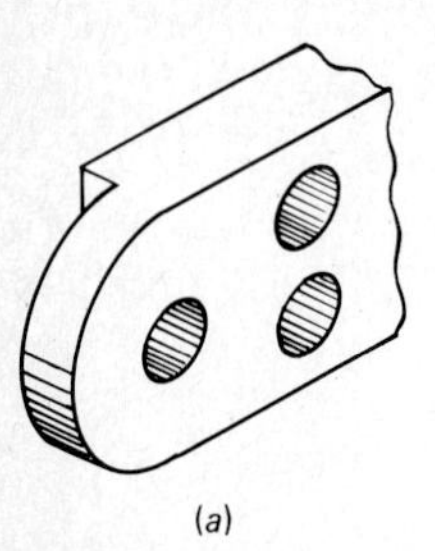

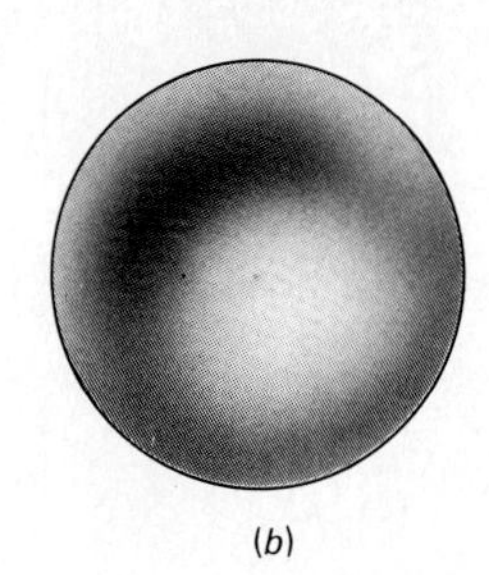

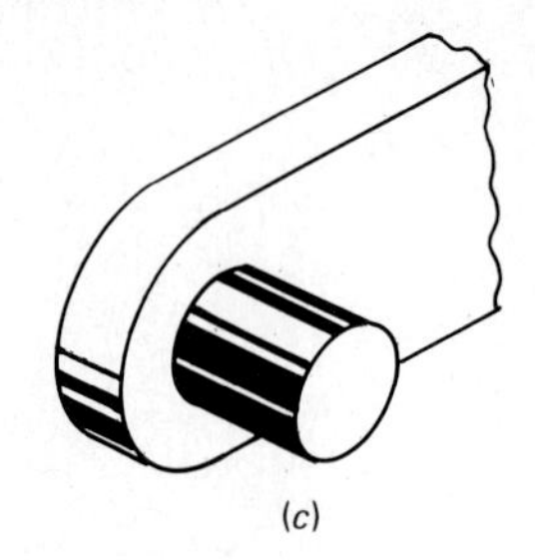

FIG. 10.32 Types of shading: (*a*) line shading, (*b*) continuous shading, and (*c*) block shading enhance the effect of pictorials.

the darkest shade and the best visual effect. See Fig. 10.32.

Continuous shading is normally performed with a soft pencil, using a blend of light to heavy tones. Solid or block shading provides deep tones and is normally used on shiny metallic objects. See Fig. 10.32.

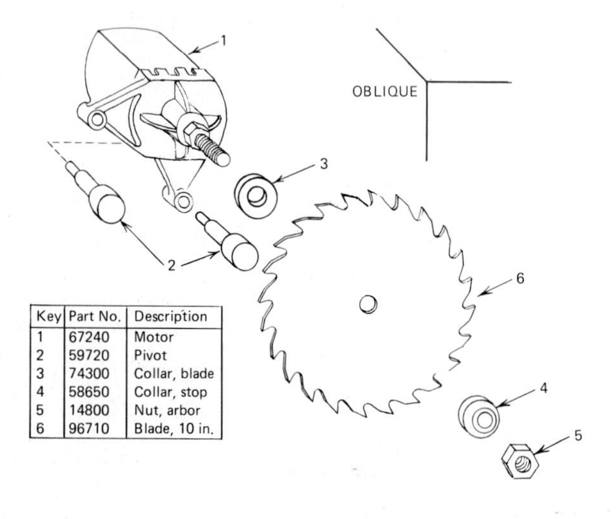

Key	Part No.	Description
1	67240	Motor
2	59720	Pivot
3	74300	Collar, blade
4	58650	Collar, stop
5	14800	Nut, arbor
6	96710	Blade, 10 in.

FIG. 10.33 An exploded oblique pictorial.

10.5 EXPLODED PICTORIALS

An exploded pictorial provides a readily understandable drawing showing how several parts fit into an assembly. These pictorials may be constructed as axonometric or oblique. Frequently, these exploded pictorials are partially dimensioned and have individual parts identified clearly. Some exploded pictorials are sectioned for further clarification of complex internal details. Figures 10.33 and 10.34 show exploded pictorials of common devices.

10.6 APPLICATIONS FROM COMPUTER GRAPHICS

The computer allows great versatility in the display of pictorials. Computer programs exist for axonometric, oblique, and perspective displays. Let us consider the possibilities for axonometric display as an example. Figure 10.35 shows two orthographic views and an isometric pictorial of a cube with one beveled edge. The geometry describing this cube can be inputted to a computer graphics system as a series of points.

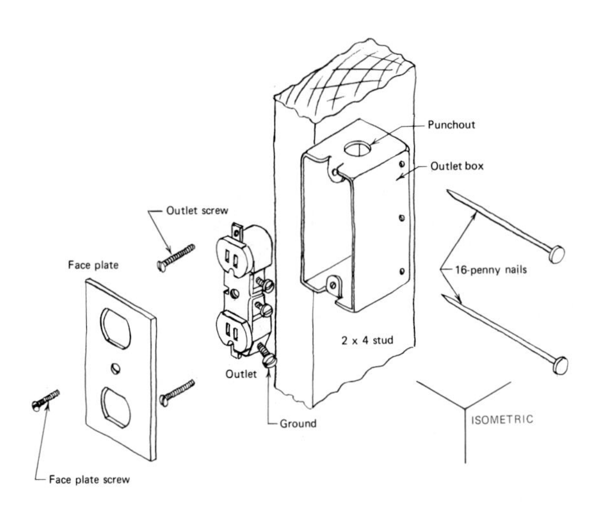

FIG. 10.34 An exploded isometric freehand pictorial.

The only required points are the endpoints of the lines which bound all the surfaces. These points are often called nodes.

Once the geometry is known, any orthographic view of the object may be obtained by means of a computer program which will produce rotations of a solid object similar to those shown in Fig. 10.5. The screen of the graphics display would be considered the projection plane, and any axonometric projection could be displayed simply by specifying the desired rotation.

One method of specifying the rotation is to specify the three angles the principal axes make with the projection plane. If the angles specified are all equal to 35°16′, an isometric drawing will be displayed. This method is impractical because for a given orienta-

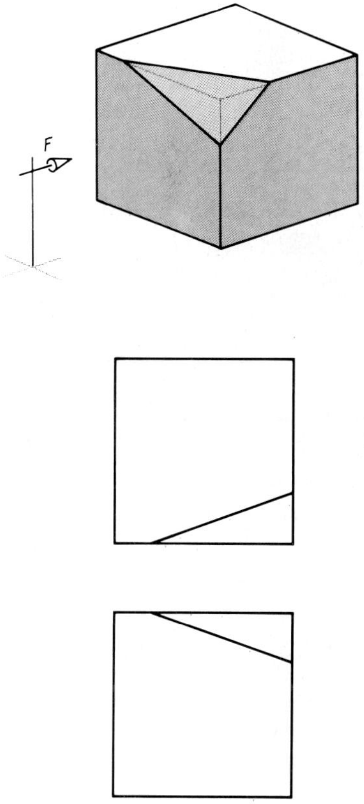

FIG. 10.35 If the geometry of a solid can be defined, it can be established as a database in a computer graphics system and displayed in any desired pictorial format.

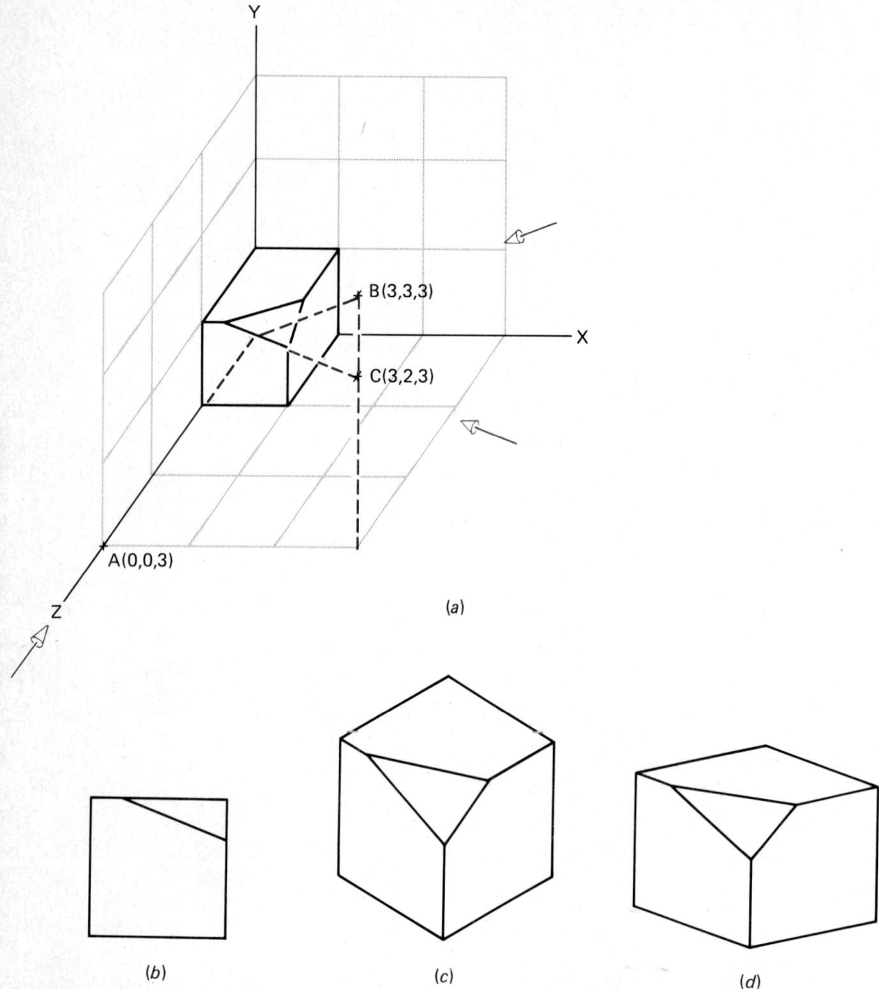

FIG. 10.36 The viewing direction, established from a coordinate in space, will produce a specific view of the object. The view is produced by transformation of the database to the given viewpoint.

tion of the object, the angles are difficult to compute.

A better method is illustrated in Fig. 10.36*a*. Imagine the object in question being positioned as shown relative to an *xyz* coordinate system. If you choose a "viewpoint" as any coordinate in space, a line of sight is established from this point to the origin. The object is then viewed orthographically relative to the viewpoint. For example, choosing point *A* (0, 0, 3) as the viewpoint will provide a front view of the object, as shown in Fig. 10.36*b*. Similarly, viewpoints *B* (3, 3, 3) and *C* (3, 2, 3) will produce the isometric and dimetric pictorials, respectively, as shown in Fig. 10.36*c* and *d*.

Similar software exists which when given a viewpoint or viewpoints will produce oblique or perspective pictorials.

The pictorials illustrated here are displayed on the computer as "wire-frame" drawings. When studying pictorials, the user must visualize a wire-frame drawing as a solid object and realize that all surfaces and interior points must be defined also, even though only the edges of the object are displayed. A great deal of research is under way to generate "solid" models on a graphics display unit. Most high-resolution graphics workstations now have a software package which will generate some form of shading to emulate the solid model.

Problems

10.1 Explain the difference between axonometric projection and axonometric drawing.

10.2 With a small sketch, illustrate the three types of axonometric drawings. Show the variation in receding axes.

10.3 Illustrate freehand the three standard positions of an isometric axis selection.

10.4 List an advantage and a disadvantage of isometric drawings.

10.5 Construct an isometric pictorial of a stop sign and post.

10.6 Using the four-center construction, draw a 50-cent piece enlarged three times.

10.7 Construct a drawing of a counterbored hole as illustrated in Fig. 10.30. Make the drawing:

(a) Freehand isometric
(b) Instrument oblique

10.8 Sketch a correct pictorial for each of the objects shown in Fig. 10.37.

10.9 Sketch a correct pictorial for each of the objects shown in Fig. 10.38.

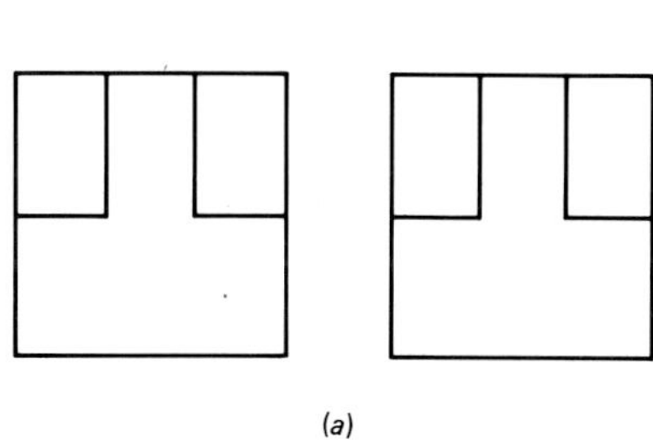
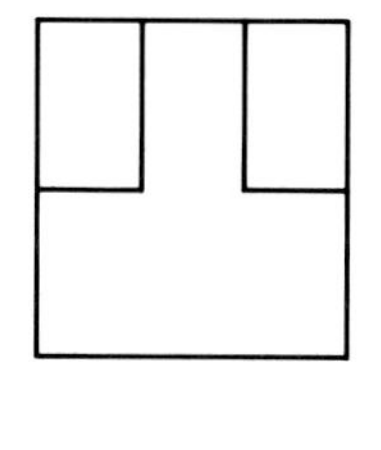

(a)

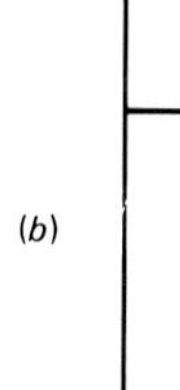

(b)

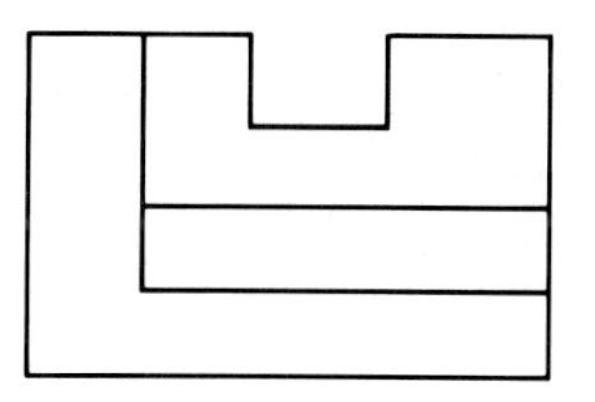

FIG. 10.37 *(c)*

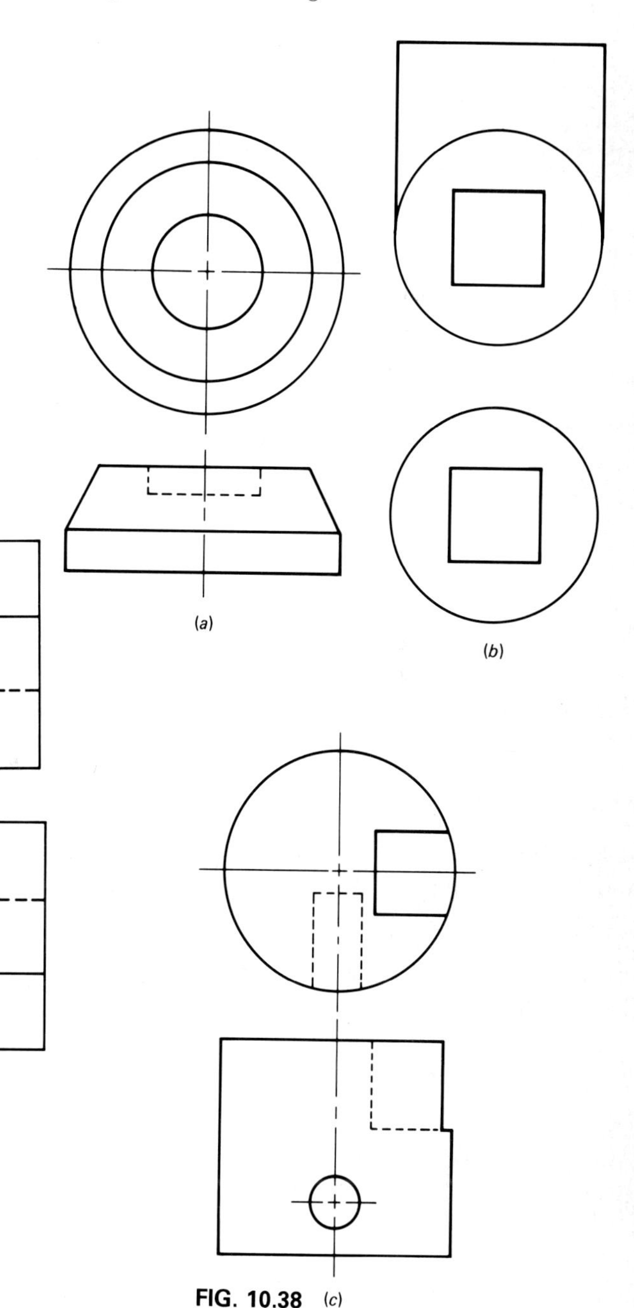

(a) *(b)*

FIG. 10.38 *(c)*

10.10 Draw an isometric pictorial of the object shown in Fig. 10.39.

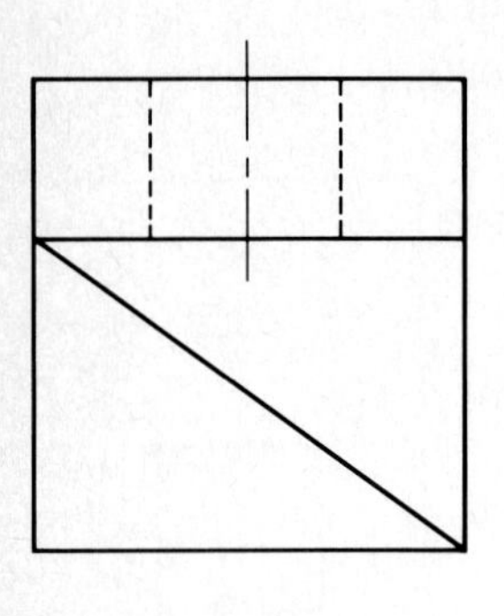

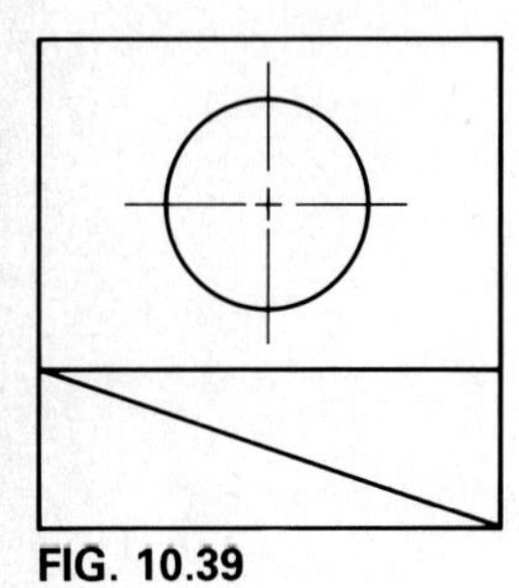

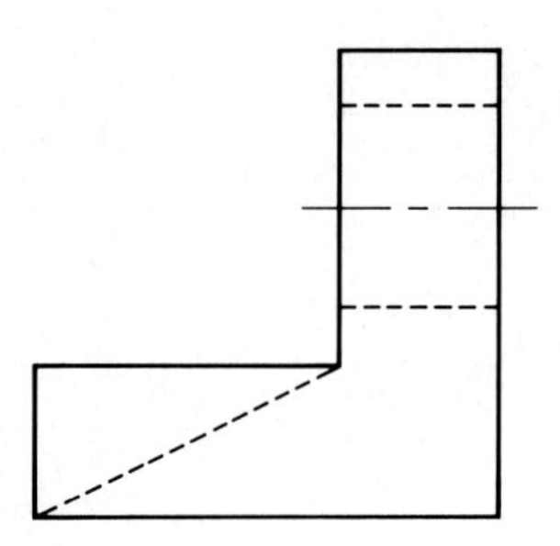

FIG. 10.39

10.11 Using reversed axes, draw Fig. 10.40 in isometric pictorial.

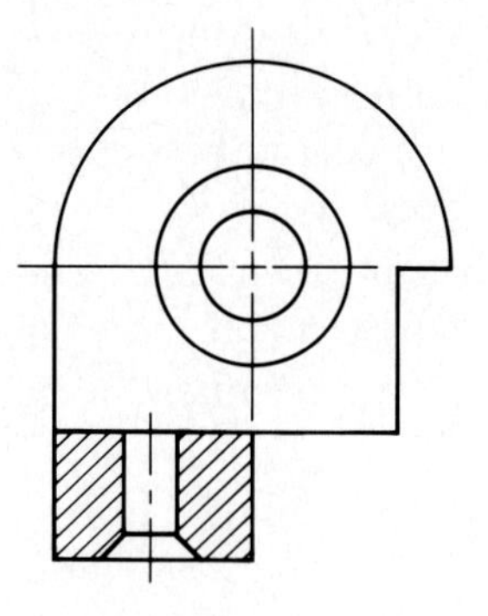

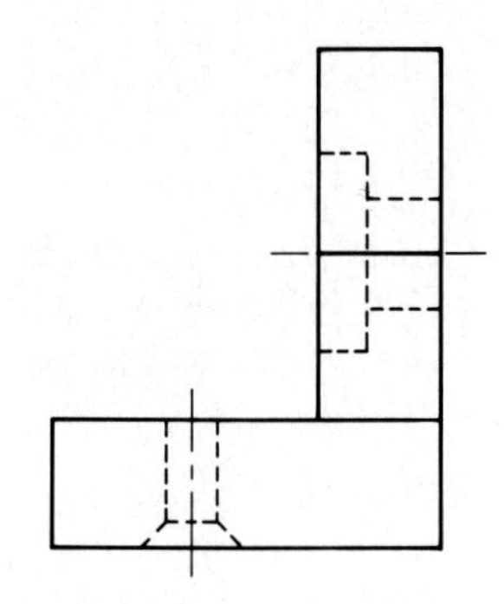

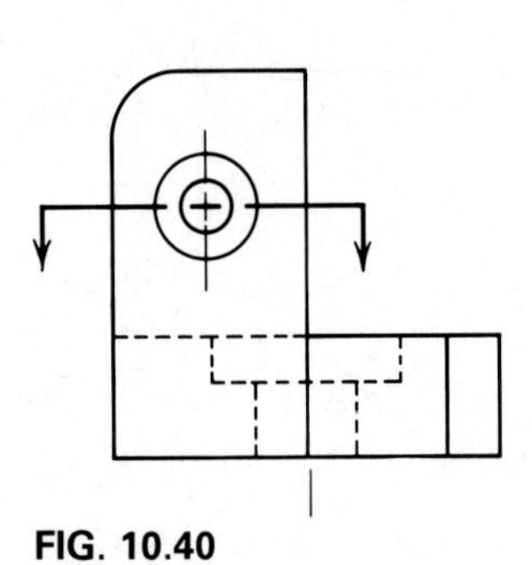

FIG. 10.40

10.12 Construct an oblique pictorial of the objects in Fig. 10.41.
(*a*) Cabinet
(*b*) Cavalier

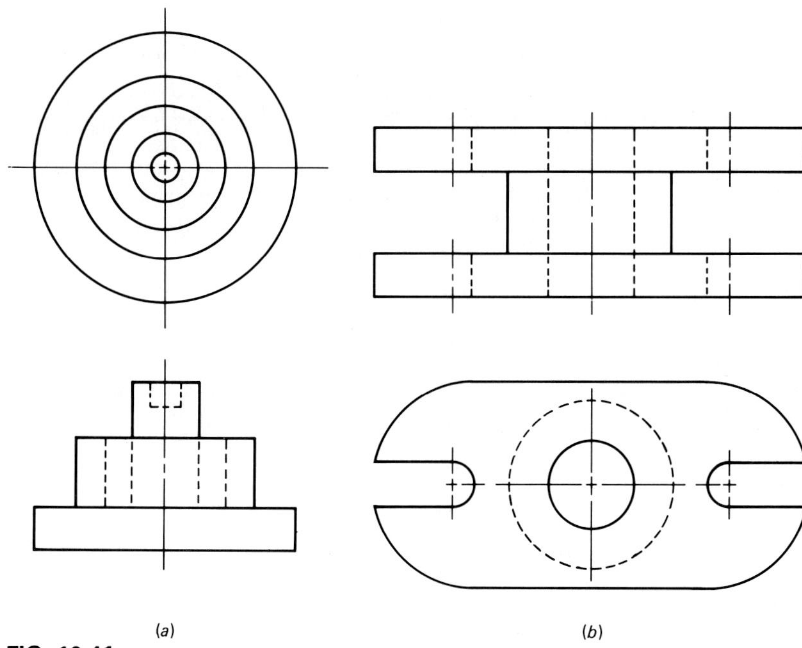

FIG. 10.41

chapter 11

Sectional Views

11.1 INTRODUCTION

Sectional views, frequently called sections, provide a method of showing the interior details of an object that are difficult to visualize from hidden lines in exterior views. Figure 11.1 shows how to visualize a section. Imagine that a cutting plane is passed through the object perpendicular to your line of sight. Then assume that the portion of the object between your eye and the imaginary cutting plane is removed to expose the interior details. Interior surfaces that have been cut by this plane are indicated by section lining (crosshatching). Whenever practical, the section view should appear in correct orthographic projection with normal exterior views, although we will present several examples where some other arrangement is more convenient.

11.2 CUTTING PLANE

The cutting plane is indicated by a broken line similar to the reference line used in orthographic projection. This plane is shown as an edge in the view adjacent to the one showing the sectional view. The sectional view in Fig. 11.1 is drawn in orthographic projection in Fig. 11.2. The cutting plane appears as an edge in the top view. Arrowheads are used to indicate the viewing direction. For a sectional view drawn in orthographic projection, no special labeling is required. If more than one section-

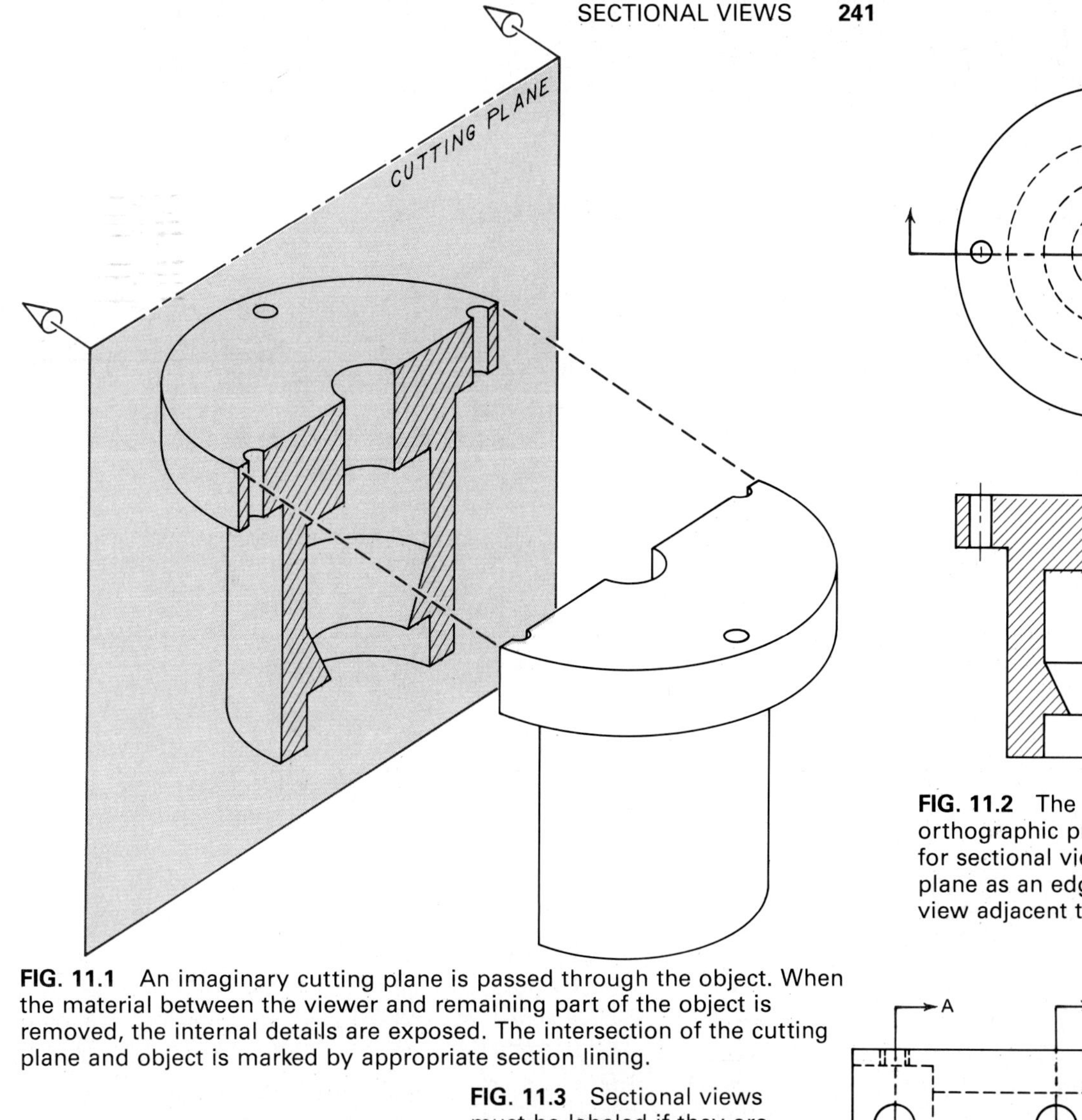

FIG. 11.1 An imaginary cutting plane is passed through the object. When the material between the viewer and remaining part of the object is removed, the internal details are exposed. The intersection of the cutting plane and object is marked by appropriate section lining.

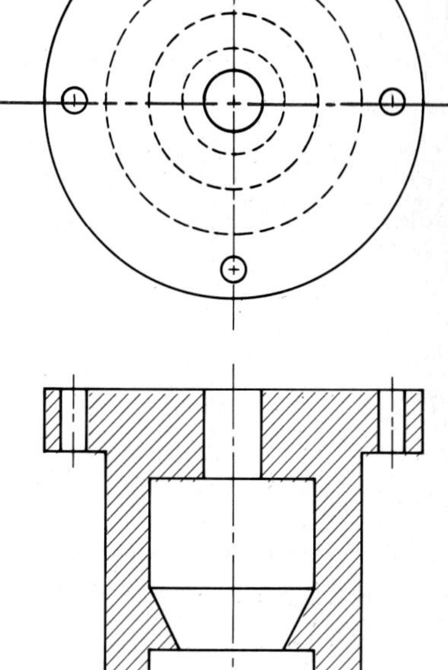

FIG. 11.2 The principles of orthographic projection are applied for sectional views. The cutting plane as an edge is indicated in the view adjacent to the sectional view.

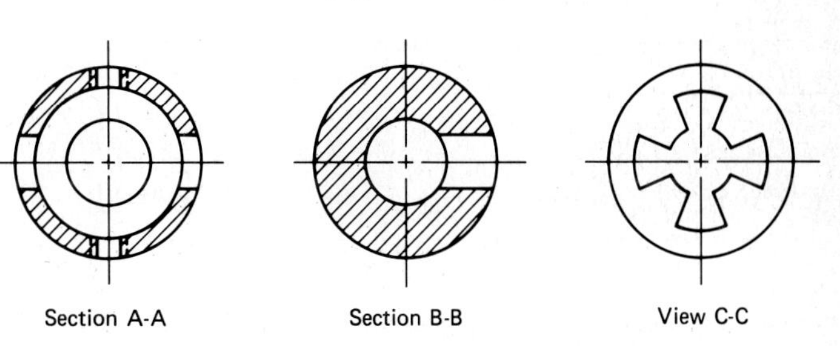

FIG. 11.3 Sectional views must be labeled if they are not in orthographic projection. Views not involving a cutting plane are labeled not as sections but as views.

al view of an object appears or if sectional views are not in orthographic projection, the sections are labeled alphabetically and if possible arranged so that the views lie alphabetically from left to right on the drawing sheet.

Figure 11.3 indicates the preferred labeling procedure for multiple sections or sections that are not in projection. Note that view C-C is not a section view since a cutting plane is not involved. However, the notation is similar and is used to clarify a view which is not in correct orthographic projection.

11.3 SECTION LINING

Figure 11.4 shows selected standard section lining symbols. Others are found in the standard from the American National Standards Institute, ANSI Y14.3—1975. If the distinction between materials is not important in a drawing, the general-use (cast-iron) symbol is usually used. The general-use symbol may be used when working with a single part (made entirely of the same material).

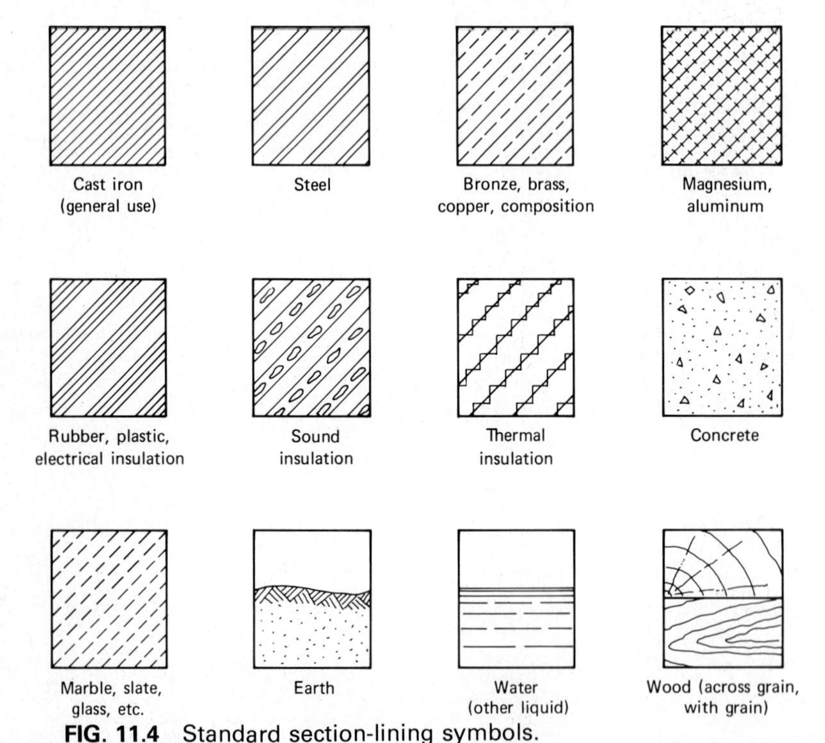

FIG. 11.4 Standard section-lining symbols.

FIG. 11.5 Appropriate section lining for two pieces of the same material which abut.

In a section of an assembly (a drawing showing the relationship between two or more parts), different symbols may be used to identify each part clearly if the parts are made of different materials. If some or all of the parts are made from the same material, which is often the case, the general-use symbol may be used with the angle of lining changed to distinguish between parts. Figure 11.5 shows appropriate crosshatching for two parts in a section made from the same material.

11.4 FULL SECTIONS

A full section is obtained by extending the cutting plane through the object, as shown in Fig. 11.6*a*. If the object is symmetrical, the cutting plane follows the center line. The cutting plane may be omitted if its position is obvious (Fig. 11.6*b*), or it may be shown if that would enhance the drawing.

Conventional sectioning practice is to omit hidden lines be-

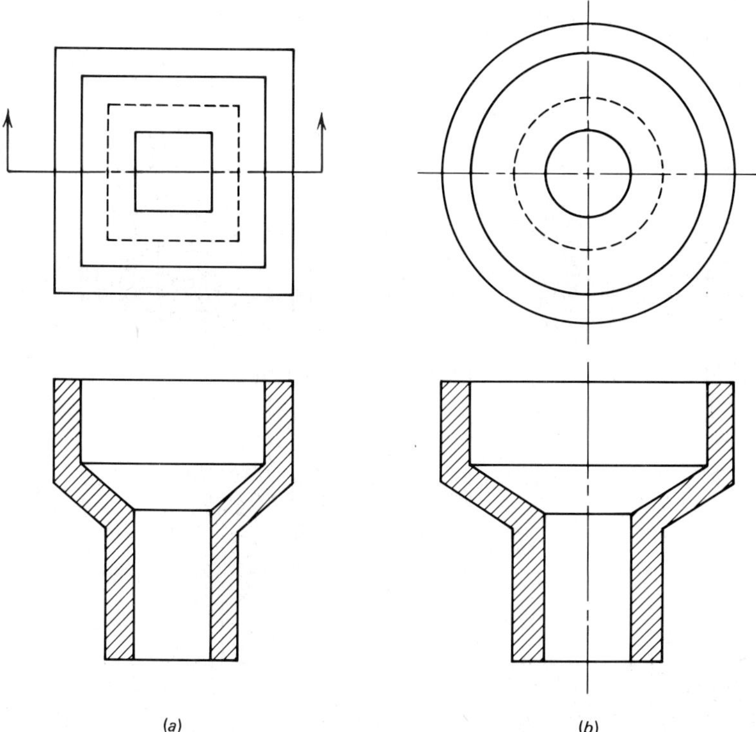

FIG. 11.6 Full section. Cutting plane line may be omitted if position is obvious.

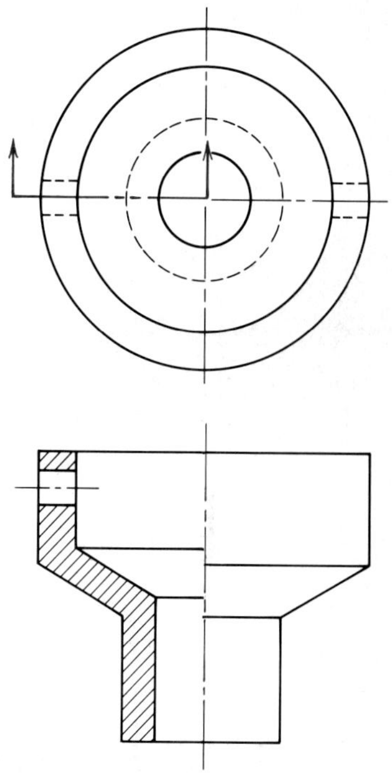

FIG. 11.7 Half section. The section and external view are divided by a centerline.

hind the cutting plane unless they are necessary to make the drawing clearer. Even when shown, hidden lines should be used sparingly in sections. Only the critical hidden lines should appear; others should be omitted.

11.5 HALF SECTIONS

The half section is used when you wish to show both internal detail and the exterior view of an object. This type of section is most effective when the object is symmetric or nearly so. Therefore, one half of the object illustrates internal detail, and the other half shows a normal external projection. The half section is visualized by imagining that one quarter of the object has been removed. Figure 11.7 shows the position of the cutting plane in the top view and the resulting half section in the front view. The section view and the external view are separated by a center line. The hidden lines in the external-view portion have been omitted according to conventional practice, although they could be added if needed to clarify the drawing.

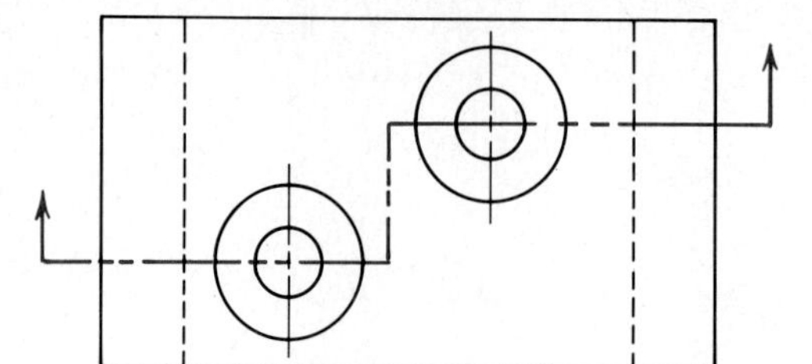

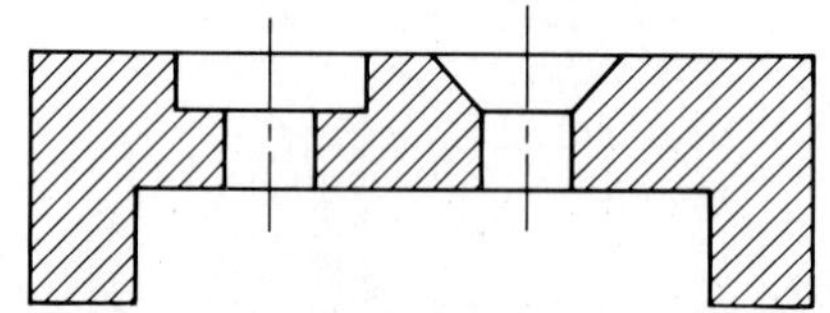

FIG. 11.8 Offset section. Cutting plane line must be shown on adjacent view.

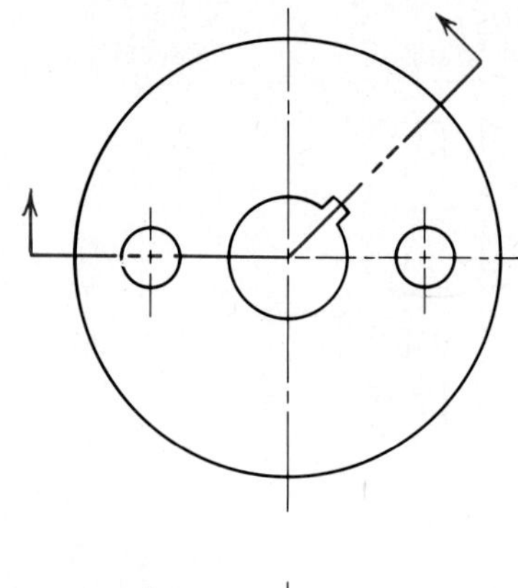

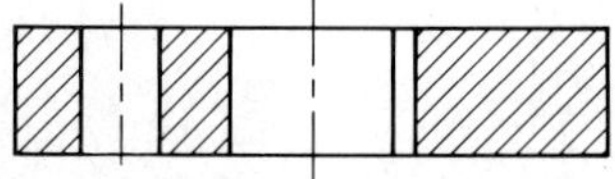

FIG. 11.9 Aligned section. The aligned portion is rotated perpendicular to the viewing direction.

11.6 OFFSET SECTIONS

The offset section allows internal details that are not in the same plane to be shown in a single section. The cutting plane goes through the object but is stepped or offset so that details in different planes can be shown. The offset is generally drawn at 90° to the viewing direction. However, this "break" in the cutting plane is not shown in the section view. The cut material is lined as though the offset were not present. Therefore, an offset cutting plane must always be shown on the drawing, since the presence of the offset cannot be detected in the section view. Figure 11.8 shows the appropriate use of an offset section.

11.7 ALIGNED SECTIONS

The aligned section, which is illustrated in Fig. 11.9, is similar to the offset section in that the cutting plane does not pass through the object in a straight line. The difference is that the aligned section is placed at some angle other than 90°. The features cut by the cutting plane are rotated into a plane perpendicular to the viewing direction. In Fig. 11.9, the aligned section allows the keyway to appear more clearly in the front sectional view.

11.8 REMOVED SECTIONS

If space is not available or if for other practical reasons a section view cannot be drawn in orthographic projection, a removed section can be drawn. If the removed section is placed on a sep-

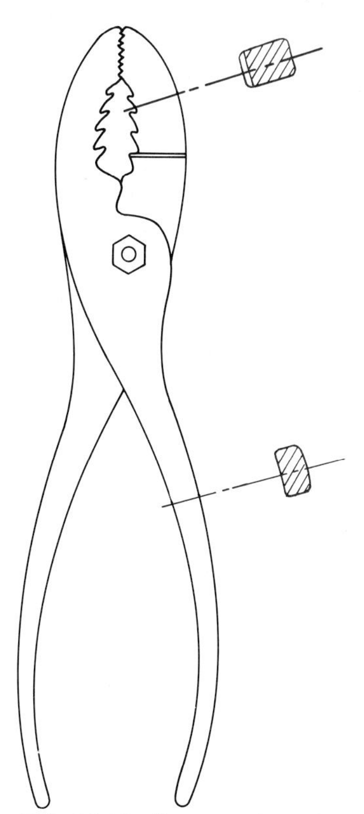

FIG. 11.10 Removed sections.

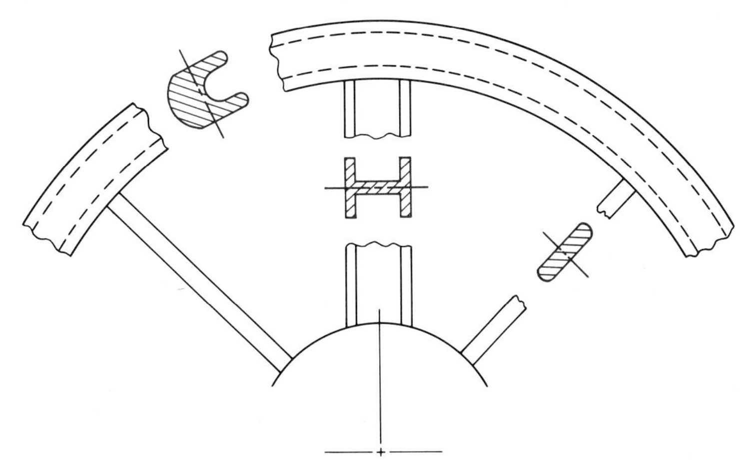

FIG. 11.11 Revolved sections.

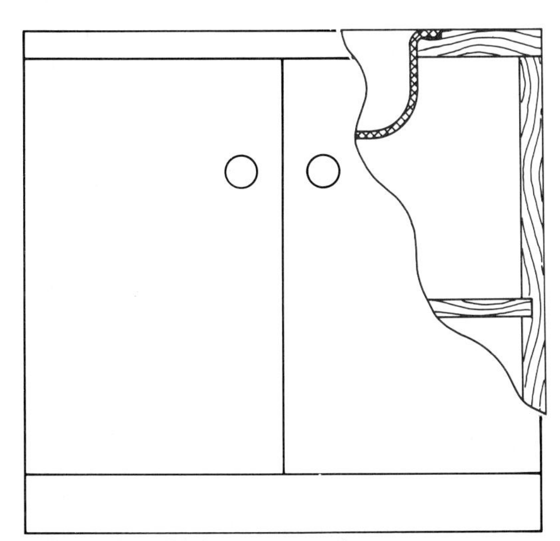

FIG. 11.12 Broken-out section.

arate sheet, a note is placed near the cutting plane (with a leader pointing to the cutting plane) explaining where the section will be found. Likewise, a note should appear with the section view so that the cutting plane can be located. Most applications of the removed section are of the type shown for the pliers in Fig. 11.10, where the sections of the handle and jaw are conveniently located.

11.9 REVOLVED SECTIONS

Elongated items such as spokes of wheels, shafts, etc., can be shown in section simply by passing a cutting plane through the object perpendicular to its axis and then revolving it 90° without translation so that the cut face appears in the same view. The cutting plane is not shown. Break lines often are used to separate the visible portion of the object from the revolved section, leaving a small gap for clarity. Figure 11.11 illustrates uses of the revolved section.

11.10 BROKEN-OUT SECTION

Often, only part of the object must be revealed by a section view, and this can be accomplished in a single orthographic view with a broken-out section. Figure 11.12 shows a broken-out section of a cabinet revealing the side, the top, a shelf, and a sink. A break line separates the broken-out portion from the remainder of the drawing.

11.11 AUXILIARY SECTIONS

If a sectional view drawn in orthographic projection appears in any view other than a principal view, it is called an auxiliary sec-

FIG. 11.13 Auxiliary section.

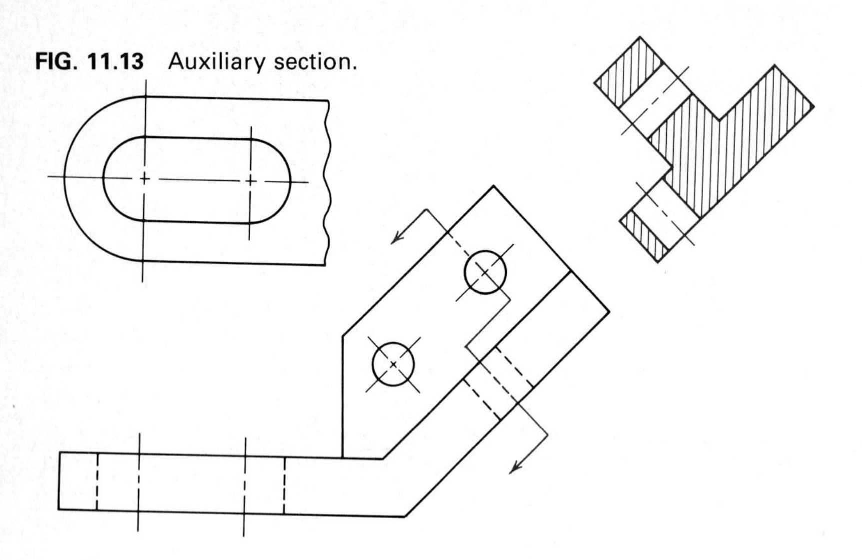

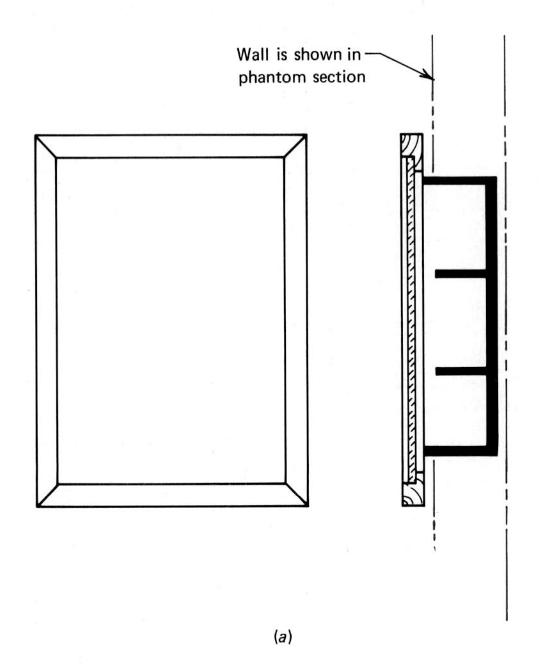

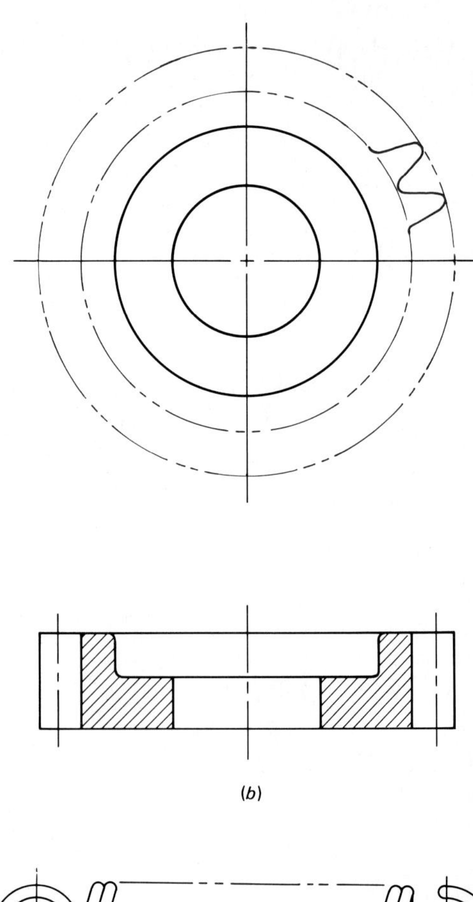

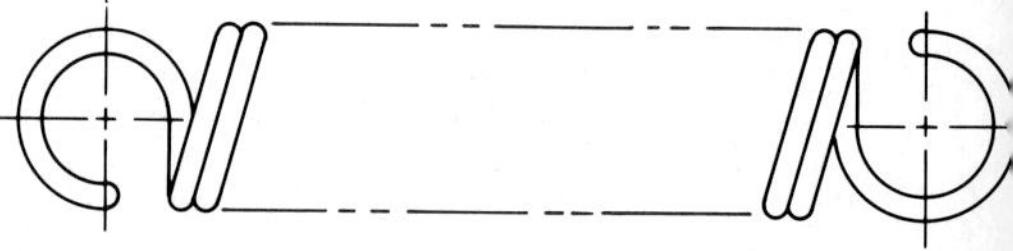

FIG. 11.14 Three applications for the phantom section.

tion. The auxiliary offset section in Fig. 11.13 clearly shows the T-shaped flange and hole locations.

11.12 PHANTOM SECTIONS

If a part is to be positioned relative to another part, it may be appropriate to show the parts together in a sectional view without omitting the external part. To accomplish this, a phantom section may be used. As shown in Fig. 11.14*a*, the internal part can be clearly seen inside the phantom external part drawn with phantom lines.

Phantom lines are also used to represent alternative positions of objects such as doors and levers as well as to represent repeated segments of a part, such as gear teeth and the spring that are shown in Fig. 11.14*b* and *c*.

11.13 CONVENTIONAL REPRESENTATION APPLIED TO SECTIONS

The concept of conventional representation was introduced in Chap. 10 and will be extended here to include items specifically related to section views.

Briefly, conventional practice allows violations of true orthographic projection in the interest of time (cost) and ease of understanding. These deviations are widely used and must become part of the engineer's knowledge.

The conventional practices covered in Chap. 10 include line precedence, rotation of holes and other features into the viewing plane, and simplified drawing of small intersections. These practices generally apply to section views as well as external views.

In instances where the cutting plane passes through relatively thin elements of an object, such as ribs and lugs, the section lining is omitted to avoid giving the impression of a massive object. See Fig. 11.15*a*. Alternatively, spaced section lining may be used for the same purpose. The spaced lining shown in Fig. 11.15*b* is necessary to distinguish between the rib that is cut and the rib that is viewed behind the cutting plane in the section.

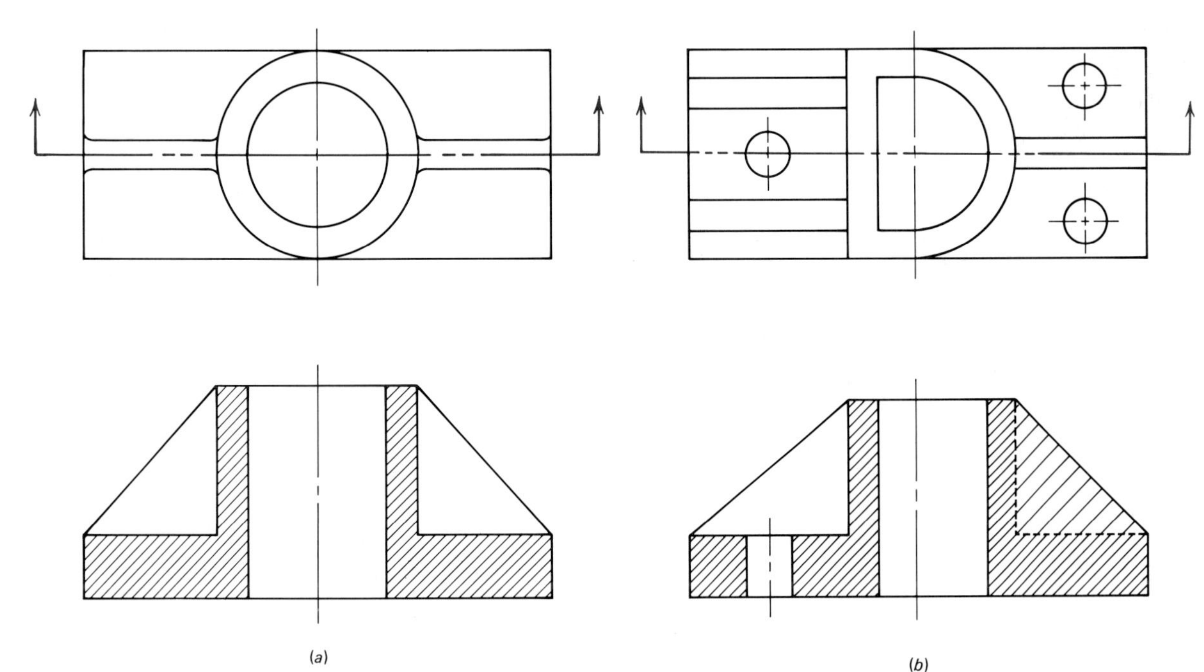

FIG. 11.15 Conventional practices for ribs in section.

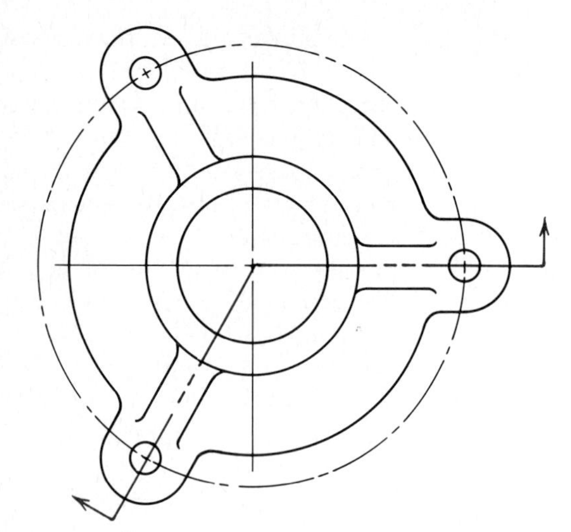

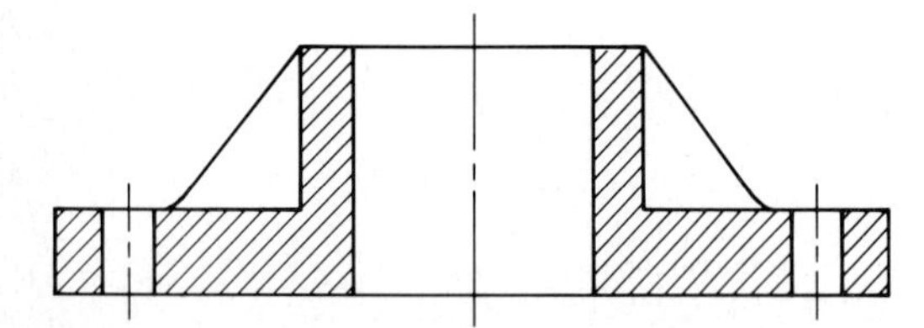

FIG. 11.16 Conventional practice for lugs and ribs in section. As with an aligned section, the features are rotated into a plane perpendicular to the viewing direction.

In Fig. 11.16, the rib and lug have been revolved to the plane of the section view, and the ribs are not lined. Note also the use of the bolt circle, which may be used to dimension the location of the bolt holes relative to the center of the object.

The section view in Fig. 11.17 illustrates another conventional practice related to sections. Here the cap screws and shaft, although severed by the cutting plane, are not lined. Generally, items such as bolts, keys, shafts, screws, and pins that are cut longitudinally by the cutting plane are not lined in the section view. If they are cut by the plane perpendicular to their axes, these parts are section-lined in the usual way.

When drawing or reading drawings of complex objects, you should become familiar with the standards which outline the con-

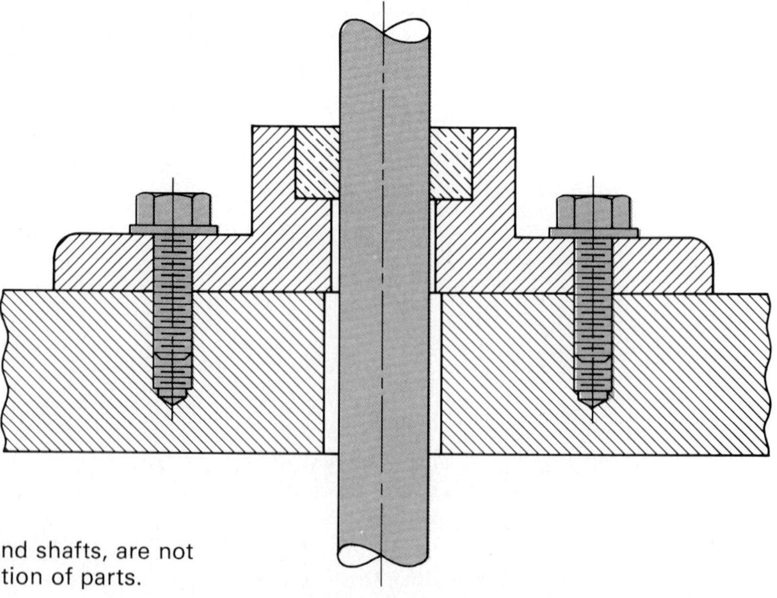

FIG. 11.17 Smaller objects, such as fasteners and shafts, are not section-lined in order to maintain a clear separation of parts.

ventional practices in detail. One of these standards is ANSI Y14.3—1975.

11.14 APPLICATIONS FROM COMPUTER GRAPHICS

Orthographic section drawings can be created with a computer-graphics terminal in the same way as can exterior views. The only difference is the required section lining, which can be generated as individual lines or as a special character in the system software. As with any drawing created on a graphics terminal, a hard copy can be obtained from an appropriate pen plotter, electrostatic plotter, or other type of hard-copy unit.

One function of engineering that is becoming more automated through the aid of computers is the drafting function. Many systems on the market today perform the tasks of drawing electronically. The input to these systems may be a light pen, digitizer board, thumbwheel, joystick, and so on. The drafter sits in front of the screen and simply enters the symbols and connects them to form the drawing. See Fig. 11.18. The engineer becomes involved with these drafting systems in a supervisory manner or in the development of the software that carries out the commands of the terminal operator (drafter).

A much more complex problem involves having the geometry of a solid residing in the computer as a numeric database and programming the computer to determine and display an arbitrary section view. Consider the object shown in Fig. 11.8. To have a computer determine the offset section, the cutting plane must be defined geometrically. The section view is then the intersection of the plane and the known object. Finally, the cut surfaces must be lined to complete the section.

FIG. 11.18 A computer-graphics terminal using appropriate software can produce sectional views of complex objects. (Courtesy of Control Data Corporation.)

Problems

11.1 Define each of the following types of sections and draw or sketch a simple example to illustrate the use of each type:

- (*a*) Full section
- (*b*) Half section
- (*c*) Offset section
- (*d*) Aligned section
- (*e*) Removed section
- (*f*) Revolved section
- (*g*) Broken-out section
- (*h*) Auxiliary section
- (*i*) Phantom section

11.2 Explain the similarities between the conventional practice of rotating a feature into the plane of a full section and the use of an aligned section.

11.3 List one section where the cutting plane must be shown, one where it is sometimes shown, and one where it is never shown.

11.4 Describe the conventional practices that apply to fasteners shown in sectional views.

11.5 Draw or sketch each of the objects in Fig. 11.19 on a sheet of paper, converting one of the given orthographic views to an appropriate section view.

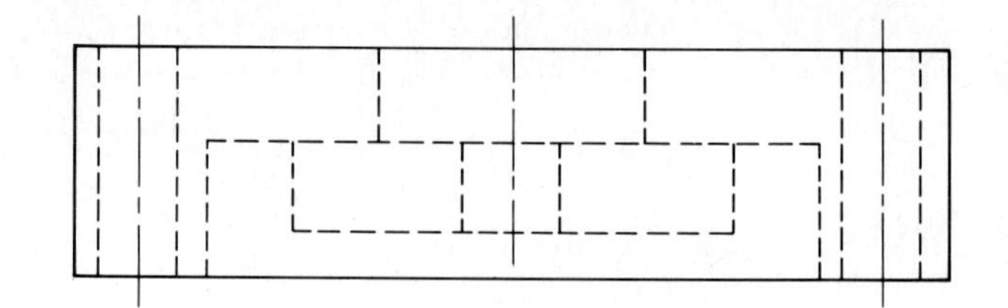

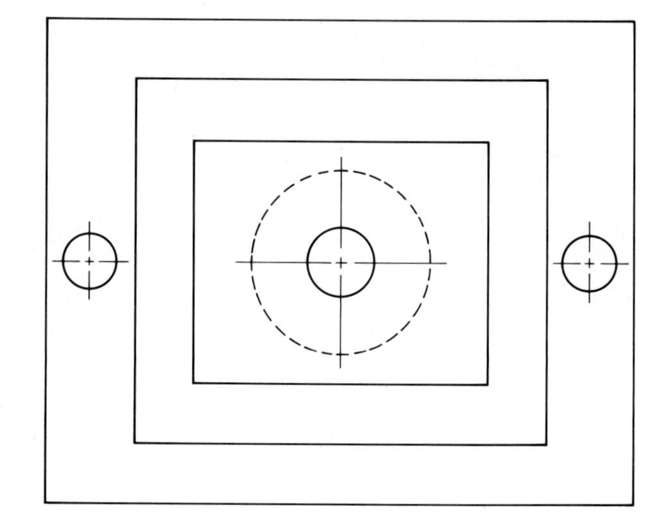

(*a*)

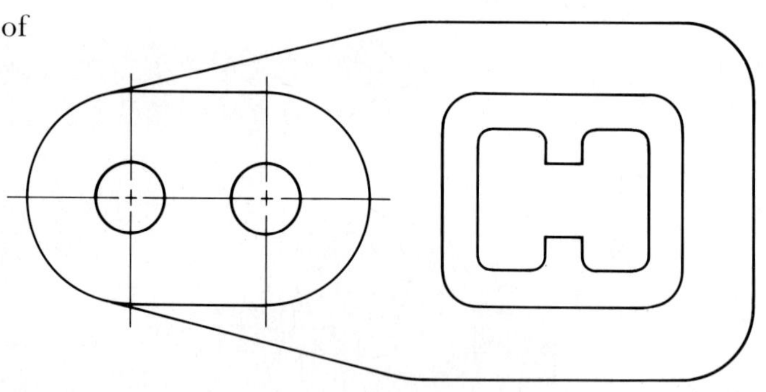

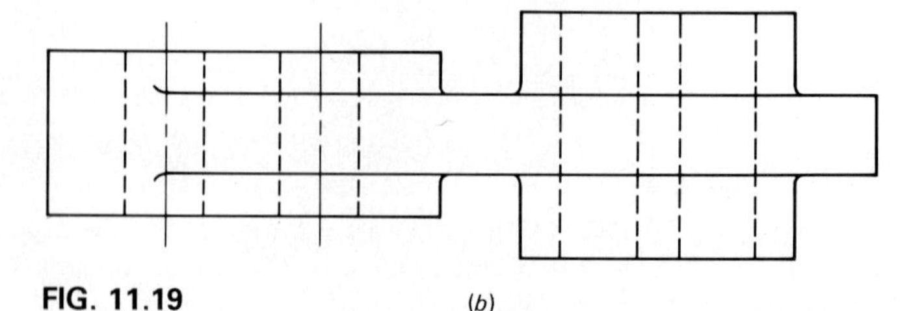

FIG. 11.19 (*b*)

11.6 Sketch the given top view of Fig. 9.8 but substitute a full section for the given front view.

11.7 Draw or sketch the H and F views of Fig. 11.20. Add an appropriate revolved section of the knurled handle and a removed cross section to illustrate the contour of the support ribs for the end wrench.

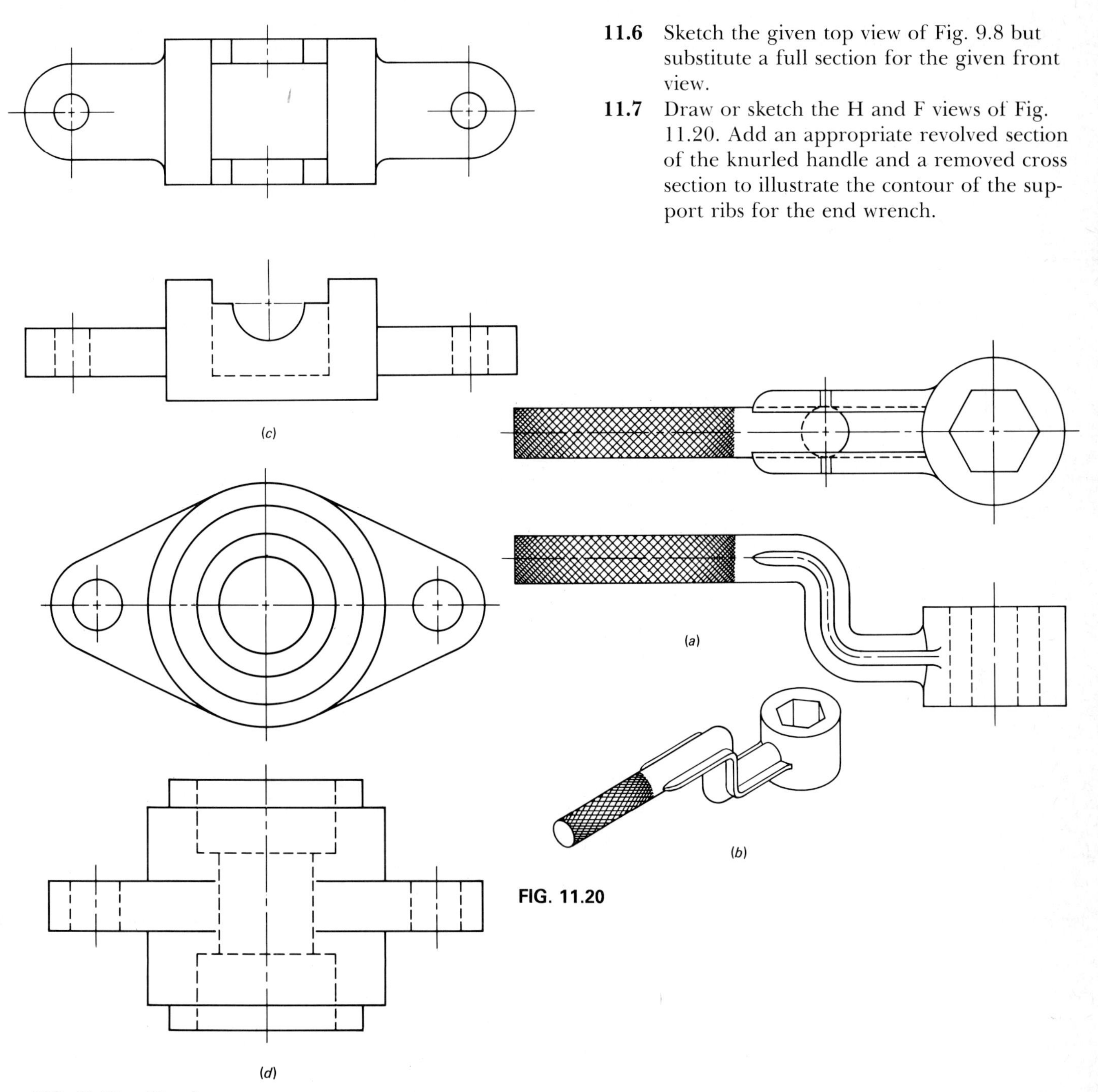

FIG. 11.20

FIG. 11.19. (*Cont.*)

11.8 Draw or sketch the front view of Fig. 11.21. Using conventional practices, substitute an appropriate profile view.

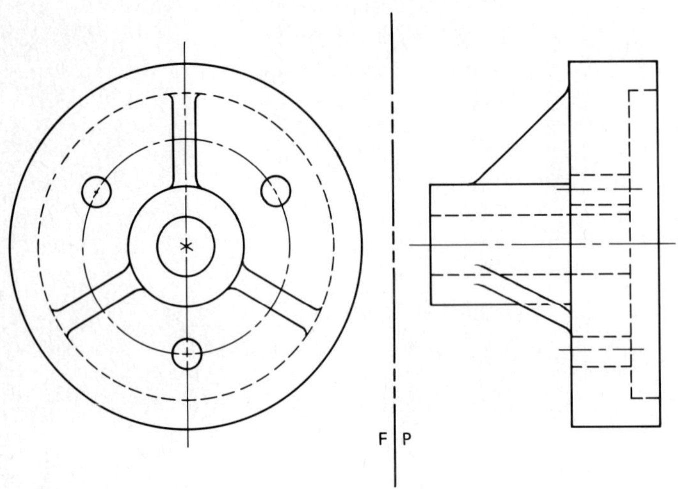

FIG. 11.21

11.9 Draw or sketch the necessary orthographic views of the objects shown in Fig. 11.22. Use appropriate sections to enhance understanding of the drawings.

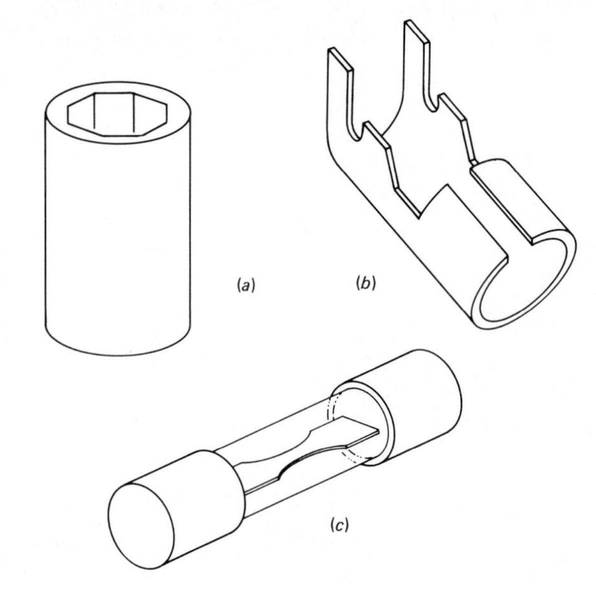

FIG. 11.22

chapter 12

Dimensioning

12.1 INTRODUCTION

With a knowledge of orthographic projection, it is possible to delineate a three-dimensional object by constructing an appropriate multiview. To complete the description of the object, dimensions must be added to the drawing. Dimensions include such items as locations and sizes of features, materials to be used, and surface finish. To dimension a drawing accurately, you must be aware of the users of the drawing. For example, those who make the patterns, set up the machines for manufacture, assemble the parts, and use the product may employ the drawing to effect their tasks. Also, you must consider how the part mates with other parts and how the part is to be used, i.e., the function of the part.

Agencies such as the American National Standards Institute (ANSI) have established standards for dimensioning. "Engineering Drawing and Related Practices—Dimensioning and Tolerancing," which is designated ANSI Y14.5M—1982, replaces ANSI Y14.5—1973. The dimensioning concepts discussed here will conform to the standards set forth in Y14.5M—1982. You are encouraged to refer to this standard for specific details and additional dimensioning concepts.

12.2 DIMENSIONING FUNDAMENTALS

Even the most complicated part can be broken down into simple geometric shapes, including prisms, cylinders, cones, spheres, and portions of these figures.

Each simple form must be dimensioned to indicate its size and location from some datum (reference point, line, or surface).

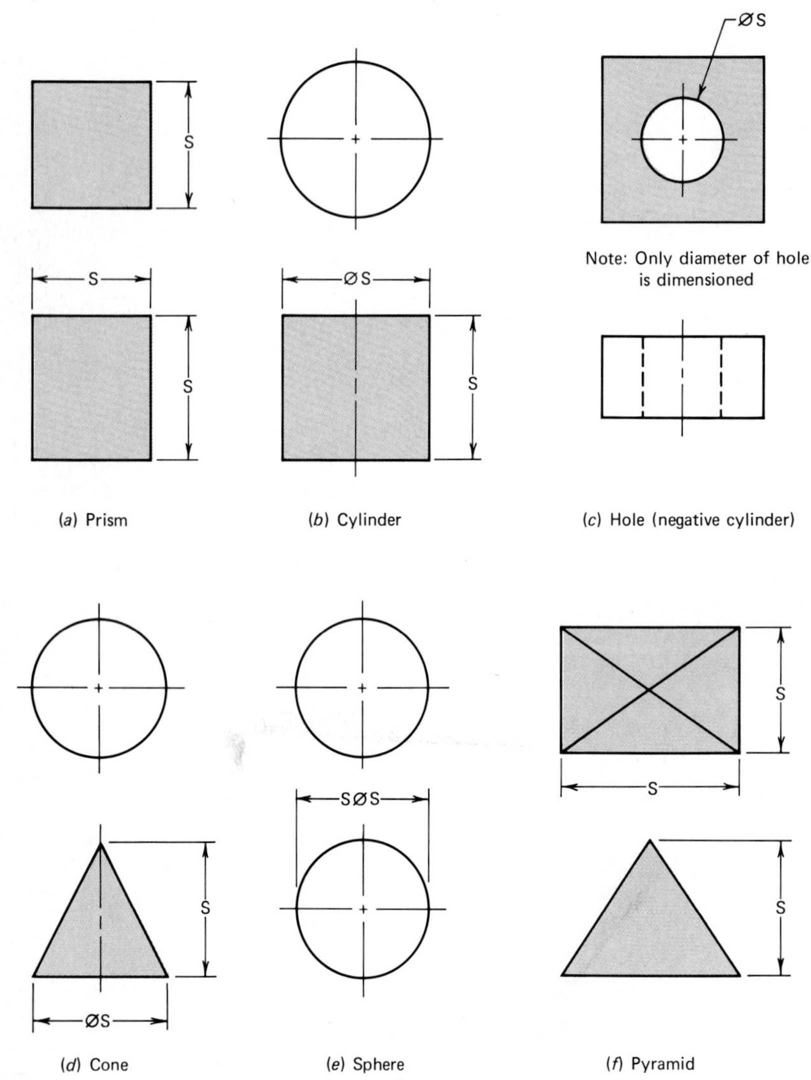

FIG. 12.1 Dimensioning geometric shapes.

12.2.1 Size Dimensioning

The recommended practice for dimensioning the size of some simple geometric shapes is illustrated in Fig. 12.1. S has been used to designate a size dimension. Refer to Fig. 12.1 and consider the following:

1. A prism is dimensioned by specifying two of its dimensions in a principal view and the third in an adjacent view.

2. A cylinder is sized by showing both diameter and length in the rectangular view. Note that the symbol ∅ precedes the size dimension S and refers to diameter.

3. A hole, also called a negative cylinder, is dimensioned by giving the diameter in the circular view as a note with a leader pointing to the hole. If a hole does not go through (blind hole), the depth may be specified in the same view with the diameter or on a rectangular view.

4. The diameter and height of a cone are placed in the view showing the cone as a triangular shape.

5. Only one dimension is required for a sphere. The diameter is preceded by the symbol S∅, which stands for spherical diameter.

6. A pyramid is dimensioned by showing two dimensions in the view where the base is seen as a rectangle.

12.2.2 Location Dimensioning

The positions of features (holes, slots, locating pins, and so on) must be accurately located to ensure proper functioning of a part, particularly if it must mate or function with other parts. We begin by selecting reference lines

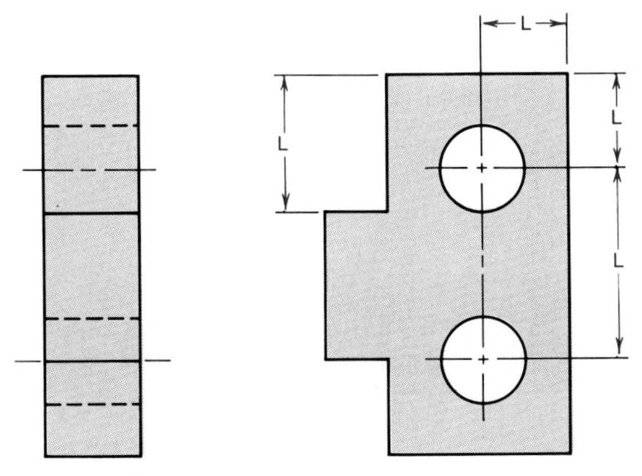

FIG. 12.2 Location dimensions.

or surfaces (datums) from which to position features. This selection is influenced by manufacturing and quality-control practices. For example, a hole can be located more accurately from a flat machined surface than from a rough-cast surface.

Features may be located with dimensions from surface to surface, from center line to surface, or from center line to center line. Figure 12.2 illustrates a simple bracket where the symbol L is used to denote a location dimension. Note that size dimensions have been omitted although the location dimension for the notch could also be interpreted as a size dimension. The designer has decided to locate the notch and hole from the top edge of the part while locating the second hole relative to the first hole. This procedure ensures that the two holes are placed correctly relative to each other even if the distance from hole to edge is slightly incorrect. Note that the location dimensions involving center lines are placed in the view where the features appear circular.

12.3 DIMENSIONING PRACTICE

You should perform the following steps sequentially to produce a properly dimensioned drawing:

1. Consider the necessary orthographic views to describe the object. In most cases the same views will be adequate for dimensioning purposes.

2. Visualize the simple geometric shapes that make up the object and place size dimensions on each shape.

3. Choose surfaces or center lines that will serve as datums and locate each feature relative to these datums.

4. Place the overall dimensions for the object.

5. Add title information and other necessary notes.

12.3.1 Contour Rule

It is natural to look for the size and location dimensions of a fea-

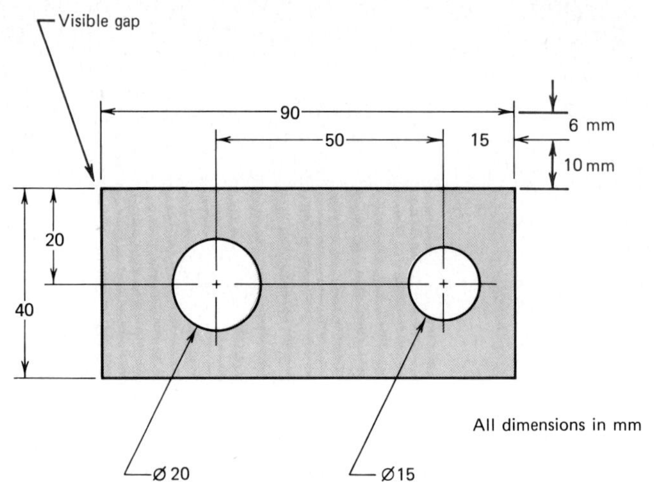

FIG. 12.3 Spacing of dimensions.

ture in the view where it appears in its most characteristic shape. Therefore, it is logical to place dimensions where one would expect to find them. We will refer to this as the contour rule. Thus, a hole is located and sized in the view where it is seen as a circle, and a slot is dimensioned where one can see through it, not where it appears as hidden lines. Occasionally dimensions are referred to a hidden line (the depth of a blind hole, for example), but generally it is preferable to dimension visible features.

12.3.2 Spacing of Dimensions

The basic rule of placing dimensions is not to crowd. Figure 12.3 illustrates the correct spacing to use. Leave a minimum of 10 mm from the object to the first dimension line and about 6 mm between parallel dimension lines. Dimensions will normally be placed outside the views. It is permissible to place a dimension inside a view if this will enhance clarity significantly. This procedure should be regarded as the exception rather than the rule.

12.3.3 Dimension Lines, Extension Lines, and Leaders

Dimensions are placed on a view by means of dimension lines and extension lines or by means of a leader from a note to the applicable feature. Extension lines are used to show continuations of visible lines on the object. They are lighter weight than the object lines and begin with a small gap next to the object lines. Dimension lines are also lightweight; they are drawn between extension lines parallel to the object line they reference. Arrowheads are used at the ends of a dimension line terminating at an extension line or center line, as can be seen in Fig. 12.4. Arrowheads are drawn so that they are about three times as long as they are wide. Normally, dimension lines are broken somewhere near the center for placement of the dimension.

Avoid having extension lines cross each other or cross dimension lines. In situations where crossing is unavoidable, there should be no break in the lines, except when an extension line crosses an arrowhead, in which case a break is preferable. If smaller location or size dimensions are placed near the object and larger dimensions are placed progressively outward, most line crossings can be avoided.

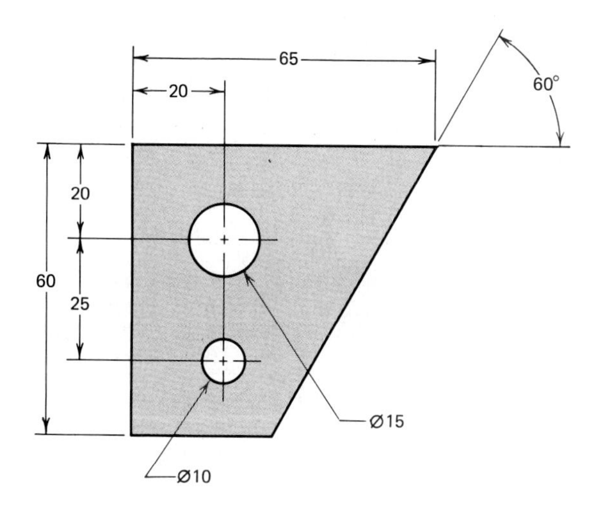

FIG. 12.4 Dimension lines, extension lines, and leaders.

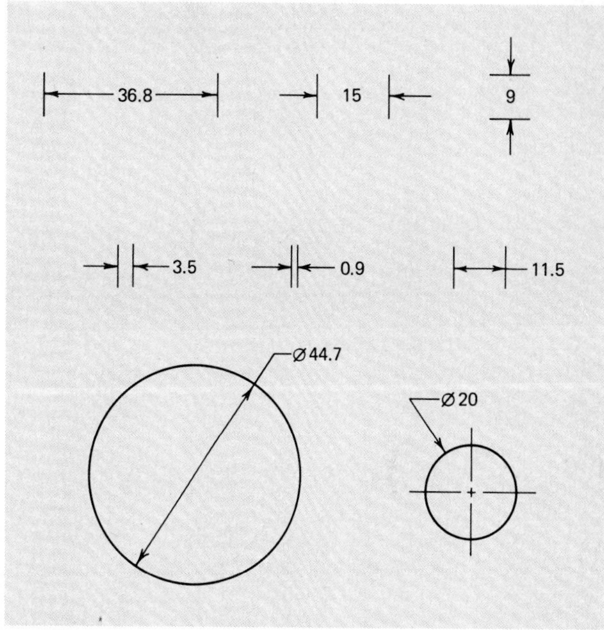

FIG. 12.5 Millimeter dimensions.

Leaders are lightweight lines that terminate on one end with an arrowhead pointing at a feature and on the other end with a short horizontal line directed toward a note or specification. As in Fig. 12.4, when a leader is directed toward a circle or arc, the leader should be radial. All line segments of a leader should be straight.

The angle in Fig. 12.4 is dimensioned by means of extension lines from the object and a dimension line that is an arc centered at the apex of the angle, with arrowheads at each end. Note that where object lines or center lines are drawn at right angles and are not dimensioned, a 90° angle is implied.

12.3.4 Dimensioning Units

Most drawings are dimensioned in decimal units, although specialized drawings such as structural steel drawings use fractional values. We will direct our attention to the decimal values in this text.

Many drawings of machine parts are now prepared using millimeters, although some industries and companies still prefer a decimal inch system and certainly older drawings will have decimal inch dimensions. Thus, it is necessary for you to be familiar with both types.

Several conventions regarding millimeter dimensioning have been adopted that are derived from the international agreements that define Système International d'Unites (SI) units. They are as follows (refer to Fig. 12.5 for examples):

1. A zero must precede the decimal point for values less than 1.

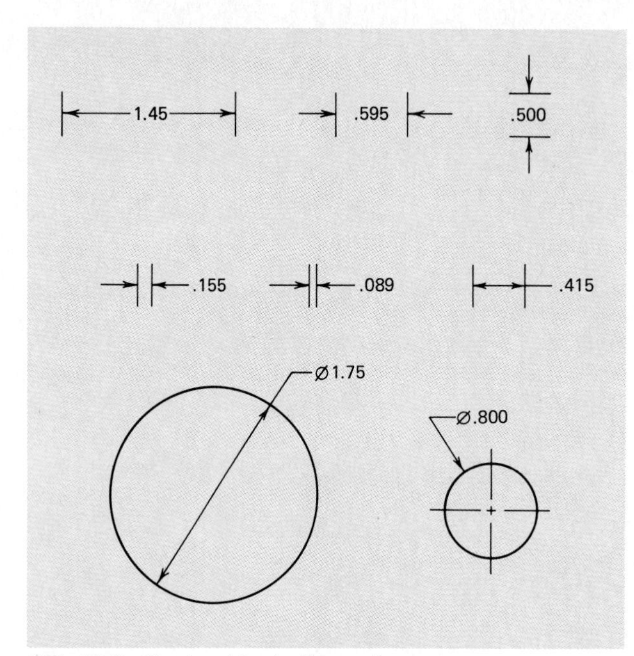

FIG. 12.6 Decimal inch dimensions.

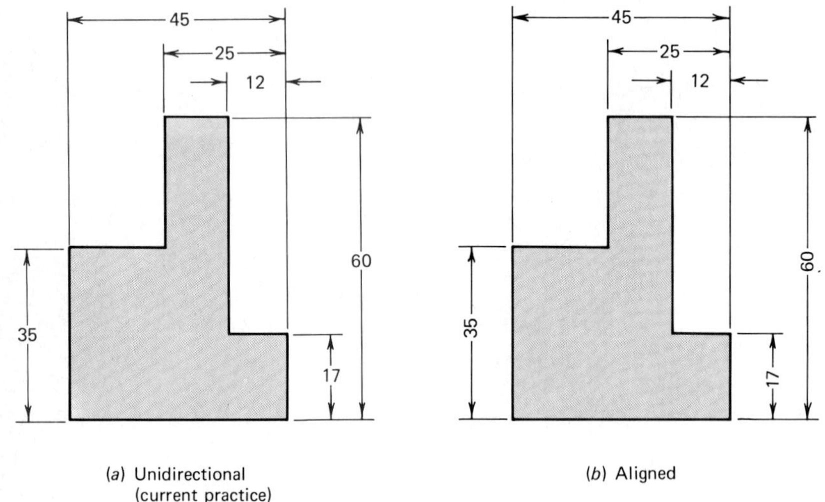

FIG. 12.7 Dimension orientation.

2. No decimal point or following zero is used for values that are whole numbers (12, not 12. or 12.0).

3. The last number of a decimal fraction should not be a zero (12.3, not 12.30). An exception will be noted later in the discussion of limit dimensions.

4. Spaces rather than commas are used to separate large or small numbers into groups of three (22 050 and 0.065 42 rather than 22,050 and 0.065,42).

The standards for dimensioning with decimal inches differ somewhat from those for millimeters. These standards are as follows (see Fig. 12.6 for examples):

1. No zero precedes the decimal point for values less than 1.

2. Zeros should be added in order to express the dimension to the same number of decimal places as the tolerance.

12.3.5 Additional Practices

Current practice requires that dimensions and notes be placed on a drawing so that they are readable from the bottom of the drawing. This is consistent with what was previously called the unidirectional system. The other option is called the aligned system. In the latter case, dimensions are written to align with the dimension line and be readable from the bottom or the right side of the drawing. Refer to Fig. 12.7 for examples of both methods of dimensioning.

Reference dimensions may be placed on a drawing to provide technically unnecessary but helpful information. For example, an overall dimension may be given as a reference dimension even though intermediate dimensions specify the part size completely. The reference dimension is denoted by enclosing it in parentheses. Without an indication that such a dimension is a reference dimension, the object will be overdimensioned. An example of the correct use of reference dimensions is shown in Fig. 12.8.

Figure 12.8 also introduces use of the symbol X to specify repetitive features. In this case, two holes of the same diameter are required, so 2 X ∅20 means that there are two holes of diameter 20 mm. Alternative notations are ∅20 2 PLACES or ∅20 2 HOLES. In this context × means times or places but in other situations × can mean "by", as in 20 × 50, which means 20 by 50 mm. Care is needed to avoid confusion. It is appropriate to use 3 × 20 × 50 as a note to mean three identical features of size 20 by 50 mm.

12.4 DIMENSIONING FEATURES

Procedures for dimensioning simple features such as cylinders and prisms were introduced when size dimensioning was discussed earlier in this chapter. Additional features that require unique methods are examined here.

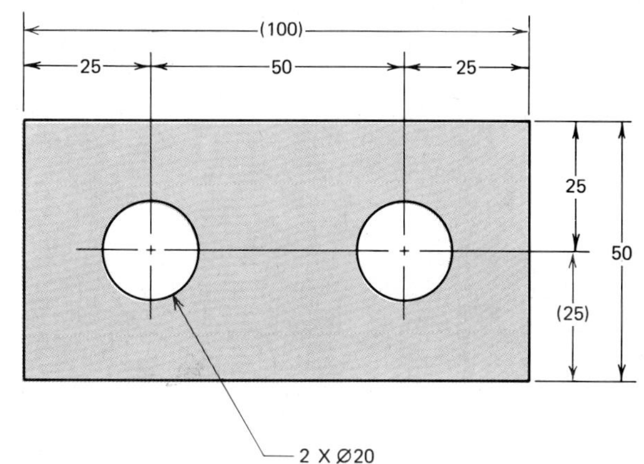

FIG. 12.8 Reference dimensions.

12.4.1 Symbols

Current standards employ symbols which replace many notes that previously were required in order to make the drawing understandable universally. Figure 12.9 defines some of the symbols in current use.

SYMBOL	MEANING
∅	Diameter
S∅	Spherical diameter
R	Radius
CR	Common radius
SR	Spherical radius
$\frown$ 25	Arc dimension
3X	Number of times/places
(15)	Reference dimension
<u>75</u>	Dimension not to scale
⌴	Counterbore/spotface
∨	Countersink
↧	Depth/deep
□	Square shape
√	Surface texture

FIG. 12.9 Dimensioning symbols.

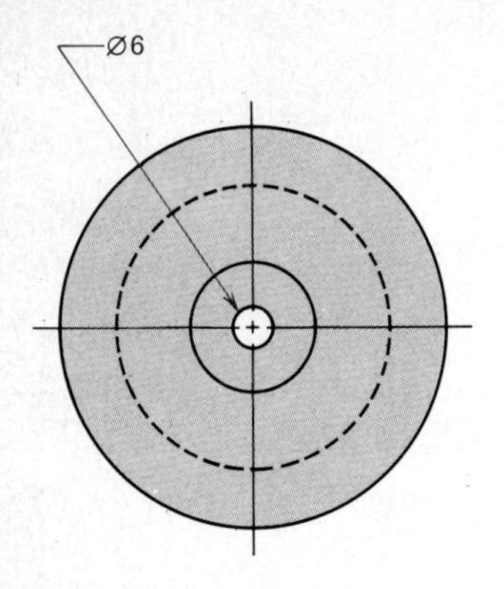

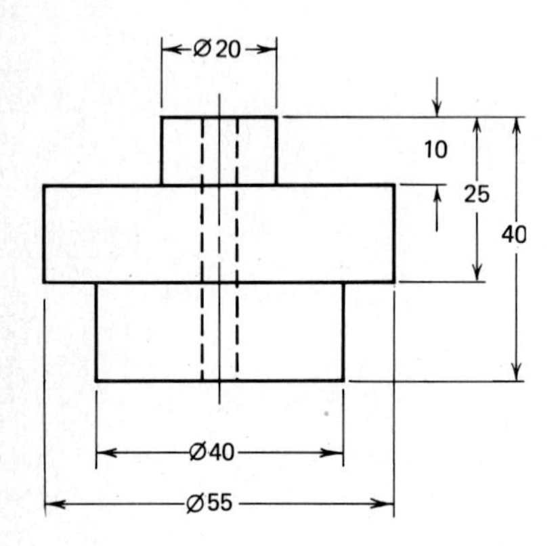

FIG. 12.10 Dimensioning diameters.

12.4.2 Diameters and Radii

The procedures for dimensioning positive and negative cylinders were indicated in Sec. 12.2.1. A part having several diameters is dimensioned in Fig. 12.10.

The radius is denoted by a leader and arrowhead pointing at the arc. The arrowhead never is placed at the center of the radius. The space available determines where the dimension is placed. Some radii have no center position shown. A small cross is placed at the center of radius if the center location must be shown and is to be dimensioned. The concept of a spherical radius should be readily apparent. See Fig. 12.11 for examples of radius and spherical radius.

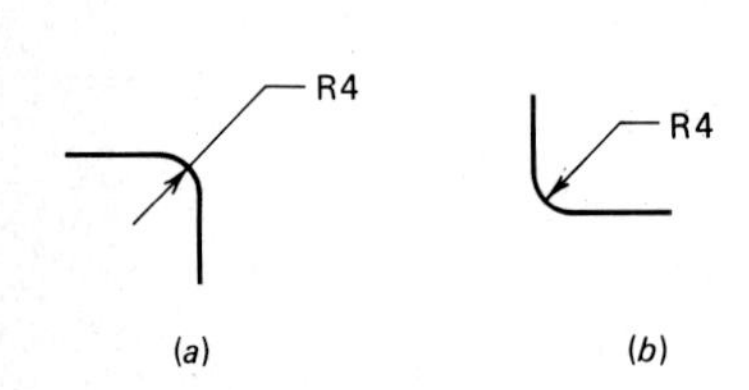

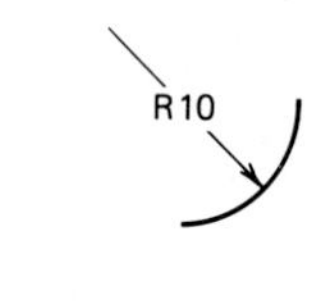

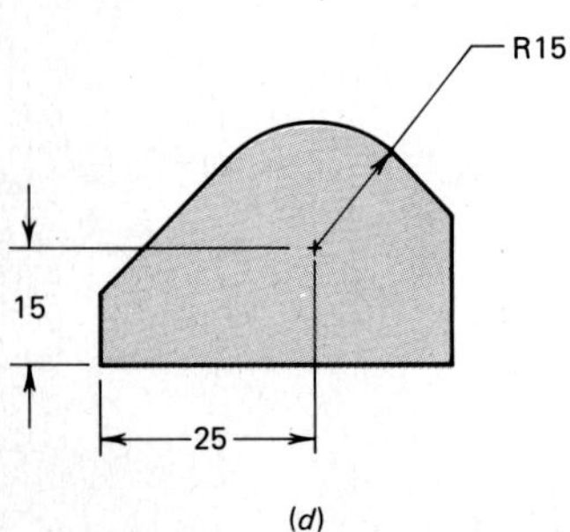

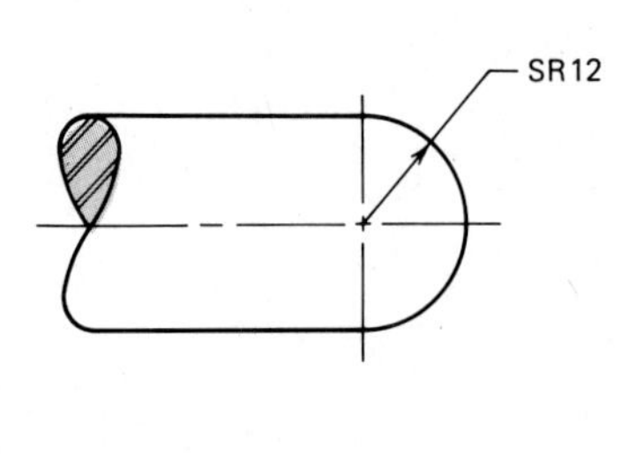

FIG. 12.11 Dimensioning radii.

12.4.3 Angles, Arcs, and Chords

When an angle is dimensioned, extensions of the radii are drawn, and the angle is placed in dimension lines that follow the surface. For an arc, the arc symbol is used in curved dimension lines, but the extension lines are parallel. The chord is shown by a straight dimension line and parallel extension lines. Each of these is shown in Fig. 12.12.

12.4.4 Partially and Fully Rounded Ends

Figure 12.13 illustrates how an object with rounded ends should be dimensioned. The overall

60°
30
28

(*a*) Angle (*b*) Arc (*c*) Chord

FIG. 12.12 Dimensioning angles, arcs, and chords.

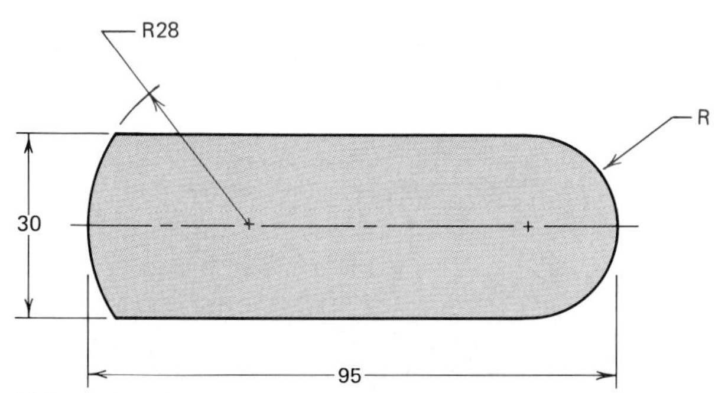

FIG. 12.13 Dimensioning objects with rounded ends.

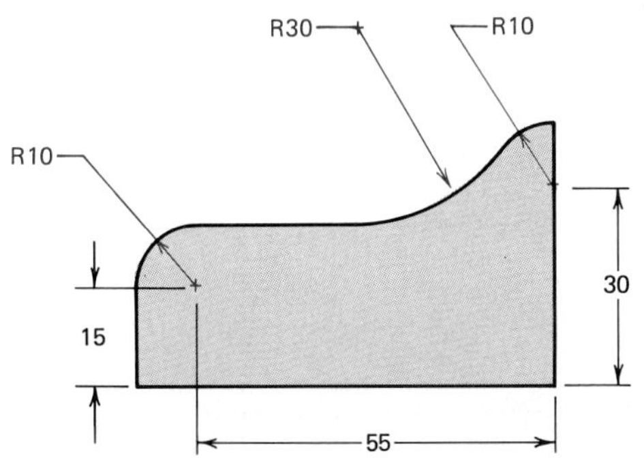

FIG. 12.14 Dimensioning outlines having circular arcs.

length and width of the object are shown. The right end is fully rounded, and so the radius is indicated with the symbol *R* but is not dimensioned. The left end is partially rounded, and so the radius must be given.

12.4.5 Outlines with Arcs

If an outline is made up with a series of arcs, it is necessary to locate a sufficient number of arc centers and to dimension each radius. If the arc centers are not explicitly located, they are found from the tangency points of the radii. Figure 12.14 describes an object with a surface composed of arcs.

12.4.6 Outlines That Are Irregular

An outline such as the one illustrated in Fig. 12.15 may not be definable by a series of circular arcs. This irregular surface can be adequately described by dimensioning some points along the surface. It is assumed that a smooth curve is to be faired between the points. The more points defined, the smaller the possibility of error in the surface, but of course there is a practical limit.

Alternatively, an *xy* coordinate system can be established with the origin conveniently chosen perhaps at the lower left point on the object. Then it is only necessary to mark and number the points along the irregular surface and provide a table of *xy* coordinate values for each point elsewhere on the drawing.

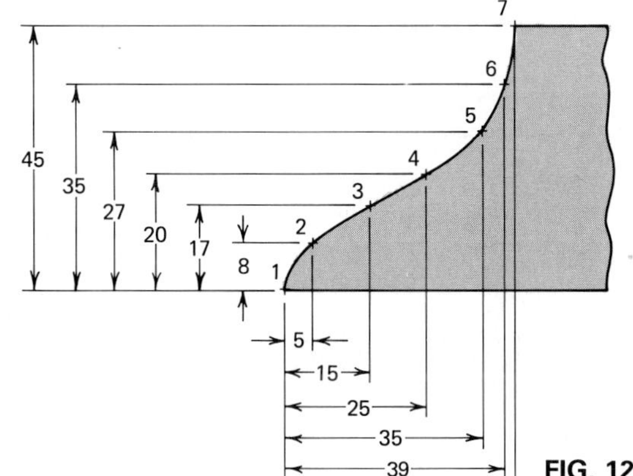

TABLE

Pt.	X	Y
1	0	0
2	5	8
3	15	17
4	25	

FIG. 12.15 Dimensioning irregular outlines.

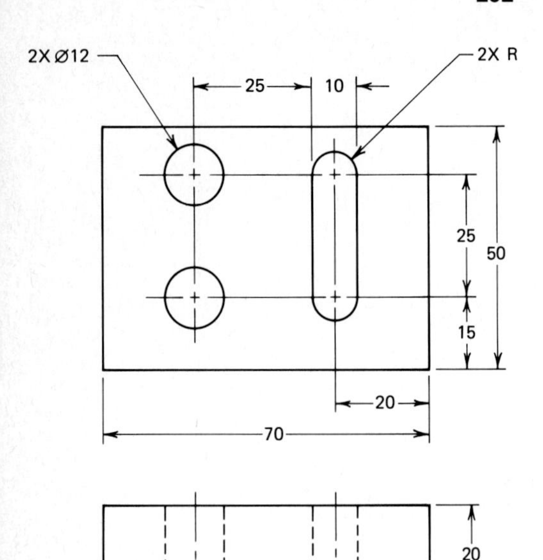

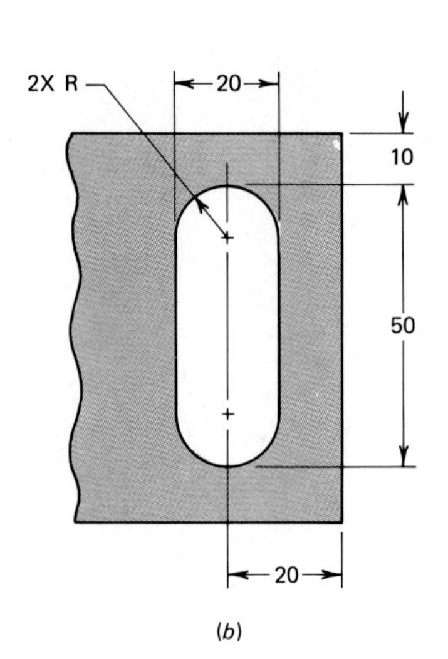

FIG. 12.16 Dimensioning slotted holes.

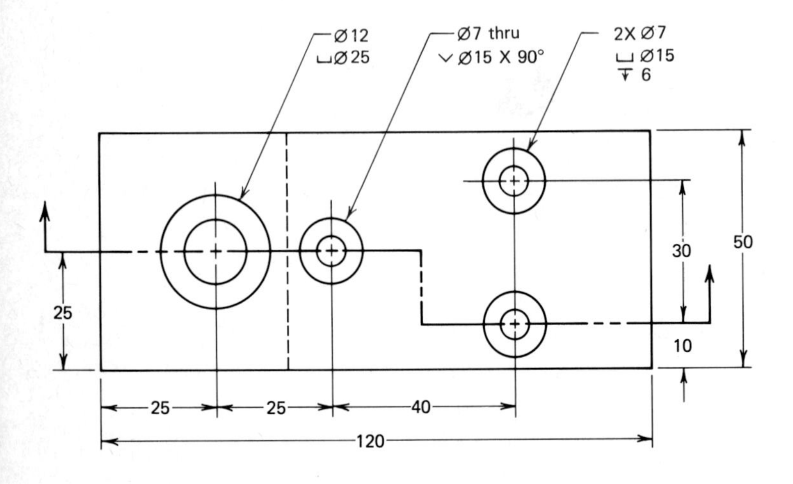

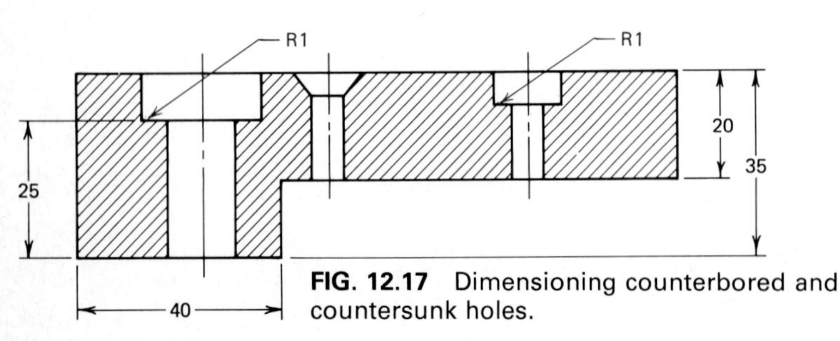

FIG. 12.17 Dimensioning counterbored and countersunk holes.

12.4.7 Slotted Holes

A slotted hole may be dimensioned by giving its width and the locations of its center lines, as shown in Fig. 12.16*a*, or by locating the position of the longitudinal center line at one end and giving the length and width, as shown in Fig. 12.16*b*. In both cases the radii of the ends of the slots are noted but are not dimensioned. Recall that the notation $2 \times R$ means a radius in two places.

12.4.8 Counterbored Holes

A counterbored hole is often used to allow the head of a fastener to be recessed below the surface of a part. It is specified by giving the diameter of the hole followed by the counterbore symbol ⌴ with the diameter of the counterbore. The depth of the counterbore can be given as part of the note, as shown for the small holes in Fig. 12.17. The counterbore dimension can also be shown in the rectangular view of the hole, as indicated for the large hole in Fig. 12.17. The method used depends on which dimension is critical. Also, the corner radius should be given in the rectangular view of the counterbore.

12.4.9 Countersunk Holes

Flathead screws can be made flush with the surface if the hole is countersunk. The countersunk hole is specified by giving the diameter of the hole followed by the countersunk symbol ⌵ and the maximum diameter and

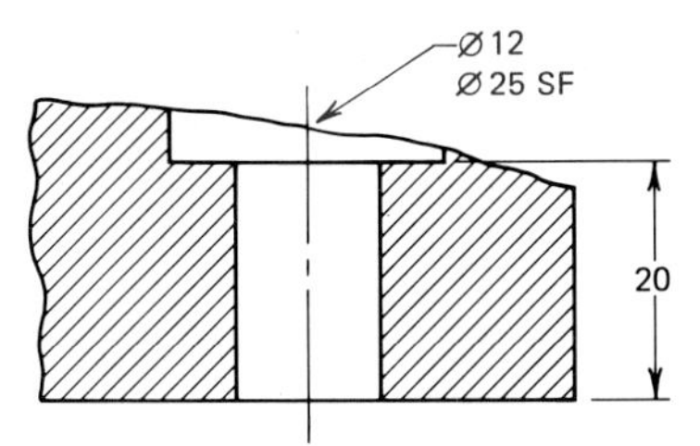

FIG. 12.18 Dimensioning a spot face.

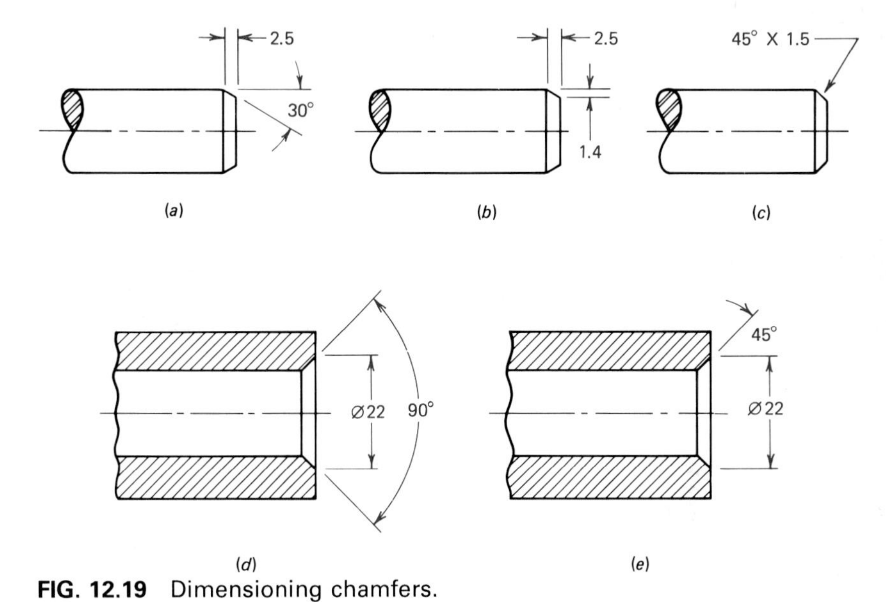

FIG. 12.19 Dimensioning chamfers.

angle of the countersink. An angle of 82° is the most common, although 90° or other angles sometimes are specified. See Fig. 12.17 for an example.

12.4.10 Spotface

The spotface is used to clean up a surface in order to provide a flat bearing area around a hole on an otherwise unfinished surface. The hole diameter is given, followed by the maximum diameter of the spotface and the symbol. The depth of a spotface may be specified if some reference surface from which to measure is available. The remaining material may also be specified, as shown in Fig. 12.18. If no depth is given, the implication is that the spotface is just deep enough to provide the required flat bearing area of the specified diameter.

12.4.11 Chamfers

Chamfers may be either external, as on the end of a bolt, or internal. The chamfer cleans up the edge of the cylinder or hole, making the assembly easier by providing a small guide for engaging the parts. Figure 12.19 describes alternatives for dimensioning chamfers. Dimensioning by a note (Figure 12.19*c*) is acceptable only in the case of the 45° chamfer, where the linear dimensions are equal.

12.4.12 Keyseats

Keys are used to prevent rotational motion between a shaft and the part (hole) through which it passes. Keyseats for square or flat keys are dimensioned as shown in Fig. 12.20. Note that the depth of the keyseat is given from the opposite side of the shaft or hole. Length should be specified in a related view if the keyseat does not extend from end to end of the part.

12.4.13 Bolt Circles

Sometimes several holes are centered at a fixed radius from a point on a part. This condition can be dimensioned by giving the diameter of a bolt circle. The bolt

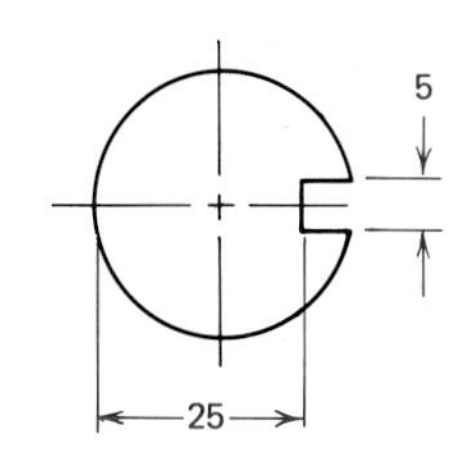

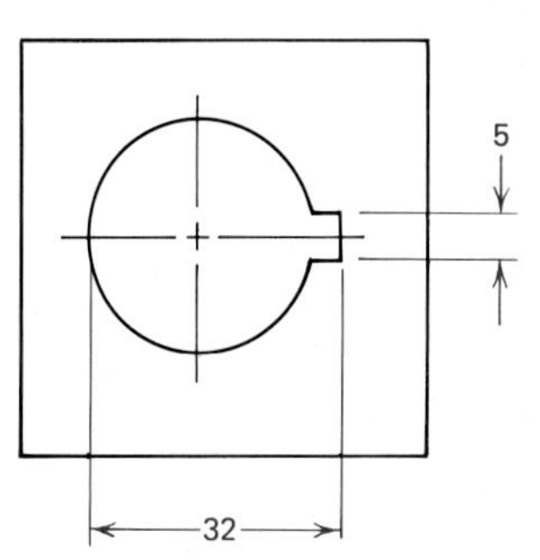

FIG. 12.20 Dimensioning keyseats.

circle is shown by a center line passing through the centers of the holes. Figure 12.21 describes the angular position of the holes by the note "equally spaced." If the holes are not equally spaced, the angular position of each hole should be given. Alternatively, the *xy* coordinates of each hole center may be specified.

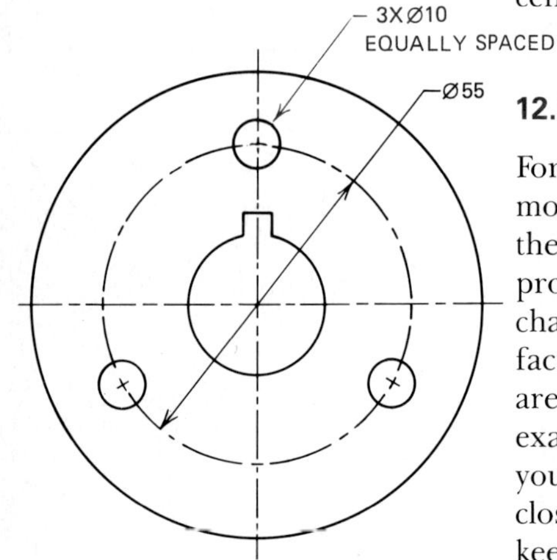

FIG. 12.21 Dimensioning bolt circles.

12.5 TOLERANCING

For a firm to be competitive in modern manufacturing practice, their parts must be mass-produced and therefore interchangeable. The costs of manufacturing increase rapidly as parts are made closer and closer to exact dimensions. As a designer, you must decide how close is close enough for the application, keeping in mind that the more variation in dimension you can allow, the lower the cost of production. Tolerancing is the practice of placing limits on part dimensions on a drawing after careful consideration of the often divergent requirements of interchangeability and cost. Experience plays a large role in determining the requirements for the fit of mating parts. Discussion of techniques for selecting a particular fit is better left to courses in machine design. We can, however, learn to apply standard fits to the dimensions on a drawing. This will be the thrust of the remaining material in Sec. 12.5. The standards available include ASA (American Standards Association) B4.1—1967 (inch) and ANSI B4.2—1978 (metric). Examples of inch standards are found in Appendix C, and examples of metric standards are presented in Appendix B.

12.5.1 Definitions

Shaft. A piece or part that is held or constrained in position.

Hole. A piece or part that holds or constrains a shaft.

Nominal size. The approximate or common size that is used to refer to a dimension. For example, a shaft may be nominally referred to as 1 in or 25 mm even though it may vary in measured diameter.

Basic size. The numeric value used to describe the theoretically exact dimension (size, location, and so on) to which acceptable variations are applied. The basic size contains decimal digits consistent with the size variation. A shaft with a nominal size of 25 mm may have a basic size of 25.00 mm.

Actual size. The measured dimension of the part.

Maximum material condition (MMC). The dimension that corresponds to the maximum acceptable amount of material in the part: the largest shaft or the smallest hole.

Least material condition (LMC). The dimension that corresponds to the minimum acceptable amount of material in the part: the smallest shaft or the largest hole.

Tolerance. The difference between the maximum and minimum limits of a part dimension.

Allowance. The difference between the maximum material conditions of mating parts: the smallest hole minus the maximum shaft. The allowance can be positive, negative, or zero and thus must be stated with a sign.

Refer to Fig. 12.22 for an illustration of these definitions.

12.5.2 Fits

Fit generally describes the looseness or tightness observed between mating parts after application of the allowance and tolerances. Fit is classified as clearance, transition, or interference.

A clearance fit ensures that mating pieces will always have clearance between the hole and shaft throughout their size variations. Thus, the allowance will be positive, and this condition clearly identifies a clearance fit.

An interference fit occurs when there is always interference between the hole and the shaft. The smallest shaft is larger than the largest hole for this fit. The allowance will certainly be negative, but this is not a sufficient condition to define an interference fit.

Transition fit refers to a condition where there is clearance between the shaft and hole over a portion of the range of size limits and interference over the remainder of the range. Since the allowance defines the tightest fit, this occurs when there is interference, and therefore the allowance is negative. A negative allowance suggests either interference or transition fits but does not distinguish between them. The maximum clearance must be checked. If it is positive, the fit is transition; if it is negative, the fit is interference.

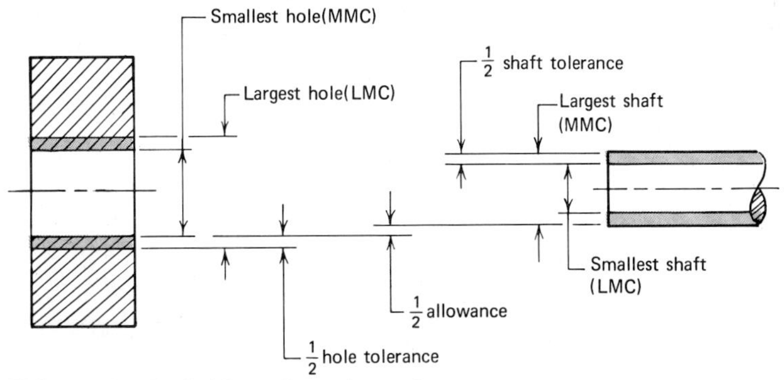

FIG. 12.22 Definitions for tolerancing.

12.5.3 Application of Tolerances and Allowances

Three equations are used to determine the limit dimensions from a given basic size, shaft and hole tolerances, and allowance:

Maximum shaft = minimum shaft + shaft tolerance

Maximum hole = minimum hole + hole tolerance

Allowance = minimum hole − maximum shaft

The given values plus these equations are not sufficient to define the limits for the hole and shaft uniquely. A specification of how the basic size relates to the limits of the shaft or hole also must be found. Two systems for defining this connection are commonly used: the basic shaft system and the basic hole system. Once this specification is made, mathematically, you have three equations in seven variables with four of the variables known, leaving three to be found.

The basic hole system assigns the basic size to the MMC of the hole. Thus, the minimum hole

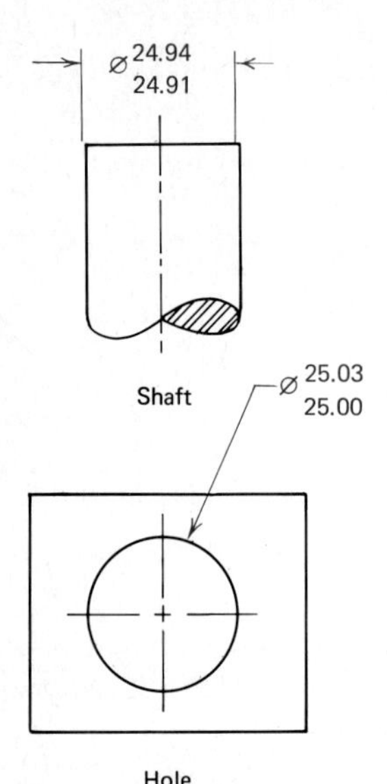

FIG. 12.23 Limit dimensionsing. Basic hole system with unilateral specifications.

becomes known in our set of equations. Take, for example, a design that requires a 25-mm (nominal size) shaft-hole combination. Suppose the tolerances of both the hole and the shaft are 0.03 mm and we seek a clearance fit with an allowance of +0.06 mm. The basic hole system assigns the minimum hole the size of 25.00 (the basic size).

Our equations become

$$\text{Maximum shaft} = \text{minimum shaft} + 0.03$$

$$\text{Maximum hole} = 25.00 + 0.03$$

$$+ 0.06 = 25.00 - \text{maximum shaft}$$

We can immediately solve the second equation for the maximum hole size of 25.00 + 0.03 = 25.03 mm. Also, the maximum shaft is found from the third equation: maximum shaft = 25.00 − 0.06 = 24.94 mm. Now the first equation allows us to calculate the minimum shaft as 24.94 − 0.03 = 24.91 mm.

The limit dimensions are now entirely determined and all that remains is for them to be placed on the drawing. Referring to Fig. 12.23, you can see that the maximum value is always placed above the minimum value. The pair of values is preceded by the diameter symbol ∅. No line is used between the values, as may be seen in drawings based on earlier standards.

The basic shaft system of limit calculations assigns the basic size to the MMC of the shaft. Let us use the conditions from the previous example but apply the basic shaft system to the computations. The same fundamental equations are still valid. Applying them with the basic size equated to the maximum shaft, we have

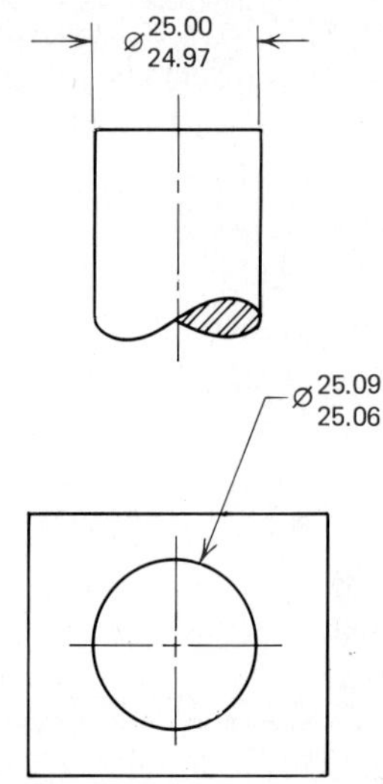

FIG. 12.24 Limit dimensioning. Basic shaft system with unilateral specifications.

$$25.00 = \text{minimum shaft} + 0.03$$

$$\text{Maximum hole} = \text{minimum hole} + 0.03$$

$$+0.06 = \text{minimum hole} - 25.00$$

Solution of these equations gives

Maximum shaft = 25.00 mm
Minimum shaft = 24.97 mm
Maximum hole = 25.09 mm
Minimum hole = 25.06 mm

These values are placed on the drawings as in Fig. 12.24

The basic hole system is the most widely used because it allows standard-size drills, for example, to be used to produce the hole and because the shaft can be turned to the specified size. The basic shaft system is useful when several parts must be fitted over the same shaft. A constant-diameter shaft can be manufactured with holes of varying sizes placed in the mating parts providing the desired type of fit for each.

The examples just considered employed the unilateral dimensioning practice, where the entire tolerance is applied on one side of the basic size. Another concept is bilateral tolerancing, where the basic size becomes the average of the MMC and the LMC of the hole for the basic hole system and of the shaft for the basic shaft system.

For example, for a nominal size of 1 in, hole and shaft tolerances of .002, and an allowance of −.003, we wish to calculate the limit dimensions by means of the basic hole system. We now have an additional equation that basic size = (maximum hole + minimum hole)/2. The full set of four equations to be solved is as follows:

1.000 = (maximum hole + minimum hole)/2

Maximum shaft = minimum shaft + .002

Maximum hole = minimum hole + .002

−.003 = minimum hole − maximum shaft

The first and third equations can be solved for the hole limits, yielding maximum hole = 1.001 and minimum hole = .999. From the allowance equation we get

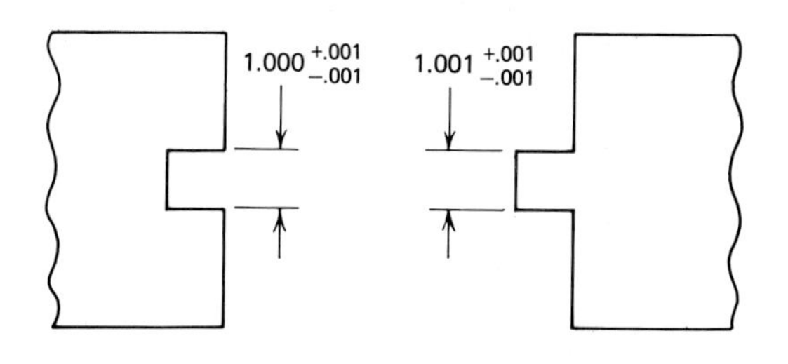

(*a*) Hole (*b*) Shaft

FIG. 12.25 Expressing bilateral tolerances on a drawing.

maximum shaft = 1.002, and from the second equation we get minimum shaft = 1.000. The average shaft is then 1.001. These values are placed on the drawing in Fig. 12.25. Examination of the maximum clearance gives 1.001 − 1.000 = +.001. Therefore, the fit is classified as transition.

12.5.4 Standard Fits for Inch Applications

ASA B4.1—1967 establishes fits which may be used to compute limit dimensions for mating parts. Tables extracted from this standard are included in Tables C.15 to C.19. Each type of fit is referenced by two letters plus one class number. These designations are not intended to appear on drawings; they only supply a mechanism of description for educational purposes. The actual limits should be used on drawings. The classes of fit are as follows:

RC1-RC9: running and sliding fits

LC1-LC11: locational clearance fits

LT1-LT6: locational transition fits

LN1-LN3: locational interference fits

FN1-FN5: force and shrink fits

Each class of fit, RC3, for example, is intended to provide similar performance throughout the range of sizes for that fit. The running and sliding fits have a built-in lubrication allowance. RC1 and RC2 are intended for sliding fits, while RC3 through RC9 range from precision running fits to loose running fits as the class number increases.

The locational fits provide a method of calculation of locations of mating parts from clearance fits, where easy assembly of parts is required, to interference fits, where accuracy of location is the overriding requirement.

The force or shrink fits are designed so that pressures between the shaft and hole are maintained throughout the size range. These fits are intended for drive applications. Shrink fits are produced by heating the part with the hole and cooling the shaft so that assembly requires either no force or less force than would be necessary otherwise. Once the parts return to ambient temperatures, they are tightly bound together by the stresses that are developed.

The tables are designed to allow the designers to apply values algebraically to the basic size in order to obtain the limit dimensions for the shaft and hole. Careful examination reveals that the increment to be added to the basic size to obtain the smallest hole is always zero; this indicates that the tables are built from the basic hole system.

As an example, compute the limit dimensions for a 3/4-in shaft and hole designed for an RC4 fit. The basic size of .7500 falls in the range of 0.71-1.19 in Table C15, yielding table values of $\begin{matrix}+1.2\\0\end{matrix}$ for the hole and $\begin{matrix}-0.8\\-1.6\end{matrix}$ for the shaft. These values are in thousandths of an inch.

Thus the minimum hole is .7500 + 0 = .7500, the maximum hole is .7500 + .0012 = .7512, the minimum shaft is .7500 + (−.0016) = .7484, and the maximum shaft is .7500 + (−.0008) = .7492.

The limits are:

Hole: .7500 − .7512
Shaft: .7484 − .7492

Table C15 also gives the limits of clearance as

$$\begin{matrix}0.8\\2.8\end{matrix}$$

The minimum clearance is the allowance and is .0008, while the maximum clearance is .0028.

Although the tables are based on the basic hole system, they can be readily adapted to the basic shaft system as needed. Since the unilateral basic shaft system specifies that the basic size should be the maximum shaft value, it is only necessary to add an increment to each of the four limits so that the increment for the maximum shaft is zero. To illustrate, compute the limit dimension for the previous example, using the basic shaft system. The table provides for the 3/4-in RC4 fit the following values:

+1.2	−0.8
0	−1.6

If 0.8 is added to each value in the table, the increment for the maximum shaft becomes zero, as desired. These new values are

+2.0	0
+0.8	−0.8

The required limits are now

Hole: .7508 − .7520
Shaft: .7492 − .7500 (basic size)

The tolerances of each part and the allowance between parts remain the same, and so the same fit is obtained as before.

12.5.5 Standard Fits for Metric Applications

The definitions in Sec. 12.5.1 were sufficient to compute limit dimensions and to use standard tables for inch applications. A few additional definitions are needed to understand the tables of standard metric fits from ANSI B4.2—1978.

Deviation. The size minus the basic size.

Upper deviation. The maximum size minus the basic size.

Lower deviation. The minimum size minus the basic size.

Fundamental deviation. The upper or lower deviation, whichever is closest to the basic size. Designated by a letter: uppercase for a hole, lowercase for a shaft.

International tolerance grade. A number that establishes the amount of the part variation. Larger numbers refer to larger part variation.

Tolerance zone. This combines the fundamental deviation and the international tolerance grade to represent the tolerance and its position relative to the basic size.

It is easy to see how these definitions are used by studying a couple of examples.

Fundamental deviation
International tolerance grade
Hole 20 H 11
Basic size
Tolerance zone

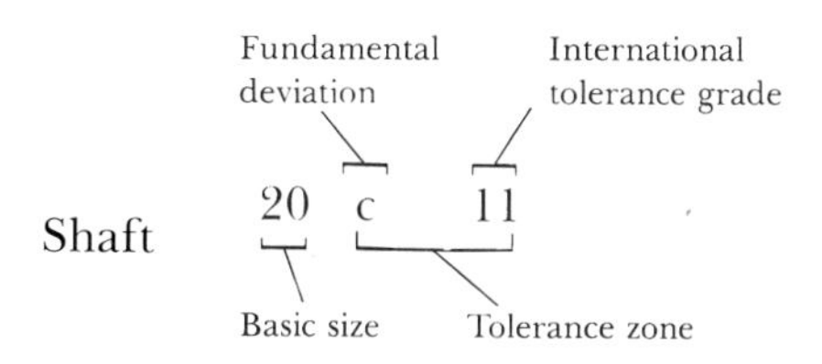

The limits of the hole represented here can be shown on a drawing as

$$20\text{H}11 \quad \text{or} \quad 20\text{H}11\left(\begin{matrix}20.130\\20.000\end{matrix}\right)$$

$$\text{or} \quad \begin{matrix}20.130\\20.000\end{matrix}\ (20\text{H}11)$$

where the numbers in parentheses are for reference. For a person beginning to use standard metric fits, the latter alternative is probably the most descriptive and informative. The first alternative

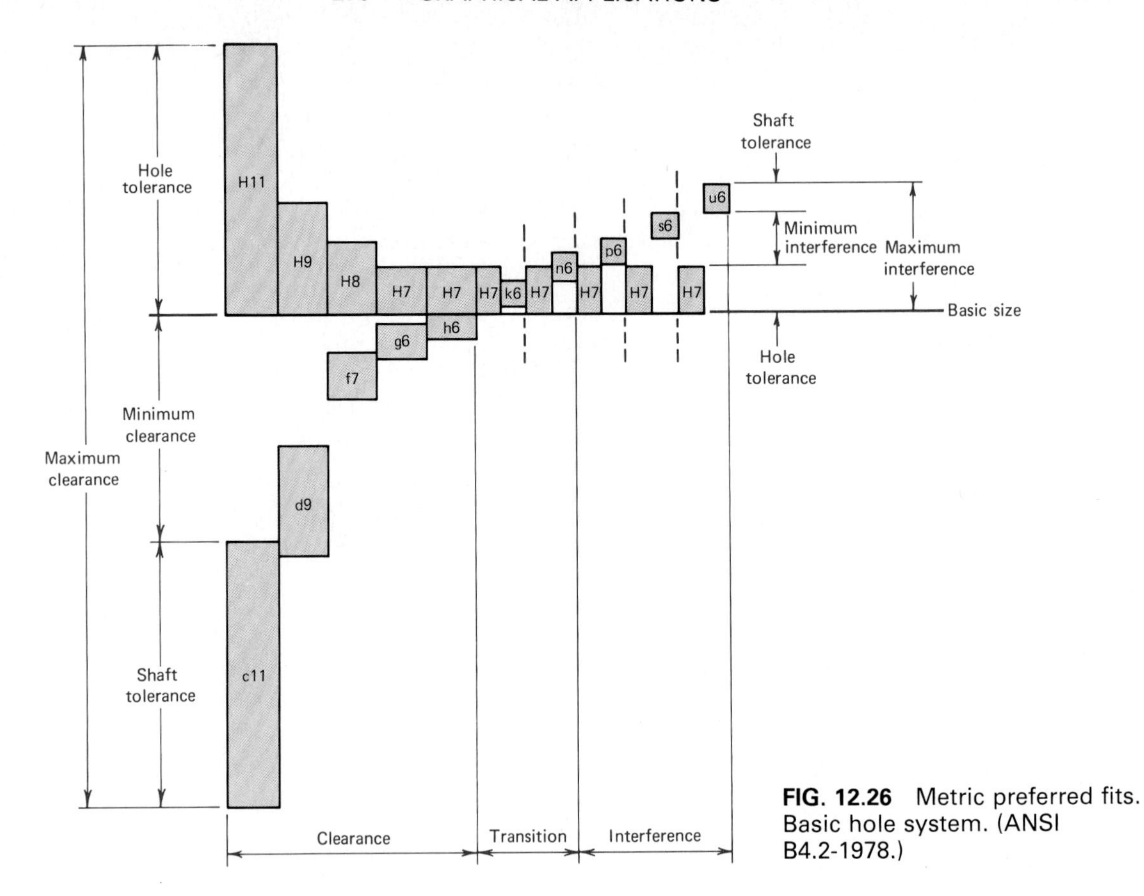

FIG. 12.26 Metric preferred fits. Basic hole system. (ANSI B4.2-1978.)

would be used by those experienced with the metric fits.

Although there are many possible metric fits, a small group of preferred fits has been defined for use. Figure 12.26 shows graphically the preferred fits using the basic hole system. The fundamental deviation for each of the holes is H for the basic hole system. This simply means that the smallest hole is equal in size to the basic size. Note that in Fig. 12.26, H11 through H7 all lie on the basic size line. Consider the fit H11/c11. The hole and shaft tolerances are both relatively large, and the minimum clearance provides a clearance fit. At the other end of the scale, the fit H7/u6 provides rather tight tolerances and an interference fit.

Unlike the case of inch-fit standards, where the basic hole system tables must be modified for the basic shaft system, there exists a separate but comparable metric fit standard for the basic shaft system. Figure 12.27 graphically shows this case. Here the fundamental deviation for each shaft is h. Note that h11 through

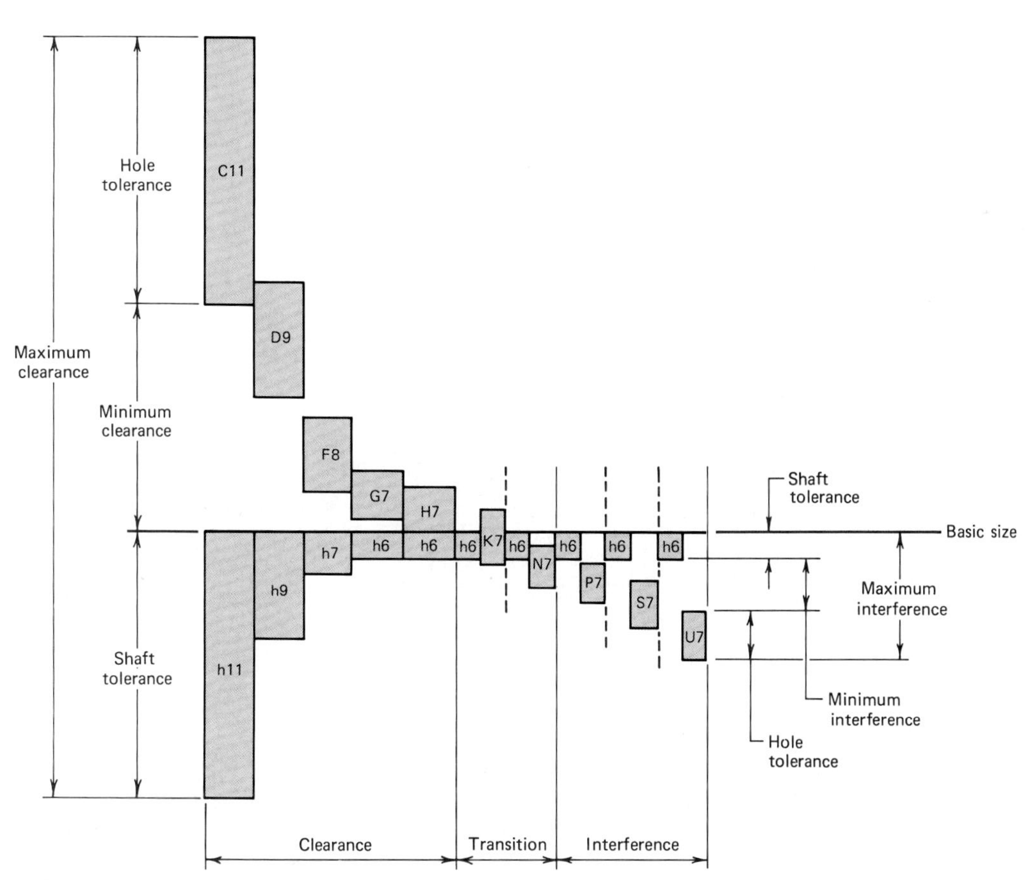

FIG. 12.27 Metric preferred fits. Basic shaft system. (ANSI B4.2-1978.)

h6 all touch the basic size line. It should be understood that H11/c11 and C11/h11 provide equivalent fits for the basic hole and basic shaft systems, respectively.

A description of each of the preferred fits is given in Fig. 12.28. From this figure, if you wanted to design a sliding fit, you would use H7/g6 for the basic hole system and G7/h6 for the basic shaft system.

Tables B.17 to B.19 are used to find preferred limit dimensions. Table B.17 (a duplicate of Fig. 12.28) is used to define a fit. Table B.18 can be used to find the hole limits as a deviation from the basic size. Table B.19 provides the shaft limits. Values in Tables B.18 and B.19 are in micrometers.

Example 12.1

Determine the limit dimensions for a 10-mm hole and shaft for a sliding fit.

Basic hole system: From Fig. 12.28 or Table B.17, we have learned that we require a fit of H7/g6. From Tables B.18 and B.19 in Appendix B, we find the following, after applying the deviations to the basic size:

HOLE	SHAFT	FIT
10.015	9.995	0.029
10.000	9.986	0.005

	ISO SYMBOL		DESCRIPTION	
	Hole Basis	Shaft Basis		
Clearance fits	H11/c11	C11/h11	*Loose running* fit for wide commercial tolerances or allowances on external members.	More clearance
	H9/d9	D9/h9	*Free running* fit not for use where accuracy is essential, but good for large temperature variations, high running speeds, or heavy journal pressures.	
	H8/f7	F8/h7	*Close running* fit for running on accurate machines and for accurate location at moderate speeds and journal pressures.	
	H7/g6	G7/h6	*Sliding* fit not intended to run freely, but to move and turn freely and locate accurately.	
	H7/h6	H7/h6	*Locational clearance* fit provides snug fit for locating stationary parts; but can be freely assembled and disassembled.	
Transition fits	H7/k6	K7/h6	*Locational transition* fit for accurate location, a compromise between clearance and interference.	More interference
	H7/n6	N7/h6	*Locational transition* fit for more accurate location where greater interference is permissible.	
Interference fits	H7/p6[1]	P7/h6	*Locational interference* fit for parts requiring rigidity and alignment with prime accuracy of location but without special bore pressure requirements.	
	H7/s6	S7/h6	*Medium drive* fit for ordinary steel parts or shrink fits on light sections, the tightest fit usable with cast iron.	
	H7/u6	U7/h6	*Force* fit suitable for parts which can be highly stressed or for shrink fits where the heavy pressing forces required are impractical.	

[1]Transition fit for basic sizes in range from 0 through 3 mm.

FIG. 12.28 Description of preferred fits. (ANSI B4.2-1978.)

The column headed "Fit" gives the maximum and minimum clearances in millimeters which can be calculated from the known dimensions. Recall that the minimum clearance is the allowance.

Basic shaft system: Likewise we have the fit, G7/h6. From Tables B.18 and B.19 in Appendix B, we can develop the following:

HOLE	SHAFT	FIT
10.020	10.000	0.029
10.005	9.991	0.005

As expected, the maximum and minimum clearances are identical for the basic hole and basic shaft systems.

If your required basic size is not contained in the tables in this text, refer to ANSI B4.2—1978.

12.5.6 Additional Considerations

Angular tolerances may be expressed either in decimal degrees or in degrees, minutes, and seconds. Use of either unilateral or bilateral limit expressions is permissible. See Fig. 12.29 for an example of unilateral limit dimensions for an angle.

Tolerance buildup between features should be closely monitored when you place dimensions on a drawing. The acceptability of a given tolerance level is largely dictated by the function of the part. Three methods—chain, baseline, and direct dimensions—are explained here.

In Fig. 12.30*a*, chain dimensioning allows feature *V* to be 75 ± 0.15 mm from surface *S*. The tolerances are simply summed to obtain the overall tolerance. If it is desired that the distance from S to V be limited to a variation of ± 0.05 mm, baseline dimensioning as shown

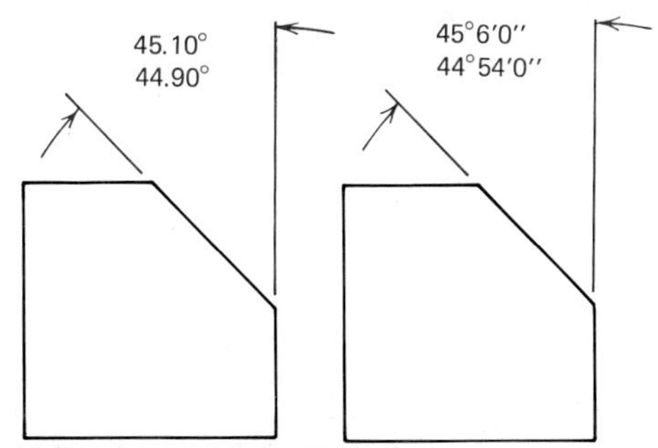

FIG. 12.29 Limit dimensions for angles.

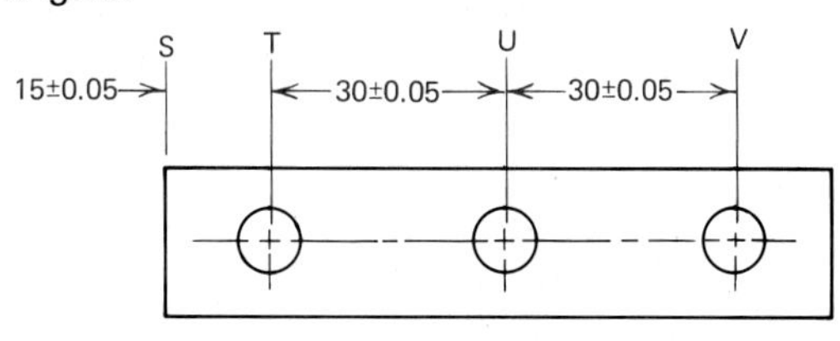

(*a*) Chain

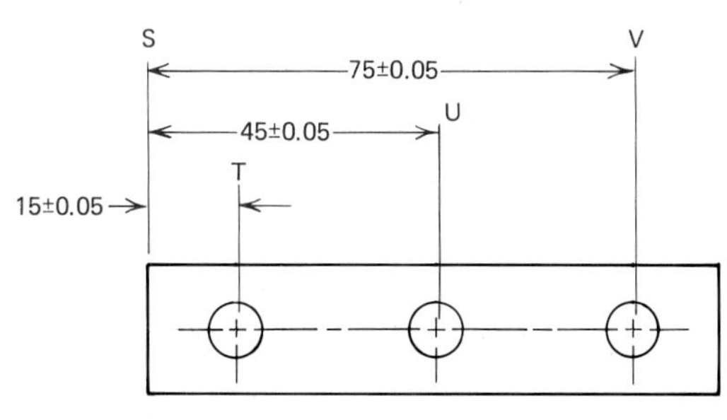

(*b*) Base line

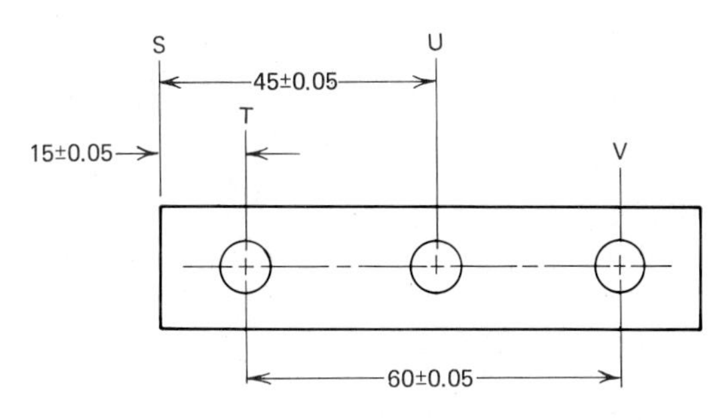

(*c*) Direct

FIG. 12.30 Controlling tolerance buildup.

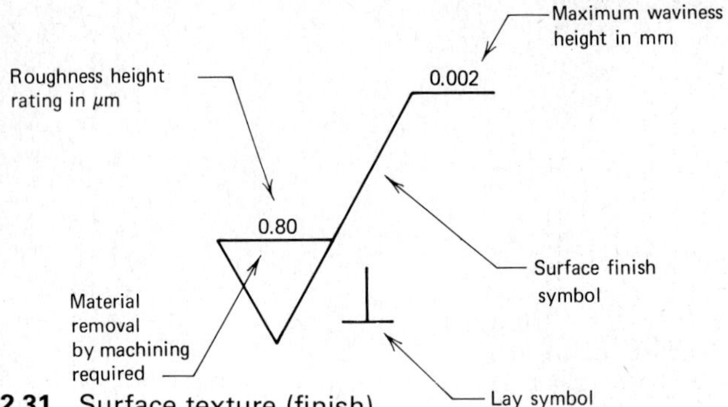

FIG. 12.31 Surface texture (finish) symbol.

in Fig. 12.30*b* can be employed. This method still allows the distance from *T* to *V* to vary by ± 0.10 mm. If the function of the part requires that the distance between *T* and *V* be critically controlled, direct dimensioning as seen in Fig. 12.30*c* should be used.

Surface texture or finish may be important to the operation of mating parts and to their service life. All surfaces have slight irregularities, which are described by terms such as roughness, waviness, lay, and flaws.

Roughness results from production methods and is a measure of the deviation from a mean surface line. Roughness height is expressed in micrometers (μm) or microinches (μin).

Waviness is a larger variation caused by strains, vibration, warping, and so on. It is expressed in millimeters or inches (peak to valley).

Lay describes the predominant direction of marks left by machine tools.

Flaws are random irregularities in the surface, such as cracks, checks, or scratches.

Surface texture is specified on a drawing by means of the surface finish symbol or a general note. The surface finish symbol either touches the surface or is placed on a leader whose arrowhead touches the affected surface. An example of a surface finish symbol is given in Fig. 12.31.

The many details of the specification of surface finish are beyond the scope of this text. You are referred to the standard that describes surface texture symbols, ANSI Y14.36—1978, and to machine design texts and handbooks.

12.6 GEOMETRIC DIMENSIONING AND TOLERANCING

The intent of a designer is to make the engineering drawing unambiguous. Thus, it is important that dimensions and tolerances be specifically stated and held to during manufacture for each part of the design. When a part must mate with other parts, it is important that the engineering drawing contain the necessary information to ensure interchangeability of mating parts during assembly. A good engineering drawing will not be subject to controversy and guesswork when it is interpreted.

Figure 12.32 shows a drawing of a part dimensioned according to the traditional coordinate plus and minus system. Let us look at some possible interpretations of this drawing which may or may not cause problems in assembly.

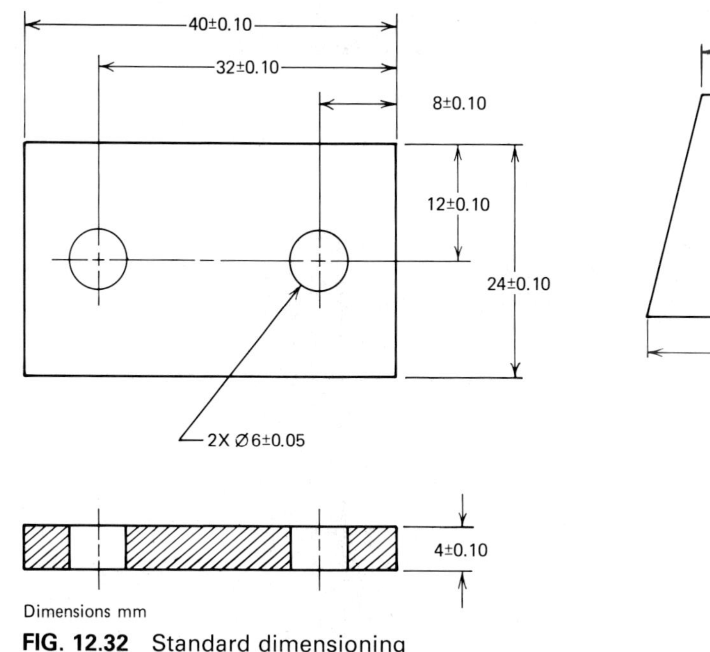

FIG. 12.32 Standard dimensioning using coordinates.

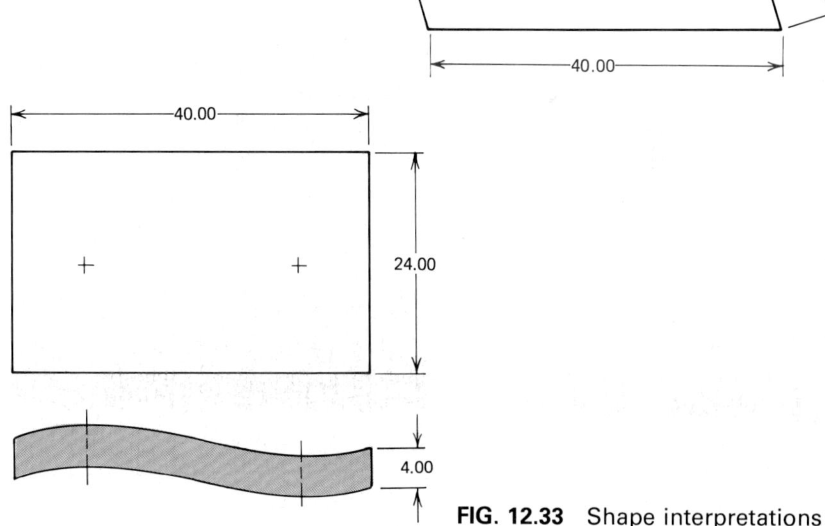

FIG. 12.33 Shape interpretations from coordinate dimensions.

Since we know from the drawing the form of the part, it appears that if it is manufactured according to the dimensions and tolerances given, the part will be acceptable. However, the drawing does not reveal the function of the part, opening the possibility for parts to be unusable even though they are made according to the drawing.

Figure 12.33 reveals some possible interpretations of the form of the part. The differences are exaggerated for clarity. Would the trapezoidal shape in Fig. 12.33*a* be acceptable? It does fit the limit requirements set forth in the drawing. Is a rectangular shape with 90° corners needed? If so, the interpretation shown in Fig. 12.33*b*, although possible, would cause the part to be unacceptable. Must the part fit flush with another flat surface? The interpretation illustrated in Fig. 12.33*c* would obviously prohibit a

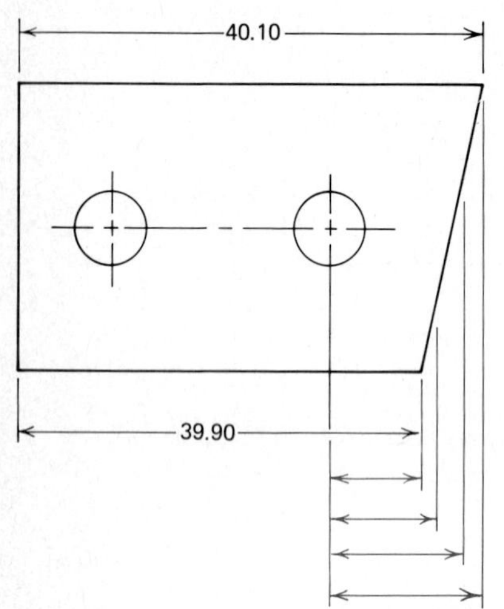

FIG. 12.34 Position interpretation.

proper fit without some physical adjustment during assembly.

Let us look at some possibilities for the location of the holes. Assume that you are part of the quality-control group and must verify the location of the holes and accept or reject the part according to the results of your measurements compared with the requirements stated on the drawing. Figure 12.34 shows one dilemma you might face. Where do you take your measurement from to determine whether the first hole is properly located with respect to the right-hand edge? The drawing (Fig. 12.32) requires that the center of the hole be located 8 ± 0.10 from the right-hand edge. It is not difficult to see that with permissible tolerances on the overall length of the part, the verification of the location of the hole centers from the edges of the part is guesswork at best.

Suppose that the holes in the part are to fit over two pins exactly 24 mm apart and 6 mm in diameter, as shown in Fig. 12.35*a*. For the permissible variation in the center locations and hole diameters, situations may arise where the part would not fit. Figure 12.35*b* shows one such situation where hole centers are 24.20 mm apart, which is within the drawing requirements. Even if both holes were drilled out to the maximum permissible diameter (6.05 mm), the part would not fit.

You can visualize many more possibilities for variations in the part which according to the drawing are permissible but which would render the part unusable for a specified function. The examples just discussed point out the need for the designer to convey on the drawing a unique object which when manufactured will satisfy the form and function requirements of the design. The sophistication of modern engineering designs requires better methods of communicating requirements. These better methods are provided by the system of geometric dimensioning and tolerancing.

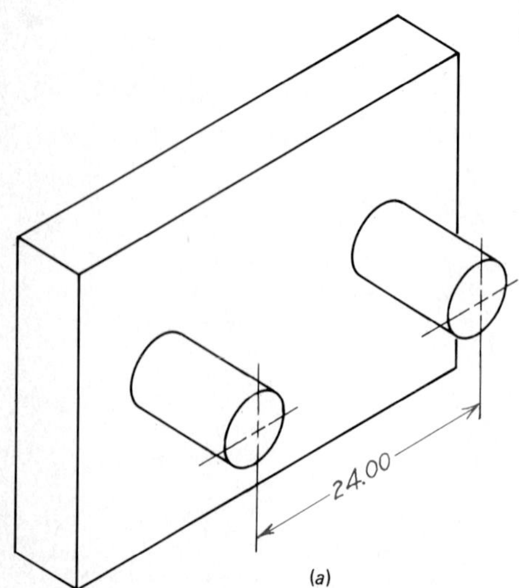

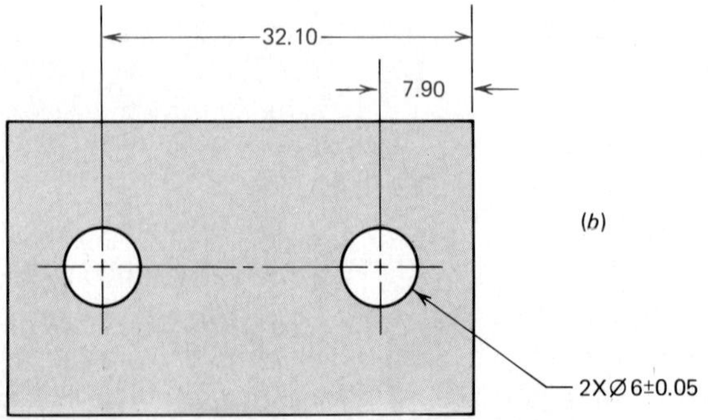

FIG. 12.35 Potential difficulties in matchup of mating parts.

12.6.1 Definitions, Characteristics, and Symbols

Geometric dimensioning and tolerancing is a system of drawing which supplements the traditional techniques with specifications which permit uniformity and convenience in the interpretation of drawings. It is essential that this system be used when

1. Part features are critical to its function
2. Standard interpretation of the part drawing is not satisfactory
3. Consistency in manufacturing and quality control methods is required
4. Interchangeability of parts is required

The characteristics and symbols are shown in Fig. 12.36. We will illustrate the purpose and use of many of these with examples in the following sections. First a brief discussion of the use of these symbols is presented.

	Type of tolerance	Characteristic	Symbol
Individual features	Form	Straightness	—
		Flatness	▱
		Circularity (roundness)	○
		Cylindricity	⌭
Individual or related features	Profile	Profile of a line	⌒
		Profile of a surface	⌓
Related features	Orientation	Angularity	∠
		Perpendicularity	⊥
		Parallelism	//
	Location	Position	⌖
		Concentricity	◎
	Runout	Circular runout	↗
		Total runout	⌰
Other symbols		Maximum material condition (MMC)	Ⓜ
		Regardless of feature size (RFS)	Ⓢ
		Basic (exact) dimension	[XX.XX]
		Datum identification	[–A–]
		Feature control	[⊥ | Ø0.1 Ⓜ | C]

FIG. 12.36 Geometric tolerancing definitions and symbols.

The *maximum material condition* Ⓜ is defined in Sec. 12.5.1. It is the smallest hole or the largest shaft. This concept is very important in geometric tolerancing because it allows a relation between the form and function of a specific part feature to be prescribed.

Regardless of feature size Ⓢ means that the tolerance of form must be met regardless of the specific size of the part feature. This is a more restrictive principle than that of MMC, for it does not allow a relation between form and function to exist.

Basic dimension is a theoretical value that describes the exact or perfect geometry of a feature. It must be associated with a permissible variation (tolerance) for the feature.

Datum identifications are points, lines, or surfaces which are assumed to be exact and are used as references to locate the features on a part or to establish the form of a feature.

The *feature symbol* is a frame which contains a geometric characteristic symbol, datum references, tolerance, and the MMC if applicable. The specific format is as follows.

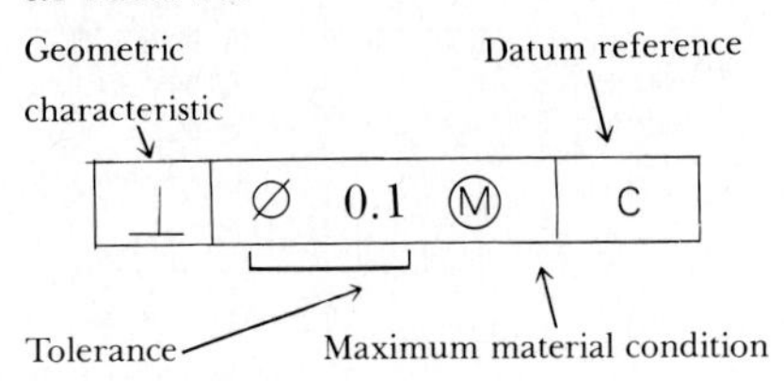

Specifically, this feature control symbol states that the feature must be perpendicular to datum surface C within a cylindrical tolerance of 0.1 at MMC. We will explain how to put together these feature symbols for specific feature requirements.

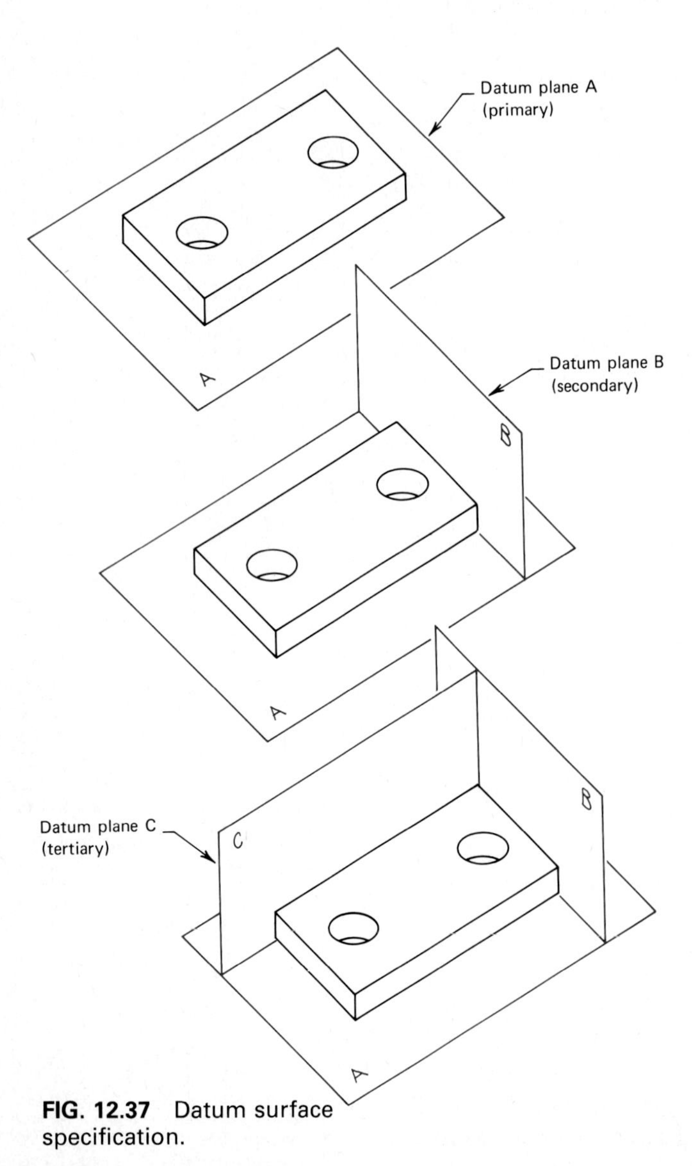

FIG. 12.37 Datum surface specification.

12.6.2 Datums

Any of the types of tolerance listed in Fig. 12.36 which involve related features—profile, orientation, location, and runout tolerances—will have one or more datum references. Most applications will require the establishment of datum planes. Datum points, lines, and cylindrical surfaces are also possible but will not be discussed here. Figure 12.37 shows the establishment of three theoretically exact datum planes for the object in Fig. 12.32. In general, the largest or most important surface is selected as the reference for the primary datum plane. This requires the manufactured part to have a minimum of three points of contact with this datum plane. The designer has the choice of selecting these planes, since it is the design function which will be conveyed on the drawing by the principles of geometric dimensioning and tolerancing.

If necessary, secondary and tertiary datum planes can be established. The three datum planes are mutually perpendicular and provide a framework in which to determine whether features lie within the specified tolerance. Figure 12.38 indicates

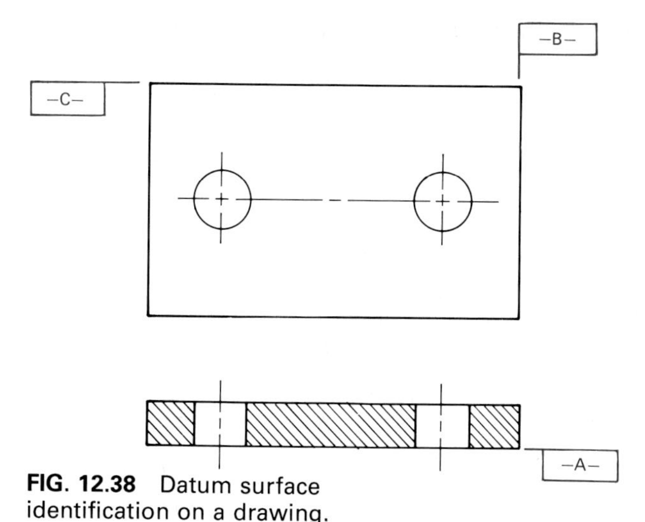

FIG. 12.38 Datum surface identification on a drawing.

how the datum planes are denoted on a drawing.

12.6.3 Form Tolerances

Straightness means that an element of a part feature must be a straight line. The application of a straightness tolerance specifies a zone within which all points on the elements of the part feature must lie. See Fig. 12.39 for an example.

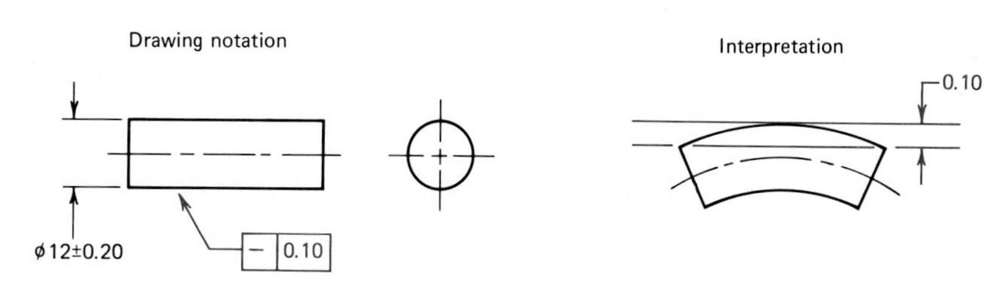

Note: The straightness tolerance must be less than the size tolerance.

FIG. 12.39 Straightness specification.

Flatness indicates that all elements of a surface are in a plane. A flatness tolerance calls for a tolerance zone which lies between two parallel planes. Figure 12.40 gives an example of flatness tolerance.

You are encouraged to look into ANSI Y14.5M to note further details of the straightness and flatness requirements as well as information on the form tolerances for circularity and cylindricity.

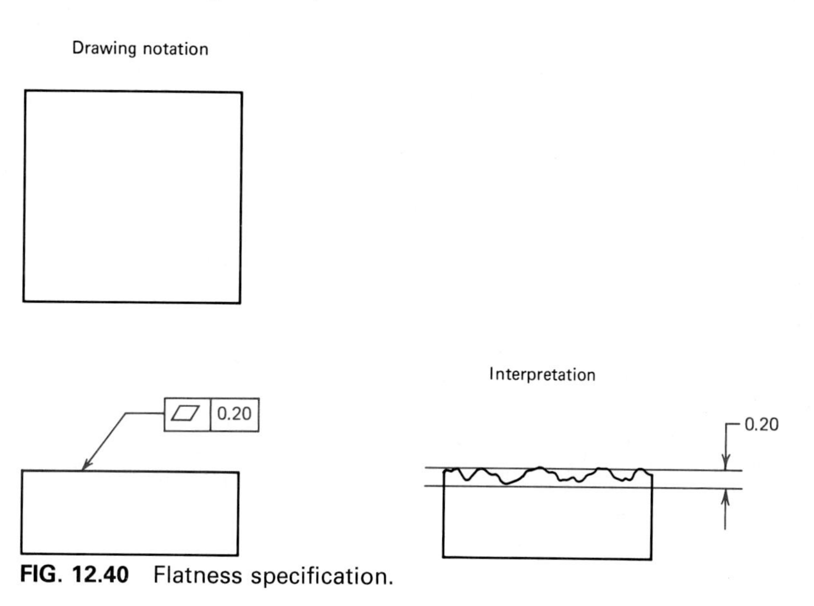

FIG. 12.40 Flatness specification.

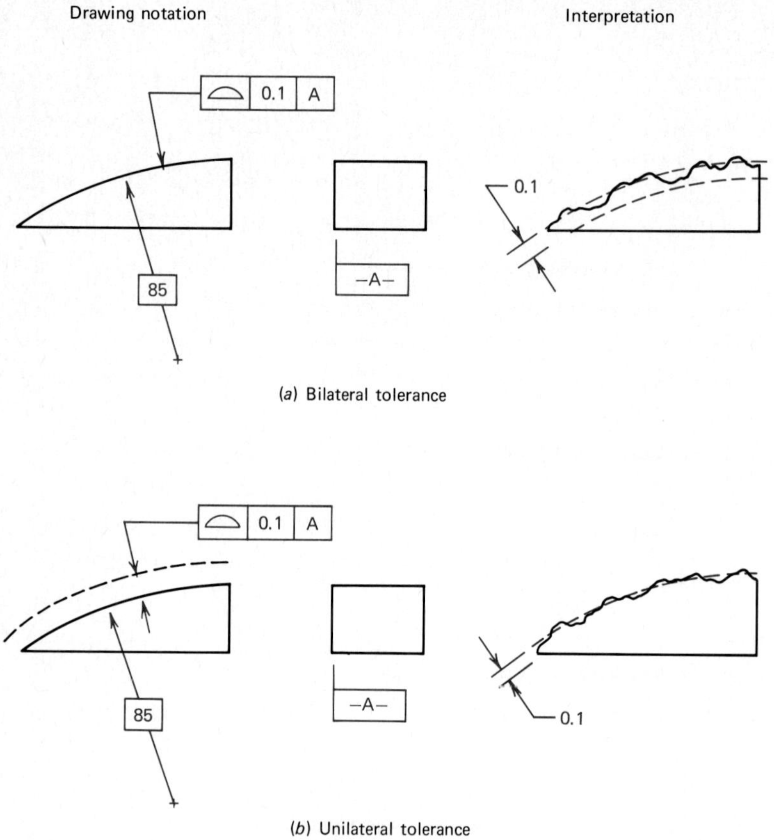

FIG. 12.41 Surface profile control.

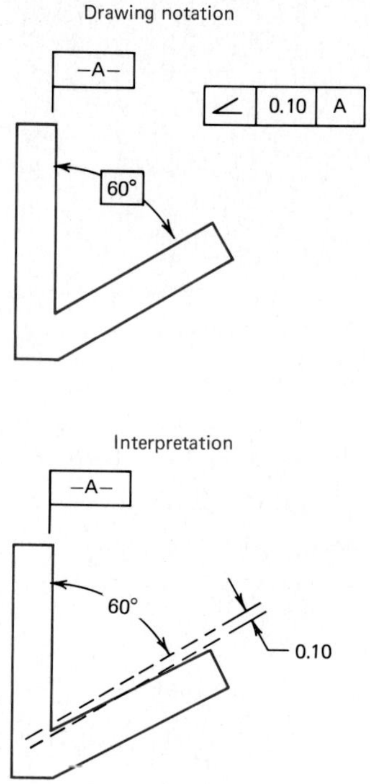

FIG. 12.42 Angularity control.

12.6.4 Profile Tolerances

A *surface profile* tolerance is a three-dimensional zone which restricts a particular feature of a part. The tolerance zone can be defined as unilateral or bilateral. See Fig. 12.41 for an example of control of a surface.

A *line profile* tolerance is a two-dimensional zone established along the length of a particular part feature. In this case, it is not desired to control an entire surface but only one line on the surface.

12.6.5 Orientation Tolerances

Angularity tolerance states that a surface or axis must lie within a specified tolerance zone in relation to a datum plane. Angularity applies to surfaces at angles other than 90° from the datum. In Fig. 12.42, the dimensioned surface must lie between two parallel planes 0.10 mm apart and at an angle of 60° with datum plane *A*.

Perpendicularity specifies the condition of a surface that is at a 90° angle with a datum plane.

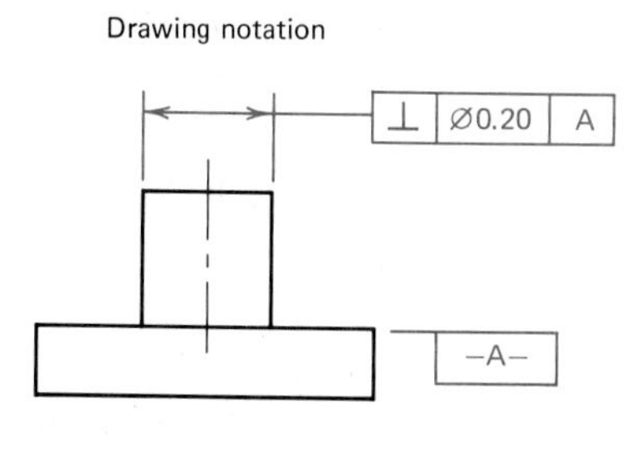

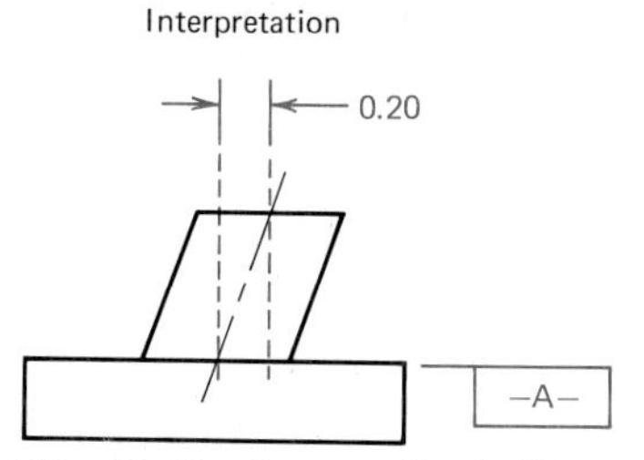

FIG. 12.43 Perpendicularity control.

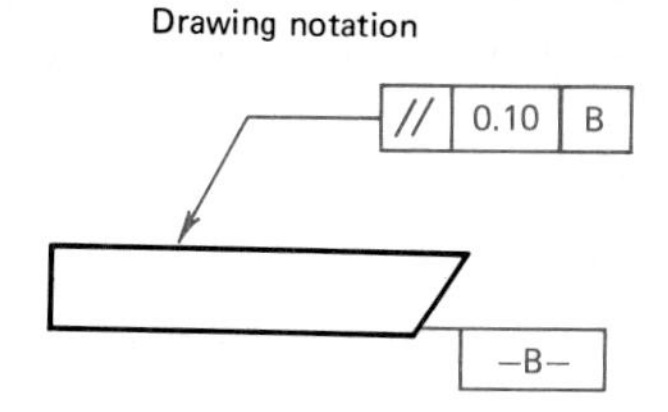

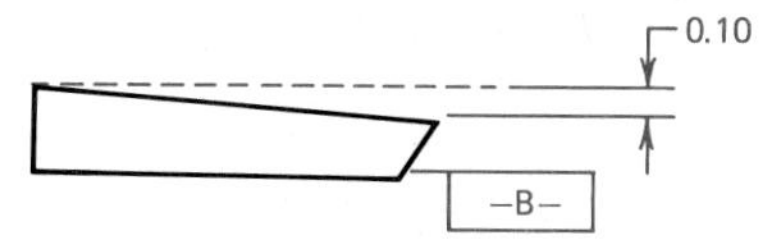

FIG. 12.44 Parallelism control.

Figure 12.43 illustrates an application of perpendicularity tolerance.

Parallelism is a condition which controls the position of a surface which is to be parallel to a datum plane within a specified limit. Figure 12.44 shows an application of the parallelism requirement.

12.6.6 Runout Tolerances

Total runout is a method which permits control of a surface relative to a datum axis. The surfaces that are controlled may be constructed at 90° to the datum axis or around the datum axis. Figure 12.45*a* shows the runout control on a cylindrical surface constructed around the datum axis. Figure 12.45*b* shows the control placed on a surface perpendicular to the datum axis. In general, there are three methods of controlling coaxial features that are

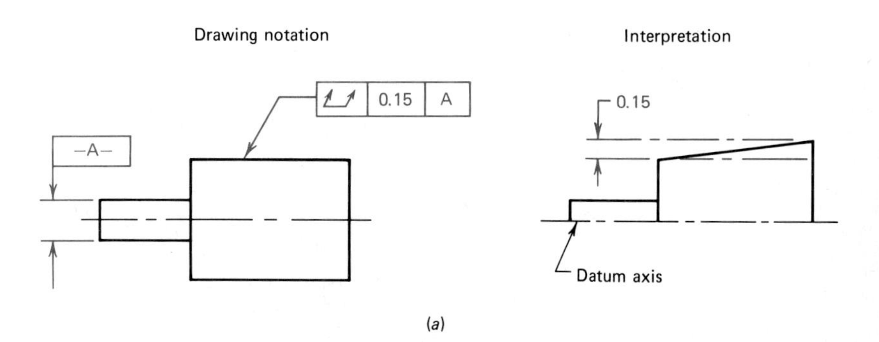

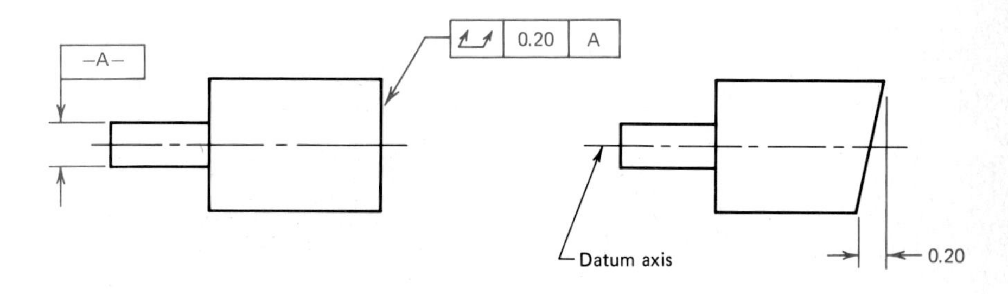

FIG. 12.45 Runout control.

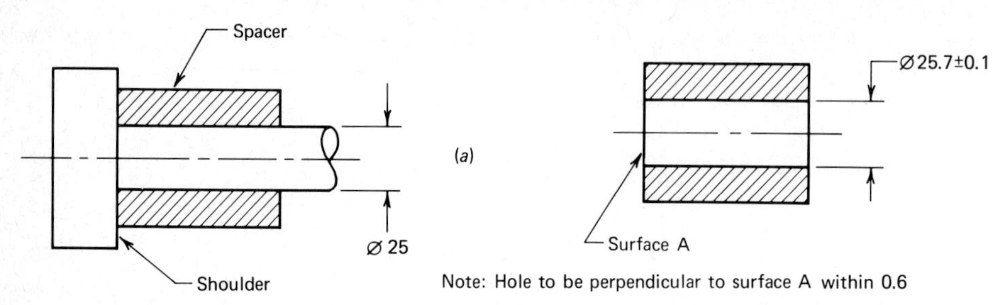

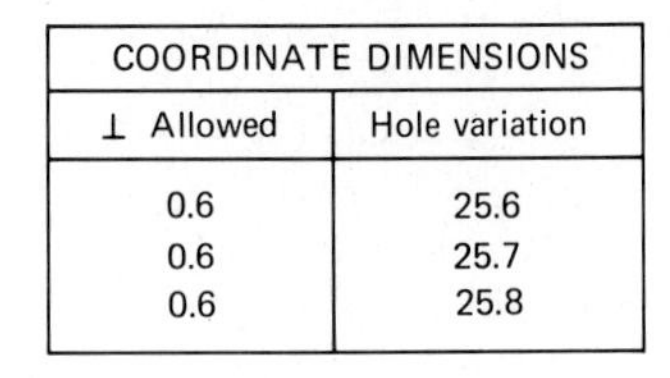

COORDINATE DIMENSIONS	
⊥ Allowed	Hole variation
0.6	25.6
0.6	25.7
0.6	25.8

(c)

FIG. 12.46 Controlling geometric characteristics with coordinate dimensions.

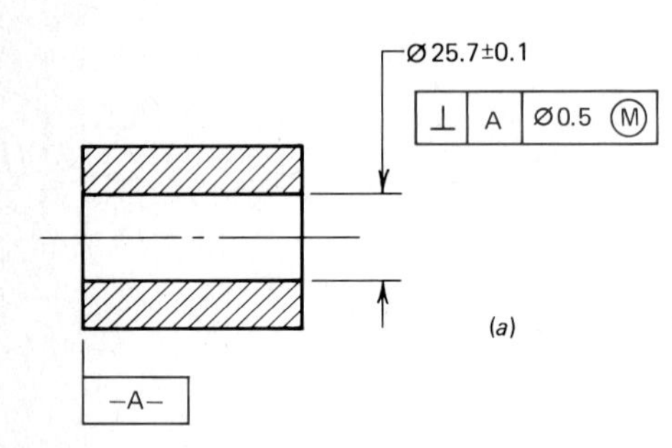

APPLICATION OF MMC	
⊥ Allowed	Hole variation
0.6	25.6
0.7	25.7
0.8	25.8

(b)

FIG. 12.47 Controlling geometric characteristics with application of MMC.

functionally related: circularity or runout, position tolerance, and concentricity tolerance. The ANSI Y14.5M standard should be consulted for specific details on controlling coaxial features.

12.6.7 Application of the MMC

Suppose that we want to design a spacer to fit over a shaft as closely as possible and lie flush with the shoulder, as shown in Fig. 12.46*a*. The shaft diameter of 25 mm is fixed. There is a need then to control the surface that will lie flush to the shoulder and the diameter of the hole. Conventionally, we might do this as indicated in Fig. 12.46*b*. The table in Fig. 12.46*c* shows the combined control placed on the spacer by coordinate dimensions. We note that the MMC of the hole is 25.6 and represents the so-called worst condition for assembly because it would have the minimum clearance between hole and shaft. We also note that if the hole size increases toward its maximum diameter (least material condition), increased perpendicularity tolerance will be tolerated and will not jeopardize the integrity of the spacer. A relationship can be described between the two design criteria; however, the conventional coordinate dimensions do not accommodate the relationship.

Consider the situation where the deviation from MMC of the size tolerance (diameter) is applied directly to the orientation tolerance (perpendicularity), thus allowing more flexibility during manufacture. Figure 12.47*a* shows the dimensioning specifications, and the table in Fig. 12.47*b* shows the combined control placed on the spacer. This transfer of size tolerance to form tolerance is a fundamental rule of geometric dimensioning and tolerancing.

We can extend this relationship to the situation of zero tolerance at MMC. This situation, another fundamental rule of geometric dimensioning and tolerancing, requires perfect form at the MMC. In other words, no feature may extend beyond the boundary established by the MMC. Figure 12.48 illustrates the effect of zero tolerance at the MMC. Note the increased flexibility in allowable combinations of size and form tolerances permitted by the application of zero tolerance.

There are situations where one does not want to permit any relationship to exist between size and form. In this case the symbol Ⓢ must be specified in the feature frame to indicate that the speci-

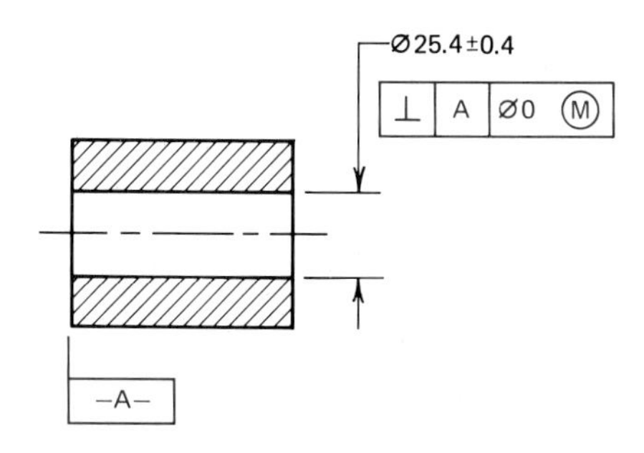

Zero tolerance at MMC	
⊥ Allowed	Hole variation
0	25
0.1	25.1
0.2	25.2
0.3	25.3
0.4	25.4
0.5	25.5
0.6	25.6
0.7	25.7
0.8	25.8

FIG. 12.48 Application of zero tolerance at MMC to control geometric characteristics.

fied form tolerance applies regardless of feature size.

12.6.8 Tolerances of Location

The example illustrated in Fig. 12.35 points out the need to locate center distances between such features as holes and slots. The position tolerancing procedure enables us to do this. Position tolerancing and MMC are always considered in relation to each other. Figure 12.49 illustrates the relation.

The coordinate system of dimensioning allows the position of the center of a hole to vary within a square. See Fig. 12.50*a* and mentally check the extremes to verify the square tolerance zone. This means that a hole center can

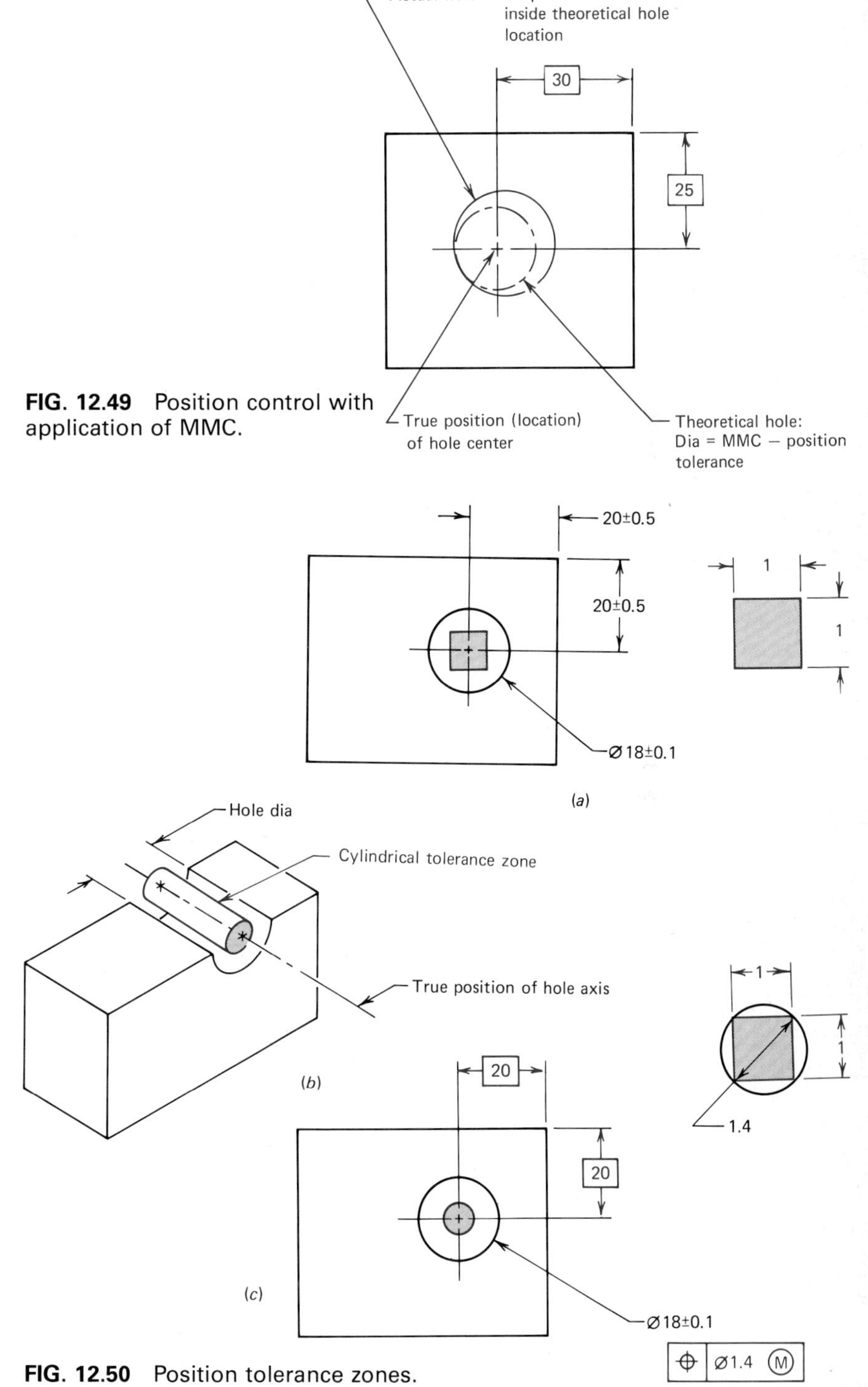

FIG. 12.49 Position control with application of MMC.

FIG. 12.50 Position tolerance zones.

be located farther from the desired true position if it varies along the diagonal of the square rather than along the coordinate axes. Application of position tolerance defines a cylindrical tolerance zone for the axis of a hole (Fig. 12.50*b*). This allows for a circular tolerance zone on the drawing in Fig. 12.50*c*. Note that the diameter of the circular tolerance zone is equal to the length of the diagonal of the square, thus increasing manufacturing flexibility without sacrificing design conditions.

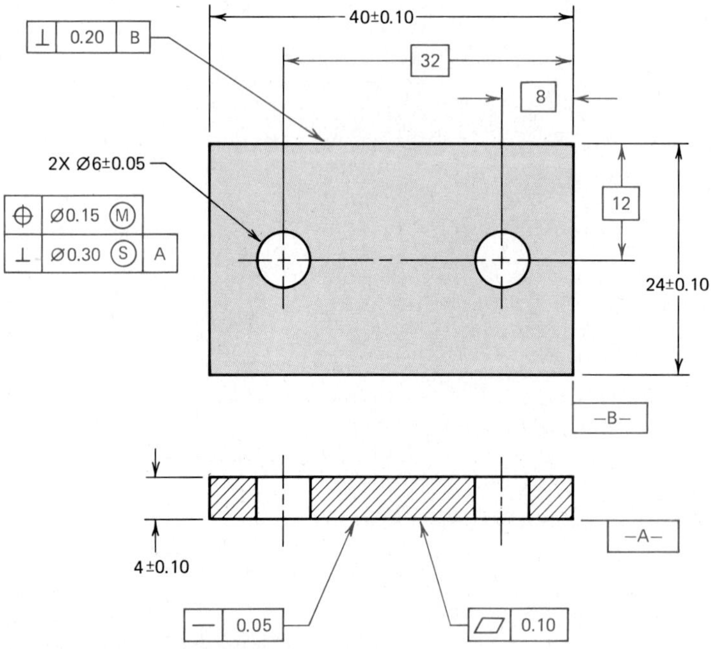

FIG. 12.51 Geometric dimensioning and tolerancing applied to ensure design conditions.

12.6.9 Geometric Tolerancing and Dimensioning Example

Referring back to the part dimensioned conventionally in Fig. 12.32, let us assume that we know the following design criteria.

1. The bottom surface of the plate must be flat within 0.10.
2. The surfaces from which the hole centers are located must be perpendicular within 0.20 at MMC.
3. The holes must be located within 0.15 of true position at MMC.
4. The plate must be straight within 0.05.
5. The hole surfaces must be perpendicular to the plate within 0.30 regardless of feature size.

Figure 12.51 is a completely dimensioned drawing of the object, using geometric tolerancing and dimensioning to accomplish the design objectives.

12.7 APPLICATIONS FROM COMPUTER GRAPHICS

Dimensioning a part automatically with a computer-graphics system is possible to a certain extent. If we remember that the geometry of an object is stored in computer memory in numeric terms, we can see that dimensions can be calculated by the computer if the endpoints of the dimensions are specified. Figure 12.52 illustrates a dimensional drawing produced with a computer and plotter. In this case, the part geometry exists in three dimensions in computer memory, with

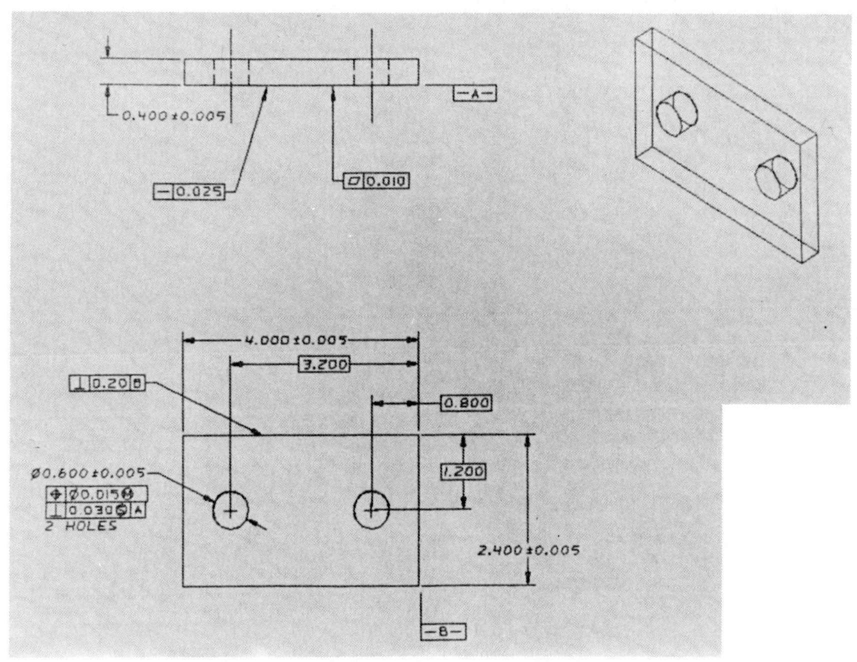

FIG. 12.52 Dimensioning with interactive computer graphics.

two dimensions being displayed in each orthographic view. The geometry was input to the computer in actual size, with the computer calculating a proper scale factor to display the multiview conveniently on the screen.

To dimension the object, the terminal operator accesses the portion of the software that performs the dimensioning. This is one part of a complete two-dimensional drafting package which allows the object to be displayed, enlarged, rotated, and so forth, on the display monitor. When the operator signals the computer for a dimension, a prompt is issued asking for endpoints. The endpoints are signaled using a light pen, digitizer board, or keypad. The computer then calculates the dimension and prompts for a location. When the location is signaled, the computer draws extension lines, dimension lines, and arrowheads if used and then labels the numeric value of the dimension appropriately. Baseline dimensions will be placed in relation to each other automatically according to the standards.

Tolerances can be placed with a dimension but must be inputted by the designer, since the computer generally will not contain the software necessary to compute the limit dimensions. Certain common fits may be an exception. Geometric tolerancing and dimensioning may also be incorporated on a computer drawing as shown in Fig 12.52.

Problems

Note: Several of the dimensioning problems are given with a scale tha may be used to determine proper dimensions. The units for the scales given are mm.

12.1 Sketch necessary orthographic views for parts *a* through *f* of Fig. 9.30 and dimension using correct basic dimensioning standards. Use S for size dimensions and L for location dimensions.

12.2 Sketch necessary views for Fig. 9.32 and dimension using correct basic dimensioning standards. Use S for size dimensions and L for location dimensions.

12.3 Draw or sketch the cable hold-down clamp in Fig. 12.53*a* or the heat sink in Fig. 12.53*b* to the assigned scale. Dimension each part.

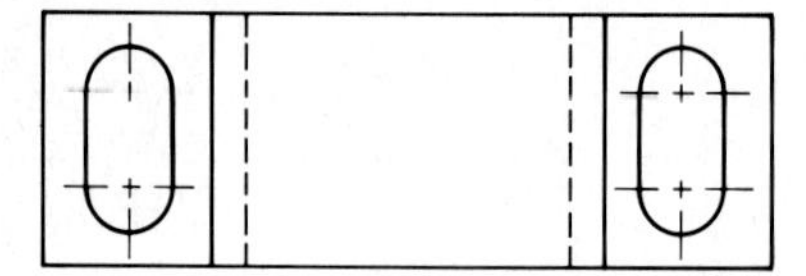

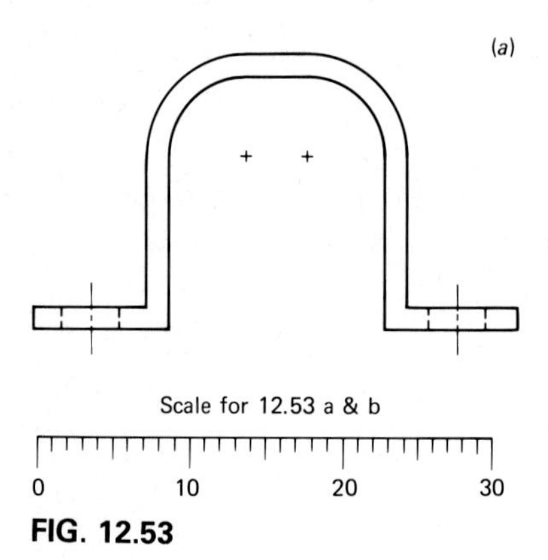

FIG. 12.53

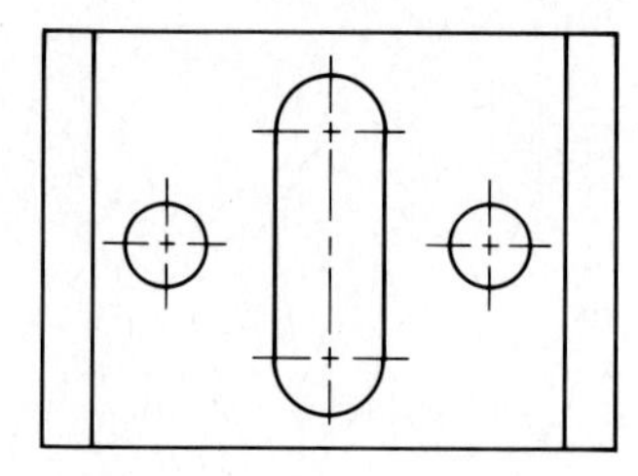

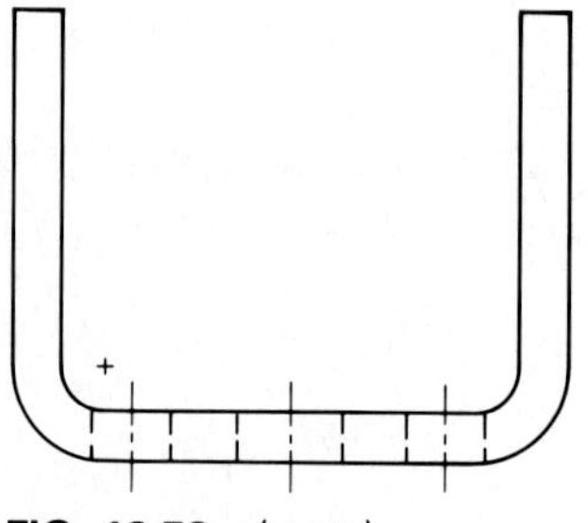

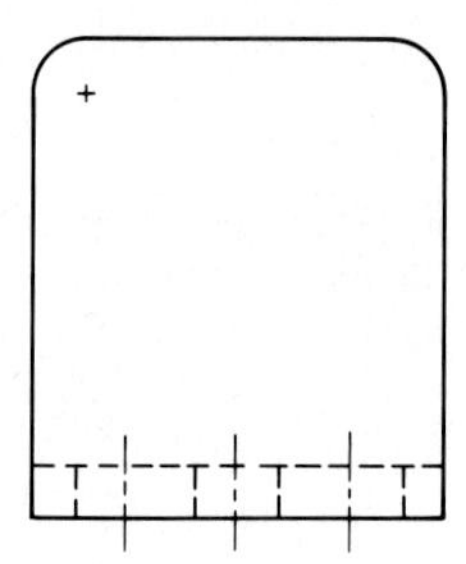

FIG. 12.53 (*cont.*)

12.4 Draw or sketch, with proper dimensions, the switch arm. Use the scale assigned. See Fig. 12.54.

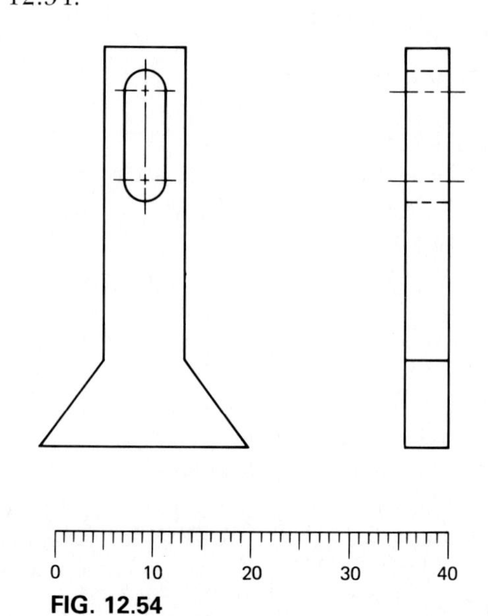

FIG. 12.54

12.5 An adjustable slider is depicted in Fig. 12.55. Sketch or draw to a scale assigned by your instructor and dimension completely.

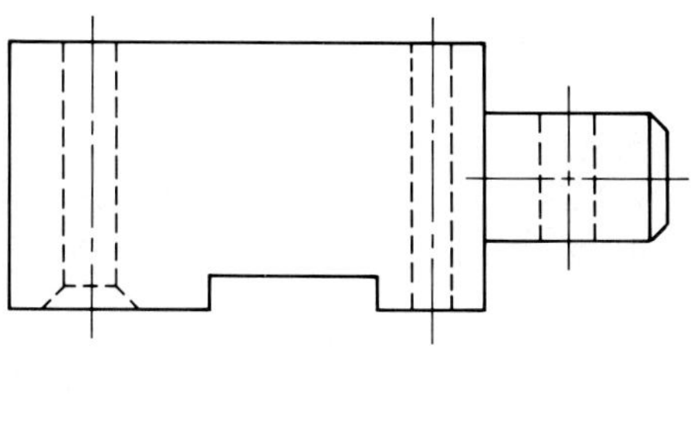

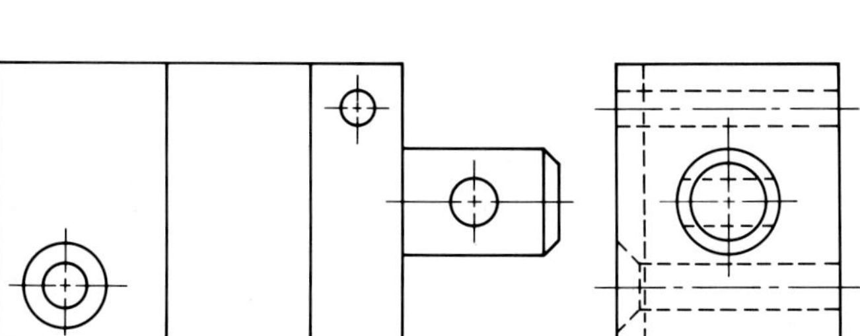

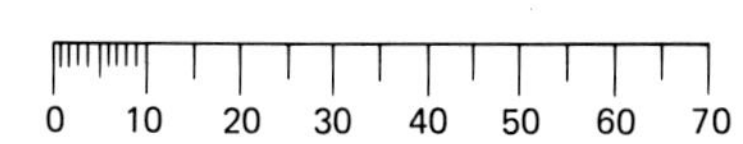

FIG. 12.55

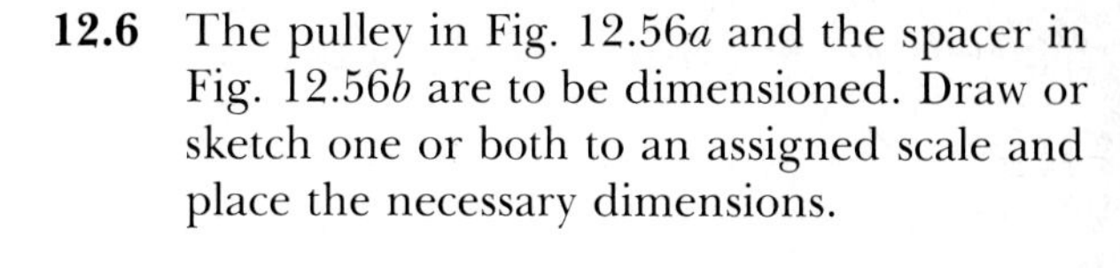

12.6 The pulley in Fig. 12.56*a* and the spacer in Fig. 12.56*b* are to be dimensioned. Draw or sketch one or both to an assigned scale and place the necessary dimensions.

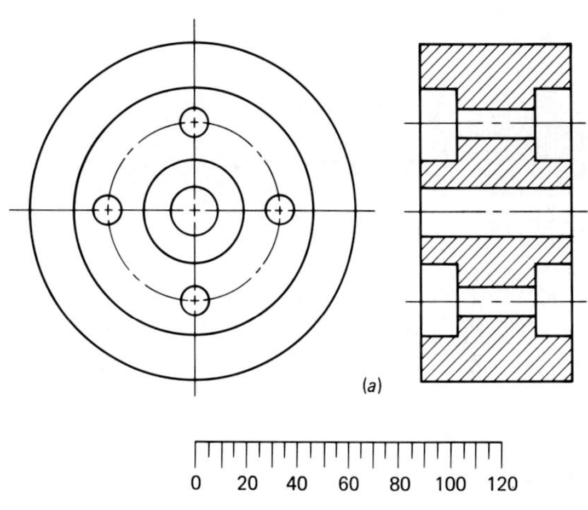

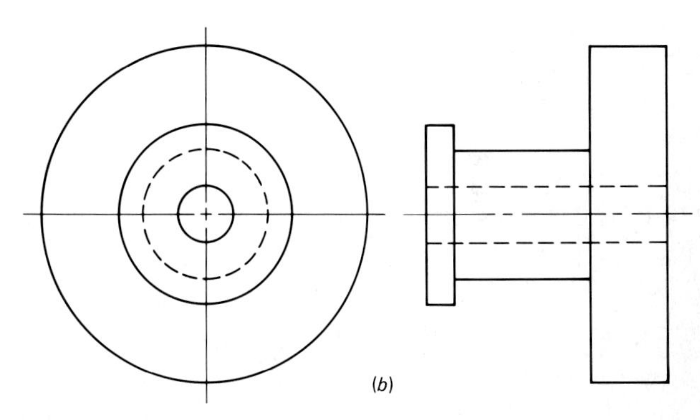

FIG. 12.56

12.7 Sketch or draw the necessary views of the arbor in Fig. 12.57 to a scale assigned and dimension.

12.8 Produce a correctly dimensioned drawing or sketch of the crank arm shown in Fig. 12.58. Use an appropriate scale for your paper size.

12.9 Draw or sketch the vise base shown in Fig. 12.59. Place the necessary dimensions. Use the scale assigned by your instructor.

12.10 For the mating shaft and hole in Fig. 12.60, calculate the limit dimensions that should be filled in for the following conditions. Use the basic hole system.

(*a*) Nominal size = 0.45 in, RC3 fit
(*b*) Nominal size = 0.10 in, RC1 fit
(*c*) Nominal size = 0.25 in, RC5 fit
(*d*) Nominal size = 2.00 in, RC8 fit
(*e*) Nominal size = 1.00 in, FN1 fit
(*f*) Nominal size = 3.50 in, FN5 fit

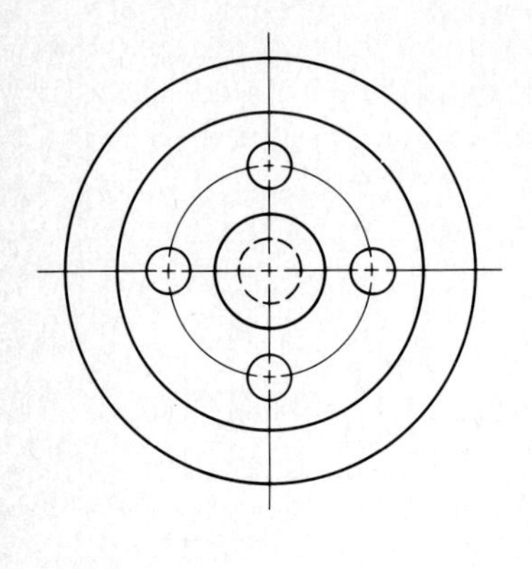

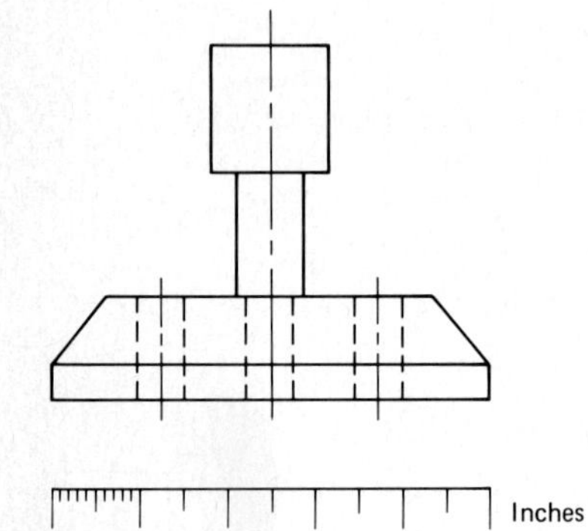

0 1 2 3 4 5 Inches

FIG. 12.57

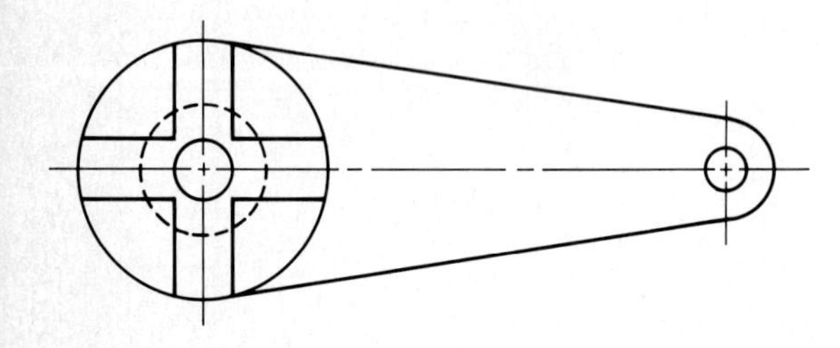

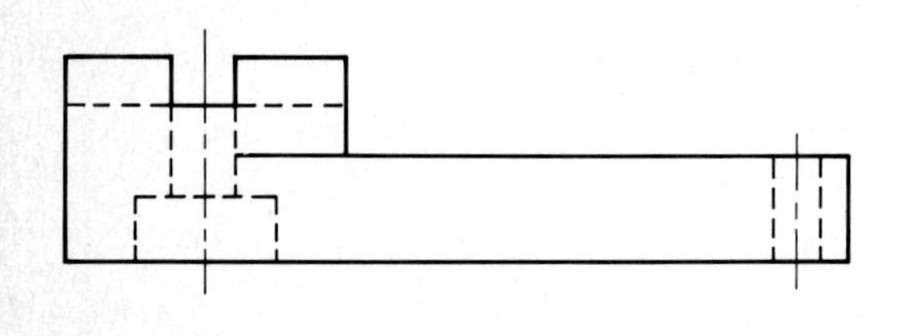

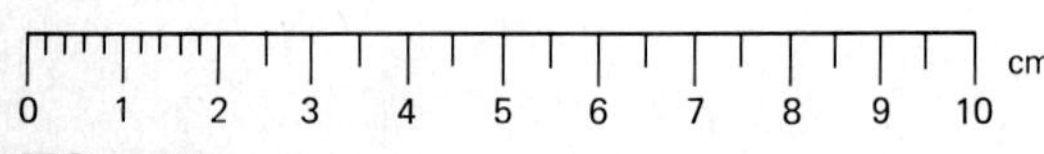

FIG. 12.58

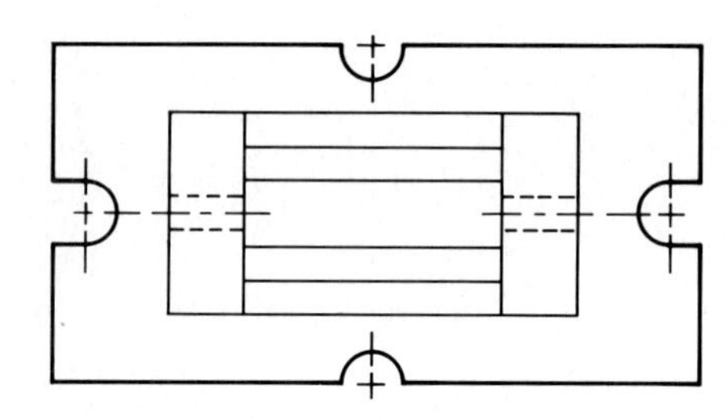

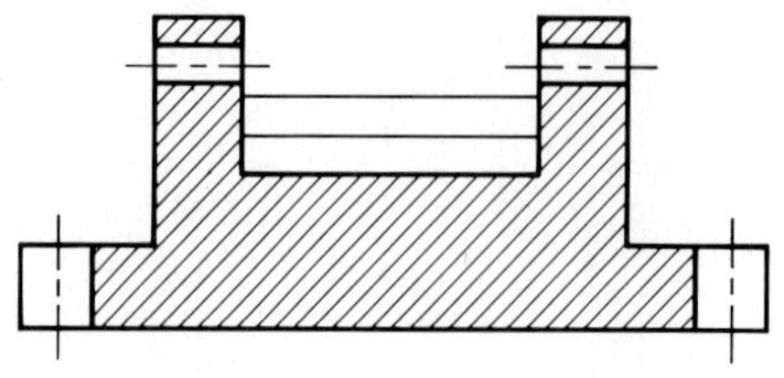

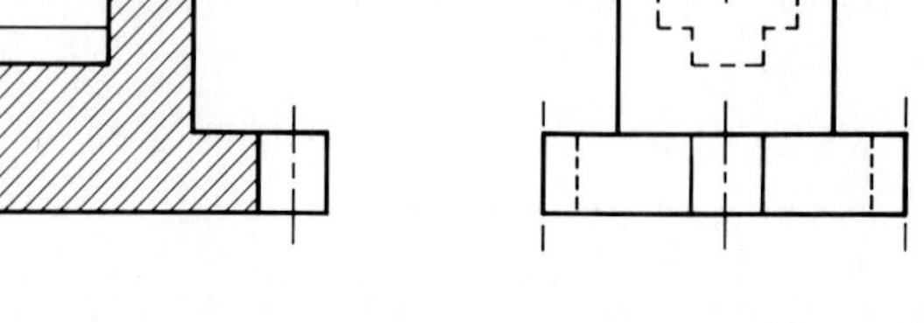

0 10 20 30 40 50 60 70 80 mm

FIG. 12.59

(*g*) Nominal size = 0.25 in, FN2 fit
(*h*) Nominal size = 2.25 in, FN4 fit
(*i*) 40 H8/f7
(*j*) 10 H11/c11
(*k*) 2 H7/s6
(*l*) 20 H7/u6
(*m*) 100 H9/d9
(*n*) 50 H7/g6
(*o*) 6 H11/c11
(*p*) 8 H7/s6

12.11 In Fig. 12.61, bearing *C* has an interference fit on shaft *B*, which in turn maintains a clearance fit with block *A*. Shaft *B* is kept from turning in block *A* by a Woodruff key. For this assembly, compute the limit dimensions of the shaft, the inner race of the bearing, and the hole in the block using the basic shaft system for:

(*a*) Nominal shaft size = 0.5 in, RC7 and FN4 fits
(*b*) Nominal shaft size = 2.0 in, RC8 and FN3 fits
(*c*) Nominal shaft size = 0.2 in, RC2 and FN1 fits
(*d*) Nominal shaft size = 1.0 in, RC5 and FN3 fits
(*e*) Nominal shaft size = 25 mm, G7/h6 and S7/h6 fits
(*f*) Nominal shaft size = 100 mm, D9/h9 and U7/h6 fits
(*g*) Nominal shaft size = 2 mm, F8/h7 and S7/h6 fits
(*h*) Nominal shaft size = 10 mm, C11/h11 and U7/h6 fits

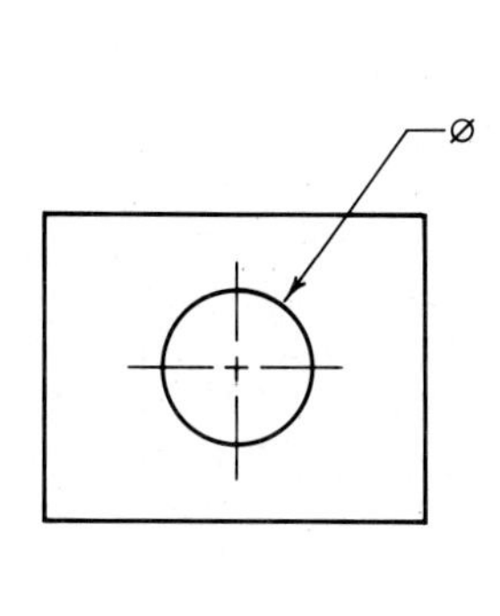

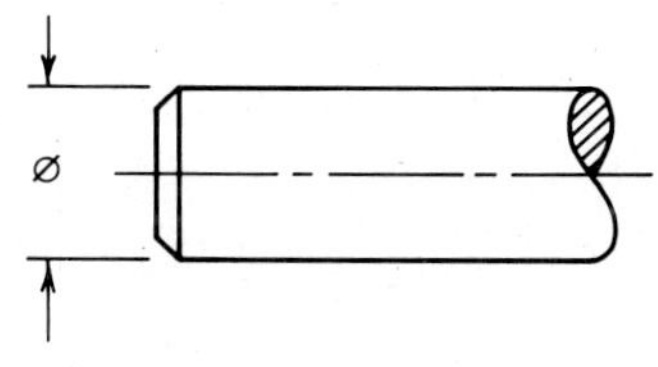

FIG. 12.60

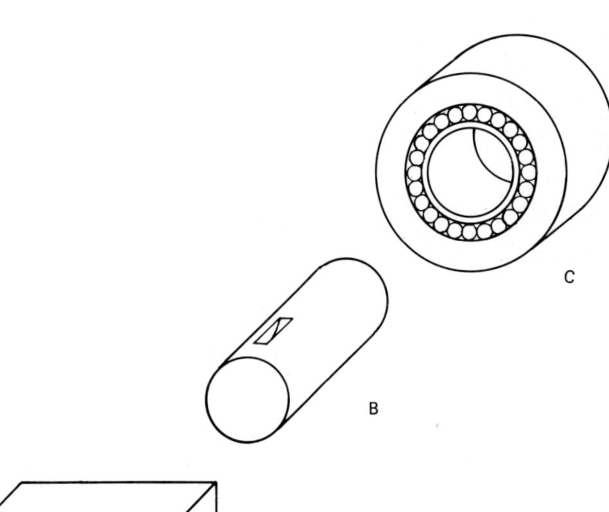

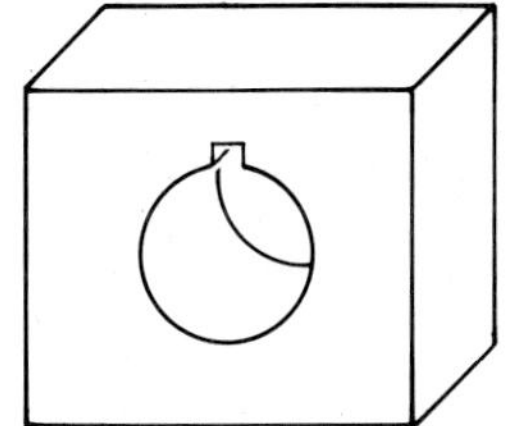

FIG. 12.61

12.12 Draw a feature control symbol for each of the following forms. Sketch an appropriate object and illustrate the tolerance zone for each symbol.

(*a*) A surface straight to within 0.05
(*b*) A surface flat within 0.07
(*c*) A surface perpendicular to datum *B* within 0.10
(*d*) A 45° angle relative to datum *A* within 0.25
(*e*) A surface parallel to datum *C* within 0.15
(*f*) A hole perpendicular to a plate within a cylindrical tolerance zone of 0.07. Establish an appropriate datum on your sketch.

12.13 The part shown in Fig. 12.62 is dimensioned with the traditional coordinate method. The following functional requirements must be met:

(*a*) All nonspecified tolerances are 0.2 applied bilaterally.
(*b*) The hole must be perpendicular to surface *A* within 0.3.
(*c*) The hole must be located true position within 0.1 at MMC.

Sketch or redraw the part and dimension, using geometric tolerancing and dimensioning procedures.

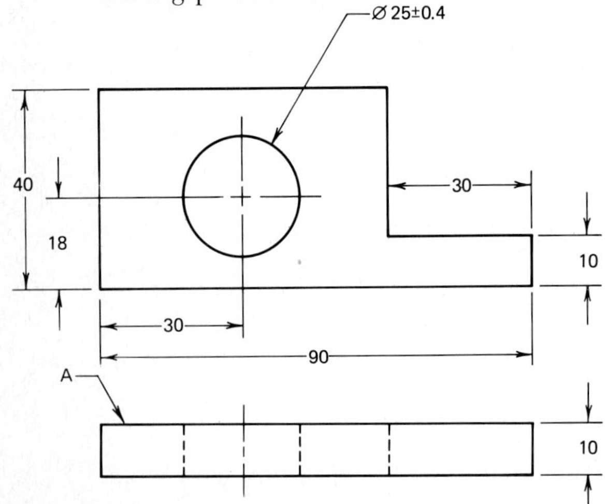

FIG. 12.62

12.14 The part shown in Fig. 12.63 is dimensioned with the traditional coordinate method. The following functional requirements must be met:

(*a*) All tolerances are 0.4 applied bilaterally.
(*b*) The cylinder must be perpendicular to surface *A* within a cylindrical tolerance zone of 0.10 at MMC.
(*c*) Surface *B* must be parallel to *A* within 0.05.
(*d*) Surface *B* must be flat within 0.15.

Sketch or redraw the part and dimension, using geometric tolerancing and dimensioning procedures.

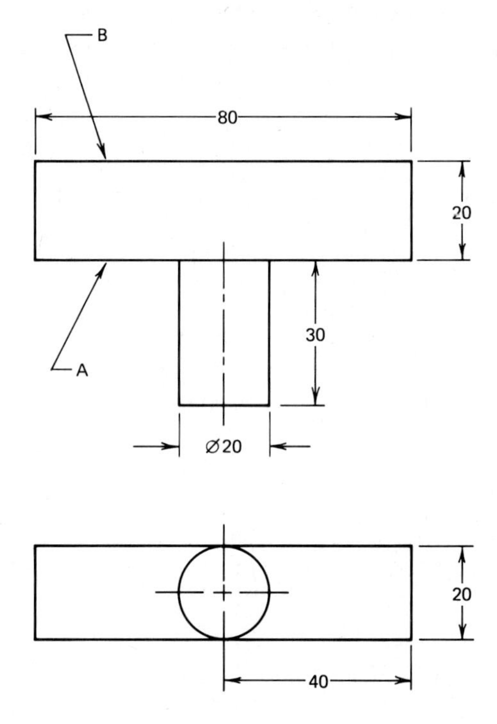

FIG. 12.63

12.15 The collar in Fig. 12.64*a* is to be designed to fit over a shaft whose virtual condition (extreme limit of the shaft for assembly) is 19.6 mm, as shown. The collar is to seat against the plate shown. Develop a table showing the allowable perpendicularity of the hole in the collar compared with the actual hole size for each of the following conditions on the collar:

(*a*) Dimensioned as in Fig. 12.64*b*
(*b*) Dimensioned as in Fig. 12.64*c*
(*c*) Dimensioned as in Fig. 12.64*d*

Make tables similar to those in Sec. 12.6.

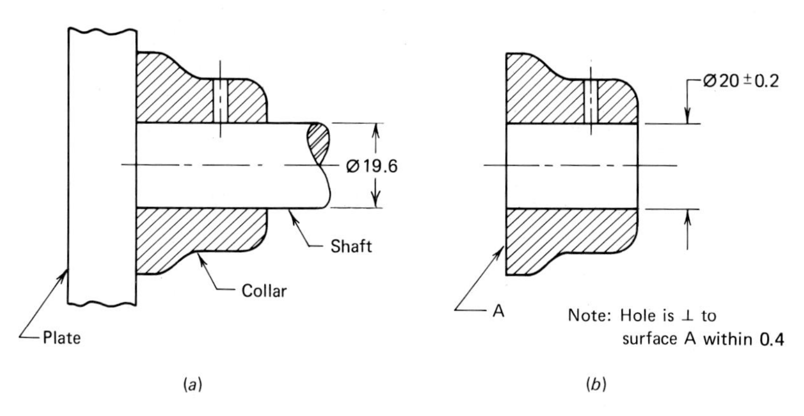

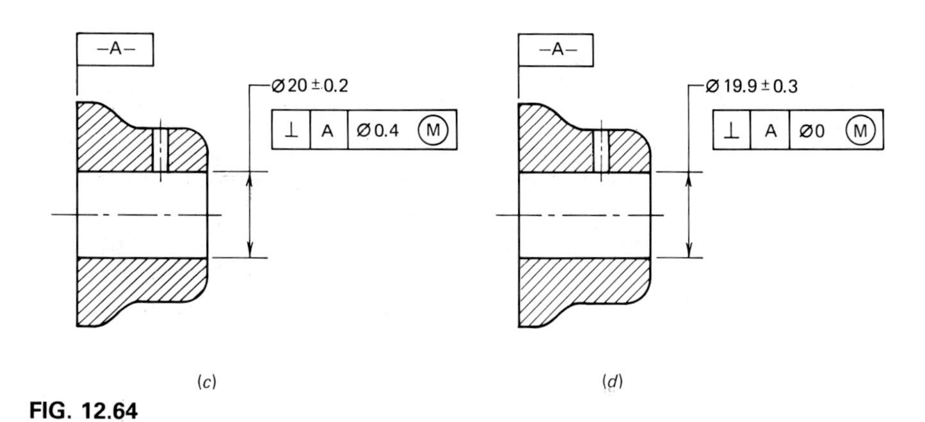

FIG. 12.64

chapter 13

Fasteners

13.1 INTRODUCTION

Engineers encounter thousands of fasteners in designing products. In most design applications the engineer chooses among different fasteners on the basis of geometric, structural, and economic considerations. In this chapter, we will discuss several of the more commonly used fasteners manufactured according to both English and metric standards.

Fasteners may be classified as threaded, locking, or fixed. Threaded fasteners include bolts, studs, and the various forms of screws, such as machine, set, and cap. Locking fasteners may be used separately or in conjunction with other fasteners. Examples include nuts, washers, cotter pins,

FIG. 13.1 Some of the many fasteners available today.

taper pins, straight pins, keys, springs, and splines. Fixed fasteners in common use include welds, solders, brazing, rivets, and a multitude of adhesives.

13.2 DEVELOPMENT OF STANDARD THREADS

Threaded fasteners have the screw thread as the securing element. The first use of screws occurred over 2000 years ago, but extensive use of threads did not occur until the Middle Ages. The increase in usage came about because of the development of machine tools to make the threads. No attempt was made during the Middle Ages to standardize the screw thread. The hole was threaded and measured, and then a screw was prepared to fit the particular hole. The earliest attempts at standardization occurred in England around the middle of the nineteenth century, but these did not become universally adopted.

The first effort at thread standardization in the United States came at the end of the Civil War, when the Franklin Institute developed the United States thread. This effort was satisfactory for industrial needs until World War II, when the Allies found previous standards were not adequate to ensure interchangeability of parts for the war effort. The solution to this need was the Unified screw thread. Since this development, many organizations have contributed to the establishment and publication of standards. Currently, the American National Standards Institute (ANSI) publications are the most widely accepted standards.

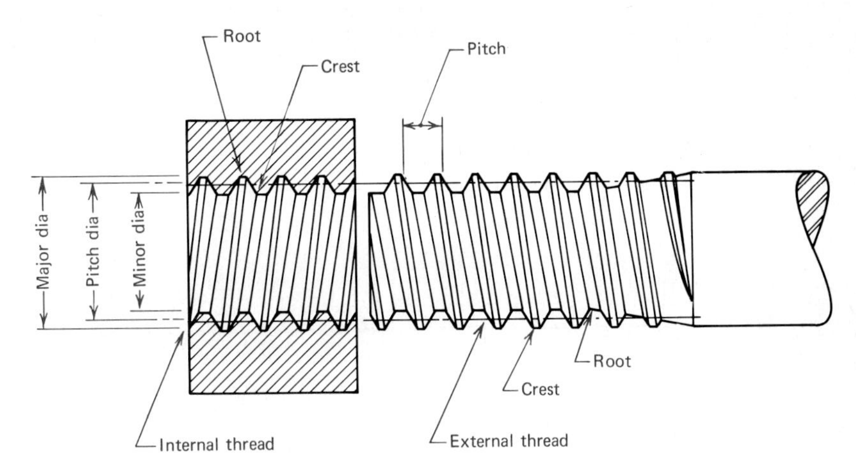

FIG. 13.2 Thread nomenclature.

13.3 SCREW THREAD DEFINITIONS AND NOMENCLATURE

With reference to Fig. 13.2, the following terms describe the screw thread.

1. A *screw thread* is a uniform wedge-shaped section in the form of a helix on the external or internal surface of a cylinder (straight thread) or a cone (taper thread).

2. If a thread traverses a path in a clockwise and receding direction when viewed axially, it is a *right-hand thread*. Any thread is assumed to be right-hand unless specified LH (left-hand).

3. The largest diameter of the screw thread is called the *major diameter*.

4. The smallest diameter is called the *minor diameter*.

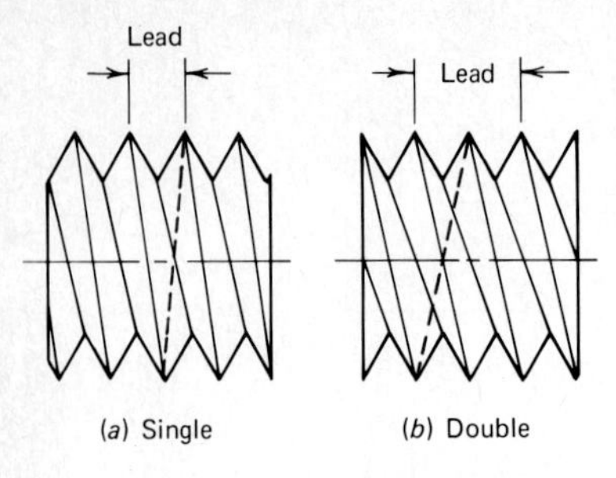

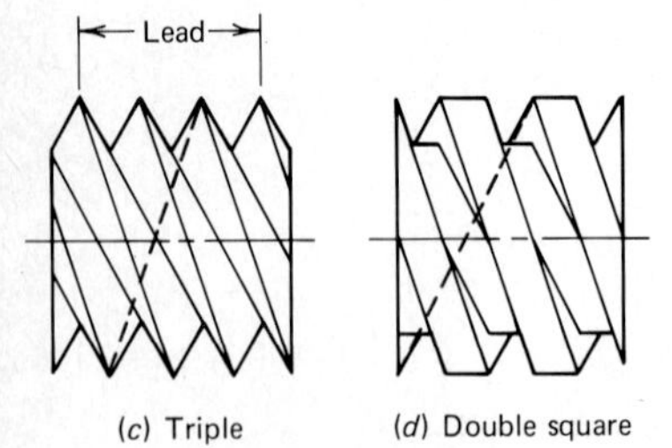

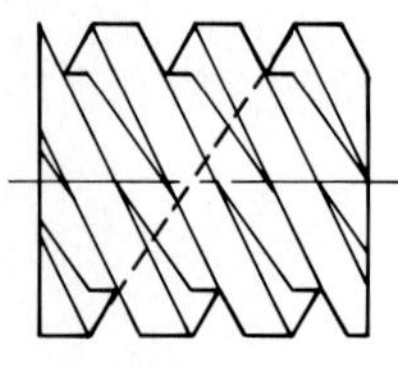

FIG. 13.3 Thread lead.

5. The *pitch* is the distance between corresponding points on a thread; it is measured parallel to the thread axis.

6. The *pitch diameter* is the diameter of an imaginary cylinder whose surface cuts the threads at a point where the width of the thread and the open space (groove) are equal.

7. On an external thread, the point that is farthest from the axis is called the *crest*. On an internal thread, the crest is the point nearest the axis.

8. The converse of the crest is the *root*.

9. The *lead* is the distance a threaded section moves axially in one revolution. Lead equals pitch for a single thread. The lead is equal to two pitches for a double thread (see also Fig. 13.3).

10. The *depth* of a thread is equal to one-half the difference between the major and minor diameters.

13.4 AVAILABLE THREAD FORMS

Threads may be used to transmit power and motion as well as to secure. A large number of thread forms are available to accomplish these functions. Figure 13.4 illustrates common thread forms.

1. The *sharp V* thread has pointed crests and roots which are difficult to manufacture and maintain. Usage is limited, but this thread form is often portrayed on drawings because of its simplicity.

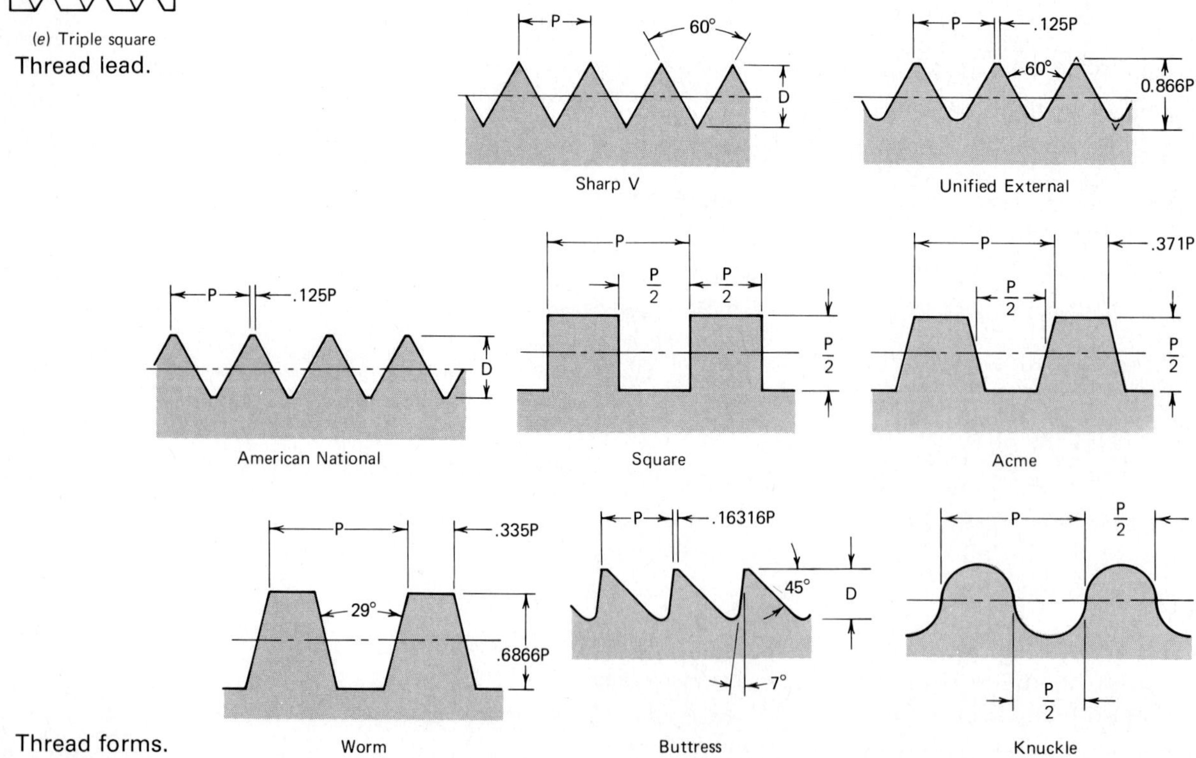

FIG. 13.4 Thread forms.

2. The *Unified* and *American National* thread forms are similar to the sharp V but have slightly rounded crests and roots. The name Unified was chosen by the United States, Canada, and Great Britain when these countries agreed on standardized fasteners.

3. Power transmission is the principal use of the *square thread*. The automobile jack represents one familiar application. The thread is often rounded slightly to reduce manufacturing problems.

4. The square thread has largely been replaced by the *acme thread*. This thread has the advantages of strength, ease of manufacture, and ease of disengagement. The *worm thread* is very similar to the acme thread.

5. The *buttress thread* combines the strength of the V thread and the efficiency of the square thread. It is used in large artillery pieces, jacks, and other devices which require the transmission of power in one direction only.

6. One of the most commonly encountered threads is the *knuckle thread*, which is found on glass jars and light bulbs. This thread is produced by molding the fastener from glass, plastic, sheet metal, or other material instead of cutting or rolling the thread, which are the standard processes for metal thread manufacturing.

13.4.1 Thread Series

The current standards for screw threads show several combinations of number of threads per unit length of material and diameter. These thread series have evolved from specific design requirements and the attempt to standardize fasteners throughout the world.

There are two unit systems, metric and English, which specify standard thread series. In the metric system, only one series is specified with a range from a 1.6-mm diameter with a pitch of 0.35 mm to a 100-mm diameter with a pitch of 6 mm. The pitch is the reciprocal of the number of threads per unit length.

There are several designations of thread series in the English system, in contrast to the one series designation in the metric system. We will not present here all the available thread series in the English system. Instead, we will concentrate on the most common series, the Unified screw thread. Refer to the ANSI standards for details on other thread series. A second difference between the metric and English systems of thread series is that the English system specifies the number of threads per inch instead of pitch.

The coarse thread series are designated UNC (Unified Coarse) or NC (National Coarse). This series allows rapid assembly and is used for materials with low tensile strength. The coarse thread resists stripping, which is a problem particularly with internal threads.

The fine thread series, designated UNF or NF, is used on thin-walled materials where short engagement or vibration is likely.

The extra-fine thread, UNEF or NEF, is used where high-frequency vibrations are encountered and very short engagements are possible.

The eight-thread series provides an alternative to the coarse and fine series. Its designation is 8 UN or 8 N. The 8 refers to threads per inch.

The twelve-thread series provides a medium-fine thread pitch for fastener diameters greater than 1.5 in. The designation is 12 UN or 12 N.

The sixteen-thread series provides a fine thread pitch for large-diameter fasteners. Common applications include retaining nuts and adjusting collars. The designation is 16 UN or 16 N.

Tables B.2 and C.2 illustrate the differences in these thread series. The Unified screw thread standard has other permissible thread series, but for most applications the ones described here will be satisfactory.

13.4.2 Classes of Fit

Two classes of fit are available for metric threads; three classes are available for Unified threads. The class of fit is determined by the amount of tolerance (variation from the stated size) permitted during manufacture.

For metric threads, the close fit is specified if desired; otherwise, the general-purpose fit is assumed. In terms of tolerance, the close fit has less permissible variation in size than the general-purpose fit and is therefore more costly to produce.

The three classes of fit for Unified screw threads are 1A or 1B, 2A or 2B, and 3A or 3B. External threads carry the A designation, and internal threads carry the B designation. The permissible variation in size decreases as the class of fit number increases. Class 1A and 1B threads are used for quick operations in which fits are not critical. Class 2A and 2B threads are used for general-purpose applications, while 3A and 3B threads are used for close tolerance requirements and when vibration is a factor.

13.4.3 Specification of Threads

In general, the thread specification is part of a total fastener specification. We will begin by outlining the requirements for specifying metric and Unified threads.

The metric thread is designated by size, pitch, and tolerance class as follows:

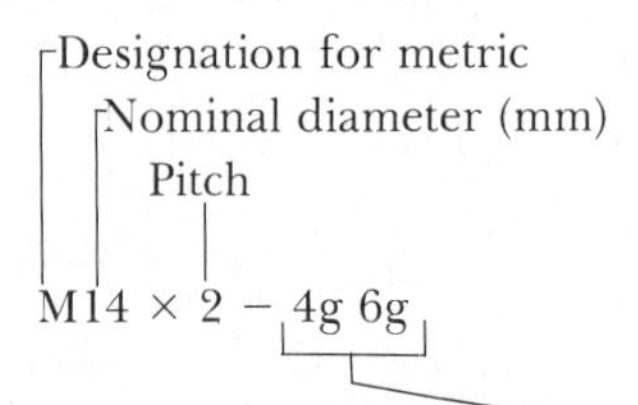

The letter M is a metric thread symbol indicating that the thread specification is based on the International Organization for Standardization (ISO); 14 × 2 is the designation for nominal diameter and pitch of the thread. M14 × 2 is called the basic designation or specification of the metric thread. For most applications the basic designation is satisfactory; it represents a general-purpose thread to be used with normal engagement.

For applications which have specific thread manufacturing requirements, the tolerance class designation shown below must be included.

Tolerance grade
Tolerance position
4g 6g
Major diameter tolerance symbol
Pitch diameter tolerance symbol

The ISO recommends the use of three tolerance grades: 4, 6, and 8. The numbers represent the tolerance magnitude; thus, 4 would represent close tolerance requirements, while 8 would reflect a large tolerance range.

The tolerance position determines the material limits (allowance) for the thread. Use of uppercase and lowercase letters enables one to make a distinction between internal and external threads. Lowercase letters are used for external threads and indicate the following tolerance positions:

e Large allowance
g Small allowance
h Zero allowance

Similarly, uppercase letters represent internal threads as follows:

G Small allowance
H Zero allowance

If the tolerance symbols for both the major and pitch diameters are the same, only one of them is written in the thread specification.

The following examples will serve to illustrate metric thread specification.

M10 × 1.5	Basic designation, general-purpose thread
M24 × 3–6g8g	Thread tolerance class included, external thread
M6 × 1–4H6H	Internal thread
M8 × 1.25–6g	Equal tolerance classes, external thread
M24 × 3–6H–LH	Equal tolerance classes, internal left-hand thread

When only the basic designation is used, the type of fastener makes it obvious whether the thread will be external or internal. For example, a bolt thread will be external, while the thread for the corresponding nut will be internal.

Table B.2 lists the basic designations for metric threads.

A Unified thread is designated by the nominal diameter, threads per inch of length, thread form, thread series, and thread class symbol, as follows:

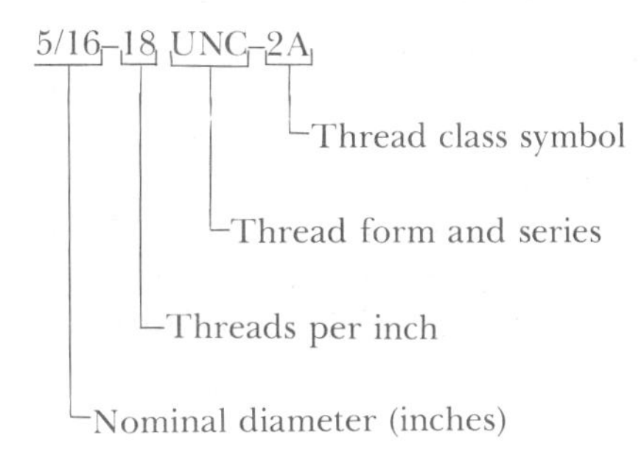

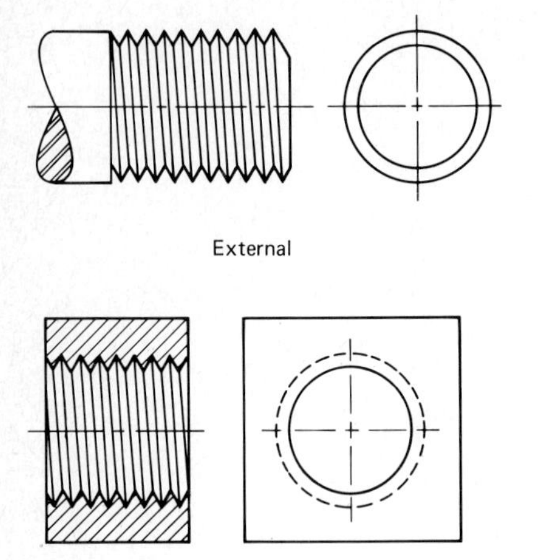

FIG. 13.5 Detailed thread representation.

In this example, we have a 5/16 (0.3125 fractional equivalent) nominal diameter thread. It is a Unified Coarse (UNC) thread with 18 threads per inch. The thread class symbol 2A represents a general-purpose (2) external (A) thread.

Several other examples are shown to further illustrate the designation for Unified threads.

1/4–28 UNF–3A	Fine form, close tolerance, external
0.8750–20 UNEF–3B	Extra-fine form, special-purpose, internal
1.0–8 UNC–1B–LH	Coarse, liberal tolerance, internal, left-hand

Closer interpretation of the metric and Unified specifications reveals that the metric grade 6 is close to the Unified general-purpose designation 2A or 2B. If the tolerance position is to be specified for a Unified thread, it is done in the form of a note, such as

0.750–10 UNC–2A
MAJOR DIAMETER 0.7353–0.7482
PITCH DIAMETER 0.6773–0.6832

This note corresponds to the uppercase and lowercase letters used in the designation of tolerance position for metric threads. In normal situations this note is not used, just as the tolerance class designation is not always placed on a metric thread designation. A designer should always consult the appropriate standard if there is a question about the thread specification.

13.4.4 Thread Delineation on a Drawing

The ultimate use of a drawing dictates the method of presenting the threaded fasteners. A detailed drawing such as Fig. 13.2 or Fig. 13.5 is appropriate for a textbook or some other illustrative purpose. These drawings are time-consuming and expensive and do not necessarily enhance the interpretation of the specifications for threaded fasteners. Therefore, in practice, threads are portrayed by one of the two methods shown in Fig. 13.6. The schematic representation shown

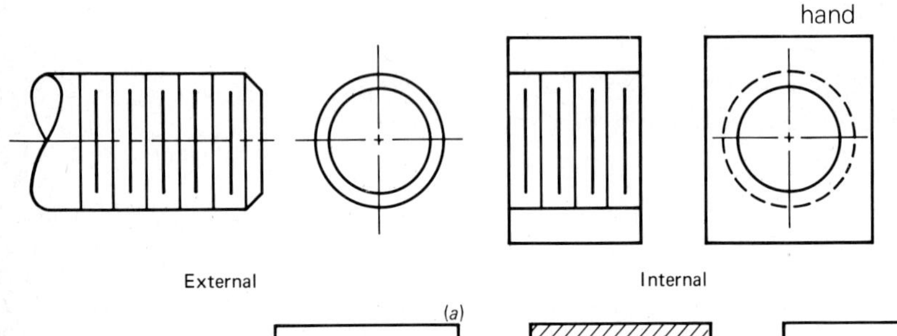

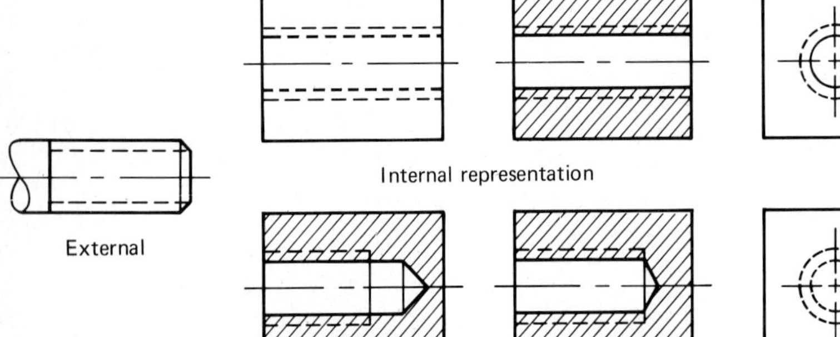

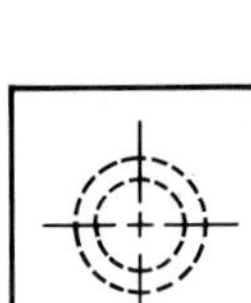

FIG. 13.6 Conventional thread representation (*a*) Schematic thread delineation. (*b*) Simplified method.

in Fig. 13.6*a* is used frequently in assembly drawings and by organizations that prefer a realistic representation of thread. The simplified representation in Fig. 13.6*b* is used widely because of its ease of drawing. Figure 13.7 illustrates the drawing technique for the construction of a schematic representation. The technique is not complicated and produces a realistic result.

13.5 THREADED FASTENERS

As stated earlier, the specification of a threaded fastener begins with the unit system to be used: metric or English. We will limit the discussion to the Unified series only in the English system.

There is a definite movement toward using metric units exclusively in engineering. This movement has been under way for many years, but some areas have been slow to convert. Until recently, engineers have had very few metric standards to work with, and this has hindered the changeover. Engineers today must become familiar with both metric and Unified fasteners in order to function in the industry. Several types of threaded fasteners from both the metric and English systems are included in Apps. B and C.

13.5.1 Bolts and Nuts

A bolt is constructed with a head on one end and threads on the other. The bolt is passed through holes in the materials to be joined and has a nut threaded

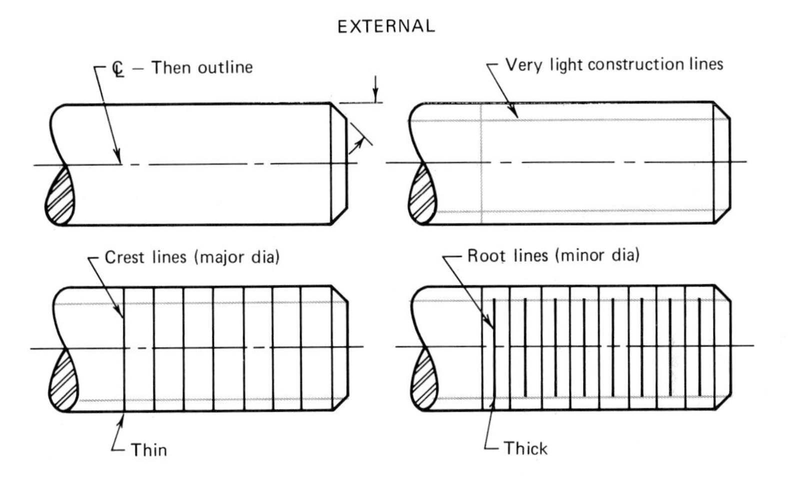

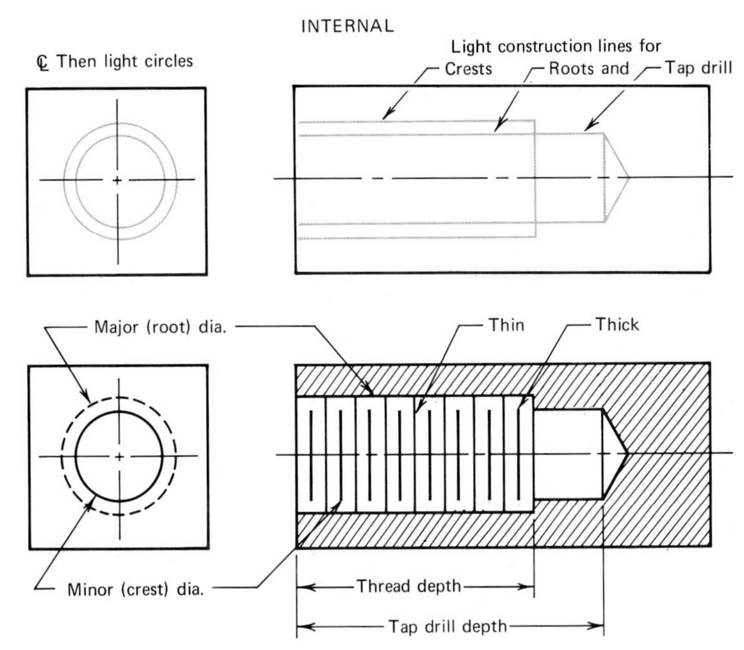

FIG. 13.7 Construction procedures for producing a schematic thread representation.

on the end. Washers may be used on either end or on both ends of a bolt. The heads on bolts are normally round, square, or hexagonal. Connections using bolts and nuts are illustrated in Fig. 13.8*a* and *g*.

A metric bolt is specified by the bolt type, basic thread designation, length, and other notes which call out special features or manufacturing requirements.

HEX BOLT, M24 × 3 × 120, STAINLESS STEEL

ROUND HEAD SHORT SQUARE NECK BOLT M10 × 1.5 × 60

A metric nut is designated by name, basic thread designation, and notes for special features. The nut specification to correspond to the stainless-steel bolt above is

HEX NUT, STYLE 1, M24 × 3, STAINLESS STEEL

The specification for a bolt using the Unified thread series consists of the thread designation, length, name, and notes for special features.

5/8–11 UNC × 3-3/4 HEX BOLT, STAINLESS STEEL

1.5–6 UNC × 8 SQUARE BOLT, STEEL

The corresponding nut for the Unified hex bolt above would be

5/8–11 UNC HEX NUT, STAINLESS STEEL

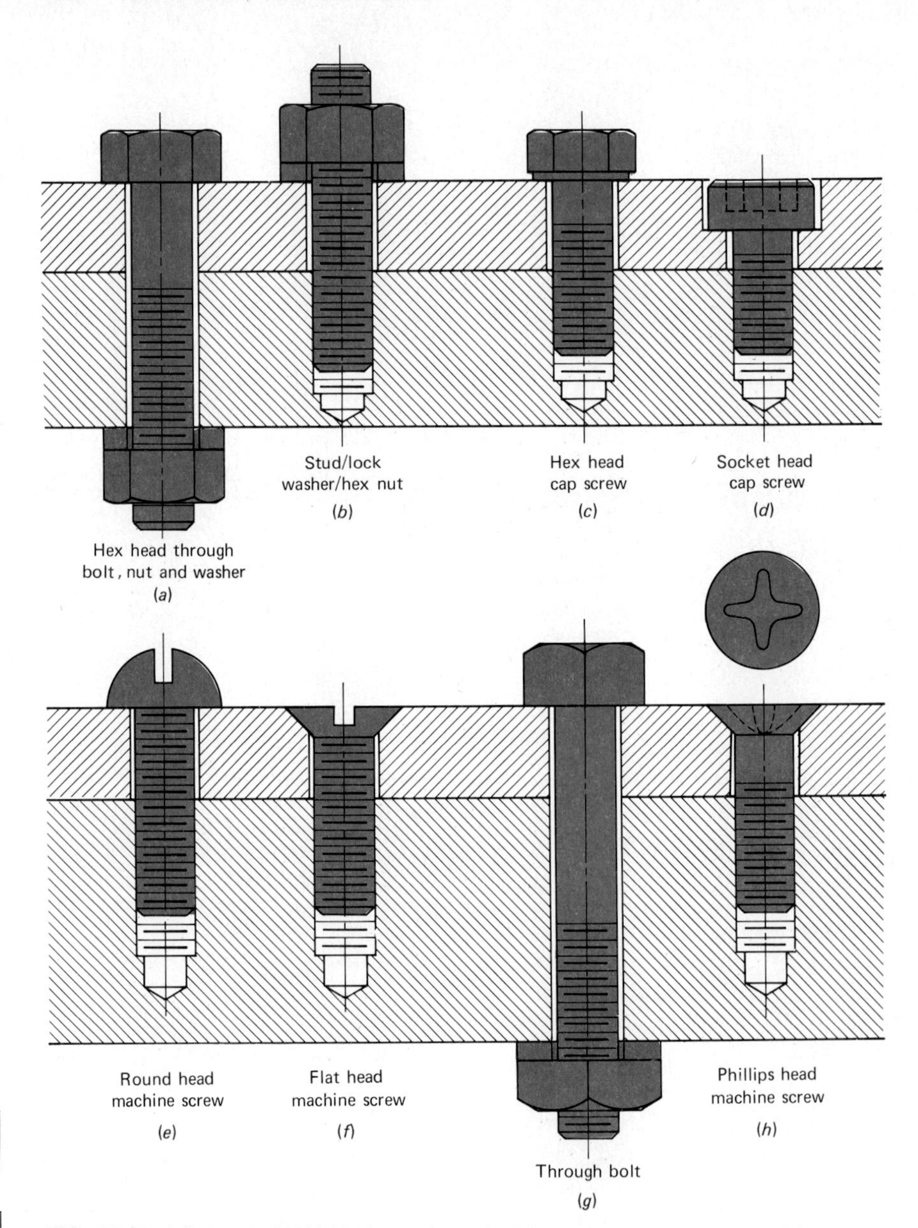

FIG. 13.8 (*a*) through (*h*) Various threaded fasteners in assembly. (i) Several fastener head types with associated tool.

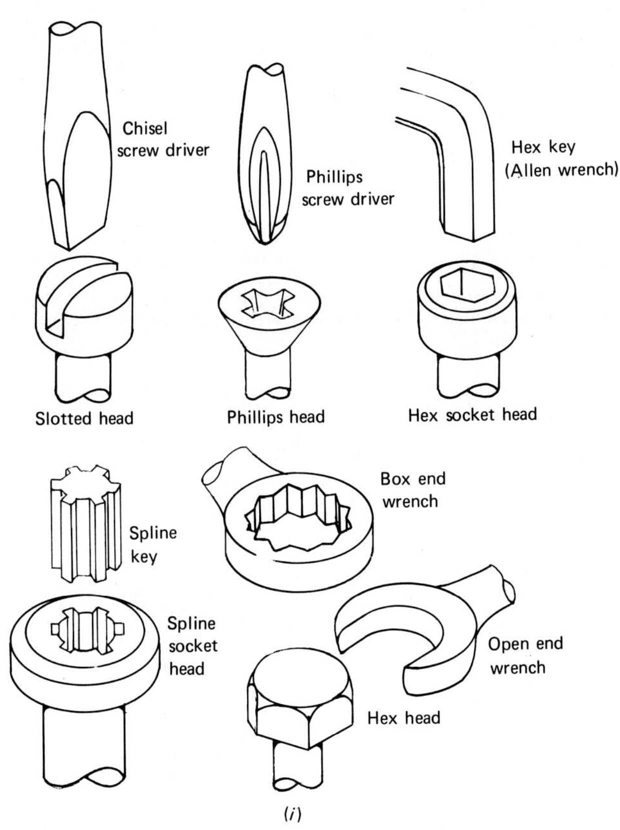

FIG. 13.8 *(con't.)*

13.5.2 Studs

A special type of fastener called a stud is threaded on both ends. See Fig. 13.8*b*. In normal use the stud is passed through a hole in one piece and into a tapped (threaded) hole in a second piece. A nut is then threaded onto the exposed thread and tightened. The length of the thread on the nut end should be long enough to prevent binding before the nut is drawn up tightly.

Studs are generally ordered for specific applications and thus are considered a nonstandard item. The complete specification is given on a detail drawing. This detail will include the thread lengths if it is a stud like the one shown in Fig. 13.8*b*. When one refers to a specific stud, a designation similar to that for a bolt is used.

A metric stud is referred to by name, basic thread designation, length, and notes for special features, as follows:

DOUBLE END STUD, M14 × 2 × 120, STEEL

A Unified stud is referred to by thread specification, length, and notes for special features, as follows:

0.75–10 UNC–2A × 3 STUD, STEEL

13.5.3 Cap Screws

A cap screw, like a stud, passes through a hole in one piece and is threaded into a hole in the second piece. The head of the cap screw is an integral part of the screw. Five types of heads are standard: hex, flat, round, fillister, and socket. See Fig. 13.8*c* and *d* for two examples.

A metric cap screw is designated by name and head type, basic thread designation, length, and notes for special features, as follows:

SOCKET HEAD CAP SCREW, M10 × 1.5 × 35, STAINLESS STEEL

A Unified cap screw is called out by thread designation, length, name, head type, and notes for special features, as follows:

1/2–13 UNC–2A × 2½ FLAT HEAD CAP SCREW, STEEL

0.5–13 UNC–2A × 2.5 FILLISTER HEAD CAP SCREW, STEEL

13.5.4 Machine Screws

A machine screw functions like a screw or a bolt but is typically used without a nut. Nine standard heads are available, with the most popular having either a straight slot or a Phillips slot (cross). Examples of machine screws are shown in Fig. 13.8*e*, *f*, and *h*.

A metric machine screw is designated similar to a cap screw.

HEX HEAD MACHINE SCREW, M6 × 1 × 50

A Unified machine screw designation is arranged as follows:

0.375–16 UNC–2A × 1.5 ROUND HEAD MACHINE SCREW

Figure 13.8*i* illustrates several fastener head types along with the associated tools used with the fasteners.

13.5.5 Set Screws

A set screw interacts between two pieces by threading into a hole in the outer piece and then bearing against the inner piece, as illustrated in Fig. 13.9. Set screws are usually square-headed or headless, with several socket shapes. The set screw can serve as a fastener to hold two pieces together, to restrict but not prevent relative motion between two pieces, or as a locking device to prohibit relative motion. Set screws can be found in most households on doorknobs and towel bars.

Metric set screws are designated by type of point, name, thread designation, and notes for special features, as follows:

CONE POINT HEX SOCKET SET SCREW, M10 × 1.5 × 40, STEEL

Unified set screws are called out by thread designation, length, type of point, name, and notes for special features, as follows:

0.5–13 UNC–2A × 1 FLAT POINT HEX SOCKET SET SCREW, STEEL

Without additional discussion, the reader is referred to Fig. 13.10 for several additional types of threaded fasteners that are in common use.

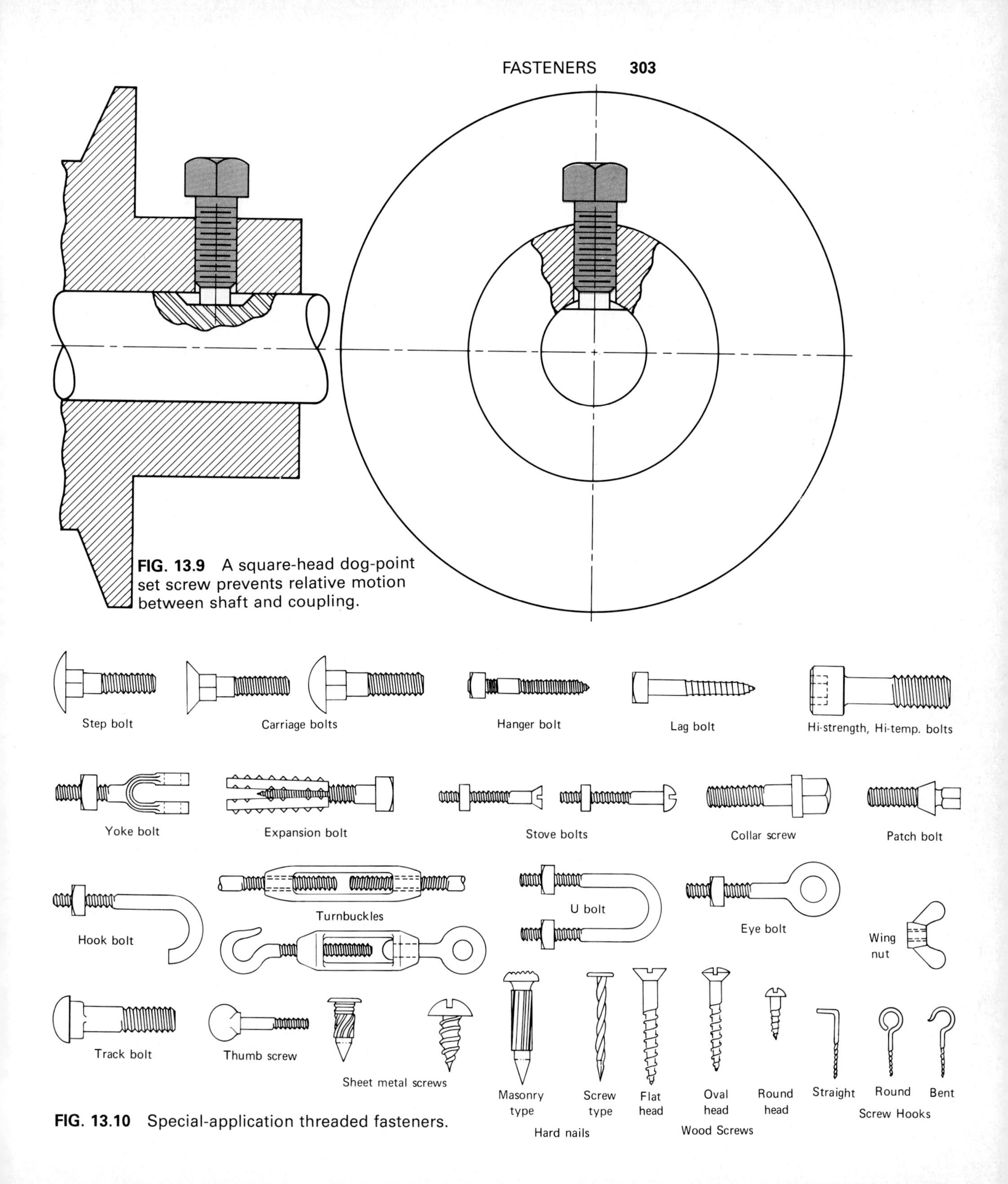

FIG. 13.9 A square-head dog-point set screw prevents relative motion between shaft and coupling.

FIG. 13.10 Special-application threaded fasteners.

13.6 LOCKING DEVICES

The set screw is only one of a wide range of locking devices that are used to prohibit relative motion. Among these locking devices are castle nuts, jam nuts, cotter pins, straight pins, taper pins, keys, springs, and splines. Each has specific applications, and manufacturers' catalogs should be consulted for the applications and the method of specification. A few examples are shown in Apps. B and C.

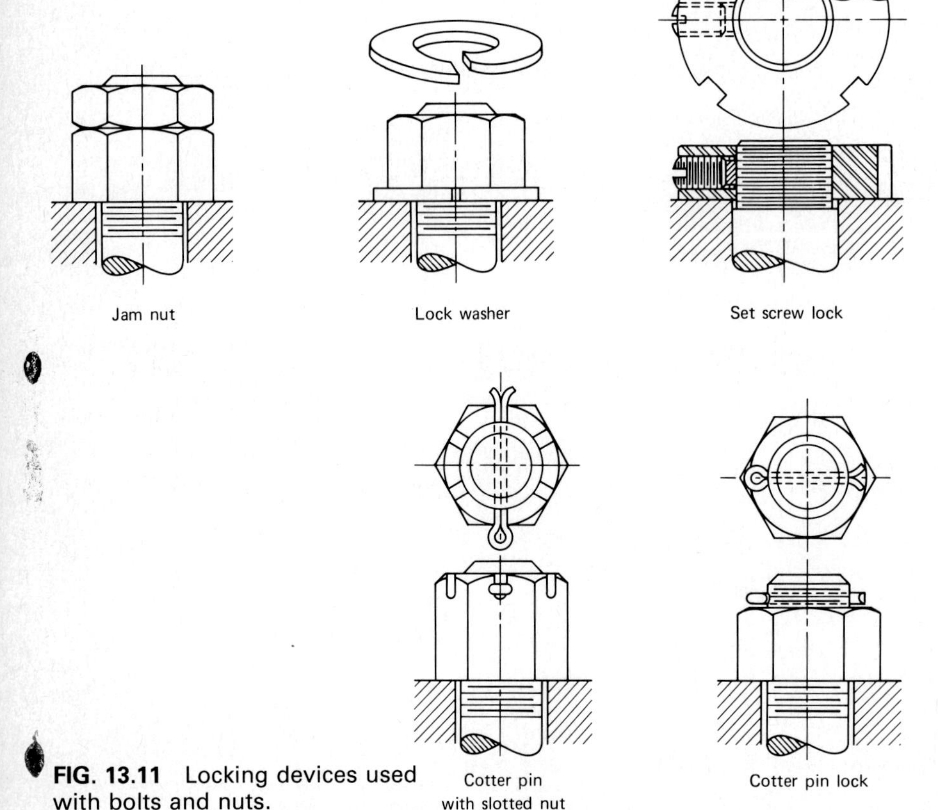

FIG. 13.11 Locking devices used with bolts and nuts.

13.6.1 Nuts

Nuts were mentioned in the discussion of bolts. Several types of nuts are shown in Fig. 13.11, along with devices used to prevent nuts from working loose. Specifications for several nut types can be found in Apps. B and C. In Fig. 13.11, note the use of washers, cotter pins, set screws, and double nuts as devices to prevent loosening.

13.6.2 Washers

Plain (flat) and lock washers are classified as light, medium, heavy, or extra-heavy. Washers are specified by the diameter of the fastener over which the washer must fit and by the classification. For example,

1/2 MEDIUM LOCK WASHER

7/16 EXTRA HEAVY PLAIN WASHER

PLAIN WASHER, 8 mm, REGULAR

would be acceptable specifications. Examples of complete specifications for some washers are presented in Apps. B and C.

13.6.3 Pins

Straight, taper, and cotter pins are used to secure castle and slotted nuts, to prevent parts from coming loose, to hold hubs to shafts, and for simple adjustments in mating parts. Straight pins and cotter pins are specified by diameter and length. Taper pins are usually specified by a manufacturer's number. See

Apps. B and C for specifications of pins. Figure 13.12 shows an application of a taper pin to secure a pulley to a shaft. Some examples of specifications are as follows:

PIN, STRAIGHT, 0.5 × 1.5, STEEL

PIN, STRAIGHT, 6 mm × 50 mm, STEEL

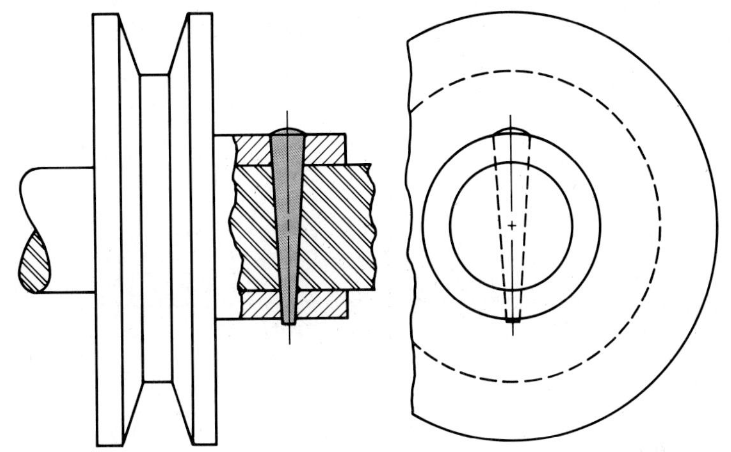

FIG. 13.12 A taper pin can be used to secure a pulley to a shaft.

13.6.4 Keys

Keys are devices that prohibit relative motion between a shaft and pulleys, gears, and so forth. The shaft will have a groove, called a keyseat, cut into it, and the pulley or gear will have a mating keyway. See Fig. 13.13.

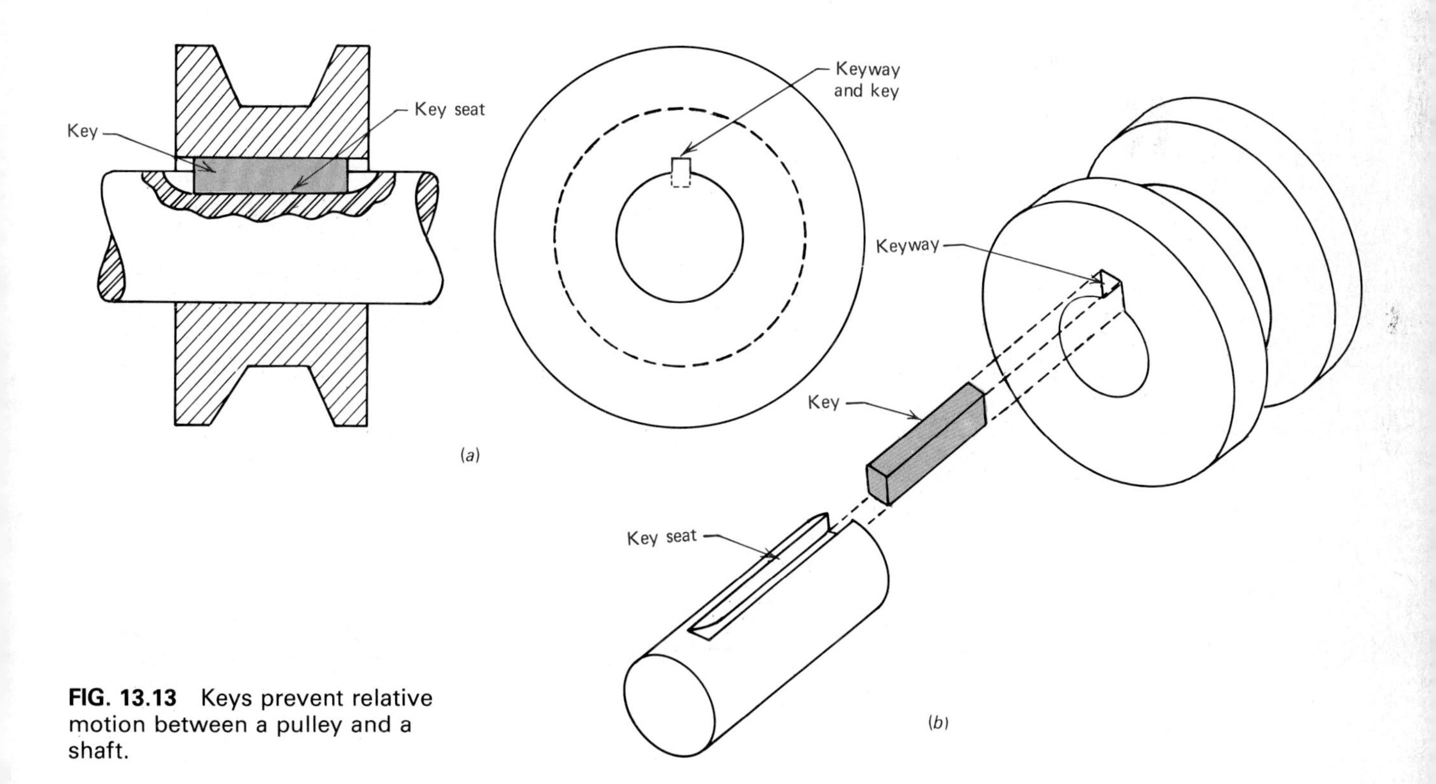

FIG. 13.13 Keys prevent relative motion between a pulley and a shaft.

The key then lies partially in the keyway and keyseat, effectively locking the shaft and pulley or gear together. Keys are identified by name and number, such as Pratt and Whitney #5 and Woodruff #405, or by shape, including square, flat, and taper. Examples may be found in Apps. B and C and in Fig. 13.14.

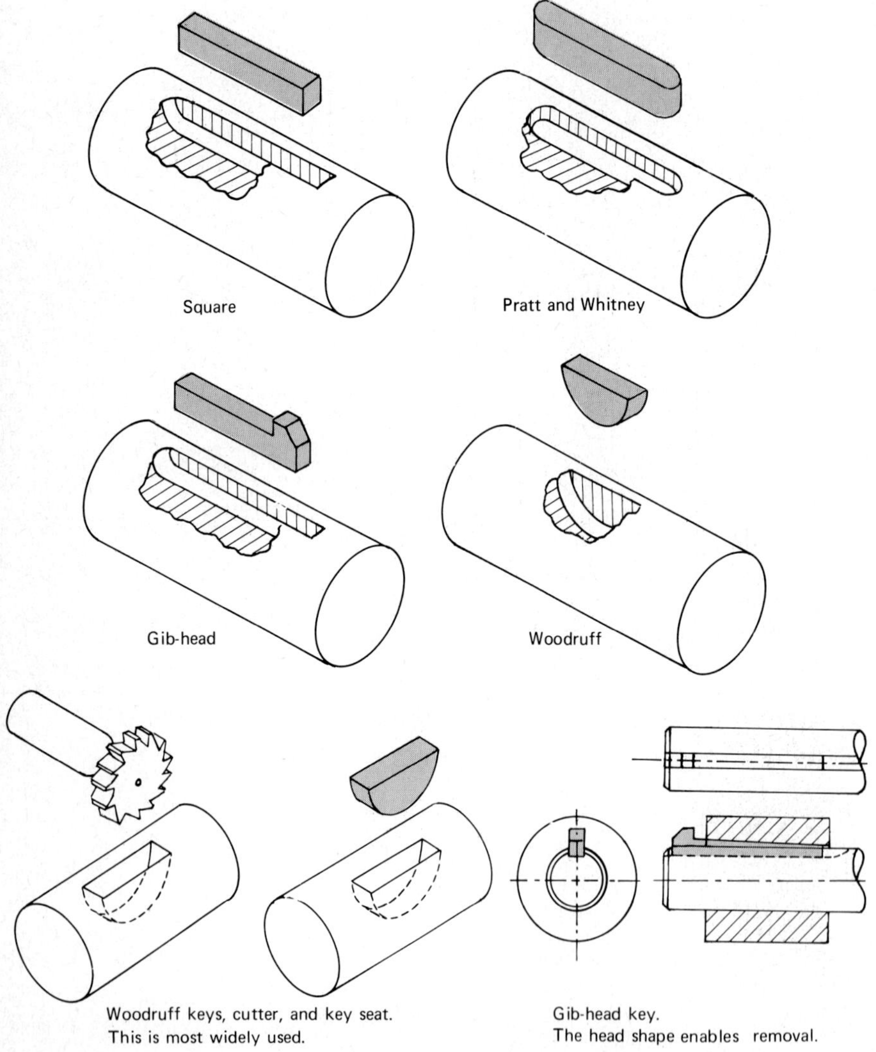

FIG. 13.14 Several key and corresponding keyway shapes.

13.6.5 Springs

Springs are identified by basic shape, either helical or spiral. A complete spring specification will also include the following:

1. Material (usually a carbon steel)
2. Dimensions
   (a) Free height: length of spring under zero load
   (b) Outside diameter: measured in free condition
   (c) Wire size
   (d) Number of coils
3. Style of ends
4. Travel: amount of compression or extension

The three major classifications of springs are compression, torsion, and extension. See Fig. 13.15. Compression springs are wound with the coils separated so that the load is resisted by shortening the spring. Extension springs resist tension loads by becoming longer. Note that the coils are wound tightly in the free condition. Torsion springs resist loads applied perpendicular to the axis of the spring. Torsion springs may be right-hand or left-hand and may have open or closed coils.

13.6.6 Splines

A spline is produced by cutting a series of groove or keyways longitudinally on a cylindrical shaft. Either internal or external splines may be formed. Figure 13.16 illustrates the spline.

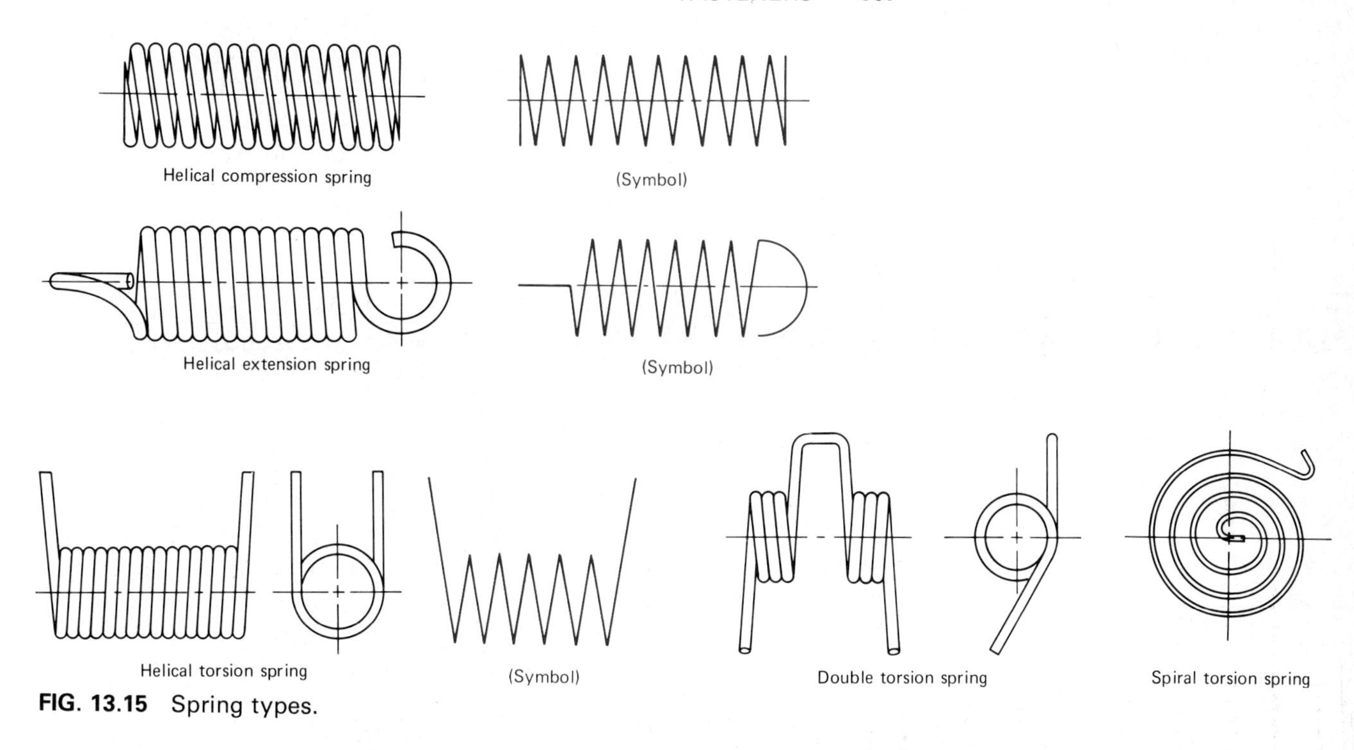

FIG. 13.15 Spring types.

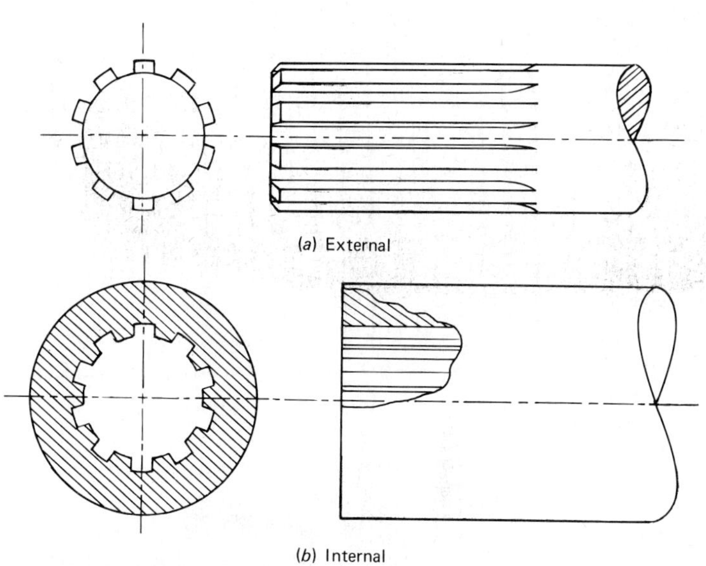

FIG. 13.16 Splines.

13.7 FIXED FASTENERS

A number of design situations require a fixed fastener, that is, a fastener that need not be removed once it is in place. Examples include nails, adhesives, welds, rivets, brazing, and soldering. Adhesives, welds, soldering, and rivets will be discussed briefly here.

13.7.1 Adhesives

The past two decades have seen adhesives become a common type of fastener. Adhesives have been recognized as engineering materials and have become important in production engineering. They are classified according to rheological properties (flow characteristics): thermoplastic, elastomeric, and thermohardened. Thermoplastic adhesives have poor heat resistance, which is a disadvantage in terms of strength, but an advantage in terms of sealing characteristics. Thermoplastics have good peeling resistance over a wide range of temperatures. Elastomeric adhesives are made from a large group of substances which have the elastic properties of rubber. Elastomers are resilient, heat-resistant, and high-strength. Thermosetting, or thermohardened, adhesives have the highest heat resistance of the three classes of adhesives. They have good strength properties but tend to be brittle.

Adhesives may be grouped into categories on the basis of whether they are liquid, paste, or dried. Liquid adhesives are subgrouped according to the solvent employed. Water-emulsified adhesives are inexpensive and fireproof but are difficult to dry except on porous materials. Volatile solvents used instead of water allow for a wider range of application, especially with nonporous materials. Pastes are very easy to apply, fill voids efficiently, and have excellent adhering qualitites. Dry adhesives come as powders, films, or tapes. These adhesives pass through the liquid state during the curing process. They are easy to handle and store.

A great deal of research is being conducted for the purpose of improving existing adhesives and finding new ones. One area of research is attempting to develop adhesives that will expand slightly rather than contract. This will improve adhering qualities and strength properties.

13.7.2 Soldering and Welding

Soldering is a process which joins two metals with a third metal, which is called a solder. The solder, usually a thin layer of a lead-tin alloy, is heated to a molten state and placed between the two metals to be joined. For best results the surfaces of the metals to be joined should be free of surface oxides. The oxides may be removed by a cleaner called a flux.

Welding is a process which also joins two metals in much the same way. The best adhesive for joining two metals would be a third metal of the same material.

This is what a weld accomplishes, but the process is difficult. Most metals are produced from a molten state with a prescribed cooling and heat-treating process to develop the required properties for the specific application. Welding basically reverses that process by creating localized spots of molten metal on the surfaces of the two metals, which then solidify upon cooling to create the joint.

Three welding processes are in common use: arc, gas, and resistance. In arc welding, the areas of the metals to be joined are brought to a very high temperature, over 3300 K, by creating an electric arc between the metal and a welding rod. Both the rod and the metal melt and then fuse during cooling.

Gas welding involves essentially the same process as does arc welding except that the heat is created by burning a mixture of gases, usually oxygen and acetylene. Beads of molten metal are formed by melting the welding rod and the metals to be joined.

Resistance welding, like arc and gas welding, develops very high temperatures to melt the surface of the metals that are to be joined. The metals are placed together under tremendous mechanical load, and then an electric current is passed through the metals. The high electric resistance creates the heating necessary to melt the contact surfaces of the metals. The weld is completed by cooling while still under the high loading.

Welds are classified by the type of joint. The butt joint aligns the metals side to side. The lap joint places one metal surface on top of the other, and the corner joint places the end of one piece against the side of the other. Each of these classes has different types of welds that can create the desired joint effectively. For example, in Fig. 13.17, a butt joint can be created with a bevel, V, square, J, or U weld. Note in Fig. 13.17 that the weld is designated by a symbol rather than by draw-

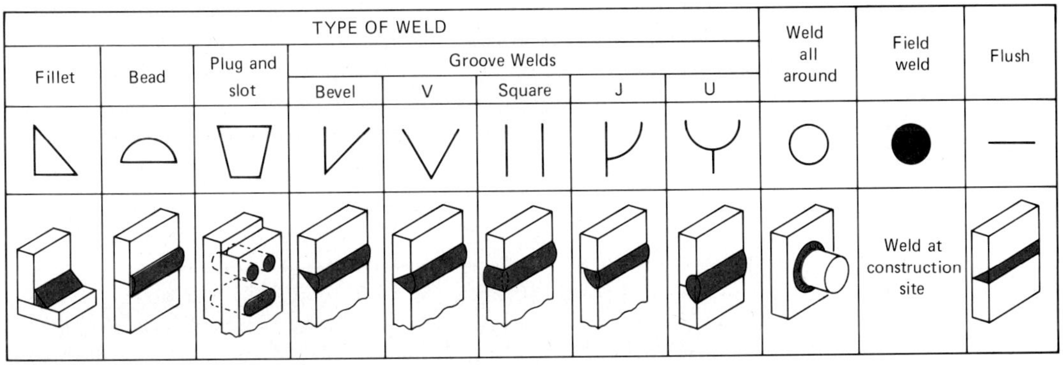

TYPE OF WELD								Weld all around	Field weld	Flush
Fillet	Bead	Plug and slot	Groove Welds							
			Bevel	V	Square	J	U			
									Weld at construction site	

(Graphic representation of each symbol)

FIG. 13.17 The types of welds shown with the corresponding symbol used on drawings.

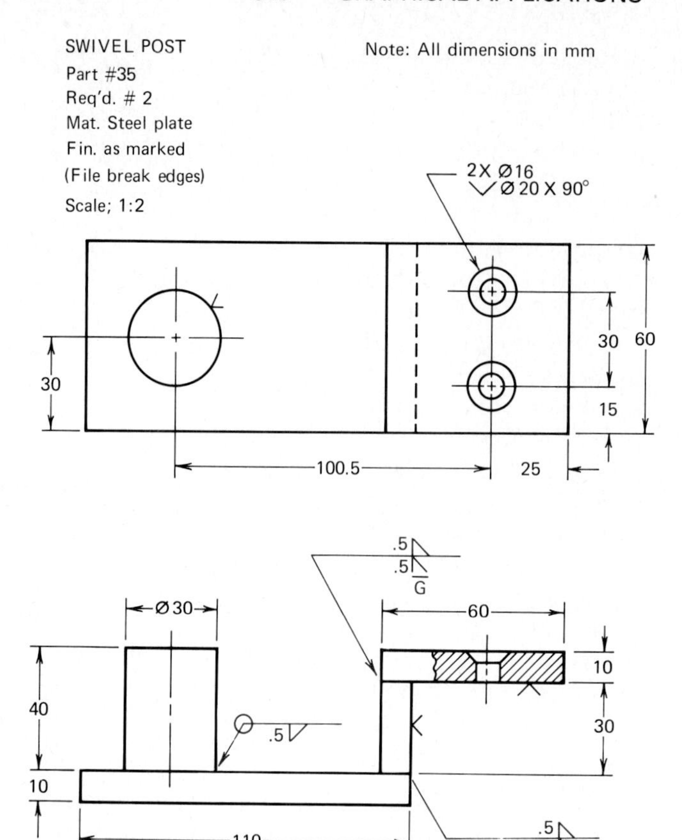

FIG. 13.18 A drawing using welding specifications.

ing the joint. Figure 13.18 shows an engineering drawing with the welding requirements placed on the drawing in symbol form. Look at the symbol and refer back to Fig. 13.17 to see the shape of weld that is called for.

13.7.3 Rivets

Rivets are fasteners which form the connection between two pieces of material in much the same way as a bolt and nut. Rivets are usually made of wrought iron or steel. They are short round bars with a head on one end. The rivet is forced through matching holes in the two materials to be joined. A head is formed on the other end by hammering, thus making the fastener fixed rather than removable like the nut and bolt.

Rivets are classified as field or shop, depending on where they are installed. Rivets are used to join other materials such as belts, clothing, and jewelry.

Standard shapes of small and large rivets are shown in Fig. 13.19.

13.8 APPLICATIONS FROM COMPUTER GRAPHICS

In many of the figures in this chapter you can see the complexity of representing the various fasteners on a drawing. The schematic and simplified thread representations were developed primarily to reduce the task of drafting. However, these representations are not satisfactory for representing fasteners in pictorial.

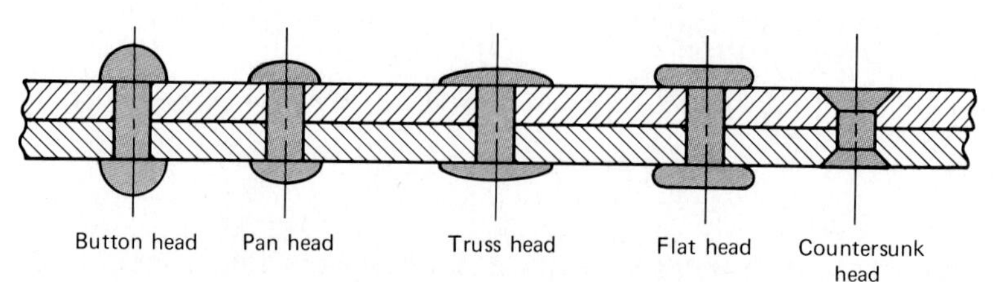

FIG. 13.19 A few of the types of rivets in common use.

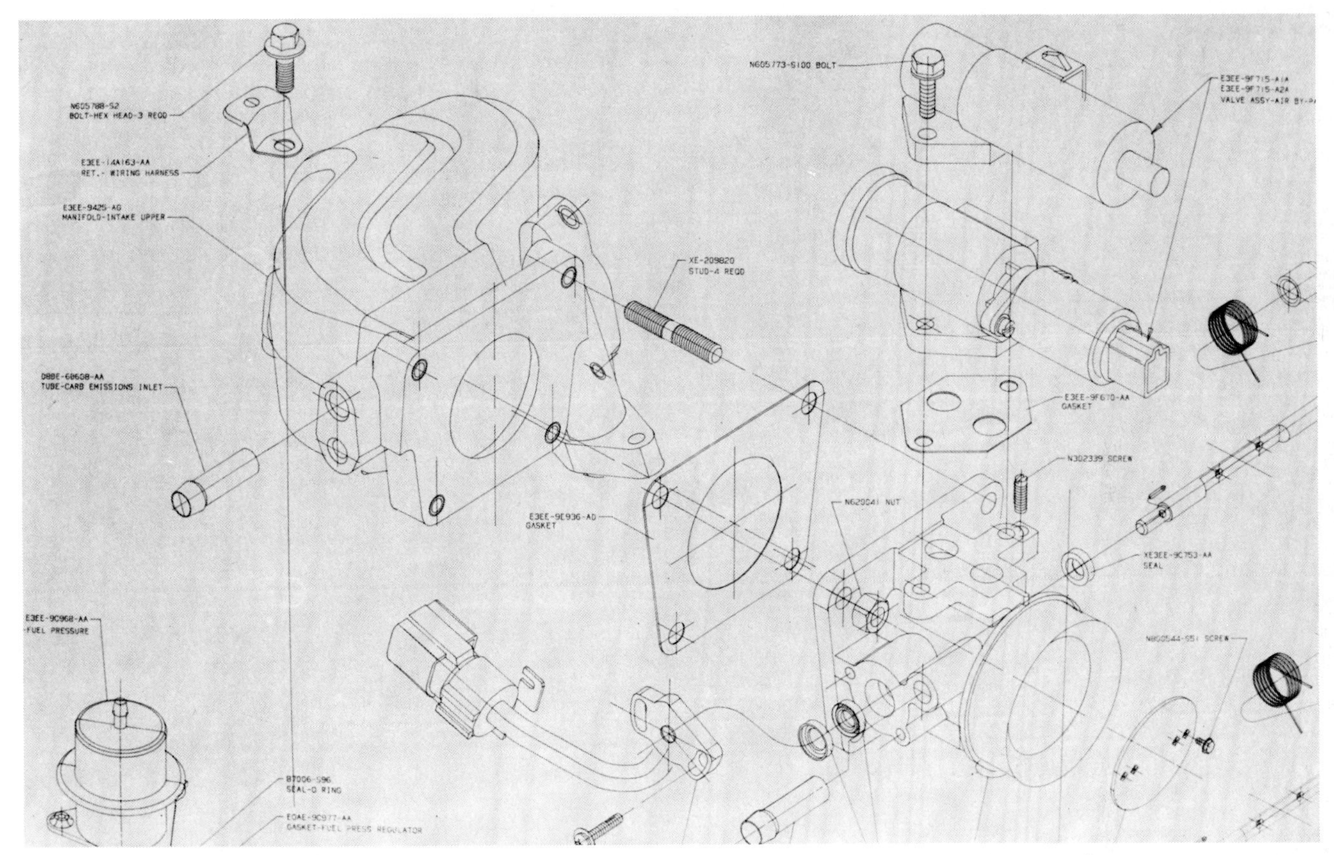

FIG. 13.20 A pictorial generated from a computer database. (*Courtesy of Ford Motor Company.*)

The exploded pictorial in Fig. 13.20 shows fasteners that are drawn by a computer-controlled plotter. Each fastener is drawn in a few seconds and will be produced in exactly the same way on any subsequent plot. This speed and consistency cannot be attained by a drafter.

Once the geometry of a fastener is defined in three dimensions and placed in the computer as a database, the fastener can be drawn to any scale and can be represented from any viewing direction. Furthermore, such variables as thread length and overall fastener length can be specified, and the resulting fastener can be created from a standard database for that particular type of fastener. Thus, the computer not only speeds the drafting process for fasteners but enhances visualization of how the fastener is incorporated into the final assembly.

Problems

13.1 Make a freehand sketch similar to Fig. 13.8. Name each fastener. Write an appropriate metric or English specification for each fastener.

13.2 Sketch a shaft with several sectioned adjustable spacers similar to those in Fig. 13.21, illustrating the following set screws. Include an appropriate specification.

At *A*: square-head, cup-point
At *B*: slotted headless, cone point
At *C*: socket-head, oval point
At *D*: square-head, dog-point
At *E*: socket-head, flat-point

13.3 Sketch a V pulley and shaft similar to the one shown in Fig. 13.12, showing a taper pin in its functional position.

13.4 Sketch a single view of the hanger device shown in Fig. 13.22, using the following fasteners. Write appropriate specifications.

At *A*: hex bolt and nut with lock washer
At *B*: flat-head cap screw
At *C*: eye bolt secured with two jam nuts

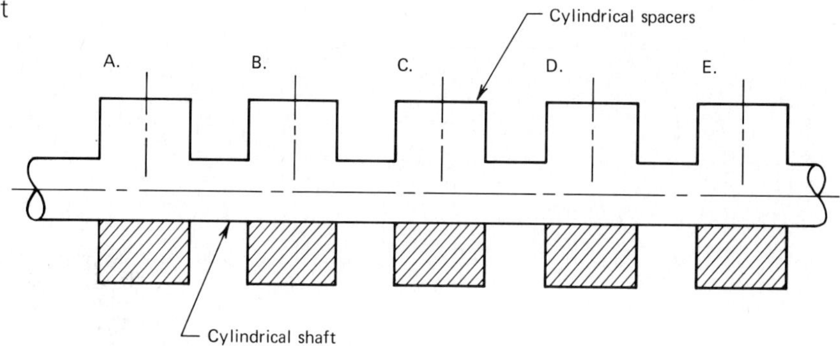

FIG. 13.21

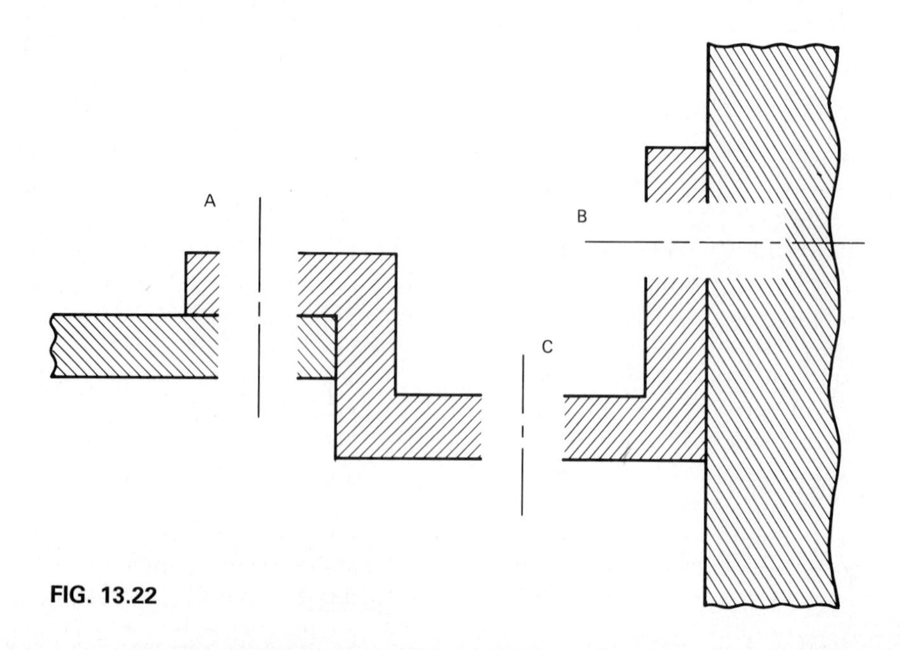

FIG. 13.22

13.5 With a single-view, sectioned, freehand sketch, illustrate how two pieces of strap steel can be riveted together. Use at least three different types of rivets or rivet heads.

13.6 Sketch the pulley-gear layout shown in Fig. 13.23. Complete the assembly by showing the following fasteners properly oriented in the mechanism. Write appropriate specifications.

At *A*: square-head, cup-point set screw
At *B*: socket-head dog-point set screw
At *C*: square key
At *D*: flat washer with a square nut
At *E*: Woodruff key
At *F*: flat-head cap screw
At *G*: semifinished hex bolt with lock washer and nut (spotfaced bearing surfaces on base)

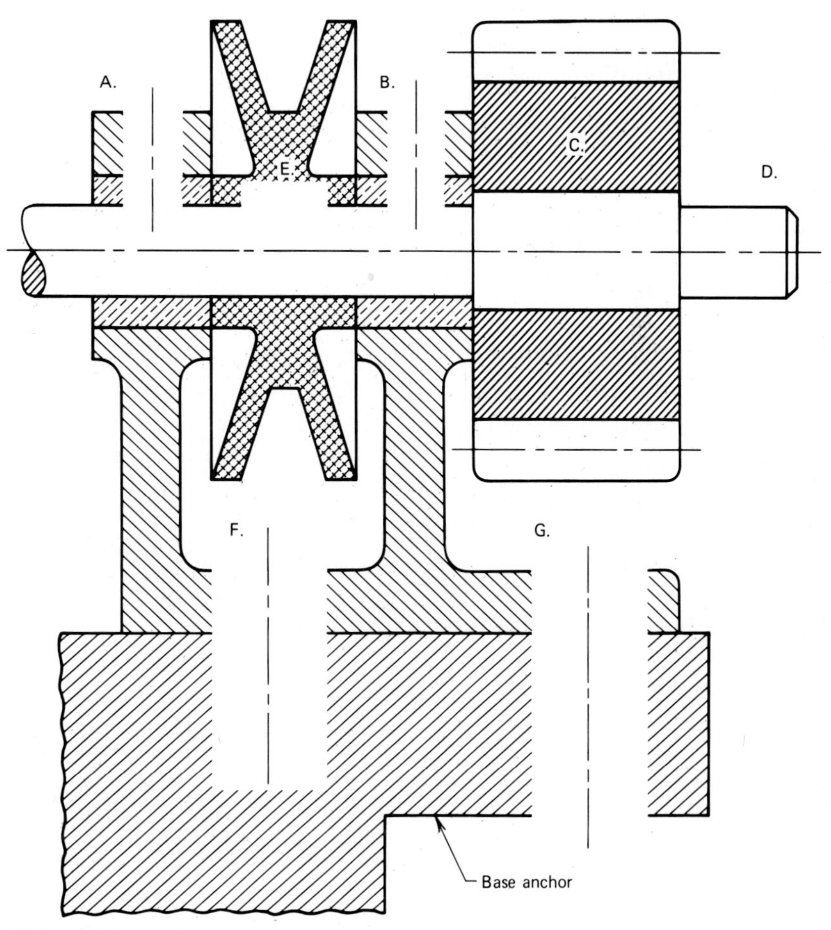

FIG. 13.23

13.7 Sketch or draw (and complete) the two partial views shown in Fig. 13.24. Complete the views enough to show appropriate fasteners of your choice in their functional assembled positions. Assume that the pivot plate and handle arm are cast parts and that counterbored, countersunk, and spotfaced holes are used where safety and security are needed. Carefully consider the use of pins, Woodruff keys, cap screws, machine screws, studs, and various jam nuts and washers. Write appropriate specifications.

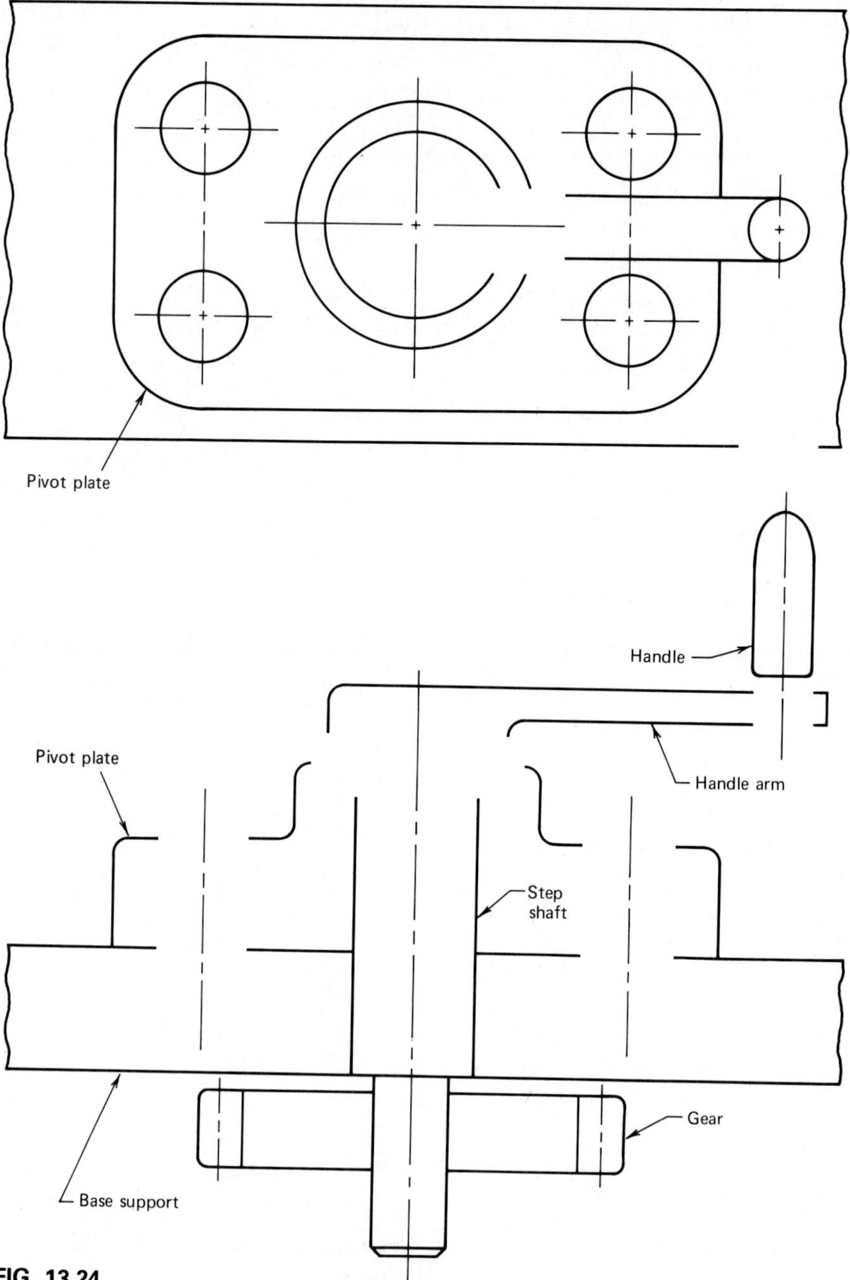

FIG. 13.24

chapter 14

Design Drawings

14.1 THE DESIGN PROCESS

Design is the distinguishing feature of engineering. The efforts of engineering are directed toward the development of devices, systems, and processes to satisfy human needs. To carry out the development efforts effectively and efficiently, engineers follow a design process which logically and systematically governs the design work. A nine-step design process is listed below.

1. Identification of a need
2. Definition of the problem
3. Search
4. Criteria and constraints
5. Alternative solutions
6. Analysis
7. Decision
8. Specification
9. Communication

There are many other design processes with a varying number of steps listed in the literature. Close inspection will show that although the steps are somewhat different, the procedures and resultant effect are the same.

The common thread in this process is the set of design drawings. Also called engineering drawings or working drawings, these are used to convey information among the various personnel involved in the design and manufacturing areas. Most design activities involve three-dimensional objects directly or indirectly; this indicates the importance of one's having the ability to describe geometry clearly and concisely.

In an industrial setting, identification of need is generally done within the marketing division. The description of the need will most likely contain sketches. In some cases the need may be perceived as a variation of an existing product, and thus existing design drawings may be reviewed and modifications proposed that convey the need. During the stage of defining the problem, sketches and layout drawings are commonly used. The definition of the problem takes place at the project management level; if the problem is large in scope, a special project group will be established to carry out the design.

The search phase involves climbing the "learning curve," that is, gathering knowledge about the problem and the factors affecting the problem. The amount of time spent in this phase depends on prior experience with the problem area. The engineer will be involved with reading drawings of similar designs as part of this learning effort.

Establishment of the criteria and constraints involves some geometric considerations, for example, length, volume, shape, and clearance with related objects. Again, sketches and layout drawings are an integral part of the establishment of criteria and constraints.

Within the time frame and cost constraints of the design, various alternative solutions are developed. These alternatives may be in the form of sketches or, if time permits, formal drawings. Each alternative solution contains sufficient characteristics to be considered in a preliminary screening of the solutions against the criteria and constraints. The most promising alternatives are prepared in greater detail for the analysis stage.

Analysis involves the determination of the properties of the alternative solutions. It includes the development of physical and mathematical models which are tested against the laws of nature and the criteria and constraints that have been specified. Much analysis work today is performed with the aid of a computer. If computer graphics techniques are used, you can see the thread of graphics extending through the analysis. Figure 14.1 illustrates analysis with a computer-graphics system.

With all information at hand, project management will make the decision on the final design. Once the decision is made, specification and communication of the design are required for manufacturing, assembly, testing, and marketing. During the specification stage, a complete set of design drawings is developed. This set becomes the official operating document for the design. Any changes that are incorporated during manufacturing, assembly, testing, and redesign must be approved at the project management level and incorporated into the set of design drawings.

The various types of drawings used in the design process are of

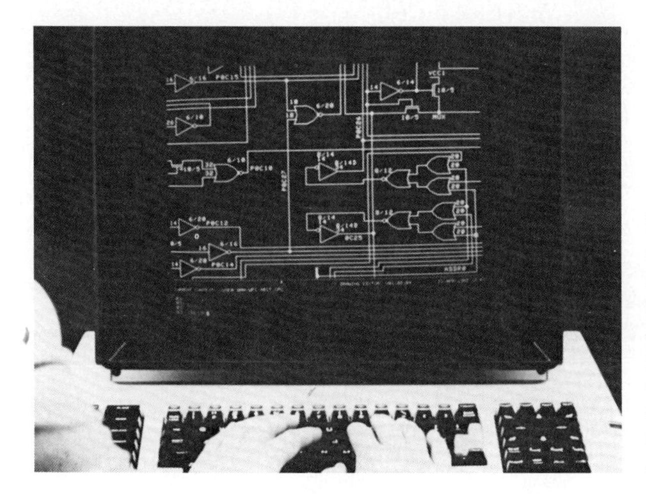

FIG. 14.1 The computer plays an important role in engineering design. Analysis of the performance of an electronic circuit is one example. (*Courtesy of Control Data Corporation.*)

such importance to the engineer that each will be described and illustrated in this chapter.

14.2 IDEA SKETCH

An idea sketch is a rapidly constructed single-view, multiview, or pictorial freehand delineation of a concept. It provides little if any detail of dimensions, motion, or other characteristics of the concept. Figures 14.2 through 14.4 show idea sketches using a single-view, multiview, and pictorial rendition, respectively, of three different mechanical log splitters.

14.3 CONCEPT SKETCH

A concept sketch is a more complete delineation than is an idea sketch. It is generally freehand and combines multiview and pictorial drawings. It is common to provide numerous explanatory notes that clarify the basic idea. It is difficult to describe precisely what should be included in a con-

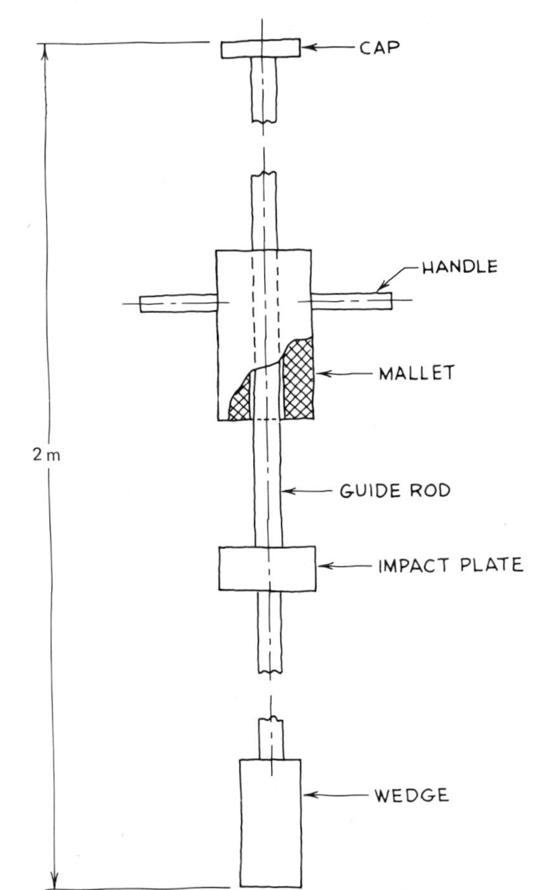

FIG. 14.2 Idea sketch: single view of a log splitter.

cept sketch. Certainly it should include the form and function of the concept, although minor details and exact specifications are generally not included. Figures 14.5 and 14.6 illustrate two pictorial concepts of one part of the mechanical log splitter shown in Fig. 14.4

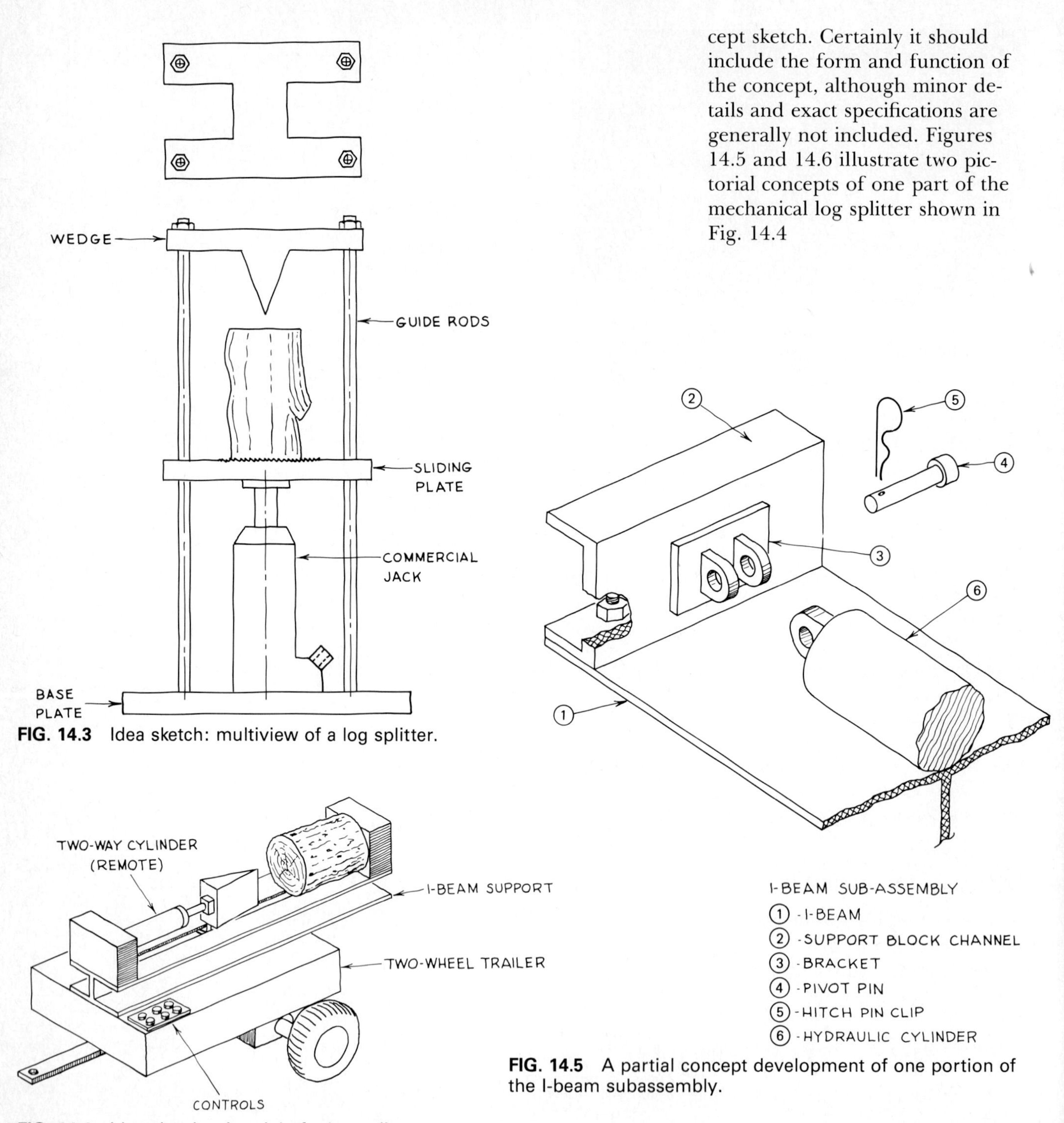

FIG. 14.3 Idea sketch: multiview of a log splitter.

FIG. 14.4 Idea sketch: pictorial of a log splitter.

FIG. 14.5 A partial concept development of one portion of the I-beam subassembly.

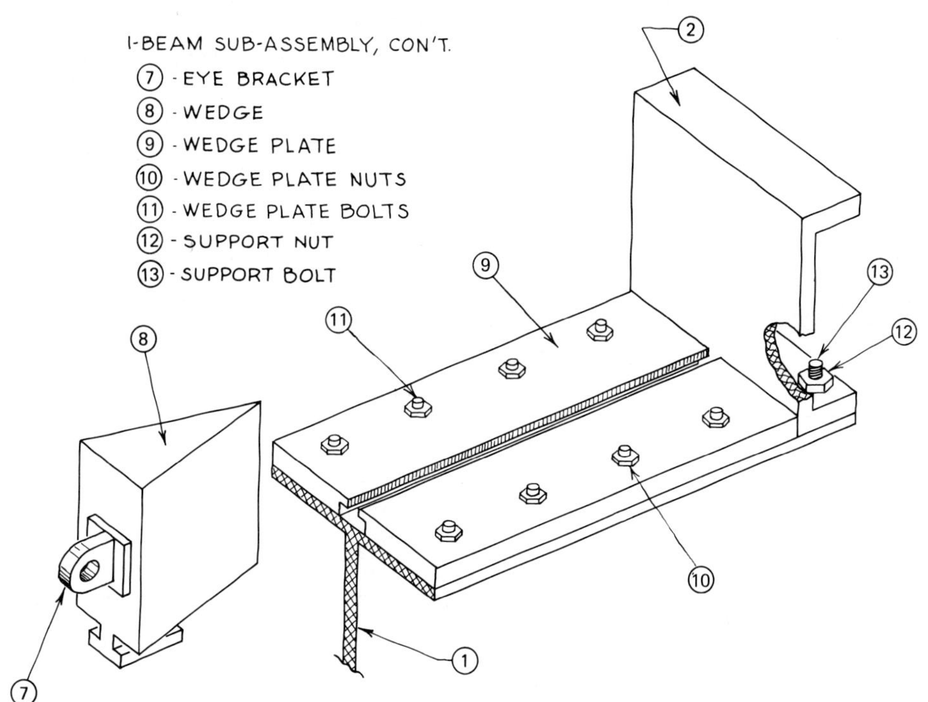

FIG. 14.6 A partial concept development of another portion of the I-beam subassembly.

14.4 LAYOUT DRAWING

A layout is a very accurate instrument drawing to scale, which shows the location of critical points, clearance distances, and other critical design geometry. It is the skeleton or foundation for the design of the individual parts, and it predetermines the ultimate functional characteristics of the device. The layout drawing controls the form of the design by placing constraints on the geometry.

Consider the sequence of events in the design of a house. First the size, location, and position of the house must be established on the lot plan. The footings are located, the floor plan is established, and finally, the vertical design is performed. These drawings are the architectural layouts that are a prerequisite to detailed house plans. The layouts serve as a guide for adding plumbing, gas lines, electric wiring, heating ducts, cabinets, closets, and so on. These details can be added only after the basic functional considerations have been established.

In mechanical design, layouts establish the location of centers of wheels, gears, shafts, and other rotating parts. Center lines for connecting rods, pistons, and other moving parts are located according to the laws of kinematics. Clearances can be established and checked on a layout drawing.

To determine cylinder length
for fixed wedge travel.

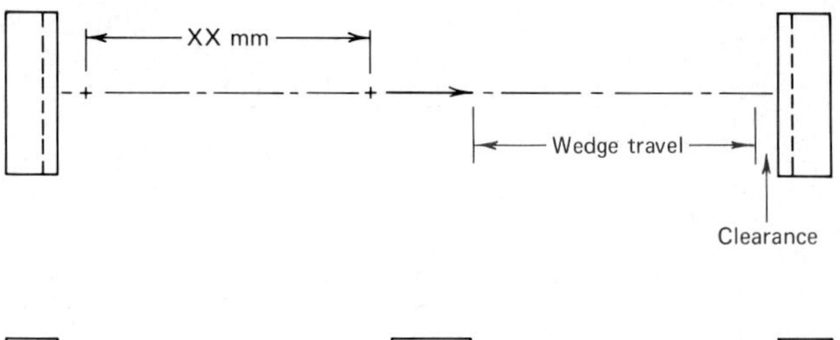

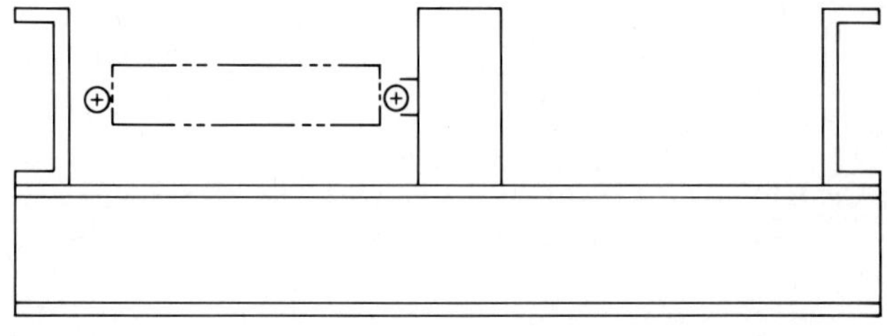

Scale 1:x

FIG. 14.7 A layout drawing is used to determine the cylinder requirements.

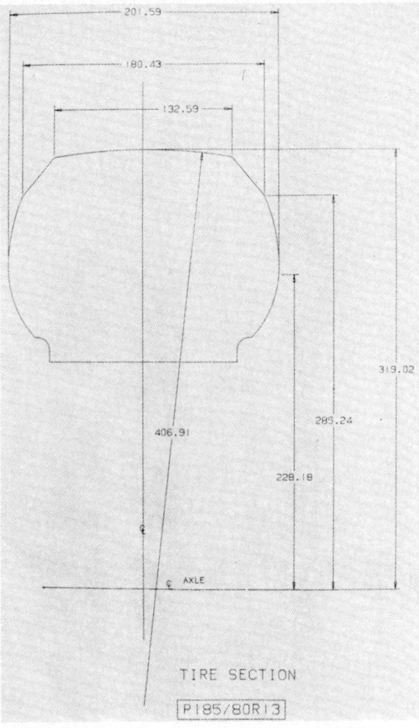

FIG. 14.8 Layout drawing of the tire section for an automobile. (*Courtesy of Ford Motor Co.*)

All this information is necessary for the analysis which will determine the shape, size, and materials to establish the final design and determine whether the design satisfies the constraints.

Figure 14.7 is a layout of the I-beam subassembly of the mechanical log splitter, showing the relationship between wedge travel, wedge size, and clearance distance necessary to split a log without part interference. Figure 14.8 is an example of a layout generated with a computer-aided design (CAD) system.

14.5 ASSEMBLY DRAWING

An assembly is generally an orthographic multiview drawing of a complete design, for example, a pencil, radio, automobile, or space shuttle. If the design is very complex, the assembly will show only functional relationships between various components (called subassemblies) of the design. Each component will be identified on the assembly by letter, number, or name. If an assembly has relatively few parts, each part is identified by letter, number, or name. When letters or numbers are used, they are inscribed inside a circle (balloon) connected to the part by a leader.

Individual parts within an assembly are not dimensioned even though the drawing is to scale. Only overall dimensions and critical points that are needed for clearances of moving parts and assembly instructions are specified.

Assembly drawings are frequently used to show alternative positions of component parts. Alternative positions are drawn with either dotted lines or phantom lines.

The major function of the assembly is to locate all parts in their relative positions in the design. This enables a design engineer, test technician, or assembler to visualize the total design quickly. Proper identification of individual parts and subassemblies enables one to tie together a complete set of design drawings.

Figure 14.9 shows the assembly of the log splitter. Because the log splitter is somewhat complex, the assembly uses balloon identification to call out three subassemblies. Overall dimensions are also included; however, most of the detail is left for the subassemblies. Compare the details in Fig. 14.9 with those shown in the pictorial idea sketch in Fig. 14.4.

14.5.1 Subassemblies

As indicated previously, the number and nature of individual parts together with the complexity of a design dictate the number of subassemblies to use. Simple objects, such as a ball-point pen, do not need a subassembly; each individual part can be identified on the assembly drawing. The mechanical log splitter shown in Fig. 14.9 was sectioned into three subassemblies: I beam, controls, and trailer. It is possible to divide subassemblies further. This is quite common for complex de-

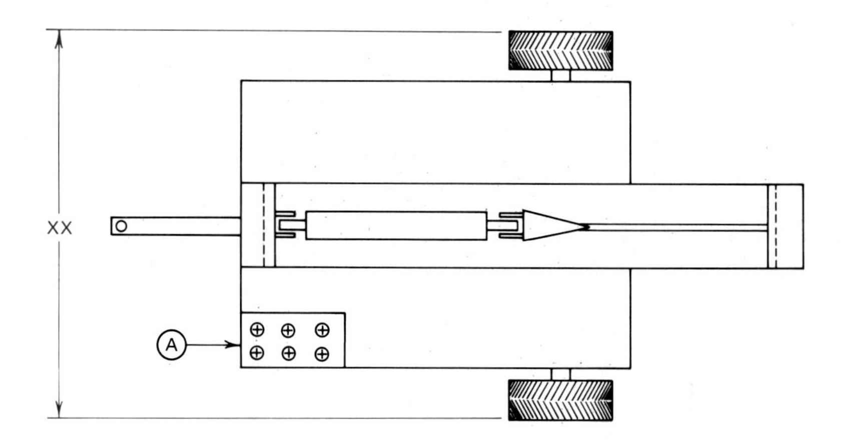

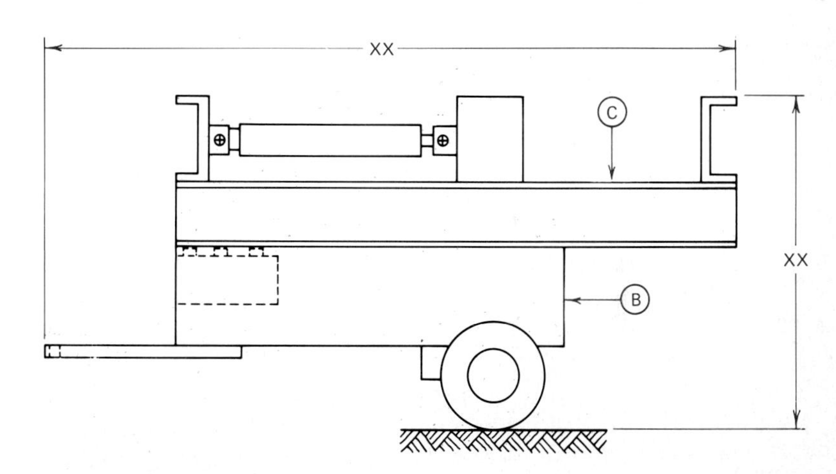

Log Splitter Assembly

(A) - Control subassembly

(B) - Trailer subassembly

(C) - I-beam subassembly

FIG. 14.9 Orthographic assembly of the logsplitter.

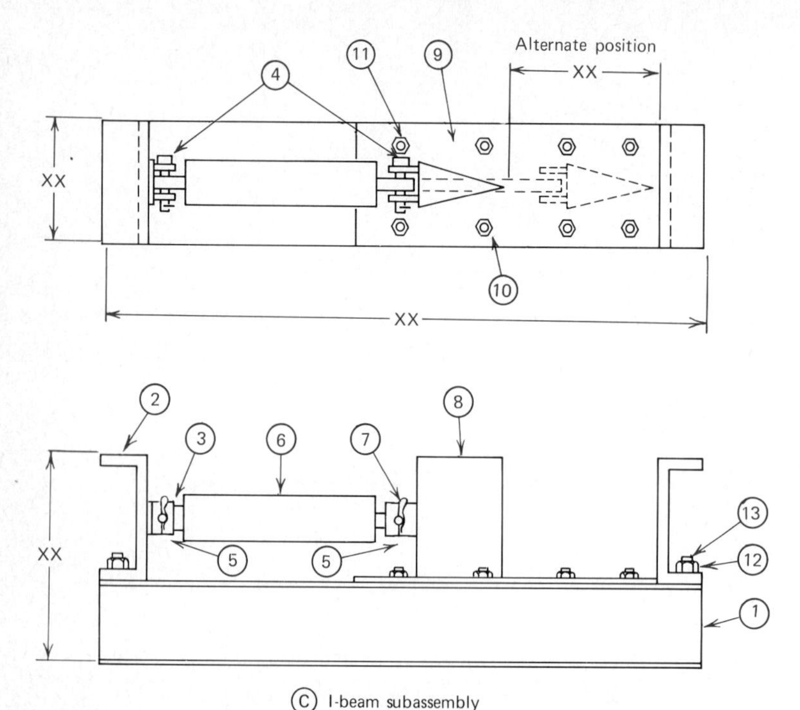

FIG. 14.10 Subassembly Ⓒ. See Fig. 14.9 for entire assembly. See Fig 14.13 for part indentification.

FIG. 14.11 Exploded pictorial displayed on a computer graphics terminal. (*Courtesy of Control Data Corporation.*)

signs such as those for automobiles and airplanes.

Figure 14.10 shows an orthographic drawing of the I-beam subassembly of the mechanical log splitter. Note the use of numbers in the balloon identification for individual parts of the subassembly. This subassembly drawing is identified with the letter C, corresponding to the notation on the assembly drawing of Fig. 14.9.

The overriding criteria for setting up assemblies and subassemblies is to demonstrate the form and functional relationships of individual parts in a design. A little time spent doing this can enhance the communication phase of the design effort.

14.5.2 Pictorial Assemblies

Often it is convenient to demonstrate how all the various parts of an assembly fit together. The pictorial assembly frequently is helpful in visualizing the design, particularly for people who are not experienced in reading orthographic drawings. Figures 14.4 through 14.6 show examples of pictorial assemblies and subassemblies.

An excellent method of showing the assembly, installation, and maintenance requirements of a design to the consumer is the exploded pictorial. Common examples of the use of exploded pictorials are the drawings provided to purchasers of outdoor barbecue grills, tricycles, bicycles, and children's toys. This technique is also helpful for industrial assemblies, as shown in Fig. 14.11.

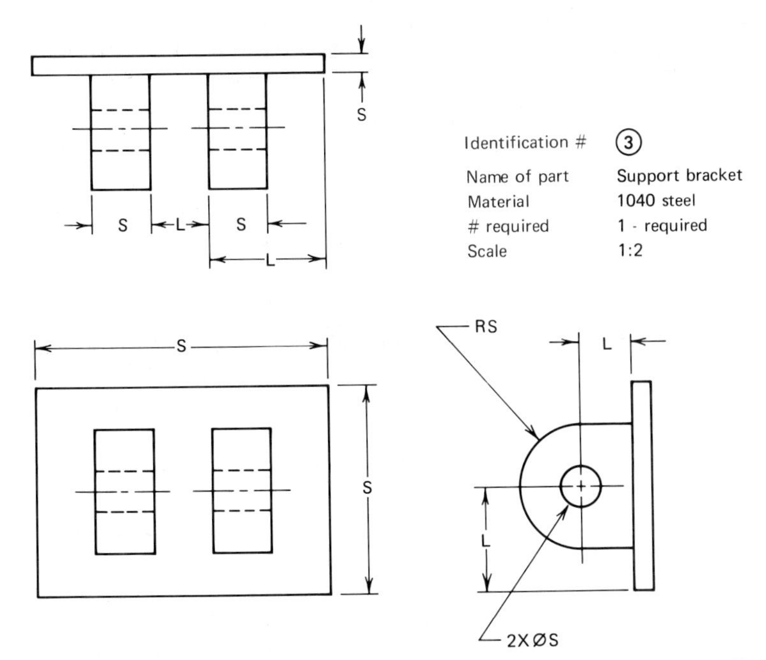

FIG. 14.12 Detail drawing of part number ③ of subassembly Ⓒ.

14.6 DETAIL DRAWING

A detail is an orthographic drawing of a single part. It contains the necessary views in correct projection along with the complete dimensions and specifications required for manufacturing. In general, the production area receives only the detail drawing to be used in manufacture. Therefore, it is imperative that all necessary information be included.

Detail drawings are made of all parts of the design except those which are considered standard. (Standard parts can be ordered from a catalog or purchased from a vendor directly.) Many industries maintain a parts inventory which may be accessed for standard parts. Obviously, it is economical to use as many standard parts as possible, since no investment need be made in design and development of those parts. Items such as bolts, nuts, keys, pins, motors, resistors, and capacitors should be standard parts. The companies that produce standard parts will usually make drawings of the parts available upon request.

Figure 14.12 is a detail drawing of the support bracket (part no. 3) on the I-beam assembly of the mechanical log splitter. It contains appropriate, properly dimensioned orthographic views, together with the identification number, part name, material, number required per assembly, and scale. Note how the identification number ties the detail drawing to the subassembly of Fig. 14.10, which in turn relates to the complete assembly of Fig. 14.9. This careful identification process is very important in maintaining accurate documentation of the design. If a design change is affected, the identification process enables the change to be included in all aspects of the design, using the set of design drawings as reference.

14.7 PARTS LIST (BILL OF MATERIALS)

In order to ensure that all parts, including standard parts, are ordered or manufactured and brought to a central assembly point at approximately the same time, a parts list is prepared and included with the set of design drawings. The parts list includes identification numbers, part name, number required per assembly, and material along with notes indicating whether the part is standard or whether special circumstances exist for manufacture.

The order in which parts appear on the list depends generally on their size and importance in the overall assembly. Main castings or forgings are listed first, machined parts next, and standard parts last. The parts are listed in numeric or alphabetic order corresponding to the subassembly or assembly drawing.

If a parts list is started at the bottom of a drawing, the order of items will be from the bottom upward. If it is started at the top of a drawing, the numeric or alphabetic order will be from the top down.

Figure 14.13 shows the parts list for the I-beam subassembly of the mechanical log splitter.

ID#	PART NAME	#REQ'D.	MAT'L SPECS AND GENERAL NOTES
1	I-beam	1	ASTM A36, W 12 X 65 X 72 (1)
2	Support channel	2	ASTM A36, C12 X 30 X 12
3	Support bracket	1	AISI C 1040 Cold-drawn steel
4	Pivot pin	2	Straight, 15 mm X 80 mm, steel
5	Hitch pin clip	2	3 mm x 40 mm, stock # HPC–3
6	Hydraulic cylinder	1	Parker-Hannifin, Std Double-acting 50 mm
7	Eye bracket	1	AISI C1040 Cold-drawn steel
8	Wedge	1	AISI C1040 Cold-drawn steel
9	Wedge plate	2	AISI C1040 Cold-drawn steel
10	Machine nuts	8	Hex nut, M10 X 1.5, steel
11	Machine bolts	8	Hex bolt, M10 X 1.5 X 50, steel
12	Machine nuts	6	Hex nut, M16 X 2, steel
13	Machine bolts	6	Hex bolt, M16 X 2 X 60, steel

(1) Note: ASTM A36, W12 X 65 is the latest std available (1980) non-metric.

FIG. 14.13 Parts list for the I-beam subassembly.

14.8 OTHER CONSIDERATIONS FOR DESIGN DRAWINGS

We have described the types of drawings that are needed in the preliminary design, final design, and manufacturing functions. The set of design drawings that constitute the formal records generally include assembly, subassemblies, and details. The idea and concept sketches along with layout drawings are used to arrive at the final design but are not included in the final set of design drawings. They are, however, kept on file for future redesign efforts and possible patent considerations.

Two other aspects of design drawings must be discussed here: title blocks and reproduction processes.

14.8.1 Title Blocks

Title blocks are generally located along the bottom or in the lower right-hand corner of a drawing. The information normally included in a title block is company name and address, name of design depicted in the drawing, drawing identification number, name or initials of the person who prepared the drawing, name or initials of the person who checked the drawing, approving authority (project engineer), date of preparation, dates of changes, scale, and so forth. Important drawing information such as general tolerances, material finish, heat treatment, and painting specifications may be in the title block but in general will appear as notes on the drawing.

Figure 14.14 illustrates two standard company title blocks. The Tektronix, Inc. title block is computer-generated, while the Fisher Controls title block was preprinted on standard paper size. In general, the title blocks are on the paper before the part drawing is begun.

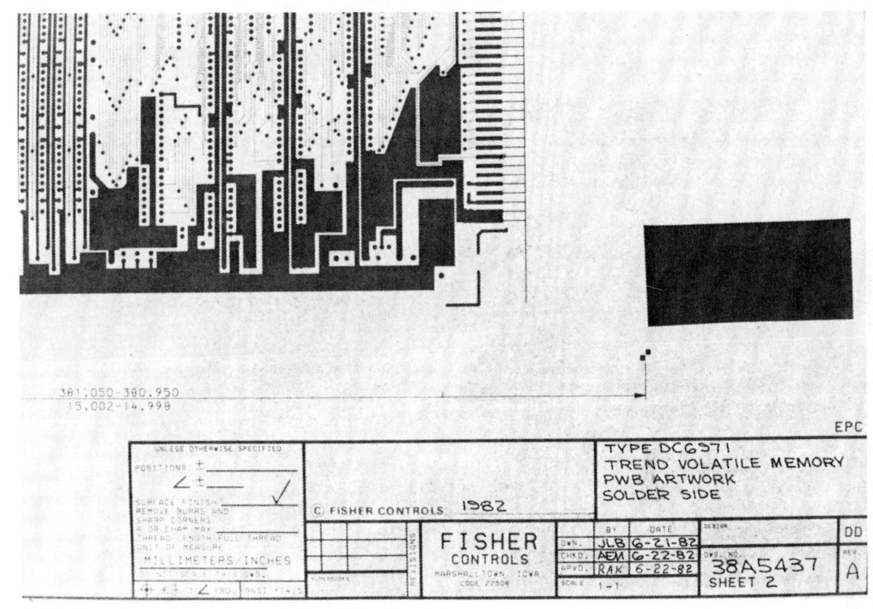

FIG. 14.14a A standard title block with entries made by hand.

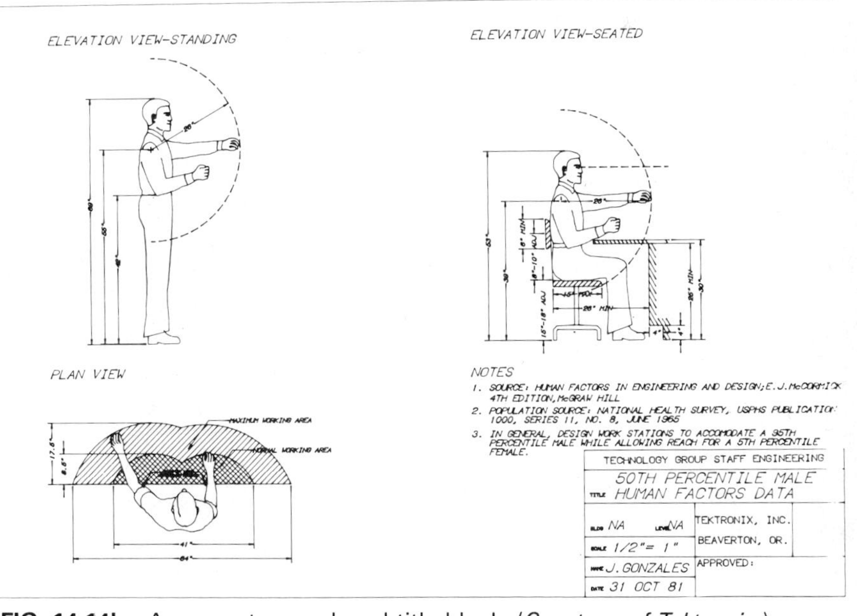

FIG. 14.14b A computer-produced title block. (*Courtesy of Tektronix.*)

14.8.2 Drawing Reproduction

In order to distribute drawings to the appropriate persons, many copies of a specific drawing must be made. Idea and concept sketches have a limited distribution and most likely will be copied quickly by means of such reproduction processes as electrostatic (Xerox, for example) or thermofax.

The set of design drawings are normally constructed on a transparent or translucent medium such as tracing paper, vellum, or Mylar. A process called dry diazo is the most common method of reproducing these drawings. A paper or film that is light-sensitive can be exposed from the drawing. Black, blue, or red lines can be produced with hot ammonia vapor on different types of paper. A very common selection is blue lines on white paper—hence the name blue-line print. When many copies are needed, it is convenient to produce an intermediate print called a sepia. The sepia has brown lines on a translucent sheet and is used to make additional prints. Figure 14.15 shows a typical machine used for producing copies of working drawings.

Computer-generated drawings can be done using ink on any media placed in the plotter. A typical application is to generate a computer drawing using ink on Mylar or vellum and use that as the original for diazo blue-line prints.

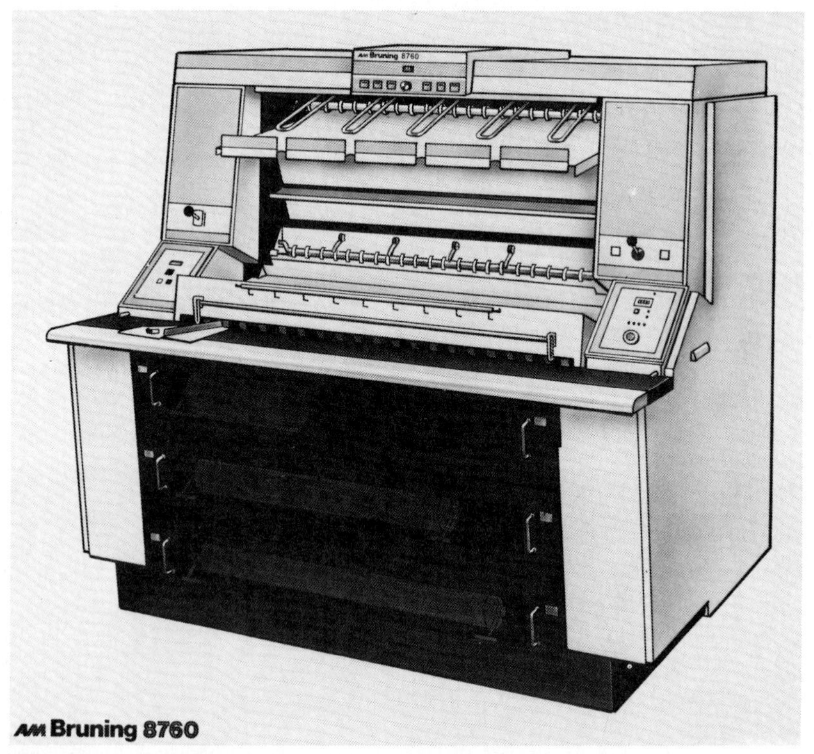

FIG. 14.15 A modern machine for reproducing drawings. (*Courtesy of Bruning.*)

14.9 DESIGN DRAWINGS AND THE COMPUTER

The text up to this point has stressed the fundamentals of engineering graphics, specifically, the how and why of doing things so that others will correctly interpret and understand your work. The impact of computer-graphics techniques on engineering graphics has been briefly introduced so that you may begin to appreciate the potential of the computer in the area of engineering graphics. The computer is a tool that can assist your engineering efforts greatly, but you must know, for example, how a drawing should be presented to meet accepted standards.

This chapter has drawn on material from all the previous chapters to describe how to produce a set of design drawings that completely specify a design effort. The fundamentals and manual procedures presented, coupled with on-the-job experience, should allow you to be productive.

The principal reason why computers have been added to the engineering work environment is to improve productivity in design, drafting, analysis, manufacturing, and a host of other applications. The computer will allow you to do more work in a shorter period of time and do it better than you were capable of doing previously. In the area of CAD, drafting, and manufacturing, increases in productivity of 2:1 to 20:1 over manual methods have been recorded. This represents a strong incentive for industry to invest in the required equipment and user training. Larger industries, such as the aircraft and automobile industries, have led the way, but smaller industries are rapidly adding computer-graphics systems to their inventory of tools. There is little doubt that most industries will be using computer graphics in the near future.

Part Four of this book has been designed to begin your training in understanding and using computer graphics. Your instruction will continue throughout your academic career and through training on the job as long as you are a productive engineer.

Problems

14.1 Assume that you have been given the idea sketch for a mechanical-hydraulic log splitter, as illustrated in Fig. 14.4. Using your own ideas and creative design ability, develop the following:

(*a*) A complete set of design drawings for the I-beam support subassembly. The set should include:

i. Concept sketches of the subassembly (not necessarily those illustrated in the text)
ii. Layout drawing
iii. Subassembly drawing
iv. Detail drawing for each nonstandard part
v. Bill of materials for subassembly

(*b*) A complete set of design drawings for the two-wheel trailer subassembly. The set should include:

i. Concept sketches of the subassembly
ii. Subassembly drawing
iii. Detail drawing for each nonstandard part
iv. Bill of materials for subassembly

(*c*) A complete set of design drawings for the controls subassembly. The set should include:

i. Concept sketches to define necessary functions
ii. Block diagrams to show the interconnection of the major functions
iii. A schematic to illustrate the necessary electric circuits and connections
iv. A wiring diagram to be used for assembly

14.2 Assume that you are in the employ of a company that is planning to manufacture the mechanical-hydraulic log splitter illustrated in Fig. 14.9. You are asked to design a trailer hitch which will allow the I-beam assembly to remain level while in operation (trailer unhitched from towing vehicle). The simple bar hitch illustrated in Fig. 14.9 can be changed or modified to whatever extent is needed to produce a cost-effective solution. You are to provide a complete set of design drawings that include the following:

(*a*) Concept sketches
(*b*) Layout drawing showing range of leveling available (assume a height for hitch attachment point to towing vehicle)
(*c*) Assembly drawing
(*d*) Detail drawings for all nonstandard parts
(*e*) Bill of materials

14.3 Prepare a number of detail drawings for the mechanical-hydraulic log splitter illustrated in Fig. 14.9. Your instructor will assign specific parts. For each item, develop a concept sketch of the part (freehand) and then a detail drawing.

14.4 Figure. 14.16 illustrates a simple hinge for an interior door. Prepare a set of working drawings for this assembly, including the following:

(*a*) Layout to show operational characteristics
(*b*) Detail drawings to include production dimensions
(*c*) Sketch showing hinge in position on door

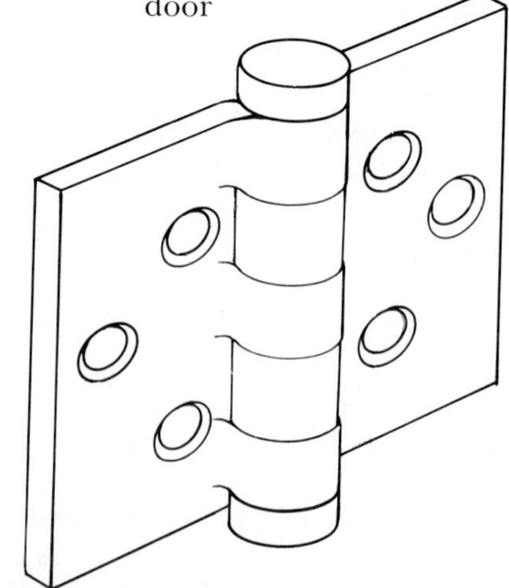

FIG. 14.16

14.5 Starting with concept sketches, develop a design for a small four-wheel wagon. Once the ideas have been defined, make a set of working drawings that include the following:

(*a*) Concept sketches
(*b*) Layout drawing
(*c*) Assembly drawing
(*d*) Detail drawings
(*e*) Bill of materials

14.6 Complete the requirements listed in Prob. 14.5, this time for a wheelbarrow.

14.7 Figure 14.17*a* is a pictorial representation of a three-point hitch used to manipulate various farm implements from a tractor. In designing a hitch, one of the key considerations is the geometry. The geometry of the hitch is best seen in the profile view (Fig. 14.17*b*). Only one half of the hitch is needed to study the geometry. The names of the various hitch parts are also indicated in Fig. 14.17*b*. Establishment of the geometry for a particular configuration can be done with a layout drawing, as shown in Fig. 14.17*c*. The design constraints in addition to the dimensions shown in Fig. 14.17*c* are as follows:

1. The driver starts in the horizontal position shown when the end of the arm is at its lowest point. Driver pivots about (0, 0).
2. The attachment point (*A*) of the link to the arm can be varied along the arm. Measured from (*x*, *y*), this distance is called *L*.
3. The regions designated area 1 and area 2 are, respectively, the areas within which the end of the arm (*S*) must start and end for proper hitch movement.
4. The point (*x*, *y*) may lie anywhere in or on the 10″ × 10″ pivot block area. The arm pivots about (*x*, *y*).

Find the lengths of the arm, link, and angle through which the driver must rotate to move point *S* (where load is attached) from area 1 to area 2 for the conditions given below. Use a layout drawing at a scale of 1″ = 5″.

(*a*) $(x, y) = (0, -10)$ and $L = 50$ percent

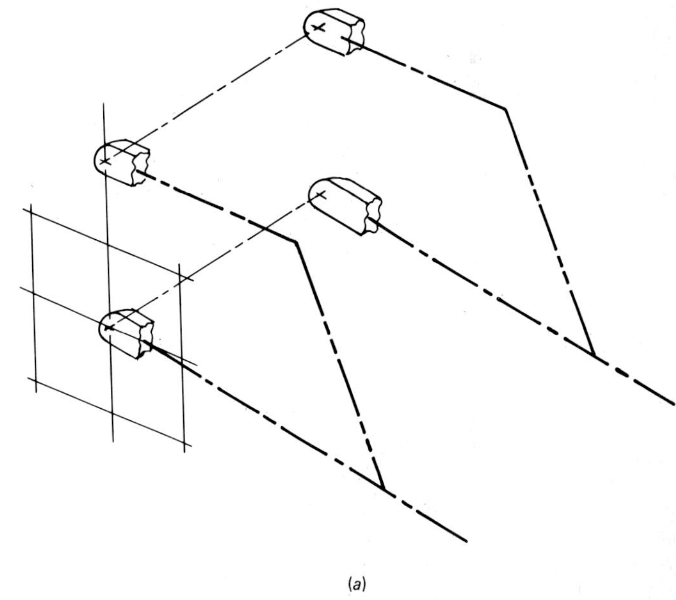
(*a*)

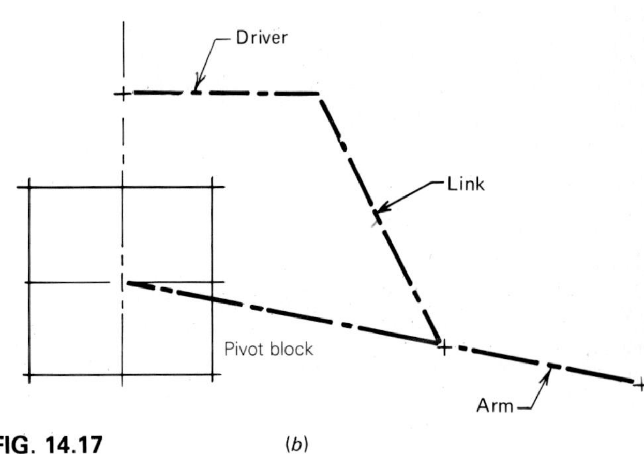

(*b*)

FIG. 14.17

arm length, 60 percent arm length, and 80 percent arm length.

(*b*) Do part *a* for $(x, y) = (-5, -5)$

(*c*) Do part *a* for $(x, y) = (2.5, -12.5)$

(*d*) Do part *a* for different locations of areas 1 and 2 as specified by your instructor.

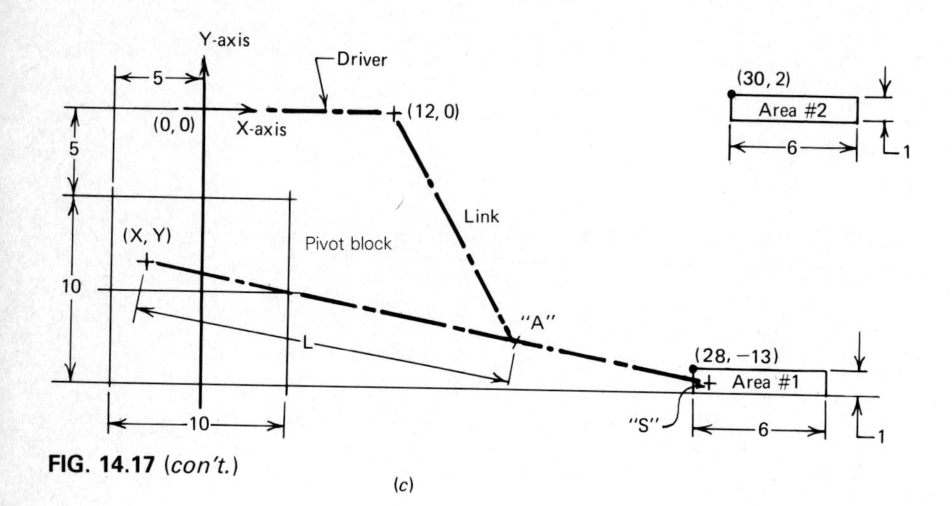

FIG. 14.17 (*con't.*)

PART FOUR

INTRODUCTION TO COMPUTER GRAPHICS

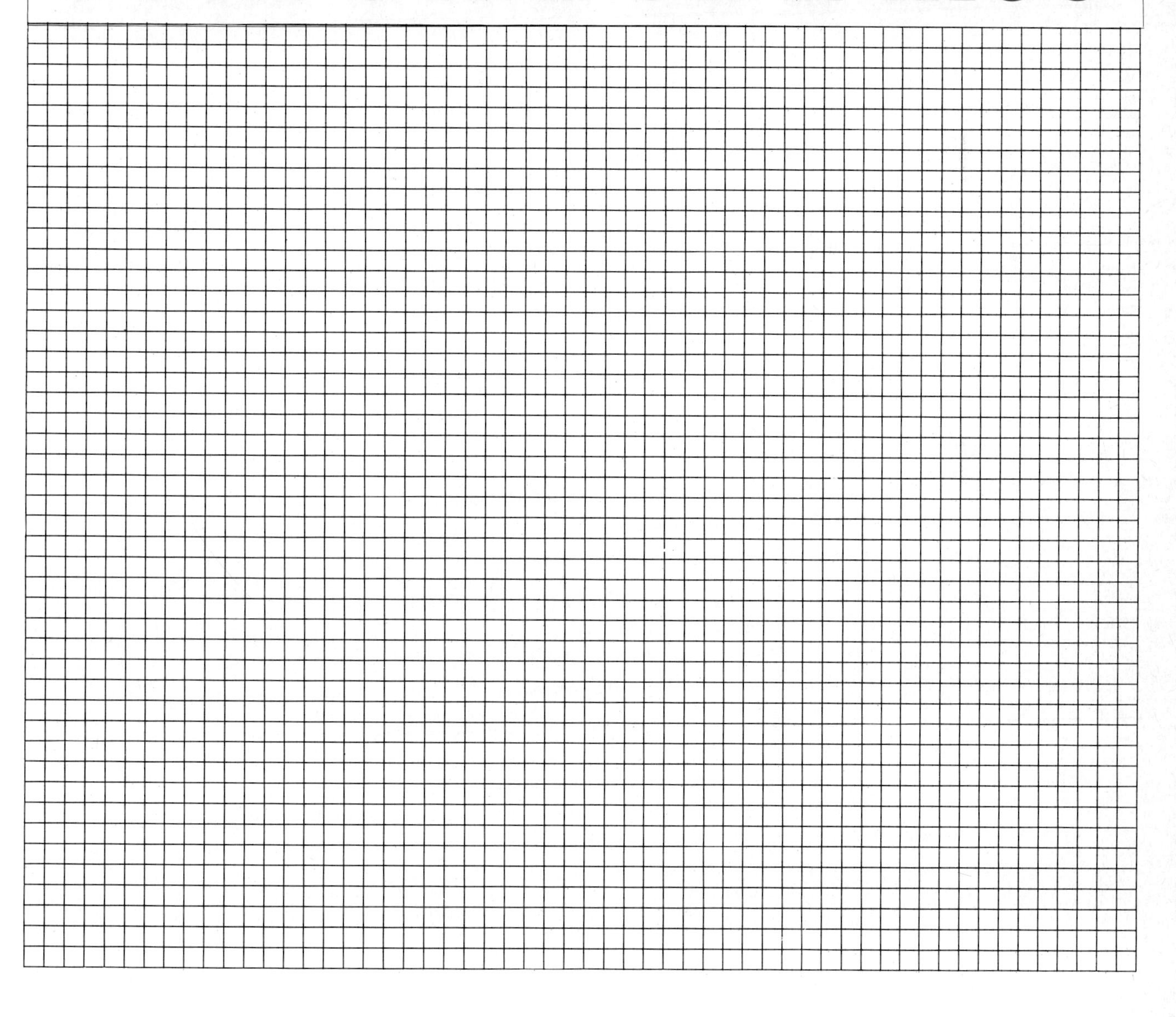

chapter 15

Survey of Computer Graphics

15.1 INTRODUCTION

Computer graphics affects our lives daily. In the medium of television, commercials, newscasts, weathercasts, and credits reflect the application of computer graphics. People who live in regions where warnings of severe weather are broadcast have seen a band of print move across a portion of the screen superimposed over the regular programming. Commercials use animation techniques and changing colors to attract attention to a message being conveyed to the viewer.

The techniques of computer graphics form the basis of video games. Other familiar examples include the flight simulators used to train pilots, the interactive airline scheduling terminals seen at airline ticket counters and the interactive student course scheduling used at many colleges and universities.

Why has the development of computer graphics had such a profound effect on engineering and most other functions of modern industry? Part of the answer lies in the very nature of computer graphics, that is, the capability of displaying pictures on a screen. The human mind has the ability to interpret a picture in one step. If the data that constitute the picture were displayed on the screen in the form of words and numbers, the difficulty of interpretation would be obvious. Our eyes would scan the words and numbers essentially one at a time, and then we would have to translate the information to a concept that could be visualized and understood. Look at the examples shown in Fig. 15.1*a* and *b*. Which is easier to study and comprehend quickly?

Humans are naturally suited to interacting with information in an image rather than a "printed page" form. The computer graphics system allows us to function at a high level of productivity. Computer graphics systems are thus becoming a very significant part of the engineer's tool kit.

If we consider an interactive system, the engineer and computer graphics terminal become a working team, each supplementing the other's efforts and producing a highly productive problem-solving combination. We will list some characteristics of a computer and some of the human (engineer) to see how the two can blend into an effective team.

Characteristics of a computer:

1. Large capacity for storing data (time-independent)

2. Excellent performance on repetitive tasks

3. Good capability for numeric analysis

4. Poor capability to detect significant information

5. No intuitive analysis

6. Rapid production of output (electronic, mechanical)

Characteristics of humans:

1. Good ability to detect significant information

2. Good intuitive analysis

3. Slow production of output (manual)

4. Time-dependent capacity for storing data

5. Poor tolerance for repetitive tasks

6. Poor capability for numeric analysis

We can see that the characteristics are mutually supporting. We can also see that with a computer graphics system, the display of pictures enhances the engineer's logic and intuition and permits better and faster decisions in the overall engineering effort (Fig. 15.2).

WORD DESCRIPTION

An ice scraper is in the shape of a rectangular prism 150 X 80 X 8 mm. One end is beveled from zero thickness to maximum thickness in a length of 50 mm. The other end is semicircular with a 20 mm diameter hole through. The center of the hole is 110 mm from the beveled end and 40 mm from either side of the scraper

(*a*)

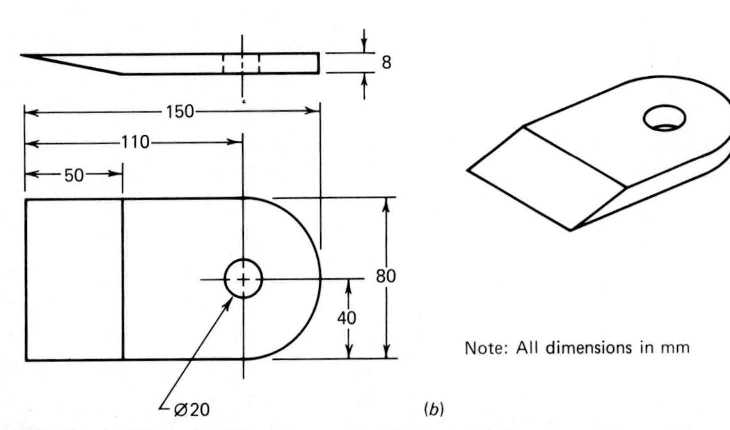

FIG. 15.1 Which provides a clearer representation: (*a*) the word description or (*b*) the drawing?

FIG. 15.2 The engineer and a computer graphics workstation form a productive team. (*Reproduced with permission of Hewlett-Packard Company.*)

In modern industry, the computer supports all the design and manufacturing functions. The concept of a common database has effectively integrated these engineering functions and requires the engineer to be aware of the ramifications of changes anywhere in the design and manufacturing process. Figure 15.3 shows schematically the kind of support provided by the computer. Note that all computers used in the design and manufacturing sequence would be networked and would have access to the common database.

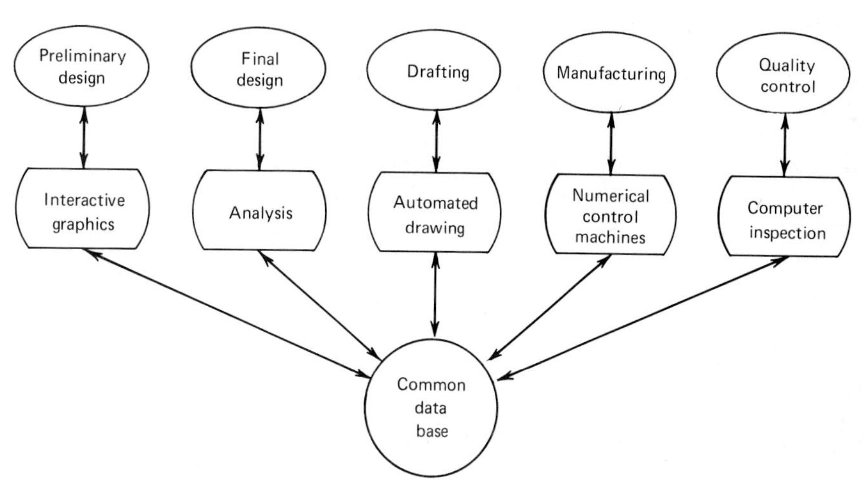

FIG. 15.3 A common database supports all engineering functions in a modern industry.

Consider the database and resulting database management requirements of the shell of the space shuttle orbiter shown in Fig. 15.4. The performance requirements of the shuttle led to the unique concept of covering its skin with 8000 tiles. The tiles were to assist in heat dissipation during the reentry portion of a mission. Once the geometry of the surface of the shuttle had been determined, it was set up as a database [imagine it as hundreds of thousands of (x, y, z) coordinates representing points on the surface]. Then each of the 8000 tiles was designed to conform to a specific section of the geometry of the shuttle's surface. The database served as the verification means by which the shape of the tiles was finalized.

Aerodynamics and structural personnel worked with this geometry database in the determination of loads and the resulting stresses and deformations on the vehicle during the mission. Using the same database, they applied the principles of thermodynamics to determine temperature distributions during flight, particularly during reentry. The results of this analysis were also provided to the materials engineers to assist in the selection of tile material and fastening substance.

Remember that all this engineering effort took place with computer assistance. Any changes to the database suggested by analyses were incorporated and made instantly available to all groups that had access to the database. In many cases the groups were subcontractors located in different areas of the country. The only database management control necessary was a computer communication line. It is obvious that this method of managing the design and manufacturing operations can be greatly superior to a set of engineering drawings. Changes are instantaneous and not subject to misunderstanding that can arise from mailing delays, slow communication between contractor and subcontractor, and the sheer volume of paper necessary to keep up with the changes.

In Part Four of this book, we develop some aspects of comput-

FIG. 15.4 This space shuttle display on a CRT was generated from a geometric database. (*Courtesy of Control Data Corporation.*)

er graphics which engineers need as part of their tool kit. In this chapter, we describe briefly the characteristics of the hardware and software currently used for engineering applications. Chapter 16 describes the fundamentals of computer-aided drafting. Three-dimensional computer graphics is presented in Chap. 17. Specific engineering applications in computer-aided design (CAD) is discussed in Chap. 18. As you proceed through the material on computer graphics, keep in mind the graphics principles you have studied. These principles will remain fundamental to engineering when you are using the computer. The methods of displaying graphics are changing because of the development of computer graphics, but the importance of graphics fundamentals to an engineer's ability to perform productively has not diminished.

15.2 COMPUTER GRAPHICS SYSTEMS

In order to establish the concept of a computer graphics system it is necessary to describe the components of a digital computer and then add the elements that make up the system we call computer graphics. This is essentially how modern-day computer graphics systems are conceived and built.

15.2.1 Computers

All computers are composed of the same basic elements. A central processing unit (CPU) consists of (1) an arithmetic unit which performs the work of adding, shifting, and so on, (2) a control unit that manages the computer operations, such as deciding the sequence of operations and sending signals to other units at the proper time, and (3) a storage unit (memory) where information may be stored and retrieved as required. Attached to the CPU are various devices that allow the user to input and output information on command (I/O).

The evolution of digital computers has been rapid and dramatic. The first digital computers were large electromechanical machines that could perform arithmetic operations in approximately 10^{-1} s. Today's mainframe computers occupy a small fraction of the space needed by their predecessors and perform arithmetic operations in a time on the order of 10^{-9} s. The development of integrated electronic circuitry has been the major contributor to the reduction in computer size and the corresponding increase in performance.

The digital computer stores and manipulates data in binary form, a series of 0s and 1s. One bit of data is a single 0 or 1. However, for practical operations, a group of 8 bits, or 1 byte, is commonly used to represent an item of data such as a letter of the alphabet or a number. The byte is operated on as a unit by the computer system.

Digital computers are often classified according to the amount of information that can be han-

FIG. 15.5*a* An early computer developed by Hollerith in the latter part of the 19th century. *(Courtesy of International Businees Machines Corporation.)*

FIG. 15.5*b* A modern digital computer, occupying little more space than the Hollerith machine (Fig. 15.5*a*), but having several orders of magnitude greater capability.

dled at one time. This amount is called a computer word and must be an even multiple of the byte length. For example, most programmable hand calculators and some smaller desk-top personal computers have an 8-bit or 1-byte word length; historically, they were classified as microcomputers. Computers which have 16 bits, or 2 bytes, per computer word were called minicomputers. Finally, machines which handle 32 or more bits per computer word were referred to as mainframe systems. These exclusive classifications are no longer possible today. There are computers which have a design computer word length such as 8 bits but have some of the capability of a 16-bit machine in handling data. Several new personal computers have 32-bit capability. A new class of computers, called supercomputers, has evolved that uses 32-bit or more computer word lengths. They use the new concepts of parallel and vector processing to effect tremememdous increases in computing power. Further information can be found in computer architecture books and vendor catalogs.

Because a computer handles data in binary form, it must have programs (software) available which take the instructions from the user and translate them into

FIG. 15.5*c* The silicon chip containing thousands of electronic circuits is the key factor in the growth of digital computing. (*Courtesy of International Business Machines Corporation.*)

machine language that can be understood by the computer. The term compiler is used to describe programs which translate user-oriented, high-level interpretive languages such as FORTRAN, Pascal, and BASIC into machine-understood code. These and other high-level languages enable us to communicate easily with a computer. It would be a most difficult communication task to write programs in machine language.

Computers are accessed for information in two modes: batch and interactive. In the batch mode, the computer reads in the previously prepared program and data on cards, executes the program, and provides a printout of the results, all without intervention from the user. The interactive mode allows one to sit at a terminal, enter the program and data, and receive the output on a cathode-ray tube (CRT). In this instance, the user can control directly what the computer does by observing the program operation and output and making adjustments with a command language. This mode of operation allows great flexibility for the user and is most important if one wants to use a computer graphics program. Refer to Fig. 15.2.

15.2.2 Computer Graphics

A computer graphics system requires two items in addition to the basic elements described in the previous section. First, a device is needed that is capable of displaying graphic output such as lines and text. Two such devices are a graphics CRT and a plotter. Second, a program (software) is needed which allows the user to describe in some logical manner the "picture" that is to be generated on the display device. This software will translate graphics commands into corresponding machine language and will control the display device in the generation of graphic output.

The development of computer graphics can be traced back to the mid-1950s with the use of CRTs for documentation of engineering data such as parts lists.

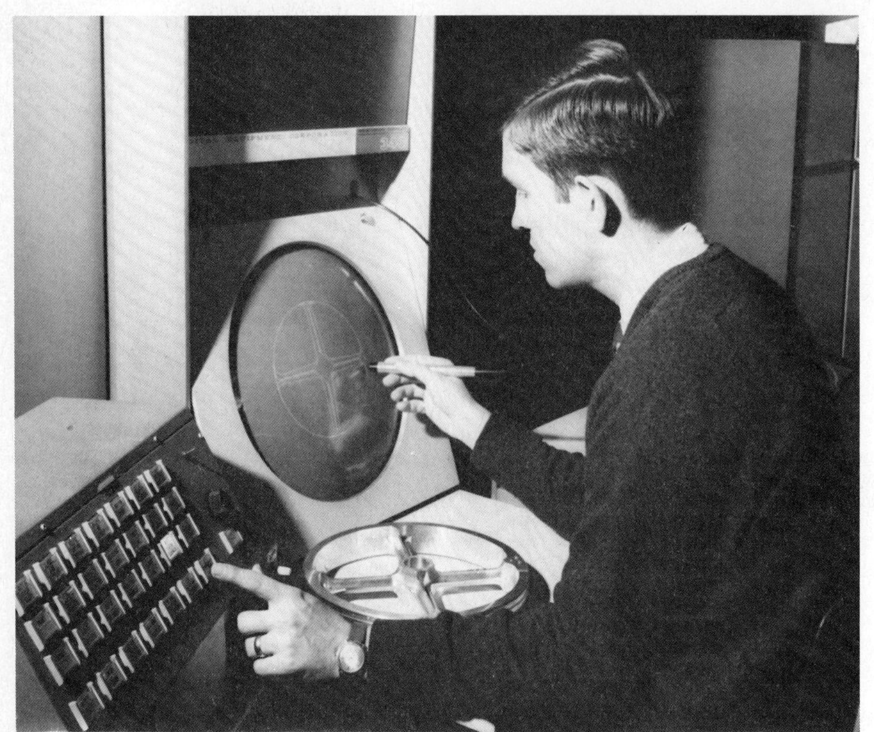

FIG. 15.6 An early-model CRT developed around 1960. (*Courtesy of Lockhead-Georgia Corporation, a division of Lockhead Aircraft Corporation.*)

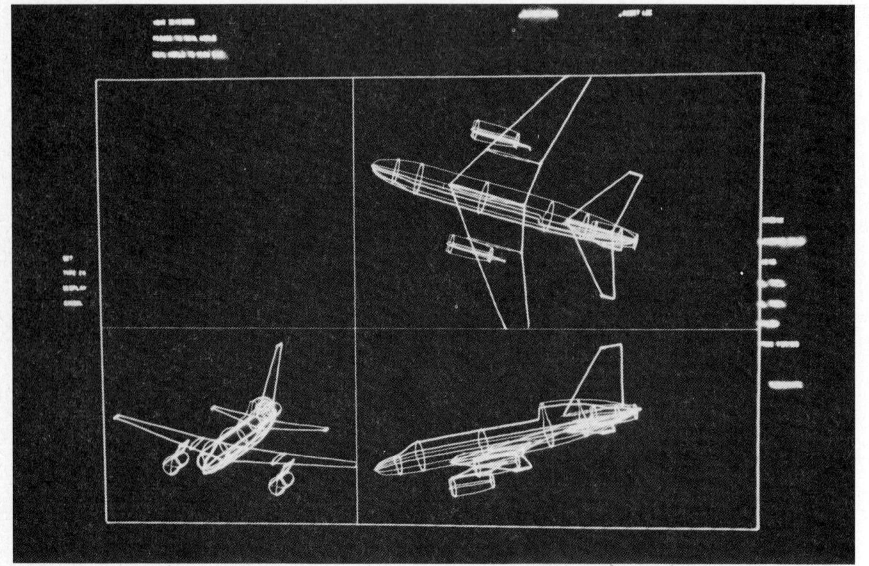

FIG. 15.7 Multiple displays on a single screen enhance the visualization of the design. This model represents the first step in establishing a database for the new aircraft. (*Courtesy of Lockhead-Georgia Corporation, a division of Lockhead Aircraft Corporation.*)

During the late 1950s, computer programs were developed which would define the shape of a part and the required machine cutter path to manufacture the part on a numerically controlled (N/C) machine. During the late 1950s, the development of the light pen occurred. This was the first real interactive computer graphics system. The next important invention was the digitizer tablet, which allowed geometry to be inputted electronically to a CRT from a peripheral device.

Perhaps the single most significant development in computer graphics technology was the "Sketchpad" system developed in the early 1960s by Ivan Sutherland of the Massachusetts Institute of Technology. This system could create and alter drawings interactively, and it became the basis for the rapid growth of computer graphics during the next several years.

The aerospace and automobile industries led the way in the development of computer graphics applications. General Motors' DAC-1, McDonnell-Douglas's CADD, and Lockheed's CADAM systems were developed in the mid-1960s and are still in general industrial use today. The development of the storage tube in the mid-1960s opened the potential of computer graphics to many additional applications because of the relatively low cost. "Turnkey," or complete, computer graphics workstations became popular.

In the mid-1970s, the development of raster graphics further broadened the areas of application of computer graphics. At a comparatively low cost, raster graphics could provide color and selective erasing which the storage tube could not. Computer graphics thus became available to the general public.

Since the mid-1970s, the growth rate of computer graphics has been phenomenal. At this time the growth rate is estimated to be about 40 percent per year. From an engineering point of view, it appears that by the late-1980s upward of 90 percent of engineering drawings will be computer-produced and one-third of manufacturing operations will be computer-controlled to some extent. The use of the computer in engineering design is expanding rapidly. In the forseeable future many parts will be designed, tested, and manufactured untouched by humans.

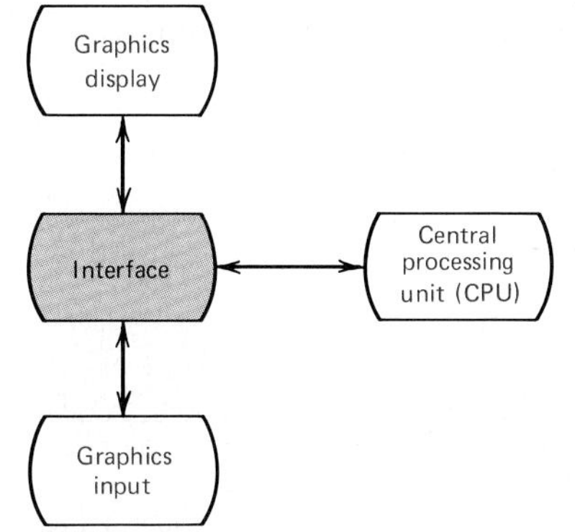

FIG. 15.8 A schematic representation of the hardware components of a computer graphics system.

15.2.3 A Complete Computer Graphics System

The hardware configuration for a computer graphics system can be represented in schematic form, as shown in Fig. 15.8. As stated in Sec. 15.2.1, the CPU contains memory, arithmetic logic, and a controller. The graphics display may consist of one or more of several devices, which are discussed in Secs. 15.3 and 15.4. The graphics input devices are described in Sec. 15.4. The interface is needed for the communication links between the components of the system. The interface consists of both hardware components and software. However, many of the technical aspects of the interface are beyond the scope of this discussion.

The physical configuration of a computer graphics system varies considerably. The components shown in Fig. 15.8 are essentially housed in a single device (Fig. 15.2). These "stand-alone" devices are used for such applications as drafting and preliminary component design. Such systems are becoming more powerful and versatile and are quite capable of performing complex engineering tasks.

For very large design and analysis tasks, the computer graphics system may consist of several workstations. Each workstation would have input and display capability, but the CPU would be remotely located, often hundreds of miles away. In many cases several workstations may share the same CPU or host computer. It is

FIG. 15.9 This workstation can stand alone for many engineering tasks and also be linked with a remote host computer. (*Courtesy of Control Data Corporation.*)

quite common to have a computer graphics workstation which has stand-alone capability and access to a host computer. See Fig. 15.9.

15.3 DISPLAY SCREENS

The most significant feature of a computer graphics system is the interactive display device, a CRT. This device enables the user to construct and visualize two- and three-dimensional quantities very rapidly. Three major types of CRTs are used: the refresh line-drawing display, the direct-view storage tube, and the raster scan display.

With a refresh line-drawing display, lines (vectors) are drawn from point to point by an electron beam impinging on a phosphor screen. On this type of display screen, lines are redrawn (refreshed) about 30 times per second. Any slower refresh rate would make the screen appear to flicker. Since the line-drawing display draws one line at a time, much as a drafter would do on paper, there is a limit to the complexity of objects that can be handled by the system without noticeable flicker. Many vector refresh systems are capable of around $1.5(10^4)$ m/s of beam speed. Translated into a refresh rate of 30 times per second, this would allow 500 m of lines to be drawn in each refresh cycle. If alphanumeric characters (formed by several very short vectors) are included, the length available for lines on the object decreases correspondingly. The refresh display can be changed by altering the display memory, thus permitting selective erasing of portions of the screen. An important characteristic of a refresh display is its ability to generate animation (apparent motion of the screen picture). This is accomplished by changing the refresh memory contents by small increments for each refresh cycle.

The direct-view storage tube (DVST) is also a line-drawing system, but it differs from the refresh system in that long-persistence phosphors are used. Once it is drawn on the screen, the picture may remain there for up to an hour before fading. These systems are capable of handling very complex objects and providing a clear, sharp display. Resolutions are very high (around 10^6 displayable points on the screen). The drawing speed (upwards of 100 m/s) is much slower than that of the refresh system, but the amount of information displayed is not a limitation as it is with the refresh system. The DVST requires that the total screen be erased before a new image can be drawn. This means that any change in an image, however minor, requires that a new image be drawn. Therefore, the DVST is not suitable for animation.

The raster scan display creates an image by controlling the intensity of a matrix of dots (pixels) as it scans the screen line by line. This is the same basis used for displaying a television picture. This system allows for a full range of colors and the creation of images that appear shaded for a more three-dimensional effect. Applications of the raster scan display include flight simulators, film animation, technical illustration, printing and plotting (used by the publishing industry for textbook production), and image processing.

Table 15.1 provides a brief comparision of the three CRT systems. Selection of a particular system depends on the intended use and the cost.

15.4 INPUT-OUTPUT DEVICES

Input-output devices provide the means of communicating with the computer and obtaining results in a permanent form. Input devices include the keyboard, card reader, magnetic tape reader, light pen, digitizer tablet, joystick, mouse, trackball, and dial. Output devices include printers and several types of plotters. The display screen (CRT) is also an output device; however, because of

TABLE 15.1
Comparison of Display Characteristics of Three CRT Types

CRT TYPE		
VECTOR REFRESH	**STORAGE TUBE (DVST)**	**RASTER**
1. Limited color possible	Single color	Full color
2. High resolution	High resolution	Low to medium resolution
3. Displays lines	Displays lines	Displays lines and surfaces
4. Dynamic (animation)	No dynamics	Dynamic
5. Image refreshed 30 times per second	Image remains on screen	Image refreshed 30 times per second
6. Complex drawings may flicker	Complex drawings displayed easily	Resolution may limit complexity

its highly interactive nature, it was considered separately in Sec. 15.3.

15.4.1 Input Devices

Every input device takes user data and prepares it for processing by the computer system. Devices such as card readers and magnetic tape readers are common to most computer systems. Computer graphics systems use various means of obtaining user data and preparing the data for the graphics software, including light pens, digitizer tablets, joysticks, trackballs, and dials. Essentially, the input device sends a graphics command to the computer. This command may be to establish a point on the display screen, draw a line, display an alphanumeric character, erase a character or line, and so forth.

Light pens are used to point at objects on a refreshed display screen. See Fig. 15.10. A light pen operates by detecting brightness on a very small segment of the display screen and identifying the object at which the pen is pointing. The identity of the object becomes the data to the graphics software, where user instructions will carry out the desired change in the object. Light pens work only on refreshed and raster scan display screens and in general are poor in resolution compared with other input devices. The principal advantage is that they provide a convenient way for the user to perform graphics tasks interactively.

FIG. 15.10 Using a light pen with a computer graphics terminal. Entire objects may be moved on the screen to desired locations. (*Courtesy of IMLAC.*)

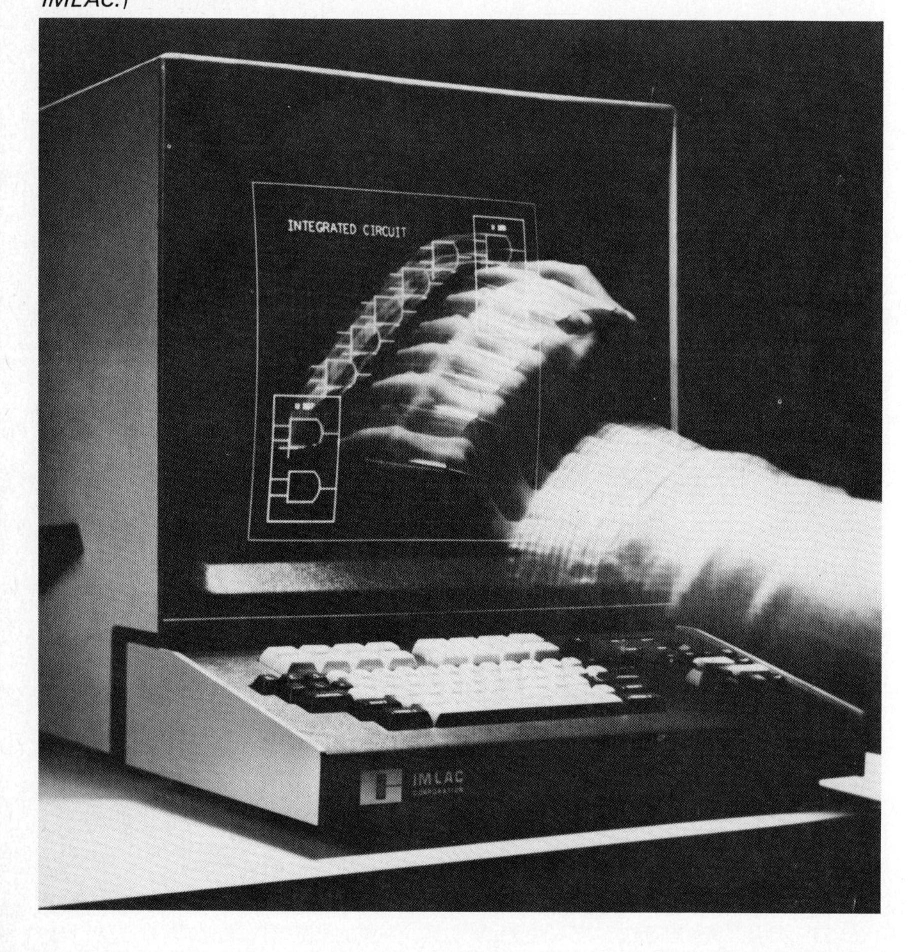

The digitizer tablet, like the light pen, is a convenient device for converting user data to acceptable input for the computer. Essentially a flat surface like a table, the digitizer tablet measures the *xy* coordinates of a point with respect to a reference frame on the tablet and sends the coordinates to the computer. The graphics software can convert these coordinates into any convenient representation of length for display on the screen. Typical resolution of digitizer tablets is between 40 and 400 units per cm of length. User data are transmitted to the tablet by a stylus, an instrument with a sharp point for accurate surface contact. Some digitizer tablets use a cross-hair device for selecting points on the surface. Figure 15.11 shows a digitizer tablet.

An alternative to the light pen and digitizer tablet is a device which controls the cursor on the display screen. The cursor is the symbol which represents the current active point on the display screen. One of these cursor-control devices is the joystick. Operated the same way as "stick" or column control on airplanes, the joystick manipulates the cursor across and up and down the display screen. See Fig. 15.12. The trackball is a spherical device which is rotated with the palm of the hand. The third commonly used device is a dial which simply moves the cursor in the direction in which the dial is rotated. When two dials are placed on the keyboard, one for x direction cursor control and the other for y direction control, the combination is referred to as thumbwheels.

The keyboard remains a significant means with which to input data. However, the keyboard is not as efficient an input device for most graphics users as are some of the devices just discussed. The keyboard remains the most efficient device for entering general alphanumeric data. Because engineering efforts require alphanumeric data as well as graphics data, the keyboard is a part of nearly all computer

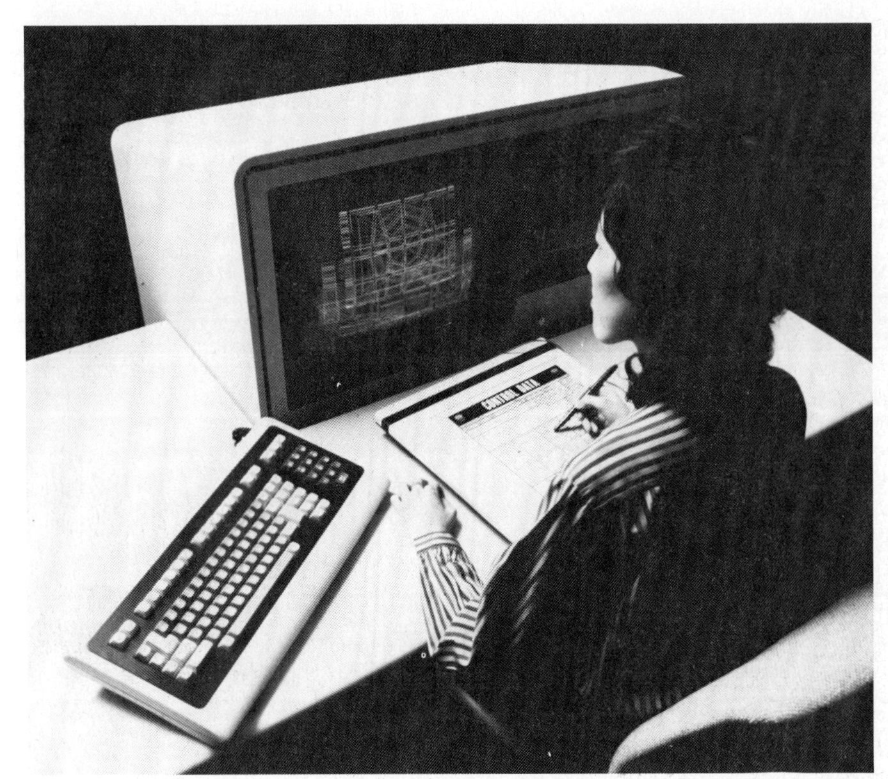

FIG. 15.11 The digitizer tablet enables information to be transported from the tablet through the digitizing pen to the CRT. (*Courtesy of Control Data Corporation.*)

FIG. 15.12 The joystick is one means of controlling the position of the graphics cursor on the CRT.

graphics systems. Certain functional keyboards are used frequently with other input devices. For example, in conjunction with a joystick, a keyboard containing cursor-control keys will permit the user to move the cursor large distances rapidly (joystick) and very small distances with accurate control (keyboard).

The future will bring voice data-entry systems. These systems have the advantage of allowing the user to keep the eyes and hands free for other tasks; this should promote higher productivity because of fewer errors.

15.4.2 Output Devices

While the display screen serves as the hub of the computer graphics workstation, in every engineering application there is a need for permanent results to be maintained. Devices that produce a permanent record are used to develop engineering reports, presentations, and archival documents. The ability to produce consistent quality work in a relatively short period of time is a significant advantage of computer-generated graphics over manual drafting—an advantage that is even more significant if many small changes are needed, since the permanent record (database) can be changed quickly and new permanent results can be produced immediately.

We are all familiar with the alphanumeric printer, as shown in Fig. 15.13. This device cannot, however, produce permanent graphic output. Instead, a plotter

FIG. 15.13 A line printer for alphanumeric information.

or hard-copy unit is needed. Examples of such devices are the flatbed, drum, and electrostatic plotters; thermal and dot-matrix plotters; and photographic film.

The flatbed plotter, which is shown in Fig. 15.14, moves a pen in an *xy* frame of reference. The position of the pen is controlled by an electrical signal through a driving motor. Multicolor plots are possible because the plotter can be programmed to select different pens during the plotting operation.

The drum plotter operates on the same principle as does the flatbed plotter except that the *xy* position of the pen is controlled by movement of both the pen holder and the paper. Figure 15.15 shows a typical drum plotter which has the advantage of using much less floor space than do similar-purpose flatbed plotters.

Electrostatic plotters operate on the same principle as does a raster scan display screen. The data to be plotted are converted into a series of dots which are then inputted to the plotter. As the special paper moves through the plotter, a static electric charge is imparted to the paper where the image exists, much the way dots are lit on a raster display screen. The visible image is created by applying a toner to the charged area of the plot. The paper is treated to hold the electric charge for an indefinite period of time. Some electrostatic plotters produce the entire image at once, using a photoconductive plate.

FIG. 15.14 A flatbed plotter. (*Courtesy of Western Graphtec.*)

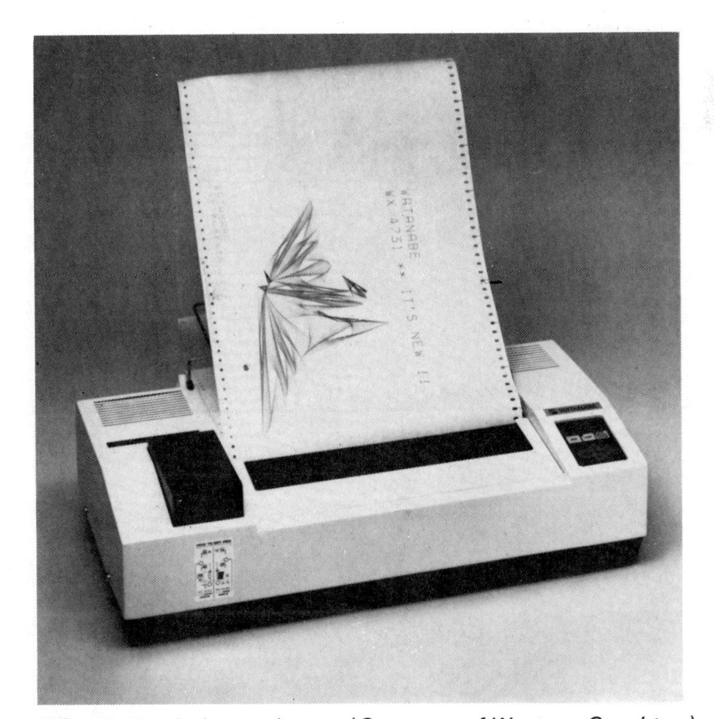

FIG. 15.15 A drum plotter. (*Courtesy of Western Graphtec.*)

With the plate process, it is possible to create images with various shades of gray. Otherwise, color plots are not possible with standard electrostatic plotters, although a new concept of laser generation of the image allows for color plots. Electrostatic plotters produce images that are similar to raster display images. Thus, resolution is not as sharp as it is with the pen plotters. Figure 15.16 shows one type of electrostatic plotter.

A less common method of producing graphics images on paper is the thermal plotting process. This method applies heat to temperature-sensitive paper. The disadvantages of this method are that the original paper must be kept refrigerated and that the imprinted copies tend to deteriorate with time.

A common hard-copy process used with raster devices is the dot-matrix printer. This device produces the image from the display screen by using small needles to impact a ribbon against the paper to form dots.

A camera with the appropriate film represents another means of obtaining hard copy. Under satisfactory lighting conditions, a picture of the display screen is easy to obtain. If Polaroid film is used, the hard copy is produced almost instantaneously. Other types of film produce the permanent hard copy in the time it takes to develop the film. Use of a color raster

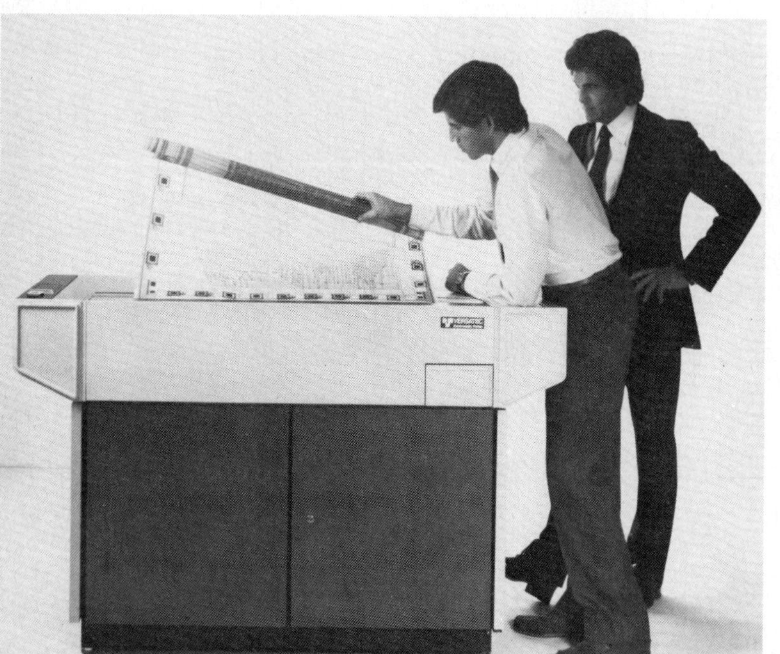

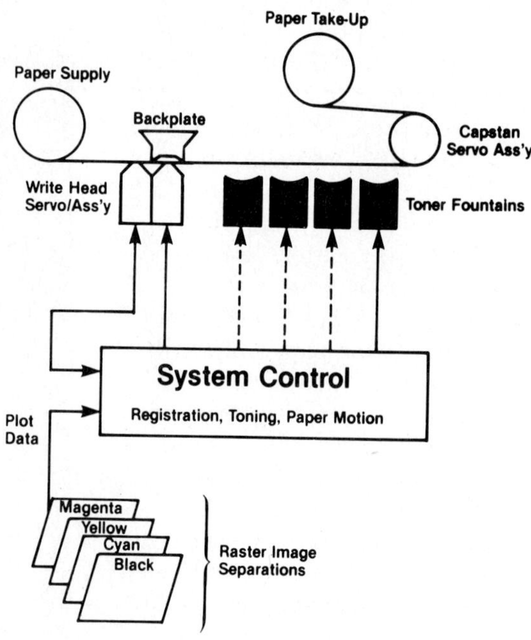

FIG. 15.16 The operation and control of an electrostatic color plotter. (*Courtesy of Versatec.*)

scan display and color slide film produces a quality set of slides that can be used for any professional presentation.

A great deal of output information is kept within computer storage facilities. Objects which have been designed using computer-assisted techniques and computer graphics have their databases stored on magnetic tape or disk. While not in image form as with other output devices, the graphics database can be retrieved quickly and converted to image form. Many of the microcomputers use floppy disks as a storage medium. These small devices store vast amounts of data which can be quickly loaded into the main computer memory for processing.

15.5 SOFTWARE

Software is the means by which the user interfaces with the computer graphics system. The applications of computer graphics have expanded greatly in recent years, as a result of which, software packages have been developed which can be used on some systems but not on others. When you prepare to use a particular computer graphics system, it will take you from a few hours to several months to learn the capabilities of the system and develop expertise in using the software efficiently and effectively.

Three aspects having to do with the use of software although important can be only briefly covered here.

15.5.1 Geometry Generation

For the user of a computer graphics system to specify the data for an object, a line of communication between the user and the system must be established. This communcation interface involves graphics commands and a programming language.

Graphics commands are the functions used to set up the display screen or other output device and construct the image. For example, commands such as CLEARSCREEN, WINDOW, and VIEWPORT define the condition of the screen initially, the region of the screen on which the image will be displayed, and the transformation from object coordinates to screen coordinates. These commands are examples only and may vary in notation from system to system; however, all systems include these commands. Once the display screen has been initialized, the image is produced with graphics commands similar to MOVE (X,Y) (position the screen cursor at *x*, *y*), DRAW (X,Y) (draw a line from current position to new position *x*, *y*), and TEXT 'INFO' (place the text information 'INFO' at the current cursor location). Many other graphics commands are available on most systems. Some of these will be discussed in later chapters of this book.

What methods are available to organize the graphics commands for an image in a logical sequence and input them to the system, thus creating a graphics

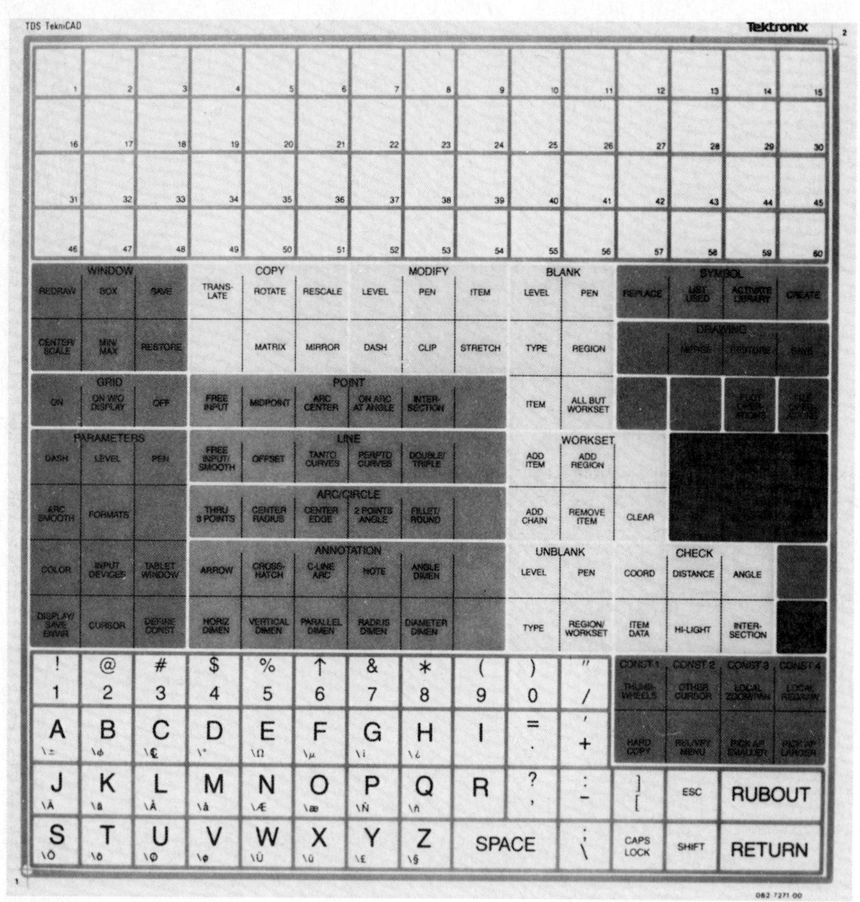

FIG. 15.17 A menu of graphics commands which can be activated with a digitizing pen. (*Courtesy of Tektronix.*)

database? One method involves the use of an input device such as a keyboard or digitizer tablet. In this case the data are input in raw form and the graphics commands are generated during the operation and stored as a file. The file is then available to run at any time to create the image on the screen.

A second method is to use a menu, as shown in Fig. 15.17. Such a menu presents a complete set of the graphics commands to the user, and each command is accessible through the digitizer tablet to which the menu is affixed. The system software contains the special functions which respond to the user's request from the menu. One option on all menus is a command to save for future use what has been created on the display screen.

A third method requires combining the graphics commands with a high-level programming language such as FORTRAN or Pascal and creating a program which is very similar to those which produce alphanumeric results. It is necessary to include the appropriate commands so that the system understands that you are creating graphic images, not alphanumeric data.

An area of concern to users of computer graphics systems is the portability of software. In general you would not expect a graphics package that generates geometry to run on several different systems without modification. In many cases the programmer must start over when trying to adapt a graphics package to a different system. The incompatibility lies not only in the software but in hardware items such as display units.

15.5.2 Viewing the Model

Once the geometry has been established as a database, it can be verified by viewing it from different directions or by changing the geometry to reflect different viewing properties. One example of this is different ways of representing hidden surfaces such as dashed or solid boundaries or complete suppression of the hidden surface.

If the model is two-dimensional, the viewer can investigate changes by such means as dragging and rubberbanding. These techniques, described in Chap. 16, allow different shapes and display conditions to be viewed and judged instantaneously.

For three-dimensional models, the most basic software for viewing the model produces a wire-frame diagram which shows only the outlines of the surfaces. This is produced by establishing points on the surface and connecting the points with lines. The software for wire-frame models becomes more complex as the objects become more complex and as the display requirements for hidden surfaces are included. Realistic three-dimensional images are possible with a shading software package. Through the use of color and projection techniques, a two-dimensional screen display of an object can appear quite realistic. See Fig. 15.18.

15.5.3 Multipurpose Software Packages

As an engineer, you will be primarily a user of computer graphics software rather than a producer of the software. In many cases you may be in a position to select a computer graphics system and compatible software. You will find a massive array of special-purpose and general-purpose hardware and software available in the marketplace.

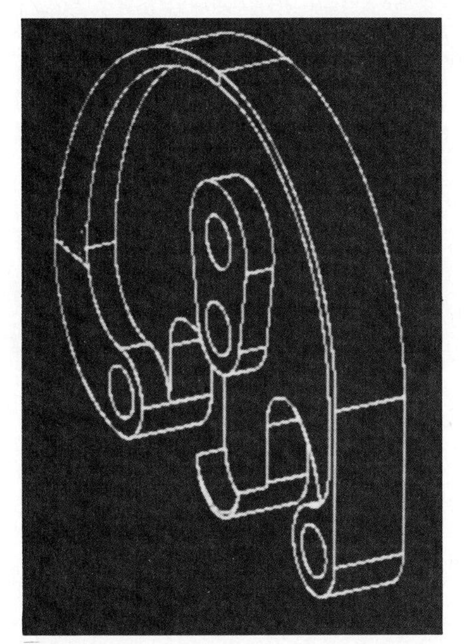

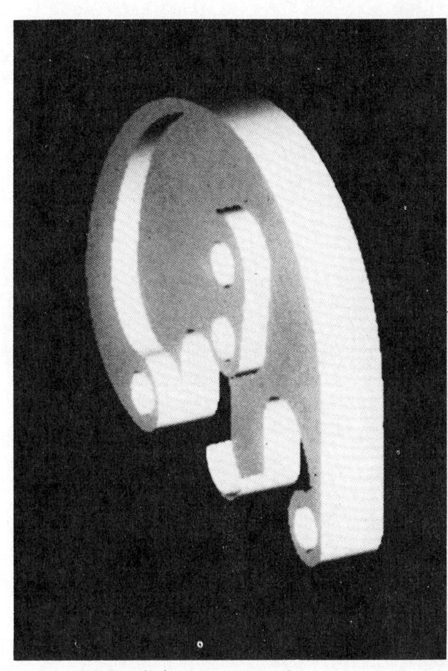

FIG. 15.18 Representation of geometry on a CRT: (*a*) the wire frame model; and (*b*) a shaded version of *a*. (*Courtesy of Control Data Corporation.*)

The first step is that of listing your specific needs. What do you want the computer graphics system to do for you and your organization? Once this is established, you can be sure that there is a "turnkey" system available to help you. A turnkey system is a combination of hardware and software which can be installed and brought online very quickly. The software has been proved, and after an appropriate training period your operators will be productive on the system.

There are general-purpose software packages designed to run on several different systems. For example, a package which analyzes structural shapes may be used by aircraft engineers on wings and fuselages, by civil engineers on bridge structures, by mechanical engineers on automobile shapes, and so on. Most likely some time will be required to adapt the package from one system to another even though it is advertised as portable. Again, a training period will be necessary for those who will be using the package.

In rare instances you may choose to produce your own software package for a specific set of needs. For example, you may want to develop a two-dimensional drawing package to be used in your drafting department. Before attempting this, look at some commercial packages carefully and determine whether the investment in time and programming personnel is economically feasible. Computer graphics software must support more functions than software written for alphanumeric output. It is therefore more complex and may not warrant engineering and programming time if general-purpose software is commercially available.

Problems

15.1 For a computer graphics system on your campus, document the following:

- *(a)* The input devices that the system accepts
- *(b)* The type of display screen (e.g., color raster)
- *(c)* The resolution of the display screen
- *(d)* The type of permanent output device available
- *(e)* A list and brief description of commercial software packages available on the system which are used by students in your department

15.2 There is a great deal of terminology associated with computers and computer graphics, most of which will not be discussed in this text. Look at computer graphics texts and find and record definitions or brief descriptions of the following terms:

- *(a)* Byte, bit, and computer word
- *(b)* Direct-view storage tube (DVST)
- *(c)* Display controller
- *(d)* Frame buffer
- *(e)* Painter's algorithm (hidden surface removal)
- *(f)* Warnock algorithm (hidden surface removal)
- *(g)* Displacement joystick, or force-operated joystick
- *(h)* Laser printer
- *(i)* Modem, or acoustic coupler
- *(j)* Baud rate

chapter 16

Computer-Aided Drafting

16.1 INTRODUCTION

In Chap. 15, you learned about equipment that can produce graphic output with the aid of a computer. The equipment uses programs (software) especially developed for the task and data (models) to produce the necessary graphics. Various modeling techniques are available, each providing advantages in certain types of applications.

Models may be categorized as 2-D, 2½-D, and 3-D. A 2-D model is basically the same as an orthographic drawing. It represents a 3-D object with a multiview drawing. On the computer, a 2-D model is merely a series of lines, text, and symbols that can be displayed, printed, or plotted. This is generally the model that computer-aided drafting systems utilize. You can think of the screen as equivalent to a sheet of paper and the light pen, digitizing tablet, or keyboard as replacing the pencil, template, drafting machine, and eraser. You create a drawing of an object on the computer using the same principles of drafting that are used in drawing by hand. The 2-D computer model is stored electronically, thereby allowing you to quickly and easily modify the model for new designs or engineering changes; of course, you can obtain a paper copy whenever needed.

The so-called 2½-D model is actually a 3-D model but not a general one. The model is created by producing a 2-D model and then providing the third dimension through software, resulting in a 3-D model. Thus, the only objects that can be properly represented with the 2½-D model are those which have a constant cross section. The procedure is similar to an extrusion in the metal or plastic industries. A common application of the 2½-D model is the foundation of a building, where the plan form can be defined and the vertical dimension added as a constant value. Other examples include structural members such as the channel section shown in Fig. 16.1*a*. Although a computer-aided system would not be used for the design, another example of a 2½-D object is the name sculpture, typically made of wood, which can be purchased for a desk (see Fig. 16.1*b*).

The 3-D model is the most general of the model types, allowing objects that vary in cross section to be described. A 2-D drawing can still be displayed if appropriate viewing directions such as top, front, profile, and so forth, are chosen. Pictorial displays are also possible with the 2½-D and 3-D models. This chapter discusses the characteristics and creation of 2-D models; 3-D models are covered in more detail in Chap. 17.

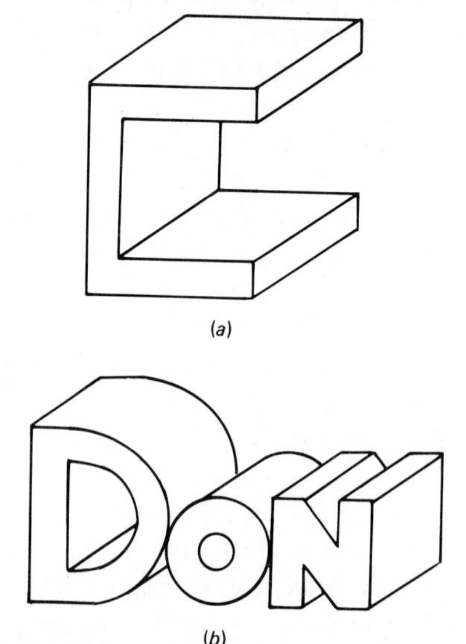

(*a*)

(*b*)

FIG. 16.1 Pictorial representation of 2½-D models.

16.2 CREATING A 2-D MODEL FROM A SIMPLE GRAPHICS PACKAGE

In this chapter, we have somewhat arbitrarily divided 2-D model creation in terms of the type of graphics package used. This section will describe model building using a simple graphics package that allows you to move the graphics cursor on the screen without drawing, draw a line from point to point, place text on the drawing, and so on. An even simpler package could be described that requires you to define the conditions of each pixel on the screen to create the picture. This procedure is so cumbersome that it is not useful for creating engineering drawings; thus, it will not be discussed further.

Section 16.3 will focus on the methods used to prepare a 2-D model using a more sophisticated package of the type that may be found in many 2-D drafting systems.

You have seen in Chap. 15 that there are line-drawing and raster systems, storage tube and refresh terminals, and a variety of input devices that can be combined into a graphics workstation. In addition, there are many different graphics languages in use. Because of this variety, we cannot discuss in detail each possible combination. Instead, we will try to describe in a generic way the fundamental operations required to produce a 2-D model. The language used does not necessarily match any known system; it was chosen to illustrate the important operations. You should study the material to learn about the process and terminology rather than to learn a specific graphics language. The specifics that apply to your local hardware and software will be introduced by your instructor as required.

To get started on many graphics terminals, you must select a window and a viewport. We will describe these operations in greater detail later, but for now, think of the window as the space within which your drawing will appear and the viewport as the area of the screen where the window (drawing) will be located. Let us assume for our current purposes that the window corresponds to the screen and that the viewport is the entire screen. In fact, this is often the system default condition.

16.2.1 Fundamental Commands

Typical fundamental commands used to operate a simple graphics package are:

MOVE (X, Y): Move from current position to coordinates X, Y without drawing a line.

DRAW (X, Y): Draw a line from current position to coordinates X, Y.

RMOVE ($\Delta X, \Delta Y$): Move ΔX units in the X direction and ΔY units in the Y direction from the current position. No line is drawn.

RDRAW ($\Delta X, \Delta Y$): Draw from current position to final coordinates of $X_{initial} + \Delta X$ and $Y_{initial} + \Delta Y$.

TEXT (SAMPLE): Write text "SAMPLE" beginning at current position.

These commands may have different names, such as POSITION instead of MOVE, VECTOR instead of DRAW, and so on. Sometimes MOVE and RMOVE are combined into one command and the system must decide whether you want to make an absolute or a relative move by checking to see if you used a number or a signed (+/−) number, respectively.

Other typical commands give you the ability to select text size, text font, and text orientation with respect to the horizontal. If the system uses color, you may select writing color and background (screen) color. Some type of CLEAR operation will certainly be available on all systems.

Although these fundamental commands are sufficient to create a drawing on the screen, to use them we must define the coordinate system. Here we must distinguish between user coordinates and screen coordinates. With user coordinates, the object is oriented while building the model. Thus, you define the user coordinates to conform to the size of

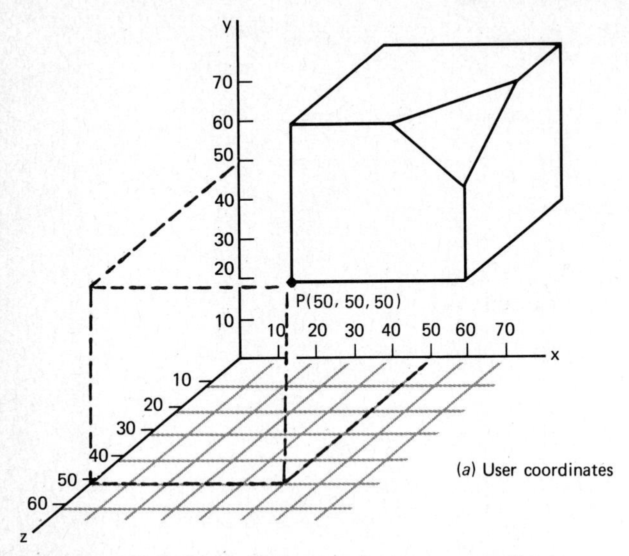

(a) User coordinates

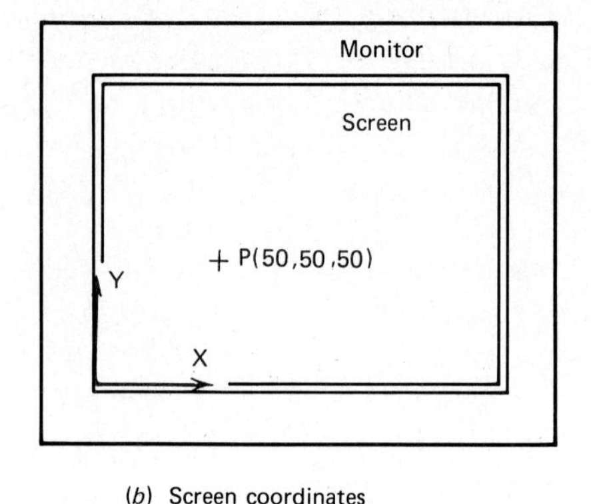

(b) Screen coordinates

FIG. 16.2 Definition of user and screen coordinates.

the object and the units you wish to use. Screen coordinates are dictated by the system you are using. They are usually defined with the origin at the lower left-hand corner of the screen with X increasing to the right and Y increasing upward. See Fig. 16.2. Some systems, however, place the origin at the upper left or at some other location. Typical screen coordinate XY ranges are 256 by 256, 130 by 100, 767 by 479, and 4096 by 4096 and are related to the resolution of the CRT. The conversion from user coordinates to screen coordinates is controlled by the window and viewport specifications.

With this brief background, we will write a series of commands that create the drawing of the "SIMPLE OBJECT" shown with user coordinates in Fig. 16.3*a*. We will use the generic commands previously defined, except for RMOVE and RDRAW.

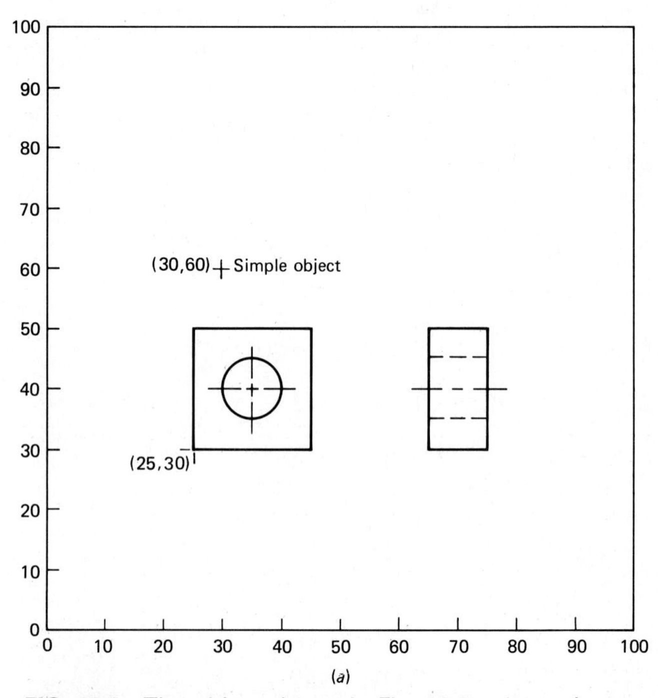

(*a*)

FIG. 16.3 The object shown in Fig. 16.3*a*, drawn in user coordinates, is programmed in Fig. 16.3*b*. In this example the screen coordinates are equal to the user coordinates.

```
CLEAR  (Clear screen)
MOVE (25,30)  (Move to lower left-hand corner of front view)
DRAW (25,50)(45,50)(45,30)(25,30)  (Draw front view outline)
MOVE (65,30)  (Move to lower left-hand corner of profile view)
DRAW (65,50)(75,50)(75,30)(65,30)  (Draw profile view outline)
MOVE (63,40) ┐
DRAW (68,40) │
MOVE (69,40) │
DRAW (71,40) ├ (Draw center line in profile view)
MOVE (72,40) │
DRAW (77,40) ┘
MOVE (28,40) ┐
DRAW (33,40) │
MOVE (34,40) │
DRAW (36,40) │
MOVE (37,40) │
DRAW (42,40) ├ (Draw center lines in front view)
MOVE (35,33) │
DRAW (35,38) │
MOVE (35,39) │
DRAW (35,41) │
MOVE (35,42) │
DRAW (35,47) ┘
MOVE (65,45) ┐
DRAW (67,45) │
MOVE (69,45) │
DRAW (71,45) │
MOVE (73,45) │
DRAW (75,45) │
MOVE (65,35) ├ (Draw hidden lines in profile view)
DRAW (67,35) │
MOVE (69,35) │
DRAW (71,35) │
MOVE (73,35) │
DRAW (75,35) │
MOVE (40,40) ┘
DRAW (39.92,40.87)(39.70,41.71)(39.33,42.50) ┐
DRAW (38.83,43.21)(38.21,43.83)(37.50,44.33) │
DRAW (36.71,44.70)(35.87,44.92)(35.00,45.00) │
DRAW (34.13,44.92)(33.29,44.70)(32.50,44,33) │
DRAW (31.79,43.83)(31.17,43.21)(30.67,42.50) │
DRAW (30.30,41.71)(30.08,40.87)(30.00,40.00) │ (Draw circle in
DRAW (30.08,39.13)(30.30,38.29)(30.67,37.50) ├  front view,
DRAW (31.17,36.79)(31.79,36.17)(32.50,35.67) │  using 10° increments)
DRAW (33.29,35.30)(34.13,35.08)(35.00,35.00) │
DRAW (35.87,35.08)(36.71,35.30)(37.50,35.67) │
DRAW (38.21,36.17)(38.83,36.79)(39.33,37.50) │
DRAW (39.70,38.29)(39.92,39.13)(40.00,40.00) ┘
MOVE (30,60)          ┐
TEXT (SIMPLE OBJECT)  ┘ (Place text above drawing)
```

(*b*)

These commands are easy to use but are more difficult to visualize when you are interpreting previously written commands because you can quickly lose where you are on the screen. With the possible exception of the first absolute MOVE to establish a position, all the other MOVE and DRAW commands can be replaced with RMOVE and RDRAW with appropriate changes in the arguments.

The particular order in which elements of the object were drawn is arbitrary. We have assumed that several DRAW commands can be compressed into one command. This is typical of many systems. Thus, DRAW (10,20) (20,30) is equivalent to DRAW (10,20) followed by DRAW (20,30). The circle was drawn with 10° increments, which is again an arbitrary choice but represents a compromise between drawing an extremely rough circle with few points and using many points (1° increments, for example) which would require 10 times the number of DRAW commands.

It is apparent that this method of creating a drawing on a CRT is cumbersome and extremely time-consuming when the drawing is complex. More sophisticated graphics packages combine many of these primitive commands into groups that draw a circle, for example. This will be discussed in more detail in Sec. 16.3.

Depending on the system, the picture may be drawn as the commands are typed in, but more likely you will create a file of commands which, when executed with a run command, will draw the picture. As each command is executed in turn, the drawing is displayed. The file can be saved for future use as it was written or for editing at a later time. Also, a paper copy of the commands or the drawing can be made via a printer or plotter, respectively.

16.2.2 Display Controls

Many computer graphics systems provide the user with a means of controlling how much of a drawing (if less than all) is to be displayed and where on the screen it should be displayed. This is done through the window and viewport concepts, which were mentioned previously. Let us discuss window and viewport in greater detail. Again, each system has its own unique set of commands, but each accomplishes about the same task. Many use a form of SET WINDOW (arguments) and SET VIEWPORT (arguments) to establish the window and viewport, respectively.

The term window refers to the portion of the drawing you wish to display. Let us assume that the square, circle, and triangle in Fig. 16.4 represent the entire drawing we have produced. By setting the window (simply a range of X and Y user–coordinate values) as the large outside rectangle, you are asking that the entire drawing be displayed. On the other hand, if the window depicted by the dashed rectangle is set, only the

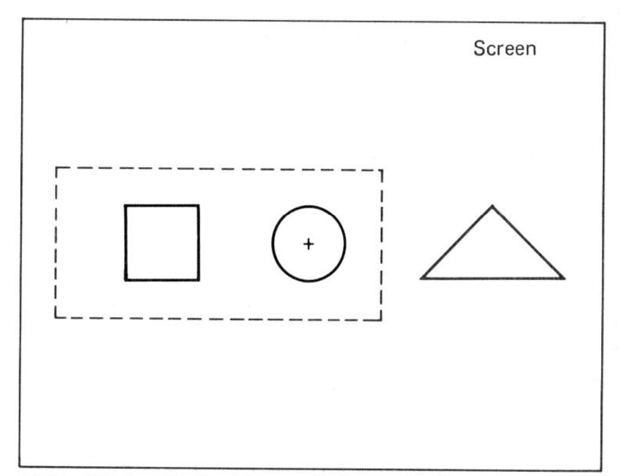

FIG. 16.4 Selection of the window, the portion of a drawing to be displayed on the screen.

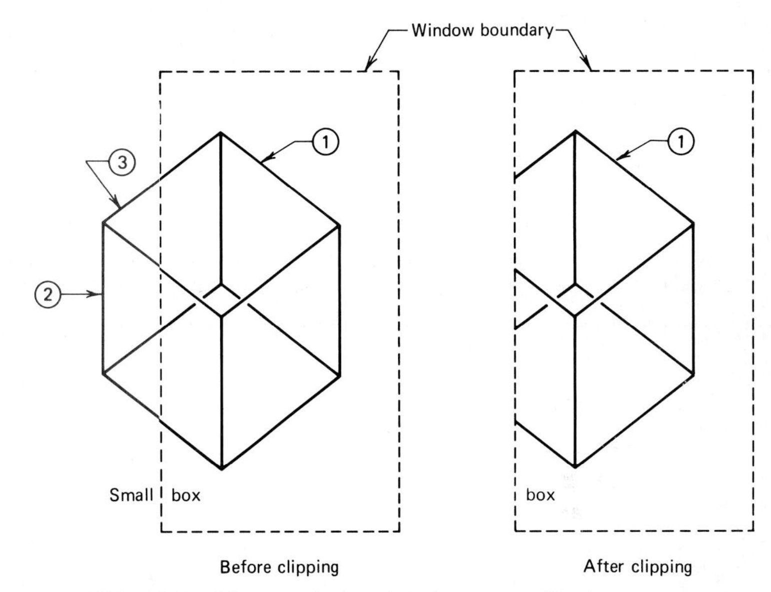

FIG. 16.5 After a window has been specified, the computer "clips" the portion of the drawing outside the window. The clipped section is not displayed but remains as a part of the database of the object so it can be displayed at another time if desired.

square and the circle will appear in the display; the triangle will not. Remember that we are concerned here with displaying all or part of a drawing. Even though we choose a window that shows only a part, the 2-D model still exists; if the window is increased to include the entire drawing, the entire model will appear in the display. Therefore, at each window setting the computer must decide what parts of the model are within the window boundaries, delete (clip) those outside the window from the display data (remember that this does not delete data from the 2-D model), and display the remaining parts.

Clipping the unwanted display elements is accomplished either by software (program) or by a special-purpose hardware device. Consider the model of the small box shown in Fig. 16.5. If the window indicated by the dashed line is selected, the clipping software or device should produce the display elements at the right. The dashed window boundary is not displayed. The clipping algorithm must handle three types of elements in this case. Examples of each element type are designated 1, 2, and 3 in Fig. 16.5. Element type 1 is in this case a line which is entirely within the window, and so it should be displayed. Type 2 is completely outside the window, and so it should not be displayed. In type 3, the element is partially inside and partially outside the window. The algorithm must determine the point of intersection of the boundary and the element. Then the element can be broken into two segments, one to be displayed and one not. In a more general

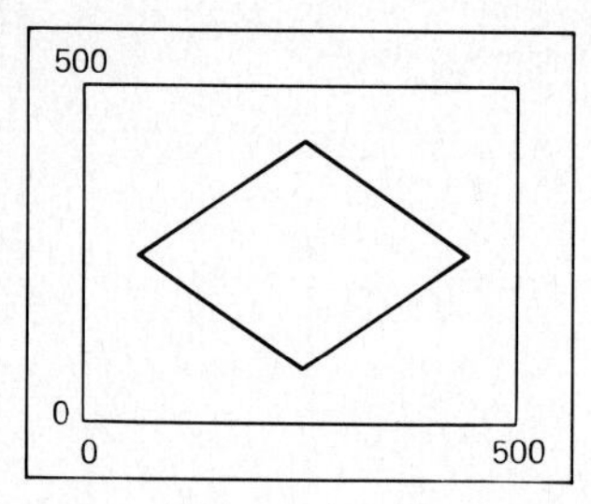

Viewport: $0 \leq X \leq 500$, $0 \leq Y \leq 500$

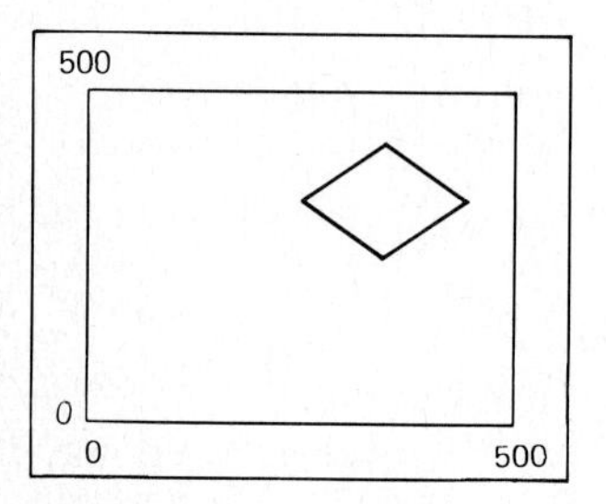

Viewport: $250 \leq X \leq 500$, $250 \leq Y \leq 500$

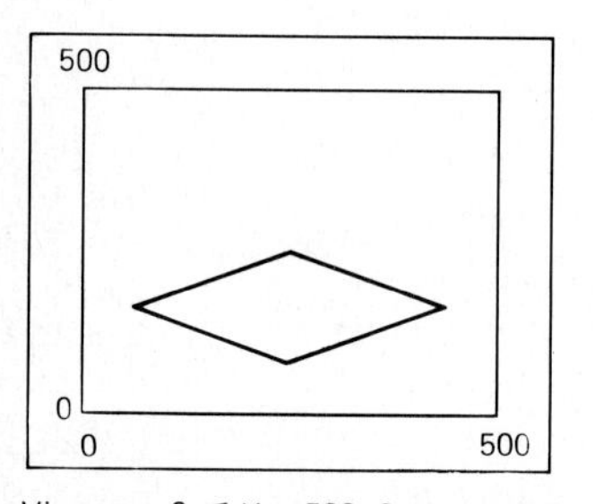

Viewport: $0 \leq X \leq 500$, $0 \leq Y \leq 250$

FIG. 16.6 Effect of changing the viewport.

case a line may extend through two boundaries of the window, and so two intersection points must be found.

The viewport defines a space on the screen where display material is to be placed. The X and Y ranges defining the viewport are given in screen coordinates. If we take as an example a screen that has an X coordinate range of 0 to 500 and a similar Y coordinate range of 0 to 500, and if the viewport is set so that $250 \leq X \leq 500$ and $250 \leq Y \leq 500$, the display will appear in the upper-right quadrant. Figure 16.6 illustrates some possible viewport specifications and shows the viewport effect on the same model. Note that if the ratio of height to width (aspect ratio) of the viewport is changed, the display is distorted. Certainly the window and viewport specifications must be coordinated. Remember that the window is specified in terms of a range of screen coordinates. If you don't wish to distort the display of the object, the aspect ratios of the window and viewport must be equal.

Multiple viewports for displaying more than one model at a time are also possible. A common occurrence of this is when you wish to display a menu on one portion of the screen, your drawing on another portion, and perhaps a prompt or system information on a third. Figure 16.7 shows a possible arrangement.

Other display controls are also available on most systems. Some type of translation is usually possible where an element in the drawing or the entire drawing can be moved on the screen. This is accomplished by the system using the MOVE and DRAW commands and the viewport or window settings. Rotation is also normally available where a coordinate conversion is accom-

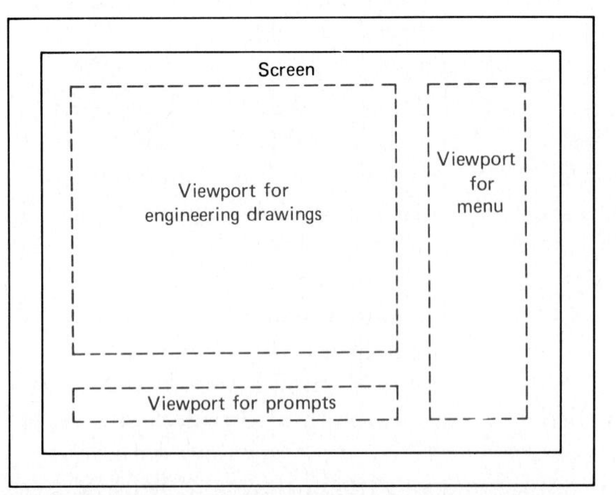

FIG. 16.7 The computer screen can be divided into sections to form convenient working areas.

plished by the software so that the arguments of the primitive commands are changed, resulting in the object's being drawn with a new orientation.

16.3 CREATING A 2-D MODEL FROM A DRAFTING PACKAGE

By a drafting package, we refer to a set of programs that combine primitive graphics commands so that the user finds it quicker and more convenient to draw lines or circles, place text, dimension objects, and so forth. This package must be designed so that drawings can be created, changed, stored, and plotted with a minimum of effort by the user. There is a wide range of drafting packages on the market. Each package is configured for a certain combination of hardware devices. For example, one package may use a keyboard, cathode-ray tube (CRT), digitizing board, and plotter for input/output while another may operate with on-screen menus and a light pen rather than the digitizing board. One typical system using these approaches is shown in Fig. 16.8.

Rather than try to describe in detail any specific system now in existence (which may be out of date as you read this), let us define some of the functions a drafting package should be able to perform. Some of these functions or capabilities are listed below in alphabetic order. A brief explanation of each is given.

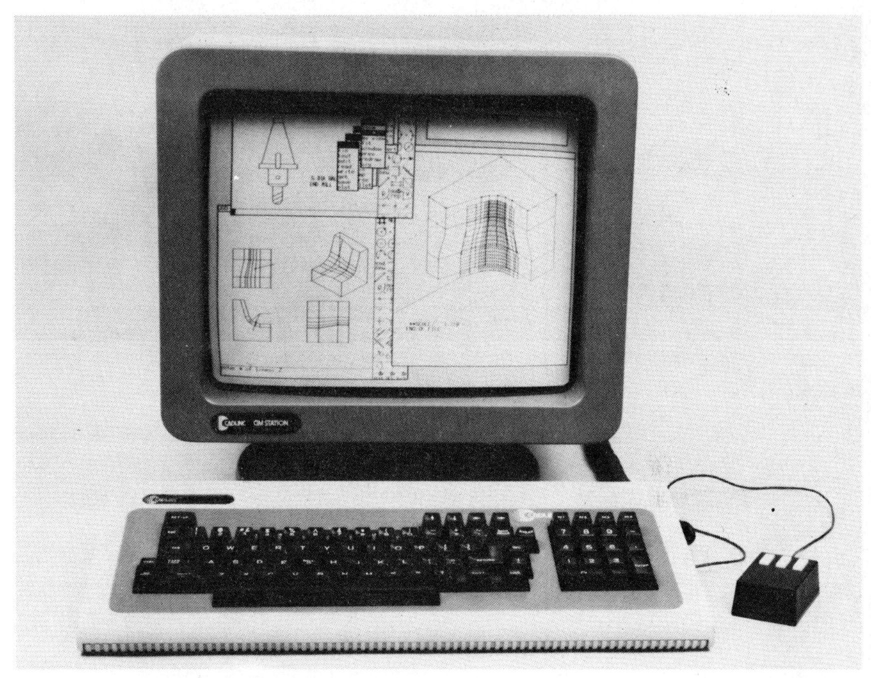

FIG. 16.8 A typical computer graphics workstation showing a keyboard, CRT and puck. (*Courtesy of CADLINC.*)

Arc. An arc of a circle is placed on the screen from generally three types of input characteristics. The arc may be defined by specifying the center of the arc plus the two ends of the arc (Fig. 16.9*a*). It may also be specified by placing the ends of the arc plus one additional point along the arc (Fig. 16.9*b*) or by defining the center point, the radius, and the starting and ending angles with respect to the horizontal (Fig. 16.9*c*).

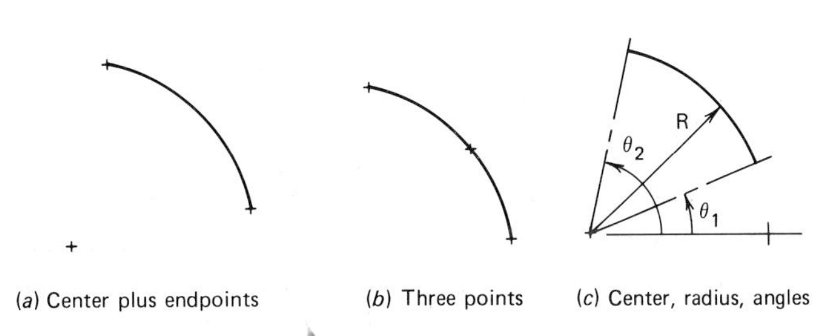

(*a*) Center plus endpoints (*b*) Three points (*c*) Center, radius, angles

FIG. 16.9 Alternative methods of specifying circular arcs.

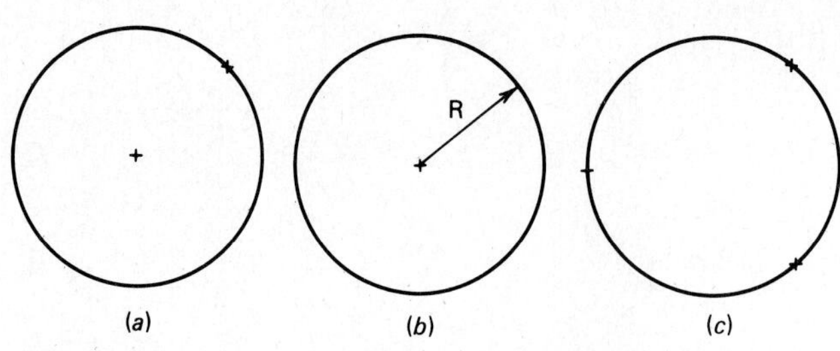

FIG. 16.10 Alternative methods of generating a circle.

Circle. A circle is drawn on the screen with a specified diameter and position. Usually three choices are available for specification: center and one point on the circle (Fig. 16.10*a*), center and radius (Fig. 16.10*b*), and three points on the circle (Fig. 16.10*c*).

Curve. (open or closed). An arbitrary curve is generated that passes through a series of points specified by the user. Some curve-fitting techniques (B-spline, for example) are used to derive the equation of a smooth curve through the points that are then drawn. The open-curve option fits a curve through the points defined and may require a point prior to the desired starting point and one beyond the endpoint to be used to establish the slope at the starting and ending points (Fig. 16.11*a*). The closed-curve option assumes that the first and last points specified should also be connected (Fig. 16.11*b*).

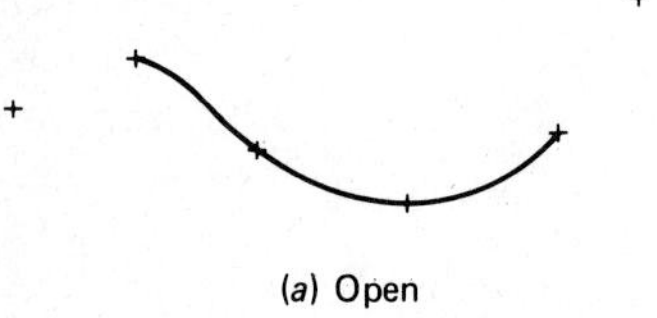

(*a*) Open

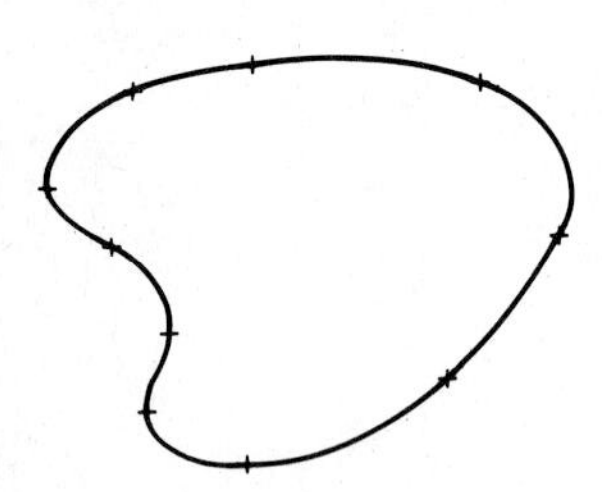

(*b*) Closed

FIG. 16.11 Open and closed curves.

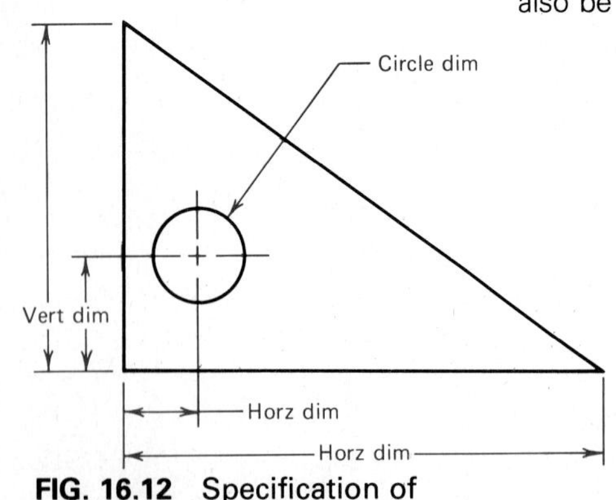

FIG. 16.12 Specification of dimensions.

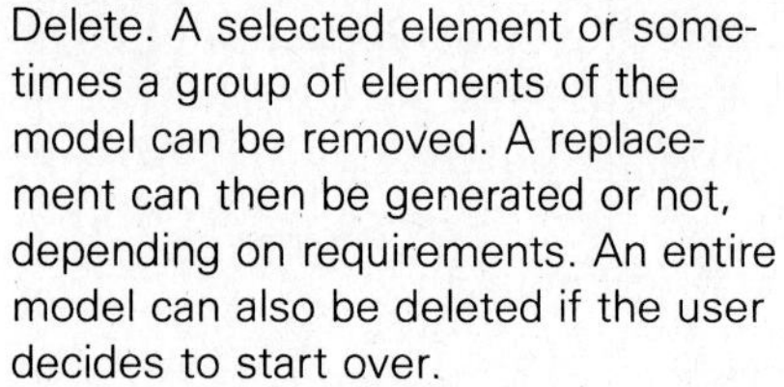

Delete. A selected element or sometimes a group of elements of the model can be removed. A replacement can then be generated or not, depending on requirements. An entire model can also be deleted if the user decides to start over.

Dimensioning. Semiautomatic or automatic dimensioning is used to place dimensions on a drawing. In semiautomatic dimensioning, the user may be required to specify where extension lines should be placed and where the dimension text should be, and the value of the dimension. Automatic dimensioning uses the coordinate values of the element being dimensioned to calculate the dimension value. The user also specifies whether the dimension should be horizontal, vertical, or parallel to a chosen element. A leader is used to dimension the radius of an arc, and either a leader or a circle diagonal is used for the diameter of a circular element. See Fig. 16.12.

Layer. The information that constitutes the complete drawing can be broken into categories, with each category being placed on a different layer. Then each layer can be displayed independently or in any combination, depending on the needs of the user. A typical example is that of placing the dimensions on a different layer from the one containing the line drawing. Then the user has the option of displaying only the line drawing or the complete drawing with the dimensions. All layers together make up the complete 2-D model.

Leader. A leader is placed on the model by specifying the position of the arrowhead and then the other end. The order can be reversed in some systems, but some order is necessary to get the arrowhead on the proper end.

Line. A line is drawn on the screen according to the wishes of the user. The line may be specified to be solid, dashed, or a center line. The line is usually defined by giving its endpoints and may be specified by requiring that the line be horizontal, vertical, or at a specific angle. This is especially necessary when using a light pen or digitizing tablet, where exact placement of the endpoints is difficult (Fig. 16.13).

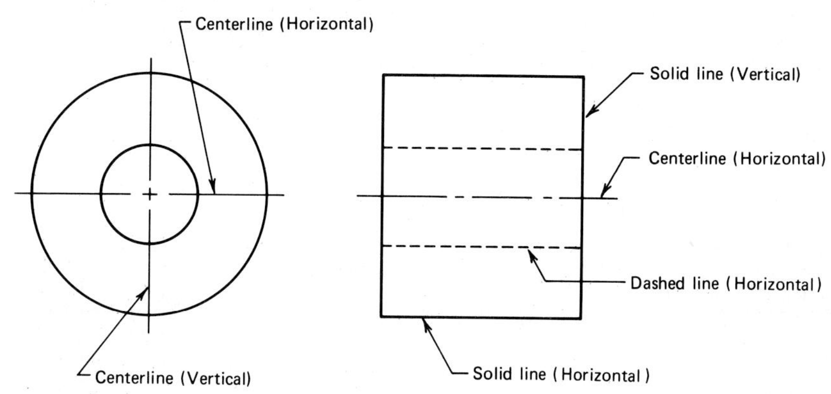

FIG. 16.13 Production of various types of lines on a computer-produced drawing.

Modify. An element can be selected for some change. For example, a dimension value may have to be corrected; a line moved, lengthened, or shortened or its type changed from solid to dashed; or text may have to be updated because of an engineering design change.

Print. A paper copy (hard copy) can be produced. The term print often refers to printed text but can also be used to create a copy of the graphic display on the screen via a hard-copy unit.

Plot. A finished drawing (in a single color or in multiple colors) can be produced on a flatbed or drum pen plotter.

Redraw. After a series of creation or modification steps has been completed, the user can require that all elements of the layer(s) be redrawn on the screen after it has been cleared. This gets rid of unwanted clutter and identifies the exact status of the model.

Rotate. An element or group of elements can be selected to be rotated about a specified axis through a given angle.

Save/Recall. Once a model has been completed or the user must stop work on a project temporarily, the model can be stored under a filename on a disk, tape, and the like by using a save command. The file is recovered by using recall and the name of the desired file.

Section. Crosshatching may be placed on a section view by identifying the outline of the space to be lined and by defining the angle with respect to the horizontal and the spacing of the crosshatching (see Figure 16.14). Specialized types of crosshatching may also be options.

Text. Characters (letters, numerals, and symbols) may be put onto the screen. Various options may be available, such as font (type of letter, for example, italics) and size. The user may also specify that a character string be horizontal, vertical, or at some angle (perhaps parallel to a given line).

Translate. An element or an entire object may be moved from one location on the screen to another. Some systems provide a "dragging" operation that allows the user to cause the object, once it is identified, to appear to move across the screen following a light pen or a stylus on a digitizer tablet. Dragging requires a refresh CRT, since a storage tube type must be fully erased and the entire image redrawn, which is generally too slow to show apparent motion.

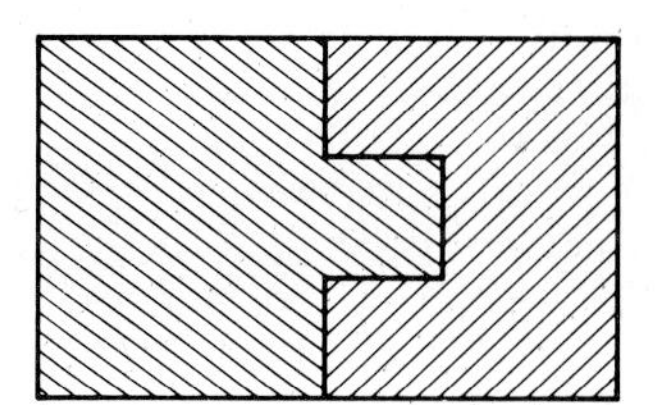

FIG. 16.14 Section lining.

Viewport. Viewport defines the area of the screen (in screen coordinates) where the user wishes the image to appear. See Sec. 16.2.2 for details.

Window. Window identifies the portion of the model to be viewed. It is given in user coordinates. See Sec. 16.2.2 for details.

Zoom. The user can request that the image on the screen be enlarged by a specified factor, for example, 2 (double the original size). Obviously, a portion of the original image, assuming the viewport is not changed, can no longer be displayed. Usually the user points out a position on the original image with a light pen or tablet stylus which should become the center of the enlarged (or reduced) image.

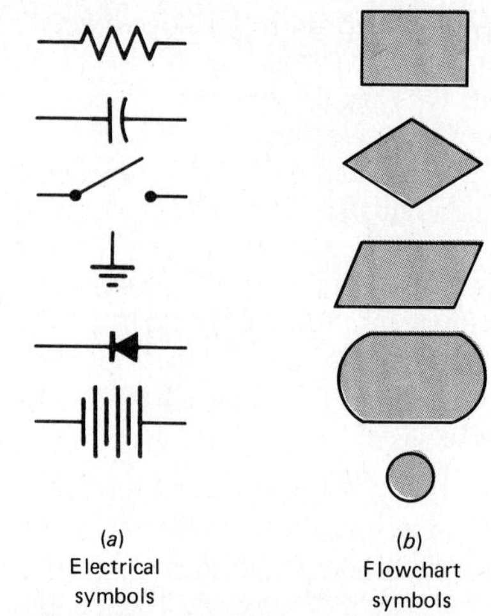

FIG. 16.15 Special symbols for application drawings. These symbols are part of the menu and are called with a single command.

This list of operations is not intended to be exhaustive but rather representative of typical systems. Simple 2-D drafting packages may have only a portion of these operations and may require many keystrokes to accomplish a specific operation. Sophisticated packages probably will have more operations than have been listed and tend to require fewer keystrokes by the user per operation. Also, the particular name used for each operation varies from package to package.

Because of the nature of this text, we have emphasized packages used to produce engineering drawings. Some of these packages provide special symbols or allow the user to create special symbols to be used for electrical schematics, piping drawings, welding drawings, and flowcharts, for example. Thus, rather than requiring the user to laboriously draw these symbols from the line or circle options whenever needed, the symbol need be done only once and stored as an entity for later use. A special menu may even be built with the symbols required for a specialized drawing. Figure 16.15 shows typical symbols used for electrical schematics and flowcharts.

To help illustrate how a 2-D drafting package might work, let us define a hypothetical system consisting of a CRT, microcomputer, keyboard, digitizing tablet and puck, plotter, and floppy disk drive. A menu that attaches to the tablet is the primary interactive element of the package, a portion of which is shown in Fig. 16.16.

We will describe the operations users might follow in creating a dimensioned drawing of the ob-

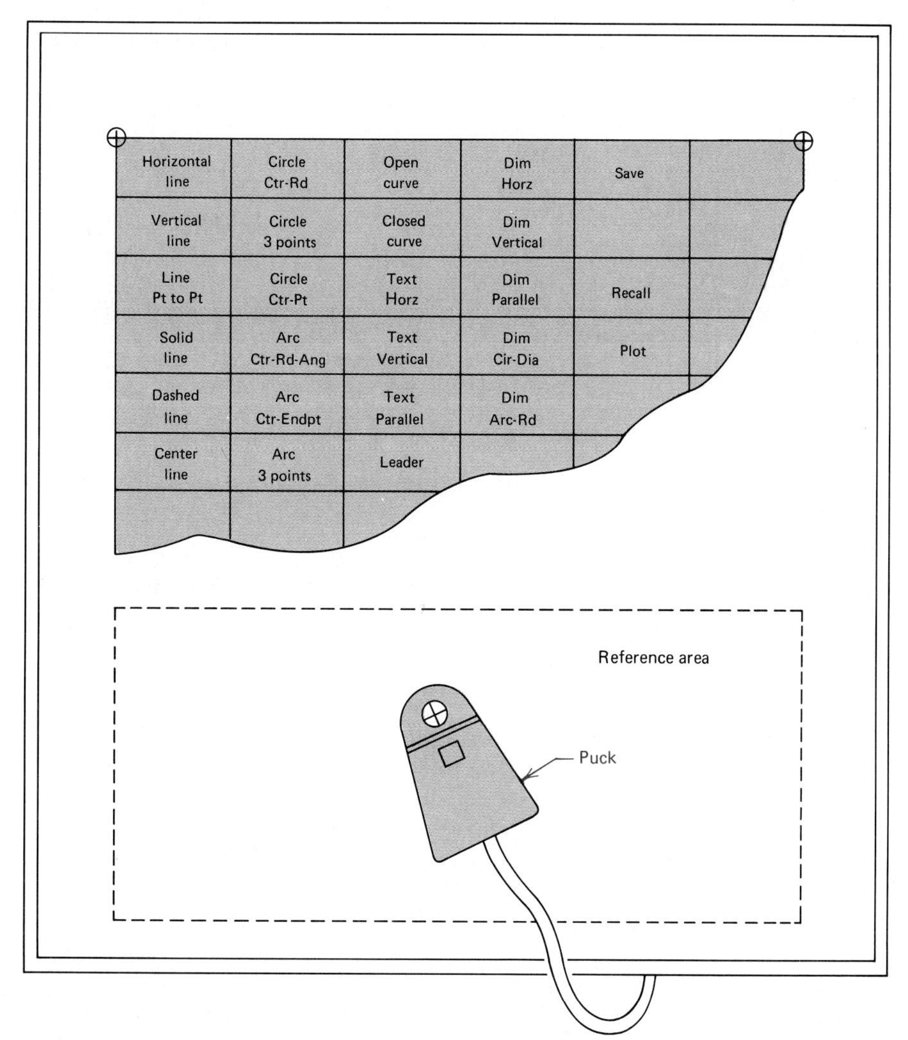

FIG. 16.16 A digitizing puck and 2-D drafting menu.

ject shown in Fig. 16.3. This is the object we previously modeled using more primitive commands. We will assume that when the puck is placed on the area of the tablet surrounded by the dashed line, the cursor follows the puck movements on the screen. Thus, the dashed box can be thought of as a representation of the screen surface. When the button on the puck is pressed, the position of the cursor on the screen is recorded. When the puck is in the menu region, a particular command is selected by placing the cross hairs of the puck over the desired menu item and pressing the button on the puck. For ease of reference, we will use the

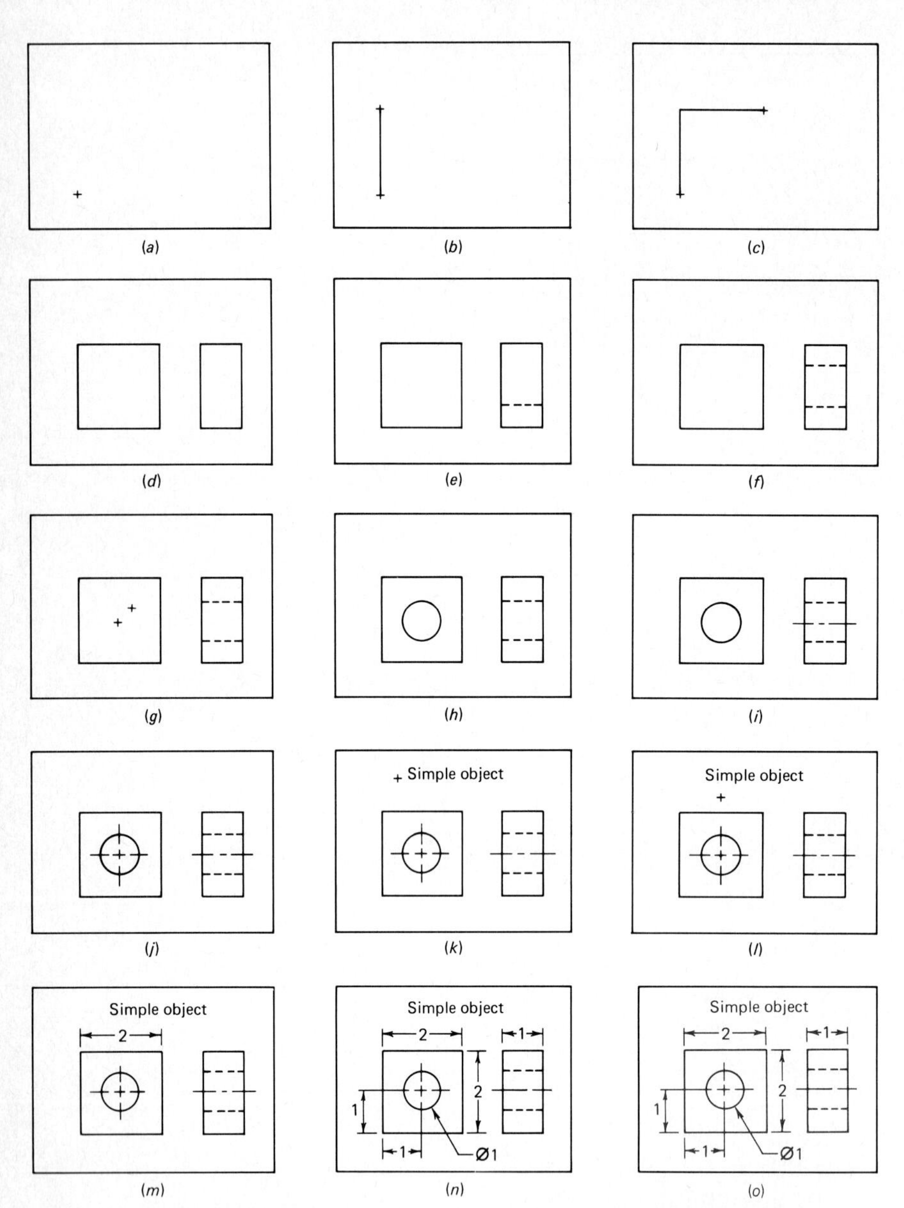

FIG. 16.17 A step-by-step procedure for producing a multiview drawing with a 2-D drafting package.

word digitize to mean the act of selecting a tablet position with the puck and pressing the button on the puck.

We will further assume that the package uses a rubberband type of line drawing. That is, while in the line-drawing mode, once one end of the line is digitized, a line from that point to the current cursor position is drawn until the other end of the line is digitized. Therefore, you get to effectively "try" a line to see how it looks before deciding to keep it.

With these preliminaries out of the way, let us write the steps you might follow to build the 2-D model using our hypothetical 2-D drafting package and specified workstation. Refer to Fig. 16.17 as you read through the steps to see what is being produced on the screen.

1. Load the 2-D drafting package from the floppy disk and perform all initialization actions to activate the package.

2. Place the menu parallel to the top of the tablet, as shown in Fig. 16.16. Digitize the upper corners of the menu to establish its position on the tablet.

3. Digitize two opposite corners of the area on the tablet you wish to have correspond to the screen area.

4. Begin the line work by digitizing SOLID LINE (often the default line type) followed by VERTICAL LINE. Place the puck in the cursor-control area of the tablet and digitize the lower end of the leftmost line (Fig. 16.17*a*). Move the puck to the upper end of the same line and digitize (Fig. 16.17*b*). With the VERTICAL LINE, only the vertical coordinate of the second point is used; the horizontal coordinate is set equal to the first point.

5. Select HORIZONTAL LINE (we assume that the SOLID LINE option will remain in effect until changed), establish the initial point, digitize, establish the final line point, and digitize to get Fig. 16.17*c*.

6. Finish the remaining two boundary lines of the front view in a similar fashion.

7. Create the profile view outline, following the previously described procedures (Fig. 16.17*d*).

8. Select DASHED LINE followed by HORIZONTAL LINE to begin to place the hidden lines in the right profile view. Locate and digitize each endpoint of the lower hidden line (Fig. 16.17*e*). Select HORIZONTAL LINE again and generate the other hidden line (Fig. 16.17*f*).

9. Select SOLID LINE and then CIRCLE CTR-POINT. Digitize the center of the circle in the front view and then digitize any point on the circle itself (Fig. 16.17*g*). The circle will then be drawn (Fig. 16.17*h*).

10. Select CENTERLINE followed by HORIZONTAL LINE in preparation for placing the center line in the profile view. Digitize each endpoint of the line (Fig. 16.17*i*). Select HORIZONTAL LINE again and place the horizontal center line in the front view. Select VERTICAL LINE and place the vertical center line (Fig. 16.17*j*).

11. Select TEXT HORIZONTAL and then locate and digitize the lower left-hand corner of the text string. Type in the string on the keyboard. Press RETURN to end the text entry (Fig. 16.17*k*).

12. Dimension the width of the part by selecting DIMENSION HORIZONTAL. Digitize the top corners of the front view. Select TEXT HORIZONTAL which while in the dimensioning mode expects a position for the text to be digitized (Fig. 16.17*l*). Type in the value of the dimension and press RETURN to end the dimensioning operation. The extension lines, dimension lines, arrowheads, and text will be drawn (Fig. 16.17*m*).

13. Repeat the procedure in step 12 with appropriate horizontal or vertical options for all required linear dimensions. Use DIMENSION CIRCLE-DIA, LEADER, and HORIZONTAL TEXT to dimension the hole (Fig. 16.17*n*).

14. Obtain a hard copy of the drawing by selecting PLOT, which could be a color plot as in Fig. 16.17*o*.

15. Save the drawing as a file by selecting SAVE and by typing in a filename.

Problems

16.1 For the object specified from Fig. 16.18, do the portion of the problem assigned by your instructor. Use proportions from the figure.

(a) Using the graphics language (MOVE, DRAW, and the like) in this chapter, write the series of steps needed to produce the necessary orthographic views of the object.

(b) Using the graphics language in this chapter, write the series of steps needed to produce a completely dimensioned detail drawing of the object.

(c) Using the graphics language appropriate for your local computer graphics system, write the series of steps required to produce the necessary orthographic views of the object. Obtain the program listings and plots as directed by your instructor.

(d) Using the graphics language appropriate for your local computer graphics system, write the series of steps required to create a completely dimensioned detail drawing of the object. Obtain program listings and plots as directed by your instructor.

16.2 For the object specified from Fig. 16.18, do the portion of the problem assigned by your instructor. Use proportions from the figure.

(a) Using the commands for the 2-D drafting package described in the text, write the series of steps needed to produce the necessary orthographic views of the object.

(b) Using the commands for the 2-D drafting package described in the text, write the series of steps needed to produce a completely dimensioned detail drawing of the object.

(c) Using the 2-D drafting package on your local system, produce the necessary orthographic views of the object. Plot the results as assigned by your instructor.

(d) Using the 2-D drafting package on your local system, create a completely dimensioned detail drawing of the object. Plot the results as assigned by your instructor.

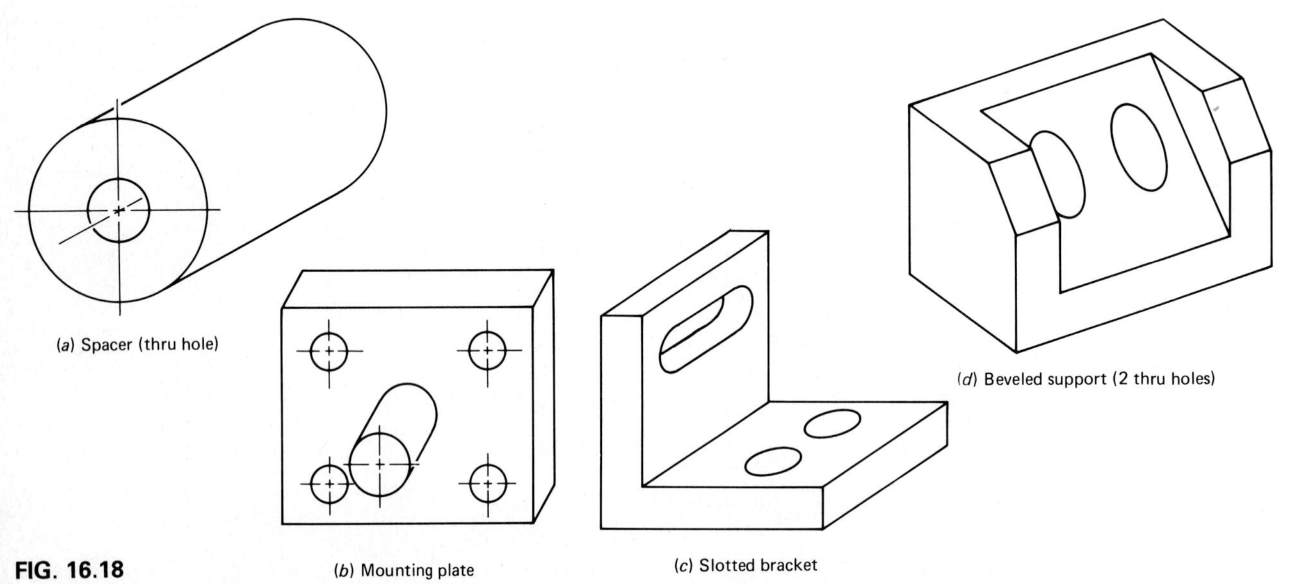

(*a*) Spacer (thru hole)

(*b*) Mounting plate

(*c*) Slotted bracket

(*d*) Beveled support (2 thru holes)

FIG. 16.18

chapter 17

3-D Computer Graphics

17.1 INTRODUCTION

Computer modeling of complex objects generally requires 3-D modeling techniques. Even objects as simple as a rectangular pyramid do not lend themselves to the 2-D methods described in Chap. 16 if the objective of the modeling is to create a common database from which drawings can be made, stresses can be analyzed, numerically controlled metal cutting tools can be driven, and so on. Even the 2½-D techniques are inadequate, since a pyramid does not have a constant cross section.

The geometric model is at the center of the common database that was discussed in Chap. 15. Without a complete and accurate geometric model, all the power of the common database cannot be used. In this chapter, we will discuss some of the 3-D modeling techniques, including the wire-frame model, surface models, and solid modeling.

The wire-frame model is the simplest of these models to create and has the associated benefits of little computer time expended, small amount of memory required, and minimal engineering time involved. The wire-frame model provides information about surface discontinuities but contains no information about the interior of the object or about most of the details of the surface. Even with these deficiencies, the wire-frame model is very useful, and a significant amount of this chapter will be devoted to it. Figure 17.1 shows a wire-frame model as seen on the display screen.

The wire-frame model is not adequate to describe the hood of an automobile, for example. This complex surface geometry must be represented by means of one of the surface modeling methods. Here the details of the surface can be accurately described. The surface model can become the input to a program that will analyze the stresses resulting from specified loads. It can also be used to drive an NC machine which produces a die set (or progressive dies) that can be used to manufacture the part. Figure 17.2 shows the use of patches to represent the surface of an object on a continuous basis.

Even though the surface model

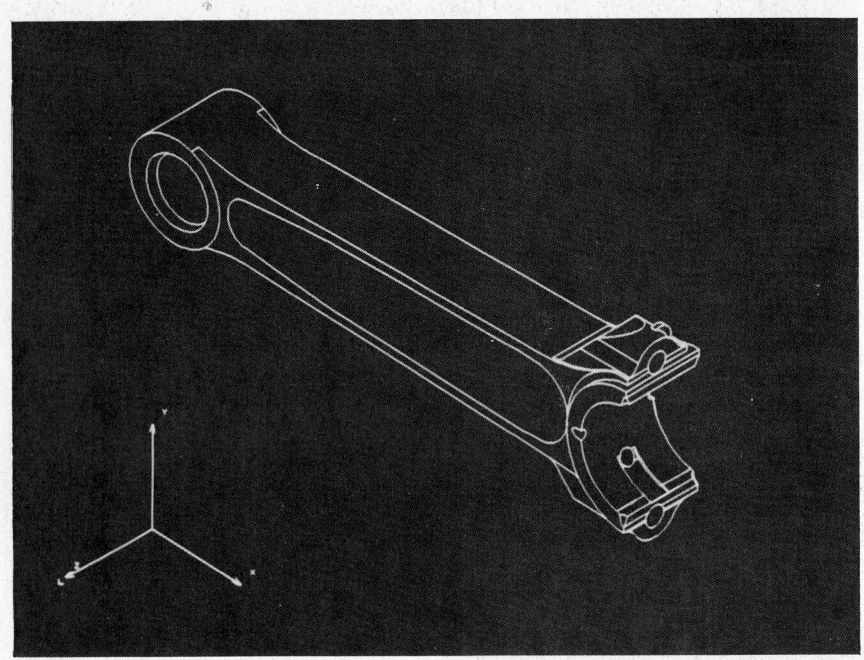

FIG. 17.1 Wire-frame pictorials of parts can be readily generated on the computer once the geometry of the part is defined. (*Courtesy of Control Data Corporation.*)

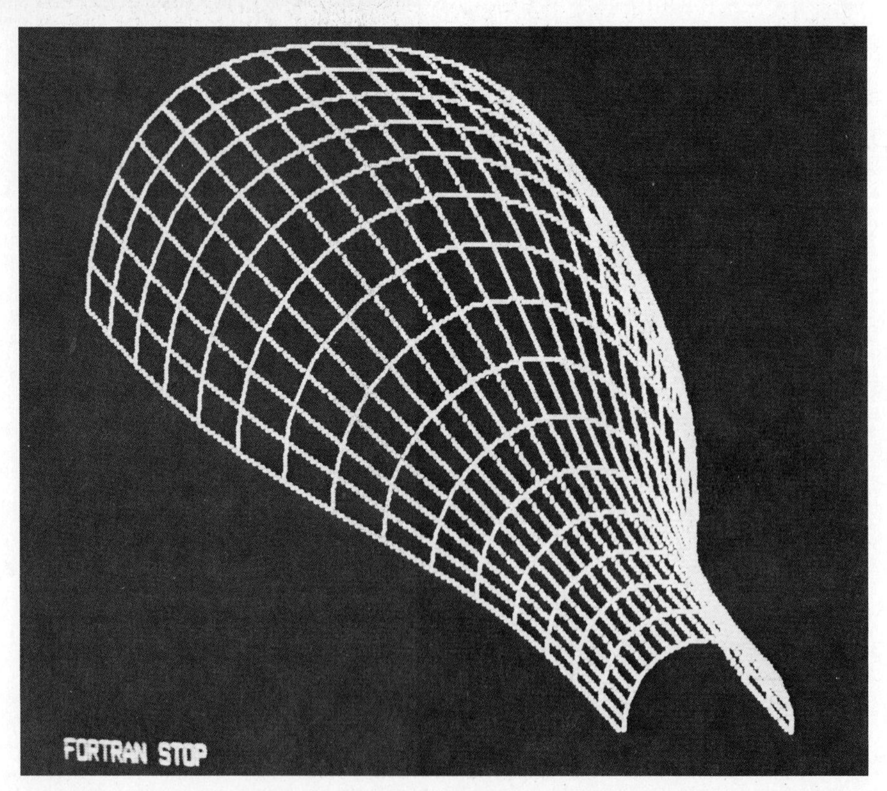

FIG. 17.2 Models of complex surfaces can be modeled using a patch technique.

FIG. 17.3 Solid modeling can produce realistic pictures of objects. (*Courtesy of Control Data Corporation.*)

provides an advantage over the wire-frame model at some cost of computer time and memory, it is not directly useful in describing the interior of a solid. The solid model approach generally combines primitives such as simple prisms, cylinders, spheres, and cones to build more complex shapes. With solid modeling, sectional views can be computer-generated. Also, important calculations involving mass, moment of inertia, and the like can be readily performed. A model created by means of solid modeling is shown in Fig. 17.3. An introduction to the concepts of surface and solid modeling is given later in this chapter.

As you study some of the concepts from 3-D computer graphics, keep in mind that we are trying to describe an object of three dimensions on a two-dimensional surface. In this case we are using the computer screen for the display medium instead of the drawing paper that was used in earlier parts of this book. We are using the computer as the method of producing the display of the objects rather than pencil, dividers, triangles, and the like. We are using vector and analytic geometry concepts as the theoretical basis for describing the geometry rather than the principles of orthographic projection and descriptive geometry. The goal is the same, but the tools are different.

17.2 FUNDAMENTALS OF 3-D COMPUTER GRAPHICS

A 3-D object can be represented in the computer mathematically by describing the coordinates of the various points on the object and defining how these points are connected. This method results in a wire-frame model. An alternative method is solids modeling, which allows section views and the like to be generated. Both methods will be discussed in this chapter, with emphasis on the wire-frame model.

We will introduce the mathematics fundamentals necessary to build wire-frame models and show how they may be displayed on the screen. We will also relate computer-generated views with those previously described in the descriptive geometry material in Chaps. 4 through 8.

17.2.1 Vector Concepts

We will use a traditional cartesian coordinate system as the basis for the analysis. For convenience, we will define a right-handed system of x, y, and z, as seen in Fig. 17.4. A right-handed system means that the x, y, and z axes are chosen so that when you point the fingers of your right hand in the direction of the positive x axis and curl them toward your palm in the direction of the positive y axis, your thumb will point toward the positive z axis.

The position of a point P in this coordinate system is given as (x, y, z) where x, y, and z are the distances along the x, y, and z axes, respectively, from the origin of the system.

An alternative way of describing the location of point P is to begin at the origin and first travel some distance along the x axis, then travel a distance parallel to the y axis, and finally move parallel to the z axis to point P. This process is depicted in Fig. 17.5.

For convenience in writing this sequence, we define a vector as a quantity having both magnitude and direction. This is shown graphically as a line terminating in an arrowhead, where the length of the line represents the magnitude of the vector and its orientation represents the direc-

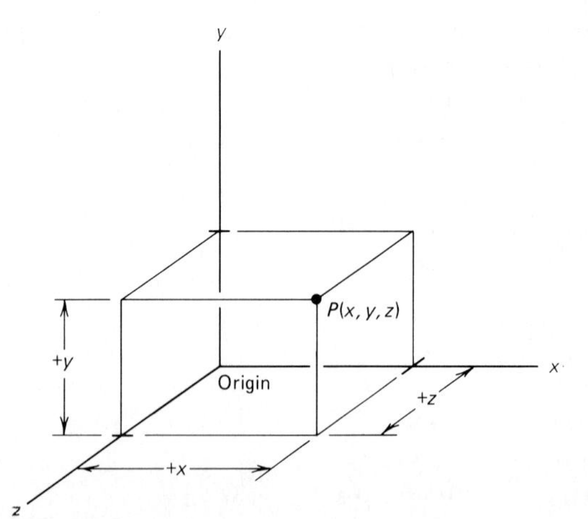

FIG. 17.4 Cartesian coordinate systems.

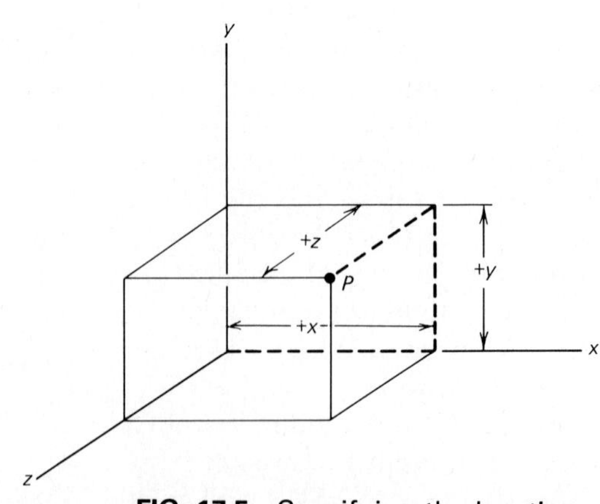

FIG. 17.5 Specifying the location of a point.

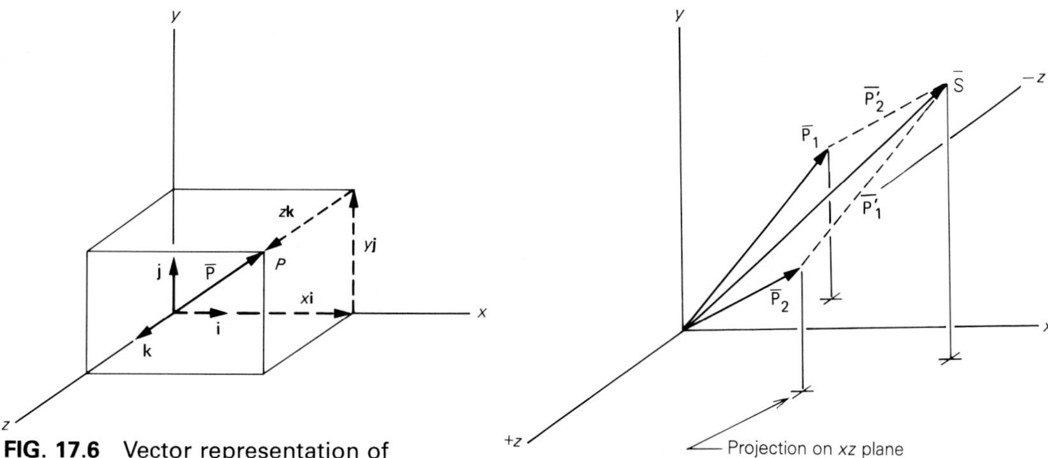

FIG. 17.6 Vector representation of a point in space.

FIG. 17.7 Vector addition.

tion. The end with the arrowhead is the terminal end (terminus), and the opposite end is the origin.

A vector having a length of 1 is called a unit vector. We will denote the unit vectors along the x, y, and z axes as $\mathbf{i}$, $\mathbf{j}$, and $\mathbf{k}$, respectively. Referring to Fig. 17.6, we can consider each of the three segments traveled from the origin to point P to be vectors. The vector along the x axis has a magnitude of x and a direction the same as $\mathbf{i}$; it can be written $x\mathbf{i}$. Likewise, we obtain $y\mathbf{j}$ and $z\mathbf{k}$ for the other segments. If P is thought of as the terminal end of a vector $\overline{\mathrm{P}}$ from the origin, we have

$$\overline{\mathrm{P}} = x\mathbf{i} + y\mathbf{j} + z\mathbf{k} \quad (17.1)$$

where + refers to vector addition.

Vectors can be added and subtracted in general. Subtraction can be considered to be addition where a negative coefficient exists.

In Fig. 17.7, vectors $\overline{\mathrm{P}}_1$ and $\overline{\mathrm{P}}_2$ are added to produce vector $\overline{\mathrm{S}}$. This is done graphically by translating $\overline{\mathrm{P}}_2$ parallel to itself such that its origin coincides with the terminus of $\overline{\mathrm{P}}_1$ (results in $\overline{\mathrm{P}}_2'$). The sum is then the vector from the origin to the terminus of $\overline{\mathrm{P}}_2'$. Mathematically this is written

$$\overline{\mathrm{S}} = \overline{\mathrm{P}}_1 + \overline{\mathrm{P}}_2 = (x_1\mathbf{i} + y_1\mathbf{j} + z_1\mathbf{k}) + (x_2\mathbf{i} + y_2\mathbf{j} + z_2\mathbf{k}) \quad (17.2)$$

where (x_1, y_1, z_1) and (x_2, y_2, z_2) are the coordinates of the terminal ends of $\overline{\mathrm{P}}_1$ and $\overline{\mathrm{P}}_2$, respectively. The sum $\overline{\mathrm{S}}$ is found by independently adding the coefficients of $\mathbf{i}$, $\mathbf{j}$, and $\mathbf{k}$ to give

$$\overline{\mathrm{S}} = (x_1 + x_2)\mathbf{i} + (y_1 + y_2)\mathbf{j} + (z_1 + z_2)\mathbf{k} \quad (17.3)$$

Therefore, the addition of vectors is not unlike algebraic addition as long as you work with each unit vector separately.

Multiplication of vectors is a somewhat different matter. There are two distinct vector multiplications. The first is the vector dot product (or scalar product) symbolized by $\overline{\mathrm{P}}_1 \bullet \overline{\mathrm{P}}_2$ that results in a scalar quantity. The second is the cross product (or vector product) $\overline{\mathrm{P}}_1 \times \overline{\mathrm{P}}_2$ that produces a vector that is normal to both $\overline{\mathrm{P}}_1$ and $\overline{\mathrm{P}}_2$.

The dot product may be evaluated as

$$\overline{\mathrm{P}}_1 \bullet \overline{\mathrm{P}}_2 = |\overline{\mathrm{P}}_1||\overline{\mathrm{P}}_2| \cos\theta \qquad 0 \le \theta \le \pi \quad (17.4)$$

where $|\overline{\mathrm{P}}|$ represents the magnitude of a vector and θ is the angle between $\overline{\mathrm{P}}_1$ and $\overline{\mathrm{P}}_2$. See Fig. 17.8. An alternative method of evaluation is

$$\overline{\mathrm{P}}_1 \bullet \overline{\mathrm{P}}_2 = x_1x_2 + y_1y_2 + z_1z_2 \quad (17.5)$$

which is generally more convenient to use. Note here that the dot product can be determined from Eq. (17.5), and then the angle between the vectors can be found from Eq. (17.4). This is a simple method for determining the angle between arbitrary vectors.

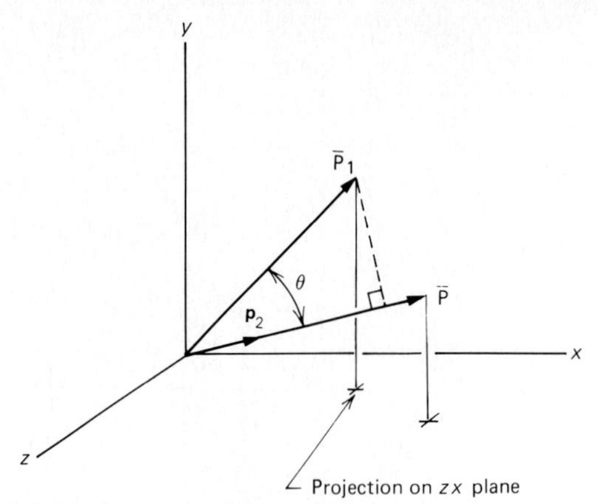

FIG. 17.8 Angle between vectors.

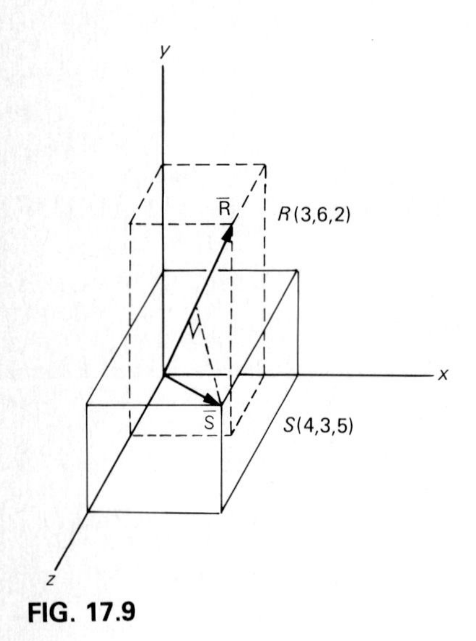

FIG. 17.9

A very useful property of the dot product in graphics applications is that the projection of vector $\bar{P}_1$ on vector $\bar{P}_2$ is given by $\bar{P}_1 \bullet \mathbf{p}_2$, where $\mathbf{p}_2$ is a unit vector in the direction of $\bar{P}_2$ (Fig. 17.8).

A unit vector along the direction of any general vector is given by that vector divided by its magnitude. Therefore,

$$\mathbf{p}_2 = \frac{\bar{P}_2}{|\bar{P}_2|} = \frac{x_2\mathbf{i} + y_2\mathbf{j} + z_2\mathbf{k}}{(x_2^2 + y_2^2 + z_2^2)^{1/2}} \quad (17.6)$$

We can then write

$$\bar{P}_1 \bullet \mathbf{p}_2 = (x_1\mathbf{i} + y_1\mathbf{j} + z_1\mathbf{k}) \bullet \frac{(x_2\mathbf{i} + y_2\mathbf{j} + z_2\mathbf{k})}{(x_2^2 + y_2^2 + z_2^2)^{1/2}}$$
$$= \frac{x_1x_2 + y_1y_2 + z_1z_2}{(x_2^2 + y_2^2 + z_2^2)^{1/2}} \quad (17.7)$$

It follows that the x component of $\bar{P}_1$ is simply $\bar{P}_1 \bullet \mathbf{i}$ and that $\bar{P}_1 \bullet \mathbf{j}$ and $\bar{P}_1 \bullet \mathbf{k}$ are the y and z components, respectively. It is also true that if the dot product of two vectors is zero, the vectors are perpendicular [from Eq. (17.4), $\cos\theta = 0$ and thus $\theta = \pi/2$].

Example 17.1
For vectors $\bar{R} = 3\mathbf{i} + 6\mathbf{j} + 2\mathbf{k}$ and $\bar{S} = 4\mathbf{i} + 3\mathbf{j} + 5\mathbf{k}$, determine
(*a*) The dot product $\bar{R} \bullet \bar{S}$
(*b*) The angle between the vectors
(*c*) The component of $\bar{S}$ in the direction of $\bar{R}$

Solution
Refer to Fig. 17.9.

(*a*) From Eq. (17.5),

$$\bar{R} \bullet \bar{S} = (3\mathbf{i} + 6\mathbf{j} + 2\mathbf{k}) \bullet (4\mathbf{i} + 3\mathbf{j} + 5\mathbf{k})$$
$$= 3(4) + 6(3) + 2(5)$$
$$= 40$$

(*b*) Then from Eq. (17.4),

$$\bar{R} \bullet \bar{S} = |\bar{R}||\bar{S}| \cos\theta = 40$$
$$|\bar{R}| = (3^2 + 6^2 + 2^2)^{1/2} = (49)^{1/2} = 7$$
$$|\bar{S}| = (4^2 + 3^2 + 5^2)^{1/2}$$
$$= (50)^{1/2}$$
$$= 7.071$$

so

$$\cos\theta = \frac{40}{7(7.071)} = 0.8081$$
$$\theta = \cos^{-1}(0.8081)$$
$$= 36.09°$$

(*c*) To obtain the component of $\bar{S}$ in the direction of $\bar{R}$, we use Eq. (17.7), which gives $\bar{S} \bullet \mathbf{r}$.

$$\bar{S} \bullet \mathbf{r} = (4\mathbf{i} + 3\mathbf{j} + 5\mathbf{k}) \bullet \frac{(3\mathbf{i} + 6\mathbf{j} + 2\mathbf{k})}{|\bar{R}|}$$
$$= \frac{4(3) + 3(6) + 5(2)}{7}$$
$$= \frac{40}{7}$$
$$= 5.714$$

The cross product of two vectors $\bar{P}_1 \times \bar{P}_2$ is a vector that is perpendicular to both $\bar{P}_1$ and $\bar{P}_2$ or, expressed differently, perpendicular to a plane defined by $\overline{P}_1$ and $\overline{P}_2$. The direction of the cross product is found from the right-hand rule. To visualize this, first place the origin of $\overline{P}_2$ at the terminal end of $\overline{P}_1$. Extend your fingers along $\overline{P}_1$ and curl them in the direction of $\overline{P}_2$. Your thumb then points in the direction of the cross product. The cross product is easily represented by the use of a determinant as

$$\overline{P}_1 \times \overline{P}_2 = \begin{vmatrix} \mathbf{i} & \mathbf{j} & \mathbf{k} \\ x_1 & y_1 & z_1 \\ x_2 & y_2 & z_2 \end{vmatrix} \quad (17.8)$$

FIG. 17.10 Cross product.

The determinant is then evaluated as

$$\overline{P}_1 \times \overline{P}_2 = \mathbf{i}\begin{vmatrix} y_1 z_1 \\ y_2 z_2 \end{vmatrix} - \mathbf{j}\begin{vmatrix} x_1 z_1 \\ x_2 z_2 \end{vmatrix} + \mathbf{k}\begin{vmatrix} x_1 y_1 \\ x_2 y_2 \end{vmatrix}$$
$$= (y_1 z_2 - y_2 z_1)\mathbf{i} - (x_1 z_2 - x_2 z_1)\mathbf{j} + (x_1 y_2 - x_2 y_1)\mathbf{k} \quad (17.9)$$

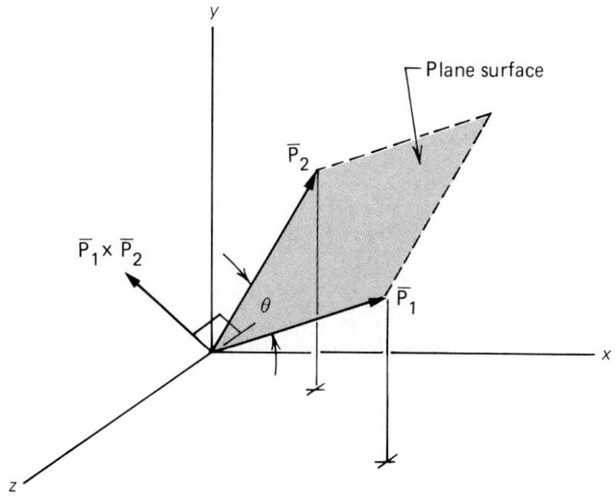

Figure 17.10 shows the result graphically. Note that $\overline{P}_2 \times \overline{P}_1 = -\overline{P}_1 \times \overline{P}_2$. Thus, if you reverse the order of the cross product, you get a vector of the same magnitude but in the opposite direction. Use the right-hand rule to convince yourself of the direction.

The magnitude of $\overline{P}_1 \times \overline{P}_2$ is

$$|\overline{P}_1 \times \overline{P}_2| = |\overline{P}_1||\overline{P}_2| \sin \theta \quad (17.10)$$

where θ is the angle between $\overline{P}_1$ and $\overline{P}_2$. This magnitude can be physically interpreted as the area of the parallelogram with sides $\overline{P}_1$ and $\overline{P}_2$.

Example 17.2

For the vectors in the previous example, $\overline{R} = 3\mathbf{i} + 6\mathbf{j} + 2\mathbf{k}$ and $\overline{S} = 4\mathbf{i} + 3\mathbf{i} + 5\mathbf{k}$, calculate (see Fig. 17.9)

(*a*) A vector normal to the plane formed by $\overline{R}$ and $\overline{S}$

(*b*) The area of the parallelogram with sides $\overline{R}$ and $\overline{S}$

(*c*) The angle between $\overline{R}$ and $\overline{S}$

Solution

(*a*) The cross product of $\overline{R}$ and $\overline{S}$ gives a vector normal to both $\overline{R}$ and $\overline{S}$ (therefore normal to the desired plane). From Eq. (17.9),

$$\overline{R} \times \overline{S} = \begin{vmatrix} \mathbf{i} & \mathbf{j} & \mathbf{k} \\ 3 & 6 & 2 \\ 4 & 3 & 5 \end{vmatrix}$$
$$= \mathbf{i}[(6)(5) - (3)(2)] - \mathbf{j}[(3)(5) - (4)(2)] + \mathbf{k}[(3)(3) - (4)(6)]$$
$$= 24\mathbf{i} - 7\mathbf{j} - 15\mathbf{k}$$

(*b*) The area of the parallelogram is represented by the magnitude $|\overline{R} \times \overline{S}|$.

$$|\overline{R} \times \overline{S}| = (24^2 + 7^2 + 15^2)^{1/2} = (850)^{1/2} = 29.15 \text{ square units}$$

(*c*) Recalling that the magnitude of $\overline{R} \times \overline{S}$ relates to the angle between $\overline{R}$ and $\overline{S}$ from Eq. (17.10), we have

$$|\overline{R} \times \overline{S}| = (850)^{1/2} = |\overline{R}||\overline{S}| \sin\theta = (3^2 + 6^2 + 2^2)^{1/2} (4^2 + 3^2 + 5^2)^{1/2} \sin\theta$$

or

$$\sin\theta = \frac{(850)^{1/2}}{(49)^{1/2}(50)^{1/2}} = \left[\frac{850}{(49)(50)}\right]^{1/2} = 0.5890$$

Therefore,

$$\theta = \sin^{-1}(0.5890) = 36.09°$$

which is the same answer as previously obtained, as it should be.

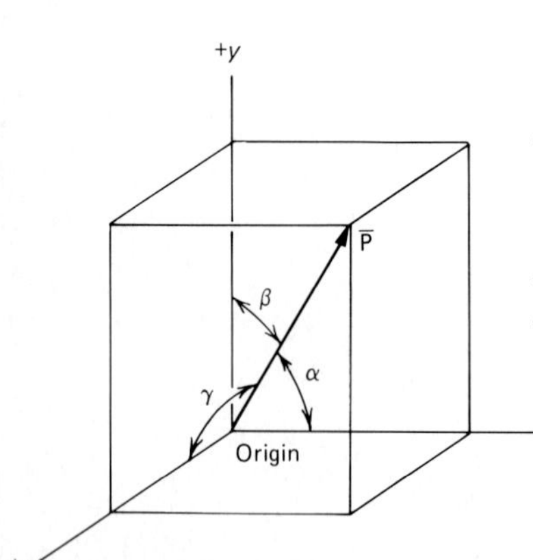

FIG. 17.11 Direction cosines.

The angle that vector $\overline{P}$ makes with each of the coordinate axes is of general interest. The angles α, β, and γ are shown in Fig. 17.11. They can be found by the following analysis.

For the vector $\overline{P}$ given by

$$\overline{P} = x\mathbf{i} + y\mathbf{j} + z\mathbf{k} \qquad (17.11)$$

a unit vector $\mathbf{p}$ along $\overline{P}$ is given by

$$\mathbf{p} = \frac{\overline{P}}{|\overline{P}|} = \frac{x\mathbf{i} + y\mathbf{j} + z\mathbf{k}}{|\overline{P}|} = \frac{x}{|P|}\mathbf{i} + \frac{y}{|P|}\mathbf{y} + \frac{z}{|P|}\mathbf{k} \qquad (17.12)$$

Then note that the angle between $\overline{P}$ and the x axis, denoted α, is given by

$$\cos\alpha = \frac{x \text{ component of } \overline{P}}{|\overline{P}|} = \frac{x}{|P|}$$

Similar relationships hold for β and γ. Thus

$$\mathbf{p} = \cos\alpha\mathbf{i} + \cos\beta\mathbf{j} + \cos\gamma\mathbf{k} \qquad (17.13)$$

If $\overline{P}$ is interpreted as a line from the origin to a point P, cos α, cos β, and cos γ are called the direction cosines of the line.

Example 17.3

Determine the angle that vector $\overline{A} = 10\mathbf{i} + 5\mathbf{j} + 12\mathbf{k}$ makes with each coordinate axis.

Solution

We must first find a unit vector along $\overline{A}$ as

$$\mathbf{a} = \frac{10\mathbf{i} + 5\mathbf{j} + 12\mathbf{k}}{[10^2 + 5^2 + 12^2]^{1/2}} = \frac{10}{16.40}\mathbf{i} + \frac{5}{16.40}\mathbf{j} + \frac{12}{16.40}\mathbf{k} = 0.6098\mathbf{i} + 0.3049\mathbf{j} + 0.7317\mathbf{k}$$

From Eq. (17.13),

$$\mathbf{a} = \cos\alpha\mathbf{i} + \cos\beta\mathbf{j} + \cos\gamma\mathbf{k}$$

Therefore

$$\cos\alpha = 0.6098$$
$$\cos\beta = 0.3049$$
$$\cos\gamma = 0.7317$$

and

$$\alpha = \cos^{-1}(0.6098) = 52.42°$$
$$\beta = \cos^{-1}(0.3049) = 72.25°$$
$$\gamma = \cos^{-1}(0.7317) = 42.97°$$

This may be interpreted as follows. The angle between $\overline{A}$ and the x axis is 52.42°, the angle between $\overline{A}$ and the y axis is 72.25°, and the angle between $\overline{A}$ and the z axis is 42.97°.

A summary of useful vector relationships is presented below. In these equations, **i**, **j**, and **k** are the unit vectors in the x, y, and z directions, respectively; $\overline{P}_i = x_i\mathbf{i} + y_i\mathbf{j} + z_i\mathbf{k}$; c and k are scalar constants; and $\mathbf{p}_i$ is a unit vector in the direction of $\overline{P}_i$.

$$\overline{P}_1 + \overline{P}_2 = \overline{P}_2 + \overline{P}_1 = (x_1 + x_2)\mathbf{i} + (y_1 + y_2)\mathbf{j} + (z_1 + z_2)\mathbf{k} \quad (17.14)$$

$$-\overline{P}_1 = -x_1\mathbf{i} - y_1\mathbf{j} - z_1\mathbf{k} \quad (17.15)$$

$$k\overline{P}_1 = kx_1\mathbf{i} + ky_1\mathbf{j} + kz_1\mathbf{k} \quad (17.16)$$

$$|\overline{P}_1| = (x_1^2 + y_1^2 + z_1^2)^{1/2} \quad (17.17)$$

$$\overline{P}_1 + (\overline{P}_2 + \overline{P}_3) = (\overline{P}_1 + \overline{P}_2) + \overline{P}_3 \quad (17.18)$$

$$c(k\overline{P}_1) = ck\overline{P}_1 \quad (17.19)$$

$$(c + k)\overline{P}_1 = c\overline{P}_1 + k\overline{P}_1 \quad (17.20)$$

$$k(\overline{P}_1 + \overline{P}_2) = k\overline{P}_1 + k\overline{P}_2 \quad (17.21)$$

$$\overline{P}_1 \cdot \overline{P}_2 = \overline{P}_2 \cdot \overline{P}_1 = |\overline{P}_1||\overline{P}_2| \cos\theta = x_1x_2 + y_1y_2 + z_1z_2 \quad (17.22)$$

$$\overline{P}_1 \cdot (\overline{P}_2 + \overline{P}_3) = \overline{P}_1 \cdot \overline{P}_2 + \overline{P}_1 \cdot \overline{P}_3 \quad (17.23)$$

$$(k\overline{P}_1) \cdot \overline{P}_2 = \overline{P}_1 \cdot (k\overline{P}_2) = k(\overline{P}_1 \cdot \overline{P}_2) \quad (17.24)$$

$$\overline{P}_1 \times \overline{P}_2 = \begin{vmatrix} \mathbf{i} & \mathbf{j} & \mathbf{k} \\ x_1 & y_1 & z_1 \\ x_2 & y_2 & z_2 \end{vmatrix} \quad (17.25)$$

$$\overline{P}_1 \times \overline{P}_2 = -\overline{P}_2 \times \overline{P}_1 \quad (17.26)$$

$$\overline{P}_1 \times (\overline{P}_2 + \overline{P}_3) = \overline{P}_1 \times \overline{P}_2 + \overline{P}_1 \times \overline{P}_3 \quad (17.27)$$

$$(k\overline{P}_1) \times \overline{P}_2 = \overline{P}_1 \times (k\overline{P}_2) = k(\overline{P}_1 \times \overline{P}_2) \quad (17.28)$$

$$\mathbf{i} \times \mathbf{j} = \mathbf{k} \quad (17.29)$$
$$\mathbf{j} \times \mathbf{k} = \mathbf{i} \quad (17.30)$$
$$\mathbf{k} \times \mathbf{i} = \mathbf{j} \quad (17.31)$$

$$\overline{P}_1 \cdot (\overline{P}_2 \times \overline{P}_3) = \begin{vmatrix} x_1 & y_1 & z_1 \\ x_2 & y_2 & z_2 \\ x_3 & y_3 & z_3 \end{vmatrix} \quad (17.32)$$

17.2.2 Analytic Geometry Concepts

The vector concepts we have just introduced can be used to derive the equations of lines and planes in space. In particular, we will show how to obtain the equation of a line passing through a given point in space along a certain direction and that of a line passing through two given points. We also will look at a plane through a given point perpendicular to a known vector as well as at a plane through three points in space.

The equation of a line through a given point along a known direction will be derived in parametric form. The point is called P_1, and the direction is represented as a unit vector **v**. In Fig. 17.12, a position vector $\overline{P}_1$ to

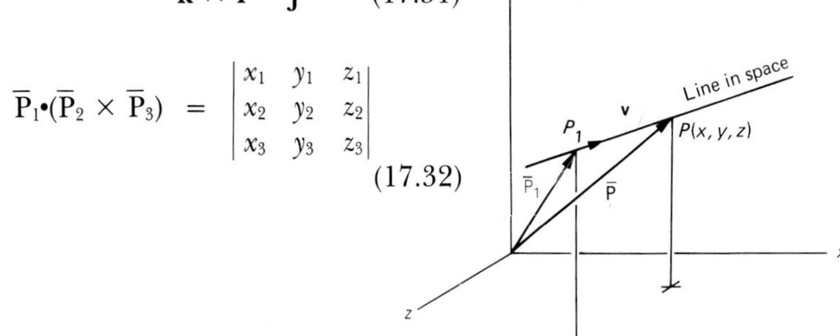

FIG. 17.12 Line through a given point in specified direction (**v**).

point P_1 and a vector $\overline{P}$ to a general point on the line P are established. The relative position vector from P_1 to P is then $m\mathbf{v}$, where m is a scalar parameter. Then

$$\begin{aligned}\overline{P} &= \overline{P}_1 + m\mathbf{v} \\ &= x\mathbf{i} + y\mathbf{j} + z\mathbf{k} \\ &= x_1\mathbf{i} + y_1\mathbf{j} + z_1\mathbf{k} \\ &\qquad + m(v_x\mathbf{i} + v_y\mathbf{j} + v_z\mathbf{k}\end{aligned} \tag{17.33}$$

where x, y, z = coordinates of a general position on the line

x_1, y_1, z_1 = coordinates of a known point on the line

v_x, v_y, v_z = x, y, z coefficients, respectively, of the known unit vector

Example 17.4

Find in parametric form the equation of the line that passes through point (1, 2, 4) in the direction of

$$\frac{1}{\sqrt{3}}\mathbf{i} + \frac{1}{\sqrt{3}}\mathbf{j} + \frac{1}{\sqrt{3}}\mathbf{k}$$

Solution

Equation (17.33) gives

$$\begin{aligned}\overline{P} &= x\mathbf{i} + y\mathbf{j} + z\mathbf{k} \\ &= (1)\mathbf{i} + (2)\mathbf{j} + (4)\mathbf{k} \\ &\qquad + m\left(\frac{\mathbf{i}}{\sqrt{3}} + \frac{\mathbf{j}}{\sqrt{3}} + \frac{\mathbf{k}}{\sqrt{3}}\right)\end{aligned}$$

Then writing in component form yields

$$x = 1 + \frac{m}{\sqrt{3}}$$
$$y = 2 + \frac{m}{\sqrt{3}}$$
$$z = 4 + \frac{m}{\sqrt{3}}$$

Varying parameter m gives points on the line. When $m = 0$, $x = 1$, $y = 2$, $z = 4$, which is the known point. When $m = -\sqrt{3}$, $x = 0$, which is the yz plane. For the same value of m,

$$y = 2 + \frac{-\sqrt{3}}{\sqrt{3}} = 1$$

and

$$z = 4 + \frac{-\sqrt{3}}{\sqrt{3}} = 3$$

Thus (0, 1, 3) is the point where the line passes through the yz plane.

If the direction of the line is not known but two points on the line are known, we can still find the equation of the line in parametric form. In Fig. 17.13, P_1 and P_2 are the known points on the line and P is a general point on the line. Observe that

$$\begin{aligned}\overline{P} &= \overline{P}_1 + \overline{P_1P} \\ &= \overline{P}_1 + m\,(\overline{P_1P_2}) \\ &= \overline{P}_1 + m\,(\overline{P}_2 - \overline{P}_1) \\ &= \overline{P}_1 - m\overline{P}_1 + m\overline{P}_2 \\ &= (1 - m)\overline{P}_1 + m\overline{P}_2\end{aligned} \tag{17.34}$$

which gives the equation of the line in parametric form with m as the parameter.

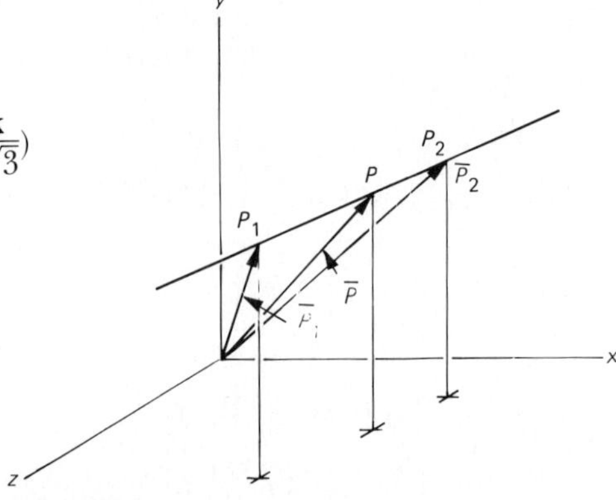

FIG. 17.13 Line through two specified points.

Example 17.5
Find the equation of the line passing through (3, 4, 3) and (−1, 4, 6).

Solution
From Eq. (17.34) we have

$$\overline{\mathrm{P}} = (1 - m)\overline{\mathrm{P}}_1 + m\overline{\mathrm{P}}_2$$

or

$$x\mathbf{i} + y\mathbf{j} + z\mathbf{k} = (1 - m)(3\mathbf{i} + 4\mathbf{j} + 3\mathbf{k}) + m(-\mathbf{i} + 4\mathbf{j} + 6\mathbf{k})$$

Therefore

$$\begin{aligned} x &= (1 - m)(3) - m \\ &= 3 - 3m - m \\ &= 3 - 4m \\ y &= (1 - m)(4) + 4m \\ &= 4 - 4m + 4m \\ &= 4 \\ z &= (1 - m)(3) + 6m \\ &= 3 - 3m + 6m \\ &= 3 + 3m \end{aligned}$$

Note that y = constant, which means that the line is parallel to the xz plane. If the xz plane is horizontal, the line is a horizontal line.

Next we will consider the equation of a plane that passes through a given point in space and is normal to a given vector. If P represents a general point on the plane and P_1 is a specified point through which the plane must pass, the vector $\overline{\mathrm{P}_1\mathrm{P}} = \overline{\mathrm{P}} - \overline{\mathrm{P}}_1$ must lie on the plane (see Fig. 17.14). If $\mathbf{n}$ is a unit vector normal to the desired plane, it must be normal to $\overline{\mathrm{P}} - \overline{\mathrm{P}}_1$. Thus, the dot product of $\mathbf{n}$ and $\overline{\mathrm{P}} - \overline{\mathrm{P}}_1$ must be zero. We have

$$\mathbf{n} \cdot (\overline{\mathrm{P}} - \overline{\mathrm{P}}_1) = 0 \quad (17.35)$$

or

$$\mathbf{n} \cdot \overline{\mathrm{P}} = \mathbf{n} \cdot \overline{\mathrm{P}}_1$$

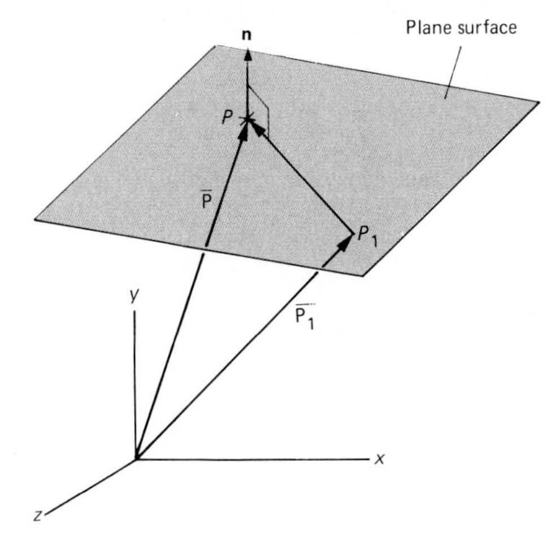

FIG. 17.14 Plane through a point and normal to a vector.

which is the equation of the required plane in vector form. Since $\mathbf{n}$ appears on both sides of the equation, and recalling that a unit vector is derived from a general vector by dividing by the magnitude of that vector, any normal vector can be substituted for the unit vector and Eq. (17.35) will still be valid.

Example 17.6
Derive the equation of a plane through point (1, 3, 5) such that the plane is normal to vector $\mathbf{i} + 2\mathbf{j} + \mathbf{k}$.

Solution

The desired unit vector

$$\mathbf{n} = \frac{\mathbf{i} + 2\mathbf{j} + \mathbf{k}}{\sqrt{6}}$$

Then from Eq. (17.35), we see that

$$\frac{1}{\sqrt{6}}(\mathbf{i} + 2\mathbf{j} + \mathbf{k}) \cdot (x\mathbf{i} + y\mathbf{j} + z\mathbf{k}) = \frac{1}{\sqrt{6}}(\mathbf{i} + 2\mathbf{j} + \mathbf{k}) \cdot (\mathbf{i} + 3\mathbf{j} + 5\mathbf{k})$$

Thus

$$x + 2y + z = 1 + 6 + 5$$

or

$$x + 2y + z = 12$$

Instead of knowing the vector direction normal to a plane, we may know three points on the plane. Assuming that these three points are all different and are not colinear (lying on the same straight line), we can find the equation of a unique plane.

From Fig. 17.15, observe that since $\overline{P_1P_2}$ and $\overline{P_2P_3}$ are vectors that lie on the plane, their cross product will produce a vector that is normal to the plane. If we then take another vector in the plane, for example, $\overline{P_1P}$, and produce a dot product with the previously described normal vector, we should get a result of zero (normal vector and vector lying on the plane are perpendicular, resulting in a zero dot product). Thus

$$\overline{P_1P} \cdot (\overline{P_1P_2} \times \overline{P_2P_3}) = 0 \quad (17.36)$$

or

$$(\overline{P} - \overline{P}_1) \cdot [(\overline{P}_2 - \overline{P}_1) \times (\overline{P}_3 - \overline{P}_2)] = 0$$

Written in determinant form, Eq. (17.36) becomes [see Eq. (17.32)]

$$\begin{vmatrix} x - x_1 & y - y_1 & z - z_1 \\ x_2 - x_1 & y_2 - y_1 & z_2 - z_1 \\ x_3 - x_2 & y_3 - y_2 & z_3 - z_2 \end{vmatrix} = 0 \quad (17.37)$$

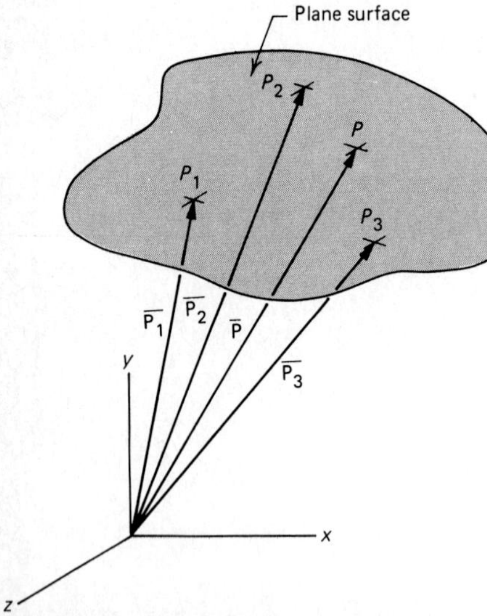

Example 17.7

Find the equation of the plane shown in Fig. 17.16.

Solution

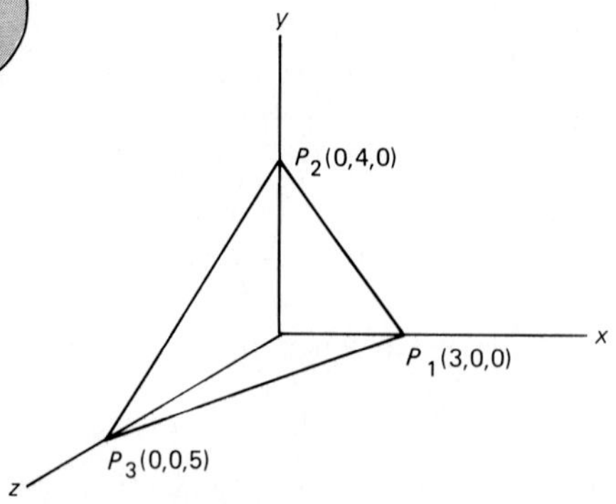

Three points on the plane are

$$(x_1, y_1, z_1) = (3, 0, 0)$$
$$(x_2, y_2, z_2) = (0, 4, 0)$$
$$(x_3, y_3, z_3) = (0, 0, 5)$$

Using Eq. (17.37) we have

$$\begin{vmatrix} x - 3 & y - 0 & z - 0 \\ 0 - 3 & 4 - 0 & 0 - 0 \\ 0 - 0 & 0 - 4 & 5 - 0 \end{vmatrix} = \begin{vmatrix} x-3 & y & z \\ -3 & 4 & 0 \\ 0 & -4 & 5 \end{vmatrix} = 0$$

Evaluation yields

$$(x - 3)(20 - 0) - y(-15 - 0) + z(12 - 0) = 0$$

or

$$20x - 60 + 15y + 12z = 0$$

or

$$20x + 15y + 12z = 60$$

FIG. 17.15 Specifying a plane in terms of three points on the plane.

FIG. 17.16

17.2.3 Creation of a 3-D Wire-Frame Model

We will describe here the data necessary for a wire-frame picture of an object to be drawn by a computer. Essentially all that is required is that a series of points be described between which lines are to be drawn. We will restrict our discussion here to straight lines between the points, although it should be clear that other types of curves can be generated to connect points (recall, however, that a curved line is obtained by a series of short straight-line segments on the display screen). Some techniques of viewing the model will be covered in Sec. 17.2.4.

The computer program that is to display a wire-frame model must at a minimum have a series of points and instructions that show which points must connect to which other points. Let us look at a simple polyhedron, in this case a triangular pyramid, as an example. Figure 17.17 shows this object placed in a cartesian coordinate system. To simplify our considerations, we have placed the object so that all coordinate values are positive.

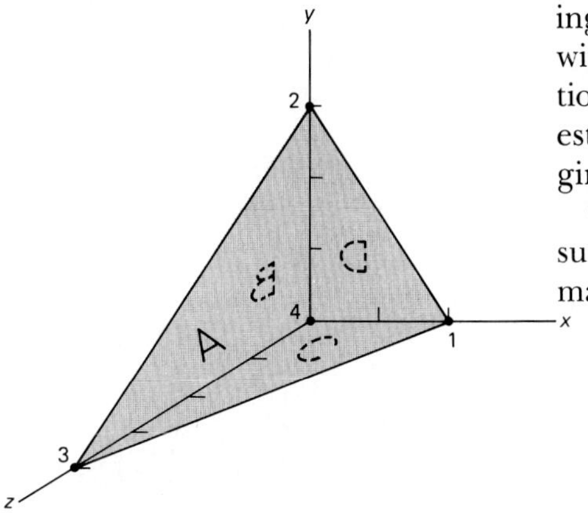

The corner points of this body will be referred to as nodes. For this pyramid, four such nodes are required. The coordinate values for these nodes are

NODE	
1	(2, 0, 0)
2	(0, 3, 0)
3	(0, 0, 4)
4	(0, 0, 0)

The object also has four planar surfaces marked *A*, *B*, *C*, and *D* on the figure. Surfaces which are not visible as the object is drawn are marked with dashed line symbols. These planes can be used to define how points should be connected by straight lines where the lines represent the edges of the object. We will want to consider the removal of hidden lines later, and because of this we must establish a consistent way of defining each plane. Specifically for a wire-frame drawing, this convention is unnecessary, but we must establish a pattern from the beginning.

Our standard for defining each surface will be to list the nodes making up each surface. This will

FIG. 17.17 Wire-frame model of a triangular pyramid.

be done in a counterclockwise sense when viewing the object from the outside. For example, surface *A* can be defined as being made up of nodes 1-2-3. Assuming a closed surface, node 3 must connect to node 1. In similar fashion, surface *B* is made up of 2-4-3; surface *C*, 1-3-4; and surface D, 1-4-2.

The software available for each local system will determine the specific requirements for data input for a wire-frame model. The discussion here cannot hope to cover all possibilities but is intended to demonstrate the type of information required by the computer and a simple way of supplying that information.

Example 17.8

Determine the necessary coordinates of the nodes and the definitions of the surfaces for a wire-frame model of the object shown in Fig. 17.18.

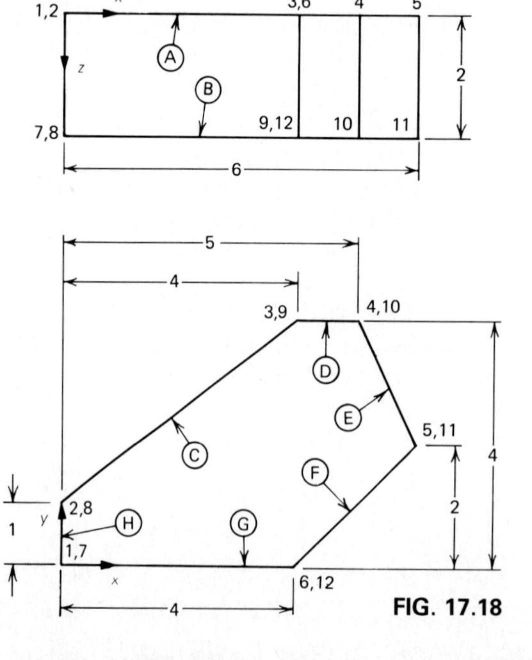

FIG. 17.18

Solution

Select an origin for a cartesian coordinate system such that the object nodes have zero or positive coordinate values. Then identify the nodes by numbers in the two views. This requires careful visualization of the object. The coordinates of each node can now be written as follows:

NODE	
1	0, 0, 0
2	0, 1, 0
3	4, 4, 0
4	5, 4, 0
5	6, 2, 0
6	4, 0, 0
7	0, 0, 2
8	0, 1, 2
9	4, 4, 2
10	5, 4, 2
11	6, 2, 2
12	4, 0, 2

Each planar surface must be identified on the object. We have chosen to use letter designations (see Fig. 17.18). Using the counterclockwise convention, the planes are as follows:

PLANE	
A	1-2-3-4-5-6
B	7-12-11-10-9-8
C	8-9-3-2
D	9-10-4-3
E	10-11-5-4
F	12-6-5-11
G	7-1-6-12
H	1-7-8-2

17.2.4 Viewing a 3-D Wire-Frame Mode

We have defined a wire-frame model of an object in a cartesian coordinate system. We can now develop a method for viewing the model on a cathode-ray tube (CRT) that is similar to the orthographic projection methods we learned earlier. If we imagine viewing the model by looking orthographically (projection rays parallel as though the viewer were a great distance from the object) along the z axis toward the origin, we will see the view of an object as depicted in Fig. 17.19. The resulting image on the screen is what we have called a front view. Note that this image is simply a plot of the xy coordinates of each node and the connecting lines. As in the case of orthographic projection, the coordinate normal to the view (z in this case) cannot be seen. We can use this idea to devise a technique for viewing the model from any point in space that we choose.

Imagine that we want to view an object from a specific point in space. We will refer to this point as a viewpoint. It has coordinates of x_{vp}, y_{vp}, z_{vp} in the coordinate system in which the model of the object is defined. For ease of use, we will call this coordinate system the object system. The object system is fixed to the object. Figure 17.20 shows the viewpoint in the object system, and we imagine that we will be looking from the viewpoint toward the origin, thus establishing a viewing direction.

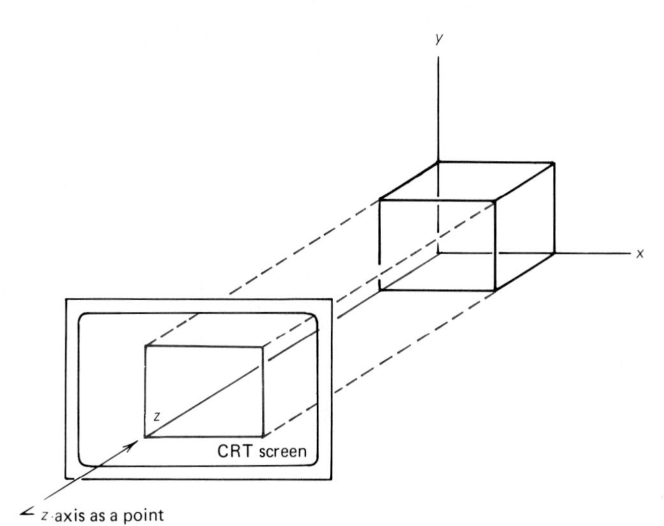

FIG. 17.19 Viewing a wire-frame model.

Our original viewing direction was along the z axis of the object system. We now define a viewing coordinate system such that we will always be looking along the z axis of the viewing system toward the origin of the system (the origins of the object and viewing systems coincide). To distinguish the viewing system from the object system, we will call the axes of the viewing system x_v, y_v, and z_v. Thus, in Fig. 17.19, the x_v, y_v, and z_v axes were exactly aligned with x, y, and z axes of the object.

To observe the object in the viewing system, we must have a relationship between each point of the object in the two coordinate systems. The coordinate

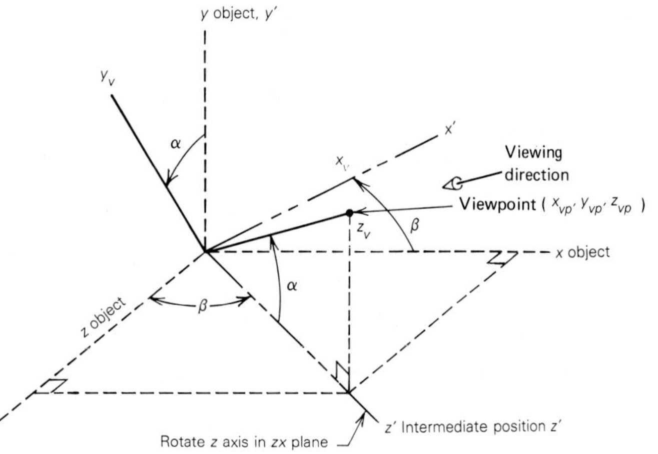

FIG. 17.20 The viewing direction is achieved by two rotations.

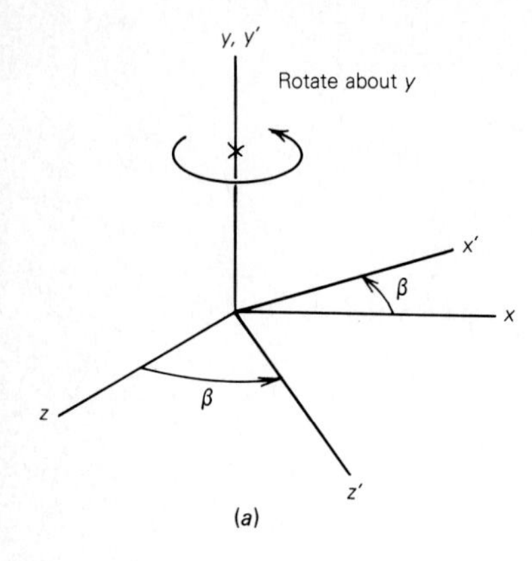

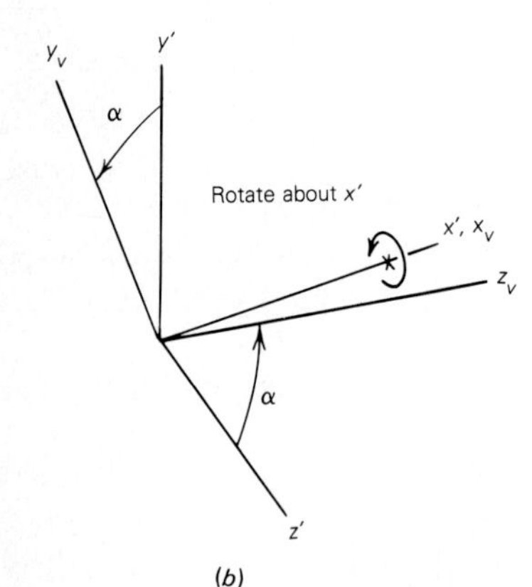

FIG. 17.21 Rotations are needed to define the viewing axis system.

conversion can be visualized as two rotations of the viewing system which is initially aligned with the object system and is then rotated through angles β and α to align the z_v axis with the desired viewing direction (see Fig. 17.20). For clarity, each rotation is performed separately in Fig. 17.21. The first rotation through an angle β is about the y axis to obtain a intermediate x', y', z' system (Fig. 17.21*a*). Then a second rotation is performed through an angle α about the x' axis such that the z_v axis passes through the viewpoint.

We can determine the relationship between coordinates in the object system and the viewing system by using the two-angle rotation process. First, we can find the relationship between the coordinates of the object in the object system (x, y, z) and the intermediate system (x', y', z'). The relationship will include the rotation angle β, as described in Fig. 17.21*a*. Next we can find the relationship between the coordinates in the intermediate system (x', y', z') and the final viewing system (x_v, y_v, z_v) which will include the rotation angle α as shown in Fig. 17.21*b*.

Consider a point P in the intermediate system having coordinates (x', y', z'). Figure 17.22*a* shows this point in relation to the intermediate system with y' shown as a point. We wish to find the coordinates of P in the object system (x, y, z). This is accomplished by projecting point P onto the x and z axes. The coordinates of P are found as follows (see Fig. 17.22*b*:

$$
\begin{aligned}
x &= PT + OA \\
y &= y' \\
z &= AR - RT
\end{aligned}
$$

Substituting the trigonometric relationships for PT, OA, AR, RT in terms of x', z', and β yields the first set of transformation equations.

$$
\begin{aligned}
x &= x' \cos \beta + z' \sin \beta \\
y &= y' \\
z &= z' \cos \beta - x' \sin \beta
\end{aligned}
$$

or

$$
\begin{aligned}
x &= x' \cos \beta + z' \sin \beta \\
y &= y' \\
z &= -x' \sin \beta + z' \cos \beta
\end{aligned}
\tag{17.38}
$$

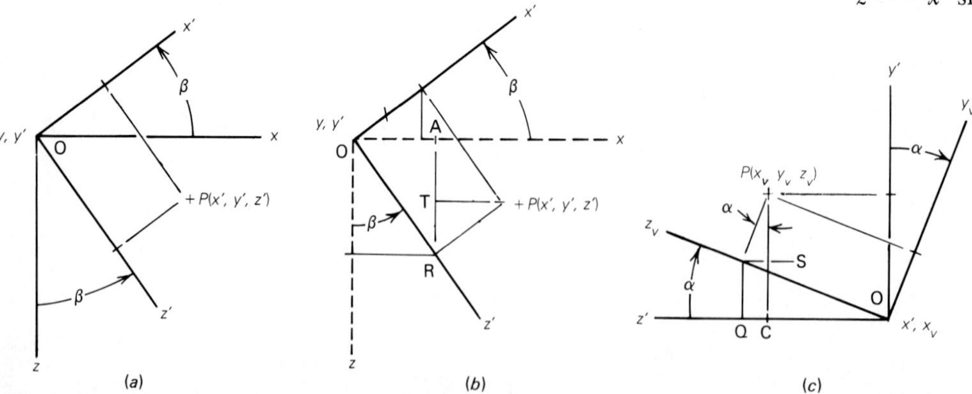

FIG. 17.22 Definition of the viewing coordinates.

Using Cramer's rule, we can solve for x', y', z' in terms of x, y, z, and β.

$$x' = \frac{\begin{vmatrix} x & 0 & \sin\beta \\ y & 1 & 0 \\ z & 0 & \cos\beta \end{vmatrix}}{\begin{vmatrix} \cos\beta & 0 & \sin\beta \\ 0 & 1 & 0 \\ -\sin\beta & 0 & \cos\beta \end{vmatrix}}$$

$$= \frac{x\begin{vmatrix} 1 & 0 \\ 0 & \cos\beta \end{vmatrix} - 0\begin{vmatrix} y & 0 \\ z & \cos\beta \end{vmatrix} + \sin\beta\begin{vmatrix} y & 1 \\ z & 0 \end{vmatrix}}{\cos\beta\begin{vmatrix} 1 & 0 \\ 0 & \cos\beta \end{vmatrix} - 0\begin{vmatrix} 0 & 0 \\ -\sin\beta & \cos\beta \end{vmatrix} + \sin\beta\begin{vmatrix} 0 & 1 \\ -\sin\beta & 0 \end{vmatrix}}$$

$$= \frac{x\cos\beta - z\sin\beta}{\cos^2\beta + \sin^2\beta}$$

$$= x\cos\beta - z\sin\beta$$

A similar computation will yield z'. The resulting set of transformed coordinates are

$$\begin{aligned} x' &= x\cos\beta - z\sin\beta \\ y' &= y \\ z' &= x\sin\beta + z\cos\beta \end{aligned} \qquad (17.39)$$

The second step of the transformation, the rotation through angle α about the x' axis, is shown in Fig. 17.22*c*. Following the same procedures as before, we can project point P onto the y' and z' axes and obtain

$$\begin{aligned} x' &= x_v \\ y' &= PS + SC \\ z' &= OQ - QC \end{aligned}$$

which yield upon substitution of the appropriate trigonometric relationships

$$\begin{aligned} x' &= x_v \\ y' &= y_v\cos\alpha + z_v\sin\alpha \\ z' &= z_v\cos\alpha - y_v\sin\alpha \end{aligned} \qquad (17.40)$$

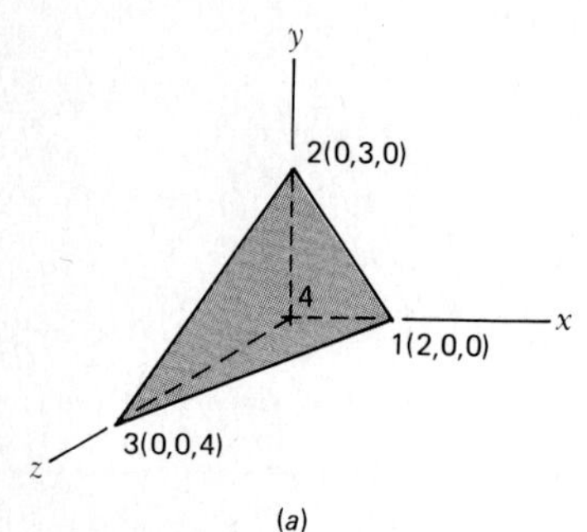

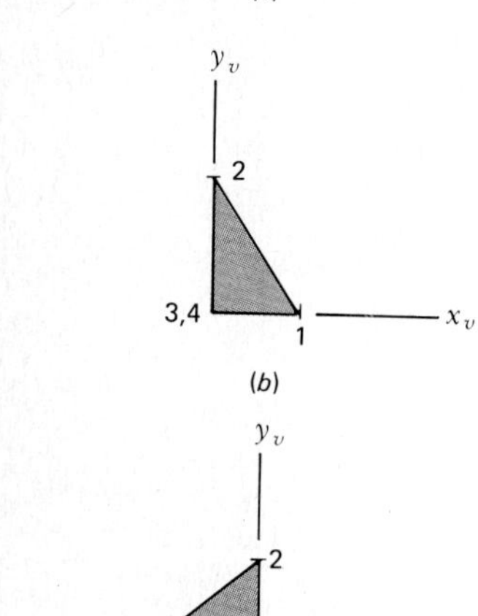

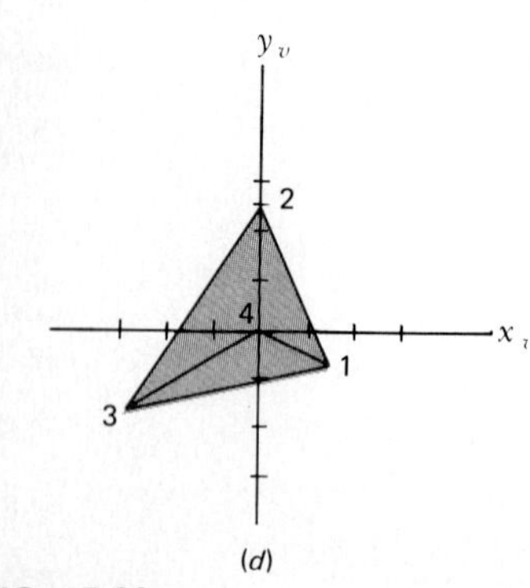

FIG. 17.23

Again using Cramer's rule, x_v, y_v, z_v can be solved for in terms of x', y', z' and α.

$$
\begin{aligned}
x_v &= x' \\
y_v &= y'\cos\alpha - z'\sin\alpha \\
z_v &= y'\sin\alpha + z'\cos\alpha
\end{aligned} \tag{17.41}
$$

The reader should carry out the algebra to verify these expressions.

Finally, substitution of Eq. (17.39) into Eq. (17.41) yields

$$
\begin{aligned}
x_v &= x\cos\beta - z\sin\beta \\
y_v &= y\cos\alpha - x\sin\alpha\sin\beta \\
&\quad - z\sin\alpha\cos\beta \\
z_v &= y\sin\alpha + x\cos\alpha\sin\beta \\
&\quad + z\cos\alpha\cos\beta
\end{aligned} \tag{17.42}
$$

which relate the object coordinates x, y, z with the viewing coordinates x_v, y_v, z_v.

From a desired viewpoint, the angles α and β can be calculated and substituted into Eq. (17.42) set, which provides the coordinates of each point on an object in the viewing coordinate system. Refer to Fig. 17.20 and note that

$$\sin\beta = \frac{x_{vp}}{(x_{vp}^2 + z_{vp}^2)^{1/2}} \tag{17.43}$$

$$\sin\alpha = \frac{y_{vp}}{(x_{vp}^2 + y_{vp}^2 + z_{vp}^2)^{1/2}} \tag{17.44}$$

For the case of

$x_{vp} = z_{vp} = 0$

Eq. (17.43) is indeterminate. However, this condition defines the horizontal (top) view for which $\beta = 0$, in order to be consistent with the front view.

Example 17.9

Calculate the coordinates of each of the four nodes in the triangular pyramid in Fig. 17.23*a* if the viewpoint is (*a*) (0, 0, 5), (*b*) (5, 0, 0) and (*c*) (5, 5, 5). Plot the resulting figures.

Solution

The object coordinates of each node are:

NODE	
1	(2, 0, 0)
2	(0, 3, 0)
3	(0, 0, 4)
4	(0, 0, 0)

For the viewpoint (0, 0, 5), from Eqs. (17.43) and (17.44) (or inspection)

$$\sin\beta = \frac{0}{(0 + 5^2)^{1/2}} = 0$$

$$\sin\alpha = \frac{0}{(0 + 0 + 5^2)^{1/2}} = 0$$

Thus $\alpha = \beta = 0°$ and Eqs. (17.42) yield

$$
\begin{aligned}
x_v &= x\cos 0° - z\sin 0° = x \\
y_v &= y\cos 0° - x\sin 0°\sin 0° \\
&\quad - z\sin 0°\cos 0° \\
&= y \\
z_v &= y\sin 0° + x\cos 0°\sin 0° \\
&\quad + z\cos 0°\cos 0° \\
&= z
\end{aligned}
$$

which means that the viewing and object systems are superimposed as we would have expected. Plotting the (x_v y_v) values gives the desired picture (see Fig. 17.23*b*).

For part (b), viewpoint (5, 0, 0)

$$\sin\beta = \frac{5}{(5^2 + 0^2)^{1/2}} = 1$$

$$\beta = 90°$$

$$\sin\alpha = \frac{0}{(5^2 + 0^2 + 0^2)^{1/2}} = 0$$

$$\alpha = 0°$$

Then

$$x_v = x \cos 90° - z \sin 90°$$
$$= -z$$
$$y_v = y \cos 0° - x \sin 0° \sin 90° - z \sin 0° \cos 90°$$
$$= y$$
$$z_v = y \sin 0° + x \cos 0° \sin 90° + z \cos 0° \cos 90°$$
$$= x$$

The plot is shown in Fig. 17.23*c* and is the right profile view. For viewpoint (5, 5, 5) we have

$$\sin \beta = \frac{5}{(5^2 + 5^2)^{1/2}} = \frac{5}{\sqrt{50}} = \frac{1}{\sqrt{2}}$$
$$\beta = 45°$$
$$\sin \alpha = \frac{5}{(5^2 + 5^2 + 5^2)^{1/2}} = \frac{1}{\sqrt{3}}$$
$$\alpha = 35.264°$$

The viewing coordinates from Eq. (17.42) are

$$x_v = x \cos 45° - z \sin 45°$$
$$= 0.7071x - 0.7071z$$
$$y_v = y \cos 35.264° - x \sin 35.264° \sin 45° - z \sin 35.264° \cos 45°$$
$$= 0.8165y - 0.4082x - 0.4082z$$
$$z_v = y \sin 35.264° + x \cos 35.264° \sin 45° + z \cos 35.264° \cos 45°$$
$$= 0.5774y + 0.5774x + 0.5774z$$

The coordinates of node 1 in the viewing coordinate system are

NODE 1

$$x_v = 0.7071(2) - 0.7071(0) = 1.4142$$
$$y_v = 0.8165(0) - 0.4082(2) - 0.4082(0) - 0.8164$$
$$z_v = 0.5774(0) + 0.5774(2) + 0.5774(0) = 1.1548$$

LIKEWISE FOR NODES 2, 3, AND 4

2	(0, 2.4495, 1.7322)
3	(−2.8284, −1.6328, 2.3096)
4	(0, 0, 0)

By plotting (x_v, y_v), since you are looking along the z_v axis, the result is as in Fig. 17.23*d*. This then is an isometric view of the pyramid.

The viewpoint concept allows us to "see" an object from any point in space but does not address the problem of rotating the image on the screen when viewed from the spatial point. Once the node points are found in the viewing coordinate system, you need only rotate the z_v axis (other rotations are already handled through α and β). Thus, before plotting the image, a rotation conversion can be performed on x_v and y_v (z_v would not change). Scaling (change of image size) and translation (movement of the image on the screen) conversions can also be done before plotting. Details of these operations are left to the reader.

Now that we know how to determine what we will see from a given viewpoint, we must reverse the situation. That is, we must be able to find the viewpoint that allows us to see some specified feature of the object. The vector analysis previously introduced can be conveniently used to find a viewpoint (viewing direction) that will show true length of a line, the true shape of a plane, and the like. These vector techniques parallel descriptive geometry methods (the solution view idea) and can be easily used in computer graphics to define the necessary viewpoint from which to view the object.

The solution view for the true length of a line is the image on a projection plane parallel to the line. This means that the line of sight is normal to the line. Two points on the line in Fig. 17.24, P_1 and P_2, define a vector along the line as $\overline{P_1P_2} = \overline{P_2} - \overline{P_1}$. If the vector cross product between $\overline{P_1P_2}$ and any other vector is calculated, the result is a vector normal to $\overline{P_1P_2}$ (as well as the other vector). Since $\overline{P_1P_2}$ can be crossed with any other vector, this means that there are an infinite number of normal vectors to $\overline{P_1P_2}$ that together define a plane normal to $\overline{P_1P_2}$. Thus, if our viewpoint is placed anywhere along these normal vectors, we will see the object line in true length.

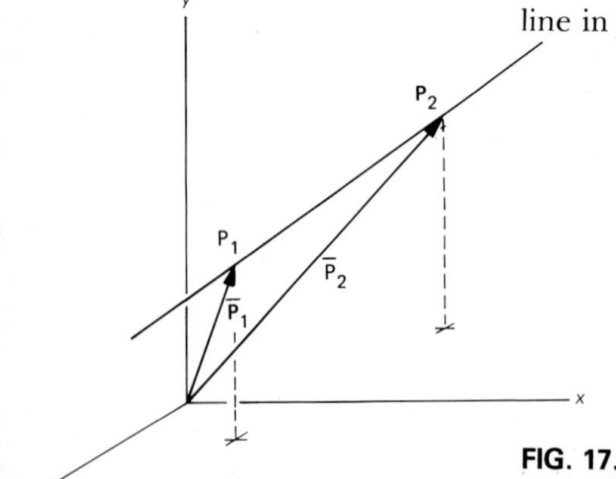

FIG. 17.24 Finding a vector normal to a given line.

Example 17.10

Find a vector normal to the line passing through points (3, 5, 2) and (4, 6, −4). Determine a viewpoint suitable to view the line in true length.

Solution

Take $P_1 = (3, 5, 2)$ and $P_2 = (4, 6, -4)$. The relative position vector from P_1 to P_2 is $\overline{P_1P_2} = \overline{P_2} - \overline{P_1} = (4-3)\mathbf{i} + (6-5)\mathbf{j} + (-4-2)\mathbf{k}$ or $\mathbf{i} + \mathbf{j} - 6\mathbf{k}$. Since any other vector can be used to perform the cross product, we can choose the absolute position vector to point P_1 which is $3\mathbf{i} + 5\mathbf{j} + 2\mathbf{k}$. The cross product $\overline{P_1P_2} \times \overline{P_1}$ gives a vector normal to the line as

$$\overline{N} = \begin{vmatrix} \mathbf{i} & \mathbf{j} & \mathbf{k} \\ 1 & 1 & -6 \\ 3 & 5 & 2 \end{vmatrix} = 32\mathbf{i} - 20\mathbf{j} + 2\mathbf{k}$$

As a check, $\overline{N} \cdot \overline{P_1P_2}$ should be zero.

$$\overline{N} \cdot \overline{P_1P_2} = 32(1) - 20(1) + 2(-6) = 0$$

Therefore, if we chose a viewpoint anywhere along vector $\overline{N}$, the view created will show the line $\overline{P_1P_2}$ in true length. We could simply choose (32, −20, 2) as a viewpoint or alternatively any scalar multiple of it such as (16, −10, 1).

The true shape of a plane is seen on a projection plane oriented parallel to the edge view of the object plane. Therefore, the line of sight is normal to the object plane. From vector analysis, if two vectors, say $\overline{P_1P_2}$ and $\overline{P_2P_3}$,

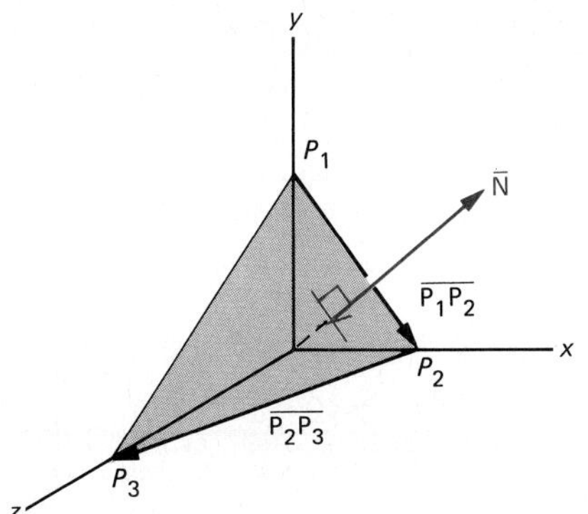

FIG. 17.25 Viewpoint to see a plane in true shape.

lie on the object plane, their cross product will produce a vector normal to that plane.

Example 17.11
If points $P_1 = (0, 4, 0)$, $P_2 = (3, 0, 0)$, and $P_3 = (0, 0, 5)$ define a plane, find a vector normal to the plane. Define a viewpoint which will show the plane in true shape.

Solution
From Fig. 17.25, we see that

$$\overline{P_1P_2} = (3 - 0)\mathbf{i} + (0 - 4)\mathbf{j} + (0)\mathbf{k} = 3\mathbf{i} - 4\mathbf{j}$$
$$\overline{P_2P_3} = (0 - 3)\mathbf{i} + (0)\mathbf{j} + (5 - 0)\mathbf{k} = -3\mathbf{i} + 5\mathbf{k}$$

Since these vectors lie in the plane $P_1P_2P_3$, their cross product creates a vector normal to the plane

$$\overline{N} = \overline{P_2P_3} \times \overline{P_1P_2} = \begin{vmatrix} \mathbf{i} & \mathbf{j} & \mathbf{k} \\ -3 & 0 & 5 \\ 3 & -4 & 0 \end{vmatrix} = 20\mathbf{i} + 15\mathbf{j} + 12\mathbf{k}$$

Had we taken $\overline{P_1P_2} \times \overline{P_2P_3}$, we would have gotten $-20\mathbf{i} - 15\mathbf{j} - 12\mathbf{k}$. This merely means that we can view the plane in true shape from either side. An appropriate viewpoint is (20, 15, 12) from our normal vector calculation.

From descriptive geometry, the inclination (slope angle, slope, or grade) of a line is seen when the line is in true length in an elevation view. Thus, we must ensure both conditions at once. As you recall, the vector cross product yields a vector normal to both vectors in the product. Therefore, if one vector is along the line and the other vector is vertical, we can obtain the true length and an elevation view simultaneously. The simplest vertical vector in the x, y, z system we have been using is the unit vector $\mathbf{j}$, assuming that the y axis is vertical. Any vector normal to the y axis will lie in the zx plane. Figure 17.26*a* illustrates a vector $\overline{P}$ in the zy plane and the unit vector $\mathbf{j}$. You can see that in this special case the normal vector to the plane established by $\mathbf{j}$ and $\overline{P}$ will be a vector $\overline{N}$ along the x axis. If the vector $\overline{P}$ rotates about the y axis, it will form a cone, and so the $\overline{N}$ vector will not always coincide with the x axis but will always be in the xz plane.

Example 17.12
Find a viewpoint that will allow you to view the inclination of a line described by endpoints $P_1(10, 5, 3)$ and $P_2(5, 3, 4)$.

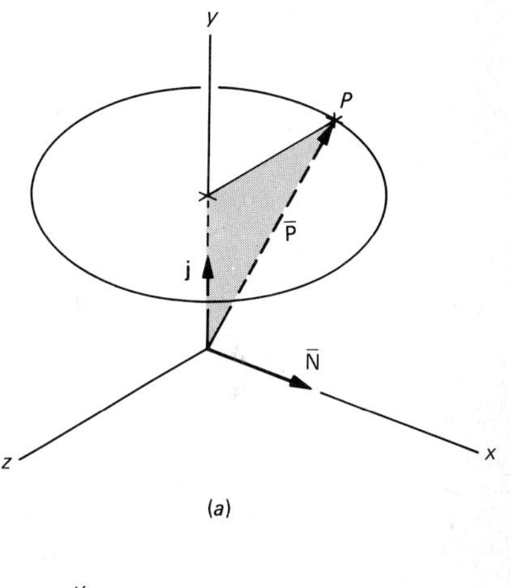

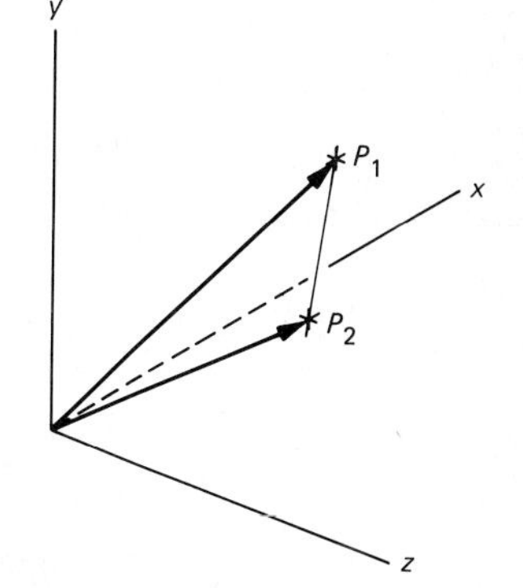

FIG. 17.26 Viewpoint to see the inclination of a line.

Solution

A vector along the line is given by (see Fig. 17.26*b*):

$$(10 - 5)\mathbf{i} + (5 - 3)\mathbf{j} + (3 - 4)\mathbf{k} = 5\mathbf{i} + 2\mathbf{j} - \mathbf{k}$$

A vector normal to $5\mathbf{i} + 2\mathbf{j} - \mathbf{k}$ and normal to the y axis is

$$\bar{\mathrm{I}} = (5\mathbf{i} + 2\mathbf{j} - \mathbf{k}) \times \mathbf{j} = \begin{vmatrix} \mathbf{i} & \mathbf{j} & \mathbf{k} \\ 5 & 2 & -1 \\ 0 & 1 & 0 \end{vmatrix} = \mathbf{i}(1) - \mathbf{j}(0) + \mathbf{k}(5) = \mathbf{i} + 5\mathbf{k}$$

The y component of this vector is zero, indicating that the vector is horizontal or that a viewpoint derived from it puts the observer in the xz plane looking at the origin. As a check, form the dot product between the vector along the line and $\bar{\mathrm{I}}$. The result should be zero if the two are normal.

$$(5\mathbf{i} + 2\mathbf{j} - \mathbf{k}) \cdot (\mathbf{i} + 0\mathbf{j} + 5\mathbf{k}) = 5 + 0 - 5 = 0$$

A suitable viewpoint is then (1, 0, 5).

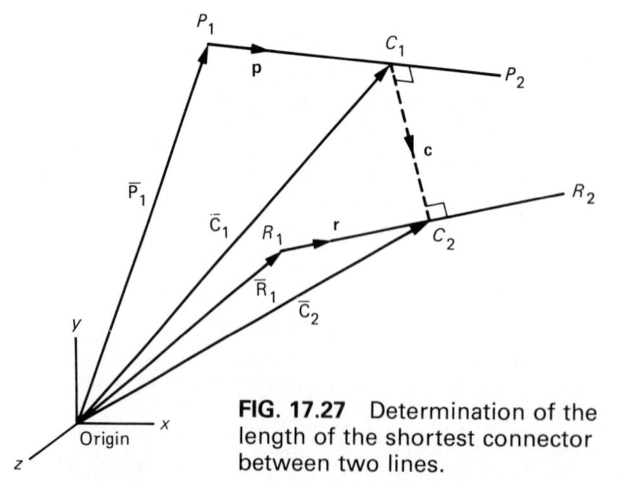

FIG. 17.27 Determination of the length of the shortest connector between two lines.

The common perpendicular (shortest connector) between two skew lines can be seen in a view where one of the lines appears as a point in accordance with graphics theory. The common perpendicular is seen in true length and is perpendicular to the line that is not seen as a point. Thus, if our purpose is to see the common perpendicular in true length, it is only necessary to look in a direction along a vector representing either of the two skew lines.

Furthermore, if we want to calculate the length of the common perpendicular, we can do so using vector analysis. In Fig. 17.27, let $\mathbf{p}$ be a unit vector along $\overline{P_1P_2}$, $\mathbf{r}$ be a unit vector along $\overline{R_1R_2}$, and $\mathbf{c}$ be a unit vector along the common perpendicular $\overline{C_1C_2}$ ($\mathbf{c}$ is unknown at the beginning).

Then

$$\bar{C}_1 = \bar{P}_1 + |\overline{P_1C_1}|\mathbf{p} = \bar{R}_1 + |\overline{R_1C_2}|\mathbf{r} - |\overline{C_1C_2}|\mathbf{c}$$

If we take the scalar product with $\mathbf{c}$, we get

$$\bar{P}_1 \cdot \mathbf{c} + |\overline{P_1C_1}|\mathbf{p} \cdot \mathbf{c} = \bar{R}_1 \cdot \mathbf{c} + |\overline{R_1C_2}|\mathbf{r} \cdot \mathbf{c} - |\overline{C_1C_2}|\mathbf{c} \cdot \mathbf{c}$$

Note that for the common perpendiculars, $\mathbf{p}$ and $\mathbf{r}$ are each perpendicular to $\mathbf{c}$; therefore, $\mathbf{p} \cdot \mathbf{c} = 0$ and $\mathbf{r} \cdot \mathbf{c} = 0$. We have left

$$\bar{P}_1 \cdot \mathbf{c} = \bar{R}_1 \cdot \mathbf{c} - |\overline{C_1C_2}|$$

or

$$|\overline{C_1C_2}| = (\bar{R}_1 - \bar{P}_1) \cdot \mathbf{c}$$

Although we do not know $\mathbf{c}$ in general, we can recognize that $\mathbf{c}$ is perpendicular to $\mathbf{p}$ and $\mathbf{r}$ and is therefore proportional to $\mathbf{p} \times \mathbf{r}$.

The unit vector **c** can then be written

$$\mathbf{c} = \frac{\mathbf{p} \times \mathbf{r}}{|\mathbf{p} \times \mathbf{r}|}$$

The length of the shortest connector $|\overline{C_1C_2}|$ becomes

$$|\overline{C_1C_2}| = \left| \frac{(\overline{R}_1 - \overline{P}_1) \cdot (\mathbf{p} \times \mathbf{r})}{|\mathbf{p} \times \mathbf{r}|} \right| \tag{17.45}$$

where the absolute value sign is used to ensure a positive result.

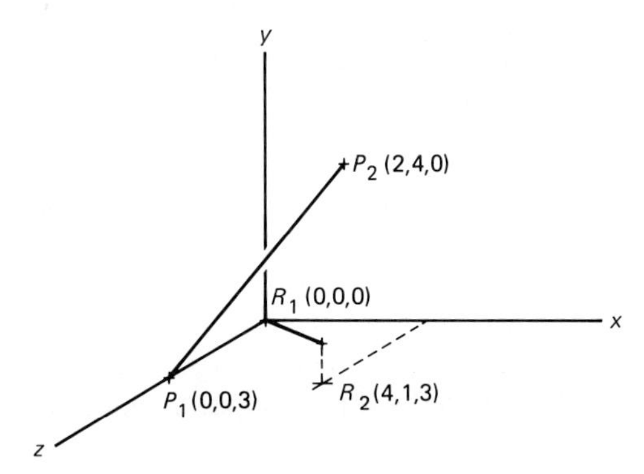

FIG. 17.28

Example 17.13

Find the length of the shortest connector between the lines $\overline{P_1P_2}$ and $\overline{R_1R_2}$ depicted in Fig. 17.28. Describe how this connector could be viewed in true length.

Solution

The unit vector along $\overline{P_1P_2}$ is

$$\mathbf{p} = \frac{(2-0)\mathbf{i} + (4-0)\mathbf{j} + (0-3)\mathbf{k}}{(2^2 + 4^2 + 3^2)^{1/2}} = \frac{2\mathbf{i} + 4\mathbf{j} - 3\mathbf{k}}{\sqrt{29}}$$

Likewise, the unit vector along $\overline{R_1R_2}$ is

$$\mathbf{r} = \frac{4\mathbf{i} + \mathbf{j} + 3\mathbf{k}}{(4^2 + 1^2 + 3^2)} = \frac{4\mathbf{i} + \mathbf{j} + 3\mathbf{k}}{\sqrt{26}}$$

Then

$$\mathbf{p} \times \mathbf{r} = \frac{1}{\sqrt{(29)(26)}} \begin{vmatrix} \mathbf{i} & \mathbf{j} & \mathbf{k} \\ 2 & 4 & -3 \\ 4 & 1 & 3 \end{vmatrix}$$

$$= \frac{\mathbf{i}(12 + 3) - \mathbf{j}(6 + 12) + \mathbf{k}(2 - 16)}{\sqrt{(29)(26)}}$$

$$= \frac{15\mathbf{i} - 18\mathbf{j} - 14\mathbf{k}}{\sqrt{(29)(26)}}$$

and

$$\frac{\mathbf{p} \times \mathbf{r}}{|\mathbf{p} \times \mathbf{r}|} = \frac{\dfrac{15\mathbf{i} - 18\mathbf{j} - 14\mathbf{k}}{\sqrt{(29)(26)}}}{\dfrac{(15^2 + 18^2 + 14^2)^{1/2}}{\sqrt{(29)(26)}}}$$

$$= \frac{15\mathbf{i} - 18\mathbf{j} - 14\mathbf{k}}{(745)^{1/2}}$$

The length of the shortest connector becomes (noting $\overline{R}_1 = 0$ and $\overline{P}_1 = 3\mathbf{k}$)

$$\text{Length} = \left| \frac{(0 - 3\mathbf{k}) \cdot (15\mathbf{i} - 18\mathbf{j} - 14\mathbf{k})}{(745)^{1/2}} \right|$$

$$= \frac{3(14)}{(745)^{1/2}} = 1.53 \text{ units}$$

Several viewpoints are possible that allow you to see the shortest connector in true length. The simplest is probably (4, 1, 3), since looking from R_2 toward R_1 is along one of the skew lines. You could also look along unit vector **p** and use a viewpoint of (2, 4, −3).

17.3 DISPLAYING REALISTIC IMAGES

As we have discussed previously, the geometric model is the heart of a CAD/CAM system. From it, many good things can come, such as engineering drawings, stress analysis, and the like. The wire-frame representation is limited in its usefulness for certain further processing. Surface models allow us to completely define the surface of an object more complex than a polyhedron described by the wire-frame model. The solid model extends our capabilities further, providing information necessary for section views, calculation of volume and mass, and so on.

As a user of CAD/CAM systems, you must see what you are creating by means of an image. The wire-frame image can be enhanced for some uses by eliminating hidden lines or surfaces or specifying that hidden lines be shown as dashed lines, following traditional drawing practice. Depending on the type of display device available, 3-D objects can be made to seem more realistic by varying the intensity of lines on the object or using shades of gray or color to create an image that is very picturelike. These more realistic images can be extremely helpful in detecting errors in a geometric model, for example, or in illustrating what the object will eventually look like before any material cutting.

17.3.1 Hidden Line and Surface Removal

An image of a geometric model can be more clearly displayed by removal of lines that cannot be seen through opaque surfaces. The trade-off for this more realistic image is additional computer time to make the necessary calculations and possibly a reduction in dynamic image display, depending on the capability of the computer. An example of the visual effect of removing hidden lines from a wire-frame model display is shown in Fig. 17.29.

The wire-frame model can be used to illustrate some possible techniques of removing hidden lines or surfaces from the display. Note that these hidden lines are only identified by a program that then instructs the display device not to draw them. The lines are

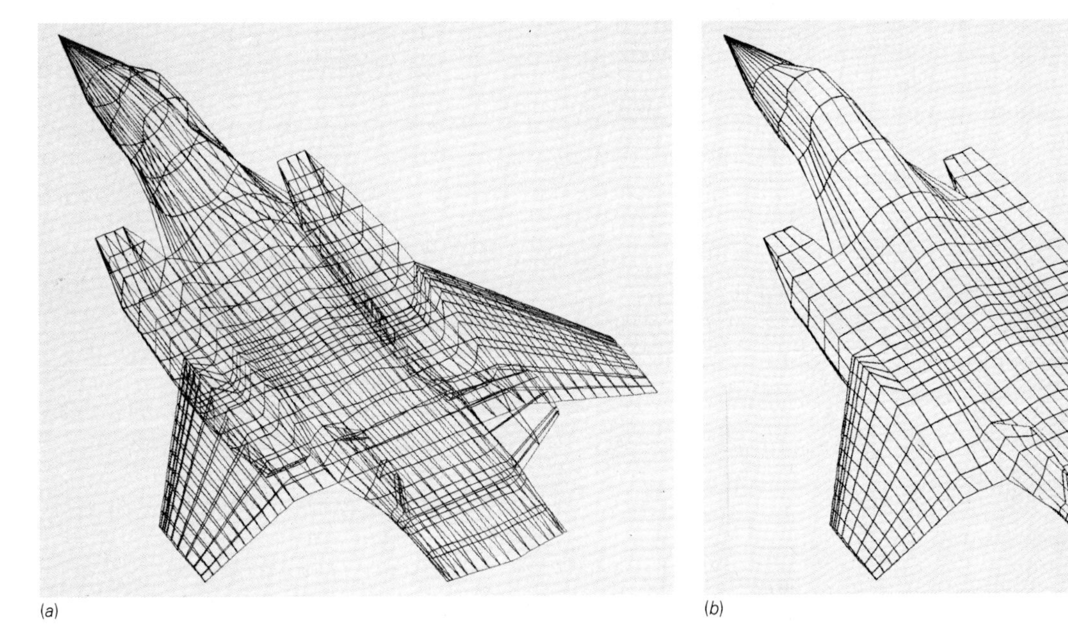

FIG. 17.29 Effect of hidden line removed from the display image. (*a*) Display with hidden lines drawn. (*b*) Display with hidden lines removed. (*Courtesy of Ames Research Center, NASA*)

not removed from the geometric database.

If the wire-frame description is of a simple object such as a convex polyhedron, a simple concept is available to detect whether each surface on the object is visible. This is true because each surface is entirely visible or entirely hidden or is seen as an edge. Recall that if the cross product of two vectors is computed, the result is a vector normal to each of the two original vectors. Also, two vectors lying on a surface that are not colinear are sufficient to define a planar surface. Thus, if you select two adjacent edges of a planar surface on the object as vectors so that you are moving in a counterclockwise sense when the object is viewed from the *outside*, the cross product will yield an outward vector normal to the surface. You need only determine from your viewpoint whether the surface normal points toward you (surface is visible), away from you (surface is hidden), or perpendicular to your sight line (surface is seen as an edge).

Then if the object is thought of as a series of surfaces, the visible surfaces can be drawn on the screen as solid lines and the hidden surfaces can be drawn with dashed lines or not drawn at all. Consider the following example to determine how the hidden surface method can be used.

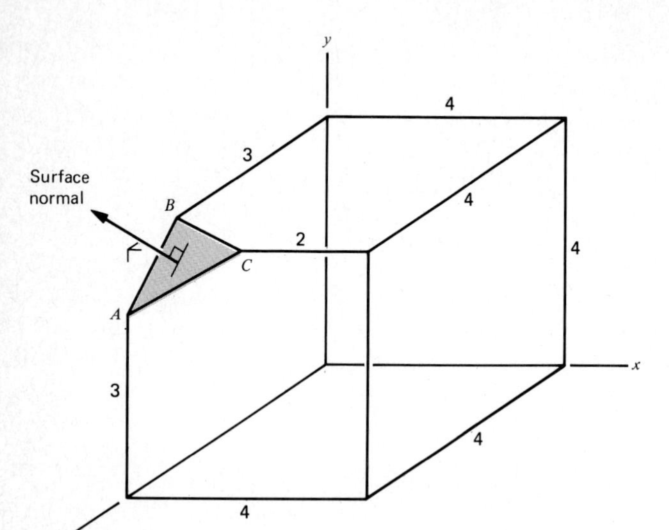

FIG. 17.30 Determining the visibility of a surface from a given viewpoint.

Example 17.14

For the object in Fig. 17.30, determine whether surface ABC is visible or hidden if the viewpoint is (0, 0, 10) or (10, 0, 0).

Solution

First determine the coordinates of points A, B, and C.

$$A = (0, 3, 4)$$
$$B = (0, 4, 3)$$
$$C = (2, 4, 4)$$

Then consider vectors parallel to lines AC and CB (boundaries of plane ABC). Take in a counterclockwise sense as viewed from outside the object.

$$\overline{\text{AC}} = (2 - 0)\mathbf{i} + (4 - 3)\mathbf{j} + (4 - 4)\mathbf{k} = 2\mathbf{i} + \mathbf{j}$$

$$\overline{\text{CB}} = (0 - 2)\mathbf{i} + (4 - 4)\mathbf{j} + (3 - 4)\mathbf{k} = -2\mathbf{i} - \mathbf{k}$$

The cross product of $\overline{\text{AC}}$ and $\overline{\text{CB}}$ yields an outward surface normal to plane ABC.

$$\overline{\text{AC}} \times \overline{\text{CB}} = \begin{vmatrix} \mathbf{i} & \mathbf{j} & \mathbf{k} \\ 2 & 1 & 0 \\ -2 & 0 & -1 \end{vmatrix} = \mathbf{i}(-1) - \mathbf{j}(-2) + \mathbf{k}(+2) = -\mathbf{i} + 2\mathbf{j} + 2\mathbf{k}$$

To determine whether this normal vector points toward or away from the viewer, you can use the form of the dot product as $\overline{\text{X}} \cdot \overline{\text{Y}} = |\overline{\text{X}}|\,|\overline{\text{Y}}| \cos \alpha$, where α is the angle between the vectors $\overline{\text{X}}$ and $\overline{\text{Y}}$. If a vector from the origin to the viewpoint is dotted with the normal vector from the surface, an angle α less than 90° indicates a visible surface, while an angle greater than 90° means that the surface is hidden from that viewpoint.

(*a*) For viewpoint (0, 0, 10), the vector from the origin to the viewpoint is 10**k**. When this is dotted with the surface normal,

$$10\mathbf{k} \cdot (-\mathbf{i} + 2\mathbf{j} + 2\mathbf{k}) = 10(1^2 + 2^2 + 2^2)^{1/2} \cos \alpha_1$$

or

$$20 = 10(9)^{1/2} \cos \alpha_1$$
$$\alpha_1 = \cos^{-1} \frac{20}{10(3)} = 48.19°$$

Since α_1 is the angle between the surface normal and the vector to the viewpoint and is less than 90°, the surface is visible from (0, 0, 10).

(*b*) For viewpoint (10, 0, 0), the vector from the origin to the viewpoint is 10**i**. Dot with the surface normal to get

$$10\mathbf{i} \cdot (-\mathbf{i} + 2\mathbf{j} + 2\mathbf{k}) = 10(9)^{1/2} \cos \alpha_2$$
$$-10 = 10(3) \cos \alpha_2$$
$$\alpha_2 = \cos^{-1} \left(-\frac{1}{3}\right) = 109.47°$$

The resulting angle α_2 shows that ABC is not visible from viewpoint (10, 0, 0).

We must extend our thinking somewhat for more complex objects where a portion of an object may hide a part of another surface rather than the entire surface. A feature as simple as a notch (see Fig. 17.31) will cause the elementary hidden surface routine to fail for certain viewpoints. As you can see from the viewpoint depicted, the shaded surface would be judged to be visible from the elementary technique previously described although it is partially hidden. Other viewpoints of the same object can be handled correctly by the simple technique just described, such as viewing along the y axis.

Since the elementary hidden surface technique can correctly identify "back" surfaces which cannot be seen at all, it is sometimes applied first to eliminate some surfaces from further visibility considerations. Many schemes are available to handle increasingly complex situations. We will describe only one additional technique for establishing the priority of one plane surface over another from the standpoint of visibility.

If you project a view of the wire-frame model from the chosen viewpoint by drawing the xy coordinate points and connections after completing the necessary rotations, each pair of planes constituting the object should have one of the following relationships. The planes may not overlap at all (Fig. 17.32*a*), one plane may entirely contain another (Fig. 17.32*b*), or the two planes may overlap partially (Fig. 17.32*c*).

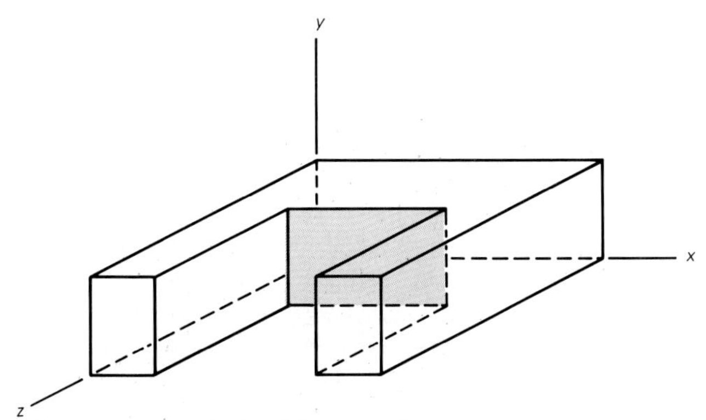

FIG. 17.31 Partially hidden surface.

The idea is then to take a test plane, that is, a plane on the object that is potentially visible, and compare it with every other plane on the object one at a time. The test plane will survive all tests and be visible, will be found to be totally hidden, or will be partially visible. In a single check, if the test plane does not overlap the comparison plane, the test plane remains potentially visible until all other planes are checked.

If the planes appear as in Fig. 17.32*b*, an xy point on the enclosed plane must be chosen and the z value for it must be compared with the z value for the same x and y on the enveloping plane. Then, if the z coordinate of the test plane is greater than it is for the comparison plane, the test plane is closer to the observer and remains potentially visible. On the other hand, if z for the

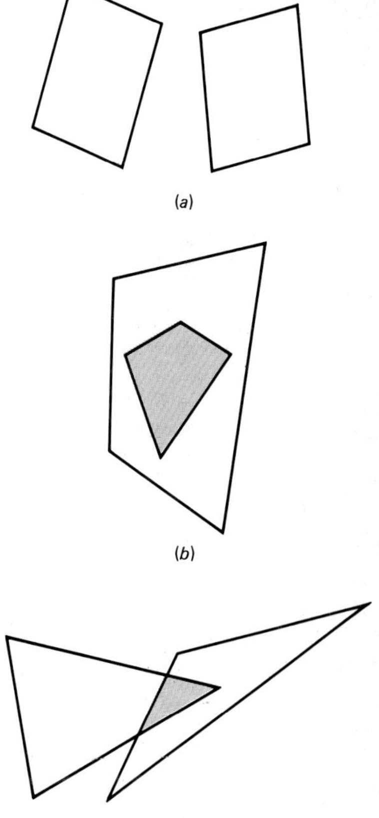

FIG. 17.32 Testing visibility of plane surfaces.

test plane is less than that for the comparison plane, the test plane is either entirely hidden or partially hidden (depending on the relative size of the two planes; one plane can hide the other, but conversely, the other plane can only partially hide the first). In any case, a conclusion can be drawn.

The partial overlap situation requires that the common area (on the 2-D projection) be found. Then the decision about which plane is nearest the observer is again based on the value of z for a given x and y. Typically, the apparent intersection points are found and z's are compared there very much as was done in the visibility tests considered earlier in our discussion of descriptive geometry (plane intersections).

17.3.2 Surface Models

A wire-frame model, as discussed previously, appears to be composed of plane surfaces. In fact, we referred to "surface removal" in Sec. 17.3.1. From the point of view of the observer, the boundary lines of the surface shown in the wire-frame model are sufficient to describe a plane surface in the observer's mind. The computer model, however, consists only of the boundary lines, and nothing is known about points between lines. If some technique is used to define the surface at every point inside a boundary, the result will be a surface model.

A surface model is often built by describing the boundaries of the surface and by defining an analytic function that provides surface values inside the boundaries. The simplest case is a planar surface, which is readily described analytically, as was done in Sec. 17.2.2 (Fig. 17.33*a*). If a straight line is caused to remain parallel to itself while moving about two circles of the same size, a right circular cylinder can be generated. Any point on the surface is defined by a function (Fig. 17.33*b*). The planar surface and a portion of the cylinder can be combined to create a fillet, as shown in Fig. 17.33*c*. A general ruled surface requires a straight line to follow two different curved lines in space to create the

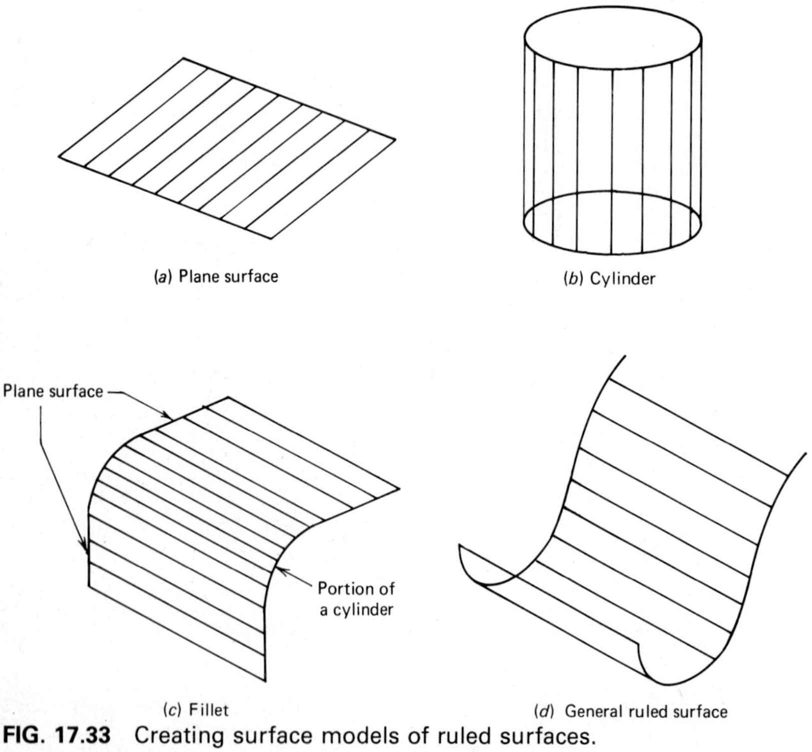

(*a*) Plane surface

(*b*) Cylinder

Plane surface

Portion of a cylinder

(*c*) Fillet

(*d*) General ruled surface

FIG. 17.33 Creating surface models of ruled surfaces.

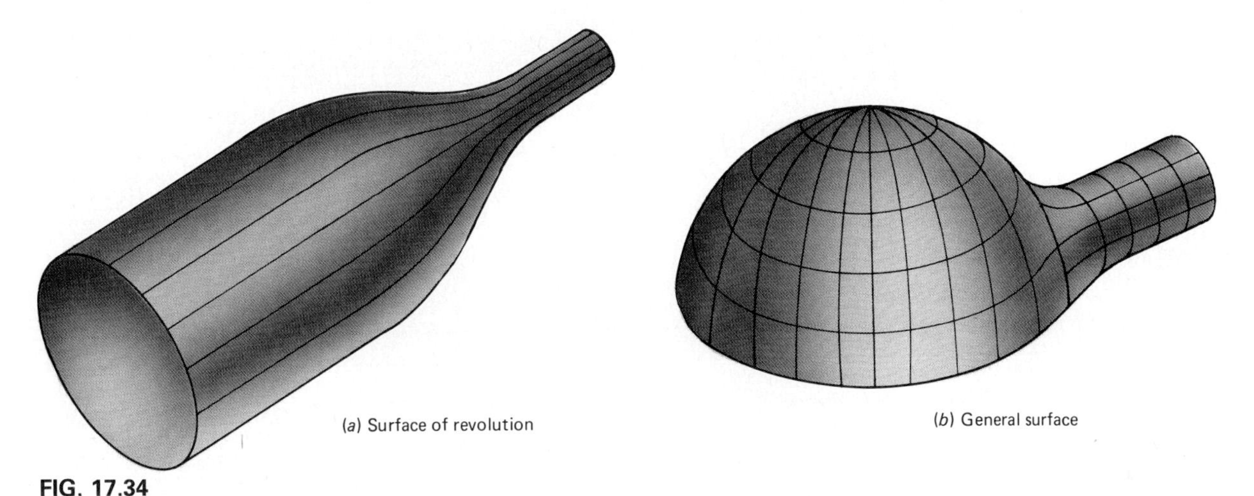

FIG. 17.34

surface, as illustrated in Fig.17.33*d*.

Surfaces that contain no straight lines can also be created. A surface of revolution is defined by revolving an arbitrary curve about an axis, perhaps resulting in the bottle shape shown in Fig. 17.34*a*. A general surface that cannot be defined by a single function is built up of patches (cubic patch or *B* surface are examples) that approximate the surface within each patch boundary by matching surface coordinates, slopes, and the like at the boundaries. An example of such a surface is shown in Fig. 17.34*b*.

Although the surface model may be displayed to look like a wire-frame model with or without hidden lines removed, it can also be displayed by assigning a different shade of gray or different color to each patch, as shown in Fig. 17.35*a*, where a light source is assumed to provide a 3-D effect. Furthermore, a smoothing technique can be employed to get rid of the sharp boundaries, as shown in Fig. 17.35*b*.

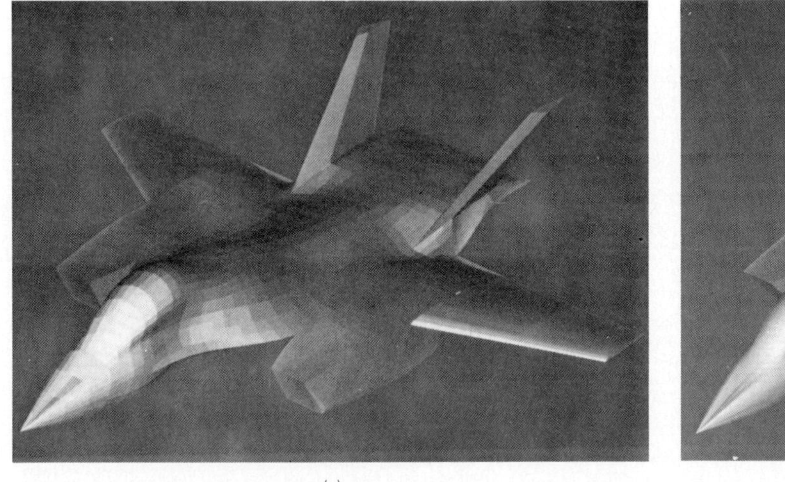

(a)

(b)

FIG. 17.35 (*a*) Shaded panels used to display a surface model. (*b*) The surface model after smoothing. (*Courtesy of Ames Research Center, NASA.*)

FIG. 17.36 Surface shading is a significant aid in detecting modeling errors. (*Courtesy of Ames Research Center, NASA.*)

Shading can be extremely helpful in visualizing the computer model, as illustrated in Fig. 17.36, where errors in the model definition become readily apparent. In general, one would probably not be able to pick out the errors in a wire-frame display of the same model.

It should be clear that once a surface model is defined, the database can be used to prepare instructions for a numerically controlled machine that can produce the surface. The surface of revolution is most applicable to turning applications because of its rotational symmetry. The general surface can be cut with an appropriate milling machine. In any case, once the machine program is done, the tool path can be displayed for visual checking before any material is cut.

17.3.3 Solid Models

As mentioned earlier in this chapter, the deficiency of the surface model in the absence of some further enhancement is that the interior of the model remains undefined. In order to draw a section view of a solid, it is necessary to know in detail how the entire solid is defined. Also, to calculate important parameters describing a solid such as its mass, center of mass, moment of inertia, and the like, more information must be available than only the surface geometry.

One approach used is that of creating the surface model and then defining analytically the mass of the solid contained within the boundaries. A second method is to define a set of solid primitives that may be combined to produce a more complicated object. This method is often referred to as solid modeling or volume modeling. We will discuss the use of primitives in more detail.

The number and type of primitives available varies somewhat in each solid modeling package. The simplest systems contain at least rectangular prisms, right circular cylinders, spheres, cones (full and truncated), and wedges. Additional primitives in sophisticated packages include toruses, ellipsoids, nonright cylinders, pyramids of various types, and some sort of general-form solid. See Fig. 17.37 for examples of primitives.

Each primitive can be scaled to the proper size, translated to the position required, and rotated to the correct angle to be added to or subtracted from other primitives to make up a complex object. In terms of a right cylinder, for example, this primitive could represent a solid object (positive cylinder) or a hole in another solid (negative cylinder). Some systems also allow primitives to be distorted. For example, a right cylinder primitive may have its centerline defined to follow an arc that creates a circular cross section at each station but a solid whose end planes are not parallel.

Once the solid model is finished, a plane may be passed through the object at an appropriate location to create a section view which can then be automatically crosshatched, resulting in a section view showing significant interior detail. Certainly, analysis routines can be used to calculate various properties, such as mass, once a material density is specified.

FIG. 17.37 Representative primitives used in solid modeling. (*Courtesy of Control Data Corporation.*)

Example 17.15
Describe how the extension bar for a socket wrench shown in Fig. 17.38 may be generated using simple solid primitives.

Solution
This simple object can be broken into four distinct solid components. Three of the components are to be added together to cre-

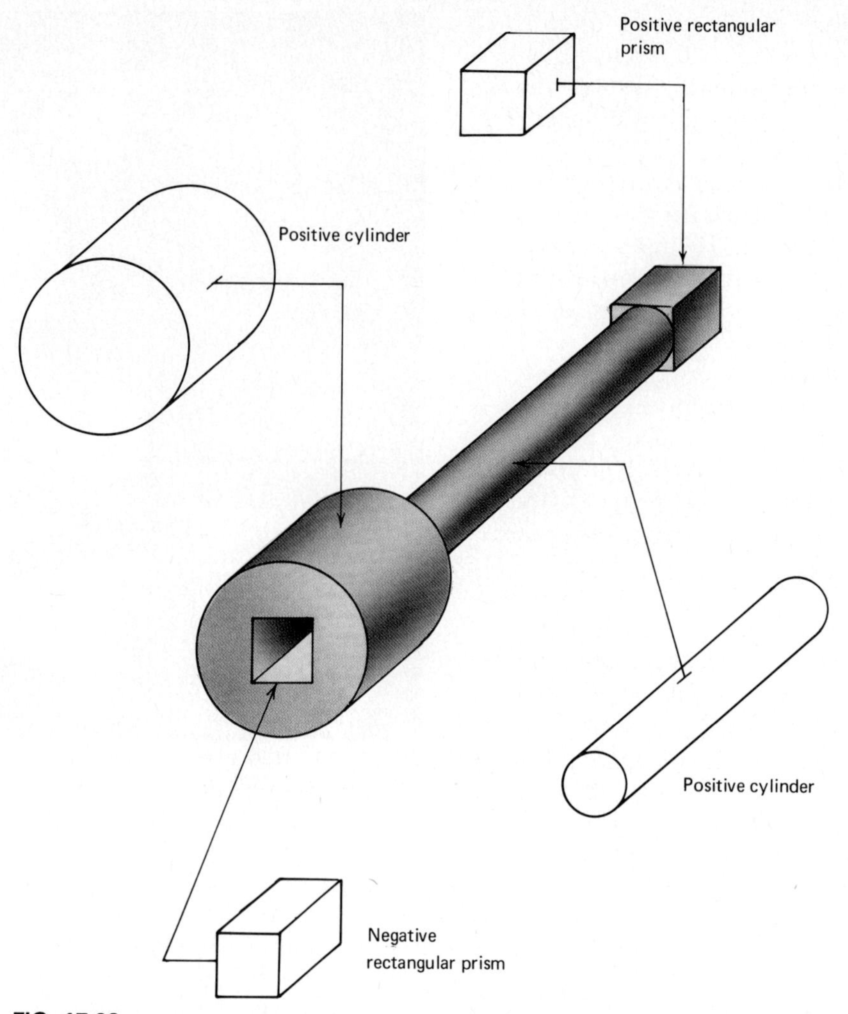

FIG. 17.38

ate the basic solid, and the fourth is to be subtracted to provide the square hole. Only two primitives are required. A right circular cylinder primitive is used twice in two different diameters, lengths, and positions to produce the left and center portions. A rectangular prism can be sized, rotated, and located to provide the right end. The same rectangular prism, sized slightly larger to provide clearance and located within the left cylinder, is subtracted from the solid, resulting in the square hole. Once the solid is finished, a plane can be passed vertically through the longitudinal axis to produce a section view of the extension bar.

Problems

17. 1 For vector $\overline{A} = 7\mathbf{i} + 4\mathbf{j} + 2\mathbf{k}$ and $\overline{B} = 3\mathbf{j} + 5\mathbf{k}$, sketch in an *xyz* coordinate system. Determine (using vectors):
- (*a*) Dot product
- (*b*) Cross product
- (*c*) Unit vector normal to the plane formed by $\overline{A}$ and $\overline{B}$
- (*d*) Component of $\overline{A}$ along $\overline{B}$ (sketch on diagram)
- (*e*) Component of $\overline{B}$ along $\overline{A}$
- (*f*) Area of the parallelogram formed by $\overline{A}$ and $\overline{B}$ (sketch)
- (*g*) Angle between $\overline{A}$ and the *y* axis
- (*h*) Angle between $\overline{B}$ and the *x* axis

17. 2 Vectors $\overline{OE} = 4\mathbf{i} - 2\mathbf{j} + 3\mathbf{k}$ and $\overline{OF} = 2\mathbf{i} + \mathbf{j} - \mathbf{k}$ form two sides of a triangular plane sector where *0* is the origin. Sketch this plane sector in *xyz* coordinates. Calculate (using vectors):
- (*a*) Area of the plane sector
- (*b*) Length of $\overline{EF}$
- (*c*) Angle between $\overline{EF}$ and $\overline{OF}$
- (*d*) Area of the projection of *OEF* on the *xy* plane

17. 3 Find the equation in parametric form of a line passing through points (2, 5, 3) and (12, 2, 1) using vector methods.

17. 4 Two points on a line are (1, 10, 12) and (−1, 12, 10). From vector analysis, determine the equation of this line in parametric form.

17. 5 Derive the equation of a horizontal line passing through the apex of a 1000-m broadcast tower in the direction N30°E. Use vector methods to find the parametric form. Assume the origin to be at the base of the tower and place the *x* axis to the north with the *z* axis upward.

17. 6 Determine the equation of the plane depicted by the triangular sector defined in Prob. 17.2. Use vectors.

17. 7 Derive the equation of the plane that passes through Houston, Texas, New York and a satellite stationed 22,000 miles above Vancouver, British Columbia. Use vectors and assume the origin to be at the center of the earth with the *x* axis through 0° latitude and the *z* axis through the north pole.

17. 8 For the object in Fig. 17.39, compute the viewpoint necessary to see the true size of the slanted surface. Compute and plot the slanted surface as viewed from (100, 100, 100).

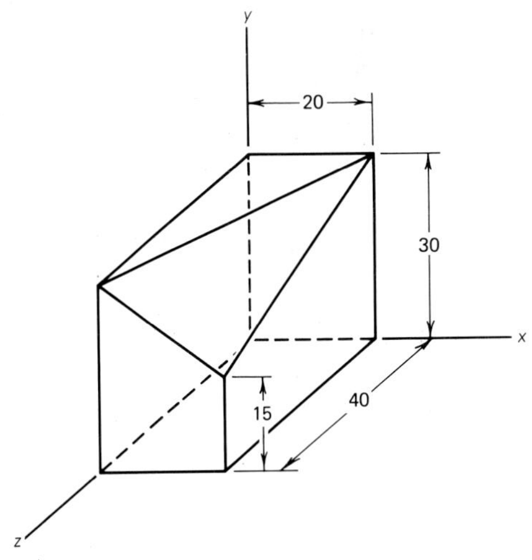

FIG. 17.39

17. 9 Determine a viewpoint suitable for measuring the inclination of a line $\overline{AB}$ passing through coordinates (−5, 7, 2) and (20, −2, 12).

17.10 Visualize your classroom. As you sit facing the front (by definition) of the room, imagine a line drawn from the center (left to right and up and down) of the front wall to the lower back right corner and a second line diagonally across the room, upper right to lower left. Using vectors, compute the shortest distance between these two lines.

17.11 Model one of the following objects or one designated by your instructor using the wire-frame method. Approximate curved surfaces by a series of planar surfaces. If possible, obtain paper copies of the top, front, and right profile views as well as an isometric pictorial of the object.

(a) Pliers
(b) Adjustable wrench
(c) Drafting pencil
(d) Bunk bed
(e) Folding chair
(f) Electric guitar
(g) Dish antenna
(h) Metal detector
(i) Table tennis paddle
(j) Personal computer with monitor
(k) Stepladder
(l) Funnel
(m) 0.5-mm lead container

chapter 18

Introduction to Engineering Design

18.1 INTRODUCTION

The steps in a design process were introduced briefly in Chap. 14. These steps are repeated here.

1. Identification of a need
2. Definition of the problem
3. Search
4. Criteria and constraints
5. Alternative solutions
6. Analysis
7. Decision
8. Specification
9. Communication

In Chap. 14 our intent was to point out what types of graphics documentation were created at each step of the process and how information is conveyed from step to step. Our purpose now is to describe how the design process can be used to solve a wide range of engineering problems. In this way, you can experience the design process and gain additional insight into how the engineer functions in an industrial setting.

In this chapter, we will observe the design process as it may be used

1. In the design of a modern commercial aircraft. The discussion will be from a corporate viewpoint. Obviously, all the details of such a design cannot be shown here, since hundreds of engineers are involved. In this first example, you will see how the design process keeps the engineering effort on a track which will produce the best solution within time and economic limitations.

2. To conceptualize solutions to a small-scale engineering problem in which a quick, effective solution is favored over a lengthy study complete with analysis of a spectrum of possible solutions. This type of problem is often assigned to one engineer or at most a team of three to five engineers and appropriate support personnel.

3. To generate an optimum solution to a specific problem. In many applications, such as building bridges, manufacturing automobiles and aircraft, storing toxic materials, and constructing computer systems, it is imperative that certain standards and design constraints be met. Thus, in a competitive situation, the industry that is first on the market with the best solution to a need has the highest potential for success. Design is a critical component in these organizations and a great deal of resources must be placed into seeking the optimum solution to problems.

FIG. 18.1 Computer-aided-design (CAD) is a large-scale activity in modern industry. Here an engineer is setting up a geometric database for subsequent analysis. (*Courtesy of Control Data Corporation.*)

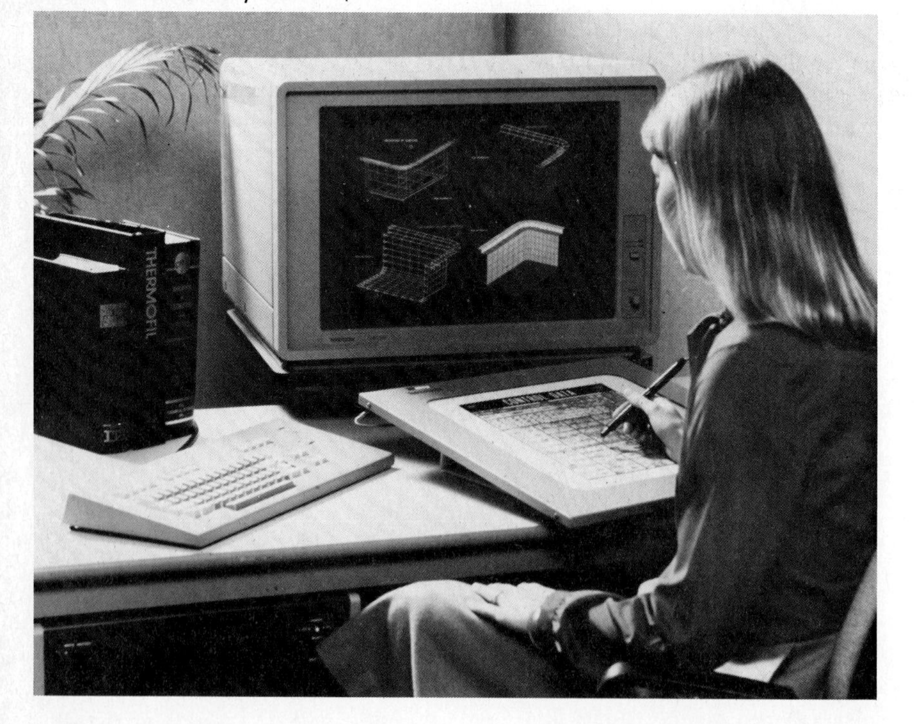

In each example, the impact of computer graphics will be discussed. Remember that the computer is a tool and note how the design engineer can take advantage of the computer. Do not think for one moment that the computer "designs." It simply responds with a vast amount of information to support the elements of the design process.

The computer has greatly increased our capacity to perform computations. Computer graphics systems have increased the capability of the designer to investigate more aspects of a given situation. It is possible today to perform analyses, decide among alternative solutions, and communicate results far quicker and with more accuracy than ever before. This, in turn, leads to better engineering and an improved quality of living.

The use of computer graphics in support of the design function is commonly referred to as computer-aided design (CAD). Modern industries today are coupling CAD with the computer-controlled manufacturing processes (computer-aided manufacturing, or CAM) through a common database. CAD/CAM techniques will become more prevalent in our industrial world in a very short time. Near the end of this chapter, the relation of the design effort with the manufacturing process will be briefly introduced.

18.2 A DESIGN EXAMPLE

Many large-scale industries such as aircraft, automobile, and construction initiate the design effort in their marketing divisions as a consequence of either a consumer request or a desire to incorporate new technology in an existing product. The decision to explore a new system, process, or device is generally made at the corporate level by persons with a great deal of experience in design. It is at this point that the general solution requirements are developed. Let us briefly trace the development of a long-range commercial airliner through a design process, commenting on the assistance provided by computer graphics during the development.

Figure 18.2 shows a typical industrial design organizational structure as it functions from recognition of a need to a final configuration ready for production. In the initial planning phase, a need is perceived for an intercontinental aircraft that can carry both passenger and cargo. The planning group will develop a concept of the anticipated routes, the amount of passengers expected on each route, and the freight to be transported. These factors are incorporated with cost and revenue data and basic aircraft parameters are then specified. No specific aircraft is proposed, but such quantities as gross takeoff weight, performance requirements for takeoff, climb, cruise and landing, and associated costs can be estimated. The results of these initial studies determine whether or not the company will pursue the effort.

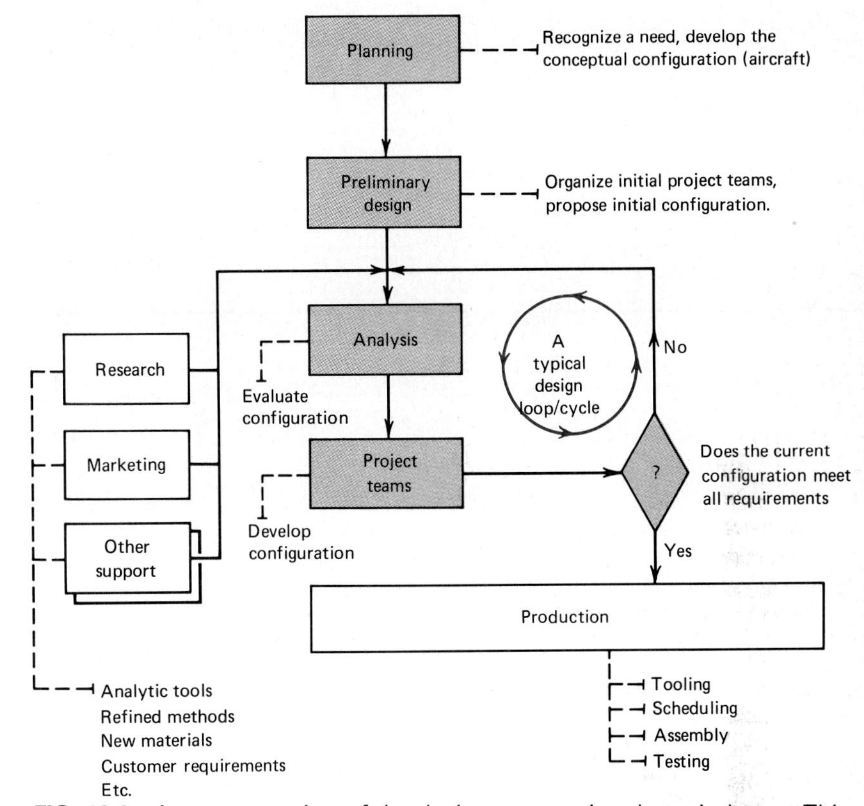

FIG. 18.2 A representation of the design process in a large industry. This particular example will be followed in the design of a new commercial aircraft.

Computer graphics in the planning stage is used in at least two ways. First, management has available a vast amount of data on existing aircraft, profit capability of existing routes, company capability for undertaking this large project, current and projected economic factors, etc. Analysts can interactively view this data in graphical form and quickly make decisions regarding the feasibility of the project. Second,

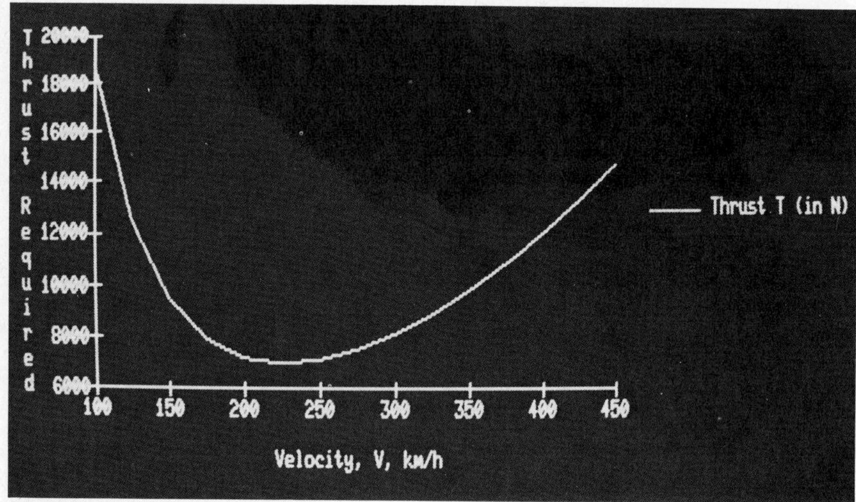

FIG. 18.3 Computer graphics enables aircraft performance factors to be viewed quickly for establishment of the overall aircraft parameters. In this plot the thrust-required curve assists in establishing an efficient cruise velocity.

the basic aircraft parameters must be established to satisfy the problem requirements. This is an interactive effort in which computer graphics can be used along with engineering experience and decision-making procedures to arrive at a conceptual aircraft.

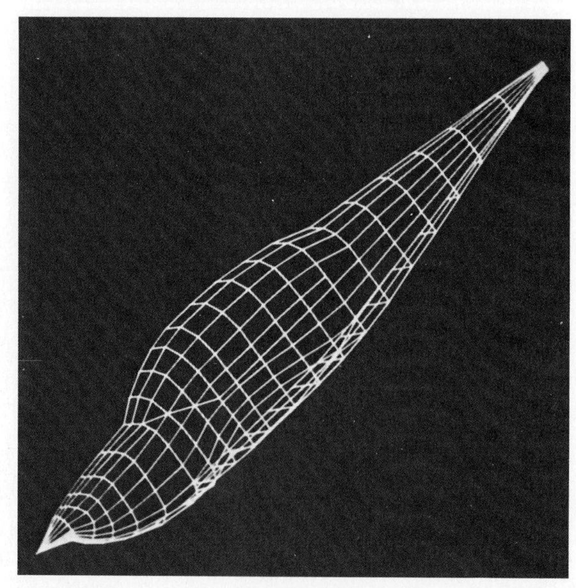

FIG. 18.4 The geometry of the fuselage is established as parts of the database. Wings, tail section, engines, and various control surfaces are also established. These can be merged into a single database for the analysis of aerodynamic and structural characteristics.

Upon approval of the project, the conceptual aircraft parameters are forwarded to preliminary design. This department will assume responsibility for developing the optimal aircraft for the given parameters. Their first task is to organize the various project and analysis groups. In order to maintain the necessary communication links on this large project, the group leaders for project and analysis work will be from the preliminary design department.

The modern design loop as shown in Fig. 18.2 begins with a proposed configuration based on the parameters specified in the planning stage. This configuration includes the major components such as fuselage, wings, tail section, and engines. The data for the aircraft geometry becomes the initial part of the common database that will be used for analysis, final design, and eventually for some of the manufacturing processes. Figure 18.4 shows a preliminary representation of part of the aircraft geometry.

Once a workable geometry is established, the analysis begins. Computer programs, called analyzers, compute mass, center of gravity, and aerodynamic characteristics such as lift and drag. Additional input is obtained from engine performance data, control systems, and structures. The aircraft is "flown" on the computer through a typical mission. A payoff function, which has been chosen by the preliminary design group, is the factor by which a

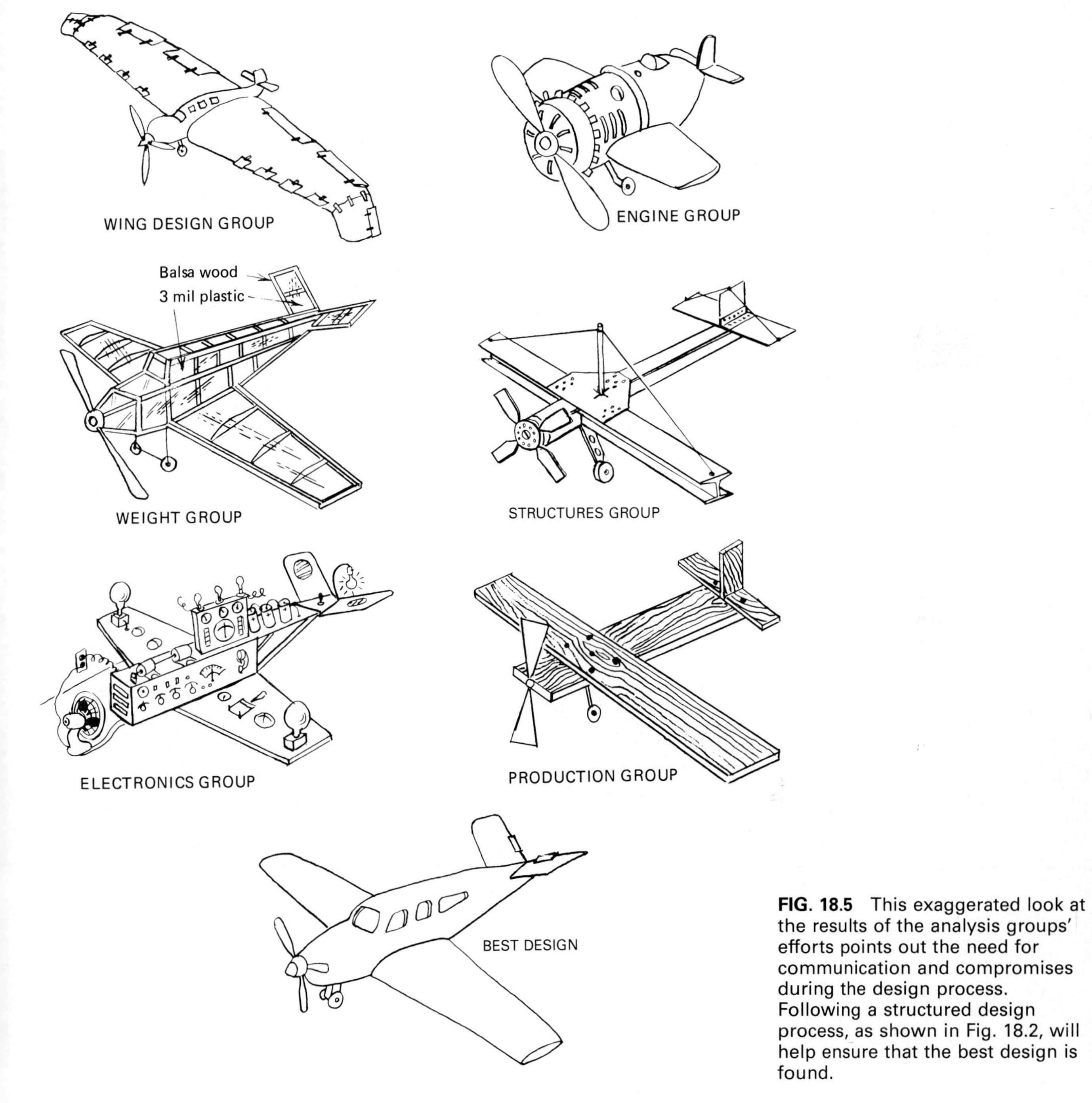

FIG. 18.5 This exaggerated look at the results of the analysis groups' efforts points out the need for communication and compromises during the design process. Following a structured design process, as shown in Fig. 18.2, will help ensure that the best design is found.

configuration is compared with other configurations. The payoff function, typically efficiency for an aircraft, is an optimum for some configuration. The designer seeks this configuration and the computer is used to investigate dozens of configurations in a fraction of the time it used to take to analyze one configuration.

It is important to realize in design that the optimum configuration for one aspect of the design is not necessarily the optimum one for another aspect. For example, the designer can begin with a nominal configuration (one which satisfies the parameters set forth in planning) and find the best configuration for aerodynamic characteristics. However, structural analysis may show that the wing shape must be altered in order to accommodate the engine mountings. Thus a tradeoff is made in order to obtain an optimum for both aerodynamic and structural requirements. Figure 18.5 is a lighthearted look at how individual analysis groups may view the aircraft design. Of course, we know that appropriate tradeoffs will result in the best overall design.

As the configuration begins to materialize as a result of the analysis and optimization procedures, the project group begins its mammoth task of finalizing the various assemblies of the aircraft. For example, project engineers who are responsible for designing the spars and ribs for the wing obtain various wing cross sections from the current database. With assistance from teams of designers and drafters operating CAD terminals, these engineers then develop a series of spars and ribs (see Fig. 18.6) which are then placed in the database for subsequent analysis by the structures group.

Additional designers will begin work on the assembly of ribs and spars to the wing covering. These designers will operate on a computer terminal from a menu containing various design and drafting operations. The common database will be used for geometry of the wing sections. For example, the designers will locate the places where rivets will be used in assembly. The size of the rivets as well as the number needed for the wing assembly will be selected. As this final design continues, the computer is storing all the information as a parts list and detail drawings which become a part of the overall database. The use of standard parts is easily incorporated, since this is a part of another database existing within the company which may include commonly used parts such as bolts, screws, plates, etc. The designer can now call for a

FIG. 18.6 The design of the individual parts is carried out with the appropriate portion of the database. Completed part designs are then added to the database. Only with the ability of the computer to store vast amounts of information in a quickly accessible format can the coordination of the design of all the parts of a complex configuration take place.

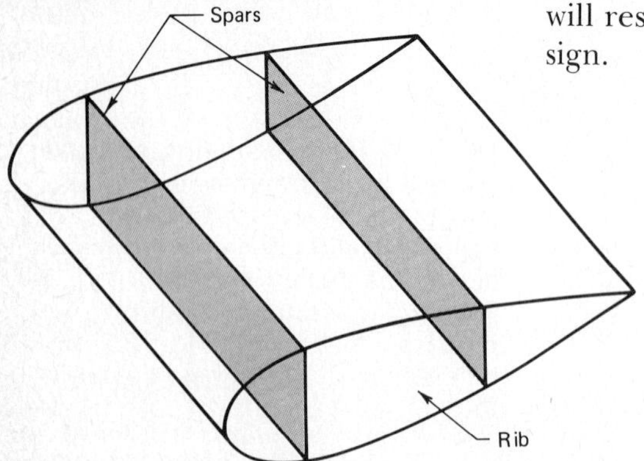

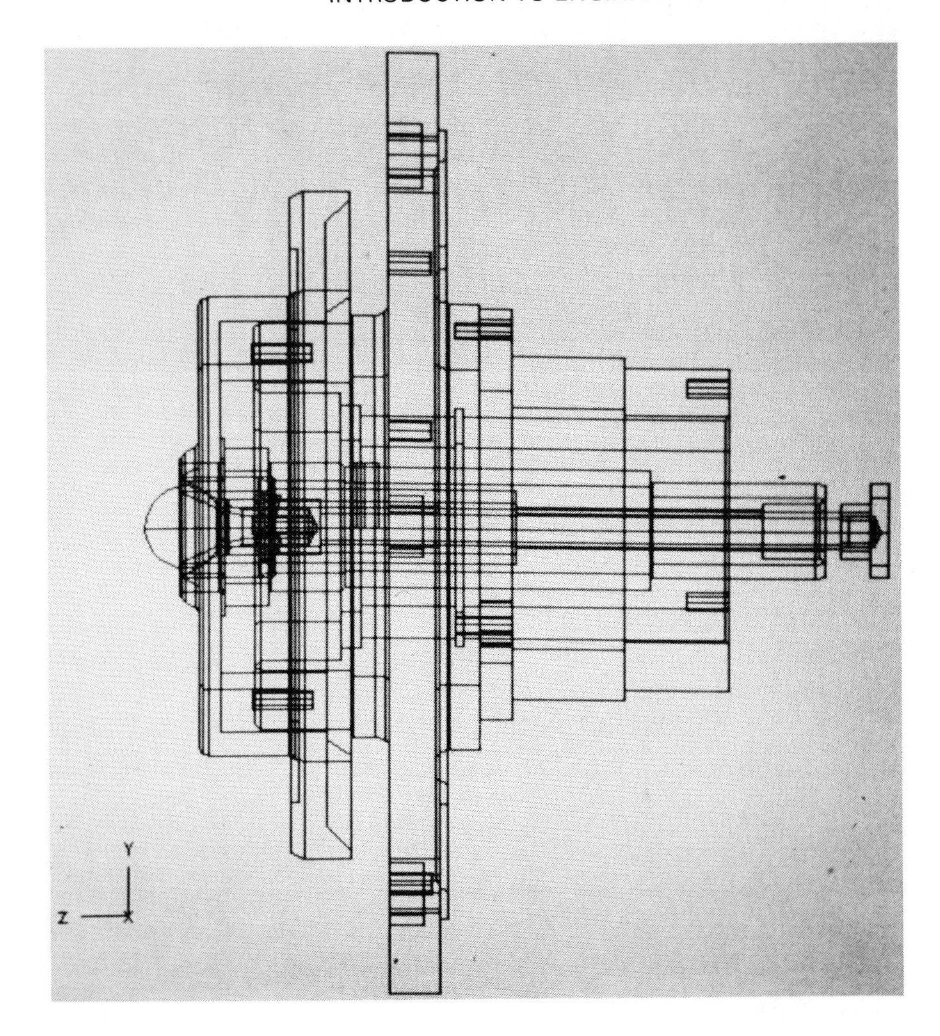

FIG. 18.7 A wire-frame assembly drawing of a mechanical part can be produced quickly on a plotter. (*Courtesy of Control Data Corporation.*)

particular type of drawing to be made on a plotter. A subassembly, detail drawing, or parts list requires only a few seconds or minutes of time to produce. See Fig. 18.7.

This brief overview of an aircraft design process does not provide you with a feeling for the magnitude of such an undertaking. From approval of a conceptual aircraft to the final database which will be used for production often takes years of time, thousands of engineers, designers, technicians, drafters, and support personnel and an investment of millions of dollars. Needless to say, any competitive edge which can be gained from new technology, shortened design time, and increased performance of the product will be supported by company management.

Before we discuss design examples of a more manageable size,

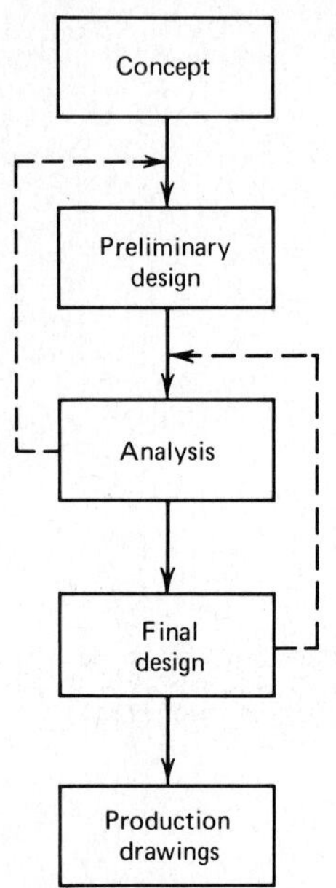

FIG. 18.8 A traditional design process showing the feedback loops. The cycle time for this process depends strongly on the reasons for going back to a previous step.

the impact of a modern design process as illustrated in Fig. 18.2 will be discussed further. Before widespread use of computer graphics and the concept of a common database, design proceeded in a somewhat direct fashion with feedback occurring somewhat slower than the design schedule. This is schematically shown in Fig. 18.8. If a problem with the proposed design was discovered during analysis, the final design would be stalled until preliminary design had corrected the problem. This time loss was expensive and affected the success of the design, particularly if the design was in competition with other designs. In addition, once the design had reached the stage of production drawings, any changes would be difficult if not impossible to make before production. These changes would be incorporated in a redesign, which in effect rendered the original design obsolete before it reached production.

Computer graphics and the common database have led to the modern design loop shown in Fig. 18.2. The activities of analysis and project engineering (final design and production drawings) are going on simultaneously. The effect is an optimum design in a shorter time period. Figure 18.9 shows how a typical analysis supported by computer graphics occurs. As the need for a particular analysis is made known, the first step is to access the common database for the required information. Called preprocessing, this step copies only the portion of the common database needed for the particular analysis and organizes this data in the format required by the analyzer. A computer graphics system is a must at the preprocessing stage in order to present vast amounts of data as geometry models, plots, charts, etc., in a short amount of time, allowing the engineer to verify the correctness.

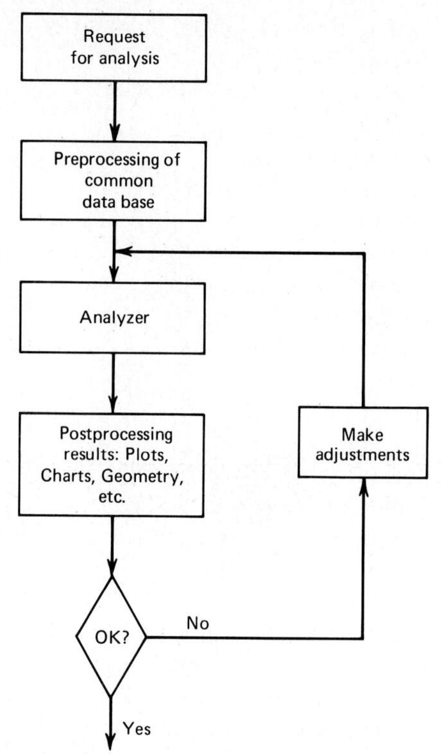

FIG. 18.9 A modern analysis loop takes advantage of the potential of the computer. The feedback cycle is timed to be much faster than is the next step that the design process requires; an optimum result can thus be obtained.

The preprocessed data is then fed to the analyzer. Most industrial analyzers are very complex computer codes which use numerical methods to accomplish the analysis. One of the most common analyzers is the finite-element code. If the geometry of the object is defined by elements as shown in Fig. 18.10*a*, then an analyzer can perform different tasks. For example, if a load is placed on the model, the displacements and stresses in the various elements can be determined, as shown in Fig. 18.10*b*. Similarly, if a heat source is placed on the model at some point, the temperature distribution throughout the model can be calculated.

Once the analysis is complete, the results, vast amounts of numerical data, can be postprocessed through a computer graphics system. This enables the engineer to quickly view and interpret the results and then make appropriate decisions. It is possible (see Fig. 18.9) that further analysis may be necessary, in which case the engineer makes the necessary adjustments to the model and database and requests a new analysis. It is most important to the overall design effort that the analysis be performed fast enough to be included in the decision-making process of the design loop (Fig. 18.2). If not, the analysis effort is at best a partial failure, since it can not be incorporated until a redesign effort is approved. Computer graphics provides the nec-

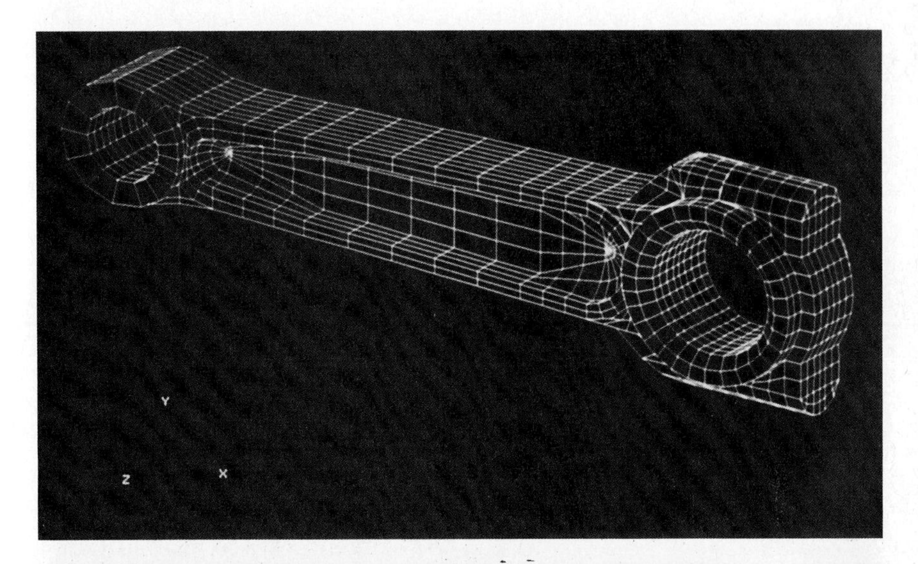

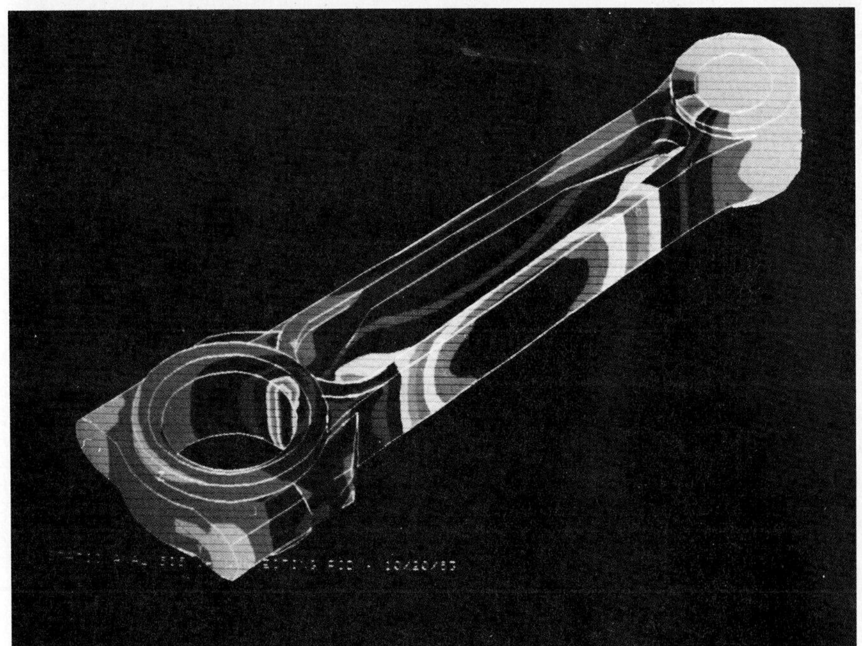

FIG. 18.10 (*a*) A finite-element model represented here as a geometric wire-frame diagram. (*b*) After loads are applied, the stresses are computed by the analyzer and displayed with shading techniques. Depending upon the results of the analysis, the geometry may be adjusted to strengthen the part. (*Courtesy of Control Data Corporation.*)

essary speed for the analysis to be an integral part of the design loop.

Note also in Fig. 18.2 the function of the research engineers and scientists. This group is responsible for developing the new analytic methods which will be incorporated into the analyzers, bringing new materials to the attention of the designers and discovering new processes for manufacturing which will lead to a better product. Although not a part of the design loop shown in Fig. 18.2, research feeds directly into the loop at the analysis function, so that new ideas can be used to advantage immediately.

18.3 CONCEPTUAL DESIGN

Section 18.2 discussed in general terms a large-scale design process. We must now investigate smaller-scale problems that will enable us to experience how the engineer works within the design process and to actually use elements of graphics, computers, and computer graphics to assist the process. Let us review the nine-step design process introduced in Chap. 14. We will give a brief description of each step and then proceed with some examples.

1. Identification of a need.
The perception of a problem or concern, which generally originates with the marketing division or planning department in an industrial organization.

2. Definition of the problem.
A concise statement, usually developed at management level, which presents a broad concept of the problem and often includes perceived solution(s).

3. Search.
The process of investigating existing solutions and conditions which add to the knowledge about the problem.

4. Criteria and constraints.
Factors which are used to evaluate alternative solutions. Criteria are desirable characteristics of a solution which allow alternative solutions to be compared on a relative basis. Constraints are quantitative limitations on the problem conditions and variables. Constraints may be in the form of limitations on the physical makeup of the solution such as size and mass, for example. They may be the restrictions imposed by the laws of nature, such as conservation of mass. In a sense, constraints may be imposed on the criteria. For example, if cost is a criterion, we may wish to keep it under $200. We can thus compare cost of alternative solutions on a relative basis as long as each alternative is less than $200 total.

5. Alternative solutions.
A variety of solutions or solution concepts which satisfy the problem definition, criteria, and constraints.

6. Analysis.
A study of alternative solutions to verify anticipated performance. The laws of nature and economics are used to evaluate models of the alternative solutions.

7. Decision.
Selection of the best solution based on analysis of the alternative solutions and evaluation against the stated criteria.

8. Specification.
Complete description of the final solution in the form of design drawings, reports, and databases.

9. Communication.
Presentation of the specifications to the groups responsible for implementation of the solution.

Inasmuch as the engineer is a problem solver, an element of creativity is necessary to yield imaginative solutions to problems. The necessity to be creative does not mean that as an engineer you must possess the qualities of a genius, but simply that you must apply your knowledge and the knowledge of others that you have access to and be able to effectively use the tools of engineering. The design process is the method by which you can maximize your creative efforts. Let us follow a problem from its inception through the phase in which alternative solutions or concepts are established; that is, let us apply the first five steps of our design process.

The interoffice memo in Fig. 18.11 briefly outlines a need within a division of a company responsible for packing and labeling cartons. Note carefully the time frame for response. The first step of the design process, identification of a need, has been clearly stated in the memo. If we work in preliminary design, we will begin our effort by trying to establish a problem definition.

In the broadest sense possible, the problem of automatically orienting cartons for labeling could begin with the actual process of filling the cartons. Figure 18.12*a* shows a problem statement in terms of an undesirable situation

FASTPAK, INC.
INTEROFFICE MEMO

Tuesday, May 15, 19__

TO: JIM WATSON, PRELIMINARY DESIGN

FROM: DAN ROGERS, VICE PRESIDENT OF ENGINEERING

I have scheduled a meeting in my office for 3:00 p.m. on Friday, May 18 to discuss the development of an automatic process for orienting cartons for a labeler in our packaging department. You are asked to present your ideas in such a manner that a decision can be made to go forward on this project or to cancel it.

In conversations with Bill Dalton in the packaging department, I have gathered the following information. You may wish to contact Bill for additional details.

1. The cartons will be in the shape of a rectangular prism with the small dimension ranging from 10-25 cm, the intermediate dimension from 30-50 cm, and the long dimension from 70-100 cm. The mass of each carton will be less than 10 kg.
2. The cartons are dropped from a chute onto one end of a 10–m conveyor. Thus the carton orientation is random.
3. The speed of the conveyor and the rate at which cartons are introduced to the conveyor can be controlled.
4. Either of the larger surfaces of the carton may be labeled.

Representatives from the planning and packaging departments will be present at the meeting.

cc: Bill Dalton

FIG. 18.11 A typical request for engineering work.

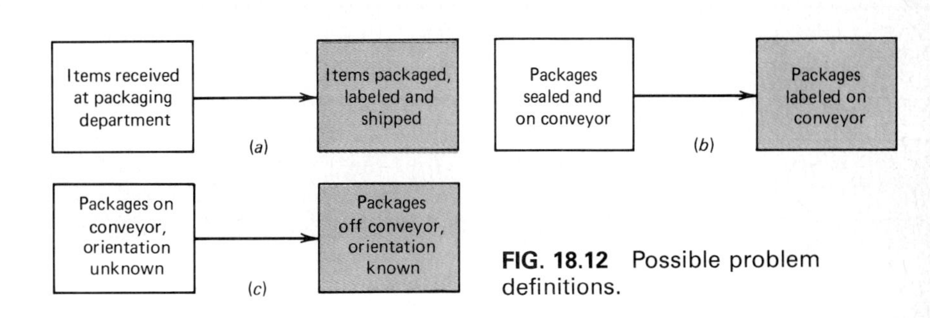

FIG. 18.12 Possible problem definitions.

on the left and the desirable situation on the right. It is the designer's task to effect the transformation from left to right.

In this case, however, we are told that the cartons are packed and sealed and dropped onto a moving conveyor. Thus, the problem statement of Fig. 18.12*a* is not compatible with the need as stated. Figure 18.12*b* restates the problem but leaves a question about how the cartons would be labeled. We therefore contact Bill Dalton, the reference indicated in the memo in Fig. 18.11, and learn that the labeler already on hand requires that the packages be stationary, so we must orient the packages correctly at some point on the conveyor and then make provisions for removing the packages for labeling. We also learn that the side to be labeled should be facing upward when it leaves the conveyor headed to the labeler. This would reduce the need for reorientation. With this new knowledge, we rewrite our problem statement as shown in Fig. 18.12*c*.

It should be emphasized that the problem definitions in Fig. 18.12*a* and *b* are not incorrect. In fact, by writing and studying many possible problem definitions, the engineer may discover a creative approach to a problem solution that will require rethinking about the original need. However, the time frame and specific requirements for the carton orientation problem make the problem statement of Fig. 18.12*c* a logical choice.

The amount of effort put forth in the search phase depends primarily on the time available. Since time is critical for this problem, we will draw on our experience in the area and use the telephone and coworkers to help fill in our "knowledge gaps" in this problem.

We begin by listing the conditions that are known or can be determined and the variables over which we can exercise control. The source of the information is indicated in parentheses.

Conditions (source)

1. Package size ranges from 10 × 30 × 70 cm to 30 × 50 × 100 cm (memo). Packages are not fragile (phone call). Mass is less than 10 kg (memo).

2. Conveyor is 10 m long (memo) by 120 cm wide (phone call).

3. Labeler requires packages to be motionless; the package will be labeled from the top (phone call).

4. The packaging department has several sources of power available including 110/220 electric and pneumatic (personal inspection).

Variables

1. Conveyor speed

2. Rate of introduction of packages to conveyor

3. Method of orienting packages

With these factors listed, we now generate a list of physical principles and devices that may be helpful in the orientation process. This list may not be complete and may be added to as we become more involved in the problem.

1. Gravity
2. Lever
3. Friction
4. Sensing devices
   (a) Photoelectric cell
   (b) Laser
5. Robotics

Other factors, such as economics and legal aspects, must also be considered. For this preliminary effort, general judgments will be made on the basis of very little data for quantitative determinations. From an economic standpoint, the simpler the solution, the more likely the economic considerations will be favorable. As much cost data as possible should be gathered, particularly on such items as sophisticated electronic components, special-purpose hardware, and labor costs for operation.

Legal factors for this application will be studied in a subsequent design (specification) phase, since the vice president's request is for a preliminary look at the feasibility of developing an automatic process for orienting cartons. Any design which will eventually impact on the general public will probably require extensive legal as well as social considerations in the preliminary phase.

Next in the design process is the establishment of the criteria and constraints. It is very easy for the engineer at this point in the design to begin thinking "solutions," since the problem constants, variables, and other factors have been recorded. However, establishing the criteria and constraints determines the range of solutions and the desirable characteristics and also provides a means of comparing alternatives and selecting the best.

The choices for the criteria are many in number. Cost, reliability, mass, ease of operation, ease of maintenance, appearance, compatability, safety features, noise level, effectiveness, durability, feasibility, and public acceptance are some general possibilities. For this preliminary design, the four or five most important criteria are sufficient. It is the prerogative of the designer to establish the relative importance.

Four criteria are selected for this problem.

1. Cost. This is frequently the most important of the criteria. To evaluate the costs associated with a design requires a good background in engineering economy and accessible data on the material, labor, and overhead costs in a particular industrial setting. For this particular problem, we would have to rely on data provided by the financial department and try to compare costs of alternative solutions on a relative basis rather than with absolute numbers.
2. Effectiveness. The final design must orient the cartons correctly in a consistent manner.
3. Reliability. The final design must function when required; that is, maintenance must be kept at a minimum.

4. Feasibility. The solution must be such that it can be implemented with reasonable effort. For example, is the technology capability within the Fastpak organization at a level high enough to maintain and operate a sophisticated solution utilizing robot technology? If not, then such a solution would not be feasible, at least at this time.

Constraints are put into quantitative form if possible. For example, although the conveyor speed is considered a variable, there will be upper and lower limits. Another call to the packaging department reveals that the conveyor speed may be set from 0.1 to 0.5 m/s, which would be a constraint on alternative solutions but would still admit a wide range of solutions which depend upon conveyor speed. Another constraint is the 10-m length of conveyor.

Constraints frequently appear as equations specifying laws of nature. Alternative solutions will be subjected to analysis, during which the performance of the alternative will be determined from the applicable laws of nature. In our carton-orientation problem, friction will most likely assist us. We will assume, for our preliminary design, that friction between the carton and conveyor belt is sufficient to prevent slipping when a small turning force is applied to the carton. We will not try to specify any of the friction equations here. We do realize that any alternative solutions prepared would be subject to analysis at the appropriate step of the design process.

We are now prepared to generate alternative solutions. All data that can be collected in the limited time is available. At this point your creativity can be most helpful. What is available to you as a designer that will enhance your creative ability and generate several good alternatives? First, the knowledge that you have gained up to this time is valuable. Most of you reading this book have not had any engineering analysis and design courses, which means that you might have some difficulty in solving complex problems at this time; but you have mathematical skills and reasoning ability which can be applied to general, conceptual problems such as the carton orientation.

Second, engineering experience in the area of application (material handling for the carton problem) is important. You do not, of course, possess this experience now. However, you have observed many phenomena which could relate to the carton orientation problem, such as escalators in an airport, for example, whereon friction between shoes and escalator surface is sufficient to prevent slipping; security devices used to spot illegal items, which require sensing devices; gates for specific flights, which is a sorting process. Many similar examples can be cited from everyday life that may whet your curiosity about applications to the carton problem.

Third, creative solutions generally are arrived at with a combination of knowledge, experience, and hard work. If you carefully study a successful engineer, or any successful person for that matter, you will discover that ideas and decision making do not come easy with them, but are arrived at after a great deal of observing, researching, and thinking. Sometimes an extra few minutes or hours spent on a problem will bear fruitful results.

Fourth, if you are a member of a team of students or professional engineers, you may conduct special sessions designed to elicit a better variety of ideas than an individual might produce. One such method is brainstorming, where an idea spoken in front of a group can trigger other ideas from different group members. Criticism of any idea is forbidden so that the group may feel relaxed and unrepressed. These sessions are usually short (10 to 15 min).

For the carton orientation problem, one alternative solution will be presented. There are many other possibilities, which are left as student exercises at the end of the chapter.

In developing an alternative solution, one must remember that the idea will be conveyed to one or more persons generally at a higher level in the organization. The solution must thus include a clear description of the form and function. The solution idea may also be sent to the analysis group, which means that pertinent data with regard to geometry, materials, and expected performance should be available. Each problem usually requires a somewhat different approach depending upon the complexity and the initial request for action. In our case, we are responding to the request outlined in Fig. 18.11.

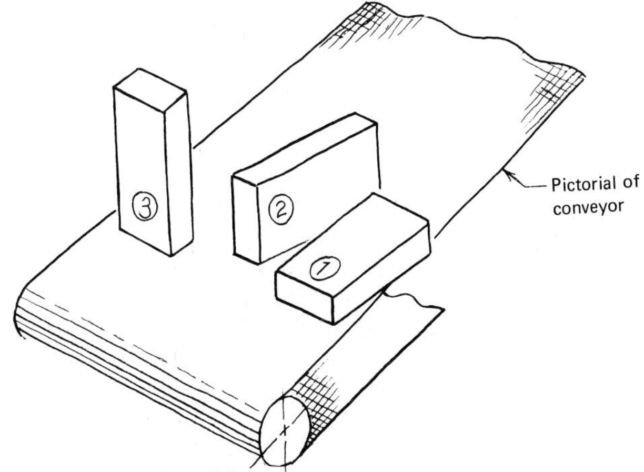

FIG. 18.13 Possible carton orientations.

Figure 18.13 shows pictorially our problem. Cartons are set on a conveyor in a random fashion. Possible orientation of the carton with respect to its sides is shown as configurations 1, 2, and 3 in Fig. 18.13. The problem at hand is to have all cartons oriented as configuration 1 when the carton is removed from the conveyor.

The alternative solution presented here consists of a three-stage process:

1. Orient cartons to configuration 1 or 2, as shown on Fig. 18.13, by installing a barrier at a 60-cm height over

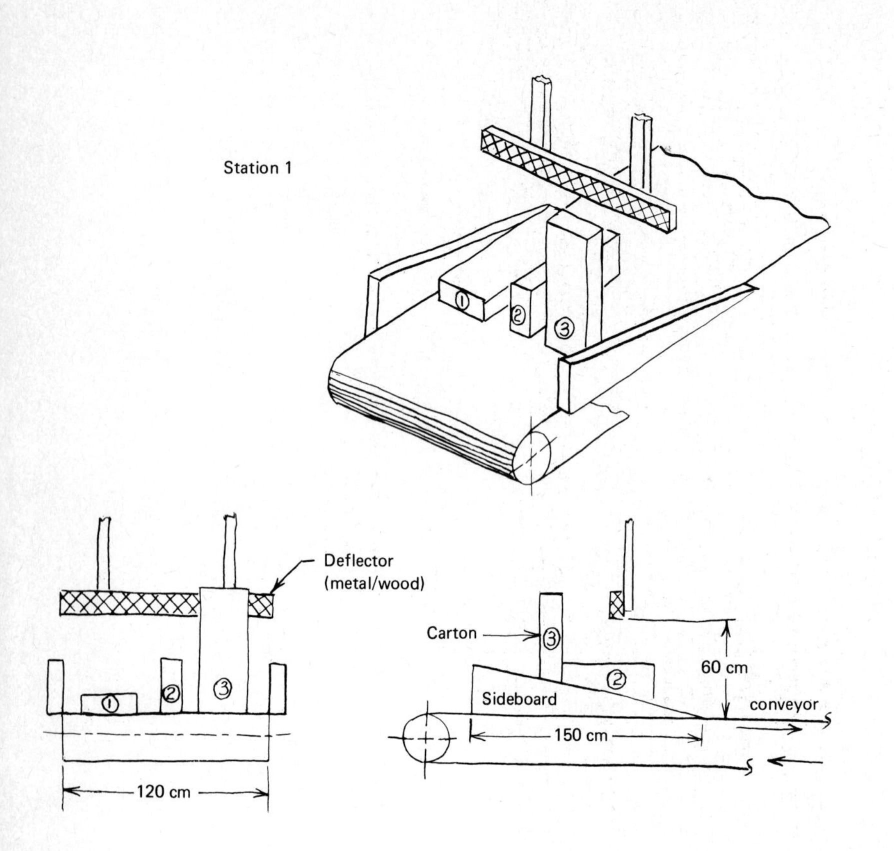

the conveyor as described in Fig. 18.14. Note the use of a multiview and a pictorial to convey both form and function.

2. Separate the cartons in configurations 1 and 2 with an adjustable metal band, as shown in Fig. 18.15.

3. Finally, the carton in the 2 position will be tipped over an edge on the right-hand side of the conveyor by a flexible band. The 1-position cartons, which are in the correct position, are simply guided off the conveyor, as shown in Fig. 18.16.

FIG. 18.14 The first stage of a possible solution to the carton-orientation problem.

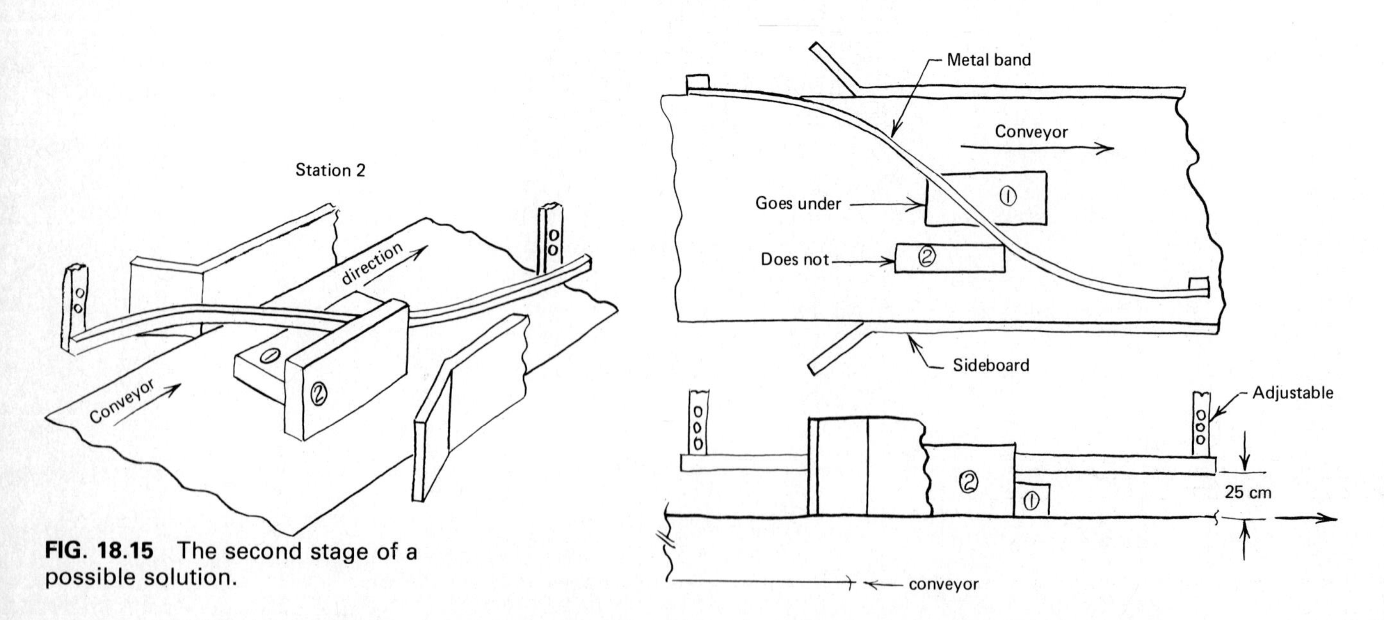

FIG. 18.15 The second stage of a possible solution.

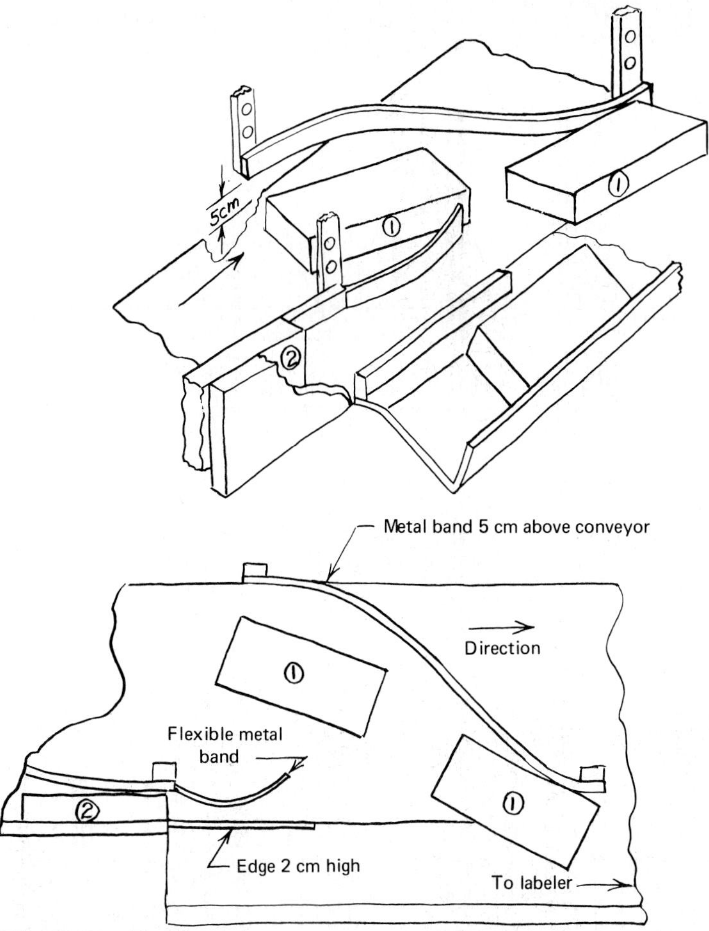

FIG. 18.16 The final carton-orientation step.

For this alternative to function effectively (criterion 2), the cartons must be put on the conveyor one at a time and spaced approximately the longest carton dimension apart. This will prevent overlapping and jamming as the cartons are sorted and guided off the conveyor.

With these sketches in hand and time to orally present these as well as other alternatives not shown here, the meeting in the vice president's office should go well.

Up to this point in the carton-oriented problem, we have not mentioned the use of a computer or computer graphics. For this particular problem, it is probably not necessary unless during the search step we can obtain some data about conveyors, material-handling equipment, and so forth. The engineer must decide when given a problem what resources can be used in the allotted time to assist in the design process.

We will not follow the carton-orientation problem through to completion of the design process, because certain aspects of the analysis are beyond the scope of this book. Instead, we will present two examples which clearly illustrate the analysis and decision-making steps of the design process.

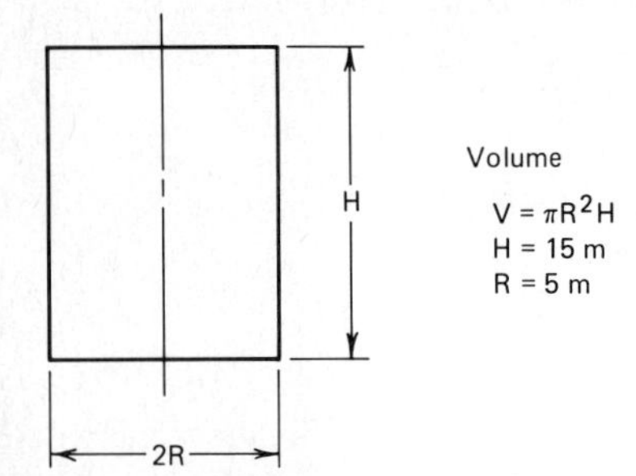

FIG. 18.17 Representation of a cylindrical water tower.

18.4 DESIGN BY ANALYSIS: EXAMPLE 1

One simplified definition of design is to create according to a plan. We have established a more detailed approach previously with the nine-step design process. To create implies an ability to achieve order from disorder. In other words, a designer must bring different ideas, objects, and principles together in a manner which satisfies the need. For most of us this creative effort is not straightforward. Consider an example which involves a volume of water contained for distribution to a residential area in a city. Suppose that we have a cylindrical-shaped water tower 15 m tall with a diameter of 10 m, as shown in Fig. 18.17. How much water can be held when the tower is full? This is a very simple *analysis* problem. We are given a system (cylindrical tower) and the operating environment (filled with water) and are asked to find a specific system response (volume). We write the expression for volume of a cylinder, substitute known quantities using consistent units, and determine the volume.

Suppose, however, that we present a more realistic situation that an engineer might encounter. Assume that the problem is stated as follows. There is a need for a water tower that has a capacity of 5000 m^3. In this case we are given the desired system response [volume of 5000 m^3], and the operating environment (filled with water) and are asked to define the system (tower). This is an example of a *synthesis* problem and there is not in general a unique solution. For example, if the tower is to be cylindrical, there are an infinite number of combinations of R and H that would yield a volume of 5000 m^3 (see Fig. 18.17). The designer may also choose other shapes. If a spherical tower is used, there is a unique solution, since the expression for volume, $V = 4/3\pi R^3$, contains a single variable. Other choices for tower shape, such as ellipsoids and right prisms, would again yield an infinite number of possibilities.

The designer needs additional conditions specified in order to select one of the possibilities. One of these conditions that is almost always imposed is a performance condition. This is an expression which defines a predetermined performance requirement of a solution. The following example will illustrate this concept.

Example 18.1

A containing tank is to be designed to hold $(5)(10^6)$ L of oil when filled. The configuration is to be in the shape of a cylinder with a hemispherical top. Material costs for the cylinder portion will run $300 per square meter; material costs for the hemisphere are $400 per square meter. Determine dimensions for the tank which result in the least material cost.

Discussion

In terms of the design steps listed in Sec. 18.1, we are at a stage where an alternative solution has been selected (step 5) and we are to analyze the given configuration and decide upon the best configuration according to the given performance condition (steps 6 and 7). Without the performance condition imposed, there is an infinite set of solutions which are valid.

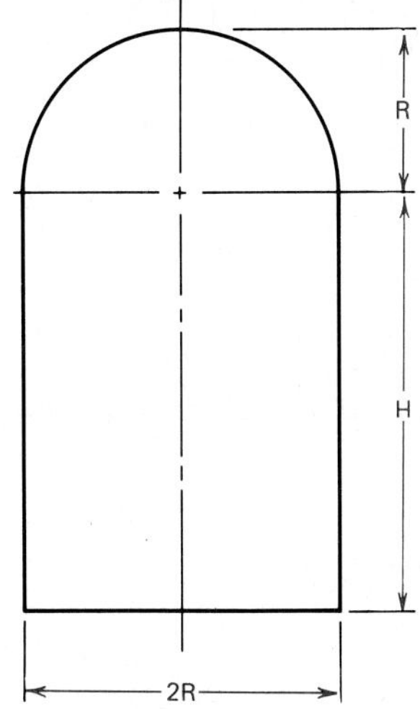

FIG. 18.18 A water tower made from a cylinder and hemisphere.

Analysis and Decision

Figure 18.18 illustrates the geometry. The variables for the problems are:

R	Radius, m
H	Height, m
V	Total volume, m^3
VH	Volume of hemisphere, m^3
VC	Volume of cylinder, m^3
AH	Surface area of hemisphere, m^2
AC	Surface area of cylinder, m^2
CC	Cost factor for cylinder, $\$/m^2$
CH	Cost factor for hemisphere, $\$/m^2$

We will first write the constraint equation for the volume,

$$\begin{aligned} V &= VC + VH \qquad (18.1) \\ &= \pi R^2 H + \frac{2\pi R^3}{3} \\ &= 5000 \end{aligned}$$

where 5000 m^3 is the equivalent of $(5)(10^6)$ L. From this equation any number of solutions to the problem can be determined. These solutions are often called nominal solutions. They satisfy the constraints but are not necessarily the best in view of the performance condition. For example, if R is chosen as 8.00 m, the corresponding height H is 19.53 m. The choices for R are, of course, not unlimited. The reader should verify that for an R equal to or greater than 13.37 m, the given configuration (cylinder capped with a hemisphere) is not possible.

We now write the cost C equation which allows us to apply the performance condition.

$$\begin{aligned} C &= (CC)(AC) + (CH)(AH) \\ &= CC(2\pi RH + \pi R^2) \\ &\qquad + CH(2\pi R^2) \\ &= CC(2\pi RH) + CC(\pi R^2) \\ &\qquad + CH(2\pi R^2) \qquad (18.2) \end{aligned}$$

The cylinder surface area includes the lateral area plus the bottom.

We can combine the volume and cost equations by solving Eq. (18.1) for H, substituting into the cost equation and simplifying the resulting expression. This will leave cost as a function of a single variable, the radius R.
From Eq. (18.1),

$$H = \frac{5000}{\pi R^2} - \frac{2R}{3}$$

Substituting into Eq. (18.2) yields

$$C = CC[2\pi R(\frac{5000}{\pi R^2} - \frac{2R}{3})] + CC(\pi R^2) + CH(2\pi R^2) \qquad (18.3)$$

At this stage each of the variables in Eq. (18.3) maintains its identity. The individual cost factors CC and CH, radius R, and required volume, 5000 m^3, are identifiable. This equation is to be minimized for the problem conditions. For our specific problem, all the factors in Eq. (18.3) are constant except for the radius.

Substituting for CC and CH and reducing, Eq. (18.3) becomes

$$C = \frac{(3)(10^6)}{R} + 700\pi R^2 \qquad (18.4)$$

Note that the contributing factors are no longer identifiable. We can now build a table of values by choosing a range for R and calculating the corresponding values of C and then plotting the results. Figure 18.19 is the resulting plot. From the graph we can read the value of R = 8.80 m for minimum material cost. The corresponding height, H = 14.69 m, is then found from Eq. (18.1).

If you have experience with differential calculus, you have recognized that cost in Eq. (18.4) can be minimized by the application of maximum/minimum procedures. This of course will yield the correct result much quicker than will the graphical procedure of Fig. 18.19. We are attempting here, however, to illustrate a design procedure, and the process we have followed can be extended to computer analysis of more complex problems for which the elementary calculus procedures cannot be used. We also get a better understanding of the effect of the problem variables on the solution by seeing the variation of cost across a range of values of radii (the design space).

If in Eq. (18.3) we replace the 5000-m^3 volume with a variable name, say K, then the equation can be used for problems which

FIG. 18.19 Determination of the optimum water-tower configuration.

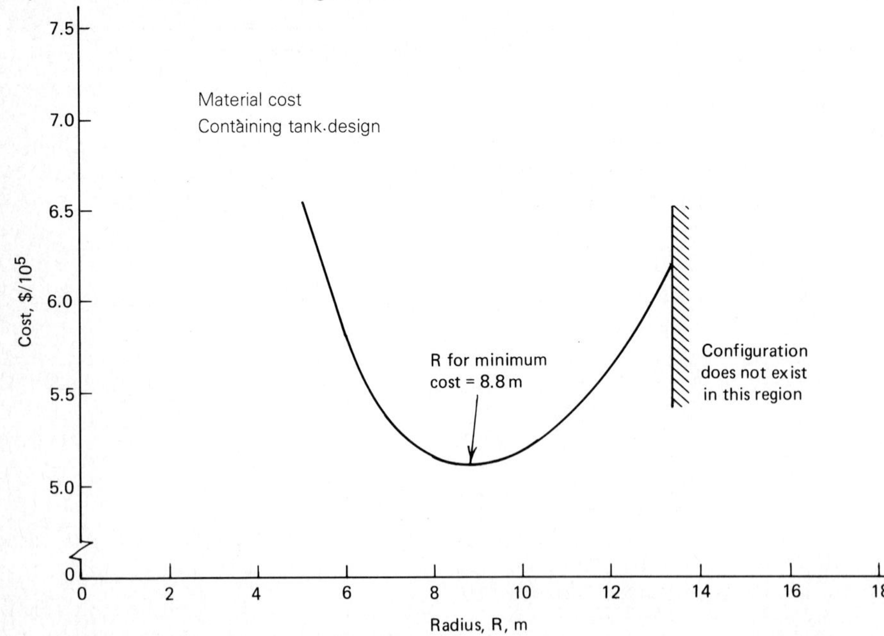

have different volume requirements and varying construction costs. It may be practical to write a computer program to evaluate Eq. (18.3), allowing the user to input CC, CH, K, and the range and increment for R. If a computer graphics plotting package is available, the user can very quickly determine many possible configurations for least cost for specified input. The program can then be used by others who are designing containing tanks of the cylinder plus hemisphere shape. Figure 18.20 shows a plot of our problem for three different construction costs for the hemisphere portion. Your knowledge of the terms in Eq. (18.3) should enable you to explain the differences in the curves in Fig. 18.20.

This example points out once again the need to understand the problem and the design space available for the solution. In Fig. 18.20, close study of the curves reveals the sensitivity of the total cost relative to the cost of the hemisphere portion. This gives the engineer a basis on which to pursue design alternatives. Inasmuch as the engineer must be able to make the correct interpretations, the computer allows for more possibilities to be studied in a shorter period of time, thereby giving the engineer more time for making decisions and creating better solutions.

The next section will further demonstrate the value of the computer and computer graphics to the design engineer.

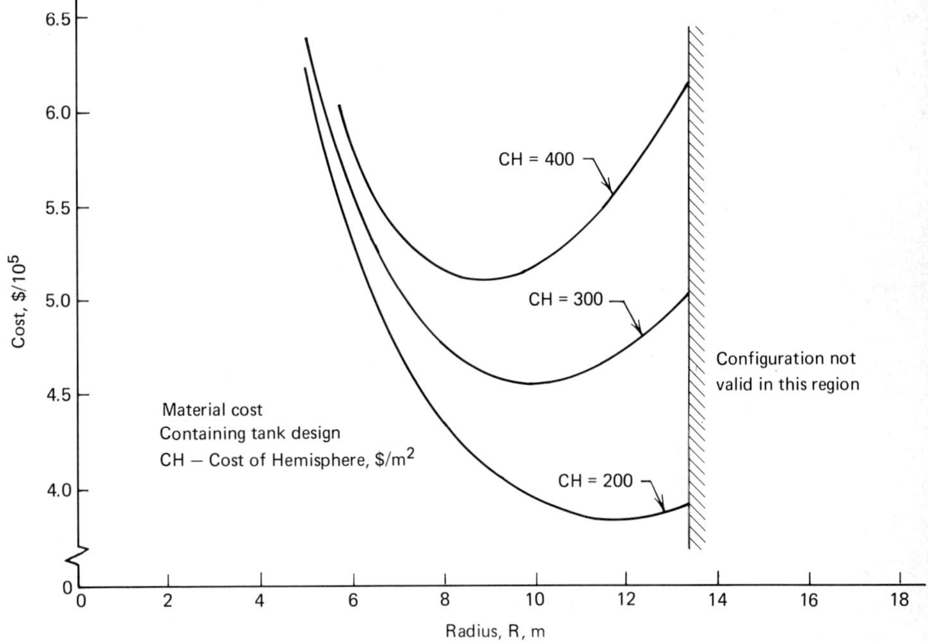

FIG. 18.20 The effect of varying one of the design parameters.

18.5 DESIGN BY ANALYSIS: EXAMPLE 2

Suppose that the crane configuration shown in Fig. 18.21 has been selected as the means to move large castings onto and off of railroad flatcars. At this stage of the design process, very little is known about the specific components of the crane. The key components of this configuration are the pin which allows rotation of the boom and the cable which controls the boom position (angle). It is critical that proper analysis be performed to determine the sizes of the cable and

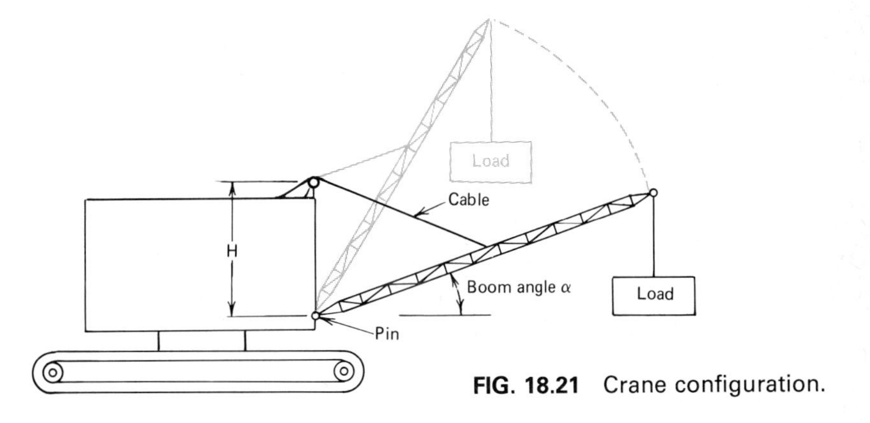

FIG. 18.21 Crane configuration.

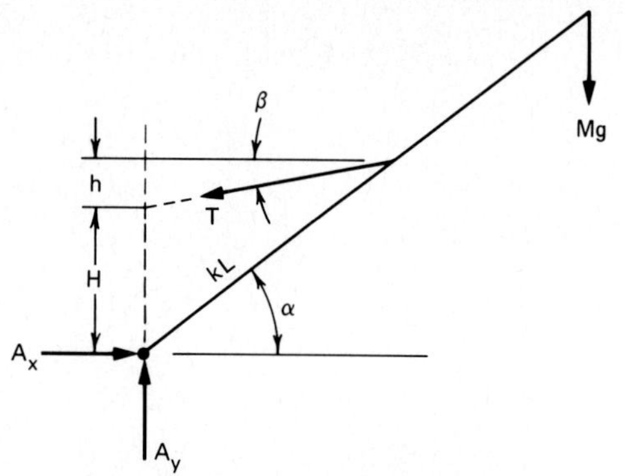

FIG. 18.22 Freebody diagram showing the boom geometry and the forces acting on the boom.

pin. Once this is accomplished, design of the remaining components can be undertaken.

Let us investigate what problem conditions and variables are involved in the determination of the cable and pin sizes. Assume that we are the analysis group and must suggest the "best" pin and cable diameters in a short time frame. We must study the problem carefully, write the conditions and variables, identify constraints, and specify the criteria for the best solution. We would then conduct the analysis and forward the results by a specified deadline.

The pin and cable are sized according to the load each supports, which in turn is a function of the load lifted by the crane. The position of the boom (boom angle) and the point of attachment of the cable along the boom will affect the load distribution. For a preliminary analysis we will neglect the mass of the boom and the cable. This will simplify our equations for this example and afford a better understanding of the relations between the variables.

Figure 18.22 shows an isolated view (called a free-body diagram) of the boom geometry and acting forces. There are nine variables in this problem. Those variables are

Ax, Ay	components of pin reaction on boom
T	tension load in cable
Mg	force due to mass M. g = 9.807 m/s^2
α	boom angle
k	constant $0.2 \leq k \leq 1$
L	length of boom
H	cable arrest height
β	angle of cable

The boom will be assumed to be in static equilibrium for our preliminary analysis. We can therefore write the three static equations of equilibrium which constrain the system.[1] From Newton's laws, the three equations of equilibrium for a coplanar noncurrent force system are

$$\begin{aligned} \rightarrow \Sigma F_x &= 0 \\ \uparrow \Sigma F_y &= 0 \\ \Sigma M_A &= 0 \end{aligned} \qquad (18.5)$$

[1] For the student who has not yet been exposed to statics in physics or engineering mechanics, a complete understanding of the equations of equilibrium is not necessary to follow this analysis segment. This will become more evident later in this section.

This will reduce the variables under our control to six (nine variables minus three equations). A geometry equation can be written relating β to the other geometry variables in Fig. 18.22. This reduces the control variables to five. To analyze the effect of all five remaining control variables on the sizes of the pin and cable is still a monumental task. If we fix four of the variables, say M, k, L and H, then one control variable, α, remains.

Example 18.2

For the crane configuration illustrated in Figs. 18.21 and 18.22, determine the maximum load that would occur for the pin and cable using the following parameters.

Boom angle α	$10° \leq \alpha \leq 80°$
Applied mass M	$(1.00)(10^4)$ kg
Boom length L	20.0 m
Cable arrest height H	4.00 m
Constant k	0.5

Discussion

Determination of the maximum load will enable us to size the pin and cable for the worst anticipated design condition. We will work with the boom angle in increments of 10° within the range given. A plot of pin load and cable load versus boom angle will provide us with the solution.

Solution

The governing equations found from Eq. (18.5) and the geometry relation are

$\rightarrow \Sigma F_x = 0$:

$$A_x - T\cos\beta = 0 \qquad (a)$$

$\uparrow \Sigma F_y = 0$:

$$A_y - T\sin\beta - Mg = 0 \qquad (b)$$

$\Sigma M_A = 0$:

$$(-T\sin\beta)(kL\cos\alpha + (T\cos\beta)(kL\sin\alpha) - Mg(L\cos\alpha) = 0$$

which reduces to

$$T = \frac{Mg\cos\alpha}{k(\cos\beta\sin\alpha - \sin\beta\cos\alpha)} = \frac{Mg\cos\alpha}{k\sin(\alpha-\beta)} \qquad (c)$$

From Fig. 18.22, the following geometry relations can be written:

$$\sin\alpha = \frac{H+h}{kL}$$

$$\tan\beta = \frac{h}{kL\cos\alpha}$$

Eliminating h and solving for β, we find

$$\beta = \tan^{-1}\left[\left(\frac{kL\sin\alpha - H}{kL\cos\alpha}\right)\right] \qquad (d)$$

A fifth equation is needed to determine the total pin load from the components A_x and A_y.

$$A = (A_x^2 + A_y^2)^{0.5} \qquad (e)$$

A table of values (Table 18.1) for α, A, and T is built using the incremental values of $\propto$ and solving the equations (*a*) to (*e*).

TABLE 18.1

α	**PIN LOAD** A, N	**CABLE TENSION** T, N
10	483 000	495 000
20	466 000	462 000
30	449 000	427 000
40	434 000	394 000
50	420 000	363 000
60	408 000	335 000
70	400 000	313 000
80	394 000	299 000

Choosing α, solving (*d*) for β, then solving (*c*), (*a*), (*b*), and (*e*) is a significant task using a calculator.

Figure 18.23 is a plot of the pin and cable loads against boom angle. It is obvious that the critical load for both the pin and cable occurs at the smallest boom angle, 10°. For the design conditions specified, the pin must thus be designed to support a load of 483 000 N and the cable a load of 495 000 N.

Example 18.3

From the results of the previous example and the following material specifications, determine the required diameters of the pin (DP) and cable (DC).

Design strength of cable = $(2)(10^8)$ Pa
Design strength of pin = $(1.2)(10^8)$ Pa

The design strengths include an appropriate factor of safety.

Solution

$$\text{Design strength} = \frac{\text{Load}}{\text{cross-sectional area}}$$

For the cable:

$$2(10^8) = \frac{495\ 000}{\frac{\pi(DC)^2}{4}}$$

$$DC = 5.61 \text{ cm}$$

For the pin:

$$(1.2)(10^8) = \frac{483\ 000}{\frac{\pi(DP)^2}{4}}$$

$$DP = 7.16 \text{ cm}$$

We now return these results to the designers so they may proceed with the overall project.

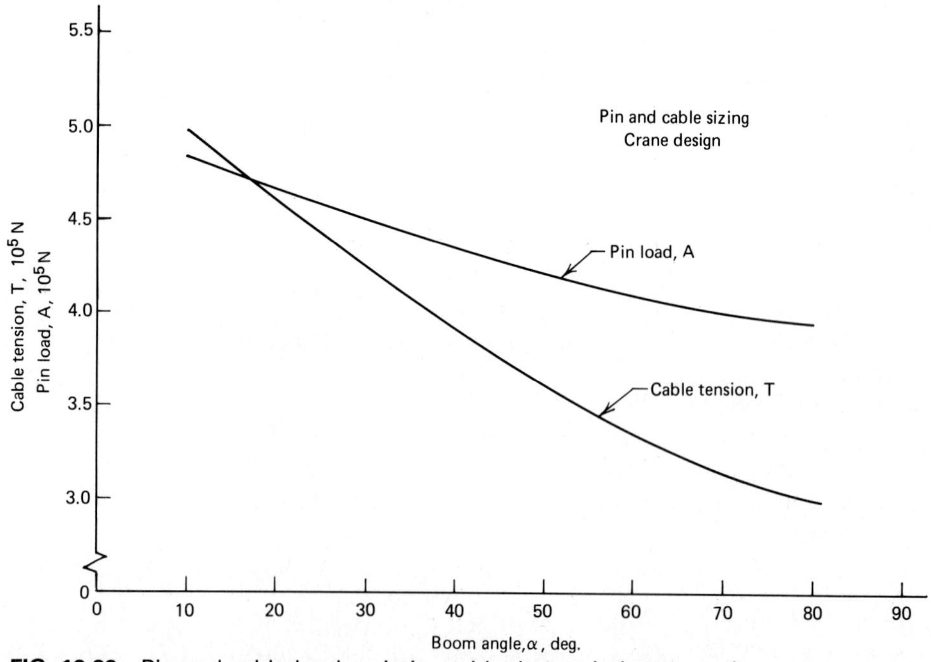

FIG. 18.23 Pin and cable load variation with change in boom angle.

Example 18.4

Determine the maximum load that the crane can lift at any boom condition within the design conditions previously specified and for the known pin and cable diameters.

Discussion

From Fig. 18.23 we can see that at a boom angle of 10°, no additional load to the 10^4 kg applied can be tolerated, since this was the critical point at which we sized our pin and cable. If we look at increasing boom angles, however, we note that the loads on the pin and cable drop. At each boom angle greater than 10°, we can say that (1) a smaller pin and cable could be used if we did not decrease the boom angle or (2) more load could be applied before the pin or cable would reach the design strength. Apparently at the largest boom angle, 80°, there is the greatest difference between the load that the pin and cable can withstand and the load that is applied. At this single point we will calculate the additional load that could be lifted.

Solution

We do not know for sure whether the pin or cable will be critical if additional load is applied at α = 80°, so we must check both cases.

Taking the cable first and using a maximum allowable load of 495 000 N and α = 80° and solving Eq. (*d*) and (*c*) in Example 18.3 for M, we have

$$\beta = 73.46°$$
$$M = 1.64\ (10^4)\ \text{kg}$$

Substituting these values into Eq. (*a*) and (*b*), solving for A_x and A_y, and then A from Eq. (*e*) in Example 18.3, we have

$$A_x = 140\ 900\ \text{N}$$
$$A_y = 635\ 400\ \text{N}$$
$$A = 650\ 800\ \text{N}$$

When the load T in the cable is critical (495 000 N), the pin load A is therefore much greater than critical at α = 80°. The maximum load that can be lifted by the crane will be determined from the condition that the pin load is critical (483 000 N) and the cable load is something less than critical.

Some reduction of Eq. (*a*) through (*e*) can be made to expedite the solution for this case.

$$\beta = 73.46°$$
$$T = \frac{(9.807)(\cos 80°)}{(0.5)(\sin 6.54°)} M = 29.90\ M$$
$$A_x = T \cos \beta = 8.512\ M$$
$$A_y = 9.807\ M + 28.66\ M = 38.47\ M$$
$$A = [(8.512M)^2 + (38.47M)^2]^{0.5}$$
$$A = 39.40\ M$$

Therefore, for critical pin load,

$$483\ 000 = 39.40\ M$$
$$M = 1.23\ (10^4)\ \text{kg}$$

Examples 18.1 through 18.4 are typical of the analysis activity that supports the design effort. The key factor is that the analysis group must respond quickly enough to be a part of the design loop (Figs. 18.2 and 18.9). In other words, if it takes a longer time than was available to determine the minimum required diameter of the pin and cable for the given design conditions, the designers will opt for a larger pin and cable than is necessary in order to maintain the design schedule. Thus the design will not be optimum and as a result will have a higher cost because larger parts will be needed to support the pin and cable on the crane. The final design may not be competitive with one in which an optimum solution was found.

The amount of computation time necessary to do the analysis for Examples 18.1 through 18.4 is not trivial but it does not take a great deal of resources. Suppose now that the following questions are asked by the designers.

1. What are the required cable and pin diameters for k = 0.3, 0.4, 0.6, 0.7?

2. What are the required cable and pin diameters for applied loads of $(2)(10^4)$ kg, $(0.5)(10^4)$ kg, and numerous other possibilities?

3. What effect does varying the boom length have on the pin and cable diameters?

4. What effect does varying the cable arrest height have on pin and cable diameters?

5. Is there a combination of the variables that would provide overall optimum (minimum) values of the pin and cable diameters for the ranges of design conditions specified?

Answers to each of these questions will require us to go through a set of calculations similar to those in Examples 18.2 through 18.4—a task that would obviously be prohibitive unless a computer were available. If an analyzer (computer program for engineering analysis) were available which would permit a user to input combinations of the variables α, M, L, H, and k and receive output in the form of loads, pin and cable diameters and graphical display of results, then very rapid turnaround could be achieved. Figure 18.24 is a com-

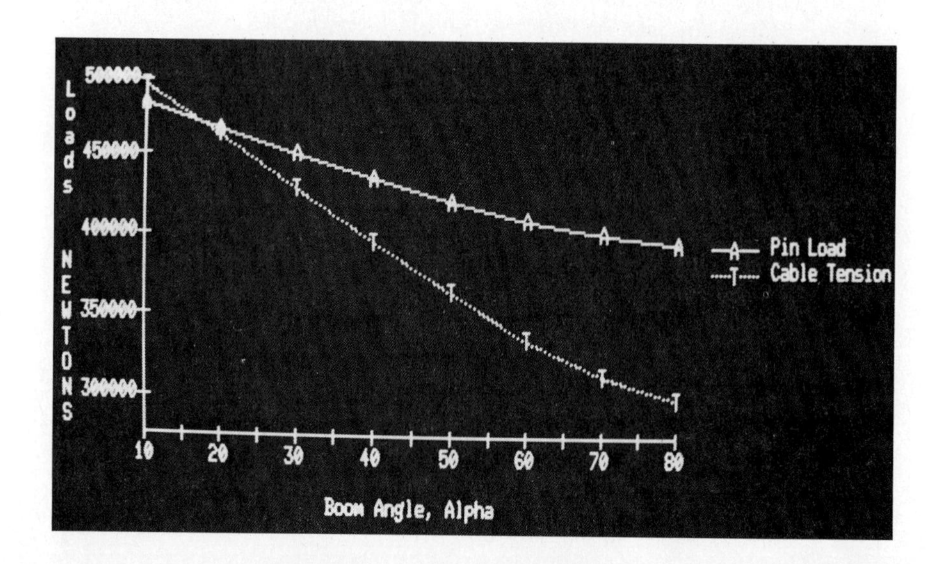

FIG. 18.24 A computer-generated plot of pin and cable load variation.

puter plot of pin and cable loads versus boom angle. This plot was obtained by inputting the parameters in Example 18.2 and requesting that the resulting table of values (Table 18.1) be plotted. The total user time involved is about 5 min from log-on to receiving a copy of the plot.

You will encounter many kinds of analyzers as you progress through your engineering program. Again remember that the computer is a tool that provides vast amounts of data in a numerical or graphical format. As the engineer, you must understand what the computer can do for you, but you must still make the decisions based on the results displayed.

Perhaps you can visualize from the previous two examples—the crane and tank problems—that engineers will be involved in the development of software (analyzers) necessary to examine design variables. The engineer is the key player needed to determine which variables are most likely to affect selected performance criteria. The engineer is also likely to develop and write subroutines that call large software packages such as numerical analyzers and plotting routines provided by vendors. It should go without saying that one objective of the educational process is to teach you the appropriate fundamentals such that upon graduation, use of software written by others is only the logical extension of a thorough understanding of that material.

18.6 INTEGRATED DESIGN AND MANUFACTURING

A major research effort is underway to use the potential of computer technology to assist in the solution of economic problems. Industrial-based nations are looking to digital technology and computer graphics to increase productivity and to develop a competitive edge in the international marketplace. The research effort is expected to result in a program called computer-integrated manufacturing (CIM).

The total design loop for a product begins with the design function but includes processing, manufacturing, assembly, quality control, and marketing, and concludes with a maintenance program and redesign effort. Each of the segments of the design loop has traditionally been identified and isolated within industry. For example, after a part is designed, the plans are sent to manufacturing, which in turn ships the finished part to assembly and so on through the loop. The link between the segments is the part drawing. Because of the time involved to send engineering changes through the system on paper (revised drawings), the part currently being marketed is likely to be different from what the design calls for. In some cases, the designed part is not the same as the manufactured part, simply because the design called for a manufacturing operation which was not possible to perform. In such a situation manufacturing adjusts the design so the part can be made. By the time the needed change is fed back through the design loop, many parts, which may be less than optimum, have been manufactured and marketed.

Attempts to improve productivity and the quality of products with technology began with automation of assembly lines in the 1940s and 1950s. This development lacked the flexibility necessary for changing to a different product. Numerically controlled machines developed in the 1950s and 1960s allowed a more rapid changeover from product to product. One fundamental problem remained: the manufacturing process remained isolated from the design function.

In recent years, the rapid development of computer-aided-design systems (CAD) has enabled the design function to incorporate more analysis in a shorter time period with the obvious result being a higher-quality design. In most industries today, however, the output of a CAD system is a set of drawings. Thus the design function still remains isolated from the remaining segments of the design loop.

CIM has as a basic premise the idea that all the functions of the design loop are interdependent. Engineers must begin to think in global terms regarding the design loop. The common thread will be the integrated database representing the product. During the design phase, the engineer will thus have access to data on manufacturing capability, material properties, and performance of similar products. A completely integrated system will shorten considerably the time between design and manufacture. The principal benefit will be a rapid response of industry to market demands with a high-quality product.

The factory of the future will most likely contain a data-processing facility which can access the product database. Software will then be produced to manufacture the product automatically. Machine centers will include robots and other computer-

FIG. 18.25 Computer-controlled manufacturing operations (CAM) are being tied closer to the design process through an integrated database. In this way, the product produced is exactly the product that was designed. (*Courtesy of Control Data Corporation.*)

controlled tools and material-handling equipment. With CIM, a product once designed can be manufactured, tested, and marketed without human intervention.

18.7 REPORTING DESIGN ACTIVITIES AND RESULTS

Throughout this chapter we have emphasized the design process up to the point where the best design is chosen. We must now complete the nine-step design process with the specification and communication steps.

Specification of the design involves the production of the necessary drawings or computer databases which are forwarded to the manufacturing function. The design drawing package was described in Chap. 14. The package includes the necessary detail drawings for manufacture as well as the assemblies and parts lists to complete the production requirements. The concept of a computer database was discussed earlier in this chapter.

All design efforts will go for naught if there are no means to communicate ongoing activities and results. The principal means of communication are the written report and oral presentation.

18.7.1 The Written Technical Report

Your engineering education will provide you with the necessary background to move into an engineering position and be productive. There are some things, e.g., company procedures and specific products, that your education will not include. There are also things that your education will include but which you may not be called upon to use in your particular job. One thing that all engineers will do regardless of company affiliation or engineering field is report writing.

As an undergraduate, you will prepare lab reports, term papers, and design reports. If you pursue graduate work, your original research requirement will be presented in a thesis. As a working engineer, progress reports, action memos, proposals, feasibility studies, and preparation of company manuals are among the many types of reports you will prepare for communication and documentation of your engineering efforts.

The first step in the preparation of a report is to establish a goal. A goal can usually be determined from the particular work assignment you have. In the case of a design problem, a reasonable goal may be: propose a solution and support your choice.

Second, a very careful plan should be made which carries out the report goal. An outline should be developed which defines the various sections of the report and details what will appear in each section. Data and other supporting evidence should be available for use in writing the report. Decisions must be made regarding tables, graphs, sketches, and drawings that are to be included in the report. A schedule must be prepared for the actual production of the report. Such things as typing arrangements, report format design, copy requirements, and the various illustration preparations must be decided upon and appropriately scheduled to meet deadline dates.

Illustrations are generally the focal point of a technical report, so quality graphics is a strong consideration. Figures may be placed within the text (narrative), at the end of a section of the report, or, in some cases, at the end of the complete report. In any event, each figure must have a number that fits into a logical sequence of numbers for all the figures in a report. When necessary, the text should refer to a figure by its number, e.g., see Fig. 3.

The actual writing of the text will be based on the plan that has been developed. To write effectively you must know the material and the terminology associated with the particular area about which you are writing. You should make effective use of a dictionary and thesaurus. After you have prepared a draft and again after the final version is prepared, have someone proofread your work carefully. Writing is not an easy task for many individuals, so plenty of lead time should be factored into your plan. Thoughts do not flow as well when you are under a short-time deadline. Rewriting will be a necessity for most of you, so allow appropriate time in the report plan.

As with any writing effort, good grammar is imperative. Coherent paragraph structure, syntactically correct sentences, and correct spelling will enhance the substance of the report. If grammar is careless, it will detract from what may be an excellent technical effort. It is a pity that much of the technical progress made is reported so badly. A little planning and careful proofing of reports will go a long way toward building a positive image.

The person and voice that are used in writing a technical report depend upon the intended use of the report.

1. First person–active voice is used in a report which represents an eyewitness account of a happening or offers a personal or group opinion about something. For example, "I (we) believe the test should be repeated upon repair of the power unit" is an opinion and would be reported in first person.

2. Second person–active voice is used for instructions in training manuals, assemblies, and job orders. "Check oil level every 30 operating hours" is an example.

3. Third person–active voice is used for dramatic effect in documents that are distribured for public use. For example, "They developed a computer model for defining a general curved surface and specifying the required mold geometry to produce the surface" is a form one may see in a newspaper account of a new development.

4. Third person–passive voice is used in reports which are read for the conclusions and the reasoning used to arrive at the conclusions. Personality does not enter into this writing viewpoint. For example, "A computer model has been developed which enables definition of a general curved surface and specifies the required mold geometry to produce the surface" is an account you may find in a technical journal, proposal, or design report.

Lengthy reports which are to be archived need an abstract for quick reference by potential readers. An abstract is a brief (one or two paragraphs) summary of the most important features of a report. Short reports, memos, and notes do not need an abstract.

A typical design report may be organized as follows:

- Cover page
- Abstract
- Table of contents
- Body
  - Introduction
  - Problem statement
  - Solution
  - Justification
- Conclusions
  - Benefits
  - Recommendations for future
- Appendix
  - Supporting evidence not in body of report

In lengthy reports, it is often convenient to include a summary section immediately following the table of contents. This makes it easier for the reader who may be judging the report content to make a decision without having to read the entire document. In most cases, the farther up the company hierarchy that a report progresses, the less of it that is read. Management will look primarily at the solution and the resulting benefits to the organization.

18.7.2 The Oral Presentation

In general, an oral presentation is made to sell an idea, educate someone, or support a written report. When you are required to make an oral presentation, you should consider carefully the preparation and rendition of the presentation.

FIG. 18.26 A student design team justifies their solution to an engineering problem.

There are five basic points to keep in mind as you prepare the presentation.

1. Know your audience. Check the size, education level, and type of group so that you may gauge the use of technical terms and depth of coverage.

2. Know the facility in which you will make the presentation. Check seating arrangements, lighting, acoustics, and speaker's area.

3. Develop your talk. Assemble facts, decide upon approach, and prepare an outline. Develop a strong opening and closing to your talk. Prepare 3×5 notes if you need them. Allow time for a lot of practice. If it is a team presentation, you must coordinate your part with all others.

4. Develop the graphics. A picture may be worth a thousand words. Use this to effect in your presentation. Decide on the type, such as slides, charts, blackboard, or models. Omit details and emphasize quality and visibility of the graphics. Practice your presentation with the graphics.

5. Obtain equipment. Find out if the room has a screen and appropriate projection equipment if you need it. If in doubt, bring your own equipment.

For the actual presentation, use the following as a guide.

1. Check the setup. Make sure that the equipment is working and that you know the time for your presentation.

2. Dress. Dress for the audience and in a professional manner.

3. Delivery. Relax and project a confident manner. Speak clearly and with appropriate loudness. If you are going to use a microphone, make sure that you practice with one beforehand to determine voice level and the way to talk into one. Maintain eye contact with your audience. Use notes but do not read from them.

4. Audience. If you lose the attention of the audience, check your voice level to determine if you can be heard fully, ascertain if the visuals can be seen properly or, finally, find out if there is a temporary distraction. Pause or repeat yourself if you feel that the audience has missed a point.

5. Questions. Time is generally reserved after a presentation for questions. Respond quickly and honestly to questions.

In all phases of the presentation, conduct yourself in a professional manner. Audiences will respond in kind.

Problems

18.1 Develop a second alternative solution to the carton-orientation example. Prepare appropriate sketches and notes for your solution. If required, prepare a brief report and/or oral presentation of the results.

18.2 Ceramic substrates (see Fig. 18.27*a*), used in the manufacture of potentiometers, must be oriented for a printing operation as shown in Fig. 18.27*b*. Initially the substrates are in random order. A feeder bowl is available to feed the parts to an orientation device. You are asked to develop and describe a concept for this orientation device through words and sketches. The device must take the supply of randomly oriented parts from the feeder bowl track and provide correctly oriented parts in a single track as shown. Part or all of the device may be within the feeder bowl if desired. If required, prepare a brief report and/or oral presentation of the results.

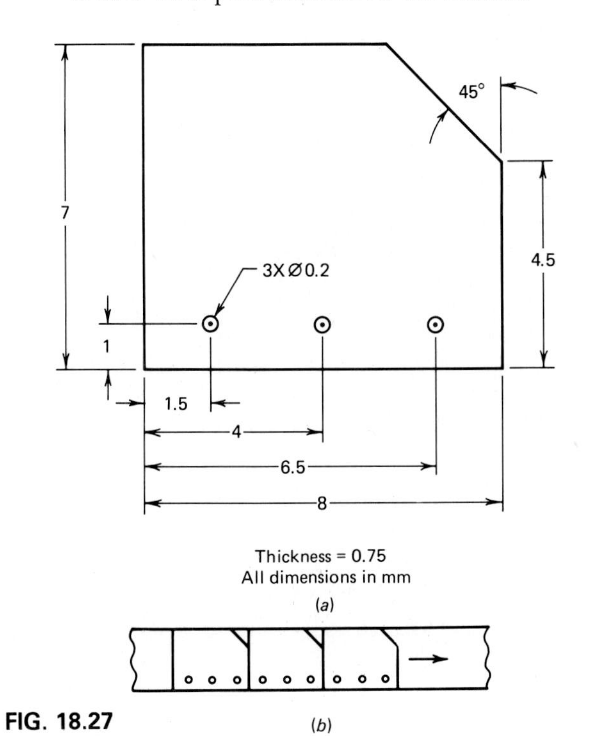

FIG. 18.27

18.3 Prepare an alternative solution using methods described in this chapter for one or more of the following. If required, prepare a brief written report and/or oral presentation.

(a) Drafting pencil for person who has lost his hands at the wrists
(b) Mechanism to improve access to the spare tire of a pickup truck
(c) Intercom system for motorcycle driver and passenger
(d) Doorbell for deaf family
(e) Convenient carrying system for student to take books and a portable personal computer to class
(f) Remote sensing system to allow a homeowner to know if a garage door is open or closed
(g) Mechanism to aid in changing a motorcycle tire when the cycle has no center stand
(h) Method for improving the traction of a rear-wheel-drive car or pickup on ice or snow
(i) Device to show by instrument panel display the engine oil level for an auto when the ignition is in the accessory position
(j) A quick and simple method of changing weights on a barbell
(k) Device for cleaning the exterior of windows up to 20 ft above the ground level while standing on the ground
(l) System to allow 25 students in a classroom to simultaneously access a central computer via their personal computers
(m) Method to reduce book losses from a campus library
(n) Device to simplify the use of dental floss by an individual
(o) System to inexpensively make a ground-floor window intruder-proof
(p) Instrument panel display to monitor the condition of the auto's battery

- (q) System to aid in making a bicycle wheel true
- (r) Auto hood lock that can be released from inside and outside the car
- (s) Folding auto ramps for convenient trunk storage
- (t) Smoke detector for deaf person
- (u) Windshield wipers that are effective during freezing conditions
- (v) Method for detering wild animals (e.g., deer) from crossing a busy highway

18.4 A company transfers packages from point to point across the country. The requirement on package size is that the girth plus the longest dimension cannot exceed 60 inches. Following the procedure of Example 18.1, prepare appropriate plots which indicate the maximum package volume that could be shipped if the package shape is

- (a) A rectangular prism with square cross section
- (b) A rectangular prism with dimensions in the ratio of 1:2:3
- (c) A cylinder (girth = circumference)

If required, prepare a brief written and/or oral presentation of the results.

18.5 Following Examples 18.2 to 18.4, determine pin and cable diameters and the maximum carrying capacity for a crane with the following parameters:

$$10° \leq \alpha \leq 70°$$
$$M = 0.85\ (10^4)\ kg$$
$$L = 22\ m$$
$$H = 4.0\ m$$
$$k = 0.75$$

Use a design strength of 1.8 (10^8) Pa for the cable and (1.1) (10^8) Pa for the pin. Even though you realize that the critical condition occurs at the smallest boom angle, carry through the computations for 10° increments on $\propto$ and construct a plot similar to Fig. 18.23.

If required, prepare a brief written report and/or oral presentation of the results.

18.6 For a fixed boom angle of 60°, and all other parameters except k the same as in Example 18.2, plot a graph of pin and cable loads versus k for $0.2 \leq k \leq 1$. Use 0.1 increments for k. State your conclusions.

If required, prepare a brief written report and/or oral presentation of the results.

18.7 For a fixed boom angle of 20°, and all other parameters except H the same as in Example 18.2, plot a graph of pin and cable loads versus H for $2 \leq H \leq 8$. Use increments of 1 for H. State your conclusions.

If required, prepare a brief written report and/or oral presentation of the results.

Appendix A

TABLE A.1
Greek Alphabet

Alpha	A	α
Beta	B	β
Gamma	Γ	γ
Delta	Δ	δ
Epsilon	E	ϵ
Zeta	Z	ζ
Eta	H	η
Theta	Θ	θ
Iota	I	ι
Kappa	K	κ
Lambda	Λ	λ
Mu	M	μ
Nu	N	ν
Xi	Ξ	ξ
Omicron	O	o
Pi	Π	π
Rho	P	ρ
Sigma	Σ	σ
Tau	T	τ
Upsilon	Υ	υ
Phi	Φ	ϕ
Chi	X	χ
Psi	Ψ	ψ
Omega	Ω	ω

TABLE A.2

QUANTITY	NAME	SYMBOL
Length	meter	m
Mass	kilogram	kg
Time	second	s
Electric current	ampere	A
Thermodynamic temperature	kelvin	K
Amount of a substance	mole	mol
Luminous intensity	candela	cd
Plane angle	radian	rad
Solid angle	steradian	sr

TABLE A.3

QUANTITY	SI UNIT SYMBOL	NAME	BASE UNITS
Frequency	Hz	hertz	s^{-1}
Force	N	newton	$kg \cdot m \cdot s^{-2}$
Pressure, stress	Pa	pascal	$kg \cdot m^{-1} \cdot s^{-2}$
Energy or work	J	joule	$kg \cdot m^2 \cdot s^{-2}$
A quantity of heat	J	joule	$kg \cdot m^2 \cdot s^{-2}$
Power, radiant flux	W	watt	$kg \cdot m^2 \cdot s^{-3}$
Electric charge	C	coulomb	$A \cdot s$
Electric potential	V	volt	$kg \cdot m^2 \cdot s^{-3} \cdot A^{-1}$
Potential difference	V	volt	$kg \cdot m^2 \cdot s^{-3} \cdot A^{-1}$
Electromotive force	V	volt	$kg \cdot m^2 \cdot s^{-3} \cdot A^{-1}$
Capacitance	F	farad	$A^2 \cdot s^4 \cdot kg^{-1} \cdot m^{-2}$
Electric resistance	Ω	ohm	$kg \cdot m^2 \cdot s^{-3} \cdot A^{-2}$
Conductance	S	siemens	$kg^{-1} \cdot m^{-2} \cdot s^3 \cdot A^2$
Magnetic flux	Wb	weber	$kg \cdot m^2 \cdot s^{-2} \cdot A^{-1}$
Magnetic flux density	T	tesla	$kg \cdot s^{-2} \cdot A^{-1}$
Inductance	H	henry	$kg \cdot m^2 \cdot s^{-2} \cdot A^{-2}$
Luminous flux	lm	lumen	$cd \cdot sr$
Illuminance	lx	lux	$cd \cdot sr \cdot m^{-2}$
Dose equivalent	Sv	sievert	$m^2 \cdot s^{-2}$
Activity (radionuclides)	Bq	becquerel	s^{-1}
Absorbed dose	Gy	gray	$m^2 \cdot s^{-2}$
Celsius temperature[1]	°C	degree Celsius	K

Note: The thermodynamic temperature (T_K) expressed in kelvins is related to Celsius temperature ($t_{°C}$) expressed in degrees Celsius by the equation

$$t_{°C} = T_K - 273.15$$

TABLE A.4

QUANTITY	UNITS	QUANTITY	UNITS
Acceleration	$m \cdot s^{-2}$	Molar entropy	$J \cdot mol^{-1} \cdot K^{-1}$
Angular acceleration	$rad \cdot s^{-2}$	Molar heat capacity	$J \cdot mol^{-1} \cdot K^{-1}$
Angular velocity	$rad \cdot s^{-1}$	Moment of force	$N \cdot m$
Area	m^2	Permeability	$H \cdot m^{-1}$
Concentration	$mol \cdot m^{-3}$	Permittivity	$F \cdot m^{-1}$
Current density	$A \cdot m^{-2}$	Radiance	$W \cdot m^{-2} \cdot sr^{-1}$
Density, mass	$kg \cdot m^{-3}$	Radiant intensity	$W \cdot sr^{-1}$
Electric charge density	$C \cdot m^{-3}$	Specific heat capacity	$J \cdot kg^{-1} \cdot K^{-1}$
Electric field strength	$V \cdot m^{-1}$	Specific energy	$J \cdot kg^{-1}$
Electric flux density	$C \cdot m^{-2}$	Specific entropy	$J \cdot kg^{-1} \cdot K^{-1}$
Energy density	$J \cdot m^{-3}$	Specific volume	$m^3 \cdot kg^{-1}$
Entropy	$J \cdot K^{-1}$	Surface tension	$N \cdot m^{-1}$
Heat capacity	$J \cdot K^{-1}$	Thermal conductivity	$W \cdot m^{-1} \cdot K^{-1}$
Heat flux density	$W \cdot m^{-2}$	Velocity	$m \cdot s^{-1}$
Irradiance	$W \cdot m^{-2}$	Viscosity, dynamic	$Pa \cdot s$
Luminance	$cd \cdot m^{-2}$	Viscosity, kinematic	$m^2 \cdot s^{-1}$
Magnetic field strength	$A \cdot m^{-1}$	Volume	m^3
Molar energy	$J \cdot mol^{-1}$	Wavelength	m

TABLE A.5

MULTIPLIER	PREFIX NAME	SYMBOL
10^{18}	exa	E
10^{15}	peta	P
10^{12}	tera	T
10^{9}	giga	G
10^{6}	*mega	M
10^{3}	*kilo	k
10^{2}	hecto	h
10^{1}	deka	da
10^{-1}	deci	d
10^{-2}	centi	c
10^{-3}	*milli	m
10^{-6}	*micro	μ
10^{-9}	nano	n
10^{-12}	pico	p
10^{-15}	femto	f
10^{-18}	atto	a

*Most often used.

TABLE A.6
Unit Conversions

TO CONVERT FROM	TO	MULTIPLY BY
acres	ft^2	4.356×10^4
acres	ha	4.0469×10^{-1}
acres	m^2	4.0469×10^3
angstroms	cm	1×10^{-8}
angstroms	in	3.9370×10^{-9}
barrels (petroleum, US)	gal (US liquid)	4.2×10^1
bushels (US)	ft^3	1.2445
bushels (US)	m^3	3.5239×10^{-2}
circular mils	cm^2	5.0671×10^{-6}
circular mils	in^2	7.8540×10^{-7}
cubic centimeters	in^3	6.1024×10^{-2}
cubic centimeters	L	1×10^{-3}
cubic centimeters	oz (US fluid)	3.3814×10^{-2}
cubic feet	bushels (US)	8.0356×10^{-1}
cubic feet	gal (US liquid)	7.4805
cubic feet	L	2.8317×10^1
cubic feet	m^3	2.8317×10^{-2}
cubic inches	cm^3	1.6387×10^1
cubic inches	L	1.6387×10^{-2}
cubic inches	oz (US fluid)	5.5411×10^{-1}
cubic meters	bushels (US)	2.8378×10^1
cubic meters	ft^3	3.5315×10^1
cubic meters	gal (US liquid)	2.6417×10^2
cubic yards	bushels (US)	2.1696×10^1
cubic yards	gal (US liquid)	2.0197×10^2
cubic yards	L	7.6455×10^2
cubic yards	m^3	7.6455×10^{-1}
fathoms	ft	6
feet	in	1.2×10^1
feet	m	3.048×10^{-1}
furlongs	ft	6.6×10^2
gallons (US liquid)	in^3	2.31×10^2
gallons (US liquid)	L	3.7854
gallons (US liquid)	m^3	3.7854×10^{-3}
gallons (US liquid)	oz (US fluid)	1.28×10^2

TABLE A.6 continued
Unit Conversions

TO CONVERT FROM	TO	MULTIPLY BY
gallons (US liquid)	pt (US liquid)	8
gallons (US liquid)	qt (US liquid)	4
hectares	acres	2.4711
hectares	ft^2	1.0764×10^5
hectares	m^2	1×10^4
hours	s	3.6×10^3
inches	cm	2.54
inches	mils	1×10^3
kilograms	lbm	2.2046
kilograms	slugs	6.8522×10^{-2}
kilometers	mi	6.2137×10^{-1}
kilometers	nmi (nautical mile)	5.3966×10^{-1}
liters	bushels (US)	2.8378×10^{-2}
liters	ft^3	3.5315×10^{-2}
liters	gal (US liquid)	2.6417×10^{-1}
meters	Å	1×10^{10}
meters	ft	3.2808
microns	m	1×10^{-6}
miles	ft	5.28×10^3
miles	furlongs	8
miles	km	1.6093
miles	nmi (nautical mile)	8.6898×10^{-1}
nautical miles	mi	1.1508
newtons	lbf	2.2481×10^{-1}
ounces (US fluid)	cm^3	2.9574×10^1
ounces (US fluid)	gal (US liquid)	7.8125×10^{-3}
ounces (US fluid)	in^3	1.8047
ounces (US fluid)	L	2.9574×10^{-2}
pounds mass	kg	4.5359×10^{-1}
pounds force	N	4.4482
radians	°	5.7296×10^1
slugs	kg	1.4594×10^1
slugs	lbm	3.2174×10^1
square centimeters	ft^2	1.0764×10^{-3}
square centimeters	in^2	1.5500×10^{-1}

TABLE A.6 continued
Unit Conversions

TO CONVERT FROM	TO	MULTIPLY BY
square feet	acre	2.2957×10^{-5}
square feet	cm^2	9.2903×10^{2}
square feet	ha	9.2903×10^{-6}
square feet	m^2	9.2903×10^{-2}
square meters	ft^2	1.0764×10^{1}
square meters	in^2	1.5500×10^{3}
square miles	acres	6.4×10^{2}
square miles	ft^2	2.7878×10^{7}
square miles	ha	2.5900×10^{2}
square miles	km^2	2.5900
square millimeters	ft^2	1.0764×10^{-5}
square millimeters	in^2	1.5500×10^{-3}

Appendix B

TABLE B.1
Metric Twist Drill Size

ALL DIMENSIONS IN MILLIMETERS
From 0.15 to 0.20, increment = 0.01
From 0.20 to 2.50, increment = 0.05; also 0.22, 0.28, 0.32, 0.38, 0.42, 0.48
From 2.50 to 10.00, increment = 0.10
From 10.00 to 14.00, increment = 0.50; also 10.20, 10.80, 11.20, 11.80, 12.20, 12.80, 13.20, 13.80
From 14.00 to 17.50, increment = 0.25
From 17.50 to 38.00, increment = 0.50
From 38.00 to 78.00, increment = 1.00

Source: ANSI B94.11M-1979.

TABLE B.2
American National Standard Metric Screw Threads

	COARSE		FINE	
NOMINAL DIA, mm	PITCH, mm	TAP DRILL DIA, mm	PITCH, mm	TAP DRILL DIA, mm
1.6	0.35	1.25		
2	0.4	1.6		
2.5	0.45	2.05		
3	0.50	2.50		
3.5	0.6	2.90		
4	0.7	3.30		
5	0.8	4.20		
6	1	5.30		
8	1.25	6.80	1	7.00
10	1.5	8.50	1.25	8.75
12	1.75	10.30	1.25	10.50
14	2	12.00	1.5	12.50
16	2	14.00	1.5	14.50
18	2.5	15.50	1.5	16.50
20	2.5	17.50	1.5	18.50
22	2.5	19.50	1.5	20.50
24	3	21.00	2	22.00

Source: ANSI B1.13M-1979, which was taken from ISO261-1973. Both these standards list other threads, many of which are not recommended by ISO or have very specialized application.

TABLE B.2 continued
American National Standard Metric Screw Threads

	COARSE		FINE	
NOMINAL DIA, mm	PITCH, mm	TAP DRILL DIA, mm	PITCH, mm	TAP DRILL DIA, mm
27	3	24.00	2	25.00
30	3.5	26.50	2	28.00
33	3.5	29.50	2	31.00
36	4	32.00	2	33.00
39	4	35.00	2	36.00
42	4.5	37.50	3	39.00
45	4.5	40.50	3	42.00
48	5	43.00	3	45.00
52	5	47.00	3	49.00
56	5.5	50.50	4	52.00
60	5.5	54.50	4	56.00
64	6	58.00	4	60.00
68	6	62.00	4	64.00
72	6	66.00		
80	6	74.00		
90	6	84.00		
100	6	94.00		

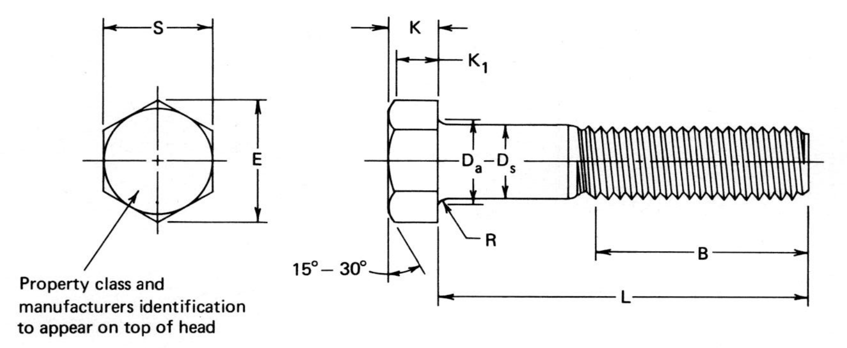

TABLE B.3

American National Standard Metric Hex Bolts

D	D_s		S		E		K		K_1	D_a	R	B (Ref)		
												Thread Length (Basic)		
NOMINAL BOLT DIA AND THREAD PITCH	BODY DIAMETER		WIDTH ACROSS FLATS		WIDTH ACROSS CORNERS		HEAD HEIGHT		WRENCH-ING HEIGHT	FILLET TRANSI-TION DIA	RADIUS OF FILLET	BOLT LENGTHS <125	BOLT LENGTHS >125 AND <200	BOLT LENGTHS >200
	Max	Min	Max	Min	Max	Min	Max	Min	Min	Max	Min			
M5 × 0.8	5.48	4.52	8.00	7.64	9.24	8.63	3.58	3.35	2.4	5.7	0.2	16	22	35
M6 × 1	6.19	5.52	10.00	9.64	11.55	10.89	4.38	3.55	2.8	6.8	0.3	18	24	37
M8 × 1.25	8.58	7.42	13.00	12.57	15.01	14.20	5.68	5.10	3.7	9.2	0.4	22	28	41
M10 × 1.5	10.58	9.42	16.00	15.57	18.48	17.59	6.85	6.17	4.5	11.2	0.4	26	32	45
M12 × 1.75	12.70	11.30	18.00	17.57	20.78	19.85	7.95	7.24	5.2	13.7	0.6	30	36	49
M14 × 2	14.70	13.30	21.00	20.16	24.25	22.78	9.25	8.51	6.2	15.7	0.6	34	40	53
M16 × 2	16.70	15.30	24.00	23.16	27.71	26.17	10.75	9.68	7.0	17.7	0.6	38	44	57
M20 × 2.5	20.84	19.16	30.00	29.16	34.64	32.95	13.40	12.12	8.8	22.4	0.8	46	52	65
M24 × 3	24.84	23.16	36.00	35.00	41.57	39.55	15.90	14.56	10.5	26.4	0.8	54	60	73
M30 × 3.5	30.84	29.16	46.00	45.00	53.12	50.55	19.75	17.92	13.1	33.4	1.0	66	72	85
M36 × 4	37.00	35.00	55.00	53.80	63.51	60.79	23.55	21.72	15.8	39.4	1.0	78	84	97
M42 × 4.5	43.00	41.00	65.00	62.90	75.06	71.71	27.05	25.03	18.2	45.4	1.2	90	96	109
M48 × 5	49.00	47.00	75.00	72.60	86.60	82.76	31.07	28.93	21.0	52.0	1.5	102	108	121
M56 × 5.5	57.20	54.80	85.00	82.20	98.15	93.71	36.20	33.80	24.5	62.0	2.0	–	124	137
M64 × 6	65.52	62.80	95.00	91.80	109.70	104.65	41.32	38.68	28.0	70.0	2.0	–	140	153
M72 × 6	73.84	70.80	105.00	101.40	121.24	115.60	46.45	43.55	31.5	70.0	2.0	–	156	169
M80 × 6	82.16	78.80	115.00	111.00	132.79	126.54	51.58	48.42	35.0	86.0	2.0	–	172	185
M90 × 6	92.48	88.60	130.00	125.50	150.11	143.07	57.74	54.26	39.2	96.0	2.0	–	192	205
M100 × 6	102.80	98.60	145.00	140.00	167.43	159.60	63.90	60.10	43.4	107.0	2.5	–	212	225

Typical Lengths:
For L < 100 mm, increment = 5 mm
For 100 < L < 200 mm, increment = 10 mm
For L > 200 mm, increment = 20 mm
Source: ANSI B18.2.3.5M-1979

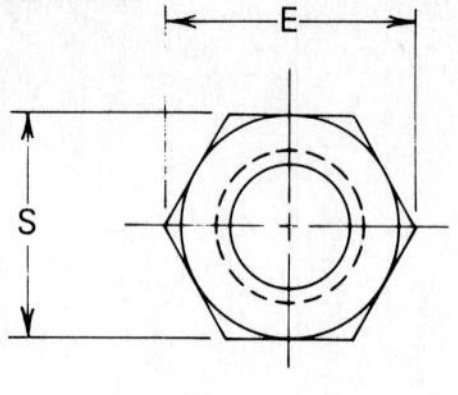

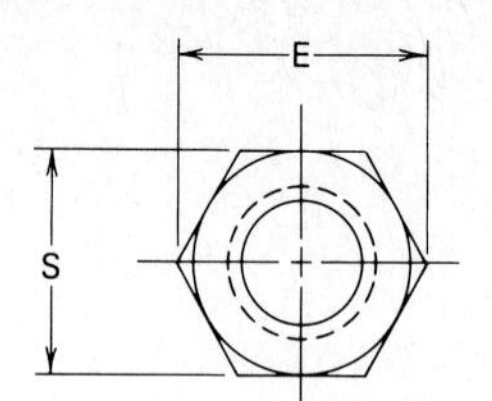

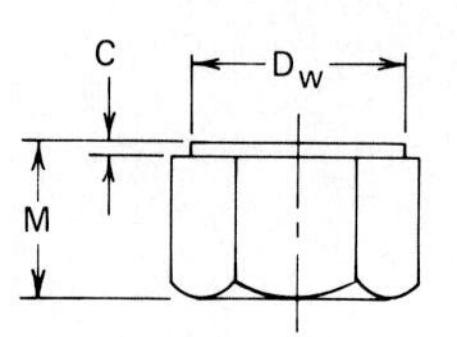

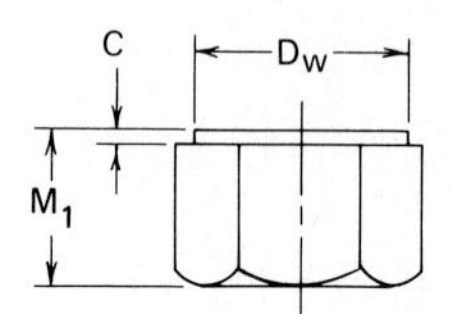

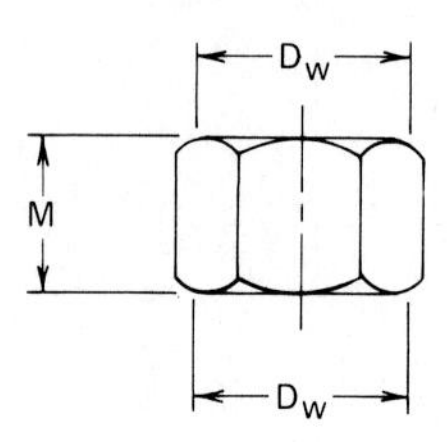

Hex nut

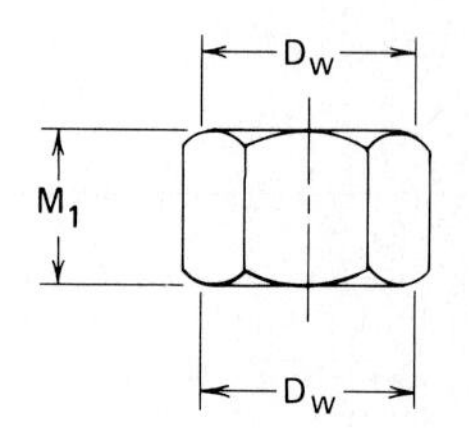

Hex jam nut

TABLE B.4

American National Standard Metric Hex Nuts

NOMINAL NUT DIA AND PITCH	S		E		D_W	C		M	
	WIDTH ACROSS FLATS		WIDTH ACROSS CORNERS		BEARING FACE DIA	WASHER FACE THICKNESS		THICKNESS	
	Max	Min	Max	Min	Min	Max	Min	Max	Min
M1.6 × 0.35	3.20	3.02	3.70	3.41	2.3	. . .	. . .	1.30	1.05
M2 × 0.4	4.00	3.82	4.62	4.32	3.1	. . .	. . .	1.60	1.35
M2.5 × 0.45	5.00	4.82	5.77	5.45	4.1	. . .	. . .	2.00	1.75
M3 × 0.5	5.50	5.32	6.35	6.01	4.6	. . .	. . .	2.40	2.15
M3.5 × 0.6	6.00	5.82	6.93	6.58	5.1	. . .	. . .	2.80	2.55
M4 × 0.7	7.00	6.78	8.08	7.66	6.0	. . .	. . .	3.20	2.90
M5 × 0.8	8.00	7.78	9.24	8.79	7.0	. . .	. . .	4.70	4.40
M6 × 1	10.00	9.78	11.55	11.05	8.9	. . .	. . .	5.20	4.90
M8 × 1.25	13.00	12.73	15.01	14.38	11.6	. . .	. . .	6.80	6.44
M10 × 1.5	16.00	15.73	18.45	17.77	14.6	. . .	. . .	8.40	8.04
M12 × 1.75	18.00	17.73	20.78	20.03	16.6	. . .	. . .	10.80	10.37
M14 × 2	21.00	20.67	24.25	23.36	19.4	. . .	. . .	12.80	12.10
M16 × 2	24.00	23.67	27.71	26.75	22.4	. . .	. . .	14.80	14.10
M20 × 2.5	30.00	29.16	34.64	32.95	27.9	0.8	0.4	18.00	16.90
M24 × 3	36.00	35.00	41.57	39.55	32.5	0.8	0.4	21.50	20.20
M30 × 3.5	46.00	45.00	53.12	50.85	42.5	0.8	0.4	25.60	24.30
M36 × 4	55.00	53.80	63.51	60.79	50.8	0.8	0.4	31.00	29.40

Source: ANSI B18.2.4.1M,
B18.2.4.3M,
and B18.2.4.5M

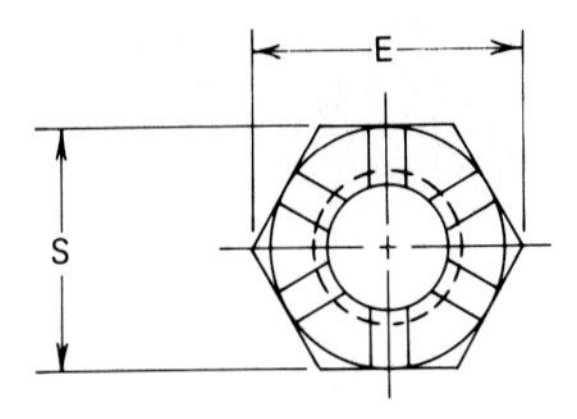

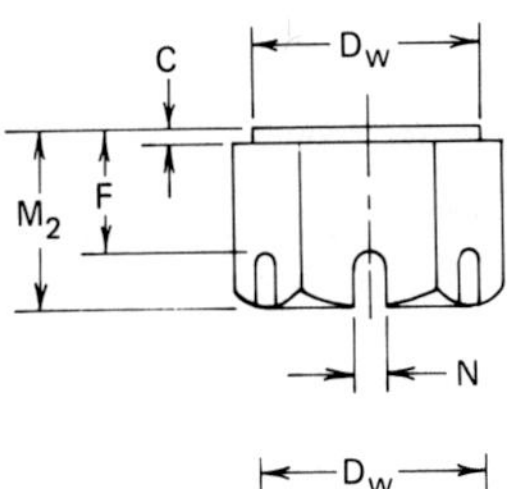

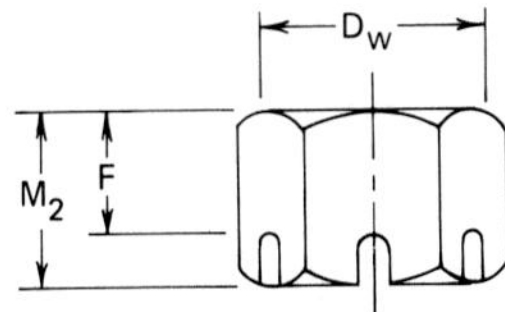

Slotted hex nut

TABLE B.4 continued

NOMINAL NUT DIA AND PITCH	M_1 THICKNESS HEX JAM		M_2 THICKNESS SLOTTED HEX		F UNSLOTTED THICKNESS		N WIDTH OF SLOT	
	Max	Min	Max	Min	Max	Min	Max	Min
M1.6 × 0.35								
M2 × 0.4								
M2.5 × 0.45								
M3 × 0.5								
M3.5 × 0.6								
M4 × 0.7								
M5 × 0.8	2.70	2.45	5.10	4.80	3.2	2.9	2.0	1.4
M6 × 1	3.20	2.90	5.70	5.40	3.5	3.2	2.4	1.8
M8 × 1.25	4.00	3.70	7.50	7.14	4.4	4.1	2.9	2.3
M10 × 1.5	5.00	4.70	9.30	8.94	5.2	4.9	3.4	2.8
M12 × 1.75	6.00	5.70	12.00	11.57	7.3	6.9	4.0	3.2
M14 × 2	7.00	6.64	14.10	13.40	8.6	8.0	4.3	3.5
M16 × 2	8.00	7.64	16.40	15.70	9.9	9.3	5.3	4.5
M20 × 2.5	10.00	9.42	20.30	19.00	13.3	12.2	5.7	4.5
M24 × 3	12.00	11.30	23.90	22.60	15.4	14.3	6.7	5.5
M30 × 3.5	16.00	14.30	28.60	27.30	18.1	16.8	8.5	7.0
M36 × 4	18.00	17.30	34.70	33.10	23.7	22.4	8.5	7.0

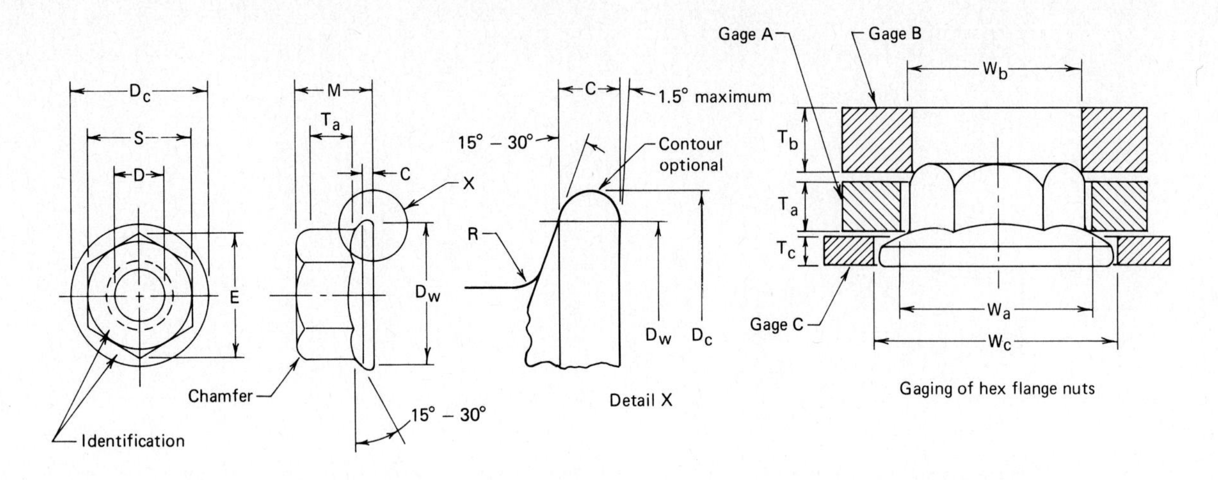

TABLE B.5

American National Standard Metric Hex Flange Nuts

	S		E		D_c	D_w	C	M	
NOMINAL NUT DIA AND THREAD PITCH	WIDTH ACROSS FLATS		WIDTH ACROSS CORNERS		FLANGE DIA	BEAR-ING CIRCLE DIA	FLANGE EDGE THICK-NESS	THICKNESS	
	Max	Min	Max	Min	Max	Min	Min	Max	Min
M5 × 0.8	8.00	7.78	9.24	8.79	11.8	9.8	1.0	5.00	4.70
M6 × 1	10.00	9.78	11.55	11.05	14.2	12.2	1.1	6.00	5.70
M8 × 1.25	13.00	12.73	15.01	14.38	17.9	15.8	1.2	8.00	7.60
M10 × 1.5	15.00	14.73	17.32	16.64	21.8	19.6	1.5	10.00	9.60
M12 × 1.75	18.00	17.73	20.78	20.03	26.0	23.8	1.8	12.00	11.60
M14 × 2	21.00	20.67	24.25	23.35	29.9	27.6	2.1	14.00	13.30
M16 × 2	24.00	23.67	27.71	26.75	34.5	31.9	2.4	16.00	15.30
M20 × 2.5	30.00	29.16	34.64	32.95	42.8	39.9	3.0	20.00	18.90

Source: ANSI B18.2.4.4M-1982

TABLE B.5 continued

R	W_a		T_a		W_b		T_b	W_c		T_c	
FLANGE TOP FILLET RADIUS	**GAGE A**				**GAGE B**			**GAGE C**			
	INSIDE DIA		**THICKNESS**		**INSIDE DIA**		**THICK-NESS**	**INSIDE DIA**		**THICKNESS**	
Max	**Max**	**Min**	**Max**	**Min**	**Max**	**Min**	**Min**	**Max**	**Min**	**Max**	**Min**
0.3	9.25	9.24	2.20	2.19	8.78	8.77	3.0	12.00	11.90	1.08	1.07
0.4	11.56	11.55	3.10	3.09	11.04	11.03	4.0	14.40	14.30	1.19	1.18
0.5	15.02	15.01	4.50	4.49	14.37	14.36	4.0	18.10	18.00	1.31	1.30
0.6	17.33	17.32	5.50	5.49	16.63	16.62	5.0	22.00	21.90	1.81	1.80
0.7	20.79	20.78	6.70	6.69	20.02	20.01	5.0	26.20	26.10	2.20	2.19
0.9	24.26	24.25	7.80	7.79	23.24	23.33	6.0	30.10	30.00	2.55	2.54
1.0	27.72	27.71	9.00	8.99	26.74	26.73	6.0	34.70	34.60	2.96	2.95
1.2	34.65	34.64	11.10	11.09	32.94	32.93	6.0	43.00	42.90	3.70	3.69

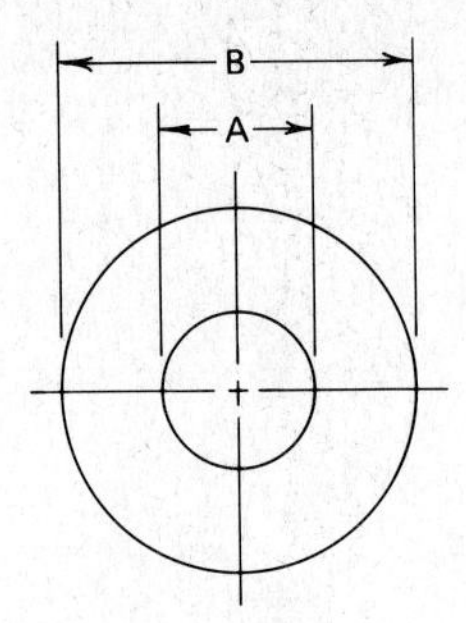

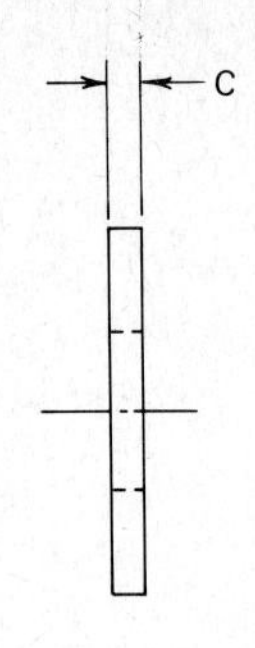

TABLE B.6
American National Standard Metric Washers

NOMINAL WASHER SIZE	WASHER SERIES	A INSIDE DIAMETER Max	A INSIDE DIAMETER Min	B OUTSIDE DIAMETER Max	B OUTSIDE DIAMETER Min	C THICKNESS Max	C THICKNESS Min
1.6	Narrow	2.09	1.95	4.00	3.70	0.70	0.50
	Regular	2.09	1.95	5.00	4.70	0.70	0.50
	Wide	2.09	1.95	6.00	5.70	0.90	0.60
2	Narrow	2.64	2.50	5.00	4.70	0.90	0.60
	Regular	2.64	2.50	6.00	5.70	0.90	0.60
	Wide	2.64	2.50	8.00	7.64	0.90	0.60
2.5	Narrow	3.14	3.00	6.00	5.70	0.90	0.60
	Regular	3.14	3.00	8.00	7.64	0.90	0.60
	Wide	3.14	3.00	10.00	9.64	1.20	0.80
3	Narrow	3.68	3.50	7.00	6.64	0.90	0.60
	Regular	3.68	3.50	10.00	9.64	1.20	0.80
	Wide	3.68	3.50	12.00	11.57	1.40	1.00
3.5	Narrow	4.18	4.00	9.00	8.64	1.20	0.80
	Regular	4.18	4.00	10.00	9.64	1.40	1.00
	Wide	4.18	4.00	15.00	14.57	1.75	1.20
4	Narrow	4.88	4.70	10.00	9.64	1.20	0.80
	Regular	4.88	4.70	12.00	11.57	1.40	1.00
	Wide	4.88	4.70	16.00	15.57	2.30	1.60
5	Narrow	5.78	5.50	11.00	10.57	1.40	1.00
	Regular	5.78	5.50	15.00	14.57	1.75	1.20
	Wide	5.78	5.50	20.00	19.48	2.30	1.60
6	Narrow	6.87	6.65	13.00	12.57	1.75	1.20
	Regular	6.87	6.65	18.80	18.37	1.75	1.20
	Wide	6.87	6.65	25.40	24.88	2.30	1.60
8	Narrow	9.12	8.90	18.80	18.37	2.30	1.60
	Regular	9.12	8.90	25.40	24.48	2.30	1.60
	Wide	9.12	8.90	32.00	31.38	2.80	2.00

TABLE B.6 continued

NOMINAL WASHER SIZE	WASHER SERIES	A INSIDE DIAMETER Max	Min	B OUTSIDE DIAMETER Max	Min	C THICKNESS Max	Min
10	Narrow	11.12	10.85	20.00	19.48	2.30	1.60
	Regular	11.12	10.85	28.00	27.48	2.80	2.00
	Wide	11.12	10.85	39.00	38.38	3.50	2.50
12	Narrow	13.57	13.30	25.40	24.88	2.80	2.00
	Regular	13.57	13.30	34.00	33.38	3.50	2.50
	Wide	13.57	13.30	44.00	43.38	3.50	2.50
14	Narrow	15.52	15.25	28.00	27.48	2.80	2.00
	Regular	15.52	15.25	39.00	38.38	3.50	2.50
	Wide	15.52	15.25	50.00	49.38	4.00	3.00
16	Narrow	17.52	17.25	32.00	31.38	3.50	2.50
	Regular	17.52	17.25	44.00	43.38	4.00	3.00
	Wide	17.52	17.25	56.00	54.80	4.60	3.50
20	Narrow	22.32	21.80	39.00	38.38	4.00	3.00
	Regular	22.32	21.80	50.00	49.38	4.60	3.50
	Wide	22.32	21.80	66.00	64.80	5.10	4.00
24	Narrow	26.12	25.60	44.00	43.38	4.60	3.50
	Regular	26.12	25.60	56.00	54.80	5.10	4.00
	Wide	26.12	25.60	72.00	70.80	5.60	4.50
30	Narrow	33.02	32.40	56.00	54.80	5.10	4.00
	Regular	33.02	32.40	72.00	70.80	5.60	4.50
	Wide	33.02	32.40	90.00	88.60	6.40	5.00
36	Narrow	38.92	38.30	66.00	64.80	5.60	4.50
	Regular	38.92	38.30	90.00	88.60	6.40	5.00
	Wide	38.92	38.30	110.00	108.60	8.50	7.00

Source: ANSI B18.22M-1981

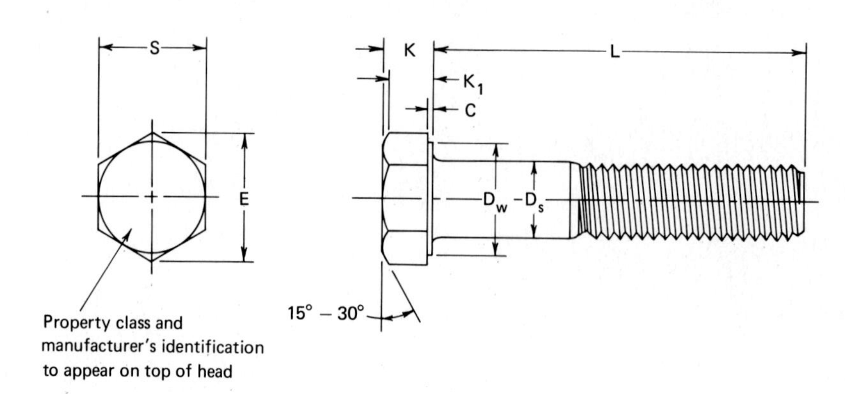

TABLE B.7

American National Standard Metric Hex Cap Screws

D	D_S		S		E		K		K_1	C		D_w	RUNOUT OF BEARING SURFACE FIM
NOM SCREW DIA AND THREAD PITCH	BODY DIAMETER		WIDTH ACROSS FLATS		WIDTH ACROSS CORNERS		HEAD HEIGHT		WRENCH-ING HEIGHT	WASHER FACE THICKNESS		WASHER FACE DIA	
	Max	Min	Max	Min	Max	Min	Max	Min	Min	Max	Min	Min	Max
M5 × 0.8	5.00	4.82	8.00	7.78	9.24	8.79	3.65	3.35	2.4	0.5	0.2	7.0	0.22
M6 × 1	6.00	5.82	10.00	9.78	11.55	11.05	4.15	3.85	2.8	0.5	0.2	8.9	0.25
M8 × 1.25	8.00	7.78	13.00	12.73	15.01	14.38	5.50	5.10	3.7	0.6	0.3	11.6	0.28
M10 × 1.5	10.00	9.78	16.00	15.73	18.48	17.77	6.63	6.17	4.5	0.6	0.3	14.6	0.32
M12 × 1.75	12.00	11.73	18.00	17.73	20.78	20.03	7.76	7.24	5.2	0.6	0.3	16.6	0.35
M14 × 2	14.00	13.73	21.00	20.67	24.25	23.35	9.09	8.51	6.2	0.6	0.3	19.6	0.39
M16 × 2	16.00	15.73	24.00	23.67	27.71	26.75	10.32	9.68	7.0	0.8	0.4	22.5	0.43
M20 × 2.5	20.00	19.67	30.00	29.16	34.64	32.95	12.88	12.12	8.8	0.8	0.4	27.7	0.53
M24 × 3	24.00	23.67	36.00	35.00	41.57	39.55	15.44	14.56	10.5	0.8	0.4	33.2	0.63
M30 × 3.5	30.00	29.67	46.00	45.00	53.12	50.85	19.48	17.92	13.1	0.8	0.4	42.7	0.78
M36 × 4	36.00	35.61	55.00	53.80	63.51	60.79	23.38	21.62	15.8	0.8	0.4	51.1	0.93
M42 × 4.5	42.00	41.38	65.00	62.90	75.06	71.71	26.97	25.03	18.2	1.0	0.5	59.8	1.09
M48 × 5	48.00	47.38	75.00	72.60	86.60	82.76	31.07	28.93	21.0	1.0	0.5	69.0	1.25
M56 × 5.5	56.00	55.26	85.00	82.20	98.15	93.71	36.20	33.80	24.5	1.0	0.5	78.1	1.47
M64 × 6	64.00	63.26	95.00	91.80	109.70	104.65	41.32	38.68	28.0	1.0	0.5	87.2	1.69
M72 × 6	72.00	71.26	105.00	101.40	121.24	115.60	46.45	43.55	31.5	1.2	0.6	96.3	1.91
M80 × 6	80.00	79.26	115.00	111.00	132.72	126.54	51.58	48.42	35.0	1.2	0.6	105.4	2.13
M90 × 6	90.00	89.13	130.00	125.50	150.11	143.07	57.74	54.26	39.2	1.2	0.6	119.2	2.41
M100 × 6	100.00	99.13	145.00	140.00	167.43	159.60	63.90	60.10	43.4	1.2	0.6	133.0	2.69

Typical Lengths: For $L<100$, increment = 5 mm.
For $100<L<200$, increment = 10 mm.
For $L>200$, increment = 20 mm.

Source: ANSI B18.2.3.1M-1979

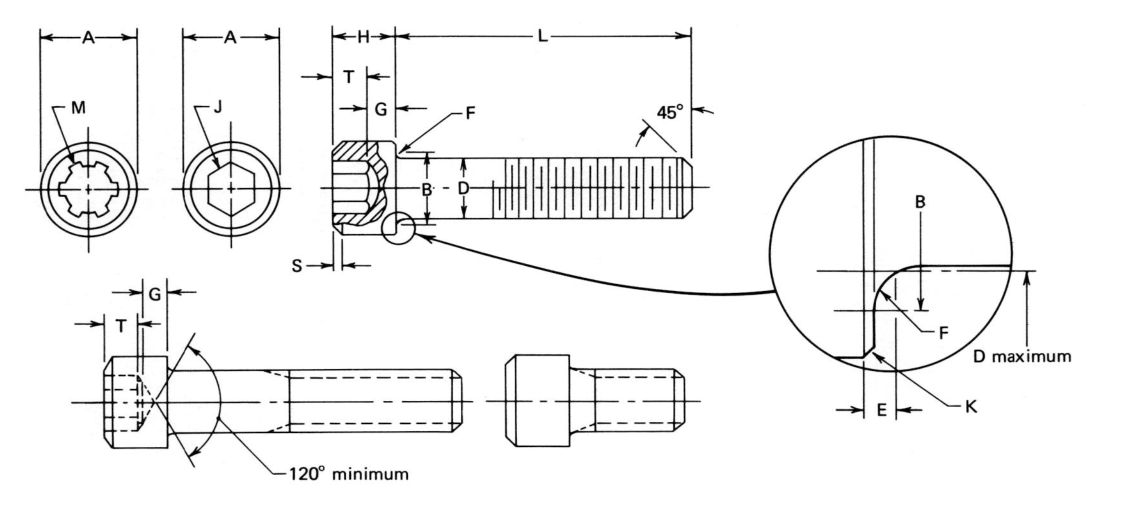

TABLE B.8
American National Standard Metric Socket Head Cap Screws

		D		A		H		S	J	M	T	G	E	F	B	B	K
NOMINAL SIZE OR BASIC SCREW DIAMETER	THD PITCH	BODY DIAMETER		HEAD DIAMETER		HEAD HEIGHT		CHAM-FER OR RADIUS	HEXAGON SOCKET SIZE	SPLINE SOCKET SIZE	KEY ENGAGE-MENT	WALL THICK-NESS	UNDER HEAD FILLET RADIUS				CHAM-FER OR RADIUS
		Max	Min	Max	Min	Max	Min	Max	Nom	Nom	Min	Min	Max	Min	Max	Min	Max
1.6	0.35	1.60	1.46	3.00	2.87	1.60	1.52	0.16	1.5	1.829	0.80	0.54	0.34	0.10	2.00	1.8	0.08
2	0.4	2.00	1.86	3.80	3.65	2.00	1.91	0.20	1.5	1.829	1.00	0.68	0.51	0.10	2.60	2.2	0.08
2.5	0.45	2.50	2.36	4.50	4.33	2.50	2.40	0.25	2.0	2.438	1.25	0.85	0.51	0.10	3.10	2.7	0.08
3	0.5	3.00	2.86	5.50	5.32	3.00	2.89	0.30	2.5	2.819	1.50	1.02	0.51	0.10	3.60	3.2	0.13
4	0.7	4.00	3.82	7.00	6.80	4.00	3.88	0.40	3.0	3.378	2.00	1.52	0.60	0.20	4.70	4.4	0.13
5	0.8	5.00	4.82	8.50	8.27	5.00	4.86	0.50	4.0	4.648	2.50	1.90	0.60	0.20	5.70	5.4	0.13
6	1	6.00	5.82	10.00	9.74	6.00	5.85	0.60	5.0	5.486	3.00	2.28	0.68	0.25	6.80	6.5	0.20
8	1.25	8.00	7.78	13.00	12.70	8.00	7.83	0.80	6.0	7.391	4.00	3.20	1.02	0.40	9.20	8.8	0.20
10	1.5	10.00	9.78	16.00	15.67	10.00	9.81	1.00	8.0		5.00	4.00	1.02	0.40	11.20	10.8	0.20
12	1.75	12.00	11.73	18.00	17.63	12.00	11.79	1.20	10.0		6.00	4.80	1.87	0.60	14.20	13.2	0.25
16	2	16.00	15.73	24.00	23.58	16.00	15.76	1.60	14.0		8.00	6.88	1.87	0.60	18.20	17.2	0.25
20	2.5	20.00	19.67	30.00	29.53	20.00	19.73	2.00	17.0		10.00	8.60	2.04	0.80	22.40	21.6	0.40
24	3	24.00	23.67	36.00	35.48	24.00	23.70	2.40	19.0		12.00	10.32	2.04	0.80	26.40	25.6	0.40
30	3.5	30.00	29.67	45.00	44.42	30.00	29.67	3.00	22.0		15.00	12.90	2.89	1.00	33.40	32.0	0.40
36	4	36.00	35.61	54.00	53.37	36.00	35.64	3.60	27.0		18.00	15.48	2.89	1.00	39.40	38.0	0.40
42	4.5	42.00	41.61	63.00	62.31	42.00	41.61	4.20	32.0		21.00	18.06	3.06	1.20	45.60	44.4	0.40
48	5	48.00	47.61	72.00	71.27	48.00	47.58	4.80	36.0		24.00	20.64	3.91	1.60	52.60	51.2	0.40

Typical lengths: L = 4, 5, 6, 8, 10, . . . , 22, 25, 30, . . . , 80, 90, 100 . . .
Source: ANSI B18.3.1-1978.

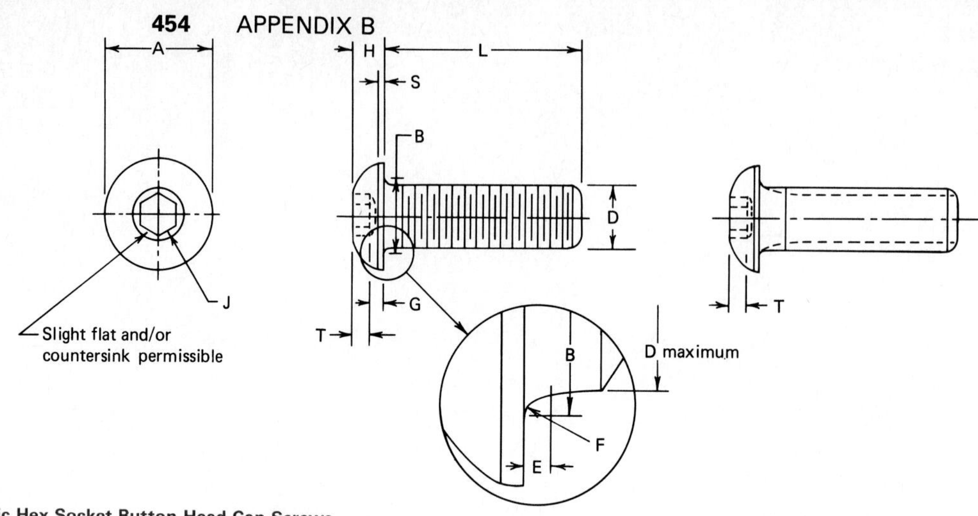

TABLE B.9
American National Standard Metric Hex Socket Button Head Cap Screws

D		A		H		S	J	T	G	B		E	F	L
										UNDER HEAD FILLET				
NOMINAL SIZE OR BASIC SCREW DIAMETER	THREAD PITCH	HEAD DIAMETER		HEAD HEIGHT		HEAD SIDE HEIGHT	HEXAGON SOCKET SIZE	KEY ENGAGE-MENT	WALL THICK-NESS	TRANSITION DIAMETER		TRANSI-TION LENGTH	JUNC-TURE RADIUS	MAXIMUM STANDARD LENGTH
		Max	Min	Max	Min	Ref	Nom	Min	Min	Max	Min	Max	Min	Nom
3	0.5	5.70	5.40	1.65	1.43	0.38	2	1.04	0.20	3.6	3.2	0.51	0.10	12
4	0.7	7.60	7.24	2.20	1.95	0.38	2.5	1.30	0.30	4.7	4.4	0.60	0.20	20
5	0.8	9.50	9.14	2.75	2.50	0.50	3	1.56	0.38	5.7	5.4	0.60	0.20	30
6	1	10.50	10.07	3.30	3.00	0.80	4	2.08	0.74	6.8	6.5	0.68	0.25	30
8	1.25	14.00	13.57	4.40	4.05	0.80	5	2.60	1.05	9.2	8.8	1.02	0.40	40
10	1.5	17.50	17.07	5.50	5.20	0.80	6	3.12	1.45	11.2	10.8	1.02	0.40	40
12	1.75	21.00	20.48	6.60	6.24	0.80	8	4.16	1.63	14.2	13.2	1.87	0.60	60
16	2	28.00	27.48	8.80	8.44	1.50	10	5.20	2.25	18.2	17.2	1.87	0.60	60

Typical lengths: L = 4, 5, 6, 8, 10, . . . , 22, 25, 30, . . . , 60.
Source: ANSI B18.3.4M-1979.

TABLE B.10
American National Standard Metric Hex Flange Screws

D	D_s		S		E		D_c	D_w	
NOM SCREW DIA AND THREAD PITCH	BODY DIAMETER		WIDTH ACROSS FLATS		WIDTH ACROSS CORNERS		FLANGE DIA	BEARING CIRCLE DIA	CIRCULAR RUNOUT OF BEARING CIRCLE FIM
	Max	Min	Max	Min	Max	Min	Max	Min	Max
M5 × 0.8	5.00	4.82	7.00	6.64	0.08	7.50	11.8	9.8	0.31
M6 × 1	6.00	5.82	8.00	7.64	9.24	8.63	14.2	12.2	0.34
M8 × 1.25	8.00	7.78	10.00	9.64	11.55	10.89	18.0	15.8	0.38
M10 × 1.5	10.00	9.78	13.00	12.57	15.01	14.20	22.3	19.6	0.43
M12 × 1.75	12.00	11.73	15.00	14.57	17.32	16.46	26.6	23.8	0.50
M14 × 2	14.00	13.73	18.00	17.57	20.78	19.85	30.5	27.6	0.55
M16 × 2	16.00	15.73	21.00	20.16	24.25	22.78	35.0	31.9	0.61
M20 × 2.5	20.00	19.67	27.00	26.16	31.18	29.56	43.0	39.9	0.76

Typical lengths: L = 10, 12, . . . , 20, 25, 30, . . . , 80, 90, 100.
Source: ANSI B18.2.3.4M-1979.

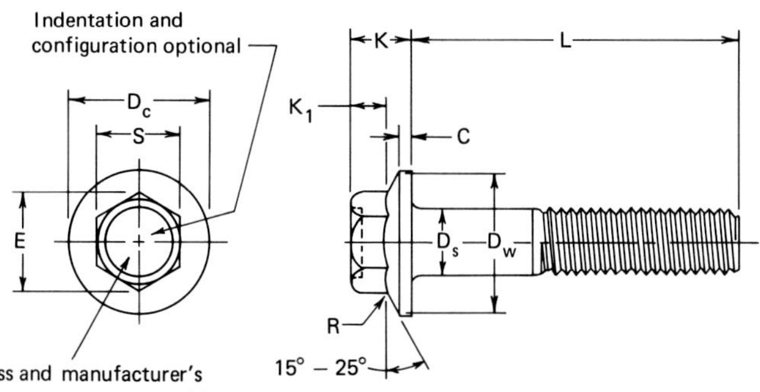

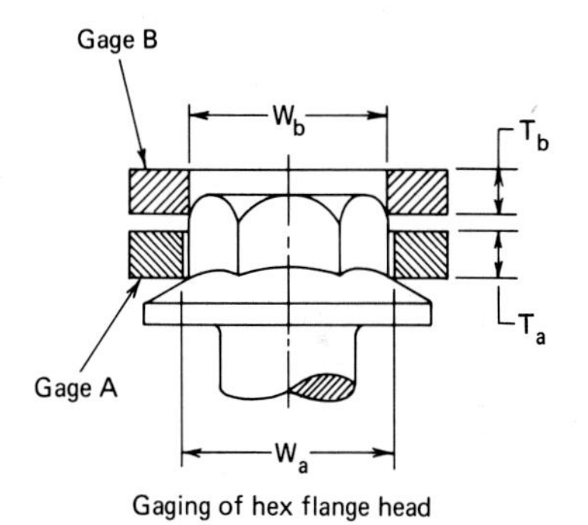

Gaging of hex flange head

TABLE B.10 continued

D	C	K	K_1	R	W_a		T_a		W_b		T_b
					GAGE A				GAGE B		
NOM SCREW DIA AND THREAD PITCH	FLANGE EDGE THICK-NESS	HEAD HEIGHT	HEX HEIGHT	FLANGE TOP FILLET RADIUS	INSIDE DIA		THICK-NESS		INSIDE DIA		THICK-NESS
	Min	Max	Min	Max	Max	Min	Max	Min	Max	Min	Min
M5 × 0.8	1.0	5.4	3.1	0.3	8.09	8.08	2.00	1.99	7.49	7.48	3
M6 × 1	1.1	6.6	3.8	0.4	9.25	9.24	2.50	2.49	8.62	8.61	3
M8 × 1.25	1.2	8.1	4.8	0.5	11.56	11.55	3.20	3.19	10.88	10.87	4
M10 × 1.5	1.5	9.2	5.4	0.6	15.02	15.01	3.60	3.59	14.19	14.18	4
M12 × 1.75	1.8	11.5	6.9	0.7	17.33	17.32	4.60	4.59	16.45	16.44	5
M14 × 2	2.1	12.8	7.7	0.8	20.79	20.78	5.10	5.09	19.84	19.83	5
M16 × 2	2.4	14.4	8.7	1.0	24.26	24.25	5.80	5.79	22.77	22.76	6
M20 × 2.5	3.0	17.1	10.2	1.2	31.19	31.18	6.80	6.79	29.55	29.54	6

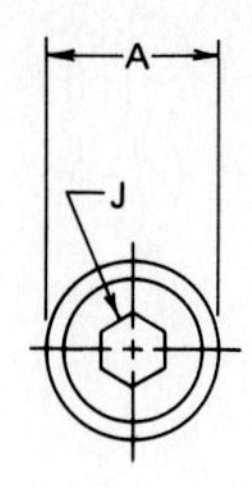

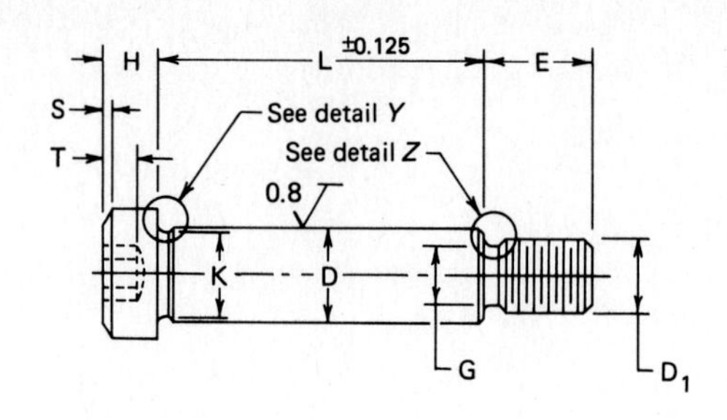

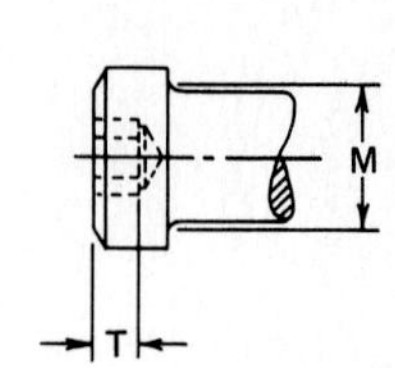

TABLE B.11
American National Standard Metric Hex Socket Head Shoulder Screws

NOMINAL SCREW SIZE OR BASIC SHOULDER DIAMETER	D	A	H	S	J	T	M
	SHOULDER DIAMETER	HEAD DIAMETER	HEAD HEIGHT	CHAM-FER OR RADIUS	HEXA-GON SOCKET SIZE	KEY ENGAGE-MENT	HEAD FIL-LET EX-TENSION DIAMETER
	Max	Max	Max	Max	Nom	Min	Max
6.5	6.487	10.00	4.50	0.6	3	2.4	7.5
8.0	7.987	13.00	5.50	0.8	4	3.3	9.2
10.0	9.987	16.00	7.00	1.0	5	4.2	11.2
13.0	12.984	18.00	9.00	1.2	6	4.9	15.2
16.0	15.984	24.00	11.00	1.6	8	6.6	18.2
20.0	19.980	30.00	14.00	2.0	10	8.8	22.4
25.0	24.980	36.00	16.00	2.4	12	10.0	27.4

Typical lengths: L = 10, 12, . . . , 20, 25, 30, . . . , 80, 90, 100.
Source: ANSI B18.3.3M-1979.

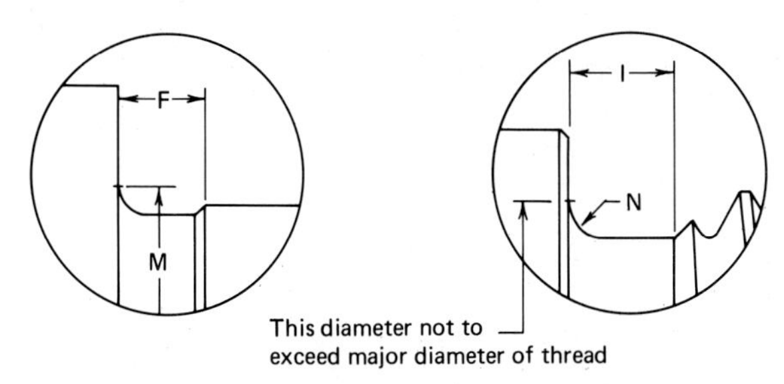

Enlarged detail *Y*

Enlarged detail *Z*

TABLE B.11 continued

NOMINAL SCREW SIZE OR BASIC SHOULDER DIAMETER	K	F	D_1		G	I	N	E
	SHOULDER NECK DIAMETER	SHOULDER NECK WIDTH	NOMINAL THREAD SIZE OR BASIC THREAD DIAMETER	THREAD PITCH	THREAD NECK DIAMETER	THREAD NECK WIDTH	THREAD NECK FILLET RADIUS	THREAD LENGTH
	Min	Max			Max	Max	Max	Max
6.5	5.92	2.5	5	0.8	3.86	2.4	0.66	9.75
8.0	7.42	2.5	6	1	4.58	2.6	0.69	11.25
10.0	9.42	2.5	8	1.25	6.25	2.8	0.80	13.25
13.0	12.42	2.5	10	1.5	7.91	3.0	0.93	16.40
16.0	15.42	2.5	12	1.75	9.57	4.0	1.03	18.40
20.0	19.42	2.5	16	2	13.23	4.8	1.30	22.40
25.0	22.42	3.0	20	2.5	16.57	5.6	1.46	27.40

TABLE B.12
American National Standard Metric Hex Socket Set Screws
Larger sizes on pages 460 and 461

D		J	L	T			C	
				FOR				
NOMINAL SIZE OR BASIC SCREW DIAMETER	THREAD PITCH	HEXAGON SOCKET SIZE Nom	NOMINAL SCREW LENGTHS	CUP AND FLAT POINTS Min	CONE AND OVAL POINTS Min	HALF DOG POINT Min	CUP POINT DIAMETER FOR TYPES I AND III Max	CUP POINT DIAMETER FOR TYPES I AND III Min
			1.5	0.6	0.6	–		
			2	0.8	0.8	0.6	0.80	0.55
1.6	0.35	0.7	2.5	1.0	1.0	0.7		
			3	1.25	1.25	1.25		
			1.5	0.6	0.6	–		
			2	0.8	0.8	–		
2	0.4	0.9	2.5	1.0	1.0	0.8	1.00	0.75
			3	1.2	1.2	1.2		
			4	1.5	1.5	1.5		
			2	0.7	0.7	–		
			2.5	1.1	1.0	0.9		
2.5	0.45	1.3	3	1.5	1.3	1.2	1.20	0.95
			4	1.8	1.8	1.8		
			2	0.6	–	–		
			2.5	1.1	0.7	–		
3	0.5	1.5	3	1.5	1.0	1.0	1.40	1.15
			4	2.1	1.5	2.0		
			5	2.1	2.1	2.1		
			2.5	1.0	–	–		
			3	1.3	1.0	1.0		
4	0.7	2	4	1.8	1.5	1.5	2.00	1.75
			5	2.3	2.0	2.0		
			6	2.3	2.3	2.3		

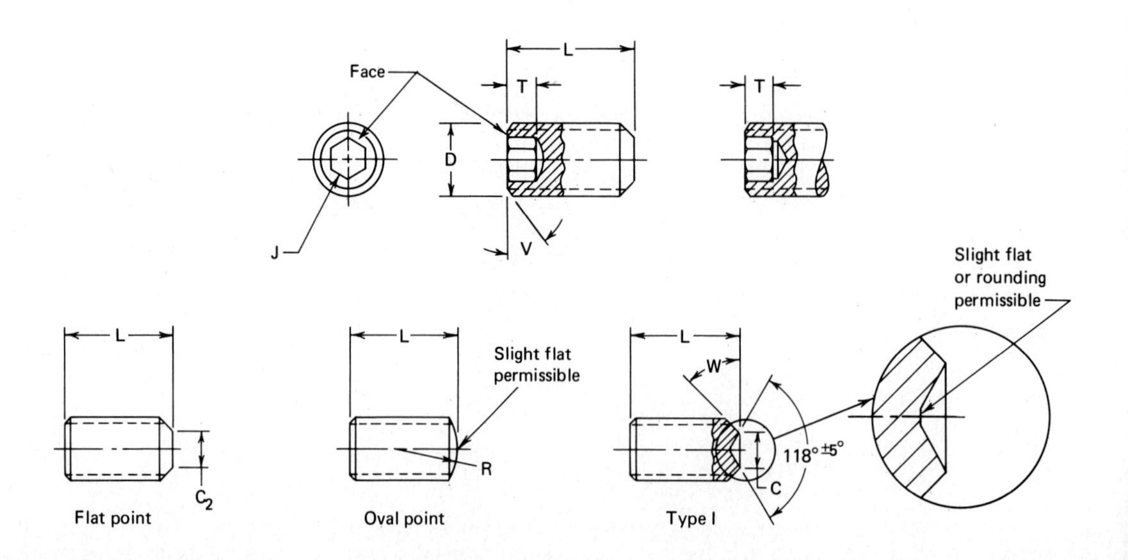

TABLE B.12 continued

C_1		C_2		R		Y	A		P		Q	
CUP POINT DIAMETER FOR TYPES II, IV AND V		FLAT POINT DIAMETER		OVAL POINT RADIUS		CONE POINT ANGLE 90° FOR THESE LENGTHS AND OVER 118° FOR SHORTER LGTHS.	FLAT OF TRUNCATION ON CONE POINT		HALF DOG POINT			
									DIAMETER		LENGTH	
Max	Min	Max	Min	Max	Min		Max	Min	Max	Min	Max	Min
0.80	0.64	0.80	0.55	1.60	1.20	3	1.16	0	0.80	0.55	0.53	0.40
1.00	0.82	1.00	0.75	1.90	1.50	3	0.2	0	1.00	0.75	0.64	0.50
1.25	1.05	1.50	1.25	2.28	1.88	4	0.25	0	1.50	1.25	0.78	0.63
1.50	1.28	2.00	1.75	2.65	2.25	4	0.3	0	2.00	1.75	0.92	0.75
2.00	1.75	2.50	2.25	3.80	3.00	5	0.4	0	2.50	2.25	1.20	1.00

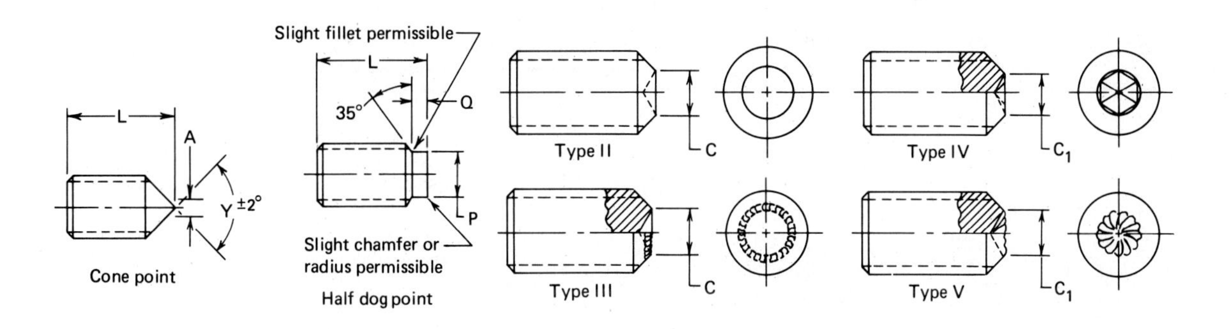

TABLE B.12 continued (see pages 458 and 459 for drawings)
American National Standard Metric Hex Socket Set Screws

D		J	L	T			C	
				FOR				
NOMINAL SIZE OR BASIC SCREW DIAMETER	THREAD PITCH	HEXAGON SOCKET SIZE	NOMINAL SCREW LENGTHS	CUP AND FLAT POINTS	CONE AND OVAL POINTS	HALF DOG POINT	CUP POINT DIAMETER FOR TYPES I AND III	
		Nom		Min	Min	Min	Max	Min
			3	1.2	–	–		
			4	2.0	1.2	–		
5	0.8	2.5	5	2.7	1.7	2.0	2.50	2.25
			6	2.7	2.0	2.5		
			8	2.7	2.7	2.7		
			4	1.8	–	–		
			5	2.5	1.8	1.5		
6	1	3	6	3.0	2.7	2.0	3.00	2.75
			8	3.0	3.0	3.0		
			5	1.8	–	–		
			6	2.5	2.3	1.8		
8	1.25	4	8	4.0	3.5	3.0	5.00	4.70
			10	4.0	4.0	4.0		
			6	2.0	–	–		
			8	3.6	3.0	2.5		
10	1.5	5	10	5.0	4.0	4.0	6.00	5.70
			12	5.0	5.0	5.0		
			8	3.0	–	–		
			10	4.5	3.8	3.5		
12	1.75	6	12	6.0	5.0	5.0	8.00	7.64
			16	6.0	6.0	6.0		
			10	3.0	–	–		
			12	4.8	3.0	3.0		
16	2	8	16	8.0	6.0	6.0	10.00	9.64
			20	8.0	8.0	8.0		
			12	–	–	–		
			16	6.0	5.0	5.0		
20	2.5	10	20	9.0	8.0	8.0	14.00	13.57
			25	10.0	10.0	10.0		
			16	5.0	–	–		
			20	8.0	7.0	6.0		
24	3	12	25	12.0	10.0	10.0	16.00	15.57
			30	12.0	12.0	12.0		

Source: ANSI B18.3.6M-1979

TABLE B.12 continued

C_1		C_2		R		Y	A		P		Q	
CUP POINT DIAMETER FOR TYPES II, IV AND V		FLAT POINT DIAMETER		OVAL POINT RADIUS		CONE POINT ANGLE 90° FOR THESE LENGTHS AND OVER 118° FOR SHORTER LGTHS.	FLAT OF TRUNCATION ON CONE POINT		HALF DOG POINT DIAMETER		HALF DOG POINT LENGTH	
Max	Min	Max	Min	Max	Min		Max	Min	Max	Min	Max	Min
2.50	2.22	3.50	3.20	4.55	3.75	6	0.5	0	3.50	3.20	1.37	1.25
3.00	2.69	4.00	3.70	5.30	4.50	8	1.5	1.2	4.00	3.70	1.74	1.50
4.00	3.65	5.50	5.20	6.80	6.00	10	2.0	1.6	5.50	5.20	2.28	2.00
5.00	4.60	7.00	6.64	8.30	7.50	12	2.5	2.0	7.00	6.64	2.82	2.50
6.00	5.57	8.50	8.14	9.80	9.00	16	3.0	2.4	8.50	8.14	3.35	3.00
8.00	7.50	12.00	11.57	12.80	12.00	20	4.0	3.2	12.00	11.57	4.40	4.00
10.00	9.44	15.00	14.57	15.80	15.00	25	5.0	4.0	15.00	14.57	5.45	5.00
12.00	11.39	18.00	17.57	18.80	18.00	30	6.0	4.8	18.00	17.57	6.49	

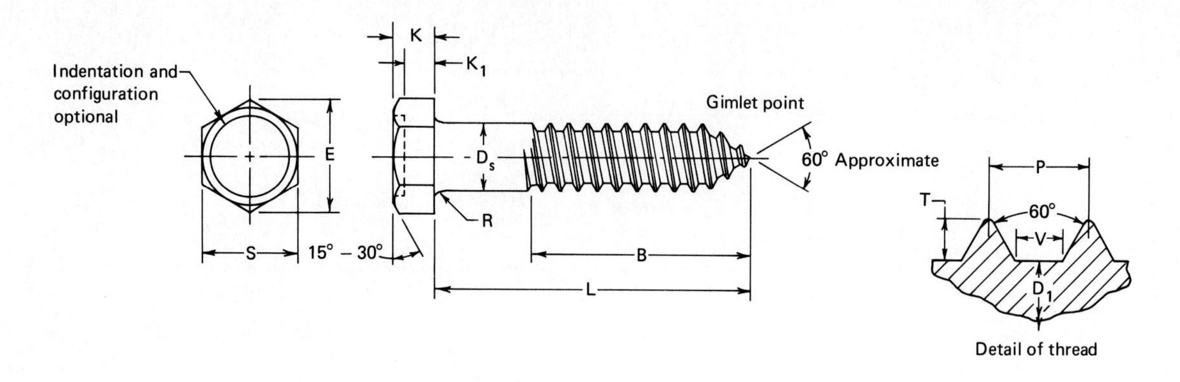

TABLE B.13
American National Standard Metric Hex Lag Screws

D	D_S		S		E		K		K_1	R		P	V	T	D_1
NOMINAL SCREW DIA, mm	BODY DIAMETER		WIDTH ACROSS FLATS		WIDTH ACROSS CORNERS		HEAD HEIGHT		WRENCH-ING HEIGHT	RADIUS OF FILLET		THREAD DIMENSIONS: THREAD PITCH	FLAT AT ROOT	DEPTH OF THREAD	ROOT DIA
	Max	Min	Max	Min	Max	Min	Max	Min	Min	Max	Min				
5	5.48	4.52	8.00	7.64	9.24	8.63	3.9	3.1	2.4	0.6	0.2	2.3	1.0	0.9	3.2
6	6.48	5.52	10.00	9.64	11.55	10.89	4.4	3.6	2.8	0.9	0.3	2.5	1.1	1.0	4.0
8	8.58	7.42	13.00	12.57	15.01	14.20	5.7	4.9	3.7	1.2	0.4	2.8	1.2	1.1	5.8
10	10.58	9.42	16.00	15.57	18.48	17.59	6.9	5.9	4.5	1.2	0.4	3.6	1.6	1.4	7.2
12	12.70	11.30	18.00	17.57	20.78	19.85	8.0	7.0	5.2	1.8	0.6	4.2	1.8	1.6	8.7
16	16.70	15.30	24.00	23.16	27.71	26.17	10.8	9.3	7.0	1.8	0.6	5.1	2.2	2.0	12.0
20	20.84	19.16	30.00	29.16	34.64	32.95	13.4	11.6	8.8	2.4	0.8	5.6	2.4	2.2	15.6
24	24.84	23.16	36.00	35.00	41.57	39.55	15.9	14.1	10.5	2.4	0.8	7.3	3.1	2.5	18.1

Typical lengths: L = 10, 12, . . . , 20, 25, 30, . . . , 80, 90, 100.
Source: ANSI B18.2.3.8M-1981.

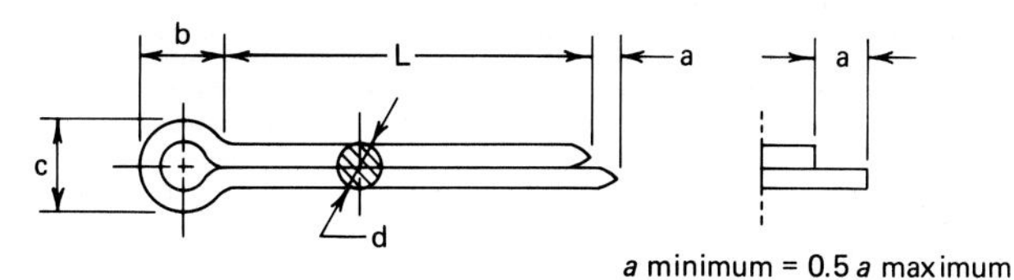

Shape of ends optional

TABLE B.14
Metric Cotter Pins (All Dimensions in millimeters)

NOMINAL SIZE[1]	d Max	d Min	a Max	b ≈	c Max	c Min	RANGE OF STANDARD LENGTHS[2] L	USE WITH: BOLT DIAMETERS OVER–TO	USE WITH: CLEVIS PIN DIAMETERS OVER–TO
0.6	0.5	0.4	1.6	2.0	1.0	0.9	4–12	–2.5	–2
0.8	0.7	0.6	1.6	2.4	1.4	1.2	5–16	2.5–3.5	2–3
1.0	0.9	0.8	1.6	3.0	1.8	1.6	6–20	3.5–4.5	3–4
1.2	1.0	0.9	2.5	3.0	2.0	1.7	8–25	4.5–5.5	4–5
1.6	1.4	1.3	2.5	3.2	2.8	2.4	8–32	5.5–7	5–6
2.0	1.8	1.7	2.5	4.0	3.6	3.2	10–40	7–9	6–8
2.5	2.3	2.1	2.5	5.0	4.6	4.0	12–50	9–11	8–9
3.2	2.9	2.7	3.2	6.4	5.8	5.1	14–63	11–14	9–12
4	3.7	3.5	4.0	8.0	7.4	6.5	18–80	14–20	12–17
5	4.6	4.4	4.0	10.0	9.2	8.0	22–100	20–27	17–23
6.3	5.9	5.7	4.0	12.6	11.8	10.3	32–125	27–39	23–29
8	7.5	7.3	4.0	16.0	15.0	13.1	40–160	39–56	29–44
10	9.5	9.3	6.3	20.0	19.0	16.6	45–200	56–80	44–69
13	12.4	12.1	6.3	26.0	24.8	21.7	71–250	80–120	69–110
16	15.4	15.1	6.3	32.0	30.8	27.0	112–280	120–170	110–160
20	19.3	19.0	6.3	40.0	38.6	33.8	160–280	170–	160–

1 Nominal size: Diameter of the cotter pin hole.
2 Standard lengths: 4, 5, 6, 8, 10, 12, 14, 16, 18, 20, 22, 25, 28, 32, 36, 40, 45, 50, 56, 63, 71, 80, 90, 100, 112, 125, 140, 160, 180, 200, 224, 250, 280
Source: ISO 1234-1976(E)

R ≈ d
d
1
50
a
l
R ≈ d
Other shape possible by agreement between the supplier and the customer

TABLE B.15
Taper Pins—Metric (Dimensions in millimeters)

d	0.6	0.8	1	1.2	1.5	2	2.5	3	4	5	6	8	10	12	16	20	25	30	40	50
a ≈	0.08	0.1	0 12	0.16	0.2	0.25	0.3	0.4	0.5	0.63	0.8	1	1.2	1.6	2	2.5	3	4	5	6.3
l FROM	4	5	6	6	8	10	10	12	14	20	25	25	30	35	40	45	50	60	80	100
l TO	8	12	16	20	25	35	35	45	55	60	90	130	160	180	200	200	200	200	200	200

Standard lengths = 2, 3, 4, 5, 6, 8, 10, 12, 14, 16, 20, 25, 30, 35, 40, 45, 50, 55, 60, 65, 70, 75, 80, 90, 100, 110, 120, 130, 140, 150, 160, 170, 180, 190, 200.
Source: ISO 2339-1974(E)

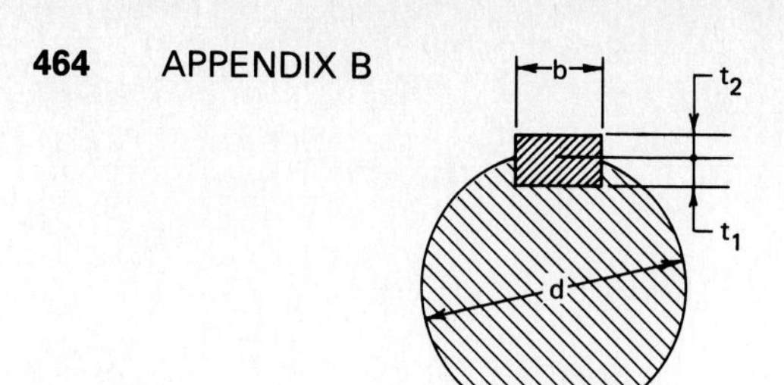

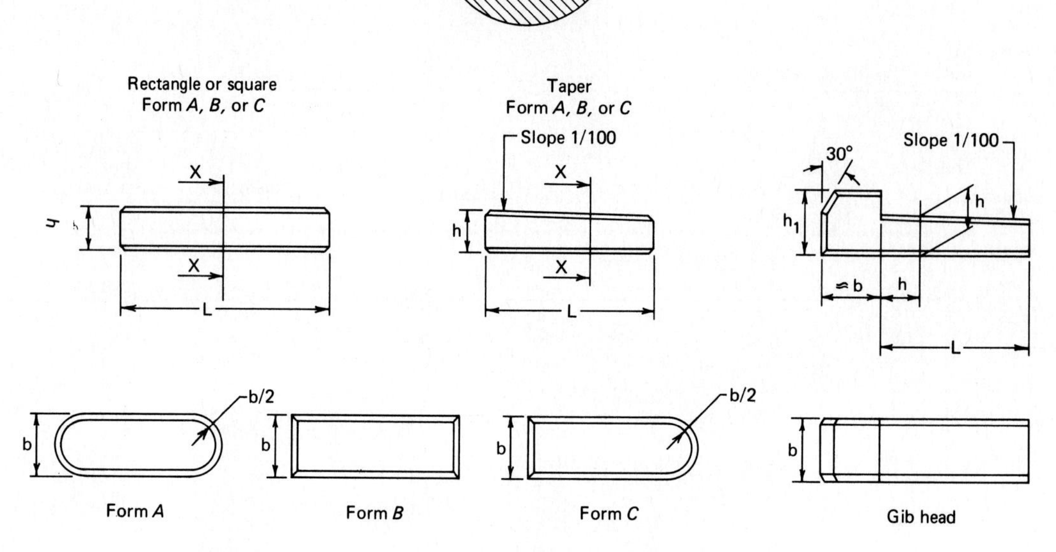

TABLE B.16
Metric Keys and Keyways—Rectangular, Square, Taper and Taper with Gib Head

DIAMETER d Over	DIAMETER d To	SECTION b×h	SHAFT t_1	HUB (PARALLEL) t_2	HUB (TAPER) t_2*	GIB HEAD h_1
6	8	2×2	1.2	1.0	0.5	–
8	10	3×3	1.8	1.4	0.9	–
10	12	4×4	2.5	1.8	1.2	7
12	17	5×5	3.0	2.3	1.7	8
17	22	6×6	3.5	2.8	2.2	10
22	30	8×7	4.0	3.3	2.4	11
30	38	10×8	5.0	3.3	2.4	12
38	44	12×8	5.0	3.3	2.4	12
44	50	14×9	5.5	3.8	2.9	14
50	58	16×10	6.0	4.3	3.4	16
58	65	18×11	7.0	4.4	3.4	18
65	75	20×12	7.5	4.9	3.9	20
75	85	22×14	9.0	5.4	4.4	22
85	95	25×14	9.0	5.4	4.4	22
95	110	28×16	10.0	6.4	5.4	25
110	130	32×18	11.0	7.4	6.4	28
130	150	36×20	12.0	8.4	7.1	32
150	170	40×22	13.0	9.4	8.1	36
170	200	45×25	15.0	10.4	9.1	40

*On gib head keys, t_2 should be measured at the end of the hub at the side where the key enters.
Source: ISO 773 and 774

TABLE B.17
American National Standard Description of Preferred Fits

	ISO SYMBOL			
	HOLE BASIS	SHAFT BASIS	DESCRIPTION	
Clearance fits	H11/c11	C11/h11	*Loose running* fit for wide commercial tolerances or allowances on external members.	More clearance ↑
Clearance fits	H9/d9	D9/h9	*Free running* fit not for use where accuracy is essential, but good for large temperature variations, high running speeds, or heavy journal pressures.	
Clearance fits	H8/f7	F8/h7	*Close running* fit for running on accurate machines and for accurate location at moderate speeds and journal pressures.	
Clearance fits	H7/g6	G7/h6	*Sliding fit* not intended to run freely, but to move and turn freely and locate accurately.	
Clearance fits	H7/h6	H7/h6	*Locational clearance* fit provides snug fit for locating stationary parts; but can be freely assembled and disassembled.	
Transition fits	H7/k6	K7/h6	*Locational transition* fit for accurate location, a compromise between clearance and interference.	
Transition fits	H7/n6	N7/h6	*Locational transition* fit for more accurate location where greater interference is permissible.	
Interference fits	H7/p6[1]	P7/h6	*Locational interference* fit for parts requiring rigidity and alignment with prime accuracy of location but without special bore pressure requirements.	
Interference fits	H7/s6	S7/h6	*Medium drive* fit for ordinary steel parts or shrink fits on light sections, the tightest fit usable with cast iron.	
Interference fits	H7/u6	U7/h6	*Force* fit suitable for parts which can be highly stressed or for shrink fits where the heavy pressing forces required are impractical.	More interference ↓

1 Transition fit for basic sizes in range from 0 through 3 mm.

TABLE B.18
American National Standard Preferred Metric Limits and Fits—Holes (Limits given in micrometers)

NOMINAL SIZES, mm	C11		D9		F8		G7		H7		H8	
	+	+	+	+	+	+	+	+	+		+	
1, 1.2, 1.6, 2, 2.5, 3	120	60	45	20	20	6	12	2	10	0	14	0
4, 5, 6	145	70	60	30	28	10	16	4	12	0	18	0
8, 10	170	80	76	40	35	13	20	5	15	0	22	0
12	205	95	93	50	43	16	24	6	18	0	27	0
16	205	95	93	50	43	16	24	6	18	0	27	0
20	240	110	117	65	53	20	28	7	21	0	33	0
25, 30	240	110	117	65	53	20	28	7	21	0	33	0
40	280	120	142	80	64	25	34	9	25	0	39	0
50	290	130	142	80	64	25	34	9	25	0	39	0
60	330	140	174	100	76	30	40	10	30	0	46	0
80	340	150	174	100	76	30	40	10	30	0	46	0
100	390	170	207	120	90	36	47	12	35	0	54	0
120	400	180	207	120	106	43	47	12	35	0	54	0
160	460	210	245	145	106	43	54	14	40	0	63	0
200	530	240	285	170	122	50	61	15	46	0	72	0

Source: ANSI B4.2-1978

TABLE B.19
American National Standard Preferred Metric Limits and Fits—Shafts (Limits given in micrometers)

NOMINAL SIZES, mm	c11		d9		f7		g6		h6		h7	
	−	−	−	−	−	−	−	−		−		−
1, 1.2, 1.6, 2, 2.5, 3	60	120	20	45	6	16	2	8	0	6	0	10
4, 5, 6	70	145	30	60	10	22	4	12	0	8	0	12
8, 10	80	170	40	76	13	28	5	14	0	9	0	15
12	95	205	50	93	16	34	6	17	0	11	0	18
16	95	205	50	93	16	34	6	17	0	11	0	21
20	110	240	65	117	20	41	7	20	0	13	0	21
25, 30	110	240	65	117	20	41	7	20	0	13	0	21
40	120	280	80	142	25	50	9	25	0	16	0	25
50	130	290	80	142	25	50	9	25	0	16	0	25
60	140	330	100	174	30	60	10	29	0	19	0	30
80	150	340	100	174	30	60	10	29	0	19	0	30
100	170	390	120	207	36	71	12	34	0	22	0	35
120	180	400	120	207	36	71	12	34	0	22	0	35
160	210	460	145	245	43	83	14	39	0	25	0	40
200	240	530	170	285	50	96	15	44	0	29	0	46

Source: ANSI B4.2-1978

TABLE B.18 continued

NOMINAL SIZES, mm	H9		H11		K7		N7		P7		S7		U7	
	+		+		+	–	–	–	–	–	–	–	–	–
1, 1.2, 1.6, 2, 2.5, 3	25	0	60	0	0	10	4	14	6	16	14	24	18	28
4, 5, 6	30	0	75	0	3	9	4	16	8	20	15	27	19	31
8, 10	36	0	90	0	5	10	4	19	9	24	17	32	22	37
12	43	0	110	0	6	12	5	23	11	29	21	39	26	44
16	43	0	110	0	6	12	5	23	11	29	21	39	26	44
20	52	0	130	0	6	15	7	28	14	35	27	48	33	54
25, 30	52	0	130	0	6	15	7	28	14	35	27	48	40	61
40	62	0	160	0	7	18	8	33	17	42	34	59	51	76
50	62	0	160	0	7	18	8	33	17	42	34	59	61	86
60	74	0	190	0	9	21	9	39	21	51	42	72	76	106
80	74	0	190	0	9	21	9	39	21	51	48	78	91	121
100	87	0	220	0	10	25	10	45	24	59	58	93	111	146
120	87	0	220	0	10	25	10	45	24	59	66	101	131	166
160	100	0	250	0	12	28	12	52	28	68	85	125	175	215
200	115	0	290	0	13	33	14	60	33	79	105	151	219	265

TABLE B.19 continued

NOMINAL SIZES, mm	h9		h11		k6		n6		p6		s6		u6	
		–		–	+	+	+	+	+	+	+	+	+	+
1, 1.2, 1.6, 2, 2.5, 3	0	25	0	60	6	0	10	4	12	6	20	14	24	18
4, 5, 6	0	30	0	75	9	1	16	8	20	12	27	19	31	23
8, 10	0	36	0	90	10	1	19	10	24	15	32	23	37	28
12	0	43	0	110	12	1	23	12	29	18	39	28	44	33
16	0	43	0	110	12	1	23	12	29	18	39	28	44	33
20	0	52	0	130	15	2	28	15	35	22	48	35	54	41
25, 30	0	52	0	130	15	2	28	15	35	22	48	35	61	48
40	0	62	0	160	18	2	33	17	42	26	59	43	76	60
50	0	62	0	160	18	2	33	17	42	26	59	43	86	70
60	0	74	0	190	21	2	39	20	51	32	72	53	106	87
80	0	74	0	190	21	2	39	20	51	32	78	59	121	102
100	0	87	0	220	25	3	45	23	59	37	93	71	146	124
120	0	87	0	220	25	3	45	23	59	37	101	79	166	144
160	0	100	0	250	28	3	52	27	68	43	125	100	215	190
200	0	115	0	290	33	4	60	31	79	50	151	122	265	236

Example

Nominal size = 16 mm; free running fit; basic hole.

Solution

From Table B.17, fit is H9/d9.

From Table B.18, hole is

$$16 \begin{matrix} +0.043 \\ +0.000 \end{matrix} \quad \text{or} \quad \begin{matrix} 16.043 \\ 16.000 \end{matrix}$$

From Table B.19, shaft is

$$16 \begin{matrix} -0.050 \\ -0.093 \end{matrix} \quad \text{or} \quad \begin{matrix} 15.950 \\ 15.907 \end{matrix}$$

Allowance = 16.000 − 15.950 = 0.050.

Loosest fit = 16.043 − 15.907 = 0.136.

Example

Nominal size = 60 mm; locational transition fit; basic shaft.

Solution

From Table B.17, fit is K7/h6.

From Table B.18, hole is

$$60 \begin{matrix} -0.021 \\ +0.009 \end{matrix} \quad \text{or} \quad \begin{matrix} 60.009 \\ 59.079 \end{matrix}$$

From Table B.19, shaft is

$$60 \begin{matrix} +0.000 \\ -0.019 \end{matrix} \quad \text{or} \quad \begin{matrix} 60.000 \\ 59.081 \end{matrix}$$

Allowance: 59.979 − 60.000 = −0.021.

Loosest fit: 60.009 − 59.981 = +0.028.

Appendix C

TABLE C.1
Twist Drill Sizes

No.	DIAMETER, in	No.	DIAMETER, in	Letter	DIAMETER, in
1	0.2280	41	0.0960	A	0.234
2	0.2210	42	0.0935	B	0.238
3	0.2130	43	0.0890	C	0.242
4	0.2090	44	0.0860	D	0.246
5	0.2055	45	0.0820	E	0.250
6	0.2040	46	0.0810	F	0.257
7	0.2010	47	0.0785	G	0.261
8	0.1990	48	0.0760	H	0.266
9	0.1960	49	0.0730	I	0.272
10	0.1935	50	0.0700	J	0.277
11	0.1910	51	0.0670	K	0.281
12	0.1890	52	0.0635	L	0.290
13	0.1850	53	0.0595	M	0.295
14	0.1820	54	0.0550	N	0.302
15	0.1800	55	0.0520	O	0.316
16	0.1770	56	0.0465	P	0.323
17	0.1730	57	0.0430	Q	0.332
18	0.1695	58	0.0420	R	0.339
19	0.1660	59	0.0410	S	0.348
20	0.1610	60	0.0400	T	0.358
21	0.1590	61	0.0390	U	0.368
22	0.1570	62	0.0380	V	0.377
23	0.1540	63	0.0370	W	0.386
24	0.1520	64	0.0360	X	0.397
25	0.1495	65	0.0350	Y	0.404
26	0.1470	66	0.0330	Z	0.413
27	0.1440	67	0.0320		
28	0.1405	68	0.0310		
29	0.1360	69	0.0292		
30	0.1285	70	0.0280		
31	0.1200	71	0.0260		
32	0.1160	72	0.0250		
33	0.1130	73	0.0240		
34	0.1110	74	0.0225		
35	0.1100	75	0.0210		
36	0.1065	76	0.0200		
37	0.1040	77	0.0180		
38	0.1015	78	0.0160		
39	0.0995	79	0.0145		
40	0.0980	80	0.0135		

Also available in sizes from 1/16 to 4 in by 64ths.

TABLE C.2
American National Standard Unified Threads

NOMINAL SIZES		BASIC MAJOR DIAMETER	THREADS PER INCH AND TAP DRILL SIZE					
Primary	Secondary		Coarse UNC	Tap drill	Fine UNF	Tap drill	Extra fine UNEF	Tap drill
0		0.0600	–	–	80	3/64		
	1	0.0730	64	53	72	53		
2		0.0860	56	50	64	50		
	3	0.0990	48	47	56	45		
4		0.1120	40	43	48	42		
5		0.1250	40	38	44	37		
6		0.1380	32	36	40	33		
8		0.1640	32	29	36	29		
10		0.1900	24	25	32	21		
	12	0.2160	24	16	28	14	32	13
1/4		0.2500	20	7	28	3	32	7/32
5/16		0.3125	18	F	24	I	32	9/32
3/8		0.3750	16	5/16	24	Q	32	11/32
7/16		0.4375	14	U	20	25/64	28	13/32
1/2		0.5000	13	27/64	20	29/64	28	15/32
9/16		0.5625	12	31/64	18	33/64	24	33/64
5/8		0.6250	11	17/32	18	37/64	24	37/64
	11/16	0.6875	...	...	...	...	24	41/64
3/4		0.7500	10	21/32	16	11/16	20	45/64
	13/16	0.8125	...	...	...	...	20	49/64
7/8		0.8750	9	49/64	14	13/16	20	53/64
	15/16	0.9375	...	...	...	...	20	57/64
1		1.0000	8	7/8	12	59/64	20	61/64
	1 1/16	1.0625	...	...	...	...	18	1
1 1/8		1.1250	7	63/64	12	1 3/64	18	1 5/64
	1 3/16	1.1875	...	...	...	...	18	1 9/64
1 1/4		1.2500	7	1 7/64	12	1 11/64	18	1 3/16
	1 5/16	1.3125	...	...	...	...	18	1 17/64
1 3/8		1.3750	6	1 7/32	12	1 19/64	18	1 5/16
	1 7/16	1.4375	...	...	...	...	18	1 3/8
1 1/2		1.5000	6	1 11/32	12	1 27/64	18	1 7/16
	1 9/16	1.5625	...	...			18	1 1/2
1 5/8		1.6250	...	...			18	1 9/16
	1 11/16	1.6875	...	...			18	1 5/8
1 3/4		1.7500	5	1 9/16				
2		2.0000	4 1/2	1 25/32				
2 1/4		2.2500	4 1/2	2 1/32				
2 1/2		2.5000	4	2 1/4				
2 3/4		2.7500	4	2 1/2				
3		3.0000	4	2 3/4				
3 1/4		3.2500	4	3				
3 1/2		3.5000	4	3 1/4				
3 3/4		3.7500	4	3 1/2				
4		4.0000	4	3 3/4				

Source: ANSI B1.1-1974

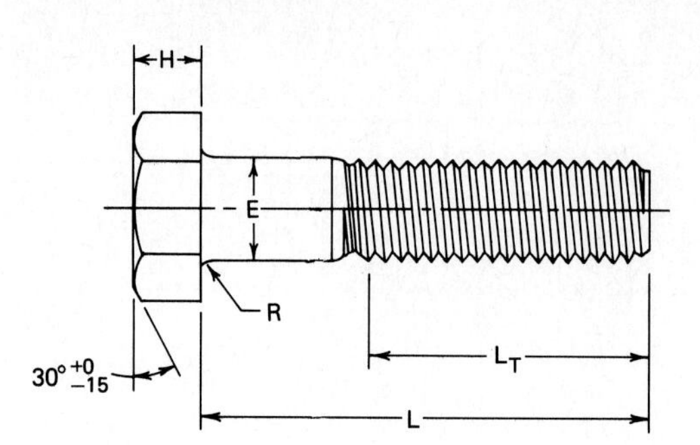

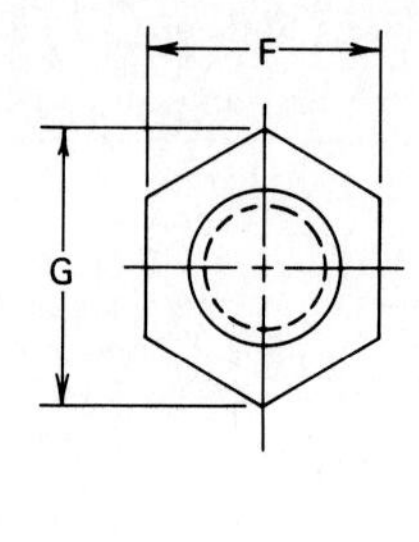

TABLE C.3
American National Standard Hex Bolts

		E	F			G		H			R		L_T	
													THREAD LENGTH FOR BOLT LENGTHS	
NOMINAL SIZE OR BASIC PRODUCT DIA		BODY DIA	WIDTH ACROSS FLATS			WIDTH ACROSS CORNERS		HEIGHT			RADIUS OF FILLET		6 IN AND SHORTER	OVER 6 IN.
		Max	Basic	Max	Min	Max	Min	Basic	Max	Min	Max	Min	Basic	Basic
1/4	0.2500	0.260	7/16	0.438	0.425	0.505	0.484	11/64	0.188	0.150	0.03	0.01	0.750	1.000
5/16	0.3125	0.324	1/2	0.500	0.484	0.577	0.552	7/32	0.235	0.195	0.03	0.01	0.875	1.125
3/8	0.3750	0.388	9/16	0.562	0.544	0.650	0.620	1/4	0.268	0.226	0.03	0.01	1.000	1.250
7/16	0.4375	0.452	5/8	0.625	0.603	0.722	0.687	19/64	0.316	0.272	0.03	0.01	1.125	1.375
1/2	0.5000	0.515	3/4	0.750	0.725	0.866	0.826	11/32	0.364	0.302	0.03	0.01	1.250	1.500
5/8	0.6250	0.642	15/16	0.928	0.906	1.083	1.033	27/64	0.444	0.378	0.06	0.02	1.500	1.750
3/4	0.7500	0.768	1 1/8	1.125	1.088	1.299	1.240	1/2	0.524	0.455	0.06	0.02	1.750	2.000
7/8	0.8750	0.895	1 5/16	1.312	1.269	1.516	1.447	37/64	0.604	0.531	0.06	0.02	2.000	2.250
1	1.0000	1.022	1 1/2	1.500	1.450	1.732	1.653	43/64	0.700	0.591	0.09	0.03	2.250	2.500
1 1/8	1.1250	1.149	1 11/16	1.688	1.631	1.949	1.859	3/4	0.780	0.658	0.09	0.03	2.500	2.750
1 1/4	1.2500	1.277	1 7/8	1.875	1.812	2.165	2.066	27/32	0.876	0.749	0.09	0.03	2.750	3.000
1 3/8	1.3750	1.404	2 1/16	2.062	1.994	2.382	2.273	29/32	0.940	0.810	0.09	0.03	3.000	3.250
1 1/2	1.5000	1.531	2 1/4	2.250	2.175	2.598	2.480	1	1.036	0.902	0.09	0.03	3.250	3.500

Typical lengths: $1 < L < 8$, increment = 0.25
$L > 8$, increment = 0.5
Source: ANSI B18.2.1-1981

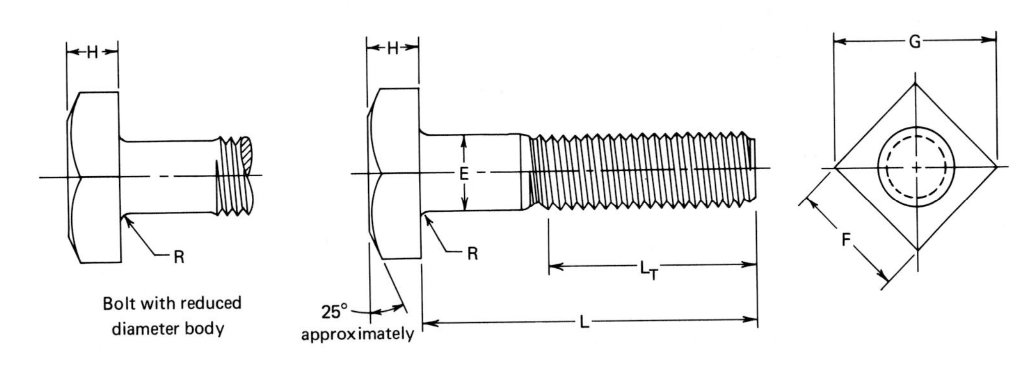

TABLE C.4
American National Standard Square Bolts

NOMINAL SIZE OR BASIC PRODUCT DIA		E	F			G		H			R		L_T	
		BODY DIA	WIDTH ACROSS FLATS			WIDTH ACROSS CORNERS		HEIGHT			RADIUS OF FILLET		THREAD LENGTH FOR BOLT LENGTHS 6 IN. AND SHORTER	THREAD LENGTH FOR BOLT LENGTHS OVER 6 IN.
		Max	Basic	Max	Min	Max	Min	Basic	Max	Min	Max	Min	Basic	Basic
¼	0.2500	0.260	⅜	0.375	0.362	0.530	0.498	11/64	0.188	0.156	0.03	0.01	0.750	1.000
5/16	0.3125	0.324	½	0.500	0.484	0.707	0.665	13/64	0.220	0.186	0.03	0.01	0.875	1.125
⅜	0.3750	0.388	9/16	0.562	0.544	0.795	0.747	¼	0.268	0.232	0.03	0.01	1.000	1.250
7/16	0.4375	0.452	⅝	0.625	0.603	0.884	0.828	19/64	0.316	0.278	0.03	0.01	1.125	1.375
½	0.5000	0.515	¾	0.750	0.725	1.061	0.995	21/64	0.348	0.308	0.03	0.01	1.250	1.500
⅝	0.6250	0.642	15/16	0.938	0.906	1.326	1.244	27/64	0.444	0.400	0.06	0.02	1.500	1.750
¾	0.7500	0.768	1⅛	1.125	1.088	1.591	1.494	½	0.524	0.476	0.06	0.02	1.750	2.000
⅞	0.8750	0.895	1 5/16	1.312	1.269	1.856	1.742	19/32	0.620	0.568	0.06	0.02	2.000	2.250
1	1.0000	1.022	1½	1.500	1.450	2.121	1.991	21/32	0.684	0.628	0.09	0.03	2.250	2.500
1⅛	1.1250	1.149	1 11/16	1.688	1.631	2.386	2.239	¾	0.780	0.720	0.09	0.03	2.500	2.750
1¼	1.2500	1.277	1⅞	1.875	1.812	2.652	2.489	27/32	0.876	0.812	0.09	0.03	2.750	3.000
1⅜	1.3750	1.404	2 1/16	2.062	1.994	2.917	2.738	29/32	0.940	0.872	0.09	0.03	3.000	3.250
1½	1.5000	1.531	2¼	2.250	2.175	3.182	2.986	1	1.036	0.964	0.09	0.03	3.250	3.500

Typical lengths: 1<L<8, increment = 0.25
L>8, increment = 0.50
Source: ANSI B18.2.1-1981

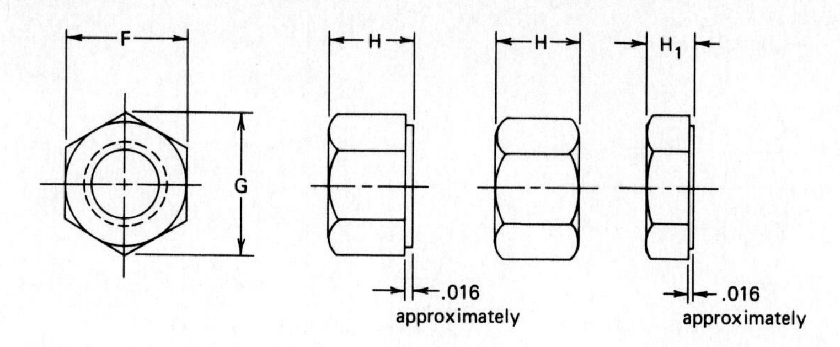

TABLE C.5

American National Standard Hex, Jam & Square Nuts

NOMINAL SIZE OR BASIC MAJOR DIA OF THREAD		F WIDTH ACROSS FLATS			G WIDTH ACROSS CORNERS		H THICKNESS HEX NUTS		
		Basic	Max	Min	Max	Min	Basic	Max	Min
1/4	0.2500	7/16	0.438	0.428	0.505	0.488	7/32	0.226	0.212
5/16	0.3125	1/2	0.500	0.489	0.577	0.557	17/64	0.273	0.258
3/8	0.3750	9/16	0.562	0.551	0.650	0.628	21/64	0.337	0.320
7/16	0.4375	11/16	0.688	0.675	0.794	0.768	3/8	0.385	0.365
1/2	0.5000	3/4	0.750	0.736	0.866	0.840	7/16	0.448	0.427
5/8	0.6250	15/16	0.938	0.922	1.083	1.051	35/64	0.559	0.535
3/4	0.7500	1 1/8	1.125	1.088	1.299	1.240	41/64	0.665	0.617
7/8	0.8750	1 5/16	1.312	1.269	1.516	1.447	3/4	0.776	0.724
1	1.0000	1 1/2	1.500	1.450	1.732	1.653	55/64	0.887	0.831
1 1/8	1.1250	1 11/16	1.688	1.631	1.949	1.859	31/32	0.999	0.939
1 1/4	1.2500	1 7/8	1.875	1.812	2.165	2.066	1 1/16	1.094	1.030
1 3/8	1.3750	2 1/16	2.062	1.994	2.382	2.273	1 11/64	1.206	1.138
1 1/2	1.5000	2 1/4	2.250	2.175	2.598	2.480	1 9/32	1.317	1.245

Source: ANSI B18.2.2-1972

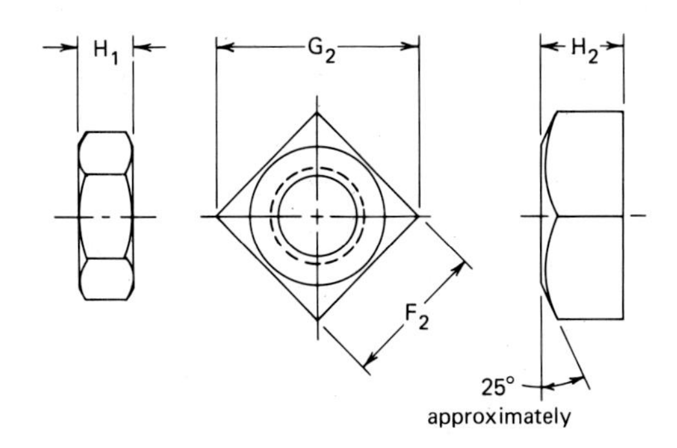

TABLE C.5 continued

NOMINAL SIZE OR BASIC MAJOR DIA OF THREAD		H_1 THICKNESS HEX JAM NUTS			F_2 WIDTH ACROSS FLATS			G_2 WIDTH ACROSS CORNERS		H_2 THICKNESS SQUARE NUTS		
		Basic	Max	Min	Basic	Max	Min	Max	Min	Basic	Max	Min
1/4	0.2500	5/32	0.163	0.150	7/16	0.438	0.425	0.619	0.584	7/32	0.235	0.203
5/16	0.3125	3/16	0.195	0.180	9/16	0.562	0.547	0.795	0.751	17/64	0.283	0.249
3/8	0.3750	7/32	0.227	0.210	5/8	0.625	0.606	0.884	0.832	21/64	0.346	0.310
7/16	0.4375	1/4	0.260	0.240	3/4	0.750	0.728	1.061	1.000	3/8	0.394	0.356
1/2	0.5000	5/16	0.323	0.302	13/16	0.812	0.788	1.149	1.082	7/16	0.458	0.418
5/8	0.6250	3/8	0.387	0.363	1	1.000	0.969	1.414	1.330	35/64	0.569	0.525
3/4	0.7500	27/64	0.446	0.398	1 1/8	1.125	1.088	1.591	1.494	21/32	0.680	0.632
7/8	0.8750	31/64	0.510	0.458	1 5/16	1.312	1.269	1.856	1.742	49/64	0.792	0.740
1	1.0000	35/64	0.575	0.519	1 1/2	1.500	1.450	2.121	1.991	7/8	0.903	0.847
1 1/8	1.1250	39/64	0.639	0.579	1 11/16	1.688	1.631	2.386	2.239	1	1.030	0.970
1 1/4	1.2500	23/32	0.751	0.687	1 7/8	1.875	1.812	2.652	2.489	1 3/32	1.126	1.062
1 3/8	1.3750	25/32	0.815	0.747	2 1/16	2.062	1.994	2.917	2.738	1 13/64	1.237	1.169
1 1/2	1.5000	27/32	0.880	0.808	2 1/4	2.250	2.175	3.182	2.986	1 5/16	1.348	1.276

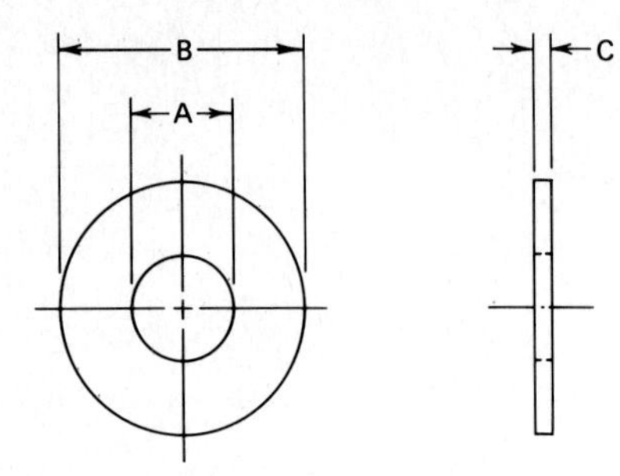

TABLE C.6
American National Standard Plain Washers-Type A

NOMINAL WASHER SIZE			INSIDE DIAMETER A			OUTSIDE DIAMETER B			THICKNESS C		
				TOLERANCE			TOLERANCE				
			Basic	Plus	Minus	Basic	Plus	Minus	Basic	Max	Min
¼	0.250	N	0.281	0.015	0.005	0.625	0;015	0.005	0.065	0.080	0.051
¼	0.250	W	0.312	0.015	0.005	0.734	0.015	0.007	0.065	0.080	0.051
5⁄16	0.312	N	0.344	0.015	0.005	0.688	0.015	0.007	0.065	0.080	0.051
5⁄16	0.312	W	0.375	0.015	0.005	0.875	0.030	0.007	0.083	0.104	0.064
⅜	0.375	N	0.406	0.015	0.005	0.812	0.015	0.007	0.065	0.080	0.051
⅜	0.375	W	0.438	0.015	0.005	1.000	0.030	0.007	0.083	0.104	0.064
7⁄16	0.438	N	0.469	0.015	0.005	0.922	0.015	0.007	0.065	0.080	0.051
7⁄16	0.438	W	0.500	0.015	0.005	1.250	0.030	0.007	0.083	0.104	0.064
½	0.500	N	0.531	0.015	0.005	1.062	0.030	0.007	0.095	0.121	0.074
½	0.500	W	0.562	0.015	0.005	1.375	0.030	0.007	0.109	0.132	0.086
9⁄16	0.562	N	0.594	0.015	0.005	1.156	0.030	0.007	0.095	0.121	0.074
9⁄16	0.562	W	0.625	0.015	0.005	1.469	0.030	0.007	0.109	0.132	0.086
⅝	0.625	N	0.656	0.030	0.007	1.312	0.030	0.007	0.095	0.121	0.074
⅝	0.625	W	0.688	0.030	0.007	1.750	0.030	0.007	0.134	0.160	0.108
¾	0.750	N	0.812	0.030	0.007	1.469	0.030	0.007	0.134	0.160	0.108
¾	0.750	W	0.812	0.030	0.007	2.000	0.030	0.007	0.148	0.177	0.122
⅞	0.875	N	0.938	0.030	0.007	1.750	0.030	0.007	0.134	0.160	0.108
⅞	0.875	W	0.938	0.030	0.007	2.250	0.030	0.007	0.165	0.192	0.136
1	1.000	N	1.062	0.030	0.007	2.000	0.030	0.007	0.134	0.160	0.108
1	1.000	W	1.062	0.030	0.007	2.500	0.030	0.007	0.165	0.192	0.136
1⅛	1.125	N	1.250	0.030	0.007	2.250	0.030	0.007	0.134	0.160	0.108
1⅛	1.125	W	1.250	0.030	0.007	2.750	0.030	0.007	0.165	0.192	0.136
1¼	1.250	N	1.375	0.030	0.007	2.500	0.030	0.007	0.165	0.192	0.136
1¼	1.250	W	1.375	0.030	0.007	3.000	0.030	0.007	0.165	0.192	0.136
1⅜	1.375	N	1.500	0.030	0.007	2.750	0.030	0.007	0.165	0.192	0.136
1⅜	1.375	W	1.500	0.045	0.010	3.250	0.045	0.010	0.180	0.213	0.153
1½	1.500	N	1.625	0.030	0.007	3.000	0.030	0.007	0.165	0.192	0.136
1½	1.500	W	1.625	0.045	0.010	3.500	0.045	0.010	0.180	0.213	0.153

Source: ANSI B 18.22.1-1965

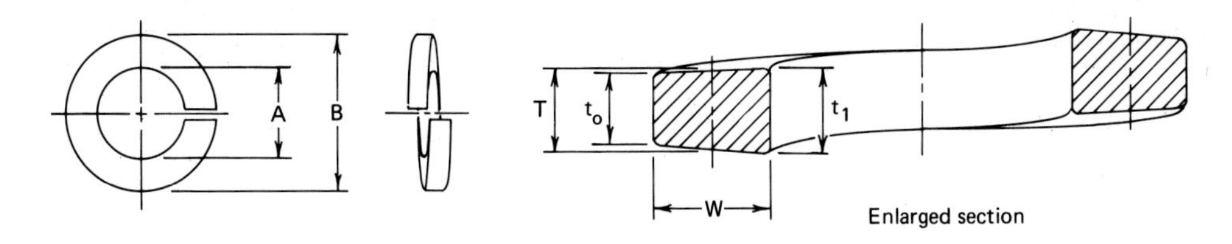

TABLE C.7
American National Standard Lock Washers

NOMINAL WASHER SIZE		A INSIDE DIAMETER Max	A INSIDE DIAMETER Min	B OUTSIDE DIAMETER Max	T MEAN SECTION THICKNESS $\frac{t_1+t_0}{2}$ Min	W SECTION WIDTH Min
1/4	0.250	0.262	0.254	0.489	0.062	0.109
5/16	0.312	0.326	0.317	0.586	0.078	0.125
3/8	0.375	0.390	0.380	0.683	0.094	0.141
7/16	0.438	0.455	0.443	0.779	0.109	0.156
1/2	0.500	0.518	0.506	0.873	0.125	0.171
9/16	0.562	0.582	0.570	0.971	0.141	0.188
5/8	0.625	0.650	0.635	1.079	0.156	0.203
11/16	0.688	0.713	0.698	1.176	0.172	0.219
3/4	0.750	0.775	0.760	1.271	0.188	0.234
13/16	0.812	0.843	0.824	1.367	0.203	0.250
7/8	0.875	0.905	0.887	1.464	0.219	0.266
15/16	0.938	0.970	0.950	1.560	0.234	0.281
1	1.000	1.042	1.017	1.661	0.250	0.297
1 1/16	1.062	1.107	1.080	1.756	0.266	0.312
1 1/8	1.125	1.172	1.144	1.853	0.281	0.328
1 3/16	1.188	1.237	1.208	1.950	0.297	0.344
1 1/4	1.250	1.302	1.271	2.045	0.312	0.359
1 5/16	1.312	1.366	1.334	2.141	0.328	0.375
1 3/8	1.375	1.432	1.398	2.239	0.344	0.391
1 7/16	1.438	1.497	1.462	2.334	0.359	0.406
1 1/2	1.500	1.561	1.525	2.430	0.375	0.422

Source: ANSI B18.21.1-1972

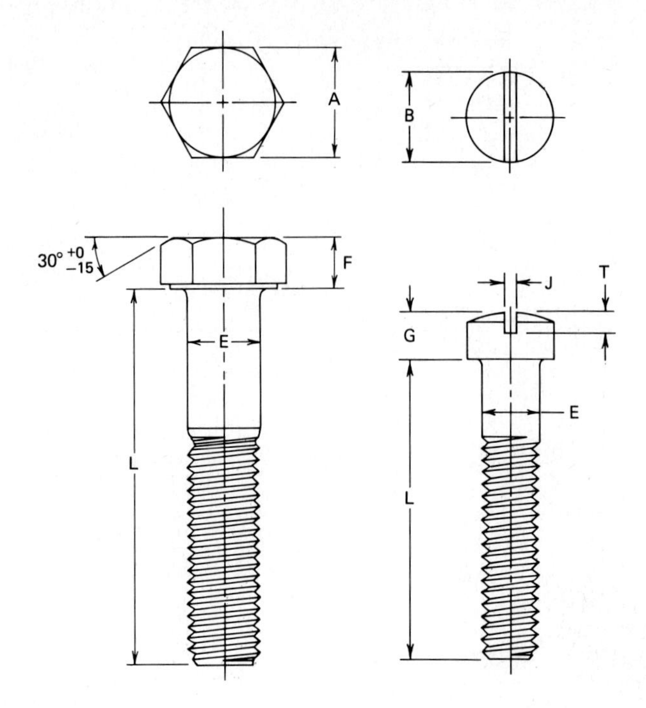

TABLE C.8

American National Standard Cap Screws

NOMINAL SIZE		E BODY DIA Max	Min	A	HEAD DIAMETER, Max B	C	D
¼	0.2500	0.2500	0.2450	0.438	0.375	0.500	0.437
5/16	0.3125	0.3125	0.3065	0.500	0.437	0.625	0.562
⅜	0.3750	0.3750	0.3690	0.562	0.562	0.750	0.625
7/16	0.4375	0.4375	0.4305	0.625	0.625	0.812	0.750
½	0.5000	0.5000	0.4930	0.750	0.750	0.875	0.812
9/16	0.5625	0.5625	0.5545	0.812	0.812	1.000	0.937
⅝	0.6250	0.6250	0.6170	0.938	0.875	1.125	1.000
¾	0.7500	0.7500	0.7410	1.125	1.000	1.375	1.250
⅞	0.8750	0.8750	0.8660	1.312	1.125	1.625	. . .
1	1.0000	1.0000	0.9900	1.500	1.312	1.875	. . .
1⅛	1.1250	1.1250	1.1140	1.688	. . .	2.062	. . .
1¼	1.2500	1.2500	1.2390	1.875	. . .	2.312	. . .
1⅜	1.3750	1.3750	1.3630	2.062	. . .	2.562	. . .
1½	1.5000	1.5000	1.4880	2.230	. . .	2.812	. . .

Typical lengths: $1<L<4$, increment = 0.25
$L>4$, increment = 0.50

Source: ANSI B 18.2.1-1981 and B 18.6.2-1972.

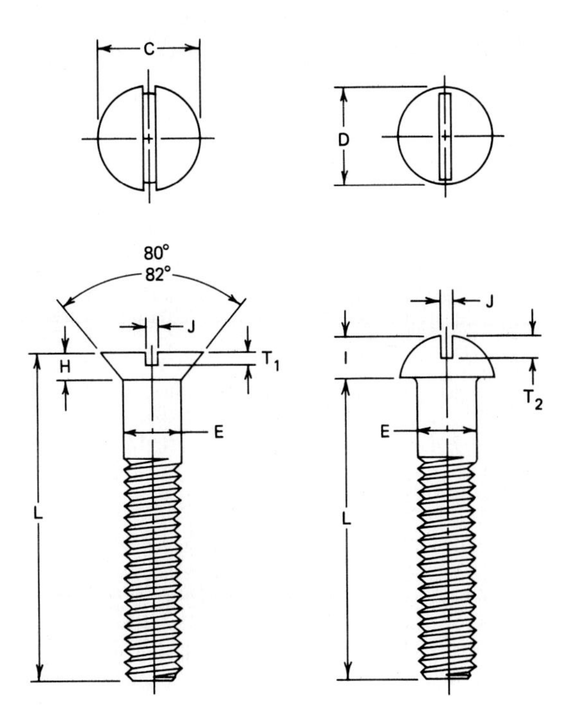

TABLE C.8 continued

NOMINAL SIZE		HEAD HEIGHT, Max				SLOT WIDTH, Max	SLOT DEPTH, Max		
		F	G	H	I	J	T	T_1	T_2
¼	0.2500	0.163	0.216	0.140	0.191	0.075	0.097	0.068	0.117
5/16	0.3125	0.211	0.253	0.177	0.245	0.084	0.115	0.086	0.151
⅜	0.3750	0.243	0.314	0.210	0.273	0.094	0.142	0.103	0.168
7/16	0.4375	0.291	0.368	0.210	0.328	0.094	0.168	0.103	0.202
½	0.5000	0.323	0.413	0.210	0.354	0.106	0.193	0.103	0.218
9/16	0.5625	0.371	0.467	0.244	0.409	0.118	0.213	0.120	0.252
⅝	0.6250	0.403	0.521	0.281	0.437	0.133	0.239	0.137	0.270
¾	0.7500	0.483	0.612	0.352	0.546	0.149	0.283	0.171	0.338
⅞	0.8750	0.563	0.720	0.423	...	0.167	0.334	0.206	...
1	1.0000	0.627	0.803	0.494	...	0.188	0.371	0.240	...
1⅛	1.1250	0.718	...	0.529	...	0.196	...	0.257	...
1¼	1.2500	0.813	...	0.600	...	0.211	...	0.291	...
1⅜	1.3750	0.878	...	0.665	...	0.226	...	0.326	...
1½	1.5000	0.974	...	0.742	...	0.258	...	0.360	...

TABLE C.9

American National Standard Machine Screws

NOMINAL SIZE		HEAD DIAMETER, Max					HEAD HEIGHT, Max					SLOT WIDTH, Max	SLOT DEPTH,[1] Max
		A	B	C	D	E	F	G	H	I	K	J	T
0000	0.0210	0.038	0.043	...	...	...	0.025	0.011	...	...	...	0.008	0.012
000	0.0340	0.059	0.064	...	...	...	0.035	0.016	...	...	...	0.012	0.017
00	0.0470	0.082	0.093	0.093	...	...	0.047	0.028	0.042	...	...	0.017	0.022
0	0.0600	0.096	0.119	0.119	0.116	...	0.055	0.035	0.056	0.044	...	0.023	0.025
1	0.0730	0.118	0.146	0.146	0.142	0.125	0.066	0.043	0.068	0.053	0.044	0.026	0.031
2	0.0860	0.140	0.172	0.172	0.167	0.125	0.083	0.051	0.080	0.062	0.050	0.031	0.037
3	0.0990	0.161	0.199	0.199	0.193	0.188	0.095	0.059	0.092	0.071	0.055	0.035	0.043
4	0.1120	0.183	0.225	0.225	0.219	0.188	0.107	0.067	0.104	0.080	0.060	0.039	0.048
5	0.1250	0.205	0.252	0.252	0.245	0.188	0.120	0.075	0.116	0.089	0.070	0.043	0.054
6	0.1380	0.226	0.279	0.279	0.270	0.250	0.132	0.083	0.128	0.097	0.093	0.048	0.060
8	0.1640	0.270	0.332	0.332	0.322	0.250	0.156	0.100	0.152	0.115	0.110	0.054	0.071
10	0.1900	0.313	0.385	0.385	0.373	0.312	0.180	0.116	0.176	0.133	0.120	0.060	0.083
12	0.2160	0.357	0.438	0.438	0.425	0.312	0.205	0.132	0.200	0.151	0.155	0.067	0.094
¼	0.2500	0.414	0.507	0.507	0.492	0.375	0.237	0.153	0.232	0.175	0.190	0.075	0.109
5/16	0.3125	0.518	0.635	0.635	0.615	0.500	0.295	0.191	0.290	0.218	0.230	0.084	0.137
⅜	0.3750	0.622	0.762	0.762	0.740	0.562	0.355	0.230	0.347	0.261	0.295	0.094	0.164
7/16	0.4375	0.625	0.812	0.812	0.863	...	0.368	0.223	0.345	0.305	...	0.094	0.170
½	0.5000	0.750	0.875	0.875	0.987	...	0.412	0.223	0.354	0.348	...	0.106	0.190
9/16	0.5625	0.812	1.000	1.000	1.041	...	0.466	0.260	0.410	0.391	...	0.118	0.214
⅝	0.6250	0.875	1.125	1.125	1.172	...	0.521	0.298	0.467	0.434	...	0.133	0.240
¾	0.7500	1.000	1.375	1.375	1.435	...	0.612	0.372	0.578	0.521	...	0.149	0.281

1. For fillister head. See standard for slot depth of other head types.

Typical lengths: $L<1$, increment $=0.125$

$1<L<4$, increment $=0.25$

$L>4$, increment $=0.50$

Source: ANSI B18.6.3-1972

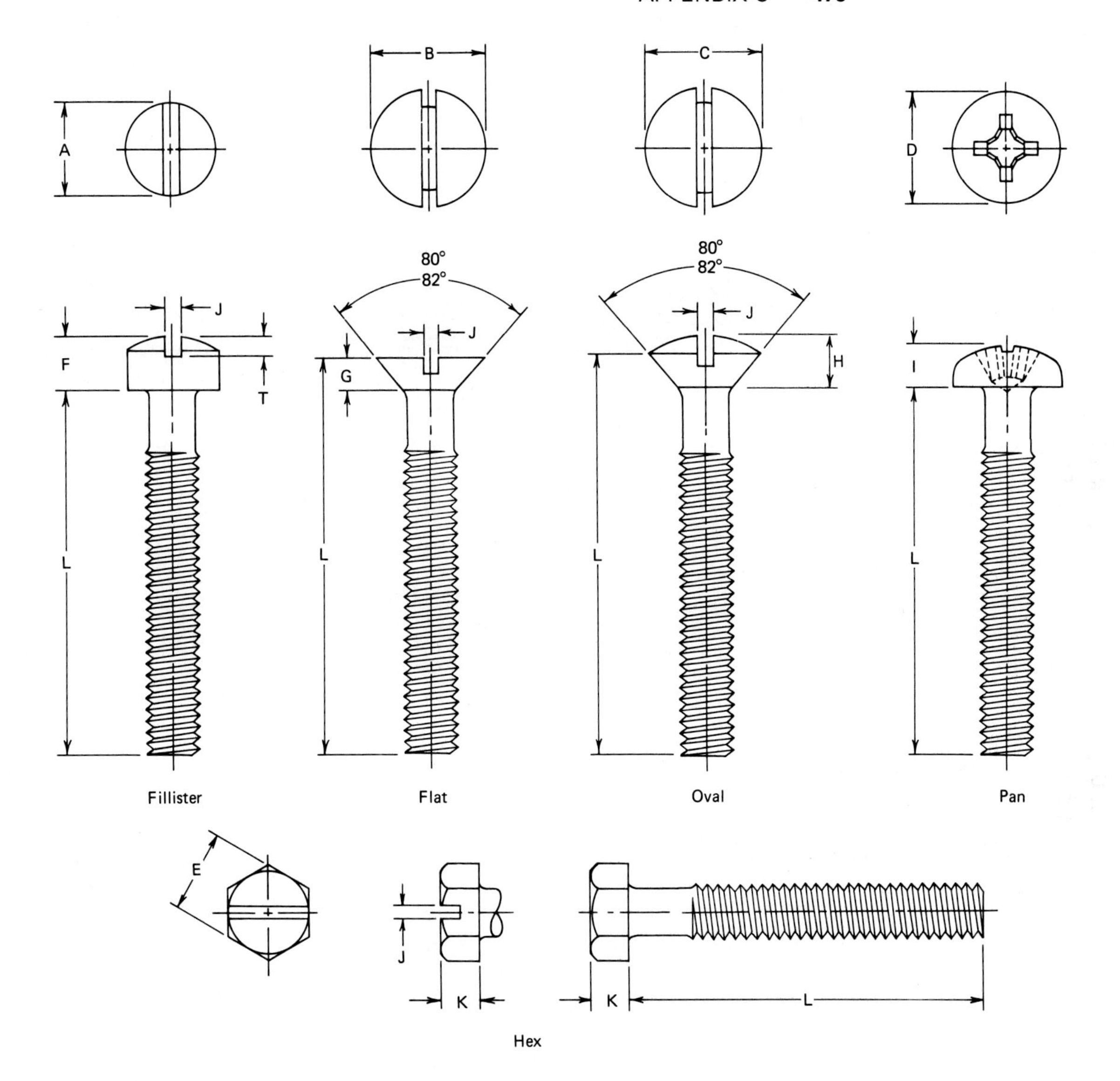
A
B
C
D
80°
82°
80°
82°
J
J
J
F
T
G
H
I
L
L
L
L
Fillister
Flat
Oval
Pan
E
J
K
K
L
Hex

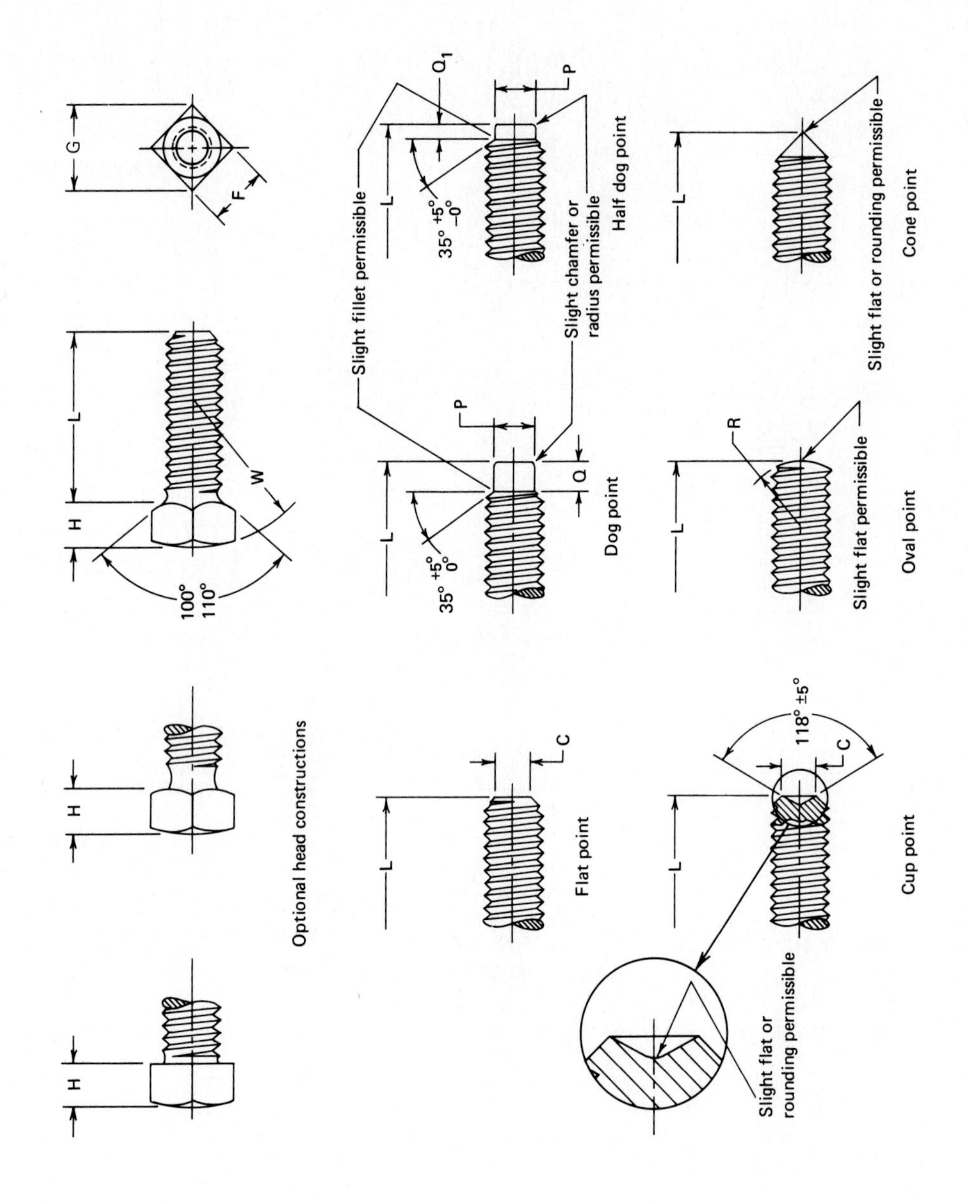

H
H
Optional head constructions
H
L
W
100°
110°
G
F
L
C
Flat point
L
C
118° ±5°
Slight flat or rounding permissible
Cup point
L
35° +5° 0°
P
Q
Dog point
Slight fillet permissible
Slight chamfer or radius permissible
L
R
Slight flat permissible
Oval point
L
35° +5° −0°
Q1
P
Half dog point
L
Slight flat or rounding permissible
Cone point

TABLE C.10

American National Standard Square Head Set Screws

		F		G		H		W	C		P		Q		Q1		R
NOMINAL SIZE OR BASIC SCREW DIAMETERS		WIDTH ACROSS FLATS		WIDTH ACROSS CORNERS		HEAD HEIGHT		HEAD RADIUS	CUP AND FLAT POINT DIAMETERS		DOG AND HALF DOG POINT DIAMETERS		POINT LENGTH				OVAL POINT RADIUS
													DOG		HALF DOG		+0.031
		Max	Min	Max	Min	Max	Min	Min	Max	Min	Max	Min	Max	Min	Max	Min	−0.000
1/4	0.2500	0.250	0.241	0.354	0.331	0.196	0.178	0.62	0.132	0.118	0.156	0.149	0.130	0.120	0.068	0.058	0.188
5/16	0.3125	0.312	0.302	0.442	0.415	0.245	0.224	0.78	0.172	0.156	0.203	0.195	0.161	0.151	0.083	0.073	0.234
3/8	0.3750	0.375	0.362	0.530	0.497	0.293	0.270	0.94	0.212	0.194	0.250	0.241	0.193	0.183	0.099	0.089	0.281
7/16	0.4375	0.438	0.423	0.619	0.581	0.341	0.315	1.09	0.252	0.232	0.297	0.287	0.224	0.214	0.114	0.104	0.328
1/2	0.5000	0.500	0.484	0.707	0.665	0.389	0.361	1.25	0.291	0.270	0.344	0.334	0.255	0.245	0.130	0.120	0.375
9/16	0.5625	0.562	0.545	0.795	0.748	0.437	0.407	1.41	0.332	0.309	0.391	0.379	0.287	0.275	0.146	0.134	0.472
5/8	0.6250	0.625	0.606	0.884	0.833	0.485	0.452	1.56	0.371	0.347	0.469	0.456	0.321	0.305	0.164	0.148	0.469
3/4	0.7500	0.750	0.729	1.060	1.001	0.582	0.544	1.88	0.450	0.425	0.562	0.549	0.383	0.367	0.196	0.180	0.562
7/8	0.8750	0.875	0.852	1.237	1.170	0.678	0.635	2.19	0.530	0.502	0.656	0.642	0.446	0.430	0.227	0.211	0.656
1	1.0000	1.000	0.974	1.414	1.337	0.774	0.726	2.50	0.609	0.579	0.750	0.734	0.510	0.490	0.260	0.240	0.750
1 1/8	1.1250	1.125	1.096	1.591	1.505	0.870	0.817	2.81	0.689	0.655	0.844	0.826	0.572	0.552	0.291	0.271	0.844
1 1/4	1.2500	1.250	1.219	1.768	1.674	0.966	0.908	3.12	0.767	0.733	0.938	0.920	0.635	0.615	0.323	0.303	0.938
1 3/8	1.3750	1.375	1.342	1.945	1.845	1.063	1.000	3.44	0.848	0.808	1.031	1.011	0.698	0.678	0.354	0.334	1.031
1 1/2	1.5000	1.500	1.464	2.121	2.010	1.159	1.091	3.75	0.926	0.886	1.125	1.105	0.760	0.740	0.385	0.365	1.125

See manufacturers catalogs for available lengths.
Source: ANSI B18.6.2-1972.

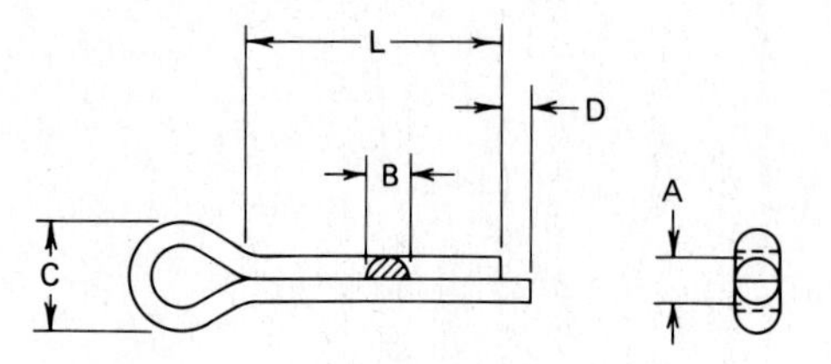

Extended prong square cut type

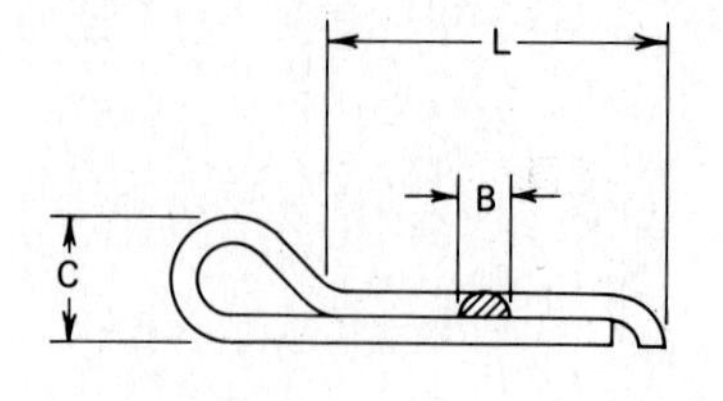

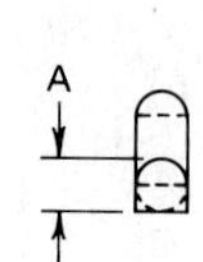

Hammer lock type

TABLE C.11
American National Standard Cotter Pins

NOMINAL SIZE OR BASIC PIN DIAMETER	A TOTAL SHANK DIAMETER		B WIRE WIDTH		C HEAD DIAMETER	D EXTENDED PRONG LENGTH	RECOMMENDED HOLE SIZE
	Max	Min	Max	Min	Min	Min	
1/32 0.031	0.032	0.028	0.032	0.022	0.06	0.01	0.047
3/64 0.047	0.048	0.044	0.048	0.035	0.09	0.02	0.062
1/16 0.062	0.060	0.056	0.060	0.044	0.12	0.03	0.078
5/64 0.078	0.076	0.072	0.076	0.057	0.16	0.04	0.094
3/32 0.094	0.090	0.086	0.090	0.069	0.19	0.04	0.109
7/64 0.109	0.104	0.100	0.104	0.080	0.22	0.05	0.125
1/8 0.125	0.120	0.116	0.120	0.093	0.25	0.06	0.141
9/64 0.141	0.134	0.130	0.134	0.104	0.28	0.06	0.156
5/32 0.156	0.150	0.146	0.150	0.116	0.31	0.07	0.172
3/16 0.188	0.176	0.172	0.176	0.137	0.38	0.09	0.203
7/32 0.219	0.207	0.202	0.207	0.161	0.44	0.10	0.234
1/4 0.250	0.225	0.220	0.225	0.176	0.50	0.11	0.266
5/16 0.312	0.280	0.275	0.280	0.220	0.62	0.14	0.312

Typical lengths: L<3, increment = 0.25
L>3, increment = 1.0
Source: ANSI B18.8.1-1972

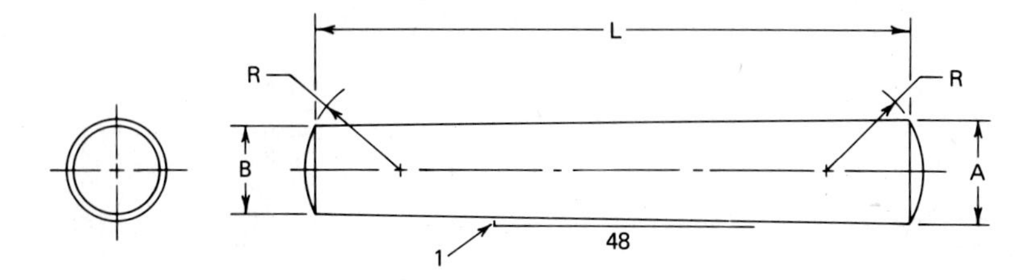

TABLE C.12
American National Standard Taper Pins

PIN SIZE AND BASIC PIN DIAMETER		A — MAJOR DIAMETER (LARGE END)				R	
		COMMERCIAL CLASS		PRECISION CLASS		END CROWN RADIUS	
		Max	Min	Max	Min	Max	Min
7/0	0.0625	0.0638	0.0618	0.0635	0.0625	0.072	0.052
6/0	0.0780	0.0793	0.0773	0.0790	0.0780	0.088	0.068
5/0	0.0940	0.0953	0.0933	0.0950	0.0940	0.104	0.084
4/0	0.1090	0.1103	0.1083	0.1100	0.1090	0.119	0.099
3/0	0.1250	0.1263	0.1243	0.1260	0.1250	0.135	0.115
2/0	0.1410	0.1423	0.1403	0.1420	0.1410	0.151	0.131
0	0.1560	0.1573	0.1553	0.1570	0.1560	0.166	0.146
1	0.1720	0.1733	0.1713	0.1730	0.1720	0.182	0.162
2	0.1930	0.1943	0.1923	0.1940	0.1930	0.203	0.183
3	0.2190	0.2203	0.2183	0.2200	0.2190	0.229	0.209
4	0.2500	0.2513	0.2493	0.2510	0.2500	0.260	0.240
5	0.2890	0.2903	0.2883	0.2900	0.2890	0.299	0.279
6	0.3410	0.3423	0.3403	0.3420	0.3410	0.351	0.331
7	0.4090	0.4103	0.4083	0.4100	0.4090	0.419	0.399
8	0.4920	0.4933	0.4913	0.4930	0.4920	0.502	0.482
9	0.5910	0.5923	0.5903	0.5920	0.5910	0.601	0.581

Typical lengths: L < 1 increment = 0.125
L > 1 increment = 0.25
Source: ANSI B18.8.2-1978

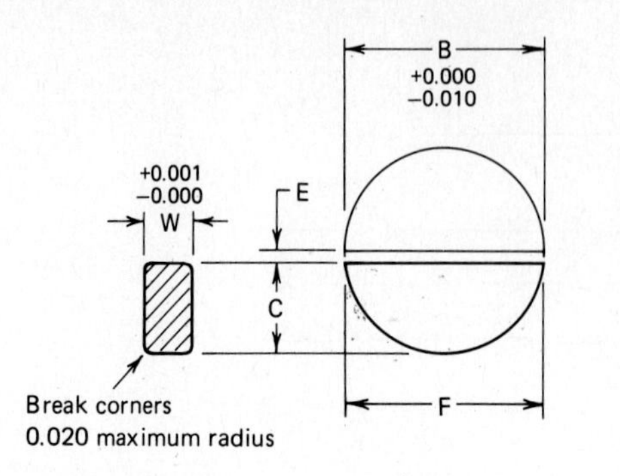

Full radius type

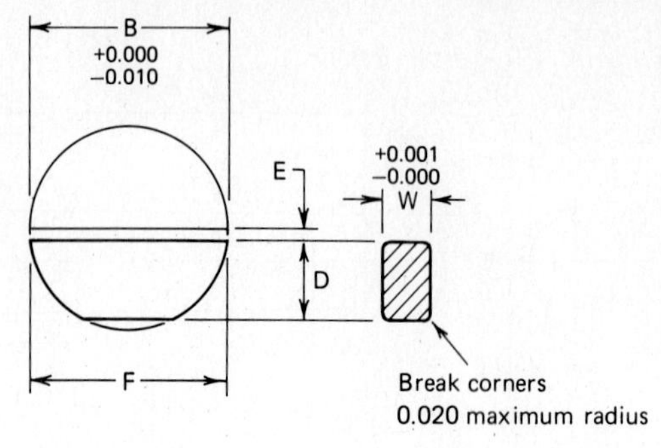

Flat bottom type

TABLE C.13
USA Standard Woodruff Keys

KEY NO.	NOMINAL KEY SIZE WXB	ACTUAL LENGTH F	HEIGHT OF KEY C Max	C Min	D Max	D Min	DISTANCE BELOW CENTER E
202	1/16 × 1/4	0.248	0.109	0.104	0.109	0.104	1/64
202.5	1/16 × 5/16	0.311	0.140	0.135	0.140	0.135	1/64
302.5	3/32 × 5/16	0.311	0.140	0.135	0.140	0.135	1/64
203	1/16 × 3/8	0.374	0.172	0.167	0.172	0.167	1/64
303	3/32 × 3/8	0.374	0.172	0.167	0.172	0.167	1/64
403	1/8 × 3/8	0.374	0.172	0.167	0.172	0.167	1/64
204	1/16 × 1/2	0.491	0.203	0.198	0.194	0.188	3/64
304	3/32 × 1/2	0.491	0.203	0.198	0.194	0.188	3/64
404	1/8 × 1/2	0.491	0.203	0.198	0.194	0.188	3/64
305	3/32 × 5/8	0.612	0.250	0.245	0.240	0.234	1/16
405	1/8 × 5/8	0.612	0.250	0.245	0.240	0.234	1/16
505	5/32 × 5/8	0.612	0.250	0.245	0.240	0.234	1/16
605	3/16 × 5/8	0.612	0.250	0.245	0.240	0.234	1/16
406	1/8 × 3/4	0.740	0.313	0.308	0.303	0.297	1/16
506	5/32 × 3/4	0.740	0.313	0.308	0.303	0.297	1/16
606	3/16 × 3/4	0.740	0.313	0.308	0.303	0.297	1/16
806	1/4 × 3/4	0.740	0.313	0.308	0.303	0.297	1/16
507	5/32 × 7/8	0.866	0.375	0.370	0.365	0.359	1/16
607	3/16 × 7/8	0.866	0.375	0.370	0.365	0.359	1/16
707	7/32 × 7/8	0.866	0.375	0.370	0.365	0.359	1/16
807	1/4 × 7/8	0.866	0.375	0.370	0.365	0.359	1/16
608	3/16 × 1	0.992	0.438	0.433	0.428	0.422	1/16
708	7/32 × 1	0.992	0.438	0.433	0.428	0.422	1/16
808	1/4 × 1	0.992	0.438	0.433	0.428	0.422	1/16
1008	5/16 × 1	0.992	0.438	0.433	0.428	0.422	1/16
1208	3/8 × 1	0.992	0.438	0.433	0.428	0.422	1/16

Source: ANSI B17.2-1967

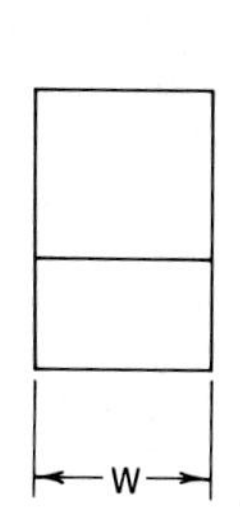

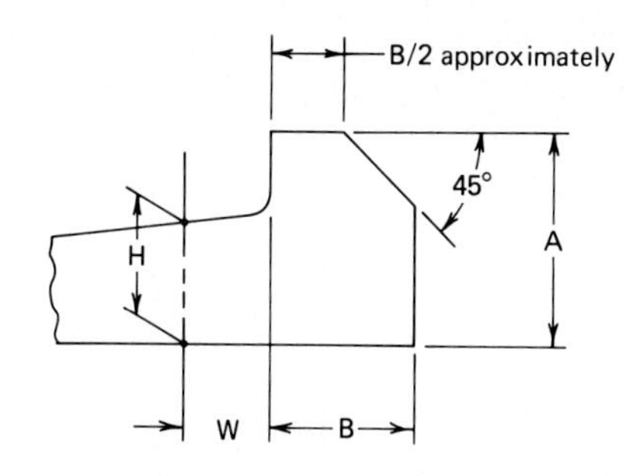

TABLE C.14
USA Standard Gib Head Keys

NOMINAL KEY SIZE WIDTH, W	SQUARE H	SQUARE A	SQUARE B	RECTANGULAR H	RECTANGULAR A	RECTANGULAR B
1/8	1/8	1/4	1/4	3/32	3/16	1/8
3/16	3/16	5/16	5/16	1/8	1/4	1/4
1/4	1/4	7/16	3/8	3/16	5/16	5/16
5/16	5/16	1/2	7/16	1/4	7/16	3/8
3/8	3/8	5/8	1/2	1/4	7/16	3/8
1/2	1/2	7/8	5/8	3/8	5/8	1/2
5/8	5/8	1	3/4	7/16	3/4	9/16
3/4	3/4	1 1/4	7/8	1/2	7/8	5/8
7/8	7/8	1 3/8	1	5/8	1	3/4

Source: ANSI B17.1-1967

TABLE C.15

Running & Sliding Fits (Limits are in thousandths of an inch)

NOMINAL SIZE RANGE, INCHES		CLASS RC 1			CLASS RC 2			CLASS RC 3			CLASS RC 4		
			STANDARD LIMITS			STANDARD LIMITS			STANDARD LIMITS			STANDARD LIMITS	
Over	To	CLEARANCE LIMITS	HOLE H5	SHAFT g4	CLEARANCE LIMITS	HOLE H6	SHAFT g5	CLEARANCE LIMITS	HOLE H7	SHAFT f6	CLEARANCE LIMITS	HOLE H8	SHAFT f7
		0.1	+0.2	−0.1	0.1	+0.25	−0.1	0.3	+0.4	−0.3	0.3	+0.6	−0.3
0	0.12	0.45	0	−0.25	0.55	0	−0.3	0.95	0	−0.55	1.3	0	−0.7
		0.15	+0.2	−0.15	0.15	+0.3	−0.15	0.4	+0.5	−0.4	0.4	+0.7	−0.4
0.12	0.24	0.5	0	−0.3	0.65	0	−0.35	1.12	0	−0.7	1.6	0	−0.9
		0.2	+0.25	−0.2	0.2	+0.4	−0.2	0.5	+0.6	−0.5	0.5	+0.9	−0.5
0.24	0.40	0.6	0	−0.35	0.85	0	−0.45	1.5	0	−0.9	2.0	0	−1.1
		0.25	+0.3	−0.25	0.25	+0.4	−0.25	0.6	+0.7	−0.6	0.6	+1.0	−0.6
0.40	0.71	0.75	0	−0.45	0.95	0	−0.55	1.7	0	−1.0	2.3	0	−1.3
		0.3	+0.4	−0.3	0.3	+0.5	−0.3	0.8	+0.8	−0.8	0.8	+1.2	−0.8
0.71	1.19	0.95	0	−0.55	1.2	0	−0.7	2.1	0	−1.3	2.8	0	−1.6
		0.4	+0.4	−0.4	0.4	+0.6	−0.4	1.0	+1.0	−1.0	1.0	+1.6	−1.0
1.19	1.97	1.1	0	−0.7	1.4	0	−0.8	2.6	0	−1.6	3.6	0	−2.0
		0.4	+0.5	−0.4	0.4	+0.7	−0.4	1.2	+1.2	−1.2	1.2	+1.8	−1.2
1.97	3.15	1.2	0	−0.7	1.6	0	−0.9	3.1	0	−1.9	4.2	0	−2.4
		0.5	+0.6	−0.5	0.5	+0.9	−0.5	1.4	+1.4	−1.4	1.4	+2.2	−1.4
3.15	4.73	1.5	0	−0.9	2.0	0	−1.1	3.7	0	−2.3	5.0	0	−2.8

TABLE C.15 continued

Running & Sliding Fits (Limits are in thousandths of an inch)

NOMINAL SIZE RANGE, INCHES		CLASS RC 5			CLASS RC 6			CLASS RC 7			CLASS RC 8			CLASS RC 9		
			STANDARD LIMITS			STANDARD LIMITS			STANDARD LIMITS			STANDARD LIMITS			STANDARD LIMITS	
Over	To	CLEARANCE LIMITS	HOLE H8	SHAFT e7	CLEARANCE LIMITS	HOLE H9	SHAFT e8	CLEARANCE LIMITS	HOLE H9	SHAFT d8	CLEARANCE LIMITS	HOLE H10	SHAFT c9	CLEARANCE LIMITS	HOLE H11	SHAFT
		0.6	+0.6	−0.6	0.6	+1.0	−0.6	1.0	+1.0	−1.0	2.5	+1.6	−2.5	4.0	+2.5	−4.0
0	0.12	1.6	0	−1.0	2.2	0	−1.2	2.6	0	−1.6	5.1	0	−3.5	8.1	0	−5.6
		0.8	+0.7	−0.8	0.8	+1.2	−0.8	1.2	+1.2	−1.2	2.8	+1.8	−2.8	4.5	+3.0	−4.5
0.12	0.24	2.0	0	−1.3	2.7	0	−1.5	3.1	0	−1.9	5.8	0	−4.0	9.0	0	−6.0
		1.0	+0.9	−1.0	1.0	+1.4	−1.0	1.6	+1.4	−1.6	3.0	+2.2	−3.0	5.0	+3.5	−5.0
0.24	0.40	2.5	0	−1.6	3.3	0	−1.9	3.9	0	−2.5	6.6	0	−4.4	10.7	0	−7.2
		1.2	+1.0	−1.2	1.2	+1.6	−1.2	2.0	+1.6	−2.0	3.5	+2.8	−3.5	6.0	+4.0	−6.0
0.40	0.71	2.9	0	−1.9	3.8	0	−2.2	4.6	0	−3.0	7.9	0	−5.1	12.8	0	−8.8
		1.6	+1.2	−1.6	1.6	+2.0	−1.6	2.5	+2.0	−2.5	4.5	+3.5	−4.5	7.0	+5.0	−7.0
0.71	1.19	3.6	0	−2.4	4.8	0	−2.8	5.7	0	−3.7	10.0	0	−6.5	15.5	0	−10.5
		2.0	+1.6	−2.0	2.0	+2.5	−2.0	3.0	+2.5	−3.0	5.0	+4.0	−5.0	8.0	+6.0	−8.0
1.19	1.97	4.6	0	−3.0	6.1	0	−3.6	7.1	0	−4.6	11.5	0	−7.5	18.0	0	−12.0
		2.5	+1.8	−2.5	2.5	+3.0	−2.5	4.0	+3.0	−4.0	6.0	+4.5	−6.0	9.0	+7.0	−9.0
1.97	3.15	5.5	0	−3.7	7.3	0	−4.3	8.8	0	−5.8	13.5	0	−9.0	20.5	0	−13.5
		3.0	+2.2	−3.0	3.0	+3.5	−3.0	5.0	+3.5	−5.0	7.0	+5.0	−7.0	10.0	+9.0	−10.0
3.15	4.73	6.6	0	−4.4	8.7	0	−5.2	10.7	0	−7.2	15.5	0	−10.5	24.0	0	−15.0

Source: ANSI B4.1-1967.

TABLE C.16

Locational Clearance Fits (Limits are in thousandths of an inch)

NOMINAL SIZE RANGE INCHES		CLASS LC 1			CLASS LC 3			CLASS LC 5		
			STANDARD LIMITS			STANDARD LIMITS			STANDARD LIMITS	
Over	To	CLEARANCE LIMITS	HOLE H6	SHAFT h5	CLEARANCE LIMITS	HOLE H8	SHAFT h7	CLEARANCE LIMITS	HOLE H7	SHAFT g6
		0	+0.25	0	0	+0.6	0	0.1	+0.4	-0.1
0	0.12	0.45	0	-0.2	1	0	-0.4	0.75	0	-0.35
		0	+0.3	0	0	+0.7	0	0.15	+0.5	-0.15
0.12	0.24	0.5	0	-0.2	1.2	0	-0.5	0.95	0	-0.45
		0	+0.4	0	0	+0.9	0	0.2	+0.6	-0.2
0.24	0.40	0.65	0	-0.25	1.5	0	-0.6	1.2	0	-0.6
		0	+0.4	0	0	+1.0	0	0.25	+0.7	-0.25
0.40	0.71	0.7	0	-0.3	1.7	0	-0.7	1.35	0	-0.65
		0	+0.5	0	0	+1.2	0	0.3	+0.8	-0.3
0.71	1.19	0.9	0	-0.4	2	0	-0.8	1.6	0	-0.8
		0	+0.6	0	0	+1.6	0	0.4	+1.0	-0.4
1.19	1.97	1.0	0	-0.4	2.6	0	-1	2.0	0	-1.0
		0	+0.7	0	0	+1.8	0	0.2	+1.2	-0.4
1.97	3.15	1.2	0	-0.5	3	0	-1.2	2.3	0	-1.1
		0	+0.9	0	0	+2.2	0	0.5	+1.4	-0.5
3.15	4.73	1.5	0	-0.6	3.6	0	-1.4	2.8	0	-1.4

Compiled From ANSI B4.1-1967

TABLE C.17

Locational Transitional Fits (Limits are in thousandths of an inch)

NOMINAL SIZE RANGE INCHES		CLASS LT 1			CLASS LT 2			CLASS LT 3		
			STANDARD LIMITS			STANDARD LIMITS			STANDARD LIMITS	
Over	To	FIT	HOLE H7	SHAFT js6	FIT	HOLE H8	SHAFT js7	FIT	HOLE H7	SHAFT k6
		-0.10	+0.4	+0.10	-0.2	+0.6	+0.2	...	...	...
0	0.12	+0.50	0	-0.10	+0.8	0	-0.2	...	...	...
		-0.15	+0.5	+0.15	-0.25	+0.7	+0.25	...	...	...
0.12	0.24	+0.65	0	-0.15	+0.95	0	-0.25	...	...	...
		-0.2	+0.6	+0.2	-0.3	+0.9	+0.3	-0.5	+0.6	+0.5
0.24	0.40	+0.8	0	-0.2	+1.2	0	-0.3	+0.5	0	+0.1
		-0.2	+0.7	+0.2	-0.35	+1.0	+0.35	-0.5	+0.7	+0.5
0.40	0.71	+0.9	0	-0.2	+1.35	0	-0.35	+0.6	0	+0.1
		-0.25	+0.8	+0.25	-0.4	+1.2	+0.4	-0.6	+0.8	+0.6
0.71	1.19	+1.05	0	-0.25	+1.6	0	-0.4	+0.7	0	+0.1
		-0.3	+1.0	+0.3	-0.5	+1.6	+0.5	-0.7	+1.0	+0.7
1.19	1.97	+1.3	0	-0.3	+2.1	0	-0.5	+0.9	0	+0.1
		-0.3	+1.2	+0.3	-0.6	+1.8	+0.6	-0.8	+1.2	+0.8
1.97	3.15	+1.5	0	-0.3	+2.4	0	-0.6	+1.1	0	+0.1
		-0.4	+1.4	+0.4	-0.7	+2.2	+0.7	-1.0	+1.4	+1.0
3.15	4.73	+1.8	0	-0.4	+2.9	0	-0.7	+1.3	0	+0.1

Source: ANSI B4.1-1967

TABLE C.16 continued

CLASS LC 7			CLASS LC 9			CLASS LC 11		
	STANDARD LIMITS			STANDARD LIMITS			STANDARD LIMITS	
CLEARANCE LIMITS	HOLE H10	SHAFT e9	CLEARANCE LIMITS	HOLE H11	SHAFT e10	CLEARANCE LIMITS	HOLE H13	SHAFT
0.6 3.2	+1.6 0	−0.6 −1.6	2.5 6.6	+2.5 0	−2.5 −4.1	5 17	+6 0	−5 −11
0.8 3.8	+1.8 0	−0.8 −2.0	2.8 7.6	+3.0 0	−2.8 −4.6	6 20	+7 0	−6 −13
1.0 4.6	+2.2 0	−1.0 −2.4	3.0 8.7	+3.5 0	−3.0 −5.2	7 25	+9 0	−7 −16
1.2 5.6	+2.8 0	−1.2 −2.8	3.5 10.3	+4.0 0	−3.5 −6.3	8 28	+10 0	−8 −18
1.6 7.1	+3.5 0	−1.6 −3.6	4.5 13.0	+5.0 0	−4.5 −8.0	10 34	+12 0	−10 −22
2.0 8.5	+4.0 0	−2.0 −4.5	5 15	+6 0	−5 −9	12 44	+16 0	−12 −28
2.5 10.0	+4.5 0	−2.5 −5.5	6 17.5	+7 0	−6 −10.5	14 50	+18 0	−14 −32
3.0 11.5	+5.0 0	−3.0 −6.5	7 21	+9 0	−7 −12	16 60	+22 0	−16 −38

TABLE C.17 continued

CLASS LT 4			CLASS LT 5			CLASS LT 6		
	STANDARD LIMITS			STANDARD LIMITS			STANDARD LIMITS	
FIT	HOLE H8	SHAFT k7	FIT	HOLE H7	SHAFT n6	FIT	HOLE H7	SHAFT n7
. . . . . .	. . . . . .	. . . . . .	−0.5 +0.15	+0.4 0	+0.5 +0.25	−0.65 +0.15	+0.4 0	+0.65 +0.25
. . . . . .	. . . . . .	. . . . . .	−0.6 −0.2	+0.5 0	+0.6 +0.3	−0.8 +0.2	+0.5 0	+0.8 +0.3
−0.7 +0.8	+0.9 0	+0.7 +0.1	−0.8 +0.2	+0.6 0	+0.8 +0.4	−1.0 +0.2	+0.6 0	+1.0 +0.4
−0.8 +0.9	+1.0 0	+0.8 +0.1	−0.9 +0.2	+0.7 0	+0.9 +0.5	−1.2 +0.2	+0.7 0	+1.2 +0.5
−0.9 +1.1	+1.2 0	+0.9 +0.1	−1.1 +0.2	+0.8 0	+1.1 +0.6	−1.4 +0.2	+0.8 0	+1.4 +0.6
−1.1 +1.5	+1.6 0	+1.1 +0.1	−1.3 +0.3	+1.0 0	+1.3 +0.7	−1.7 +0.3	+1.0 0	+1.7 +0.7
−1.3 +1.7	+1.8 0	+1.3 +0.1	−1.5 +0.4	+1.2 0	+1.5 +0.8	−2.0 +0.4	+1.2 0	+2.0 +0.8
−1.5 +2.1	+2.2 0	+1.5 +0.1	−1.9 +0.4	+1.4 0	+1.9 +1.0	−2.4 +0.4	+1.4 0	+2.4 +1.0

TABLE C.18

Locational Interference Fits (Limits are in thousandths of an inch)

NOMINAL SIZE RANGE INCHES		CLASS LN 1			CLASS LN 2			CLASS LN 3		
		Interference Limits	STANDARD LIMITS		Interference Limits	STANDARD LIMITS		Interference Limits	STANDARD LIMITS	
Over	To		HOLE H6	SHAFT n5		HOLE H7	SHAFT p6		HOLE H7	SHAFT r6
0	0.12	0 0.45	+0.25 0	+0.45 +0.25	0 0.65	+0.4 0	+0.65 +0.4	0.1 0.75	+0.4 0	+0.75 +0.5
0.12	0.24	0 0.5	+0.3 0	+0.5 +0.3	0 0.8	+0.5 0	+0.8 +0.5	0.1 0.9	+0.5 0	+0.9 +0.6
0.24	0.40	0 0.65	+0.4 0	+0.65 +0.4	0 1.0	+0.6 0	+1.0 +0.6	0.2 1.2	+0.6 0	+1.2 +0.8
0.40	0.71	0 0.8	+0.4 0	+0.8 +0.4	0 1.1	+0.7 0	+1.1 +0.7	0.3 1.4	+0.7 0	+1.4 +1.0
0.71	1.19	0 1.0	+0.5 0	+1.0 +0.5	0 1.3	+0.8 0	+1.3 +0.8	0.4 1.7	+0.8 0	+1.7 +1.2
1.19	1.97	0 1.1	+0.6 0	+1.1 +0.6	0 1.6	+1.0 0	+1.6 +1.0	0.4 2.0	+1.0 0	+2.0 +1.4
1.97	3.15	0.1 1.3	+0.7 0	+1.3 +0.7	0.2 2.1	+1.2 0	+2.1 +1.4	0.4 2.3	+1.2 0	+2.3 +1.6
3.15	4.73	0.1 1.6	+0.9 0	+1.6 +1.0	0.2 2.5	+1.4 0	+2.5 +1.6	0.6 2.9	+1.4 0	+2.9 +2.0

Source: ANSI B4.1-1967

TABLE C.19

Force and Shrink Fits (Limits are in thousandths of an inch)

NOMINAL SIZE RANGE INCHES		CLASS FN 1			CLASS FN 2			CLASS FN 3			CLASS FN 4			CLASS FN 5		
		Interference Limits	STANDARD LIMITS		Interference Limits	STANDARD LIMITS		Interference Limits	STANDARD LIMITS		Interference Limits	STANDARD LIMITS		Interference Limits	STANDARD LIMITS	
Over	To		HOLE H6	SHAFT		HOLE H7	SHAFT s6		HOLE H7	SHAFT t6		HOLE H7	SHAFT u6		HOLE H8	SHAFT x7
		0.05	+0.25	+0.5	0.2	+0.4	+0.85	...	...	...	0.3	+0.4	+0.95	0.3	+0.6	+1.3
0	0.12	0.5	0	+0.3	0.85	0	+0.6	...	...	...	0.95	0	+0.7	1.3	0	+0.9
		0.1	+0.3	+0.6	0.2	+0.5	+1.0	...	...	...	0.4	+0.5	+1.2	0.5	+0.7	+1.7
0.12	0.24	0.6	0	+0.4	1.0	0	+0.7	...	...	...	1.2	0	+0.9	1.7	0	+1.2
		0.1	+0.4	+0.75	0.4	+0.6	+1.4	...	...	...	0.6	+0.6	+1.6	0.5	+0.9	+2.0
0.24	0.40	0.75	0	+0.5	1.4	0	+1.0	...	...	...	1.6	0	+1.2	2.0	0	+1.4
		0.1	+0.4	+0.8	0.5	+0.7	+1.6	...	...	...	0.7	+0.7	+1.8	0.6	+1.0	+2.3
0.40	0.56	0.8	0	+0.5	1.6	0	+1.2	...	...	...	1.8	0	+1.4	2.3	0	+1.6
		0.2	+0.4	+0.9	0.5	+0.7	+1.6	...	...	...	0.7	+0.7	+1.8	0.8	+1.0	+2.5
0.56	0.71	0.9	0	+0.6	1.6	0	+1.2	...	...	...	1.8	0	+1.4	2.5	0	+1.8
		0.2	+0.5	+1.1	0.6	+0.8	+1.9	...	...	...	0.8	+0.8	+2.1	1.0	+1.2	+3.0
0.71	0.95	1.1	0	+0.7	1.9	0	+1.4	...	...	...	2.1	0	+1.6	3.0	0	+2.2
		0.3	+0.5	+1.2	0.6	+0.8	+1.9	0.8	+0.8	+2.1	1.0	+0.8	+2.3	1.3	+1.2	+3.3
0.95	1.19	1.2	0	+0.8	1.9	0	+1.4	2.1	0	+1.6	2.3	0	+1.8	3.3	0	+2.5
		0.3	+0.6	+1.3	0.8	+1.0	+2.4	1.0	+1.0	+2.6	1.5	+1.0	+3.1	1.4	+1.6	+4.0
1.19	1.58	1.3	0	+0.9	2.4	0	+1.8	2.6	0	+2.0	3.1	0	+2.5	4.0	0	+3.0
		0.4	+0.6	+1.4	0.8	+1.0	+2.4	1.2	+1.0	+2.8	1.8	+1.0	+3.4	2.4	+1.6	+5.0
1.58	1.97	1.4	0	+1.0	2.4	0	+1.8	2.8	0	+2.2	3.4	0	+2.8	5.0	0	+4.0
		0.6	+0.7	+1.8	0.8	+1.2	+2.7	1.3	+1.2	+3.2	2.3	+1.2	+4.2	3.2	+1.8	+6.2
1.97	2.56	1.8	0	+1.3	2.7	0	+2.0	3.2	0	+2.5	4.2	0	+3.5	6.2	0	+5.0
		0.7	+0.7	+1.9	1.0	+1.2	+2.9	1.8	+1.2	+3.7	2.8	+1.2	+4.7	4.2	+1.8	+7.2
2.56	3.15	1.9	0	+1.4	2.9	0	+2.2	3.7	0	+3.0	4.7	0	+4.0	7.2	0	+6.0
		0.9	+0.9	+2.4	1.4	+1.4	+3.7	2.1	+1.4	+4.4	3.6	+1.4	+5.9	4.8	+2.2	+8.4
3.15	3.94	2.4	0	+1.8	3.7	0	+2.8	4.4	0	+3.5	5.9	0	+5.0	8.4	0	+7.0
		1.1	+0.9	+2.6	1.6	+1.4	+3.9	2.6	+1.4	+4.9	4.6	+1.4	+6.9	5.8	+2.2	+9.4
3.94	4.73	2.6	0	+2.0	3.9	0	+3.0	4.9	0	+4.0	6.9	0	+6.0	9.4	0	+8.0

Source: ANSI B4.1-1967

Index